Elementary
Linear Algebra

Elementary Linear Algebra

NINTH EDITION

HOWARD ANTON

John Wiley & Sons, Inc.

ASSOCIATE PUBLISHER	Laurie Rosatone
ASSOCIATE EDITOR	Jennifer Battista
EDITORIAL ASSISTANT	Stacy French
FREELANCE DEVELOPMENTAL EDITOR	Anne Scanlan-Rohrer
SENIOR PRODUCTION EDITOR	Ken Santor
SENIOR DESIGNER	Dawn L. Stanley
COVER DESIGNER	David Levy
PHOTO EDITOR	Hilary Newman
ILLUSTRATION COORDINATOR	Techsetters, Inc./Sigmund Malinowski
COVER PHOTO	© John Marshall/Stone/Getty Images

This book was set in Times New Roman PS by Techsetters, Inc. and printed and bound by Von Hoffmann Press, Inc. The cover was printed by Von Hoffmann Press.

This book is printed on acid-free paper. ∞

The paper in this book was manufactured by a mill whose forest management programs include sustained yield harvesting of its timberlands. Sustained yield harvesting principles ensure that the numbers of trees cut each year does not exceed the amount of new growth.

To order books or for customer service please, call 1(800)-CALL-WILEY (225-5945).

0-471-66960-1

Printed in the United States of America

10 9 8 7 6 5 4 3

To my wife Pat
and my children
Brian, David, and Lauren
HA

This edition of *Elementary Linear Algebra*, like those that have preceded it, gives an elementary treatment of linear algebra that is suitable for students in their freshman or sophomore year. The aim is to present the fundamentals of linear algebra in the clearest possible way; pedagogy is the main consideration. Calculus is not a prerequisite, but there are clearly labeled exercises and examples for students who have studied calculus. Those exercises can be omitted without loss of continuity. Technology is also not required, but for those who would like to use MATLAB, Maple, *Mathematica*, or calculators with linear algebra capabilities, exercises have been included at the ends of the chapters that allow for further exploration of that chapter's contents.

SUMMARY OF CHANGES IN THIS EDITION

This edition contains organizational changes and additional material suggested by users of the text. Most of the text is unchanged. The entire text has been reviewed for accuracy, typographical errors, and areas where the exposition could be improved or additional examples are needed. The following changes have been made:

- Section 6.5 has been split into two sections: Section 6.5 Change of Basis and Section 6.6 Orthogonal Matrices. This allows for sharper focus on each topic.

- A new Section 4.4 Spaces of Polynomials has been added to further smooth the transition to general linear transformations, and a new Section 8.6 Isomorphisms has been added to provide explicit coverage of this topic.

- Chapter 2 has been reorganized by switching Section 2.1 with Section 2.4. The cofactor expansion approach to determinants is now covered first and the combinatorial approach is now at the end of the chapter.

- Additional exercises, including Discussion and Discovery, Supplementary, and Technology exercises, have been added throughout the text.

- In response to instructors' requests, the number of exercises that have answers in the back of the book has been reduced considerably.

- The page design has been modified to enhance the readability of the text.

Hallmark Features

- **Relationships Between Concepts:** One of the important goals of a course in linear algebra is to establish the intricate thread of relationships between systems of linear equations, matrices, determinants, vectors, linear transformations, and eigenvalues. That thread of relationships is developed through the following crescendo of theorems that link each new idea with ideas that preceded it: 1.5.3, 1.6.4, 2.3.6, 4.3.4, 5.6.9, 6.2.7, 6.4.5, 7.1.5. These theorems bring a coherence to the linear algebra landscape and also serve as a constant source of review.

- **Smooth Transition to Abstraction:** The transition from R^n to general vector spaces is often difficult for students. To smooth out that transition, the underlying geometry of R^n is emphasized and key ideas are developed in R^n before proceeding to general vector spaces.

- **Early Exposure to Linear Transformations and Eigenvalues:** To ensure that the material on linear transformations and eigenvalues does not get lost at the end of the course, some of the basic concepts relating to those topics are developed early in the text and then reviewed and expanded on when the topic is treated in more depth later in the text. For example, characteristic equations are discussed briefly in the chapter on determinants, and linear transformations from R^n to R^m are discussed immediately after R^n is introduced, then reviewed later in the context of general linear transformations.

About the Exercises

Each section exercise set begins with routine drill problems, progresses to problems with more substance, and concludes with theoretical problems. In most sections, the main part of the exercise set is followed by the *Discussion and Discovery* problems described above. Most chapters end with a set of supplementary exercises that tend to be more challenging and force the student to draw on ideas from the entire chapter rather than a specific section. The technology exercises follow the supplementary exercises and are classified according to the section in which we suggest that they be assigned. Data for these exercises in MATLAB, Maple, and *Mathematica* formats can be downloaded from **www.wiley.com/college/anton**.

Supplementary Materials for Students

Student Solutions Manual, Ninth Edition—This supplement provides detailed solutions to most theoretical exercises and to at least one nonroutine exercise of every type. (ISBN 0-471-43330-6)

Data for Technology Exercises is provided in MATLAB, Maple, and *Mathematica* formats. This data can be downloaded from **www.wiley.com/college/anton**.

Linear Algebra Solutions—Powered by JustAsk! invites you to be a part of the solution as it walks you step-by-step through a total of over 150 problems that correlate to chapter materials to help you master key ideas. The powerful online problem-solving tool provides you with more than just the answers.

Supplementary Materials for Instructors

Instructor's Solutions Manual—This new supplement provides solutions to all exercises in the text. (ISBN 0-471-44798-6)

Test Bank—This includes approximately 50 free-form questions, five essay questions for each chapter, and a sample cumulative final examination. (ISBN 0–471-44797-8)

eGrade—eGrade is an online assessment system that contains a large bank of skill-building problems, homework problems, and solutions. Instructors can automate the process of assigning, delivering, grading, and routing all kinds of homework, quizzes, and tests while providing students with immediate scoring and feedback on their work. Wiley eGrade "does the math" … and much more. For more information, visit **http://www.wiley.com/college/egrade** or contact your Wiley representative.

Web Resources—More information about this text and its resources can be obtained from your Wiley representative or from **www.wiley.com/college/anton**.

A GUIDE FOR THE INSTRUCTOR

Linear algebra courses vary widely between institutions in content and philosophy, but most courses fall into two categories: those with about 35–40 lectures (excluding tests and reviews) and those with about 25–30 lectures (excluding tests and reviews). Accordingly, I have created long and short templates as possible starting points for constructing a course outline. In the long template I have assumed that all sections in the indicated chapters are covered, and in the short template I have assumed that instructors will make selections from the chapters to fit the available time. Of course, these are just guides and you may want to customize them to fit your local interests and requirements.

The organization of the text has been carefully designed to make life easier for instructors working under time constraints: A brief introduction to eigenvalues and eigenvectors occurs in Sections 2.3 and 4.3, and linear transformations from R^n to R^m are discussed in Chapter 4. This makes it possible for all instructors to cover these topics at a basic level when the time available for their more extensive coverage in Chapters 7 and 8 is limited. Also, note that Chapter 3 can be omitted without loss of continuity for students who are already familiar with the material.

	Long Template	**Short Template**
Chapter 1	7 lectures	6 lectures
Chapter 2	4 lectures	3 lectures
Chapter 4	4 lectures	4 lectures
Chapter 5	7 lectures	6 lectures
Chapter 6	6 lectures	3 lectures
Chapter 7	4 lectures	3 lectures
Chapter 8	6 lectures	2 lectures
Total	38 lectures	27 lectures

Variations in the Standard Course

Many variations in the long template are possible. For example, one might create an alternative long template by following the time allocations in the short template and devoting the remaining 11 lectures to some of the topics in Chapters 9 and 10.

An Applications-Oriented Course

Chapter 9 contains selected applications of linear algebra that are mostly of a mathematical nature. Instructors who are interested in a wider variety of applications may want to consider the alternative version of this text, *Elementary Linear Algebra, Applications Version*, by Howard Anton and Chris Rorres. (ISBN 0-471-66959-8) That text provides applications to business, biology, engineering, economics, the social sciences, and the physical sciences.

ACKNOWLEDGEMENTS

I express my appreciation for the helpful guidance provided by the following people:

REVIEWERS AND CONTRIBUTORS

Marie Aratari, *Oakland Community College*
Nancy Childress, *Arizona State University*
Nancy Clarke, *Acadia University*
Aimee Ellington, *Virginia Commonwealth University*
William Greenberg, *Virginia Tech*
Molly Gregas, *Finger Lakes Community College*
Conrad Hewitt, *St. Jerome's University*
Sasho Kalajdzievski, *University of Manitoba*
Gregory Lewis, *University of Ontario Institute of Technology*
Sharon O'Donnell, *Chicago State University*
Mazi Shirvani, *University of Alberta*
Roxana Smarandache, *San Diego State University*
Edward Smerek, *Hiram College*
Earl Taft, *Rutgers University*
Angela Walters, *Capitol College*

Mathematical Advisors

Special thanks are due to two very talented mathematicians who read the manuscript in detail for technical accuracy and provided excellent advice on numerous pedagogical and mathematical matters.

Philip Riley, *James Madison University*
Laura Taalman, *James Madison University*

Special Contributions

The talents and dedication of many individuals are required to produce a book such as the one you now hold in your hands. The following people deserve special mention:

Jeffery J. Leader–for his outstanding work overseeing the implementation of numerous recommendations and improvements in this edition.

Chris Black, Ralph P. Grimaldi, and Marie Vanisko–for evaluating the exercise sets and making helpful recommendations.

Laurie Rosatone–for the consistent and enthusiastic support and direction she has provided this project.

Jennifer Battista–for the innumerable things she has done to make this edition a reality.

Anne Scanlan-Rohrer–for her essential role in overseeing day-to-day details of the editing stage of this project.

Kelly Boyle and Stacy French–for their assistance in obtaining pre-revision reviews.

Ken Santor–for his attention to detail and his superb job in managing this project.

Techsetters, Inc.–for once again providing beautiful typesetting and careful attention to detail.

Dawn Stanley–for a beautiful design and cover.

The Wiley Production Staff–with special thanks to Lucille Buonocore, Maddy Lesure, Sigmund Malinowski, and Ann Berlin for their efforts behind the scenes and for their support on many books over the years.

HOWARD ANTON

CONTENTS

Systems of Linear Equations and Matrices

CHAPTER CONTENTS

INTRODUCTION: Information in science and mathematics is often organized into rows and columns to form rectangular arrays, called "matrices" (plural of "matrix"). Matrices are often tables of numerical data that arise from physical observations, but they also occur in various mathematical contexts. For example, we shall see in this chapter that to solve a system of equations such as

$$5x + y = 3$$
$$2x - y = 4$$

all of the information required for the solution is embodied in the matrix

$$\begin{bmatrix} 5 & 1 & 3 \\ 2 & -1 & 4 \end{bmatrix}$$

and that the solution can be obtained by performing appropriate operations on this matrix. This is particularly important in developing computer programs to solve systems of linear equations because computers are well suited for manipulating arrays of numerical information. However, matrices are not simply a notational tool for solving systems of equations; they can be viewed as mathematical objects in their own right, and there is a rich and important theory associated with them that has a wide variety of applications. In this chapter we will begin the study of matrices.

1.1
INTRODUCTION TO SYSTEMS OF LINEAR EQUATIONS

Systems of linear algebraic equations and their solutions constitute one of the major topics studied in the course known as "linear algebra." In this first section we shall introduce some basic terminology and discuss a method for solving such systems.

Linear Equations

Any straight line in the xy-plane can be represented algebraically by an equation of the form

$$a_1 x + a_2 y = b$$

where a_1, a_2, and b are real constants and a_1 and a_2 are not both zero. An equation of this form is called a linear equation in the variables x and y. More generally, we define a ***linear equation*** in the n variables $x_1, x_2, \ldots, x_n$ to be one that can be expressed in the form

$$a_1 x_1 + a_2 x_2 + \cdots + a_n x_n = b$$

where $a_1, a_2, \ldots, a_n$, and b are real constants. The variables in a linear equation are sometimes called ***unknowns***.

EXAMPLE 1 Linear Equations

The equations

$$x + 3y = 7, \quad y = \tfrac{1}{2}x + 3z + 1, \quad \text{and} \quad x_1 - 2x_2 - 3x_3 + x_4 = 7$$

are linear. Observe that a linear equation does not involve any products or roots of variables. All variables occur only to the first power and do not appear as arguments for trigonometric, logarithmic, or exponential functions. The equations

$$x + 3\sqrt{y} = 5, \quad 3x + 2y - z + xz = 4, \quad \text{and} \quad y = \sin x$$

are *not* linear. ◆

A ***solution*** of a linear equation $a_1 x_1 + a_2 x_2 + \cdots + a_n x_n = b$ is a sequence of n numbers $s_1, s_2, \ldots, s_n$ such that the equation is satisfied when we substitute $x_1 = s_1$, $x_2 = s_2, \ldots, x_n = s_n$. The set of all solutions of the equation is called its ***solution set*** or sometimes the ***general solution*** of the equation.

EXAMPLE 2 Finding a Solution Set

Find the solution set of (a) $4x - 2y = 1$, and (b) $x_1 - 4x_2 + 7x_3 = 5$.

Solution (a)

To find solutions of (a), we can assign an arbitrary value to x and solve for y, or choose an arbitrary value for y and solve for x. If we follow the first approach and assign x an arbitrary value t, we obtain

$$x = t, \qquad y = 2t - \tfrac{1}{2}$$

These formulas describe the solution set in terms of an arbitrary number t, called a ***parameter***. Particular numerical solutions can be obtained by substituting specific values for t. For example, $t = 3$ yields the solution $x = 3$, $y = \frac{11}{2}$; and $t = -\frac{1}{2}$ yields the solution $x = -\frac{1}{2}$, $y = -\frac{3}{2}$.

If we follow the second approach and assign y the arbitrary value t, we obtain

$$x = \tfrac{1}{2}t + \tfrac{1}{4}, \qquad y = t$$

Although these formulas are different from those obtained above, they yield the same solution set as t varies over all possible real numbers. For example, the previous formulas gave the solution $x = 3$, $y = \frac{11}{2}$ when $t = 3$, whereas the formulas immediately above yield that solution when $t = \frac{11}{2}$.

Solution (b)

To find the solution set of (b), we can assign arbitrary values to any two variables and solve for the third variable. In particular, if we assign arbitrary values s and t to x_2 and x_3, respectively, and solve for x_1, we obtain

$$x_1 = 5 + 4s - 7t, \qquad x_2 = s, \qquad x_3 = t \quad \blacklozenge$$

Linear Systems

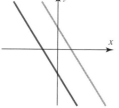

(*a*) No solution

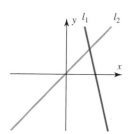

(*b*) One solution

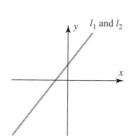

(*c*) Infinitely many solutions

Figure 1.1.1

A finite set of linear equations in the variables $x_1, x_2, \ldots, x_n$ is called a ***system of linear equations*** or a ***linear system***. A sequence of numbers $s_1, s_2, \ldots, s_n$ is called a ***solution*** of the system if $x_1 = s_1$, $x_2 = s_2, \ldots, x_n = s_n$ is a solution of every equation in the system. For example, the system

$$4x_1 - x_2 + 3x_3 = -1$$
$$3x_1 + x_2 + 9x_3 = -4$$

has the solution $x_1 = 1$, $x_2 = 2$, $x_3 = -1$ since these values satisfy both equations. However, $x_1 = 1$, $x_2 = 8$, $x_3 = 1$ is not a solution since these values satisfy only the first equation in the system.

Not all systems of linear equations have solutions. For example, if we multiply the second equation of the system

$$x + y = 4$$
$$2x + 2y = 6$$

by $\frac{1}{2}$, it becomes evident that there are no solutions since the resulting equivalent system

$$x + y = 4$$
$$x + y = 3$$

has contradictory equations.

A system of equations that has no solutions is said to be ***inconsistent***; if there is at least one solution of the system, it is called ***consistent***. To illustrate the possibilities that can occur in solving systems of linear equations, consider a general system of two linear equations in the unknowns x and y:

$$a_1 x + b_1 y = c_1 \quad (\textbf{\textit{a}}_1, \textbf{\textit{b}}_1 \text{ \textbf{not both zero}})$$
$$a_2 x + b_2 y = c_2 \quad (\textbf{\textit{a}}_2, \textbf{\textit{b}}_2 \text{ \textbf{not both zero}})$$

The graphs of these equations are lines; call them l_1 and l_2. Since a point (x, y) lies on a line if and only if the numbers x and y satisfy the equation of the line, the solutions of the system of equations correspond to points of intersection of l_1 and l_2. There are three possibilities, illustrated in Figure 1.1.1:

- The lines l_1 and l_2 may be parallel, in which case there is no intersection and consequently no solution to the system.

- The lines l_1 and l_2 may intersect at only one point, in which case the system has exactly one solution.
- The lines l_1 and l_2 may coincide, in which case there are infinitely many points of intersection and consequently infinitely many solutions to the system.

Although we have considered only two equations with two unknowns here, we will show later that the same three possibilities hold for arbitrary linear systems:

Every system of linear equations has no solutions, or has exactly one solution, or has infinitely many solutions.

An arbitrary system of m linear equations in n unknowns can be written as

$$a_{11}x_1 + a_{12}x_2 + \cdots + a_{1n}x_n = b_1$$
$$a_{21}x_1 + a_{22}x_2 + \cdots + a_{2n}x_n = b_2$$
$$\vdots \qquad \vdots \qquad \qquad \vdots \qquad \vdots$$
$$a_{m1}x_1 + a_{m2}x_2 + \cdots + a_{mn}x_n = b_m$$

where $x_1, x_2, \ldots, x_n$ are the unknowns and the subscripted a's and b's denote constants. For example, a general system of three linear equations in four unknowns can be written as

$$a_{11}x_1 + a_{12}x_2 + a_{13}x_3 + a_{14}x_4 = b_1$$
$$a_{21}x_1 + a_{22}x_2 + a_{23}x_3 + a_{24}x_4 = b_2$$
$$a_{31}x_1 + a_{32}x_2 + a_{33}x_3 + a_{34}x_4 = b_3$$

The double subscripting on the coefficients of the unknowns is a useful device that is used to specify the location of the coefficient in the system. The first subscript on the coefficient a_{ij} indicates the equation in which the coefficient occurs, and the second subscript indicates which unknown it multiplies. Thus, a_{12} is in the first equation and multiplies unknown x_2.

Augmented Matrices

If we mentally keep track of the location of the $+$'s, the x's, and the $=$'s, a system of m linear equations in n unknowns can be abbreviated by writing only the rectangular array of numbers:

$$\begin{bmatrix} a_{11} & a_{12} & \cdots & a_{1n} & b_1 \\ a_{21} & a_{22} & \cdots & a_{2n} & b_2 \\ \vdots & \vdots & & \vdots & \vdots \\ a_{m1} & a_{m2} & \cdots & a_{mn} & b_m \end{bmatrix}$$

This is called the ***augmented matrix*** for the system. (The term *matrix* is used in mathematics to denote a rectangular array of numbers. Matrices arise in many contexts, which we will consider in more detail in later sections.) For example, the augmented matrix for the system of equations

$$x_1 + x_2 + 2x_3 = 9$$
$$2x_1 + 4x_2 - 3x_3 = 1$$
$$3x_1 + 6x_2 - 5x_3 = 0$$

is

$$\begin{bmatrix} 1 & 1 & 2 & 9 \\ 2 & 4 & -3 & 1 \\ 3 & 6 & -5 & 0 \end{bmatrix}$$

REMARK When constructing an augmented matrix, we must write the unknowns in the same order in each equation, and the constants must be on the right.

The basic method for solving a system of linear equations is to replace the given system by a new system that has the same solution set but is easier to solve. This new system is generally obtained in a series of steps by applying the following three types of operations to eliminate unknowns systematically:

1. Multiply an equation through by a nonzero constant.
2. Interchange two equations.
3. Add a multiple of one equation to another.

Since the rows (horizontal lines) of an augmented matrix correspond to the equations in the associated system, these three operations correspond to the following operations on the rows of the augmented matrix:

1. Multiply a row through by a nonzero constant.
2. Interchange two rows.
3. Add a multiple of one row to another row.

Elementary Row Operations

These are called *elementary row operations*. The following example illustrates how these operations can be used to solve systems of linear equations. Since a systematic procedure for finding solutions will be derived in the next section, it is not necessary to worry about how the steps in this example were selected. The main effort at this time should be devoted to understanding the computations and the discussion.

EXAMPLE 3 Using Elementary Row Operations

In the left column below we solve a system of linear equations by operating on the equations in the system, and in the right column we solve the same system by operating on the rows of the augmented matrix.

$$x + y + 2z = 9$$
$$2x + 4y - 3z = 1$$
$$3x + 6y - 5z = 0$$

$$\begin{bmatrix} 1 & 1 & 2 & 9 \\ 2 & 4 & -3 & 1 \\ 3 & 6 & -5 & 0 \end{bmatrix}$$

Add −2 times the first equation to the second to obtain

$$x + y + 2z = 9$$
$$2y - 7z = -17$$
$$3x + 6y - 5z = 0$$

Add −2 times the first row to the second to obtain

$$\begin{bmatrix} 1 & 1 & 2 & 9 \\ 0 & 2 & -7 & -17 \\ 3 & 6 & -5 & 0 \end{bmatrix}$$

Add −3 times the first equation to the third to obtain

$$x + y + 2z = 9$$
$$2y - 7z = -17$$
$$3y - 11z = -27$$

Add −3 times the first row to the third to obtain

$$\begin{bmatrix} 1 & 1 & 2 & 9 \\ 0 & 2 & -7 & -17 \\ 0 & 3 & -11 & -27 \end{bmatrix}$$

Multiply the second equation by $\frac{1}{2}$ to obtain

$$x + y + 2z = 9$$
$$y - \tfrac{7}{2}z = -\tfrac{17}{2}$$
$$3y - 11z = -27$$

Multiply the second row by $\frac{1}{2}$ to obtain

$$\begin{bmatrix} 1 & 1 & 2 & 9 \\ 0 & 1 & -\tfrac{7}{2} & -\tfrac{17}{2} \\ 0 & 3 & -11 & -27 \end{bmatrix}$$

Add -3 times the second equation to the third to obtain

$$x + y + 2z = 9$$
$$y - \tfrac{7}{2}z = -\tfrac{17}{2}$$
$$-\tfrac{1}{2}z = -\tfrac{3}{2}$$

Add -3 times the second row to the third to obtain

$$\begin{bmatrix} 1 & 1 & 2 & 9 \\ 0 & 1 & -\tfrac{7}{2} & -\tfrac{17}{2} \\ 0 & 0 & -\tfrac{1}{2} & -\tfrac{3}{2} \end{bmatrix}$$

Multiply the third equation by -2 to obtain

$$x + y + 2z = 9$$
$$y - \tfrac{7}{2}z = -\tfrac{17}{2}$$
$$z = 3$$

Multiply the third row by -2 to obtain

$$\begin{bmatrix} 1 & 1 & 2 & 9 \\ 0 & 1 & -\tfrac{7}{2} & -\tfrac{17}{2} \\ 0 & 0 & 1 & 3 \end{bmatrix}$$

Add -1 times the second equation to the first to obtain

$$x + \tfrac{11}{2}z = \tfrac{35}{2}$$
$$y - \tfrac{7}{2}z = -\tfrac{17}{2}$$
$$z = 3$$

Add -1 times the second row to the first to obtain

$$\begin{bmatrix} 1 & 0 & \tfrac{11}{2} & \tfrac{35}{2} \\ 0 & 1 & -\tfrac{7}{2} & -\tfrac{17}{2} \\ 0 & 0 & 1 & 3 \end{bmatrix}$$

Add $-\tfrac{11}{2}$ times the third equation to the first and $\tfrac{7}{2}$ times the third equation to the second to obtain

$$x = 1$$
$$y = 2$$
$$z = 3$$

Add $-\tfrac{11}{2}$ times the third row to the first and $\tfrac{7}{2}$ times the third row to the second to obtain

$$\begin{bmatrix} 1 & 0 & 0 & 1 \\ 0 & 1 & 0 & 2 \\ 0 & 0 & 1 & 3 \end{bmatrix}$$

The solution $x = 1$, $y = 2$, $z = 3$ is now evident. ◆

EXERCISE SET 1.1

1. Which of the following are linear equations in x_1, x_2, and x_3?
   (a) $x_1 + 5x_2 - \sqrt{2}x_3 = 1$ (b) $x_1 + 3x_2 + x_1x_3 = 2$
   (c) $x_1 = -7x_2 + 3x_3$ (d) $x_1^{-2} + x_2 + 8x_3 = 5$
   (e) $x_1^{3/5} - 2x_2 + x_3 = 4$ (f) $\pi x_1 - \sqrt{2}x_2 + \tfrac{1}{3}x_3 = 7^{1/3}$

2. Given that k is a constant, which of the following are linear equations?
   (a) $x_1 - x_2 + x_3 = \sin k$ (b) $kx_1 - \tfrac{1}{k}x_2 = 9$ (c) $2^k x_1 + 7x_2 - x_3 = 0$

3. Find the solution set of each of the following linear equations.
   (a) $7x - 5y = 3$ (b) $3x_1 - 5x_2 + 4x_3 = 7$
   (c) $-8x_1 + 2x_2 - 5x_3 + 6x_4 = 1$ (d) $3v - 8w + 2x - y + 4z = 0$

4. Find the augmented matrix for each of the following systems of linear equations.
   (a) $3x_1 - 2x_2 = -1$
   $4x_1 + 5x_2 = 3$
   $7x_1 + 3x_2 = 2$
   (b) $2x_1 + 2x_3 = 1$
   $3x_1 - x_2 + 4x_3 = 7$
   $6x_1 + x_2 - x_3 = 0$
   (c) $x_1 + 2x_2 - x_4 + x_5 = 1$
   $3x_2 + x_3 - x_5 = 2$
   $x_3 + 7x_4 = 1$
   (d) $x_1 = 1$
   $x_2 = 2$
   $x_3 = 3$

5. Find a system of linear equations corresponding to the augmented matrix.

(a) $\begin{bmatrix} 2 & 0 & 0 \\ 3 & -4 & 0 \\ 0 & 1 & 1 \end{bmatrix}$

(b) $\begin{bmatrix} 3 & 0 & -2 & 5 \\ 7 & 1 & 4 & -3 \\ 0 & -2 & 1 & 7 \end{bmatrix}$

(c) $\begin{bmatrix} 7 & 2 & 1 & -3 & 5 \\ 1 & 2 & 4 & 0 & 1 \end{bmatrix}$

(d) $\begin{bmatrix} 1 & 0 & 0 & 0 & 7 \\ 0 & 1 & 0 & 0 & -2 \\ 0 & 0 & 1 & 0 & 3 \\ 0 & 0 & 0 & 1 & 4 \end{bmatrix}$

6. (a) Find a linear equation in the variables x and y that has the general solution $x = 5 + 2t$, $y = t$.

(b) Show that $x = t$, $y = \frac{1}{2}t - \frac{5}{2}$ is also the general solution of the equation in part (a).

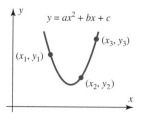

$y = ax^2 + bx + c$

(x_3, y_3)

(x_1, y_1)

(x_2, y_2)

Figure Ex-7

7. The curve $y = ax^2 + bx + c$ shown in the accompanying figure passes through the points (x_1, y_1), (x_2, y_2), and (x_3, y_3). Show that the coefficients a, b, and c are a solution of the system of linear equations whose augmented matrix is

$$\begin{bmatrix} x_1^2 & x_1 & 1 & y_1 \\ x_2^2 & x_2 & 1 & y_2 \\ x_3^2 & x_3 & 1 & y_3 \end{bmatrix}$$

8. Consider the system of equations

$$x + y + 2z = a$$
$$x \quad\;\; + \; z = b$$
$$2x + y + 3z = c$$

Show that for this system to be consistent, the constants a, b, and c must satisfy $c = a + b$.

9. Show that if the linear equations $x_1 + kx_2 = c$ and $x_1 + lx_2 = d$ have the same solution set, then the equations are identical.

10. Show that the elementary row operations do not affect the solution set of a linear system.

Discussion & Discovery

11. For which value(s) of the constant k does the system

$$x - \;\; y = 3$$
$$2x - 2y = k$$

have no solutions? Exactly one solution? Infinitely many solutions? Explain your reasoning.

12. Consider the system of equations

$$ax + by = k$$
$$cx + dy = l$$
$$ex + fy = m$$

Indicate what we can say about the relative positions of the lines $ax + by = k$, $cx + dy = l$, and $ex + fy = m$ when

(a) the system has no solutions.

(b) the system has exactly one solution.

(c) the system has infinitely many solutions.

13. If the system of equations in Exercise 12 is consistent, explain why at least one equation can be discarded from the system without altering the solution set.

14. If $k = l = m = 0$ in Exercise 12, explain why the system must be consistent. What can be said about the point of intersection of the three lines if the system has exactly one solution?

15. We could also define elementary column operations in analogy with the elementary row operations. What can you say about the effect of elementary column operations on the solution set of a linear system? How would you interpret the effects of elementary column operations?

1.2
GAUSSIAN ELIMINATION

In this section we shall develop a systematic procedure for solving systems of linear equations. The procedure is based on the idea of reducing the augmented matrix of a system to another augmented matrix that is simple enough that the solution of the system can be found by inspection.

Echelon Forms

In Example 3 of the last section, we solved a linear system in the unknowns x, y, and z by reducing the augmented matrix to the form

$$\begin{bmatrix} 1 & 0 & 0 & 1 \\ 0 & 1 & 0 & 2 \\ 0 & 0 & 1 & 3 \end{bmatrix}$$

from which the solution $x = 1$, $y = 2$, $z = 3$ became evident. This is an example of a matrix that is in *reduced row-echelon form*. To be of this form, a matrix must have the following properties:

1. If a row does not consist entirely of zeros, then the first nonzero number in the row is a 1. We call this a *leading 1*.

2. If there are any rows that consist entirely of zeros, then they are grouped together at the bottom of the matrix.

3. In any two successive rows that do not consist entirely of zeros, the leading 1 in the lower row occurs farther to the right than the leading 1 in the higher row.

4. Each column that contains a leading 1 has zeros everywhere else in that column.

A matrix that has the first three properties is said to be in *row-echelon form*. (Thus, a matrix in reduced row-echelon form is of necessity in row-echelon form, but not conversely.)

EXAMPLE 1 Row-Echelon and Reduced Row-Echelon Form

The following matrices are in reduced row-echelon form.

$$\begin{bmatrix} 1 & 0 & 0 & 4 \\ 0 & 1 & 0 & 7 \\ 0 & 0 & 1 & -1 \end{bmatrix}, \quad \begin{bmatrix} 1 & 0 & 0 \\ 0 & 1 & 0 \\ 0 & 0 & 1 \end{bmatrix}, \quad \begin{bmatrix} 0 & 1 & -2 & 0 & 1 \\ 0 & 0 & 0 & 1 & 3 \\ 0 & 0 & 0 & 0 & 0 \\ 0 & 0 & 0 & 0 & 0 \end{bmatrix}, \quad \begin{bmatrix} 0 & 0 \\ 0 & 0 \end{bmatrix}$$

The following matrices are in row-echelon form.

$$\begin{bmatrix} 1 & 4 & -3 & 7 \\ 0 & 1 & 6 & 2 \\ 0 & 0 & 1 & 5 \end{bmatrix}, \quad \begin{bmatrix} 1 & 1 & 0 \\ 0 & 1 & 0 \\ 0 & 0 & 0 \end{bmatrix}, \quad \begin{bmatrix} 0 & 1 & 2 & 6 & 0 \\ 0 & 0 & 1 & -1 & 0 \\ 0 & 0 & 0 & 0 & 1 \end{bmatrix}$$

We leave it for you to confirm that each of the matrices in this example satisfies all of the requirements for its stated form. ◆

EXAMPLE 2 **More on Row-Echelon and Reduced Row-Echelon Form**

As the last example illustrates, a matrix in row-echelon form has zeros below each leading 1, whereas a matrix in reduced row-echelon form has zeros below *and above* each leading 1. Thus, with any real numbers substituted for the $*$'s, all matrices of the following types are in row-echelon form:

$$\begin{bmatrix} 1 & * & * & * \\ 0 & 1 & * & * \\ 0 & 0 & 1 & * \\ 0 & 0 & 0 & 1 \end{bmatrix}, \quad \begin{bmatrix} 1 & * & * & * \\ 0 & 1 & * & * \\ 0 & 0 & 1 & * \\ 0 & 0 & 0 & 0 \end{bmatrix},$$

$$\begin{bmatrix} 1 & * & * & * \\ 0 & 1 & * & * \\ 0 & 0 & 0 & 0 \\ 0 & 0 & 0 & 0 \end{bmatrix}, \quad \begin{bmatrix} 0 & 1 & * & * & * & * & * & * & * & * \\ 0 & 0 & 0 & 1 & * & * & * & * & * & * \\ 0 & 0 & 0 & 0 & 1 & * & * & * & * & * \\ 0 & 0 & 0 & 0 & 0 & 1 & * & * & * & * \\ 0 & 0 & 0 & 0 & 0 & 0 & 0 & 0 & 1 & * \end{bmatrix}$$

Moreover, all matrices of the following types are in reduced row-echelon form:

$$\begin{bmatrix} 1 & 0 & 0 & 0 \\ 0 & 1 & 0 & 0 \\ 0 & 0 & 1 & 0 \\ 0 & 0 & 0 & 1 \end{bmatrix}, \quad \begin{bmatrix} 1 & 0 & 0 & * \\ 0 & 1 & 0 & * \\ 0 & 0 & 1 & * \\ 0 & 0 & 0 & 0 \end{bmatrix},$$

$$\begin{bmatrix} 1 & 0 & * & * \\ 0 & 1 & * & * \\ 0 & 0 & 0 & 0 \\ 0 & 0 & 0 & 0 \end{bmatrix}, \quad \begin{bmatrix} 0 & 1 & * & 0 & 0 & 0 & * & * & 0 & * \\ 0 & 0 & 0 & 1 & 0 & 0 & * & * & 0 & * \\ 0 & 0 & 0 & 0 & 1 & 0 & * & * & 0 & * \\ 0 & 0 & 0 & 0 & 0 & 1 & * & * & 0 & * \\ 0 & 0 & 0 & 0 & 0 & 0 & 0 & 0 & 1 & * \end{bmatrix} \blacklozenge$$

If, by a sequence of elementary row operations, the augmented matrix for a system of linear equations is put in reduced row-echelon form, then the solution set of the system will be evident by inspection or after a few simple steps. The next example illustrates this situation.

EXAMPLE 3 **Solutions of Four Linear Systems**

Suppose that the augmented matrix for a system of linear equations has been reduced by row operations to the given reduced row-echelon form. Solve the system.

(a) $\begin{bmatrix} 1 & 0 & 0 & 5 \\ 0 & 1 & 0 & -2 \\ 0 & 0 & 1 & 4 \end{bmatrix}$ (b) $\begin{bmatrix} 1 & 0 & 0 & 4 & -1 \\ 0 & 1 & 0 & 2 & 6 \\ 0 & 0 & 1 & 3 & 2 \end{bmatrix}$

(c) $\begin{bmatrix} 1 & 6 & 0 & 0 & 4 & -2 \\ 0 & 0 & 1 & 0 & 3 & 1 \\ 0 & 0 & 0 & 1 & 5 & 2 \\ 0 & 0 & 0 & 0 & 0 & 0 \end{bmatrix}$ (d) $\begin{bmatrix} 1 & 0 & 0 & 0 \\ 0 & 1 & 2 & 0 \\ 0 & 0 & 0 & 1 \end{bmatrix}$

Solution (a)

The corresponding system of equations is

$$
\begin{aligned}
x_1 & & & = & 5 \\
& x_2 & & = & -2 \\
& & x_3 & = & 4
\end{aligned}
$$

By inspection, $x_1 = 5$, $x_2 = -2$, $x_3 = 4$.

Solution (b)

The corresponding system of equations is

$$
\begin{aligned}
x_1 & & & + 4x_4 = -1 \\
& x_2 & & + 2x_4 = 6 \\
& & x_3 & + 3x_4 = 2
\end{aligned}
$$

Since x_1, x_2, and x_3 correspond to leading 1's in the augmented matrix, we call them *leading variables* or *pivots*. The nonleading variables (in this case x_4) are called *free variables*. Solving for the leading variables in terms of the free variable gives

$$
\begin{aligned}
x_1 &= -1 - 4x_4 \\
x_2 &= 6 - 2x_4 \\
x_3 &= 2 - 3x_4
\end{aligned}
$$

From this form of the equations we see that the free variable x_4 can be assigned an arbitrary value, say t, which then determines the values of the leading variables x_1, x_2, and x_3. Thus there are infinitely many solutions, and the general solution is given by the formulas

$$
x_1 = -1 - 4t, \qquad x_2 = 6 - 2t, \qquad x_3 = 2 - 3t, \qquad x_4 = t
$$

Solution (c)

The row of zeros leads to the equation $0x_1 + 0x_2 + 0x_3 + 0x_4 + 0x_5 = 0$, which places no restrictions on the solutions (why?). Thus, we can omit this equation and write the corresponding system as

$$
\begin{aligned}
x_1 + 6x_2 & & & + 4x_5 = -2 \\
& x_3 & & + 3x_5 = 1 \\
& & x_4 & + 5x_5 = 2
\end{aligned}
$$

Here the leading variables are x_1, x_3, and x_4, and the free variables are x_2 and x_5. Solving for the leading variables in terms of the free variables gives

$$
\begin{aligned}
x_1 &= -2 - 6x_2 - 4x_5 \\
x_3 &= 1 - 3x_5 \\
x_4 &= 2 - 5x_5
\end{aligned}
$$

Since x_5 can be assigned an arbitrary value, t, and x_2 can be assigned an arbitrary value, s, there are infinitely many solutions. The general solution is given by the formulas

$$
x_1 = -2 - 6s - 4t, \qquad x_2 = s, \qquad x_3 = 1 - 3t, \qquad x_4 = 2 - 5t, \qquad x_5 = t
$$

Solution (d)

The last equation in the corresponding system of equations is

$$
0x_1 + 0x_2 + 0x_3 = 1
$$

Since this equation cannot be satisfied, there is no solution to the system. ◆

Elimination Methods

We have just seen how easy it is to solve a system of linear equations once its augmented matrix is in reduced row-echelon form. Now we shall give a step-by-step *elimination* procedure that can be used to reduce any matrix to reduced row-echelon form. As we state each step in the procedure, we shall illustrate the idea by reducing the following matrix to reduced row-echelon form.

$$\begin{bmatrix} 0 & 0 & -2 & 0 & 7 & 12 \\ 2 & 4 & -10 & 6 & 12 & 28 \\ 2 & 4 & -5 & 6 & -5 & -1 \end{bmatrix}$$

Step 1. Locate the leftmost column that does not consist entirely of zeros.

$$\begin{bmatrix} 0 & 0 & -2 & 0 & 7 & 12 \\ 2 & 4 & -10 & 6 & 12 & 28 \\ 2 & 4 & -5 & 6 & -5 & -1 \end{bmatrix}$$

↑
└── **Leftmost nonzero column**

Step 2. Interchange the top row with another row, if necessary, to bring a nonzero entry to the top of the column found in Step 1.

$$\begin{bmatrix} 2 & 4 & -10 & 6 & 12 & 28 \\ 0 & 0 & -2 & 0 & 7 & 12 \\ 2 & 4 & -5 & 6 & -5 & -1 \end{bmatrix}$$

◄── The first and second rows in the preceding matrix were interchanged.

Step 3. If the entry that is now at the top of the column found in Step 1 is a, multiply the first row by $1/a$ in order to introduce a leading 1.

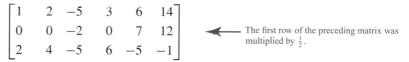

$$\begin{bmatrix} 1 & 2 & -5 & 3 & 6 & 14 \\ 0 & 0 & -2 & 0 & 7 & 12 \\ 2 & 4 & -5 & 6 & -5 & -1 \end{bmatrix}$$

◄── The first row of the preceding matrix was multiplied by $\frac{1}{2}$.

Step 4. Add suitable multiples of the top row to the rows below so that all entries below the leading 1 become zeros.

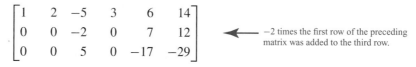

$$\begin{bmatrix} 1 & 2 & -5 & 3 & 6 & 14 \\ 0 & 0 & -2 & 0 & 7 & 12 \\ 0 & 0 & 5 & 0 & -17 & -29 \end{bmatrix}$$

◄── -2 times the first row of the preceding matrix was added to the third row.

Step 5. Now cover the top row in the matrix and begin again with Step 1 applied to the submatrix that remains. Continue in this way until the *entire* matrix is in row-echelon form.

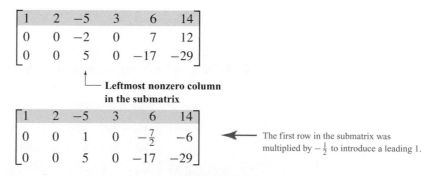

$$\begin{bmatrix} 1 & 2 & -5 & 3 & 6 & 14 \\ 0 & 0 & -2 & 0 & 7 & 12 \\ 0 & 0 & 5 & 0 & -17 & -29 \end{bmatrix}$$

↑
└── **Leftmost nonzero column**
  in the submatrix

$$\begin{bmatrix} 1 & 2 & -5 & 3 & 6 & 14 \\ 0 & 0 & 1 & 0 & -\frac{7}{2} & -6 \\ 0 & 0 & 5 & 0 & -17 & -29 \end{bmatrix}$$

◄── The first row in the submatrix was multiplied by $-\frac{1}{2}$ to introduce a leading 1.

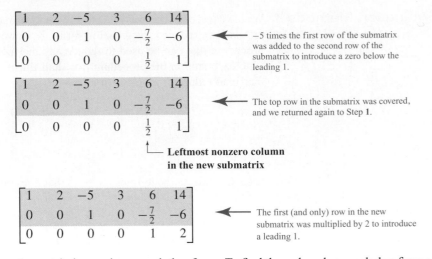

$$\begin{bmatrix} 1 & 2 & -5 & 3 & 6 & 14 \\ 0 & 0 & 1 & 0 & -\frac{7}{2} & -6 \\ 0 & 0 & 0 & 0 & \frac{1}{2} & 1 \end{bmatrix}$$

← −5 times the first row of the submatrix was added to the second row of the submatrix to introduce a zero below the leading 1.

$$\begin{bmatrix} 1 & 2 & -5 & 3 & 6 & 14 \\ 0 & 0 & 1 & 0 & -\frac{7}{2} & -6 \\ 0 & 0 & 0 & 0 & \frac{1}{2} & 1 \end{bmatrix}$$

← The top row in the submatrix was covered, and we returned again to Step **1**.

↑ **Leftmost nonzero column in the new submatrix**

$$\begin{bmatrix} 1 & 2 & -5 & 3 & 6 & 14 \\ 0 & 0 & 1 & 0 & -\frac{7}{2} & -6 \\ 0 & 0 & 0 & 0 & 1 & 2 \end{bmatrix}$$

← The first (and only) row in the new submatrix was multiplied by 2 to introduce a leading 1.

The *entire* matrix is now in row-echelon form. To find the reduced row-echelon form we need the following additional step.

Step 6. Beginning with the last nonzero row and working upward, add suitable multiples of each row to the rows above to introduce zeros above the leading 1's.

$$\begin{bmatrix} 1 & 2 & -5 & 3 & 6 & 14 \\ 0 & 0 & 1 & 0 & 0 & 1 \\ 0 & 0 & 0 & 0 & 1 & 2 \end{bmatrix}$$

← $\frac{7}{2}$ times the third row of the preceding matrix was added to the second row.

$$\begin{bmatrix} 1 & 2 & -5 & 3 & 0 & 2 \\ 0 & 0 & 1 & 0 & 0 & 1 \\ 0 & 0 & 0 & 0 & 1 & 2 \end{bmatrix}$$

← −6 times the third row was added to the first row.

$$\begin{bmatrix} 1 & 2 & 0 & 3 & 0 & 7 \\ 0 & 0 & 1 & 0 & 0 & 1 \\ 0 & 0 & 0 & 0 & 1 & 2 \end{bmatrix}$$

← 5 times the second row was added to the first row.

The last matrix is in reduced row-echelon form.

If we use only the first five steps, the above procedure produces a row-echelon form and is called ***Gaussian elimination***. Carrying the procedure through to the sixth step and producing a matrix in reduced row-echelon form is called ***Gauss–Jordan elimination***.

REMARK It can be shown that *every matrix has a unique reduced row-echelon form*; that is, one will arrive at the same reduced row-echelon form for a given matrix no matter how the row operations are varied. (A proof of this result can be found in the article "The Reduced Row Echelon Form of a Matrix Is Unique: A Simple Proof," by Thomas Yuster, *Mathematics Magazine*, Vol. 57, No. 2, 1984, pp. 93–94.) In contrast, *a row-echelon form of a given matrix is not unique*: different sequences of row operations can produce different row-echelon forms.

EXAMPLE 4   Gauss–Jordan Elimination

Solve by Gauss–Jordan elimination.

$$\begin{aligned} x_1 + 3x_2 - 2x_3 \qquad\qquad + 2x_5 \qquad\qquad &= \;\; 0 \\ 2x_1 + 6x_2 - 5x_3 - \;\; 2x_4 + 4x_5 - \;\; 3x_6 &= -1 \\ 5x_3 + 10x_4 \qquad + 15x_6 &= \;\; 5 \\ 2x_1 + 6x_2 \qquad\quad + \;\; 8x_4 + 4x_5 + 18x_6 &= \;\; 6 \end{aligned}$$

Karl Friedrich Gauss

Wilhelm Jordan

Karl Friedrich Gauss *(1777–1855)* was a German mathematician and scientist. Sometimes called the "prince of mathematicians," Gauss ranks with Isaac Newton and Archimedes as one of the three greatest mathematicians who ever lived. In the entire history of mathematics there may never have been a child so precocious as Gauss—by his own account he worked out the rudiments of arithmetic before he could talk. One day, before he was even three years old, his genius became apparent to his parents in a very dramatic way. His father was preparing the weekly payroll for the laborers under his charge while the boy watched quietly from a corner. At the end of the long and tedious calculation, Gauss informed his father that there was an error in the result and stated the answer, which he had worked out in his head. To the astonishment of his parents, a check of the computations showed Gauss to be correct!

In his doctoral dissertation Gauss gave the first complete proof of the fundamental theorem of algebra, which states that every polynomial equation has as many solutions as its degree. At age 19 he solved a problem that baffled Euclid, inscribing a regular polygon of seventeen sides in a circle using straightedge and compass; and in 1801, at age 24, he published his first masterpiece, *Disquisitiones Arithmeticae*, considered by many to be one of the most brilliant achievements in mathematics. In that paper Gauss systematized the study of number theory (properties of the integers) and formulated the basic concepts that form the foundation of the subject.

Among his myriad achievements, Gauss discovered the Gaussian or "bell-shaped" curve that is fundamental in probability, gave the first geometric interpretation of complex numbers and established their fundamental role in mathematics, developed methods of characterizing surfaces intrinsically by means of the curves that they contain, developed the theory of conformal (angle-preserving) maps, and discovered non-Euclidean geometry 30 years before the ideas were published by others. In physics he made major contributions to the theory of lenses and capillary action, and with Wilhelm Weber he did fundamental work in electromagnetism. Gauss invented the heliotrope, bifilar magnetometer, and an electrotelegraph.

Gauss, who was deeply religious and aristocratic in demeanor, mastered foreign languages with ease, read extensively, and enjoyed mineralogy and botany as hobbies. He disliked teaching and was usually cool and discouraging to other mathematicians, possibly because he had already anticipated their work. It has been said that if Gauss had published all of his discoveries, the current state of mathematics would be advanced by 50 years. He was without a doubt the greatest mathematician of the modern era.

Wilhelm Jordan *(1842–1899)* was a German engineer who specialized in geodesy. His contribution to solving linear systems appeared in his popular book, *Handbuch der Vermessungskunde* (*Handbook of Geodesy*), in 1888.

Solution

The augmented matrix for the system is

$$\begin{bmatrix} 1 & 3 & -2 & 0 & 2 & 0 & 0 \\ 2 & 6 & -5 & -2 & 4 & -3 & -1 \\ 0 & 0 & 5 & 10 & 0 & 15 & 5 \\ 2 & 6 & 0 & 8 & 4 & 18 & 6 \end{bmatrix}$$

Adding -2 times the first row to the second and fourth rows gives

$$\begin{bmatrix} 1 & 3 & -2 & 0 & 2 & 0 & 0 \\ 0 & 0 & -1 & -2 & 0 & -3 & -1 \\ 0 & 0 & 5 & 10 & 0 & 15 & 5 \\ 0 & 0 & 4 & 8 & 0 & 18 & 6 \end{bmatrix}$$

Solution

This is the system in Example 3 of Section 1.1. In that example we converted the augmented matrix

$$\begin{bmatrix} 1 & 1 & 2 & 9 \\ 2 & 4 & -3 & 1 \\ 3 & 6 & -5 & 0 \end{bmatrix}$$

to the row-echelon form

$$\begin{bmatrix} 1 & 1 & 2 & 9 \\ 0 & 1 & -\frac{7}{2} & -\frac{17}{2} \\ 0 & 0 & 1 & 3 \end{bmatrix}$$

The system corresponding to this matrix is

$$\begin{aligned} x + y + 2z &= 9 \\ y - \tfrac{7}{2}z &= -\tfrac{17}{2} \\ z &= 3 \end{aligned}$$

Solving for the leading variables yields

$$\begin{aligned} x &= 9 - y - 2z \\ y &= -\tfrac{17}{2} + \tfrac{7}{2}z \\ z &= 3 \end{aligned}$$

Substituting the bottom equation into those above yields

$$\begin{aligned} x &= 3 - y \\ y &= 2 \\ z &= 3 \end{aligned}$$

and substituting the second equation into the top yields $x = 1$, $y = 2$, $z = 3$. This agrees with the result found by Gauss–Jordan elimination in Example 3 of Section 1.1. ◆

Homogeneous Linear Systems

A system of linear equations is said to be *homogeneous* if the constant terms are all zero; that is, the system has the form

$$\begin{aligned} a_{11}x_1 + a_{12}x_2 + \cdots + a_{1n}x_n &= 0 \\ a_{21}x_1 + a_{22}x_2 + \cdots + a_{2n}x_n &= 0 \\ \vdots \qquad \vdots \qquad\qquad \vdots \qquad \vdots \\ a_{m1}x_1 + a_{m2}x_2 + \cdots + a_{mn}x_n &= 0 \end{aligned}$$

Every homogeneous system of linear equations is consistent, since all such systems have $x_1 = 0, x_2 = 0, \ldots, x_n = 0$ as a solution. This solution is called the *trivial solution*; if there are other solutions, they are called *nontrivial solutions*.

Because a homogeneous linear system always has the trivial solution, there are only two possibilities for its solutions:

- The system has only the trivial solution.
- The system has infinitely many solutions in addition to the trivial solution.

In the special case of a homogeneous linear system of two equations in two unknowns, say

$$\begin{aligned} a_1 x + b_1 y &= 0 \quad (a_1, b_1 \text{ not both zero}) \\ a_2 x + b_2 y &= 0 \quad (a_2, b_2 \text{ not both zero}) \end{aligned}$$

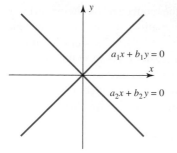

$a_1x + b_1y = 0$

$a_2x + b_2y = 0$

(*a*) Only the trivial solution

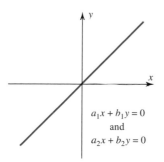

$a_1x + b_1y = 0$
and
$a_2x + b_2y = 0$

(*b*) Infinitely many solutions

Figure 1.2.1

the graphs of the equations are lines through the origin, and the trivial solution corresponds to the point of intersection at the origin (Figure 1.2.1).

There is one case in which a homogeneous system is assured of having nontrivial solutions—namely, whenever the system involves more unknowns than equations. To see why, consider the following example of four equations in five unknowns.

EXAMPLE 7   Gauss–Jordan Elimination

Solve the following homogeneous system of linear equations by using Gauss–Jordan elimination.

$$\begin{aligned}
2x_1 + 2x_2 - x_3 \phantom{{}+2x_3} + x_5 &= 0 \\
-x_1 - x_2 + 2x_3 - 3x_4 + x_5 &= 0 \\
x_1 + x_2 - 2x_3 \phantom{{}-3x_4} - x_5 &= 0 \\
x_3 + x_4 + x_5 &= 0
\end{aligned} \tag{1}$$

Solution

The augmented matrix for the system is

$$\begin{bmatrix}
2 & 2 & -1 & 0 & 1 & 0 \\
-1 & -1 & 2 & -3 & 1 & 0 \\
1 & 1 & -2 & 0 & -1 & 0 \\
0 & 0 & 1 & 1 & 1 & 0
\end{bmatrix}$$

Reducing this matrix to reduced row-echelon form, we obtain

$$\begin{bmatrix}
1 & 1 & 0 & 0 & 1 & 0 \\
0 & 0 & 1 & 0 & 1 & 0 \\
0 & 0 & 0 & 1 & 0 & 0 \\
0 & 0 & 0 & 0 & 0 & 0
\end{bmatrix}$$

The corresponding system of equations is

$$\begin{aligned}
x_1 + x_2 \phantom{{}+x_3} + x_5 &= 0 \\
x_3 + x_5 &= 0 \\
x_4 \phantom{{}+x_5} &= 0
\end{aligned} \tag{2}$$

Solving for the leading variables yields

$$\begin{aligned}
x_1 &= -x_2 - x_5 \\
x_3 &= -x_5 \\
x_4 &= 0
\end{aligned}$$

Thus, the general solution is

$$x_1 = -s - t, \qquad x_2 = s, \qquad x_3 = -t, \qquad x_4 = 0, \qquad x_5 = t$$

Note that the trivial solution is obtained when $s = t = 0$. ◆

Example 7 illustrates two important points about solving homogeneous systems of linear equations. First, none of the three elementary row operations alters the final column of zeros in the augmented matrix, so the system of equations corresponding to the reduced row-echelon form of the augmented matrix must also be a homogeneous system [see system (2)]. Second, depending on whether the reduced row-echelon form of the augmented matrix has any zero rows, the number of equations in the reduced system is the same as or less than the number of equations in the original system [compare systems (1) and (2)]. Thus, if the given homogeneous system has m equations in n unknowns with $m < n$, and if there are r nonzero rows in the reduced row-echelon form of the augmented matrix, we will have $r < n$. It follows that the system of equations corresponding to the reduced row-echelon form of the augmented matrix will have the form

$$
\begin{aligned}
\cdots\ x_{k_1} && + \Sigma(\ ) &= 0 \\
\cdots\ x_{k_2} && + \Sigma(\ ) &= 0 \\
\cdots\ \ddots && \vdots & \\
x_{k_r} + \Sigma(\ ) &= 0 &&
\end{aligned}
\tag{3}
$$

where $x_{k_1}, x_{k_2}, \ldots, x_{k_r}$ are the leading variables and $\Sigma(\ )$ denotes sums (possibly all different) that involve the $n - r$ free variables [compare system (3) with system (2) above]. Solving for the leading variables gives

$$
\begin{aligned}
x_{k_1} &= -\Sigma(\ ) \\
x_{k_2} &= -\Sigma(\ ) \\
&\ \vdots \\
x_{k_r} &= -\Sigma(\ )
\end{aligned}
$$

As in Example 7, we can assign arbitrary values to the free variables on the right-hand side and thus obtain infinitely many solutions to the system.

In summary, we have the following important theorem.

THEOREM 1.2.1

> *A homogeneous system of linear equations with more unknowns than equations has infinitely many solutions.*

REMARK Note that Theorem 1.2.1 applies only to homogeneous systems. A nonhomogeneous system with more unknowns than equations need not be consistent (Exercise 28); however, if the system is consistent, it will have infinitely many solutions. This will be proved later.

Computer Solution of Linear Systems

In applications it is not uncommon to encounter large linear systems that must be solved by computer. Most computer algorithms for solving such systems are based on Gaussian elimination or Gauss–Jordan elimination, but the basic procedures are often modified to deal with such issues as

- Reducing roundoff errors
- Minimizing the use of computer memory space
- Solving the system with maximum speed

Some of these matters will be considered in Chapter 9. For hand computations, fractions are an annoyance that often cannot be avoided. However, in some cases it is possible to avoid them by varying the elementary row operations in the right way. Thus, once the methods of Gaussian elimination and Gauss–Jordan elimination have been mastered, the reader may wish to vary the steps in specific problems to avoid fractions (see Exercise 18).

REMARK Since Gauss–Jordan elimination avoids the use of back-substitution, it would seem that this method would be the more efficient of the two methods we have considered.

It can be argued that this statement is true for solving small systems by hand since Gauss–Jordan elimination actually involves less writing. However, for large systems of equations, it has been shown that the Gauss–Jordan elimination method requires about 50% more operations than Gaussian elimination. This is an important consideration when one is working on computers.

EXERCISE SET 1.2

1. Which of the following 3×3 matrices are in reduced row-echelon form?

(a) $\begin{bmatrix} 1 & 0 & 0 \\ 0 & 1 & 0 \\ 0 & 0 & 1 \end{bmatrix}$
(b) $\begin{bmatrix} 1 & 0 & 0 \\ 0 & 1 & 0 \\ 0 & 0 & 0 \end{bmatrix}$
(c) $\begin{bmatrix} 0 & 1 & 0 \\ 0 & 0 & 1 \\ 0 & 0 & 0 \end{bmatrix}$
(d) $\begin{bmatrix} 1 & 0 & 0 \\ 0 & 0 & 1 \\ 0 & 0 & 0 \end{bmatrix}$

(e) $\begin{bmatrix} 1 & 0 & 0 \\ 0 & 0 & 0 \\ 0 & 0 & 1 \end{bmatrix}$
(f) $\begin{bmatrix} 0 & 1 & 0 \\ 1 & 0 & 0 \\ 0 & 0 & 0 \end{bmatrix}$
(g) $\begin{bmatrix} 1 & 1 & 0 \\ 0 & 1 & 0 \\ 0 & 0 & 0 \end{bmatrix}$
(h) $\begin{bmatrix} 1 & 0 & 2 \\ 0 & 1 & 3 \\ 0 & 0 & 0 \end{bmatrix}$

(i) $\begin{bmatrix} 0 & 0 & 1 \\ 0 & 0 & 0 \\ 0 & 0 & 0 \end{bmatrix}$
(j) $\begin{bmatrix} 0 & 0 & 0 \\ 0 & 0 & 0 \\ 0 & 0 & 0 \end{bmatrix}$

2. Which of the following 3×3 matrices are in row-echelon form?

(a) $\begin{bmatrix} 1 & 0 & 0 \\ 0 & 1 & 0 \\ 0 & 0 & 1 \end{bmatrix}$
(b) $\begin{bmatrix} 1 & 2 & 0 \\ 0 & 1 & 0 \\ 0 & 0 & 0 \end{bmatrix}$
(c) $\begin{bmatrix} 1 & 0 & 0 \\ 0 & 1 & 0 \\ 0 & 2 & 0 \end{bmatrix}$

(d) $\begin{bmatrix} 1 & 3 & 4 \\ 0 & 0 & 1 \\ 0 & 0 & 0 \end{bmatrix}$
(e) $\begin{bmatrix} 1 & 5 & -3 \\ 0 & 1 & 1 \\ 0 & 0 & 0 \end{bmatrix}$
(f) $\begin{bmatrix} 1 & 2 & 3 \\ 0 & 0 & 0 \\ 0 & 0 & 1 \end{bmatrix}$

3. In each part determine whether the matrix is in row-echelon form, reduced row-echelon form, both, or neither.

(a) $\begin{bmatrix} 1 & 2 & 0 & 3 & 0 \\ 0 & 0 & 1 & 1 & 0 \\ 0 & 0 & 0 & 0 & 1 \\ 0 & 0 & 0 & 0 & 0 \end{bmatrix}$
(b) $\begin{bmatrix} 1 & 0 & 0 & 5 \\ 0 & 0 & 1 & 3 \\ 0 & 1 & 0 & 4 \end{bmatrix}$
(c) $\begin{bmatrix} 1 & 0 & 3 & 1 \\ 0 & 1 & 2 & 4 \end{bmatrix}$

(d) $\begin{bmatrix} 1 & -7 & 5 & 5 \\ 0 & 1 & 3 & 2 \end{bmatrix}$
(e) $\begin{bmatrix} 1 & 3 & 0 & 2 & 0 \\ 1 & 0 & 2 & 2 & 0 \\ 0 & 0 & 0 & 0 & 1 \\ 0 & 0 & 0 & 0 & 0 \end{bmatrix}$
(f) $\begin{bmatrix} 0 & 0 \\ 0 & 0 \\ 0 & 0 \end{bmatrix}$

4. In each part suppose that the augmented matrix for a system of linear equations has been reduced by row operations to the given reduced row-echelon form. Solve the system.

(a) $\begin{bmatrix} 1 & 0 & 0 & -3 \\ 0 & 1 & 0 & 0 \\ 0 & 0 & 1 & 7 \end{bmatrix}$
(b) $\begin{bmatrix} 1 & 0 & 0 & -7 & 8 \\ 0 & 1 & 0 & 3 & 2 \\ 0 & 0 & 1 & 1 & -5 \end{bmatrix}$

(c) $\begin{bmatrix} 1 & -6 & 0 & 0 & 3 & -2 \\ 0 & 0 & 1 & 0 & 4 & 7 \\ 0 & 0 & 0 & 1 & 5 & 8 \\ 0 & 0 & 0 & 0 & 0 & 0 \end{bmatrix}$
(d) $\begin{bmatrix} 1 & -3 & 0 & 0 \\ 0 & 0 & 1 & 0 \\ 0 & 0 & 0 & 1 \end{bmatrix}$

5. In each part suppose that the augmented matrix for a system of linear equations has been reduced by row operations to the given row-echelon form. Solve the system.

(a) $\begin{bmatrix} 1 & -3 & 4 & 7 \\ 0 & 1 & 2 & 2 \\ 0 & 0 & 1 & 5 \end{bmatrix}$

(b) $\begin{bmatrix} 1 & 0 & 8 & -5 & 6 \\ 0 & 1 & 4 & -9 & 3 \\ 0 & 0 & 1 & 1 & 2 \end{bmatrix}$

(c) $\begin{bmatrix} 1 & 7 & -2 & 0 & -8 & -3 \\ 0 & 0 & 1 & 1 & 6 & 5 \\ 0 & 0 & 0 & 1 & 3 & 9 \\ 0 & 0 & 0 & 0 & 0 & 0 \end{bmatrix}$

(d) $\begin{bmatrix} 1 & -3 & 7 & 1 \\ 0 & 1 & 4 & 0 \\ 0 & 0 & 0 & 1 \end{bmatrix}$

6. Solve each of the following systems by Gauss–Jordan elimination.

(a)
$$\begin{aligned} x_1 + x_2 + 2x_3 &= 8 \\ -x_1 - 2x_2 + 3x_3 &= 1 \\ 3x_1 - 7x_2 + 4x_3 &= 10 \end{aligned}$$

(b)
$$\begin{aligned} 2x_1 + 2x_2 + 2x_3 &= 0 \\ -2x_1 + 5x_2 + 2x_3 &= 1 \\ 8x_1 + x_2 + 4x_3 &= -1 \end{aligned}$$

(c)
$$\begin{aligned} x - y + 2z - w &= -1 \\ 2x + y - 2z - 2w &= -2 \\ -x + 2y - 4z + w &= 1 \\ 3x \qquad\qquad - 3w &= -3 \end{aligned}$$

(d)
$$\begin{aligned} -2b + 3c &= 1 \\ 3a + 6b - 3c &= -2 \\ 6a + 6b + 3c &= 5 \end{aligned}$$

7. Solve each of the systems in Exercise 6 by Gaussian elimination.

8. Solve each of the following systems by Gauss–Jordan elimination.

(a)
$$\begin{aligned} 2x_1 - 3x_2 &= -2 \\ 2x_1 + x_2 &= 1 \\ 3x_1 + 2x_2 &= 1 \end{aligned}$$

(b)
$$\begin{aligned} 3x_1 + 2x_2 - x_3 &= -15 \\ 5x_1 + 3x_2 + 2x_3 &= 0 \\ 3x_1 + x_2 + 3x_3 &= 11 \\ -6x_1 - 4x_2 + 2x_3 &= 30 \end{aligned}$$

(c)
$$\begin{aligned} 4x_1 - 8x_2 &= 12 \\ 3x_1 - 6x_2 &= 9 \\ -2x_1 + 4x_2 &= -6 \end{aligned}$$

(d)
$$\begin{aligned} 10y - 4z + w &= 1 \\ x + 4y - z + w &= 2 \\ 3x + 2y + z + 2w &= 5 \\ -2x - 8y + 2z - 2w &= -4 \\ x - 6y + 3z \qquad &= 1 \end{aligned}$$

9. Solve each of the systems in Exercise 8 by Gaussian elimination.

10. Solve each of the following systems by Gauss–Jordan elimination.

(a)
$$\begin{aligned} 5x_1 - 2x_2 + 6x_3 &= 0 \\ -2x_1 + x_2 + 3x_3 &= 1 \end{aligned}$$

(b)
$$\begin{aligned} x_1 - 2x_2 + x_3 - 4x_4 &= 1 \\ x_1 + 3x_2 + 7x_3 + 2x_4 &= 2 \\ x_1 - 12x_2 - 11x_3 - 16x_4 &= 5 \end{aligned}$$

(c)
$$\begin{aligned} w + 2x - y &= 4 \\ x - y &= 3 \\ w + 3x - 2y &= 7 \\ 2u + 4v + w + 7x &= 7 \end{aligned}$$

11. Solve each of the systems in Exercise 10 by Gaussian elimination.

12. Without using pencil and paper, determine which of the following homogeneous systems have nontrivial solutions.

(a)
$$\begin{aligned} 2x_1 - 3x_2 + 4x_3 - x_4 &= 0 \\ 7x_1 + x_2 - 8x_3 + 9x_4 &= 0 \\ 2x_1 + 8x_2 + x_3 - x_4 &= 0 \end{aligned}$$

(b)
$$\begin{aligned} x_1 + 3x_2 - x_3 &= 0 \\ x_2 - 8x_3 &= 0 \\ 4x_3 &= 0 \end{aligned}$$

(c)
$$\begin{aligned} a_{11}x_1 + a_{12}x_2 + a_{13}x_3 &= 0 \\ a_{21}x_1 + a_{22}x_2 + a_{23}x_3 &= 0 \end{aligned}$$

(d)
$$\begin{aligned} 3x_1 - 2x_2 &= 0 \\ 6x_1 - 4x_2 &= 0 \end{aligned}$$

13. Solve the following homogeneous systems of linear equations by any method.

(a)
$$\begin{aligned} 2x_1 + x_2 + 3x_3 &= 0 \\ x_1 + 2x_2 &= 0 \\ x_2 + x_3 &= 0 \end{aligned}$$

(b)
$$\begin{aligned} 3x_1 + x_2 + x_3 + x_4 &= 0 \\ 5x_1 - x_2 + x_3 - x_4 &= 0 \end{aligned}$$

(c)
$$2x + 2y + 4z = 0$$
$$w \quad - y - 3z = 0$$
$$2w + 3x + y + z = 0$$
$$-2w + x + 3y - 2z = 0$$

14. Solve the following homogeneous systems of linear equations by any method.

(a)
$$2x - y - 3z = 0$$
$$-x + 2y - 3z = 0$$
$$x + y + 4z = 0$$

(b)
$$v + 3w - 2x = 0$$
$$2u + v - 4w + 3x = 0$$
$$2u + 3v + 2w - x = 0$$
$$-4u - 3v + 5w - 4x = 0$$

(c)
$$x_1 + 3x_2 \quad + x_4 = 0$$
$$x_1 + 4x_2 + 2x_3 \quad = 0$$
$$- 2x_2 - 2x_3 - x_4 = 0$$
$$2x_1 - 4x_2 + x_3 + x_4 = 0$$
$$x_1 - 2x_2 - x_3 + x_4 = 0$$

15. Solve the following systems by any method.

(a)
$$2I_1 - I_2 + 3I_3 + 4I_4 = 9$$
$$I_1 \quad - 2I_3 + 7I_4 = 11$$
$$3I_1 - 3I_2 + I_3 + 5I_4 = 8$$
$$2I_1 + I_2 + 4I_3 + 4I_4 = 10$$

(b)
$$Z_3 + Z_4 + Z_5 = 0$$
$$-Z_1 - Z_2 + 2Z_3 - 3Z_4 + Z_5 = 0$$
$$Z_1 + Z_2 - 2Z_3 \quad - Z_5 = 0$$
$$2Z_1 + 2Z_2 - Z_3 \quad + Z_5 = 0$$

16. Solve the following systems, where a, b, and c are constants.

(a)
$$2x + y = a$$
$$3x + 6y = b$$

(b)
$$x_1 + x_2 + x_3 = a$$
$$2x_1 \quad + 2x_3 = b$$
$$3x_2 + 3x_3 = c$$

17. For which values of a will the following system have no solutions? Exactly one solution? Infinitely many solutions?
$$x + 2y - 3z = 4$$
$$3x - y + 5z = 2$$
$$4x + y + (a^2 - 14)z = a + 2$$

18. Reduce
$$\begin{bmatrix} 2 & 1 & 3 \\ 0 & -2 & -29 \\ 3 & 4 & 5 \end{bmatrix}$$
to reduced row-echelon form.

19. Find two different row-echelon forms of
$$\begin{bmatrix} 1 & 3 \\ 2 & 7 \end{bmatrix}$$

20. Solve the following system of nonlinear equations for the unknown angles α, β, and γ, where $0 \le \alpha \le 2\pi$, $0 \le \beta \le 2\pi$, and $0 \le \gamma < \pi$.
$$2\sin\alpha - \cos\beta + 3\tan\gamma = 3$$
$$4\sin\alpha + 2\cos\beta - 2\tan\gamma = 2$$
$$6\sin\alpha - 3\cos\beta + \tan\gamma = 9$$

21. Show that the following nonlinear system has 18 solutions if $0 \le \alpha \le 2\pi$, $0 \le \beta \le 2\pi$, and $0 \le \gamma < 2\pi$.
$$\sin\alpha + 2\cos\beta + 3\tan\gamma = 0$$
$$2\sin\alpha + 5\cos\beta + 3\tan\gamma = 0$$
$$-\sin\alpha - 5\cos\beta + 5\tan\gamma = 0$$

22. For which value(s) of λ does the system of equations
$$(\lambda - 3)x + y = 0$$
$$x + (\lambda - 3)y = 0$$
have nontrivial solutions?

23. Solve the system

$$2x_1 - x_2 \qquad = \lambda x_1$$
$$2x_1 - x_2 + x_3 = \lambda x_2$$
$$-2x_1 + 2x_2 + x_3 = \lambda x_3$$

for x_1, x_2, and x_3 in the two cases $\lambda = 1$, $\lambda = 2$.

24. Solve the following system for x, y, and z.

$$\frac{1}{x} + \frac{2}{y} - \frac{4}{z} = 1$$
$$\frac{2}{x} + \frac{3}{y} + \frac{8}{z} = 0$$
$$-\frac{1}{x} + \frac{9}{y} + \frac{10}{z} = 5$$

25. Find the coefficients a, b, c, and d so that the curve shown in the accompanying figure is the graph of the equation $y = ax^3 + bx^2 + cx + d$.

26. Find coefficients a, b, c, and d so that the curve shown in the accompanying figure is given by the equation $ax^2 + ay^2 + bx + cy + d = 0$.

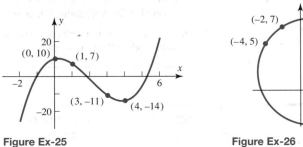

Figure Ex-25

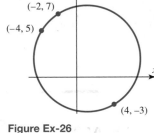

Figure Ex-26

27. (a) Show that if $ad - bc \neq 0$, then the reduced row-echelon form of

$$\begin{bmatrix} a & b \\ c & d \end{bmatrix} \quad \text{is} \quad \begin{bmatrix} 1 & 0 \\ 0 & 1 \end{bmatrix}$$

(b) Use part (a) to show that the system

$$ax + by = k$$
$$cx + dy = l$$

has exactly one solution when $ad - bc \neq 0$.

28. Find an inconsistent linear system that has more unknowns than equations.

Discussion & Discovery

29. Indicate all possible reduced row-echelon forms of

(a) $\begin{bmatrix} a & b & c \\ d & e & f \\ g & h & i \end{bmatrix}$ (b) $\begin{bmatrix} a & b & c & d \\ e & f & g & h \\ i & j & k & l \\ m & n & p & q \end{bmatrix}$

30. Consider the system of equations

$$ax + by = 0$$
$$cx + dy = 0$$
$$ex + fy = 0$$

Discuss the relative positions of the lines $ax + by = 0$, $cx + dy = 0$, and $ex + fy = 0$ when (a) the system has only the trivial solution, and (b) the system has nontrivial solutions.

31. Indicate whether the statement is always true or sometimes false. Justify your answer by giving a logical argument or a counterexample.

    (a) If a matrix is reduced to reduced row-echelon form by two different sequences of elementary row operations, the resulting matrices will be different.

    (b) If a matrix is reduced to row-echelon form by two different sequences of elementary row operations, the resulting matrices might be different.

    (c) If the reduced row-echelon form of the augmented matrix for a linear system has a row of zeros, then the system must have infinitely many solutions.

    (d) If three lines in the xy-plane are sides of a triangle, then the system of equations formed from their equations has three solutions, one corresponding to each vertex.

32. Indicate whether the statement is always true or sometimes false. Justify your answer by giving a logical argument or a counterexample.

    (a) A linear system of three equations in five unknowns must be consistent.

    (b) A linear system of five equations in three unknowns cannot be consistent.

    (c) If a linear system of n equations in n unknowns has n leading 1's in the reduced row-echelon form of its augmented matrix, then the system has exactly one solution.

    (d) If a linear system of n equations in n unknowns has two equations that are multiples of one another, then the system is inconsistent.

1.3
MATRICES AND MATRIX OPERATIONS

Rectangular arrays of real numbers arise in many contexts other than as augmented matrices for systems of linear equations. In this section we begin our study of matrix theory by giving some of the fundamental definitions of the subject. We shall see how matrices can be combined through the arithmetic operations of addition, subtraction, and multiplication.

Matrix Notation and Terminology

In Section 1.2 we used rectangular arrays of numbers, called *augmented matrices*, to abbreviate systems of linear equations. However, rectangular arrays of numbers occur in other contexts as well. For example, the following rectangular array with three rows and seven columns might describe the number of hours that a student spent studying three subjects during a certain week:

	Mon.	Tues.	Wed.	Thurs.	Fri.	Sat.	Sun.
Math	2	3	2	4	1	4	2
History	0	3	1	4	3	2	2
Language	4	1	3	1	0	0	2

If we suppress the headings, then we are left with the following rectangular array of numbers with three rows and seven columns, called a "matrix":

$$\begin{bmatrix} 2 & 3 & 2 & 4 & 1 & 4 & 2 \\ 0 & 3 & 1 & 4 & 3 & 2 & 2 \\ 4 & 1 & 3 & 1 & 0 & 0 & 2 \end{bmatrix}$$

More generally, we make the following definition.

> **DEFINITION**
>
> A **matrix** is a rectangular array of numbers. The numbers in the array are called the **entries** in the matrix.

EXAMPLE 1 Examples of Matrices

Some examples of matrices are

$$\begin{bmatrix} 1 & 2 \\ 3 & 0 \\ -1 & 4 \end{bmatrix}, \qquad \begin{bmatrix} 2 & 1 & 0 & -3 \end{bmatrix}, \qquad \begin{bmatrix} e & \pi & -\sqrt{2} \\ 0 & \frac{1}{2} & 1 \\ 0 & 0 & 0 \end{bmatrix}, \qquad \begin{bmatrix} 1 \\ 3 \end{bmatrix}, \qquad [4] \quad \blacklozenge$$

The **size** of a matrix is described in terms of the number of rows (horizontal lines) and columns (vertical lines) it contains. For example, the first matrix in Example 1 has three rows and two columns, so its size is 3 by 2 (written 3×2). In a size description, the first number always denotes the number of rows, and the second denotes the number of columns. The remaining matrices in Example 1 have sizes 1×4, 3×3, 2×1, and 1×1, respectively. A matrix with only one column is called a **column matrix** (or a **column vector**), and a matrix with only one row is called a **row matrix** (or a **row vector**). Thus, in Example 1 the 2×1 matrix is a column matrix, the 1×4 matrix is a row matrix, and the 1×1 matrix is both a row matrix and a column matrix. (The term *vector* has another meaning that we will discuss in subsequent chapters.)

REMARK   It is common practice to omit the brackets from a 1×1 matrix. Thus we might write 4 rather than [4]. Although this makes it impossible to tell whether 4 denotes the number "four" or the 1×1 matrix whose entry is "four," this rarely causes problems, since it is usually possible to tell which is meant from the context in which the symbol appears.

We shall use capital letters to denote matrices and lowercase letters to denote numerical quantities; thus we might write

$$A = \begin{bmatrix} 2 & 1 & 7 \\ 3 & 4 & 2 \end{bmatrix} \quad \text{or} \quad C = \begin{bmatrix} a & b & c \\ d & e & f \end{bmatrix}$$

When discussing matrices, it is common to refer to numerical quantities as **scalars**. Unless stated otherwise, *scalars will be real numbers*; complex scalars will be considered in Chapter 10.

The entry that occurs in row i and column j of a matrix A will be denoted by a_{ij}. Thus a general 3×4 matrix might be written as

$$A = \begin{bmatrix} a_{11} & a_{12} & a_{13} & a_{14} \\ a_{21} & a_{22} & a_{23} & a_{24} \\ a_{31} & a_{32} & a_{33} & a_{34} \end{bmatrix}$$

and a general $m \times n$ matrix as

$$A = \begin{bmatrix} a_{11} & a_{12} & \cdots & a_{1n} \\ a_{21} & a_{22} & \cdots & a_{2n} \\ \vdots & \vdots & & \vdots \\ a_{m1} & a_{m2} & \cdots & a_{mn} \end{bmatrix} \tag{1}$$

When compactness of notation is desired, the preceding matrix can be written as

$$[a_{ij}]_{m \times n} \quad \text{or} \quad [a_{ij}]$$

the first notation being used when it is important in the discussion to know the size, and the second being used when the size need not be emphasized. Usually, we shall match the letter denoting a matrix with the letter denoting its entries; thus, for a matrix B we would generally use b_{ij} for the entry in row i and column j, and for a matrix C we would use the notation c_{ij}.

The entry in row i and column j of a matrix A is also commonly denoted by the symbol $(A)_{ij}$. Thus, for matrix (1) above, we have

$$(A)_{ij} = a_{ij}$$

and for the matrix

$$A = \begin{bmatrix} 2 & -3 \\ 7 & 0 \end{bmatrix}$$

we have $(A)_{11} = 2$, $(A)_{12} = -3$, $(A)_{21} = 7$, and $(A)_{22} = 0$.

Row and column matrices are of special importance, and it is common practice to denote them by boldface lowercase letters rather than capital letters. For such matrices, double subscripting of the entries is unnecessary. Thus a general $1 \times n$ row matrix $\mathbf{a}$ and a general $m \times 1$ column matrix $\mathbf{b}$ would be written as

$$\mathbf{a} = \begin{bmatrix} a_1 & a_2 & \cdots & a_n \end{bmatrix} \quad \text{and} \quad \mathbf{b} = \begin{bmatrix} b_1 \\ b_2 \\ \vdots \\ b_m \end{bmatrix}$$

A matrix A with n rows and n columns is called a ***square matrix of order n***, and the shaded entries $a_{11}, a_{22}, \ldots, a_{nn}$ in (2) are said to be on the ***main diagonal*** of A.

$$\begin{bmatrix} a_{11} & a_{12} & \cdots & a_{1n} \\ a_{21} & a_{22} & \cdots & a_{2n} \\ \vdots & \vdots & & \vdots \\ a_{n1} & a_{n2} & \cdots & a_{nn} \end{bmatrix} \tag{2}$$

Operations on Matrices

So far, we have used matrices to abbreviate the work in solving systems of linear equations. For other applications, however, it is desirable to develop an "arithmetic of matrices" in which matrices can be added, subtracted, and multiplied in a useful way. The remainder of this section will be devoted to developing this arithmetic.

DEFINITION

Two matrices are defined to be ***equal*** if they have the same size and their corresponding entries are equal.

In matrix notation, if $A = [a_{ij}]$ and $B = [b_{ij}]$ have the same size, then $A = B$ if and only if $(A)_{ij} = (B)_{ij}$, or, equivalently, $a_{ij} = b_{ij}$ for all i and j.

12. (a) Show that if AB and BA are both defined, then AB and BA are square matrices.

(b) Show that if A is an $m \times n$ matrix and $A(BA)$ is defined, then B is an $n \times m$ matrix.

13. In each part, find matrices A, $\mathbf{x}$, and $\mathbf{b}$ that express the given system of linear equations as a single matrix equation $A\mathbf{x} = \mathbf{b}$.

(a)
$$
\begin{aligned}
2x_1 - 3x_2 + 5x_3 &= 7 \\
9x_1 - x_2 + x_3 &= -1 \\
x_1 + 5x_2 + 4x_3 &= 0
\end{aligned}
$$

(b)
$$
\begin{aligned}
4x_1 \qquad - 3x_3 + x_4 &= 1 \\
5x_1 + x_2 \qquad - 8x_4 &= 3 \\
2x_1 - 5x_2 + 9x_3 - x_4 &= 0 \\
3x_2 - x_3 + 7x_4 &= 2
\end{aligned}
$$

14. In each part, express the matrix equation as a system of linear equations.

(a)
$$
\begin{bmatrix} 3 & -1 & 2 \\ 4 & 3 & 7 \\ -2 & 1 & 5 \end{bmatrix}
\begin{bmatrix} x_1 \\ x_2 \\ x_3 \end{bmatrix}
=
\begin{bmatrix} 2 \\ -1 \\ 4 \end{bmatrix}
$$

(b)
$$
\begin{bmatrix} 3 & -2 & 0 & 1 \\ 5 & 0 & 2 & -2 \\ 3 & 1 & 4 & 7 \\ -2 & 5 & 1 & 6 \end{bmatrix}
\begin{bmatrix} w \\ x \\ y \\ z \end{bmatrix}
=
\begin{bmatrix} 0 \\ 0 \\ 0 \\ 0 \end{bmatrix}
$$

15. If A and B are partitioned into submatrices, for example,

$$
A = \left[\begin{array}{c|c} A_{11} & A_{12} \\ \hline A_{21} & A_{22} \end{array} \right]
\quad \text{and} \quad
B = \left[\begin{array}{c|c} B_{11} & B_{12} \\ \hline B_{21} & B_{22} \end{array} \right]
$$

then AB can be expressed as

$$
AB = \left[\begin{array}{c|c} A_{11}B_{11} + A_{12}B_{21} & A_{11}B_{12} + A_{12}B_{22} \\ \hline A_{21}B_{11} + A_{22}B_{21} & A_{21}B_{12} + A_{22}B_{22} \end{array} \right]
$$

provided the sizes of the submatrices of A and B are such that the indicated operations can be performed. This method of multiplying partitioned matrices is called **block multiplication**. In each part, compute the product by block multiplication. Check your results by multiplying directly.

(a) $A = \left[\begin{array}{cc|cc} -1 & 2 & 1 & 5 \\ 0 & -3 & 4 & 2 \\ \hline 1 & 5 & 6 & 1 \end{array} \right]$, $B = \left[\begin{array}{cc|c} 2 & 1 & 4 \\ -3 & 5 & 2 \\ \hline 7 & -1 & 5 \\ 0 & 3 & -3 \end{array} \right]$

(b) $A = \left[\begin{array}{ccc|c} -1 & 2 & 1 & 5 \\ 0 & -3 & 4 & 2 \\ \hline 1 & 5 & 6 & 1 \end{array} \right]$, $B = \left[\begin{array}{cc|c} 2 & 1 & 4 \\ -3 & 5 & 2 \\ 7 & -1 & 5 \\ \hline 0 & 3 & -3 \end{array} \right]$

16. Adapt the method of Exercise 15 to compute the following products by block multiplication.

(a) $\left[\begin{array}{cc|cc} 3 & -1 & 0 & -3 \\ 2 & 1 & 4 & 5 \end{array} \right] \left[\begin{array}{cc|c} 2 & -4 & 1 \\ 3 & 0 & 2 \\ \hline 1 & -3 & 5 \\ 2 & 1 & 4 \end{array} \right]$

(b) $\left[\begin{array}{c|c} 2 & -5 \\ 1 & 3 \\ 0 & 5 \\ \hline 1 & 4 \end{array} \right] \left[\begin{array}{c|ccc} 2 & -1 & 3 & -4 \\ 0 & 1 & 5 & 7 \end{array} \right]$

(c) $\left[\begin{array}{ccc|cc} 1 & 0 & 0 & 0 & 0 \\ 0 & 1 & 0 & 0 & 0 \\ 0 & 0 & 1 & 0 & 0 \\ \hline 0 & 0 & 0 & 2 & 0 \\ 0 & 0 & 0 & -1 & 2 \end{array} \right] \left[\begin{array}{cc} 3 & 3 \\ -1 & 4 \\ 1 & 5 \\ \hline 2 & -2 \\ 1 & 6 \end{array} \right]$

17. In each part, determine whether block multiplication can be used to compute AB from the given partitions. If so, compute the product by block multiplication.

Note See Exercise 15.

(a) $A = \begin{bmatrix} -1 & 2 & 1 & | & 5 \\ 0 & -3 & 4 & | & 2 \\ \hline 1 & 5 & 6 & | & 1 \end{bmatrix}$, $B = \begin{bmatrix} 2 & | & 1 & | & 4 \\ -3 & | & 5 & | & 2 \\ \hline 7 & | & -1 & | & 5 \\ 0 & | & 3 & | & -3 \end{bmatrix}$

(b) $A = \begin{bmatrix} -1 & 2 & 1 & 5 \\ 0 & -3 & 4 & 2 \\ 1 & 5 & 6 & 1 \end{bmatrix}$, $B = \begin{bmatrix} 2 & | & 1 & | & 4 \\ \hline -3 & | & 5 & | & 2 \\ 7 & | & -1 & | & 5 \\ 0 & | & 3 & | & -3 \end{bmatrix}$

18. (a) Show that if A has a row of zeros and B is any matrix for which AB is defined, then AB also has a row of zeros.

(b) Find a similar result involving a column of zeros.

19. Let A be any $m \times n$ matrix and let 0 be the $m \times n$ matrix each of whose entries is zero. Show that if $kA = 0$, then $k = 0$ or $A = 0$.

20. Let I be the $n \times n$ matrix whose entry in row i and column j is

$$\begin{cases} 1 & \text{if} \quad i = j \\ 0 & \text{if} \quad i \neq j \end{cases}$$

Show that $AI = IA = A$ for every $n \times n$ matrix A.

21. In each part, find a 6×6 matrix $[a_{ij}]$ that satisfies the stated condition. Make your answers as general as possible by using letters rather than specific numbers for the nonzero entries.

(a) $a_{ij} = 0$ if $i \neq j$ (b) $a_{ij} = 0$ if $i > j$

(c) $a_{ij} = 0$ if $i < j$ (d) $a_{ij} = 0$ if $|i - j| > 1$

22. Find the 4×4 matrix $A = [a_{ij}]$ whose entries satisfy the stated condition.

(a) $a_{ij} = i + j$ (b) $a_{ij} = i^{j-1}$ (c) $a_{ij} = \begin{cases} 1 & \text{if} \quad |i - j| > 1 \\ -1 & \text{if} \quad |i - j| \leq 1 \end{cases}$

23. Consider the function $y = f(x)$ defined for 2×1 matrices x by $y = Ax$, where

$$A = \begin{bmatrix} 1 & 1 \\ 0 & 1 \end{bmatrix}$$

Plot $f(x)$ together with x in each case below. How would you describe the action of f?

(a) $x = \begin{pmatrix} 1 \\ 1 \end{pmatrix}$ (b) $x = \begin{pmatrix} 2 \\ 0 \end{pmatrix}$ (c) $x = \begin{pmatrix} 4 \\ 3 \end{pmatrix}$ (d) $x = \begin{pmatrix} 2 \\ -2 \end{pmatrix}$

24. Let A be a $n \times m$ matrix. Show that if the function $y = f(x)$ defined for $m \times 1$ matrices x by $y = Ax$ satisfies the linearity property, then $f(\alpha w + \beta z) = \alpha f(w) + \beta f(z)$ for any real numbers α and β and any $m \times 1$ matrices w and z.

25. Prove: If A and B are $n \times n$ matrices, then $\text{tr}(A + B) = \text{tr}(A) + \text{tr}(B)$.

Discussion & Discovery

26. Describe three different methods for computing a matrix product, and illustrate the methods by computing some product AB three different ways.

27. How many 3×3 matrices A can you find such that

$$A \begin{bmatrix} x \\ y \\ z \end{bmatrix} = \begin{bmatrix} x + y \\ x - y \\ 0 \end{bmatrix}$$

for all choices of x, y, and z?

28. How many 3×3 matrices A can you find such that

$$A \begin{bmatrix} x \\ y \\ z \end{bmatrix} = \begin{bmatrix} xy \\ 0 \\ 0 \end{bmatrix}$$

for all choices of x, y, and z?

29. A matrix B is said to be a **square root** of a matrix A if $BB = A$.

(a) Find two square roots of $A = \begin{bmatrix} 2 & 2 \\ 2 & 2 \end{bmatrix}$.

(b) How many different square roots can you find of $A = \begin{bmatrix} 5 & 0 \\ 0 & 9 \end{bmatrix}$?

(c) Do you think that every 2×2 matrix has at least one square root? Explain your reasoning.

30. Let 0 denote a 2×2 matrix, each of whose entries is zero.

(a) Is there a 2×2 matrix A such that $A \neq 0$ and $AA = 0$? Justify your answer.

(b) Is there a 2×2 matrix A such that $A \neq 0$ and $AA = A$? Justify your answer.

31. Indicate whether the statement is always true or sometimes false. Justify your answer with a logical argument or a counterexample.

(a) The expressions $\mathrm{tr}(AA^T)$ and $\mathrm{tr}(A^TA)$ are always defined, regardless of the size of A.

(b) $\mathrm{tr}(AA^T) = \mathrm{tr}(A^TA)$ for every matrix A.

(c) If the first column of A has all zeros, then so does the first column of every product AB.

(d) If the first row of A has all zeros, then so does the first row of every product AB.

32. Indicate whether the statement is always true or sometimes false. Justify your answer with a logical argument or a counterexample.

(a) If A is a square matrix with two identical rows, then AA has two identical rows.

(b) If A is a square matrix and AA has a column of zeros, then A must have a column of zeros.

(c) If B is an $n \times n$ matrix whose entries are positive even integers, and if A is an $n \times n$ matrix whose entries are positive integers, then the entries of AB and BA are positive even integers.

(d) If the matrix sum $AB + BA$ is defined, then A and B must be square.

33. Suppose the array

$$\begin{bmatrix} 4 & 3 & 3 \\ 2 & 1 & 0 \\ 4 & 4 & 2 \end{bmatrix}$$

represents the orders placed by three individuals at a fast-food restaurant. The first person orders 4 burgers, 3 sodas, and 3 fries; the second orders 2 burgers and 1 soda, and the third orders 4 burgers, 4 sodas, and 2 fries. Burgers cost \$2 each, sodas \$1 each, and fries \$1.50 each.

(a) Argue that the amounts owed by these persons may be represented as a function $y = f(x)$, where $f(x)$ is equal to the array given above times a certain vector.

(b) Compute the amounts owed in this case by performing the appropriate multiplication.

(c) Change the matrix for the case in which the second person orders an additional soda and 2 fries, and recompute the costs.

1.4
INVERSES; RULES OF MATRIX ARITHMETIC

In this section we shall discuss some properties of the arithmetic operations on matrices. We shall see that many of the basic rules of arithmetic for real numbers also hold for matrices, but a few do not.

Properties of Matrix Operations

For real numbers a and b, we always have $ab = ba$, which is called the *commutative law for multiplication*. For matrices, however, AB and BA need not be equal. Equality can fail to hold for three reasons: It can happen that the product AB is defined but BA is undefined. For example, this is the case if A is a 2×3 matrix and B is a 3×4 matrix. Also, it can happen that AB and BA are both defined but have different sizes. This is the situation if A is a 2×3 matrix and B is a 3×2 matrix. Finally, as Example 1 shows, it is possible to have $AB \neq BA$ even if both AB and BA are defined and have the same size.

EXAMPLE 1 *AB* and *BA* Need Not Be Equal

Consider the matrices

$$A = \begin{bmatrix} -1 & 0 \\ 2 & 3 \end{bmatrix}, \qquad B = \begin{bmatrix} 1 & 2 \\ 3 & 0 \end{bmatrix}$$

Multiplying gives

$$AB = \begin{bmatrix} -1 & -2 \\ 11 & 4 \end{bmatrix}, \qquad BA = \begin{bmatrix} 3 & 6 \\ -3 & 0 \end{bmatrix}$$

Thus, $AB \neq BA$. ◆

Although the commutative law for multiplication is not valid in matrix arithmetic, many familiar laws of arithmetic are valid for matrices. Some of the most important ones and their names are summarized in the following theorem.

THEOREM 1.4.1

Properties of Matrix Arithmetic

Assuming that the sizes of the matrices are such that the indicated operations can be performed, the following rules of matrix arithmetic are valid.

(a) $A + B = B + A$ (Commutative law for addition)
(b) $A + (B + C) = (A + B) + C$ (Associative law for addition)
(c) $A(BC) = (AB)C$ (Associative law for multiplication)
(d) $A(B + C) = AB + AC$ (Left distributive law)
(e) $(B + C)A = BA + CA$ (Right distributive law)
(f) $A(B - C) = AB - AC$ (j) $(a + b)C = aC + bC$
(g) $(B - C)A = BA - CA$ (k) $(a - b)C = aC - bC$
(h) $a(B + C) = aB + aC$ (l) $a(bC) = (ab)C$
(i) $a(B - C) = aB - aC$ (m) $a(BC) = (aB)C = B(aC)$

To prove the equalities in this theorem, we must show that the matrix on the left side has the same size as the matrix on the right side and that corresponding entries on the two sides are equal. With the exception of the associative law in part (c), the proofs all follow the same general pattern. We shall prove part (d) as an illustration. The proof of the associative law, which is more complicated, is outlined in the exercises.

Proof (d) We must show that $A(B + C)$ and $AB + AC$ have the same size and that corresponding entries are equal. To form $A(B + C)$, the matrices B and C must have the same size, say $m \times n$, and the matrix A must then have m columns, so its size must be of the form $r \times m$. This makes $A(B + C)$ an $r \times n$ matrix. It follows that $AB + AC$ is also an $r \times n$ matrix and, consequently, $A(B + C)$ and $AB + AC$ have the same size.

Suppose that $A = [a_{ij}]$, $B = [b_{ij}]$, and $C = [c_{ij}]$. We want to show that corresponding entries of $A(B + C)$ and $AB + AC$ are equal; that is,

$$[A(B + C)]_{ij} = [AB + AC]_{ij}$$

for all values of i and j. But from the definitions of matrix addition and matrix multiplication, we have

$$[A(B + C)]_{ij} = a_{i1}(b_{1j} + c_{1j}) + a_{i2}(b_{2j} + c_{2j}) + \cdots + a_{im}(b_{mj} + c_{mj})$$
$$= (a_{i1}b_{1j} + a_{i2}b_{2j} + \cdots + a_{im}b_{mj}) + (a_{i1}c_{1j} + a_{i2}c_{2j} + \cdots + a_{im}c_{mj})$$
$$= [AB]_{ij} + [AC]_{ij} = [AB + AC]_{ij} \qquad \blacksquare$$

REMARK Although the operations of matrix addition and matrix multiplication were defined for pairs of matrices, associative laws (b) and (c) enable us to denote sums and products of three matrices as $A + B + C$ and ABC without inserting any parentheses. This is justified by the fact that no matter how parentheses are inserted, the associative laws guarantee that the same end result will be obtained. In general, *given any sum or any product of matrices, pairs of parentheses can be inserted or deleted anywhere within the expression without affecting the end result.*

EXAMPLE 2 Associativity of Matrix Multiplication

As an illustration of the associative law for matrix multiplication, consider

$$A = \begin{bmatrix} 1 & 2 \\ 3 & 4 \\ 0 & 1 \end{bmatrix}, \qquad B = \begin{bmatrix} 4 & 3 \\ 2 & 1 \end{bmatrix}, \qquad C = \begin{bmatrix} 1 & 0 \\ 2 & 3 \end{bmatrix}$$

Then

$$AB = \begin{bmatrix} 1 & 2 \\ 3 & 4 \\ 0 & 1 \end{bmatrix} \begin{bmatrix} 4 & 3 \\ 2 & 1 \end{bmatrix} = \begin{bmatrix} 8 & 5 \\ 20 & 13 \\ 2 & 1 \end{bmatrix} \quad \text{and} \quad BC = \begin{bmatrix} 4 & 3 \\ 2 & 1 \end{bmatrix} \begin{bmatrix} 1 & 0 \\ 2 & 3 \end{bmatrix} = \begin{bmatrix} 10 & 9 \\ 4 & 3 \end{bmatrix}$$

Thus

$$(AB)C = \begin{bmatrix} 8 & 5 \\ 20 & 13 \\ 2 & 1 \end{bmatrix} \begin{bmatrix} 1 & 0 \\ 2 & 3 \end{bmatrix} = \begin{bmatrix} 18 & 15 \\ 46 & 39 \\ 4 & 3 \end{bmatrix}$$

and

$$A(BC) = \begin{bmatrix} 1 & 2 \\ 3 & 4 \\ 0 & 1 \end{bmatrix} \begin{bmatrix} 10 & 9 \\ 4 & 3 \end{bmatrix} = \begin{bmatrix} 18 & 15 \\ 46 & 39 \\ 4 & 3 \end{bmatrix}$$

so $(AB)C = A(BC)$, as guaranteed by Theorem 1.4.1c. ◆

Zero Matrices

A matrix, all of whose entries are zero, such as

$$\begin{bmatrix} 0 & 0 \\ 0 & 0 \end{bmatrix}, \quad \begin{bmatrix} 0 & 0 & 0 \\ 0 & 0 & 0 \\ 0 & 0 & 0 \end{bmatrix}, \quad \begin{bmatrix} 0 & 0 & 0 & 0 \\ 0 & 0 & 0 & 0 \end{bmatrix}, \quad \begin{bmatrix} 0 \\ 0 \\ 0 \\ 0 \end{bmatrix}, \quad [0]$$

is called a *zero matrix*. A zero matrix will be denoted by 0; if it is important to emphasize the size, we shall write $0_{m \times n}$ for the $m \times n$ zero matrix. Moreover, in keeping with our convention of using boldface symbols for matrices with one column, we will denote a zero matrix with one column by **0**.

If A is any matrix and 0 is the zero matrix with the same size, it is obvious that $A + 0 = 0 + A = A$. The matrix 0 plays much the same role in these matrix equations as the number 0 plays in the numerical equations $a + 0 = 0 + a = a$.

Since we already know that some of the rules of arithmetic for real numbers do not carry over to matrix arithmetic, it would be foolhardy to assume that all the properties of the real number zero carry over to zero matrices. For example, consider the following two standard results in the arithmetic of real numbers.

- If $ab = ac$ and $a \neq 0$, then $b = c$. (This is called the *cancellation law*.)
- If $ad = 0$, then at least one of the factors on the left is 0.

As the next example shows, the corresponding results are not generally true in matrix arithmetic.

EXAMPLE 3 The Cancellation Law Does Not Hold

Consider the matrices

$$A = \begin{bmatrix} 0 & 1 \\ 0 & 2 \end{bmatrix}, \quad B = \begin{bmatrix} 1 & 1 \\ 3 & 4 \end{bmatrix}, \quad C = \begin{bmatrix} 2 & 5 \\ 3 & 4 \end{bmatrix}, \quad D = \begin{bmatrix} 3 & 7 \\ 0 & 0 \end{bmatrix}$$

You should verify that

$$AB = AC = \begin{bmatrix} 3 & 4 \\ 6 & 8 \end{bmatrix} \quad \text{and} \quad AD = \begin{bmatrix} 0 & 0 \\ 0 & 0 \end{bmatrix}$$

Thus, although $A \neq 0$, it is *incorrect* to cancel the A from both sides of the equation $AB = AC$ and write $B = C$. Also, $AD = 0$, yet $A \neq 0$ and $D \neq 0$. Thus, the cancellation law is not valid for matrix multiplication, and it is possible for a product of matrices to be zero without either factor being zero. ◆

In spite of the above example, there are a number of familiar properties of the real number 0 that *do* carry over to zero matrices. Some of the more important ones are summarized in the next theorem. The proofs are left as exercises.

THEOREM 1.4.2

Properties of Zero Matrices

Assuming that the sizes of the matrices are such that the indicated operations can be performed, the following rules of matrix arithmetic are valid.

(a) $A + 0 = 0 + A = A$

(b) $A - A = 0$

(c) $0 - A = -A$

(d) $A0 = 0; \quad 0A = 0$

Identity Matrices

Of special interest are square matrices with 1's on the main diagonal and 0's off the main diagonal, such as

$$\begin{bmatrix} 1 & 0 \\ 0 & 1 \end{bmatrix}, \quad \begin{bmatrix} 1 & 0 & 0 \\ 0 & 1 & 0 \\ 0 & 0 & 1 \end{bmatrix}, \quad \begin{bmatrix} 1 & 0 & 0 & 0 \\ 0 & 1 & 0 & 0 \\ 0 & 0 & 1 & 0 \\ 0 & 0 & 0 & 1 \end{bmatrix}, \quad \text{and so on.}$$

A matrix of this form is called an ***identity matrix*** and is denoted by I. If it is important to emphasize the size, we shall write I_n for the $n \times n$ identity matrix.

If A is an $m \times n$ matrix, then, as illustrated in the next example,

$$AI_n = A \quad \text{and} \quad I_m A = A$$

Thus, an identity matrix plays much the same role in matrix arithmetic that the number 1 plays in the numerical relationships $a \cdot 1 = 1 \cdot a = a$.

EXAMPLE 4   Multiplication by an Identity Matrix

Consider the matrix

$$A = \begin{bmatrix} a_{11} & a_{12} & a_{13} \\ a_{21} & a_{22} & a_{23} \end{bmatrix}$$

Then

$$I_2 A = \begin{bmatrix} 1 & 0 \\ 0 & 1 \end{bmatrix} \begin{bmatrix} a_{11} & a_{12} & a_{13} \\ a_{21} & a_{22} & a_{23} \end{bmatrix} = \begin{bmatrix} a_{11} & a_{12} & a_{13} \\ a_{21} & a_{22} & a_{23} \end{bmatrix} = A$$

and

$$AI_3 = \begin{bmatrix} a_{11} & a_{12} & a_{13} \\ a_{21} & a_{22} & a_{23} \end{bmatrix} \begin{bmatrix} 1 & 0 & 0 \\ 0 & 1 & 0 \\ 0 & 0 & 1 \end{bmatrix} = \begin{bmatrix} a_{11} & a_{12} & a_{13} \\ a_{21} & a_{22} & a_{23} \end{bmatrix} = A \quad \blacklozenge$$

As the next theorem shows, identity matrices arise naturally in studying reduced row-echelon forms of *square* matrices.

THEOREM 1.4.3

> *If R is the reduced row-echelon form of an $n \times n$ matrix A, then either R has a row of zeros or R is the identity matrix I_n.*

Proof   Suppose that the reduced row-echelon form of A is

$$R = \begin{bmatrix} r_{11} & r_{12} & \cdots & r_{1n} \\ r_{21} & r_{22} & \cdots & r_{2n} \\ \vdots & \vdots & & \vdots \\ r_{n1} & r_{n2} & \cdots & r_{nn} \end{bmatrix}$$

Either the last row in this matrix consists entirely of zeros or it does not. If not, the matrix contains no zero rows, and consequently each of the n rows has a leading entry of 1. Since these leading 1's occur progressively farther to the right as we move down the matrix, each of these 1's must occur on the main diagonal. Since the other entries in the same column as one of these 1's are zero, R must be I_n. Thus, either R has a row of zeros or $R = I_n$. ■

DEFINITION

If A is a square matrix, and if a matrix B of the same size can be found such that $AB = BA = I$, then A is said to be **invertible** and B is called an **inverse** of A. If no such matrix B can be found, then A is said to be **singular**.

EXAMPLE 5 Verifying the Inverse Requirements

The matrix

$$B = \begin{bmatrix} 3 & 5 \\ 1 & 2 \end{bmatrix} \quad \text{is an inverse of} \quad A = \begin{bmatrix} 2 & -5 \\ -1 & 3 \end{bmatrix}$$

since

$$AB = \begin{bmatrix} 2 & -5 \\ -1 & 3 \end{bmatrix}\begin{bmatrix} 3 & 5 \\ 1 & 2 \end{bmatrix} = \begin{bmatrix} 1 & 0 \\ 0 & 1 \end{bmatrix} = I$$

and

$$BA = \begin{bmatrix} 3 & 5 \\ 1 & 2 \end{bmatrix}\begin{bmatrix} 2 & -5 \\ -1 & 3 \end{bmatrix} = \begin{bmatrix} 1 & 0 \\ 0 & 1 \end{bmatrix} = I \quad \blacklozenge$$

EXAMPLE 6 A Matrix with No Inverse

The matrix

$$A = \begin{bmatrix} 1 & 4 & 0 \\ 2 & 5 & 0 \\ 3 & 6 & 0 \end{bmatrix}$$

is singular. To see why, let

$$B = \begin{bmatrix} b_{11} & b_{12} & b_{13} \\ b_{21} & b_{22} & b_{23} \\ b_{31} & b_{32} & b_{33} \end{bmatrix}$$

be any 3×3 matrix. The third column of BA is

$$\begin{bmatrix} b_{11} & b_{12} & b_{13} \\ b_{21} & b_{22} & b_{23} \\ b_{31} & b_{32} & b_{33} \end{bmatrix}\begin{bmatrix} 0 \\ 0 \\ 0 \end{bmatrix} = \begin{bmatrix} 0 \\ 0 \\ 0 \end{bmatrix}$$

Thus

$$BA \neq I = \begin{bmatrix} 1 & 0 & 0 \\ 0 & 1 & 0 \\ 0 & 0 & 1 \end{bmatrix} \quad \blacklozenge$$

Properties of Inverses

It is reasonable to ask whether an invertible matrix can have more than one inverse. The next theorem shows that the answer is no—*an invertible matrix has exactly one inverse.*

THEOREM 1.4.4

If B and C are both inverses of the matrix A, then $B = C$.

Proof   Since B is an inverse of A, we have $BA = I$. Multiplying both sides on the right by C gives $(BA)C = IC = C$. But $(BA)C = B(AC) = BI = B$, so $C = B$.   ∎

As a consequence of this important result, we can now speak of "the" inverse of an invertible matrix. If A is invertible, then its inverse will be denoted by the symbol A^{-1}. Thus,

$$AA^{-1} = I \quad \text{and} \quad A^{-1}A = I$$

The inverse of A plays much the same role in matrix arithmetic that the reciprocal a^{-1} plays in the numerical relationships $aa^{-1} = 1$ and $a^{-1}a = 1$.

In the next section we shall develop a method for finding inverses of invertible matrices of any size; however, the following theorem gives conditions under which a 2×2 matrix is invertible and provides a simple formula for the inverse.

THEOREM 1.4.5

The matrix

$$A = \begin{bmatrix} a & b \\ c & d \end{bmatrix}$$

is invertible if $ad - bc \neq 0$, in which case the inverse is given by the formula

$$A^{-1} = \frac{1}{ad - bc} \begin{bmatrix} d & -b \\ -c & a \end{bmatrix} = \begin{bmatrix} \dfrac{d}{ad - bc} & -\dfrac{b}{ad - bc} \\ -\dfrac{c}{ad - bc} & \dfrac{a}{ad - bc} \end{bmatrix}$$

Proof   We leave it for the reader to verify that $AA^{-1} = I_2$ and $A^{-1}A = I_2$.   ∎

THEOREM 1.4.6

If A and B are invertible matrices of the same size, then AB is invertible and

$$(AB)^{-1} = B^{-1}A^{-1}$$

Proof   If we can show that $(AB)(B^{-1}A^{-1}) = (B^{-1}A^{-1})(AB) = I$, then we will have simultaneously shown that the matrix AB is invertible and that $(AB)^{-1} = B^{-1}A^{-1}$. But $(AB)(B^{-1}A^{-1}) = A(BB^{-1})A^{-1} = AIA^{-1} = AA^{-1} = I$. A similar argument shows that $(B^{-1}A^{-1})(AB) = I$.   ∎

Although we will not prove it, this result can be extended to include three or more factors; that is,

A product of any number of invertible matrices is invertible, and the inverse of the product is the product of the inverses in the reverse order.

EXAMPLE 7   Inverse of a Product

Consider the matrices

$$A = \begin{bmatrix} 1 & 2 \\ 1 & 3 \end{bmatrix}, \qquad B = \begin{bmatrix} 3 & 2 \\ 2 & 2 \end{bmatrix}, \qquad AB = \begin{bmatrix} 7 & 6 \\ 9 & 8 \end{bmatrix}$$

Applying the formula in Theorem 1.4.5, we obtain

$$A^{-1} = \begin{bmatrix} 3 & -2 \\ -1 & 1 \end{bmatrix}, \qquad B^{-1} = \begin{bmatrix} 1 & -1 \\ -1 & \frac{3}{2} \end{bmatrix}, \qquad (AB)^{-1} = \begin{bmatrix} 4 & -3 \\ -\frac{9}{2} & \frac{7}{2} \end{bmatrix}$$

Also,

$$B^{-1}A^{-1} = \begin{bmatrix} 1 & -1 \\ -1 & \frac{3}{2} \end{bmatrix} \begin{bmatrix} 3 & -2 \\ -1 & 1 \end{bmatrix} = \begin{bmatrix} 4 & -3 \\ -\frac{9}{2} & \frac{7}{2} \end{bmatrix}$$

Therefore, $(AB)^{-1} = B^{-1}A^{-1}$, as guaranteed by Theorem 1.4.6. ◆

Powers of a Matrix

Next, we shall define powers of a square matrix and discuss their properties.

> **DEFINITION**
>
> If A is a square matrix, then we define the nonnegative integer powers of A to be
>
> $$A^0 = I \qquad A^n = \underbrace{AA \cdots A}_{n \text{ factors}} \qquad (n > 0)$$
>
> Moreover, if A is invertible, then we define the negative integer powers to be
>
> $$A^{-n} = (A^{-1})^n = \underbrace{A^{-1}A^{-1} \cdots A^{-1}}_{n \text{ factors}}$$

Because this definition parallels that for real numbers, the usual laws of exponents hold. (We omit the details.)

THEOREM 1.4.7

> ### Laws of Exponents
>
> *If A is a square matrix and r and s are integers, then*
>
> $$A^r A^s = A^{r+s}, \qquad (A^r)^s = A^{rs}$$

The next theorem provides some useful properties of negative exponents.

THEOREM 1.4.8

> ### Laws of Exponents
>
> *If A is an invertible matrix, then:*
>
> (a) A^{-1} *is invertible and* $(A^{-1})^{-1} = A$.
> (b) A^n *is invertible and* $(A^n)^{-1} = (A^{-1})^n$ *for* $n = 0, 1, 2, \ldots$.
> (c) *For any nonzero scalar k, the matrix kA is invertible and* $(kA)^{-1} = \dfrac{1}{k}A^{-1}$.

Proof

(a) Since $AA^{-1} = A^{-1}A = I$, the matrix A^{-1} is invertible and $(A^{-1})^{-1} = A$.

(b) This part is left as an exercise.

(c) If k is any nonzero scalar, results (l) and (m) of Theorem 1.4.1 enable us to write

$$(kA)\left(\frac{1}{k}A^{-1}\right) = \frac{1}{k}(kA)A^{-1} = \left(\frac{1}{k}k\right)AA^{-1} = (1)I = I$$

Similarly, $\left(\frac{1}{k}A^{-1}\right)(kA) = I$ so that kA is invertible and $(kA)^{-1} = \frac{1}{k}A^{-1}$. ■

EXAMPLE 8 Powers of a Matrix

Let A and A^{-1} be as in Example 7; that is,

$$A = \begin{bmatrix} 1 & 2 \\ 1 & 3 \end{bmatrix} \quad \text{and} \quad A^{-1} = \begin{bmatrix} 3 & -2 \\ -1 & 1 \end{bmatrix}$$

Then

$$A^3 = \begin{bmatrix} 1 & 2 \\ 1 & 3 \end{bmatrix}\begin{bmatrix} 1 & 2 \\ 1 & 3 \end{bmatrix}\begin{bmatrix} 1 & 2 \\ 1 & 3 \end{bmatrix} = \begin{bmatrix} 11 & 30 \\ 15 & 41 \end{bmatrix}$$

$$A^{-3} = (A^{-1})^3 = \begin{bmatrix} 3 & -2 \\ -1 & 1 \end{bmatrix}\begin{bmatrix} 3 & -2 \\ -1 & 1 \end{bmatrix}\begin{bmatrix} 3 & -2 \\ -1 & 1 \end{bmatrix} = \begin{bmatrix} 41 & -30 \\ -15 & 11 \end{bmatrix} \quad \blacklozenge$$

Polynomial Expressions Involving Matrices

If A is a square matrix, say $m \times m$, and if

$$p(x) = a_0 + a_1 x + \cdots + a_n x^n \tag{1}$$

is any polynomial, then we define

$$p(A) = a_0 I + a_1 A + \cdots + a_n A^n$$

where I is the $m \times m$ identity matrix. In words, $p(A)$ is the $m \times m$ matrix that results when A is substituted for x in (1) and a_0 is replaced by $a_0 I$.

EXAMPLE 9 Matrix Polynomial

If

$$p(x) = 2x^2 - 3x + 4 \quad \text{and} \quad A = \begin{bmatrix} -1 & 2 \\ 0 & 3 \end{bmatrix}$$

then

$$p(A) = 2A^2 - 3A + 4I = 2\begin{bmatrix} -1 & 2 \\ 0 & 3 \end{bmatrix}^2 - 3\begin{bmatrix} -1 & 2 \\ 0 & 3 \end{bmatrix} + 4\begin{bmatrix} 1 & 0 \\ 0 & 1 \end{bmatrix}$$

$$= \begin{bmatrix} 2 & 8 \\ 0 & 18 \end{bmatrix} - \begin{bmatrix} -3 & 6 \\ 0 & 9 \end{bmatrix} + \begin{bmatrix} 4 & 0 \\ 0 & 4 \end{bmatrix} = \begin{bmatrix} 9 & 2 \\ 0 & 13 \end{bmatrix} \quad \blacklozenge$$

Properties of the Transpose

The next theorem lists the main properties of the transpose operation.

THEOREM 1.4.9

Properties of the Transpose

If the sizes of the matrices are such that the stated operations can be performed, then

(a) $((A)^T)^T = A$

(b) $(A + B)^T = A^T + B^T$ *and* $(A - B)^T = A^T - B^T$

(c) $(kA)^T = kA^T$, *where k is any scalar*

(d) $(AB)^T = B^T A^T$

If we keep in mind that transposing a matrix interchanges its rows and columns, parts (a), (b), and (c) should be self-evident. For example, part (a) states that interchanging rows and columns twice leaves a matrix unchanged; part (b) asserts that adding and then interchanging rows and columns yields the same result as first interchanging rows and columns and then adding; and part (c) asserts that multiplying by a scalar and then interchanging rows and columns yields the same result as first interchanging rows and columns and then multiplying by the scalar. Part (d) is not so obvious, so we give its proof.

Proof (d) Let $A = [a_{ij}]_{m \times r}$ and $B = [b_{ij}]_{r \times n}$ so that the products AB and $B^T A^T$ can both be formed. We leave it for the reader to check that $(AB)^T$ and $B^T A^T$ have the same size, namely $n \times m$. Thus it only remains to show that corresponding entries of $(AB)^T$ and $B^T A^T$ are the same; that is,

$$\left((AB)^T\right)_{ij} = (B^T A^T)_{ij} \tag{2}$$

Applying Formula (11) of Section 1.3 to the left side of this equation and using the definition of matrix multiplication, we obtain

$$\left((AB)^T\right)_{ij} = (AB)_{ji} = a_{j1}b_{1i} + a_{j2}b_{2i} + \cdots + a_{jr}b_{ri} \tag{3}$$

To evaluate the right side of (2), it will be convenient to let a'_{ij} and b'_{ij} denote the ijth entries of A^T and B^T, respectively, so

$$a'_{ij} = a_{ji} \quad \text{and} \quad b'_{ij} = b_{ji}$$

From these relationships and the definition of matrix multiplication, we obtain

$$(B^T A^T)_{ij} = b'_{i1}a'_{1j} + b'_{i2}a'_{2j} + \cdots + b'_{ir}a'_{rj}$$
$$= b_{1i}a_{j1} + b_{2i}a_{j2} + \cdots + b_{ri}a_{jr}$$
$$= a_{j1}b_{1i} + a_{j2}b_{2i} + \cdots + a_{jr}b_{ri}$$

This, together with (3), proves (2). ∎

Although we shall not prove it, part (d) of this theorem can be extended to include three or more factors; that is,

The transpose of a product of any number of matrices is equal to the product of their transposes in the reverse order.

REMARK Note the similarity between this result and the result following Theorem 1.4.6 about the inverse of a product of matrices.

Invertibility of a Transpose

The following theorem establishes a relationship between the inverse of an invertible matrix and the inverse of its transpose.

THEOREM 1.4.10

> *If A is an invertible matrix, then A^T is also invertible and*
> $$(A^T)^{-1} = (A^{-1})^T \tag{4}$$

Proof   We can prove the invertibility of A^T and obtain (4) by showing that
$$A^T(A^{-1})^T = (A^{-1})^T A^T = I$$
But from part (d) of Theorem 1.4.9 and the fact that $I^T = I$, we have
$$A^T(A^{-1})^T = (A^{-1}A)^T = I^T = I$$
$$(A^{-1})^T A^T = (AA^{-1})^T = I^T = I$$
which completes the proof.   ∎

EXAMPLE 10   Verifying Theorem 1.4.10

Consider the matrices
$$A = \begin{bmatrix} -5 & -3 \\ 2 & 1 \end{bmatrix}, \qquad A^T = \begin{bmatrix} -5 & 2 \\ -3 & 1 \end{bmatrix}$$
Applying Theorem 1.4.5 yields
$$A^{-1} = \begin{bmatrix} 1 & 3 \\ -2 & -5 \end{bmatrix}, \qquad (A^{-1})^T = \begin{bmatrix} 1 & -2 \\ 3 & -5 \end{bmatrix}, \qquad (A^T)^{-1} = \begin{bmatrix} 1 & -2 \\ 3 & -5 \end{bmatrix}$$
As guaranteed by Theorem 1.4.10, these matrices satisfy (4). ◆

EXERCISE SET 1.4

1. Let
$$A = \begin{bmatrix} 2 & -1 & 3 \\ 0 & 4 & 5 \\ -2 & 1 & 4 \end{bmatrix}, \qquad B = \begin{bmatrix} 8 & -3 & -5 \\ 0 & 1 & 2 \\ 4 & -7 & 6 \end{bmatrix},$$
$$C = \begin{bmatrix} 0 & -2 & 3 \\ 1 & 7 & 4 \\ 3 & 5 & 9 \end{bmatrix}, \qquad a = 4, \qquad b = -7$$

   Show that
   (a) $A + (B + C) = (A + B) + C$ (b) $(AB)C = A(BC)$
   (c) $(a + b)C = aC + bC$ (d) $a(B - C) = aB - aC$

2. Using the matrices and scalars in Exercise 1, verify that
   (a) $a(BC) = (aB)C = B(aC)$ (b) $A(B - C) = AB - AC$
   (c) $(B + C)A = BA + CA$ (d) $a(bC) = (ab)C$

3. Using the matrices and scalars in Exercise 1, verify that
   (a) $(A^T)^T = A$ (b) $(A + B)^T = A^T + B^T$
   (c) $(aC)^T = aC^T$ (d) $(AB)^T = B^T A^T$

4. Use Theorem 1.4.5 to compute the inverses of the following matrices.

   (a) $A = \begin{bmatrix} 3 & 1 \\ 5 & 2 \end{bmatrix}$ (b) $B = \begin{bmatrix} 2 & -3 \\ 4 & 4 \end{bmatrix}$

   (c) $C = \begin{bmatrix} 6 & 4 \\ -2 & -1 \end{bmatrix}$ (d) $D = \begin{bmatrix} 2 & 0 \\ 0 & 3 \end{bmatrix}$

5. Use the matrices A and B in Exercise 4 to verify that

    (a) $(A^{-1})^{-1} = A$ (b) $(B^T)^{-1} = (B^{-1})^T$

6. Use the matrices A, B, and C in Exercise 4 to verify that

    (a) $(AB)^{-1} = B^{-1}A^{-1}$ (b) $(ABC)^{-1} = C^{-1}B^{-1}A^{-1}$

7. In each part, use the given information to find A.

    (a) $A^{-1} = \begin{bmatrix} 2 & -1 \\ 3 & 5 \end{bmatrix}$ (b) $(7A)^{-1} = \begin{bmatrix} -3 & 7 \\ 1 & -2 \end{bmatrix}$

    (c) $(5A^T)^{-1} = \begin{bmatrix} -3 & -1 \\ 5 & 2 \end{bmatrix}$ (d) $(I + 2A)^{-1} = \begin{bmatrix} -1 & 2 \\ 4 & 5 \end{bmatrix}$

8. Let A be the matrix

$$\begin{bmatrix} 2 & 0 \\ 4 & 1 \end{bmatrix}$$

    Compute A^3, A^{-3}, and $A^2 - 2A + I$.

9. Let A be the matrix

$$\begin{bmatrix} 3 & 1 \\ 2 & 1 \end{bmatrix}$$

    In each part, find $p(A)$.

    (a) $p(x) = x - 2$ (b) $p(x) = 2x^2 - x + 1$ (c) $p(x) = x^3 - 2x + 4$

10. Let $p_1(x) = x^2 - 9$, $p_2(x) = x + 3$, and $p_3(x) = x - 3$.

    (a) Show that $p_1(A) = p_2(A)p_3(A)$ for the matrix A in Exercise 9.

    (b) Show that $p_1(A) = p_2(A)p_3(A)$ for any square matrix A.

11. Find the inverse of

$$\begin{bmatrix} \cos\theta & \sin\theta \\ -\sin\theta & \cos\theta \end{bmatrix}$$

12. Find the inverse of

$$\begin{bmatrix} \frac{1}{2}(e^x + e^{-x}) & \frac{1}{2}(e^x - e^{-x}) \\ \frac{1}{2}(e^x - e^{-x}) & \frac{1}{2}(e^x + e^{-x}) \end{bmatrix}$$

13. Consider the matrix

$$A = \begin{bmatrix} a_{11} & 0 & \cdots & 0 \\ 0 & a_{22} & \cdots & 0 \\ \vdots & \vdots & & \vdots \\ 0 & 0 & \cdots & a_{nn} \end{bmatrix}$$

    where $a_{11}a_{22}\cdots a_{nn} \neq 0$. Show that A is invertible and find its inverse.

14. Show that if a square matrix A satisfies $A^2 - 3A + I = 0$, then $A^{-1} = 3I - A$.

15. (a) Show that a matrix with a row of zeros cannot have an inverse.

    (b) Show that a matrix with a column of zeros cannot have an inverse.

16. Is the sum of two invertible matrices necessarily invertible?

17. Let A and B be square matrices such that $AB = 0$. Show that if A is invertible, then $B = 0$.

18. Let A, B, and 0 be 2×2 matrices. Assuming that A is invertible, find a matrix C such that

$$\left[\begin{array}{c|c} A^{-1} & 0 \\ \hline C & A^{-1} \end{array} \right]$$

    is the inverse of the partitioned matrix

$$\left[\begin{array}{c|c} A & 0 \\ \hline B & A \end{array} \right]$$

    (See Exercise 15 of the preceding section.)

19. Use the result in Exercise 18 to find the inverses of the following matrices.

(a) $\begin{bmatrix} 1 & 1 & 0 & 0 \\ -1 & 1 & 0 & 0 \\ 1 & 1 & 1 & 1 \\ 1 & 1 & -1 & 1 \end{bmatrix}$ (b) $\begin{bmatrix} 1 & 1 & 0 & 0 \\ 0 & 1 & 0 & 0 \\ 0 & 0 & 1 & 1 \\ 0 & 0 & 0 & 1 \end{bmatrix}$

20. (a) Find a nonzero 3×3 matrix A such that $A^T = A$.

    (b) Find a nonzero 3×3 matrix A such that $A^T = -A$.

21. A square matrix A is called *symmetric* if $A^T = A$ and *skew-symmetric* if $A^T = -A$. Show that if B is a square matrix, then

    (a) BB^T and $B + B^T$ are symmetric   (b) $B - B^T$ is skew-symmetric

22. If A is a square matrix and n is a positive integer, is it true that $(A^n)^T = (A^T)^n$? Justify your answer.

23. Let A be the matrix

$$\begin{bmatrix} 1 & 0 & 1 \\ 1 & 1 & 0 \\ 0 & 1 & 1 \end{bmatrix}$$

Determine whether A is invertible, and if so, find its inverse.

   Hint   Solve $AX = I$ by equating corresponding entries on the two sides.

24. Prove:
    (a) part (b) of Theorem 1.4.1   (b) part (i) of Theorem 1.4.1
    (c) part (m) of Theorem 1.4.1

25. Apply parts (d) and (m) of Theorem 1.4.1 to the matrices A, B, and $(-1)C$ to derive the result in part (f).

26. Prove Theorem 1.4.2.

27. Consider the laws of exponents $A^r A^s = A^{r+s}$ and $(A^r)^s = A^{rs}$.

    (a) Show that if A is any square matrix, then these laws are valid for all nonnegative integer values of r and s.

    (b) Show that if A is invertible, then these laws hold for all negative integer values of r and s.

28. Show that if A is invertible and k is any nonzero scalar, then $(kA)^n = k^n A^n$ for all integer values of n.

29. (a) Show that if A is invertible and $AB = AC$, then $B = C$.

    (b) Explain why part (a) and Example 3 do not contradict one another.

30. Prove part (c) of Theorem 1.4.1.

   Hint   Assume that A is $m \times n$, B is $n \times p$, and C is $p \times q$. The ijth entry on the left side is $l_{ij} = a_{i1}[BC]_{1j} + a_{i2}[BC]_{2j} + \cdots + a_{in}[BC]_{nj}$ and the ijth entry on the right side is $r_{ij} = [AB]_{i1}c_{1j} + [AB]_{i2}c_{2j} + \cdots + [AB]_{ip}c_{pj}$. Verify that $l_{ij} = r_{ij}$.

Discussion & Discovery

31. Let A and B be square matrices with the same size.

    (a) Give an example in which $(A + B)^2 \neq A^2 + 2AB + B^2$.

    (b) Fill in the blank to create a matrix identity that is valid for all choices of A and B.
    $(A + B)^2 = A^2 + B^2 + \underline{\hspace{1.5cm}}$.

32. Let A and B be square matrices with the same size.

    (a) Give an example in which $(A + B)(A - B) \neq A^2 - B^2$.

    (b)   Let A and B be square matrices with the same size. Fill in the blank to create a matrix identity that is valid for all choices of A and B. $(A + B)(A - B) =$
_____ .

33. In the real number system the equation $a^2 = 1$ has exactly two solutions. Find at least eight different 3×3 matrices that satisfy the equation $A^2 = I_3$.

    Hint   Look for solutions in which all entries off the main diagonal are zero.

34. A statement of the form "If p, then q" is logically equivalent to the statement "If not q, then not p." (The second statement is called the ***logical contrapositive*** of the first.) For example, the logical contrapositive of the statement "If it is raining, then the ground is wet" is "If the ground is not wet, then it is not raining."

    (a)   Find the logical contrapositive of the following statement: If A^T is singular, then A is singular.

    (b)   Is the statement true or false? Explain.

35. Let A and B be $n \times n$ matrices. Indicate whether the statement is always true or sometimes false. Justify each answer.

    (a) $(AB)^2 = A^2 B^2$ (b) $(A - B)^2 = (B - A)^2$

    (c) $(AB^{-1})(BA^{-1}) = I_n$ (d) $AB \neq BA$.

36. Assuming that all matrices are $n \times n$ and invertible, solve for D.

$$ABC^T DBA^T C = AB^T$$

1.5
ELEMENTARY MATRICES AND A METHOD FOR FINDING A^{-1}

In this section we shall develop an algorithm for finding the inverse of an invertible matrix. We shall also discuss some of the basic properties of invertible matrices.

We begin with the definition of a special type of matrix that can be used to carry out an elementary row operation by matrix multiplication.

> **DEFINITION**
>
> An $n \times n$ matrix is called an ***elementary matrix*** if it can be obtained from the $n \times n$ identity matrix I_n by performing a single elementary row operation.

EXAMPLE 1 Elementary Matrices and Row Operations

Listed below are four elementary matrices and the operations that produce them.

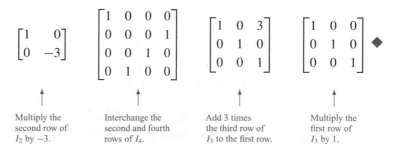

| Multiply the second row of I_2 by -3. | Interchange the second and fourth rows of I_4. | Add 3 times the third row of I_3 to the first row. | Multiply the first row of I_3 by 1. |

When a matrix A is multiplied on the *left* by an elementary matrix E, the effect is to perform an elementary row operation on A. This is the content of the following theorem, the proof of which is left for the exercises.

THEOREM 1.5.1

> ### Row Operations by Matrix Multiplication
>
> *If the elementary matrix E results from performing a certain row operation on I_m and if A is an $m \times n$ matrix, then the product EA is the matrix that results when this same row operation is performed on A.*

EXAMPLE 2 Using Elementary Matrices

Consider the matrix

$$A = \begin{bmatrix} 1 & 0 & 2 & 3 \\ 2 & -1 & 3 & 6 \\ 1 & 4 & 4 & 0 \end{bmatrix}$$

and consider the elementary matrix

$$E = \begin{bmatrix} 1 & 0 & 0 \\ 0 & 1 & 0 \\ 3 & 0 & 1 \end{bmatrix}$$

which results from adding 3 times the first row of I_3 to the third row. The product EA is

$$EA = \begin{bmatrix} 1 & 0 & 2 & 3 \\ 2 & -1 & 3 & 6 \\ 4 & 4 & 10 & 9 \end{bmatrix}$$

which is precisely the same matrix that results when we add 3 times the first row of A to the third row. ◆

REMARK Theorem 1.5.1 is primarily of theoretical interest and will be used for developing some results about matrices and systems of linear equations. Computationally, it is preferable to perform row operations directly rather than multiplying on the left by an elementary matrix.

If an elementary row operation is applied to an identity matrix I to produce an elementary matrix E, then there is a second row operation that, when applied to E, produces I back again. For example, if E is obtained by multiplying the ith row of I by a nonzero constant c, then I can be recovered if the ith row of E is multiplied by $1/c$. The various possibilities are listed in Table 1. The operations on the right side of this table are called the ***inverse operations*** of the corresponding operations on the left.

EXAMPLE 3 Row Operations and Inverse Row Operations

In each of the following, an elementary row operation is applied to the 2×2 identity matrix to obtain an elementary matrix E, then E is restored to the identity matrix by

Table 1

Row Operation on I That Produces E	Row Operation on E That Reproduces I
Multiply row i by $c \neq 0$	Multiply row i by $1/c$
Interchange rows i and j	Interchange rows i and j
Add c times row i to row j	Add $-c$ times row i to row j

applying the inverse row operation.

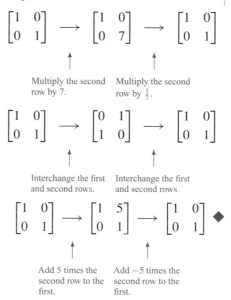

The next theorem gives an important property of elementary matrices.

THEOREM 1.5.2

Every elementary matrix is invertible, and the inverse is also an elementary matrix.

Proof If E is an elementary matrix, then E results from performing some row operation on I. Let E_0 be the matrix that results when the inverse of this operation is performed on I. Applying Theorem 1.5.1 and using the fact that inverse row operations cancel the effect of each other, it follows that

$$E_0 E = I \quad \text{and} \quad E E_0 = I$$

Thus, the elementary matrix E_0 is the inverse of E. ∎

The next theorem establishes some fundamental relationships among invertibility, homogeneous linear systems, reduced row-echelon forms, and elementary matrices. These results are extremely important and will be used many times in later sections.

THEOREM 1.5.3

Equivalent Statements

If A is an $n \times n$ matrix, then the following statements are equivalent, that is, all true or all false.

(a) *A is invertible.*
(b) *Ax = **0** has only the trivial solution.*
(c) *The reduced row-echelon form of A is I_n.*
(d) *A is expressible as a product of elementary matrices.*

Proof We shall prove the equivalence by establishing the chain of implications: $(a) \Rightarrow (b) \Rightarrow (c) \Rightarrow (d) \Rightarrow (a)$.

(a) $\Rightarrow$ (b) Assume A is invertible and let $\mathbf{x}_0$ be any solution of $A\mathbf{x} = \mathbf{0}$; thus $A\mathbf{x}_0 = \mathbf{0}$. Multiplying both sides of this equation by the matrix A^{-1} gives $A^{-1}(A\mathbf{x}_0) = A^{-1}\mathbf{0}$, or $(A^{-1}A)\mathbf{x}_0 = \mathbf{0}$, or $I\mathbf{x}_0 = \mathbf{0}$, or $\mathbf{x}_0 = \mathbf{0}$. Thus, $A\mathbf{x} = \mathbf{0}$ has only the trivial solution.

(b) $\Rightarrow$ (c) Let $A\mathbf{x} = \mathbf{0}$ be the matrix form of the system

$$
\begin{aligned}
a_{11}x_1 + a_{12}x_2 + \cdots + a_{1n}x_n &= 0 \\
a_{21}x_1 + a_{22}x_2 + \cdots + a_{2n}x_n &= 0 \\
\vdots \qquad \vdots \qquad\qquad \vdots \quad\;\; \vdots & \\
a_{n1}x_1 + a_{n2}x_2 + \cdots + a_{nn}x_n &= 0
\end{aligned}
\tag{1}
$$

and assume that the system has only the trivial solution. If we solve by Gauss–Jordan elimination, then the system of equations corresponding to the reduced row-echelon form of the augmented matrix will be

$$
\begin{aligned}
x_1 \qquad\qquad\qquad &= 0 \\
x_2 \qquad\qquad &= 0 \\
\ddots\; & \\
x_n &= 0
\end{aligned}
\tag{2}
$$

Thus the augmented matrix

$$
\begin{bmatrix}
a_{11} & a_{12} & \cdots & a_{1n} & 0 \\
a_{21} & a_{22} & \cdots & a_{2n} & 0 \\
\vdots & \vdots & & \vdots & \vdots \\
a_{n1} & a_{n2} & \cdots & a_{nn} & 0
\end{bmatrix}
$$

for (1) can be reduced to the augmented matrix

$$
\begin{bmatrix}
1 & 0 & 0 & \cdots & 0 & 0 \\
0 & 1 & 0 & \cdots & 0 & 0 \\
0 & 0 & 1 & \cdots & 0 & 0 \\
\vdots & \vdots & \vdots & & \vdots & \vdots \\
0 & 0 & 0 & \cdots & 1 & 0
\end{bmatrix}
$$

for (2) by a sequence of elementary row operations. If we disregard the last column (of zeros) in each of these matrices, we can conclude that the reduced row-echelon form of A is I_n.

(c) $\Rightarrow$ (d) Assume that the reduced row-echelon form of A is I_n, so that A can be reduced to I_n by a finite sequence of elementary row operations. By Theorem 1.5.1, each of these operations can be accomplished by multiplying on the left by an appropriate elementary matrix. Thus we can find elementary matrices $E_1, E_2, \ldots, E_k$ such that

$$
E_k \cdots E_2 E_1 A = I_n
\tag{3}
$$

By Theorem 1.5.2, $E_1, E_2, \ldots, E_k$ are invertible. Multiplying both sides of Equation (3) on the left successively by $E_k^{-1}, \ldots, E_2^{-1}, E_1^{-1}$ we obtain

$$A = E_1^{-1}E_2^{-1}\cdots E_k^{-1}I_n = E_1^{-1}E_2^{-1}\cdots E_k^{-1} \tag{4}$$

By Theorem 1.5.2, this equation expresses A as a product of elementary matrices.

(d) $\Rightarrow$ (a)   If A is a product of elementary matrices, then from Theorems 1.4.6 and 1.5.2, the matrix A is a product of invertible matrices and hence is invertible.   ■

Row Equivalence

If a matrix B can be obtained from a matrix A by performing a finite sequence of elementary row operations, then obviously we can get from B back to A by performing the inverses of these elementary row operations in reverse order. Matrices that can be obtained from one another by a finite sequence of elementary row operations are said to be *row equivalent*. With this terminology, it follows from parts (*a*) and (*c*) of Theorem 3 that an $n \times n$ matrix A is invertible if and only if it is row equivalent to the $n \times n$ identity matrix.

A Method for Inverting Matrices

As our first application of Theorem 3, we shall establish a method for determining the inverse of an invertible matrix. Multiplying (3) on the right by A^{-1} yields

$$A^{-1} = E_k \cdots E_2 E_1 I_n \tag{5}$$

which tells us that A^{-1} can be obtained by multiplying I_n successively on the left by the elementary matrices $E_1, E_2, \ldots, E_k$. Since each multiplication on the left by one of these elementary matrices performs a row operation, it follows, by comparing Equations (3) and (5), that *the sequence of row operations that reduces A to I_n will reduce I_n to A^{-1}.* Thus we have the following result:

> To find the inverse of an invertible matrix A, we must find a sequence of elementary row operations that reduces A to the identity and then perform this same sequence of operations on I_n to obtain A^{-1}.

A simple method for carrying out this procedure is given in the following example.

EXAMPLE 4   Using Row Operations to Find A^{-1}

Find the inverse of

$$A = \begin{bmatrix} 1 & 2 & 3 \\ 2 & 5 & 3 \\ 1 & 0 & 8 \end{bmatrix}$$

Solution

We want to reduce A to the identity matrix by row operations and simultaneously apply these operations to I to produce A^{-1}. To accomplish this we shall adjoin the identity matrix to the right side of A, thereby producing a matrix of the form

$$[A \mid I]$$

Then we shall apply row operations to this matrix until the left side is reduced to I; these operations will convert the right side to A^{-1}, so the final matrix will have the form

$$[I \mid A^{-1}]$$

The computations are as follows:

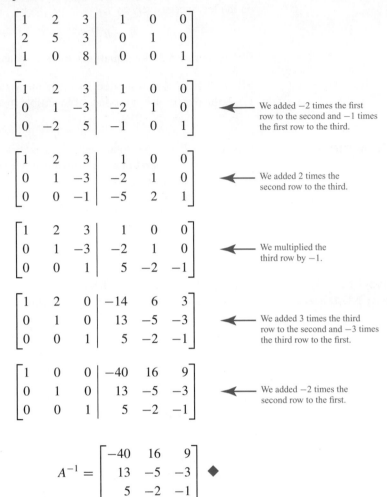

$$
\begin{bmatrix} 1 & 2 & 3 & 1 & 0 & 0 \\ 2 & 5 & 3 & 0 & 1 & 0 \\ 1 & 0 & 8 & 0 & 0 & 1 \end{bmatrix}
$$

$$
\begin{bmatrix} 1 & 2 & 3 & 1 & 0 & 0 \\ 0 & 1 & -3 & -2 & 1 & 0 \\ 0 & -2 & 5 & -1 & 0 & 1 \end{bmatrix}
$$
We added -2 times the first row to the second and -1 times the first row to the third.

$$
\begin{bmatrix} 1 & 2 & 3 & 1 & 0 & 0 \\ 0 & 1 & -3 & -2 & 1 & 0 \\ 0 & 0 & -1 & -5 & 2 & 1 \end{bmatrix}
$$
We added 2 times the second row to the third.

$$
\begin{bmatrix} 1 & 2 & 3 & 1 & 0 & 0 \\ 0 & 1 & -3 & -2 & 1 & 0 \\ 0 & 0 & 1 & 5 & -2 & -1 \end{bmatrix}
$$
We multiplied the third row by -1.

$$
\begin{bmatrix} 1 & 2 & 0 & -14 & 6 & 3 \\ 0 & 1 & 0 & 13 & -5 & -3 \\ 0 & 0 & 1 & 5 & -2 & -1 \end{bmatrix}
$$
We added 3 times the third row to the second and -3 times the third row to the first.

$$
\begin{bmatrix} 1 & 0 & 0 & -40 & 16 & 9 \\ 0 & 1 & 0 & 13 & -5 & -3 \\ 0 & 0 & 1 & 5 & -2 & -1 \end{bmatrix}
$$
We added -2 times the second row to the first.

Thus,

$$
A^{-1} = \begin{bmatrix} -40 & 16 & 9 \\ 13 & -5 & -3 \\ 5 & -2 & -1 \end{bmatrix} \blacklozenge
$$

Often it will not be known in advance whether a given matrix is invertible. If an $n \times n$ matrix A is not invertible, then it cannot be reduced to I_n by elementary row operations [part (c) of Theorem 3]. Stated another way, the reduced row-echelon form of A has at least one row of zeros. Thus, if the procedure in the last example is attempted on a matrix that is not invertible, then at some point in the computations a row of zeros will occur on the *left side*. It can then be concluded that the given matrix is not invertible, and the computations can be stopped.

EXAMPLE 5 Showing That a Matrix Is Not Invertible

Consider the matrix

$$
A = \begin{bmatrix} 1 & 6 & 4 \\ 2 & 4 & -1 \\ -1 & 2 & 5 \end{bmatrix}
$$

Applying the procedure of Example 4 yields

$$\left[\begin{array}{rrr|rrr} 1 & 6 & 4 & 1 & 0 & 0 \\ 2 & 4 & -1 & 0 & 1 & 0 \\ -1 & 2 & 5 & 0 & 0 & 1 \end{array}\right]$$

$$\left[\begin{array}{rrr|rrr} 1 & 6 & 4 & 1 & 0 & 0 \\ 0 & -8 & -9 & -2 & 1 & 0 \\ 0 & 8 & 9 & 1 & 0 & 1 \end{array}\right] \quad \longleftarrow \quad \begin{array}{l}\text{We added } -2 \text{ times the first} \\ \text{row to the second and added} \\ \text{the first row to the third.}\end{array}$$

$$\left[\begin{array}{rrr|rrr} 1 & 6 & 4 & 1 & 0 & 0 \\ 0 & -8 & -9 & -2 & 1 & 0 \\ 0 & 0 & 0 & -1 & 1 & 1 \end{array}\right] \quad \longleftarrow \quad \begin{array}{l}\text{We added the} \\ \text{second row to} \\ \text{the third.}\end{array}$$

Since we have obtained a row of zeros on the left side, A is not invertible. ◆

EXAMPLE 6 A Consequence of Invertibility

In Example 4 we showed that

$$A = \begin{bmatrix} 1 & 2 & 3 \\ 2 & 5 & 3 \\ 1 & 0 & 8 \end{bmatrix}$$

is an invertible matrix. From Theorem 3, it follows that the homogeneous system

$$\begin{aligned} x_1 + 2x_2 + 3x_3 &= 0 \\ 2x_1 + 5x_2 + 3x_3 &= 0 \\ x_1 \qquad\quad + 8x_3 &= 0 \end{aligned}$$

has only the trivial solution. ◆

EXERCISE SET 1.5

1. Which of the following are elementary matrices?

(a) $\begin{bmatrix} 1 & 0 \\ -5 & 1 \end{bmatrix}$ 　　 (b) $\begin{bmatrix} -5 & 1 \\ 1 & 0 \end{bmatrix}$ 　　 (c) $\begin{bmatrix} 1 & 0 \\ 0 & \sqrt{3} \end{bmatrix}$ 　　 (d) $\begin{bmatrix} 0 & 0 & 1 \\ 0 & 1 & 0 \\ 1 & 0 & 0 \end{bmatrix}$

(e) $\begin{bmatrix} 1 & 1 & 0 \\ 0 & 0 & 1 \\ 0 & 0 & 0 \end{bmatrix}$ 　　 (f) $\begin{bmatrix} 1 & 0 & 0 \\ 0 & 1 & 9 \\ 0 & 0 & 1 \end{bmatrix}$ 　　 (g) $\begin{bmatrix} 2 & 0 & 0 & 2 \\ 0 & 1 & 0 & 0 \\ 0 & 0 & 1 & 0 \\ 0 & 0 & 0 & 1 \end{bmatrix}$

2. Find a row operation that will restore the given elementary matrix to an identity matrix.

(a) $\begin{bmatrix} 1 & 0 \\ -3 & 1 \end{bmatrix}$ 　 (b) $\begin{bmatrix} 1 & 0 & 0 \\ 0 & 1 & 0 \\ 0 & 0 & 3 \end{bmatrix}$ 　 (c) $\begin{bmatrix} 0 & 0 & 0 & 1 \\ 0 & 1 & 0 & 0 \\ 0 & 0 & 1 & 0 \\ 1 & 0 & 0 & 0 \end{bmatrix}$ 　 (d) $\begin{bmatrix} 1 & 0 & -\frac{1}{7} & 0 \\ 0 & 1 & 0 & 0 \\ 0 & 0 & 1 & 0 \\ 0 & 0 & 0 & 1 \end{bmatrix}$

3. Consider the matrices

$$A = \begin{bmatrix} 3 & 4 & 1 \\ 2 & -7 & -1 \\ 8 & 1 & 5 \end{bmatrix}, \quad B = \begin{bmatrix} 8 & 1 & 5 \\ 2 & -7 & -1 \\ 3 & 4 & 1 \end{bmatrix}, \quad C = \begin{bmatrix} 3 & 4 & 1 \\ 2 & -7 & -1 \\ 2 & -7 & 3 \end{bmatrix}$$

Find elementary matrices E_1, E_2, E_3, and E_4 such that

(a) $E_1 A = B$ (b) $E_2 B = A$ (c) $E_3 A = C$ (d) $E_4 C = A$

4. In Exercise 3 is it possible to find an elementary matrix E such that $EB = C$? Justify your answer.

5. If a 2×2 matrix is multiplied on the left by the given matrices, what elementary row operation is performed on that matrix?

(a) $\begin{bmatrix} 0 & 1 \\ 1 & 0 \end{bmatrix}$ (b) $\begin{bmatrix} 2 & 0 \\ 0 & -3 \end{bmatrix}$ (c) $\begin{bmatrix} 1 & 0 \\ -2 & 1 \end{bmatrix}$

In Exercises 6–8 use the method shown in Examples 4 and 5 to find the inverse of the given matrix if the matrix is invertible, and check your answer by multiplication.

6. (a) $\begin{bmatrix} 1 & 4 \\ 2 & 7 \end{bmatrix}$ (b) $\begin{bmatrix} -3 & 6 \\ 4 & 5 \end{bmatrix}$ (c) $\begin{bmatrix} 6 & -4 \\ -3 & 2 \end{bmatrix}$

7. (a) $\begin{bmatrix} 3 & 4 & -1 \\ 1 & 0 & 3 \\ 2 & 5 & -4 \end{bmatrix}$ (b) $\begin{bmatrix} -1 & 3 & -4 \\ 2 & 4 & 1 \\ -4 & 2 & -9 \end{bmatrix}$ (c) $\begin{bmatrix} 1 & 0 & 1 \\ 0 & 1 & 1 \\ 1 & 1 & 0 \end{bmatrix}$

(d) $\begin{bmatrix} 2 & 6 & 6 \\ 2 & 7 & 6 \\ 2 & 7 & 7 \end{bmatrix}$ (e) $\begin{bmatrix} 1 & 0 & 1 \\ -1 & 1 & 1 \\ 0 & 1 & 0 \end{bmatrix}$

8. (a) $\begin{bmatrix} \frac{1}{5} & \frac{1}{5} & -\frac{2}{5} \\ \frac{1}{5} & \frac{1}{5} & \frac{1}{10} \\ \frac{1}{5} & -\frac{4}{5} & \frac{1}{10} \end{bmatrix}$ (b) $\begin{bmatrix} \sqrt{2} & 3\sqrt{2} & 0 \\ -4\sqrt{2} & \sqrt{2} & 0 \\ 0 & 0 & 1 \end{bmatrix}$ (c) $\begin{bmatrix} 1 & 0 & 0 & 0 \\ 1 & 3 & 0 & 0 \\ 1 & 3 & 5 & 0 \\ 1 & 3 & 5 & 7 \end{bmatrix}$

(d) $\begin{bmatrix} -8 & 17 & 2 & \frac{1}{3} \\ 4 & 0 & \frac{2}{5} & -9 \\ 0 & 0 & 0 & 0 \\ -1 & 13 & 4 & 2 \end{bmatrix}$ (e) $\begin{bmatrix} 0 & 0 & 2 & 0 \\ 1 & 0 & 0 & 1 \\ 0 & -1 & 3 & 0 \\ 2 & 1 & .5 & -3 \end{bmatrix}$

9. Find the inverse of each of the following 4×4 matrices, where k_1, k_2, k_3, k_4, and k are all nonzero.

(a) $\begin{bmatrix} k_1 & 0 & 0 & 0 \\ 0 & k_2 & 0 & 0 \\ 0 & 0 & k_3 & 0 \\ 0 & 0 & 0 & k_4 \end{bmatrix}$ (b) $\begin{bmatrix} 0 & 0 & 0 & k_1 \\ 0 & 0 & k_2 & 0 \\ 0 & k_3 & 0 & 0 \\ k_4 & 0 & 0 & 0 \end{bmatrix}$ (c) $\begin{bmatrix} k & 0 & 0 & 0 \\ 1 & k & 0 & 0 \\ 0 & 1 & k & 0 \\ 0 & 0 & 1 & k \end{bmatrix}$

10. Consider the matrix

$$A = \begin{bmatrix} 1 & 0 \\ -5 & 2 \end{bmatrix}$$

(a) Find elementary matrices E_1 and E_2 such that $E_2 E_1 A = I$.

(b) Write A^{-1} as a product of two elementary matrices.

(c) Write A as a product of two elementary matrices.

11. In each part, perform the stated row operation on

$$\begin{bmatrix} 2 & -1 & 0 \\ 4 & 5 & -3 \\ 1 & -4 & 7 \end{bmatrix}$$

by multiplying A on the left by a suitable elementary matrix. Check your answer in each case by performing the row operation directly on A.

(a)   Interchange the first and third rows.

(b)   Multiply the second row by $\frac{1}{3}$.

(c)   Add twice the second row to the first row.

12.   Write the matrix

$$\begin{bmatrix} 3 & -2 \\ 3 & -1 \end{bmatrix}$$

as a product of elementary matrices.

Note   There is more than one correct solution.

13.   Let

$$\begin{bmatrix} 1 & 0 & -2 \\ 0 & 4 & 3 \\ 0 & 0 & 1 \end{bmatrix}$$

(a)   Find elementary matrices E_1, E_2, and E_3 such that $E_3 E_2 E_1 A = I_3$.

(b)   Write A as a product of elementary matrices.

14.   Express the matrix

$$A = \begin{bmatrix} 0 & 1 & 7 & 8 \\ 1 & 3 & 3 & 8 \\ -2 & -5 & 1 & -8 \end{bmatrix}$$

in the form $A = EFGR$, where E, F, and G are elementary matrices and R is in row-echelon form.

15.   Show that if

$$A = \begin{bmatrix} 1 & 0 & 0 \\ 0 & 1 & 0 \\ a & b & c \end{bmatrix}$$

is an elementary matrix, then at least one entry in the third row must be a zero.

16.   Show that

$$A = \begin{bmatrix} 0 & a & 0 & 0 & 0 \\ b & 0 & c & 0 & 0 \\ 0 & d & 0 & e & 0 \\ 0 & 0 & f & 0 & g \\ 0 & 0 & 0 & h & 0 \end{bmatrix}$$

is not invertible for any values of the entries.

17.   Prove that if A is an $m \times n$ matrix, there is an invertible matrix C such that CA is in reduced row-echelon form.

18.   Prove that if A is an invertible matrix and B is row equivalent to A, then B is also invertible.

19.   (a)   Prove: If A and B are $m \times n$ matrices, then A and B are row equivalent if and only if A and B have the same reduced row-echelon form.

(b)   Show that A and B are row equivalent, and find a sequence of elementary row operations that produces B from A.

$$A = \begin{bmatrix} 1 & 2 & 3 \\ 1 & 4 & 1 \\ 2 & 1 & 9 \end{bmatrix}, \qquad B = \begin{bmatrix} 1 & 0 & 5 \\ 0 & 2 & -2 \\ 1 & 1 & 4 \end{bmatrix}$$

20.   Prove Theorem 1.5.1.

Discussion & Discovery

21. Suppose that A is some unknown invertible matrix, but you know of a sequence of elementary row operations that produces the identity matrix when applied in succession to A. Explain how you can use the known information to find A.

22. Indicate whether the statement is always true or sometimes false. Justify your answer with a logical argument or a counterexample.

   (a) Every square matrix can be expressed as a product of elementary matrices.

   (b) The product of two elementary matrices is an elementary matrix.

   (c) If A is invertible and a multiple of the first row of A is added to the second row, then the resulting matrix is invertible.

   (d) If A is invertible and $AB = 0$, then it must be true that $B = 0$.

23. Indicate whether the statement is always true or sometimes false. Justify your answer with a logical argument or a counterexample.

   (a) If A is a singular $n \times n$ matrix, then $A\mathbf{x} = \mathbf{0}$ has infinitely many solutions.

   (b) If A is a singular $n \times n$ matrix, then the reduced row-echelon form of A has at least one row of zeros.

   (c) If A^{-1} is expressible as a product of elementary matrices, then the homogeneous linear system $A\mathbf{x} = \mathbf{0}$ has only the trivial solution.

   (d) If A is a singular $n \times n$ matrix, and B results by interchanging two rows of A, then B may or may not be singular.

24. Do you think that there is a 2×2 matrix A such that

$$A \begin{bmatrix} a & b \\ c & d \end{bmatrix} = \begin{bmatrix} b & d \\ a & c \end{bmatrix}$$

for all values of a, b, c, and d? Explain your reasoning.

1.6 FURTHER RESULTS ON SYSTEMS OF EQUATIONS AND INVERTIBILITY

In this section we shall establish more results about systems of linear equations and invertibility of matrices. Our work will lead to a new method for solving n equations in n unknowns.

A Basic Theorem

In Section 1.1 we made the statement (based on Figure 1.1.1) that every linear system has no solutions, or has one solution, or has infinitely many solutions. We are now in a position to prove this fundamental result.

THEOREM 1.6.1

> *Every system of linear equations has no solutions, or has exactly one solution, or has infinitely many solutions.*

Proof If $A\mathbf{x} = \mathbf{b}$ is a system of linear equations, exactly one of the following is true: (a) the system has no solutions, (b) the system has exactly one solution, or (c) the system has more than one solution. The proof will be complete if we can show that the system has infinitely many solutions in case (c).

Assume that $A\mathbf{x} = \mathbf{b}$ has more than one solution, and let $\mathbf{x}_0 = \mathbf{x}_1 - \mathbf{x}_2$, where $\mathbf{x}_1$ and $\mathbf{x}_2$ are any two distinct solutions. Because $\mathbf{x}_1$ and $\mathbf{x}_2$ are distinct, the matrix $\mathbf{x}_0$ is nonzero; moreover,

$$A\mathbf{x}_0 = A(\mathbf{x}_1 - \mathbf{x}_2) = A\mathbf{x}_1 - A\mathbf{x}_2 = \mathbf{b} - \mathbf{b} = \mathbf{0}$$

If we now let k be any scalar, then

$$A(\mathbf{x}_1 + k\mathbf{x}_0) = A\mathbf{x}_1 + A(k\mathbf{x}_0) = A\mathbf{x}_1 + k(A\mathbf{x}_0)$$
$$= \mathbf{b} + k\mathbf{0} = \mathbf{b} + \mathbf{0} = \mathbf{b}$$

But this says that $\mathbf{x}_1 + k\mathbf{x}_0$ is a solution of $A\mathbf{x} = \mathbf{b}$. Since $\mathbf{x}_0$ is nonzero and there are infinitely many choices for k, the system $A\mathbf{x} = \mathbf{b}$ has infinitely many solutions. ∎

Solving Linear Systems by Matrix Inversion

Thus far, we have studied two methods for solving linear systems: Gaussian elimination and Gauss–Jordan elimination. The following theorem provides a new method for solving certain linear systems.

THEOREM 1.6.2

If A is an invertible $n \times n$ matrix, then for each $n \times 1$ matrix $\mathbf{b}$, the system of equations $A\mathbf{x} = \mathbf{b}$ has exactly one solution, namely, $\mathbf{x} = A^{-1}\mathbf{b}$.

Proof Since $A(A^{-1}\mathbf{b}) = \mathbf{b}$, it follows that $\mathbf{x} = A^{-1}\mathbf{b}$ is a solution of $A\mathbf{x} = \mathbf{b}$. To show that this is the only solution, we will assume that $\mathbf{x}_0$ is an arbitrary solution and then show that $\mathbf{x}_0$ must be the solution $A^{-1}\mathbf{b}$.

If $\mathbf{x}_0$ is any solution, then $A\mathbf{x}_0 = \mathbf{b}$. Multiplying both sides by A^{-1}, we obtain $\mathbf{x}_0 = A^{-1}\mathbf{b}$. ∎

EXAMPLE 1 Solution of a Linear System Using A^{-1}

Consider the system of linear equations

$$
\begin{aligned}
x_1 + 2x_2 + 3x_3 &= 5 \\
2x_1 + 5x_2 + 3x_3 &= 3 \\
x_1 \qquad\quad + 8x_3 &= 17
\end{aligned}
$$

In matrix form this system can be written as $A\mathbf{x} = \mathbf{b}$, where

$$A = \begin{bmatrix} 1 & 2 & 3 \\ 2 & 5 & 3 \\ 1 & 0 & 8 \end{bmatrix}, \qquad \mathbf{x} = \begin{bmatrix} x_1 \\ x_2 \\ x_3 \end{bmatrix}, \qquad \mathbf{b} = \begin{bmatrix} 5 \\ 3 \\ 17 \end{bmatrix}$$

In Example 4 of the preceding section, we showed that A is invertible and

$$A^{-1} = \begin{bmatrix} -40 & 16 & 9 \\ 13 & -5 & -3 \\ 5 & -2 & -1 \end{bmatrix}$$

By Theorem 1.6.2, the solution of the system is

$$\mathbf{x} = A^{-1}\mathbf{b} = \begin{bmatrix} -40 & 16 & 9 \\ 13 & -5 & -3 \\ 5 & -2 & -1 \end{bmatrix} \begin{bmatrix} 5 \\ 3 \\ 17 \end{bmatrix} = \begin{bmatrix} 1 \\ -1 \\ 2 \end{bmatrix}$$

or $x_1 = 1, x_2 = -1, x_3 = 2$. ◆

REMARK Note that the method of Example 1 applies only when the system has as many equations as unknowns and the coefficient matrix is invertible. This method is less efficient, computationally, than Gaussian elimination, but it is important in the analysis of equations involving matrices.

Linear Systems with a Common Coefficient Matrix

Frequently, one is concerned with solving a sequence of systems

$$A\mathbf{x} = \mathbf{b}_1, \quad A\mathbf{x} = \mathbf{b}_2, \quad A\mathbf{x} = \mathbf{b}_3, \ldots, \quad A\mathbf{x} = \mathbf{b}_k$$

each of which has the same square coefficient matrix A. If A is invertible, then the solutions

$$\mathbf{x}_1 = A^{-1}\mathbf{b}_1, \quad \mathbf{x}_2 = A^{-1}\mathbf{b}_2, \quad \mathbf{x}_3 = A^{-1}\mathbf{b}_3, \ldots, \quad \mathbf{x}_k = A^{-1}\mathbf{b}_k$$

can be obtained with one matrix inversion and k matrix multiplications. Once again, however, a more efficient method is to form the matrix

$$[A \mid \mathbf{b}_1 \mid \mathbf{b}_2 \mid \cdots \mid \mathbf{b}_k] \tag{1}$$

in which the coefficient matrix A is "augmented" by all k of the matrices $\mathbf{b}_1, \mathbf{b}_2, \ldots, \mathbf{b}_k$, and then reduce (1) to reduced row-echelon form by Gauss–Jordan elimination. In this way we can solve all k systems at once. This method has the added advantage that it applies even when A is not invertible.

EXAMPLE 2 Solving Two Linear Systems at Once

Solve the systems

$$
\begin{array}{ll}
\text{(a)} \quad
\begin{aligned}
x_1 + 2x_2 + 3x_3 &= 4 \\
2x_1 + 5x_2 + 3x_3 &= 5 \\
x_1 \qquad\ + 8x_3 &= 9
\end{aligned}
&
\text{(b)} \quad
\begin{aligned}
x_1 + 2x_2 + 3x_3 &= 1 \\
2x_1 + 5x_2 + 3x_3 &= 6 \\
x_1 \qquad\ + 8x_3 &= -6
\end{aligned}
\end{array}
$$

Solution

The two systems have the same coefficient matrix. If we augment this coefficient matrix with the columns of constants on the right sides of these systems, we obtain

$$
\left[\begin{array}{ccc|cc}
1 & 2 & 3 & 4 & 1 \\
2 & 5 & 3 & 5 & 6 \\
1 & 0 & 8 & 9 & -6
\end{array}\right]
$$

Reducing this matrix to reduced row-echelon form yields (verify)

$$
\left[\begin{array}{ccc|cc}
1 & 0 & 0 & 1 & 2 \\
0 & 1 & 0 & 0 & 1 \\
0 & 0 & 1 & 1 & -1
\end{array}\right]
$$

It follows from the last two columns that the solution of system (a) is $x_1 = 1$, $x_2 = 0$, $x_3 = 1$ and the solution of system (b) is $x_1 = 2$, $x_2 = 1$, $x_3 = -1$. ◆

Properties of Invertible Matrices

Up to now, to show that an $n \times n$ matrix A is invertible, it has been necessary to find an $n \times n$ matrix B such that

$$AB = I \quad \text{and} \quad BA = I$$

The next theorem shows that if we produce an $n \times n$ matrix B satisfying *either* condition, then the other condition holds automatically.

THEOREM 1.6.3

> *Let A be a square matrix.*
>
> (*a*) *If B is a square matrix satisfying $BA = I$, then $B = A^{-1}$.*
> (*b*) *If B is a square matrix satisfying $AB = I$, then $B = A^{-1}$.*

We shall prove part (*a*) and leave part (*b*) as an exercise.

***Proof* (a)**   Assume that $BA = I$. If we can show that A is invertible, the proof can be completed by multiplying $BA = I$ on both sides by A^{-1} to obtain

$$BAA^{-1} = IA^{-1} \quad \text{or} \quad BI = IA^{-1} \quad \text{or} \quad B = A^{-1}$$

To show that A is invertible, it suffices to show that the system $A\mathbf{x} = \mathbf{0}$ has only the trivial solution (see Theorem 3). Let $\mathbf{x}_0$ be any solution of this system. If we multiply both sides of $A\mathbf{x}_0 = \mathbf{0}$ on the left by B, we obtain $BA\mathbf{x}_0 = B\mathbf{0}$ or $I\mathbf{x}_0 = \mathbf{0}$ or $\mathbf{x}_0 = \mathbf{0}$. Thus, the system of equations $A\mathbf{x} = \mathbf{0}$ has only the trivial solution. ∎

We are now in a position to add two more statements that are equivalent to the four given in Theorem 3.

THEOREM 1.6.4

> ### Equivalent Statements
>
> *If A is an $n \times n$ matrix, then the following are equivalent.*
>
> (*a*) *A is invertible.*
> (*b*) *$A\mathbf{x} = \mathbf{0}$ has only the trivial solution.*
> (*c*) *The reduced row-echelon form of A is I_n.*
> (*d*) *A is expressible as a product of elementary matrices.*
> (*e*) *$A\mathbf{x} = \mathbf{b}$ is consistent for every $n \times 1$ matrix $\mathbf{b}$.*
> (*f*) *$A\mathbf{x} = \mathbf{b}$ has exactly one solution for every $n \times 1$ matrix $\mathbf{b}$.*

Proof   Since we proved in Theorem 3 that (*a*), (*b*), (*c*), and (*d*) are equivalent, it will be sufficient to prove that $(a) \Rightarrow (f) \Rightarrow (e) \Rightarrow (a)$.

(a) ⇒ (f)   This was already proved in Theorem 1.6.2.

(f) ⇒ (e)   This is self-evident: If $A\mathbf{x} = \mathbf{b}$ has exactly one solution for every $n \times 1$ matrix $\mathbf{b}$, then $A\mathbf{x} = \mathbf{b}$ is consistent for every $n \times 1$ matrix $\mathbf{b}$.

(e) ⇒ (a)   If the system $A\mathbf{x} = \mathbf{b}$ is consistent for every $n \times 1$ matrix $\mathbf{b}$, then in particular, the systems

$$A\mathbf{x} = \begin{bmatrix} 1 \\ 0 \\ 0 \\ \vdots \\ 0 \end{bmatrix}, \quad A\mathbf{x} = \begin{bmatrix} 0 \\ 1 \\ 0 \\ \vdots \\ 0 \end{bmatrix}, \dots, \quad A\mathbf{x} = \begin{bmatrix} 0 \\ 0 \\ 0 \\ \vdots \\ 1 \end{bmatrix}$$

are consistent. Let $\mathbf{x}_1, \mathbf{x}_2, \dots, \mathbf{x}_n$ be solutions of the respective systems, and let us form an $n \times n$ matrix C having these solutions as columns. Thus C has the form

$$C = [\mathbf{x}_1 \mid \mathbf{x}_2 \mid \cdots \mid \mathbf{x}_n]$$

As discussed in Section 1.3, the successive columns of the product AC will be

$$A\mathbf{x}_1, A\mathbf{x}_2, \dots, A\mathbf{x}_n$$

Thus

$$AC = [A\mathbf{x}_1 \mid A\mathbf{x}_2 \mid \cdots \mid A\mathbf{x}_n] = \begin{bmatrix} 1 & 0 & \cdots & 0 \\ 0 & 1 & \cdots & 0 \\ 0 & 0 & \cdots & 0 \\ \vdots & \vdots & & \vdots \\ 0 & 0 & \cdots & 1 \end{bmatrix} = I$$

By part (b) of Theorem 1.6.3, it follows that $C = A^{-1}$. Thus, A is invertible. ∎

We know from earlier work that invertible matrix factors produce an invertible product. The following theorem looks at the converse: It shows that if the product of square matrices is invertible, then the factors themselves must be invertible.

THEOREM 1.6.5

> *Let A and B be square matrices of the same size. If AB is invertible, then A and B must also be invertible.*

In our later work the following fundamental problem will occur frequently in various contexts.

A Fundamental Problem: Let A be a fixed $m \times n$ matrix. Find all $m \times 1$ matrices $\mathbf{b}$ such that the system of equations $A\mathbf{x} = \mathbf{b}$ is consistent.

If A is an invertible matrix, Theorem 1.6.2 completely solves this problem by asserting that for *every* $m \times 1$ matrix $\mathbf{b}$, the linear system $A\mathbf{x} = \mathbf{b}$ has the unique solution $\mathbf{x} = A^{-1}\mathbf{b}$. If A is not square, or if A is square but not invertible, then Theorem 1.6.2 does not apply. In these cases the matrix $\mathbf{b}$ must usually satisfy certain conditions in order for $A\mathbf{x} = \mathbf{b}$ to be consistent. The following example illustrates how the elimination methods of Section 1.2 can be used to determine such conditions.

EXAMPLE 3 Determining Consistency by Elimination

What conditions must b_1, b_2, and b_3 satisfy in order for the system of equations

$$\begin{aligned} x_1 + x_2 + 2x_3 &= b_1 \\ x_1 \quad\quad + x_3 &= b_2 \\ 2x_1 + x_2 + 3x_3 &= b_3 \end{aligned}$$

to be consistent?

Solution

The augmented matrix is

$$\begin{bmatrix} 1 & 1 & 2 & b_1 \\ 1 & 0 & 1 & b_2 \\ 2 & 1 & 3 & b_3 \end{bmatrix}$$

which can be reduced to row-echelon form as follows:

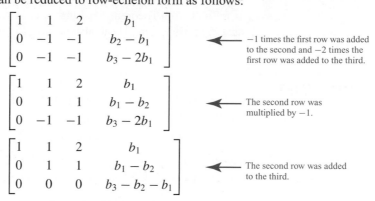

$$\begin{bmatrix} 1 & 1 & 2 & b_1 \\ 0 & -1 & -1 & b_2 - b_1 \\ 0 & -1 & -1 & b_3 - 2b_1 \end{bmatrix}$$ ← −1 times the first row was added to the second and −2 times the first row was added to the third.

$$\begin{bmatrix} 1 & 1 & 2 & b_1 \\ 0 & 1 & 1 & b_1 - b_2 \\ 0 & -1 & -1 & b_3 - 2b_1 \end{bmatrix}$$ ← The second row was multiplied by −1.

$$\begin{bmatrix} 1 & 1 & 2 & b_1 \\ 0 & 1 & 1 & b_1 - b_2 \\ 0 & 0 & 0 & b_3 - b_2 - b_1 \end{bmatrix}$$ ← The second row was added to the third.

It is now evident from the third row in the matrix that the system has a solution if and only if b_1, b_2, and b_3 satisfy the condition

$$b_3 - b_2 - b_1 = 0 \quad \text{or} \quad b_3 = b_1 + b_2$$

To express this condition another way, $A\mathbf{x} = \mathbf{b}$ is consistent if and only if $\mathbf{b}$ is a matrix of the form

$$\mathbf{b} = \begin{bmatrix} b_1 \\ b_2 \\ b_1 + b_2 \end{bmatrix}$$

where b_1 and b_2 are arbitrary. ◆

EXAMPLE 4 Determining Consistency by Elimination

What conditions must b_1, b_2, and b_3 satisfy in order for the system of equations

$$x_1 + 2x_2 + 3x_3 = b_1$$
$$2x_1 + 5x_2 + 3x_3 = b_2$$
$$x_1 \qquad + 8x_3 = b_3$$

to be consistent?

Solution

The augmented matrix is

$$\begin{bmatrix} 1 & 2 & 3 & b_1 \\ 2 & 5 & 3 & b_2 \\ 1 & 0 & 8 & b_3 \end{bmatrix}$$

Reducing this to reduced row-echelon form yields (verify)

$$\begin{bmatrix} 1 & 0 & 0 & -40b_1 + 16b_2 + 9b_3 \\ 0 & 1 & 0 & 13b_1 - 5b_2 - 3b_3 \\ 0 & 0 & 1 & 5b_1 - 2b_2 - b_3 \end{bmatrix} \tag{2}$$

In this case there are no restrictions on b_1, b_2, and b_3; that is, the given system $A\mathbf{x} = \mathbf{b}$ has the unique solution

$$x_1 = -40b_1 + 16b_2 + 9b_3, \quad x_2 = 13b_1 - 5b_2 - 3b_3, \quad x_3 = 5b_1 - 2b_2 - b_3 \tag{3}$$

for all $\mathbf{b}$. ◆

REMARK Because the system $A\mathbf{x} = \mathbf{b}$ in the preceding example is consistent for all $\mathbf{b}$, it follows from Theorem 1.6.4 that A is invertible. We leave it for the reader to verify that the formulas in (3) can also be obtained by calculating $\mathbf{x} = A^{-1}\mathbf{b}$.

EXERCISE SET 1.6

In Exercises 1–8 solve the system by inverting the coefficient matrix and using Theorem 1.6.2.

1. $\begin{aligned} x_1 + x_2 &= 2 \\ 5x_1 + 6x_2 &= 9 \end{aligned}$

2. $\begin{aligned} 4x_1 - 3x_2 &= -3 \\ 2x_1 - 5x_2 &= 9 \end{aligned}$

3. $\begin{aligned} x_1 + 3x_2 + x_3 &= 4 \\ 2x_1 + 2x_2 + x_3 &= -1 \\ 2x_1 + 3x_2 + x_3 &= 3 \end{aligned}$

4. $\begin{aligned} 5x_1 + 3x_2 + 2x_3 &= 4 \\ 3x_1 + 3x_2 + 2x_3 &= 2 \\ x_2 + x_3 &= 5 \end{aligned}$

5. $\begin{aligned} x + y + z &= 5 \\ x + y - 4z &= 10 \\ -4x + y + z &= 0 \end{aligned}$

6. $\begin{aligned} -x - 2y - 3z &= 0 \\ w + x + 4y + 4z &= 7 \\ w + 3x + 7y + 9z &= 4 \\ -w - 2x - 4y - 6z &= 6 \end{aligned}$

7. $\begin{aligned} 3x_1 + 5x_2 &= b_1 \\ x_1 + 2x_2 &= b_2 \end{aligned}$

8. $\begin{aligned} x_1 + 2x_2 + 3x_3 &= b_1 \\ 2x_1 + 5x_2 + 5x_3 &= b_2 \\ 3x_1 + 5x_2 + 8x_3 &= b_3 \end{aligned}$

9. Solve the following general system by inverting the coefficient matrix and using Theorem 1.6.2.

$$\begin{aligned} x_1 + 2x_2 + x_3 &= b_1 \\ x_1 - x_2 + x_3 &= b_2 \\ x_1 + x_2 &= b_3 \end{aligned}$$

Use the resulting formulas to find the solution if

(a) $b_1 = -1, \quad b_2 = 3, \quad b_3 = 4$ (b) $b_1 = 5, \quad b_2 = 0, \quad b_3 = 0$

(c) $b_1 = -1, \quad b_2 = -1, \quad b_3 = 3$

10. Solve the three systems in Exercise 9 using the method of Example 2.

In Exercises 11–14 use the method of Example 2 to solve the systems in all parts simultaneously.

11. $\begin{aligned} x_1 - 5x_2 &= b_1 \\ 3x_1 + 2x_2 &= b_2 \end{aligned}$

(a) $b_1 = 1, \quad b_2 = 4$

(b) $b_1 = -2, \quad b_2 = 5$

12. $\begin{aligned} -x_1 + 4x_2 + x_3 &= b_1 \\ x_1 + 9x_2 - 2x_3 &= b_2 \\ 6x_1 + 4x_2 - 8x_3 &= b_3 \end{aligned}$

(a) $b_1 = 0, \quad b_2 = 1, \quad b_3 = 0$

(b) $b_1 = -3, \quad b_2 = 4, \quad b_3 = -5$

13. $\begin{aligned} 4x_1 - 7x_2 &= b_1 \\ x_1 + 2x_2 &= b_2 \end{aligned}$

(a) $b_1 = 0, \quad b_2 = 1$

(b) $b_1 = -4, \quad b_2 = 6$

(c) $b_1 = -1, \quad b_2 = 3$

(d) $b_1 = -5, \quad b_2 = 1$

14. $\begin{aligned} x_1 + 3x_2 + 5x_3 &= b_1 \\ -x_1 - 2x_2 &= b_2 \\ 2x_1 + 5x_2 + 4x_3 &= b_3 \end{aligned}$

(a) $b_1 = 1, \quad b_2 = 0, \quad b_3 = -1$

(b) $b_1 = 0, \quad b_2 = 1, \quad b_3 = 1$

(c) $b_1 = -1, \quad b_2 = -1, \quad b_3 = 0$

15. The method of Example 2 can be used for linear systems with infinitely many solutions. Use that method to solve the systems in both parts at the same time.

(a) $\begin{aligned} x_1 - 2x_2 + x_3 &= -2 \\ 2x_1 - 5x_2 + x_3 &= 1 \\ 3x_1 - 7x_2 + 2x_3 &= -1 \end{aligned}$

(b) $\begin{aligned} x_1 - 2x_2 + x_3 &= 1 \\ 2x_1 - 5x_2 + x_3 &= -1 \\ 3x_1 - 7x_2 + 2x_3 &= 0 \end{aligned}$

In Exercises 16–19 find conditions that the b's must satisfy for the system to be consistent.

16. $\begin{aligned} 6x_1 - 4x_2 &= b_1 \\ 3x_1 - 2x_2 &= b_2 \end{aligned}$

17. $\begin{aligned} x_1 - 2x_2 + 5x_3 &= b_1 \\ 4x_1 - 5x_2 + 8x_3 &= b_2 \\ -3x_1 + 3x_2 - 3x_3 &= b_3 \end{aligned}$

18.
$$x_1 - 2x_2 - x_3 = b_1$$
$$-4x_1 + 5x_2 + 2x_3 = b_2$$
$$-4x_1 + 7x_2 + 4x_3 = b_3$$

19.
$$x_1 - x_2 + 3x_3 + 2x_4 = b_1$$
$$-2x_1 + x_2 + 5x_3 + x_4 = b_2$$
$$-3x_1 + 2x_2 + 2x_3 - x_4 = b_3$$
$$4x_1 - 3x_2 + x_3 + 3x_4 = b_4$$

20. Consider the matrices

$$A = \begin{bmatrix} 2 & 1 & 2 \\ 2 & 2 & -2 \\ 3 & 1 & 1 \end{bmatrix} \quad \text{and} \quad \mathbf{x} = \begin{bmatrix} x_1 \\ x_2 \\ x_3 \end{bmatrix}$$

(a) Show that the equation $A\mathbf{x} = \mathbf{x}$ can be rewritten as $(A - I)\mathbf{x} = \mathbf{0}$ and use this result to solve $A\mathbf{x} = \mathbf{x}$ for $\mathbf{x}$.

(b) Solve $A\mathbf{x} = 4\mathbf{x}$.

21. Solve the following matrix equation for X.

$$\begin{bmatrix} 1 & -1 & 1 \\ 2 & 3 & 0 \\ 0 & 2 & -1 \end{bmatrix} X = \begin{bmatrix} 2 & -1 & 5 & 7 & 8 \\ 4 & 0 & -3 & 0 & 1 \\ 3 & 5 & -7 & 2 & 1 \end{bmatrix}$$

22. In each part, determine whether the homogeneous system has a nontrivial solution (without using pencil and paper); then state whether the given matrix is invertible.

(a)
$$2x_1 + x_2 - 3x_3 + x_4 = 0$$
$$5x_2 + 4x_3 + 3x_4 = 0$$
$$x_3 + 2x_4 = 0$$
$$3x_4 = 0$$
$$\begin{bmatrix} 2 & 1 & -3 & 1 \\ 0 & 5 & 4 & 3 \\ 0 & 0 & 1 & 2 \\ 0 & 0 & 0 & 3 \end{bmatrix}$$

(b)
$$5x_1 + x_2 + 4x_3 + x_4 = 0$$
$$2x_3 - x_4 = 0$$
$$x_3 + x_4 = 0$$
$$7x_4 = 0$$
$$\begin{bmatrix} 5 & 1 & 4 & 1 \\ 0 & 0 & 2 & -1 \\ 0 & 0 & 1 & 1 \\ 0 & 0 & 0 & 7 \end{bmatrix}$$

23. Let $A\mathbf{x} = \mathbf{0}$ be a homogeneous system of n linear equations in n unknowns that has only the trivial solution. Show that if k is any positive integer, then the system $A^k\mathbf{x} = \mathbf{0}$ also has only the trivial solution.

24. Let $A\mathbf{x} = \mathbf{0}$ be a homogeneous system of n linear equations in n unknowns, and let Q be an invertible $n \times n$ matrix. Show that $A\mathbf{x} = \mathbf{0}$ has just the trivial solution if and only if $(QA)\mathbf{x} = \mathbf{0}$ has just the trivial solution.

25. Let $A\mathbf{x} = \mathbf{b}$ be any consistent system of linear equations, and let $\mathbf{x}_1$ be a fixed solution. Show that every solution to the system can be written in the form $\mathbf{x} = \mathbf{x}_1 + \mathbf{x}_0$, where $\mathbf{x}_0$ is a solution to $A\mathbf{x} = \mathbf{0}$. Show also that every matrix of this form is a solution.

26. Use part (a) of Theorem 1.6.3 to prove part (b).

27. What restrictions must be placed on x and y for the following matrices to be invertible?

(a) $\begin{bmatrix} x & y \\ x & x \end{bmatrix}$ (b) $\begin{bmatrix} x & 0 \\ y & y \end{bmatrix}$ (c) $\begin{bmatrix} x & y \\ y & x \end{bmatrix}$

Discussion & Discovery

28. (a) If A is an $n \times n$ matrix and if $\mathbf{b}$ is an $n \times 1$ matrix, what conditions would you impose to ensure that the equation $\mathbf{x} = A\mathbf{x} + \mathbf{b}$ has a unique solution for $\mathbf{x}$?

(b) Assuming that your conditions are satisfied, find a formula for the solution in terms of an appropriate inverse.

29. Suppose that A is an invertible $n \times n$ matrix. Must the system of equations $A\mathbf{x} = \mathbf{x}$ have a unique solution? Explain your reasoning.

30. Is it possible to have $AB = I$ without B being the inverse of A? Explain your reasoning.

31. Create a theorem by rewriting Theorem 1.6.5 in contrapositive form (see Exercise 34 of Section 1.4).

1.7
DIAGONAL, TRIANGULAR, AND SYMMETRIC MATRICES

In this section we shall consider certain classes of matrices that have special forms. The matrices that we study in this section are among the most important kinds of matrices encountered in linear algebra and will arise in many different settings throughout the text.

Diagonal Matrices

A square matrix in which all the entries off the main diagonal are zero is called a ***diagonal matrix***. Here are some examples:

$$\begin{bmatrix} 2 & 0 \\ 0 & -5 \end{bmatrix}, \qquad \begin{bmatrix} 1 & 0 & 0 \\ 0 & 1 & 0 \\ 0 & 0 & 1 \end{bmatrix}, \qquad \begin{bmatrix} 6 & 0 & 0 & 0 \\ 0 & -4 & 0 & 0 \\ 0 & 0 & 0 & 0 \\ 0 & 0 & 0 & 8 \end{bmatrix}$$

A general $n \times n$ diagonal matrix D can be written as

$$D = \begin{bmatrix} d_1 & 0 & \cdots & 0 \\ 0 & d_2 & \cdots & 0 \\ \vdots & \vdots & & \vdots \\ 0 & 0 & \cdots & d_n \end{bmatrix} \tag{1}$$

A diagonal matrix is invertible if and only if all of its diagonal entries are nonzero; in this case the inverse of (1) is

$$D^{-1} = \begin{bmatrix} 1/d_1 & 0 & \cdots & 0 \\ 0 & 1/d_2 & \cdots & 0 \\ \vdots & \vdots & & \vdots \\ 0 & 0 & \cdots & 1/d_n \end{bmatrix}$$

The reader should verify that $DD^{-1} = D^{-1}D = I$.

Powers of diagonal matrices are easy to compute; we leave it for the reader to verify that if D is the diagonal matrix (1) and k is a positive integer, then

$$D^k = \begin{bmatrix} d_1^k & 0 & \cdots & 0 \\ 0 & d_2^k & \cdots & 0 \\ \vdots & \vdots & & \vdots \\ 0 & 0 & \cdots & d_n^k \end{bmatrix}$$

EXAMPLE 1 Inverses and Powers of Diagonal Matrices

If

$$A = \begin{bmatrix} 1 & 0 & 0 \\ 0 & -3 & 0 \\ 0 & 0 & 2 \end{bmatrix}$$

then

$$A^{-1} = \begin{bmatrix} 1 & 0 & 0 \\ 0 & -\frac{1}{3} & 0 \\ 0 & 0 & \frac{1}{2} \end{bmatrix}, \quad A^5 = \begin{bmatrix} 1 & 0 & 0 \\ 0 & -243 & 0 \\ 0 & 0 & 32 \end{bmatrix}, \quad A^{-5} = \begin{bmatrix} 1 & 0 & 0 \\ 0 & -\frac{1}{243} & 0 \\ 0 & 0 & \frac{1}{32} \end{bmatrix}$$

◆

Matrix products that involve diagonal factors are especially easy to compute. For example,

$$\begin{bmatrix} d_1 & 0 & 0 \\ 0 & d_2 & 0 \\ 0 & 0 & d_3 \end{bmatrix} \begin{bmatrix} a_{11} & a_{12} & a_{13} & a_{14} \\ a_{21} & a_{22} & a_{23} & a_{24} \\ a_{31} & a_{32} & a_{33} & a_{34} \end{bmatrix} = \begin{bmatrix} d_1 a_{11} & d_1 a_{12} & d_1 a_{13} & d_1 a_{14} \\ d_2 a_{21} & d_2 a_{22} & d_2 a_{23} & d_2 a_{24} \\ d_3 a_{31} & d_3 a_{32} & d_3 a_{33} & d_3 a_{34} \end{bmatrix}$$

$$\begin{bmatrix} a_{11} & a_{12} & a_{13} \\ a_{21} & a_{22} & a_{23} \\ a_{31} & a_{32} & a_{33} \\ a_{41} & a_{42} & a_{43} \end{bmatrix} \begin{bmatrix} d_1 & 0 & 0 \\ 0 & d_2 & 0 \\ 0 & 0 & d_3 \end{bmatrix} = \begin{bmatrix} d_1 a_{11} & d_2 a_{12} & d_3 a_{13} \\ d_1 a_{21} & d_2 a_{22} & d_3 a_{23} \\ d_1 a_{31} & d_2 a_{32} & d_3 a_{33} \\ d_1 a_{41} & d_2 a_{42} & d_3 a_{43} \end{bmatrix}$$

In words, *to multiply a matrix A on the left by a diagonal matrix D, one can multiply successive rows of A by the successive diagonal entries of D, and to multiply A on the right by D, one can multiply successive columns of A by the successive diagonal entries of D.*

Triangular Matrices

A square matrix in which all the entries above the main diagonal are zero is called ***lower triangular***, and a square matrix in which all the entries below the main diagonal are zero is called ***upper triangular***. A matrix that is either upper triangular or lower triangular is called ***triangular***.

EXAMPLE 2 Upper and Lower Triangular Matrices

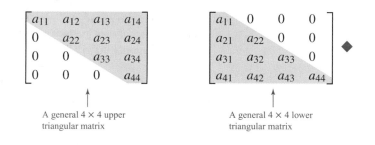

A general 4 × 4 upper triangular matrix

A general 4 × 4 lower triangular matrix

REMARK   Observe that diagonal matrices are both upper triangular and lower triangular since they have zeros below and above the main diagonal. Observe also that a *square* matrix in row-echelon form is upper triangular since it has zeros below the main diagonal.

The following are four useful characterizations of triangular matrices. The reader will find it instructive to verify that the matrices in Example 2 have the stated properties.

- A square matrix $A = [a_{ij}]$ is upper triangular if and only if the ith row starts with at least $i - 1$ zeros.

- A square matrix $A = [a_{ij}]$ is lower triangular if and only if the jth column starts with at least $j - 1$ zeros.

- A square matrix $A = [a_{ij}]$ is upper triangular if and only if $a_{ij} = 0$ for $i > j$.

- A square matrix $A = [a_{ij}]$ is lower triangular if and only if $a_{ij} = 0$ for $i < j$.

The following theorem lists some of the basic properties of triangular matrices.

THEOREM 1.7.1

(a) *The transpose of a lower triangular matrix is upper triangular, and the transpose of an upper triangular matrix is lower triangular.*

(b) *The product of lower triangular matrices is lower triangular, and the product of upper triangular matrices is upper triangular.*

(c) *A triangular matrix is invertible if and only if its diagonal entries are all nonzero.*

(d) *The inverse of an invertible lower triangular matrix is lower triangular, and the inverse of an invertible upper triangular matrix is upper triangular.*

Part (*a*) is evident from the fact that transposing a square matrix can be accomplished by reflecting the entries about the main diagonal; we omit the formal proof. We will prove (*b*), but we will defer the proofs of (*c*) and (*d*) to the next chapter, where we will have the tools to prove those results more efficiently.

***Proof* (b)**   We will prove the result for lower triangular matrices; the proof for upper triangular matrices is similar. Let $A = [a_{ij}]$ and $B = [b_{ij}]$ be lower triangular $n \times n$ matrices, and let $C = [c_{ij}]$ be the product $C = AB$. From the remark preceding this theorem, we can prove that C is lower triangular by showing that $c_{ij} = 0$ for $i < j$. But from the definition of matrix multiplication,

$$c_{ij} = a_{i1}b_{1j} + a_{i2}b_{2j} + \cdots + a_{in}b_{nj}$$

If we assume that $i < j$, then the terms in this expression can be grouped as follows:

$$c_{ij} = \underbrace{a_{i1}b_{1j} + a_{i2}b_{2j} + \cdots + a_{i(j-1)}b_{(j-1)j}}_{\substack{\text{Terms in which the row} \\ \text{number of } b \text{ is less than the} \\ \text{column number of } b}} + \underbrace{a_{ij}b_{jj} + \cdots + a_{in}b_{nj}}_{\substack{\text{Terms in which the row} \\ \text{number of } a \text{ is less than} \\ \text{the column number of } a}}$$

In the first grouping all of the b factors are zero since B is lower triangular, and in the second grouping all of the a factors are zero since A is lower triangular. Thus, $c_{ij} = 0$, which is what we wanted to prove. ■

EXAMPLE 3   Upper Triangular Matrices

Consider the upper triangular matrices

$$A = \begin{bmatrix} 1 & 3 & -1 \\ 0 & 2 & 4 \\ 0 & 0 & 5 \end{bmatrix}, \qquad B = \begin{bmatrix} 3 & -2 & 2 \\ 0 & 0 & -1 \\ 0 & 0 & 1 \end{bmatrix}$$

The matrix A is invertible, since its diagonal entries are nonzero, but the matrix B is not. We leave it for the reader to calculate the inverse of A by the method of Section 1.5 and show that

$$A^{-1} = \begin{bmatrix} 1 & -\frac{3}{2} & \frac{7}{5} \\ 0 & \frac{1}{2} & -\frac{2}{5} \\ 0 & 0 & \frac{1}{5} \end{bmatrix}$$

This inverse is upper triangular, as guaranteed by part (d) of Theorem 1.7.1. We also leave it for the reader to check that the product AB is

$$AB = \begin{bmatrix} 3 & -2 & -2 \\ 0 & 0 & 2 \\ 0 & 0 & 5 \end{bmatrix}$$

This product is upper triangular, as guaranteed by part (b) of Theorem 1.7.1. ◆

Symmetric Matrices

A square matrix A is called **symmetric** if $A = A^T$.

EXAMPLE 4 Symmetric Matrices

The following matrices are symmetric, since each is equal to its own transpose (verify).

$$\begin{bmatrix} 7 & -3 \\ -3 & 5 \end{bmatrix}, \quad \begin{bmatrix} 1 & 4 & 5 \\ 4 & -3 & 0 \\ 5 & 0 & 7 \end{bmatrix}, \quad \begin{bmatrix} d_1 & 0 & 0 & 0 \\ 0 & d_2 & 0 & 0 \\ 0 & 0 & d_3 & 0 \\ 0 & 0 & 0 & d_4 \end{bmatrix} \quad ◆$$

It is easy to recognize symmetric matrices by inspection: The entries on the main diagonal may be arbitrary, but as shown in (2), "mirror images" of entries across the main diagonal must be equal.

$$\begin{bmatrix} 1 & 4 & 5 \\ 4 & -3 & 0 \\ 5 & 0 & 7 \end{bmatrix} \tag{2}$$

This follows from the fact that transposing a square matrix can be accomplished by interchanging entries that are symmetrically positioned about the main diagonal. Expressed in terms of the individual entries, a matrix $A = [a_{ij}]$ is symmetric if and only if $a_{ij} = a_{ji}$ for all values of i and j. As illustrated in Example 4, all diagonal matrices are symmetric.

The following theorem lists the main algebraic properties of symmetric matrices. The proofs are direct consequences of Theorem 1.4.9 and are left for the reader.

THEOREM 1.7.2

If A and B are symmetric matrices with the same size, and if k is any scalar, then:

(a) A^T *is symmetric.*
(b) $A + B$ *and* $A - B$ *are symmetric.*
(c) kA *is symmetric.*

REMARK   It is not true, in general, that the product of symmetric matrices is symmetric. To see why this is so, let A and B be symmetric matrices with the same size. Then from part (d) of Theorem 1.4.9 and the symmetry, we have

$$(AB)^T = B^T A^T = BA$$

Since AB and BA are not usually equal, it follows that AB will not usually be symmetric. However, in the special case where $AB = BA$, the product AB will be symmetric. If A and B are matrices such that $AB = BA$, then we say that A and B **commute**. In summary: *The product of two symmetric matrices is symmetric if and only if the matrices commute.*

EXAMPLE 5   Products of Symmetric Matrices

The first of the following equations shows a product of symmetric matrices that *is not* symmetric, and the second shows a product of symmetric matrices that *is* symmetric. We conclude that the factors in the first equation do not commute, but those in the second equation do. We leave it for the reader to verify that this is so.

$$\begin{bmatrix} 1 & 2 \\ 2 & 3 \end{bmatrix} \begin{bmatrix} -4 & 1 \\ 1 & 0 \end{bmatrix} = \begin{bmatrix} -2 & 1 \\ -5 & 2 \end{bmatrix}$$

$$\begin{bmatrix} 1 & 2 \\ 2 & 3 \end{bmatrix} \begin{bmatrix} -4 & 3 \\ 3 & -1 \end{bmatrix} = \begin{bmatrix} 2 & 1 \\ 1 & 3 \end{bmatrix} \blacklozenge$$

In general, a symmetric matrix need not be invertible; for example, a square zero matrix is symmetric, but not invertible. However, if a symmetric matrix is invertible, then that inverse is also symmetric.

THEOREM 1.7.3

> *If A is an invertible symmetric matrix, then A^{-1} is symmetric.*

Proof   Assume that A is symmetric and invertible. From Theorem 1.4.10 and the fact that $A = A^T$, we have

$$(A^{-1})^T = (A^T)^{-1} = A^{-1}$$

which proves that A^{-1} is symmetric. $\blacklozenge$

Products AA^T and $A^T A$

Matrix products of the form AA^T and $A^T A$ arise in a variety of applications. If A is an $m \times n$ matrix, then A^T is an $n \times m$ matrix, so the products AA^T and $A^T A$ are both square matrices—the matrix AA^T has size $m \times m$, and the matrix $A^T A$ has size $n \times n$. Such products are always symmetric since

$$(AA^T)^T = (A^T)^T A^T = AA^T \quad \text{and} \quad (A^T A)^T = A^T (A^T)^T = A^T A$$

EXAMPLE 6   The Product of a Matrix and Its Transpose Is Symmetric

---

Let A be the 2×3 matrix

$$A = \begin{bmatrix} 1 & -2 & 4 \\ 3 & 0 & -5 \end{bmatrix}$$

Then

$$A^T A = \begin{bmatrix} 1 & 3 \\ -2 & 0 \\ 4 & -5 \end{bmatrix} \begin{bmatrix} 1 & -2 & 4 \\ 3 & 0 & -5 \end{bmatrix} = \begin{bmatrix} 10 & -2 & -11 \\ -2 & 4 & -8 \\ -11 & -8 & 41 \end{bmatrix}$$

$$AA^T = \begin{bmatrix} 1 & -2 & 4 \\ 3 & 0 & -5 \end{bmatrix} \begin{bmatrix} 1 & 3 \\ -2 & 0 \\ 4 & -5 \end{bmatrix} = \begin{bmatrix} 21 & -17 \\ -17 & 34 \end{bmatrix}$$

Observe that $A^T A$ and $A A^T$ are symmetric as expected. ◆

Later in this text, we will obtain general conditions on A under which $A A^T$ and $A^T A$ are invertible. However, in the special case where A is *square*, we have the following result.

THEOREM 1.7.4

If A is an invertible matrix, then $A A^T$ and $A^T A$ are also invertible.

Proof Since A is invertible, so is A^T by Theorem 1.4.10. Thus $A A^T$ and $A^T A$ are invertible, since they are the products of invertible matrices. ∎

EXERCISE SET 1.7

1. Determine whether the matrix is invertible; if so, find the inverse by inspection.

   (a) $\begin{bmatrix} 2 & 0 \\ 0 & -5 \end{bmatrix}$
   (b) $\begin{bmatrix} 4 & 0 & 0 \\ 0 & 0 & 0 \\ 0 & 0 & 5 \end{bmatrix}$
   (c) $\begin{bmatrix} -1 & 0 & 0 \\ 0 & 2 & 0 \\ 0 & 0 & \frac{1}{3} \end{bmatrix}$

2. Compute the product by inspection.

   (a) $\begin{bmatrix} 3 & 0 & 0 \\ 0 & -1 & 0 \\ 0 & 0 & 2 \end{bmatrix} \begin{bmatrix} 2 & 1 \\ -4 & 1 \\ 2 & 5 \end{bmatrix}$
   (b) $\begin{bmatrix} 2 & 0 & 0 \\ 0 & -1 & 0 \\ 0 & 0 & 4 \end{bmatrix} \begin{bmatrix} 4 & -1 & 3 \\ 1 & 2 & 0 \\ -5 & 1 & -2 \end{bmatrix} \begin{bmatrix} -3 & 0 & 0 \\ 0 & 5 & 0 \\ 0 & 0 & 2 \end{bmatrix}$

3. Find A^2, A^{-2}, and A^{-k} by inspection.

   (a) $A = \begin{bmatrix} 1 & 0 \\ 0 & -2 \end{bmatrix}$
   (b) $A = \begin{bmatrix} \frac{1}{2} & 0 & 0 \\ 0 & \frac{1}{3} & 0 \\ 0 & 0 & \frac{1}{4} \end{bmatrix}$

4. Which of the following matrices are symmetric?

   (a) $\begin{bmatrix} 2 & -1 \\ 1 & 2 \end{bmatrix}$
   (b) $\begin{bmatrix} 3 & 4 \\ 4 & 0 \end{bmatrix}$
   (c) $\begin{bmatrix} 2 & -1 & 3 \\ -1 & 5 & 1 \\ 3 & 1 & 7 \end{bmatrix}$
   (d) $\begin{bmatrix} 0 & 0 & 1 \\ 0 & 2 & 0 \\ 3 & 0 & 0 \end{bmatrix}$

5. By inspection, determine whether the given triangular matrix is invertible.

   (a) $\begin{bmatrix} -1 & 2 & 4 \\ 0 & 3 & 0 \\ 0 & 0 & 5 \end{bmatrix}$
   (b) $\begin{bmatrix} 0 & 1 & -2 & 5 \\ 0 & 1 & 5 & 6 \\ 0 & 0 & -3 & 1 \\ 0 & 0 & 0 & 5 \end{bmatrix}$

6. Find all values of a, b, and c for which A is symmetric.

$$A = \begin{bmatrix} 2 & a - 2b + 2c & 2a + b + c \\ 3 & 5 & a + c \\ 0 & -2 & 7 \end{bmatrix}$$

7. Find all values of a and b for which A and B are both not invertible.

$$A = \begin{bmatrix} a + b - 1 & 0 \\ 0 & 3 \end{bmatrix}, \qquad B = \begin{bmatrix} 5 & 0 \\ 0 & 2a - 3b - 7 \end{bmatrix}$$

8. Use the given equation to determine by inspection whether the matrices on the left commute.

   (a) $\begin{bmatrix} 1 & -3 \\ -3 & 2 \end{bmatrix} \begin{bmatrix} 4 & 1 \\ 1 & 2 \end{bmatrix} = \begin{bmatrix} 1 & -5 \\ -10 & 1 \end{bmatrix}$ (b) $\begin{bmatrix} 2 & -1 \\ -1 & 3 \end{bmatrix} \begin{bmatrix} 3 & 2 \\ 2 & 1 \end{bmatrix} = \begin{bmatrix} 4 & 3 \\ 3 & 1 \end{bmatrix}$

9. Show that A and B commute if $a - d = 7b$.

$$A = \begin{bmatrix} 2 & 1 \\ 1 & -5 \end{bmatrix}, \qquad B = \begin{bmatrix} a & b \\ b & d \end{bmatrix}$$

10. Find a diagonal matrix A that satisfies

    (a) $A^5 = \begin{bmatrix} 1 & 0 & 0 \\ 0 & -1 & 0 \\ 0 & 0 & -1 \end{bmatrix}$ (b) $A^{-2} = \begin{bmatrix} 9 & 0 & 0 \\ 0 & 4 & 0 \\ 0 & 0 & 1 \end{bmatrix}$

11. (a) Factor A into the form $A = BD$, where D is a diagonal matrix.

$$A = \begin{bmatrix} 3a_{11} & 5a_{12} & 7a_{13} \\ 3a_{21} & 5a_{22} & 7a_{23} \\ 3a_{31} & 5a_{32} & 7a_{33} \end{bmatrix}$$

    (b) Is your factorization the only one possible? Explain.

12. Verify Theorem 1.7.1b for the product AB, where

$$A = \begin{bmatrix} -1 & 2 & 5 \\ 0 & 1 & 3 \\ 0 & 0 & -4 \end{bmatrix}, \qquad B = \begin{bmatrix} 2 & -8 & 0 \\ 0 & 2 & 1 \\ 0 & 0 & 3 \end{bmatrix}$$

13. Verify Theorem 1.7.1d for the matrices A and B in Exercise 12.

14. Verify Theorem 1.7.3 for the given matrix A.

    (a) $A = \begin{bmatrix} 2 & -1 \\ -1 & 3 \end{bmatrix}$ (b) $A = \begin{bmatrix} 1 & -2 & 3 \\ -2 & 1 & -7 \\ 3 & -7 & 4 \end{bmatrix}$

15. Let A be an $n \times n$ symmetric matrix.
    (a) Show that A^2 is symmetric.
    (b) Show that $2A^2 - 3A + I$ is symmetric.

16. Let A be an $n \times n$ symmetric matrix.
    (a) Show that A^k is symmetric if k is any nonnegative integer.
    (b) If $p(x)$ is a polynomial, is $p(A)$ necessarily symmetric? Explain.

17. Let A be an $n \times n$ upper triangular matrix, and let $p(x)$ be a polynomial. Is $p(A)$ necessarily upper triangular? Explain.

18. Prove: If $A^T A = A$, then A is symmetric and $A = A^2$.

19. Find all 3×3 diagonal matrices A that satisfy $A^2 - 3A - 4I = 0$.

20. Let $A = [a_{ij}]$ be an $n \times n$ matrix. Determine whether A is symmetric.

    (a) $a_{ij} = i^2 + j^2$ (b) $a_{ij} = i^2 - j^2$

    (c) $a_{ij} = 2i + 2j$ (d) $a_{ij} = 2i^2 + 2j^3$

21. On the basis of your experience with Exercise 20, devise a general test that can be applied to a formula for a_{ij} to determine whether $A = [a_{ij}]$ is symmetric.

22. A square matrix A is called **skew-symmetric** if $A^T = -A$. Prove:

    (a)   If A is an invertible skew-symmetric matrix, then A^{-1} is skew-symmetric.

    (b)   If A and B are skew-symmetric, then so are A^T, $A + B$, $A - B$, and kA for any scalar k.

    (c)   Every square matrix A can be expressed as the sum of a symmetric matrix and a skew-symmetric matrix.

         Hint   Note the identity $A = \frac{1}{2}(A + A^T) + \frac{1}{2}(A - A^T)$.

23. We showed in the text that the product of symmetric matrices is symmetric if and only if the matrices commute. Is the product of commuting skew-symmetric matrices skew-symmetric? Explain.

    Note   See Exercise 22 for terminology.

24. If the $n \times n$ matrix A can be expressed as $A = LU$, where L is a lower triangular matrix and U is an upper triangular matrix, then the linear system $A\mathbf{x} = \mathbf{b}$ can be expressed as $LU\mathbf{x} = \mathbf{b}$ and can be solved in two steps:

    Step 1.   Let $U\mathbf{x} = \mathbf{y}$, so that $LU\mathbf{x} = \mathbf{b}$ can be expressed as $L\mathbf{y} = \mathbf{b}$. Solve this system.

    Step 2.   Solve the system $U\mathbf{x} = \mathbf{y}$ for $\mathbf{x}$.

    In each part, use this two-step method to solve the given system.

    (a) $\begin{bmatrix} 1 & 0 & 0 \\ -2 & 3 & 0 \\ 2 & 4 & 1 \end{bmatrix} \begin{bmatrix} 2 & -1 & 3 \\ 0 & 1 & 2 \\ 0 & 0 & 4 \end{bmatrix} \begin{bmatrix} x_1 \\ x_2 \\ x_3 \end{bmatrix} = \begin{bmatrix} 1 \\ -2 \\ 0 \end{bmatrix}$

    (b) $\begin{bmatrix} 2 & 0 & 0 \\ 4 & 1 & 0 \\ -3 & -2 & 3 \end{bmatrix} \begin{bmatrix} 3 & -5 & 2 \\ 0 & 4 & 1 \\ 0 & 0 & 2 \end{bmatrix} \begin{bmatrix} x_1 \\ x_2 \\ x_3 \end{bmatrix} = \begin{bmatrix} 4 \\ -5 \\ 2 \end{bmatrix}$

25. Find an upper triangular matrix that satisfies

$$A^3 = \begin{bmatrix} 1 & 30 \\ 0 & -8 \end{bmatrix}$$

Discussion & Discovery

26. What is the maximum number of distinct entries that an $n \times n$ symmetric matrix can have? Explain your reasoning.

27. Invent and prove a theorem that describes how to multiply two diagonal matrices.

28. Suppose that A is a square matrix and D is a diagonal matrix such that $AD = I$. What can you say about the matrix A? Explain your reasoning.

29. (a)   Make up a consistent linear system of five equations in five unknowns that has a lower triangular coefficient matrix with no zeros on or below the main diagonal.

    (b)   Devise an efficient procedure for solving your system by hand.

    (c)   Invent an appropriate name for your procedure.

30. Indicate whether the statement is always true or sometimes false. Justify each answer.

    (a)   If AA^T is singular, then so is A.

    (b)   If $A + B$ is symmetric, then so are A and B.

(c) If A is an $n \times n$ matrix and $A\mathbf{x} = \mathbf{0}$ has only the trivial solution, then so does $A^T\mathbf{x} = \mathbf{0}$.

(d) If A^2 is symmetric, then so is A.

CHAPTER 1
Supplementary Exercises

1. Use Gauss–Jordan elimination to solve for x' and y' in terms of x and y.
$$x = \tfrac{3}{5}x' - \tfrac{4}{5}y'$$
$$y = \tfrac{4}{5}x' + \tfrac{3}{5}y'$$

2. Use Gauss–Jordan elimination to solve for x' and y' in terms of x and y.
$$x = x'\cos\theta - y'\sin\theta$$
$$y = x'\sin\theta + y'\cos\theta$$

3. Find a homogeneous linear system with two equations that are not multiples of one another and such that
$$x_1 = 1, \quad x_2 = -1, \quad x_3 = 1, \quad x_4 = 2$$
and
$$x_1 = 2, \quad x_2 = 0, \quad x_3 = 3, \quad x_4 = -1$$
are solutions of the system.

4. A box containing pennies, nickels, and dimes has 13 coins with a total value of 83 cents. How many coins of each type are in the box?

5. Find positive integers that satisfy
$$x + y + z = 9$$
$$x + 5y + 10z = 44$$

6. For which value(s) of a does the following system have zero solutions? One solution? Infinitely many solutions?
$$x_1 + x_2 + x_3 = 4$$
$$x_3 = 2$$
$$(a^2 - 4)x_3 = a - 2$$

7. Let
$$\begin{bmatrix} a & 0 & b & 2 \\ a & a & 4 & 4 \\ 0 & a & 2 & b \end{bmatrix}$$
be the augmented matrix for a linear system. Find for what values of a and b the system has

(a) a unique solution.   (b) a one-parameter solution.

(c) a two-parameter solution.   (d) no solution.

8. Solve for x, y, and z.
$$xy - 2\sqrt{y} + 3zy = 8$$
$$2xy - 3\sqrt{y} + 2zy = 7$$
$$-xy + \sqrt{y} + 2zy = 4$$

9. Find a matrix K such that $AKB = C$ given that
$$A = \begin{bmatrix} 1 & 4 \\ -2 & 3 \\ 1 & -2 \end{bmatrix}, \quad B = \begin{bmatrix} 2 & 0 & 0 \\ 0 & 1 & -1 \end{bmatrix}, \quad C = \begin{bmatrix} 8 & 6 & -6 \\ 6 & -1 & 1 \\ -4 & 0 & 0 \end{bmatrix}$$

10. How should the coefficients a, b, and c be chosen so that the system

$$
\begin{aligned}
ax + by - 3z &= -3 \\
-2x - by + cz &= -1 \\
ax + 3y - cz &= -3
\end{aligned}
$$

has the solution $x = 1$, $y = -1$, and $z = 2$?

11. In each part, solve the matrix equation for X.

(a) $X \begin{bmatrix} -1 & 0 & 1 \\ 1 & 1 & 0 \\ 3 & 1 & -1 \end{bmatrix} = \begin{bmatrix} 1 & 2 & 0 \\ -3 & 1 & 5 \end{bmatrix}$

(b) $X \begin{bmatrix} 1 & -1 & 2 \\ 3 & 0 & 1 \end{bmatrix} = \begin{bmatrix} -5 & -1 & 0 \\ 6 & -3 & 7 \end{bmatrix}$

(c) $\begin{bmatrix} 3 & 1 \\ -1 & 2 \end{bmatrix} X - X \begin{bmatrix} 1 & 4 \\ 2 & 0 \end{bmatrix} = \begin{bmatrix} 2 & -2 \\ 5 & 4 \end{bmatrix}$

12. (a) Express the equations

$$
\begin{aligned}
y_1 &= x_1 - x_2 + x_3 \\
y_2 &= 3x_1 + x_2 - 4x_3 \\
y_3 &= -2x_1 - 2x_2 + 3x_3
\end{aligned}
\quad \text{and} \quad
\begin{aligned}
z_1 &= 4y_1 - y_2 + y_3 \\
z_2 &= -3y_1 + 5y_2 - y_3
\end{aligned}
$$

in the matrix forms $Y = AX$ and $Z = BY$. Then use these to obtain a direct relationship $Z = CX$ between Z and X.

(b) Use the equation $Z = CX$ obtained in (a) to express z_1 and z_2 in terms of x_1, x_2, and x_3.

(c) Check the result in (b) by directly substituting the equations for y_1, y_2, and y_3 into the equations for z_1 and z_2 and then simplifying.

13. If A is $m \times n$ and B is $n \times p$, how many multiplication operations and how many addition operations are needed to calculate the matrix product AB?

14. Let A be a square matrix.

(a) Show that $(I - A)^{-1} = I + A + A^2 + A^3$ if $A^4 = 0$.

(b) Show that $(I - A)^{-1} = I + A + A^2 + \cdots + A^n$ if $A^{n+1} = 0$.

15. Find values of a, b, and c such that the graph of the polynomial $p(x) = ax^2 + bx + c$ passes through the points $(1, 2)$, $(-1, 6)$, and $(2, 3)$.

16. **(For Readers Who Have Studied Calculus)** Find values of a, b, and c such that the graph of the polynomial $p(x) = ax^2 + bx + c$ passes through the point $(-1, 0)$ and has a horizontal tangent at $(2, -9)$.

17. Let J_n be the $n \times n$ matrix each of whose entries is 1. Show that if $n > 1$, then

$$(I - J_n)^{-1} = I - \frac{1}{n-1} J_n$$

18. Show that if a square matrix A satisfies $A^3 + 4A^2 - 2A + 7I = 0$, then so does A^T.

19. Prove: If B is invertible, then $AB^{-1} = B^{-1}A$ if and only if $AB = BA$.

20. Prove: If A is invertible, then $A + B$ and $I + BA^{-1}$ are both invertible or both not invertible.

21. Prove that if A and B are $n \times n$ matrices, then

(a) $\operatorname{tr}(A + B) = \operatorname{tr}(A) + \operatorname{tr}(B)$ (b) $\operatorname{tr}(kA) = k\operatorname{tr}(A)$

(c) $\operatorname{tr}(A^T) = \operatorname{tr}(A)$ (d) $\operatorname{tr}(AB) = \operatorname{tr}(BA)$

22. Use Exercise 21 to show that there are no square matrices A and B such that

$$AB - BA = I$$

23. Prove: If A is an $m \times n$ matrix and B is the $n \times 1$ matrix each of whose entries is $1/n$, then

$$AB = \begin{bmatrix} \bar{r}_1 \\ \bar{r}_2 \\ \vdots \\ \bar{r}_m \end{bmatrix}$$

where $\bar{r}_i$ is the average of the entries in the ith row of A.

24. **(For Readers Who Have Studied Calculus)** If the entries of the matrix

$$C = \begin{bmatrix} c_{11}(x) & c_{12}(x) & \cdots & c_{1n}(x) \\ c_{21}(x) & c_{22}(x) & \cdots & c_{2n}(x) \\ \vdots & \vdots & & \vdots \\ c_{m1}(x) & c_{m2}(x) & \cdots & c_{mn}(x) \end{bmatrix}$$

are differentiable functions of x, then we define

$$\frac{dC}{dx} = \begin{bmatrix} c'_{11}(x) & c'_{12}(x) & \cdots & c'_{1n}(x) \\ c'_{21}(x) & c'_{22}(x) & \cdots & c'_{2n}(x) \\ \vdots & \vdots & & \vdots \\ c'_{m1}(x) & c'_{m2}(x) & \cdots & c'_{mn}(x) \end{bmatrix}$$

Show that if the entries in A and B are differentiable functions of x and the sizes of the matrices are such that the stated operations can be performed, then

(a) $\dfrac{d}{dx}(kA) = k\dfrac{dA}{dx}$ (b) $\dfrac{d}{dx}(A+B) = \dfrac{dA}{dx} + \dfrac{dB}{dx}$

(c) $\dfrac{d}{dx}(AB) = \dfrac{dA}{dx}B + A\dfrac{dB}{dx}$

25. **(For Readers Who Have Studied Calculus)** Use part (c) of Exercise 24 to show that

$$\frac{dA^{-1}}{dx} = -A^{-1}\frac{dA}{dx}A^{-1}$$

State all the assumptions you make in obtaining this formula.

26. Find the values of a, b, and c that will make the equation

$$\frac{x^2+x-2}{(3x-1)(x^2+1)} = \frac{a}{3x-1} + \frac{bx+c}{x^2+1}$$

an identity.

Hint Multiply through by $(3x-1)(x^2+1)$ and equate the corresponding coefficients of the polynomials on each side of the resulting equation.

27. If P is an $n \times 1$ matrix such that $P^T P = 1$, then $H = I - 2PP^T$ is called the corresponding *Householder matrix* (named after the American mathematician A. S. Householder).

(a) Verify that $P^T P = 1$ if $P^T = \begin{bmatrix} \frac{3}{4} & \frac{1}{6} & \frac{1}{4} & \frac{5}{12} & \frac{5}{12} \end{bmatrix}$ and compute the corresponding Householder matrix.

(b) Prove that if H is any Householder matrix, then $H = H^T$ and $H^T H = I$.

(c) Verify that the Householder matrix found in part (a) satisfies the conditions proved in part (b).

28. Assuming that the stated inverses exist, prove the following equalities.

(a) $(C^{-1} + D^{-1})^{-1} = C(C+D)^{-1}D$ (b) $(I+CD)^{-1}C = C(I+DC)^{-1}$

(c) $(C+DD^T)^{-1}D = C^{-1}D(I+D^T C^{-1}D)^{-1}$

29. (a) Show that if $a \neq b$, then

$$a^n + a^{n-1}b + a^{n-2}b^2 + \cdots + ab^{n-1} + b^n = \frac{a^{n+1} - b^{n+1}}{a-b}$$

(b) Use the result in part (a) to find A^n if

$$A = \begin{bmatrix} a & 0 & 0 \\ 0 & b & 0 \\ 1 & 0 & c \end{bmatrix}$$

Note This exercise is based on a problem by John M. Johnson, *The Mathematics Teacher*, Vol. 85, No. 9, 1992.

CHAPTER 1
Technology Exercises

The following exercises are designed to be solved using a technology utility. Typically, this will be MATLAB, *Mathematica*, Maple, Derive, or Mathcad, but it may also be some other type of linear algebra software or a scientific calculator with some linear algebra capabilities. For each exercise you will need to read the relevant documentation for the particular utility you are using. The goal of these exercises is to provide you with a basic proficiency with your technology utility. Once you have mastered the techniques in these exercises, you will be able to use your technology utility to solve many of the problems in the regular exercise sets.

Section 1.1

T1. Numbers and Numerical Operations Read your documentation on entering and displaying numbers and performing the basic arithmetic operations of addition, subtraction, multiplication, division, raising numbers to powers, and extraction of roots. Determine how to control the number of digits in the screen display of a decimal number. If you are using a CAS, in which case you can compute with exact numbers rather than decimal approximations, then learn how to enter such numbers as π, $\sqrt{2}$, and $\frac{1}{3}$ exactly and convert them to decimal form. Experiment with numbers of your own choosing until you feel you have mastered the procedures and operations.

Section 1.2

T1. Matrices and Reduced Row-Echelon Form Read your documentation on how to enter matrices and how to find the reduced row-echelon form of a matrix. Then use your utility to find the reduced row-echelon form of the augmented matrix in Example 4 of Section 1.2.

T2. Linear Systems With a Unique Solution Read your documentation on how to solve a linear system, and then use your utility to solve the linear system in Example 3 of Section 1.1. Also, solve the system by reducing the augmented matrix to reduced row-echelon form.

T3. Linear Systems With Infinitely Many Solutions Technology utilities vary on how they handle linear systems with infinitely many solutions. See how your utility handles the system in Example 4 of Section 1.2.

T4. Inconsistent Linear Systems Technology utilities will often successfully identify inconsistent linear systems, but they can sometimes be fooled into reporting an inconsistent system as consistent, or vice versa. This typically happens when some of the numbers that occur in the computations are so small that roundoff error makes it difficult for the utility to determine whether or not they are equal to zero. Create some inconsistent linear systems and see how your utility handles them.

T5. A polynomial whose graph passes through a given set of points is called an ***interpolating polynomial*** for those points. Some technology utilities have specific commands for finding interpolating polynomials. If your utility has this capability, read the documentation and then use this feature to solve Exercise 25 of Section 1.2.

Section 1.3

T1. Matrix Operations Read your documentation on how to perform the basic operations on matrices— addition, subtraction, multiplication by scalars, and multiplication of matrices. Then perform the computations in Examples 3, 4, and 5. See what happens when you try to perform an operation on matrices with inconsistent sizes.

T2. Evaluate the expression $A^5 - 3A^3 + 7A - 4I$ for the matrix

$$A = \begin{bmatrix} 1 & -2 & 3 \\ -4 & 5 & -6 \\ 7 & -8 & 9 \end{bmatrix}$$

T3. **Extracting Rows and Columns** Read your documentation on how to extract rows and columns from a matrix, and then use your utility to extract various rows and columns from a matrix of your choice.

T4. **Transpose and Trace** Read your documentation on how to find the transpose and trace of a matrix, and then use your utility to find the transpose of the matrix A in Formula (12) and the trace of the matrix B in Example 12.

T5. **Constructing an Augmented Matrix** Read your documentation on how to create an augmented matrix $[A \mid \mathbf{b}]$ from matrices A and $\mathbf{b}$ that have previously been entered. Then use your utility to form the augmented matrix for the system $A\mathbf{x} = \mathbf{b}$ in Example 4 of Section 1.1 from the matrices A and $\mathbf{b}$.

Section 1.4 **T1.** **Zero and Identity Matrices** Typing in entries of a matrix can be tedious, so many technology utilities provide shortcuts for entering zero and identity matrices. Read your documentation on how to do this, and then enter some zero and identity matrices of various sizes.

T2. **Inverse** Read your documentation on how to find the inverse of a matrix, and then use your utility to perform the computations in Example 7.

T3. **Formula for the Inverse** If you are working with a CAS, use it to confirm Theorem 1.4.5.

T4. **Powers of a Matrix** Read your documentation on how to find powers of a matrix, and then use your utility to find various positive and negative powers of the matrix A in Example 8.

T5. Let

$$A = \begin{bmatrix} 1 & \frac{1}{2} & \frac{1}{3} \\ \frac{1}{4} & 1 & \frac{1}{5} \\ \frac{1}{6} & \frac{1}{7} & 1 \end{bmatrix}$$

Describe what happens to the matrix A^k when k is allowed to increase indefinitely (that is, as $k \to \infty$).

T6. By experimenting with different values of n, find an expression for the inverse of an $n \times n$ matrix of the form

$$A = \begin{bmatrix} 1 & 2 & 3 & 4 & \cdots & n-1 & n \\ 0 & 1 & 2 & 3 & \cdots & n-2 & n-1 \\ 0 & 0 & 1 & 2 & \cdots & n-3 & n-2 \\ \vdots & \vdots & \vdots & \vdots & & \vdots & \vdots \\ 0 & 0 & 0 & 0 & \cdots & 1 & 2 \\ 0 & 0 & 0 & 0 & \cdots & 0 & 1 \end{bmatrix}$$

Section 1.5 **T1.** Use your technology utility to verify Theorem 1.5.1 in several specific cases.

T2. **Singular Matrices** Find the inverse of the matrix in Example 4, and then see what your utility does when you try to invert the matrix in Example 5.

Section 1.6 **T1.** **Solving $A\mathbf{x} = \mathbf{b}$ by Inversion** Use the method of Example 4 to solve the system in Example 3 of Section 1.1.

T2. Compare the solution of $A\mathbf{x} = \mathbf{b}$ by Gaussian elimination and by inversion for several large matrices. Can you see the superiority of the former approach?

T3. Solve the linear system $A\mathbf{x} = 2\mathbf{x}$, given that

$$A = \begin{bmatrix} 0 & 0 & -2 \\ 1 & 2 & 1 \\ 1 & 0 & 3 \end{bmatrix}$$

Section 1.7

T1. **Diagonal, Symmetric, and Triangular Matrices** Many technology utilities provide short-cuts for entering diagonal, symmetric, and triangular matrices. Read your documentation on how to do this, and then experiment with entering various matrices of these types.

T2. **Properties of Triangular Matrices** Confirm the results in Theorem 1.7.1 using some triangular matrices of your choice.

T3. Confirm the results in Theorem 1.7.4. What happens if A is not square?

CHAPTER 2

Determinants

CHAPTER CONTENTS

INTRODUCTION: We are all familiar with functions such as $f(x) = \sin x$ and $f(x) = x^2$, which associate a real number $f(x)$ with a real value of the variable x. Since both x and $f(x)$ assume only real values, such functions are described as real-valued functions of a real variable. In this section we shall study the "determinant function," which is a real-valued function of a matrix variable in the sense that it associates a real number $f(X)$ with a square matrix X. Our work on determinant functions will have important applications to the theory of systems of linear equations and will also lead us to an explicit formula for the inverse of an invertible matrix.

2.1

DETERMINANTS BY COFACTOR EXPANSION

As noted in the introduction to this chapter, a "determinant" is a certain kind of function that associates a real number with a square matrix. In this section we will define this function. As a consequence of our work here, we will obtain a formula for the inverse of an invertible matrix as well as a formula for the solution to certain systems of linear equations in terms of determinants.

Recall from Theorem 1.4.5 that the 2×2 matrix

$$A = \begin{bmatrix} a & b \\ c & d \end{bmatrix}$$

is invertible if $ad - bc \neq 0$. The expression $ad - bc$ occurs so frequently in mathematics that it has a name; it is called the **determinant** of the matrix A and is denoted by the symbol $\det(A)$ or $|A|$. With this notation, the formula for A^{-1} given in Theorem 1.4.5 is

$$A^{-1} = \frac{1}{\det(A)} \begin{bmatrix} d & -b \\ -c & a \end{bmatrix}$$

One of the goals of this chapter is to obtain analogs of this formula to square matrices of higher order. This will require that we extend the concept of a determinant to square matrices of all orders.

Minors and Cofactors

There are several ways in which we might proceed. The approach in this section is a recursive approach: It defines the determinant of an $n \times n$ matrix in terms of the determinants of certain $(n-1) \times (n-1)$ matrices. The $(n-1) \times (n-1)$ matrices that will appear in this definition are submatrices of the original matrix. These submatrices are given a special name:

DEFINITION

If A is a square matrix, then the **minor of entry a_{ij}** is denoted by M_{ij} and is defined to be the determinant of the submatrix that remains after the ith row and jth column are deleted from A. The number $(-1)^{i+j} M_{ij}$ is denoted by C_{ij} and is called the **cofactor of entry a_{ij}**.

EXAMPLE 1 Finding Minors and Cofactors

Let

$$A = \begin{bmatrix} 3 & 1 & -4 \\ 2 & 5 & 6 \\ 1 & 4 & 8 \end{bmatrix}$$

The minor of entry a_{11} is

$$M_{11} = \begin{vmatrix} 3 & 1 & -4 \\ 2 & 5 & 6 \\ 1 & 4 & 8 \end{vmatrix} = \begin{vmatrix} 5 & 6 \\ 4 & 8 \end{vmatrix} = 16$$

The cofactor of a_{11} is

$$C_{11} = (-1)^{1+1} M_{11} = M_{11} = 16$$

Similarly, the minor of entry a_{32} is

$$M_{32} = \begin{vmatrix} 3 & 1 & -4 \\ 2 & 5 & 6 \\ 1 & 4 & 8 \end{vmatrix} = \begin{vmatrix} 3 & -4 \\ 2 & 6 \end{vmatrix} = 26$$

The cofactor of a_{32} is

$$C_{32} = (-1)^{3+2} M_{32} = -M_{32} = -26 \quad \blacklozenge$$

Note that the cofactor and the minor of an element a_{ij} differ only in sign; that is, $C_{ij} = \pm M_{ij}$. A quick way to determine whether to use $+$ or $-$ is to use the fact that the sign relating C_{ij} and M_{ij} is in the ith row and jth column of the "checkerboard" array

$$\begin{bmatrix} + & - & + & - & + & \cdots \\ - & + & - & + & - & \cdots \\ + & - & + & - & + & \cdots \\ - & + & - & + & - & \cdots \\ \vdots & \vdots & \vdots & \vdots & \vdots & \end{bmatrix}$$

For example, $C_{11} = M_{11}$, $C_{21} = -M_{21}$, $C_{12} = -M_{12}$, $C_{22} = M_{22}$, and so on.

Strictly speaking, the determinant of a matrix is a number. However, it is common practice to "abuse" the terminology slightly and use the term *determinant* to refer to the matrix whose determinant is being computed. Thus we might refer to

$$\begin{vmatrix} 3 & 1 \\ 4 & -2 \end{vmatrix}$$

as a 2×2 determinant and call 3 the entry in the first row and first column of the determinant.

Cofactor Expansions

The definition of a 3×3 determinant in terms of minors and cofactors is

$$\begin{aligned} \det(A) &= a_{11}M_{11} + a_{12}(-M_{12}) + a_{13}M_{13} \\ &= a_{11}C_{11} + a_{12}C_{12} + a_{13}C_{13} \end{aligned} \quad (1)$$

Equation (1) shows that the determinant of A can be computed by multiplying the entries in the first row of A by their corresponding cofactors and adding the resulting products. More generally, we define the determinant of an $n \times n$ matrix to be

$$\det(A) = a_{11}C_{11} + a_{12}C_{12} + \cdots + a_{1n}C_{1n}$$

This method of evaluating $\det(A)$ is called ***cofactor expansion*** along the first row of A.

EXAMPLE 2 Cofactor Expansion Along the First Row

Let $A = \begin{bmatrix} 3 & 1 & 0 \\ -2 & -4 & 3 \\ 5 & 4 & -2 \end{bmatrix}$. Evaluate $\det(A)$ by cofactor expansion along the first row of A.

Solution

From (1),

$$\det(A) = \begin{vmatrix} 3 & 1 & 0 \\ -2 & -4 & 3 \\ 5 & 4 & -2 \end{vmatrix} = 3\begin{vmatrix} -4 & 3 \\ 4 & -2 \end{vmatrix} - 1\begin{vmatrix} -2 & 3 \\ 5 & -2 \end{vmatrix} + 0\begin{vmatrix} -2 & -4 \\ 5 & 4 \end{vmatrix}$$

$$= 3(-4) - (1)(-11) + 0 = -1 \quad \blacklozenge$$

If A is a 3×3 matrix, then its determinant is

$$\det(A) = \begin{vmatrix} a_{11} & a_{12} & a_{13} \\ a_{21} & a_{22} & a_{23} \\ a_{31} & a_{32} & a_{33} \end{vmatrix}$$

$$= a_{11}\begin{vmatrix} a_{22} & a_{23} \\ a_{32} & a_{33} \end{vmatrix} - a_{12}\begin{vmatrix} a_{21} & a_{23} \\ a_{31} & a_{33} \end{vmatrix} + a_{13}\begin{vmatrix} a_{21} & a_{22} \\ a_{31} & a_{32} \end{vmatrix}$$

$$= a_{11}(a_{22}a_{33} - a_{23}a_{32}) - a_{12}(a_{21}a_{33} - a_{23}a_{31}) + a_{13}(a_{21}a_{32} - a_{22}a_{31}) \quad (2)$$

$$= a_{11}a_{22}a_{33} + a_{12}a_{23}a_{31} + a_{13}a_{21}a_{32} - a_{13}a_{22}a_{31} - a_{12}a_{21}a_{33} - a_{11}a_{23}a_{32} \quad (3)$$

By rearranging the terms in (3) in various ways, it is possible to obtain other formulas like (2). There should be no trouble checking that all of the following are correct (see Exercise 28):

$$\det(A) = a_{11}C_{11} + a_{12}C_{12} + a_{13}C_{13}$$
$$= a_{11}C_{11} + a_{21}C_{21} + a_{31}C_{31}$$
$$= a_{21}C_{21} + a_{22}C_{22} + a_{23}C_{23}$$
$$= a_{12}C_{12} + a_{22}C_{22} + a_{32}C_{32}$$
$$= a_{31}C_{31} + a_{32}C_{32} + a_{33}C_{33}$$
$$= a_{13}C_{13} + a_{23}C_{23} + a_{33}C_{33} \quad (4)$$

Note that in each equation, the entries and cofactors all come from the same row or column. These equations are called the *cofactor expansions* of $\det(A)$.

The results we have just given for 3×3 matrices form a special case of the following general theorem, which we state without proof.

THEOREM 2.1.1

Expansions by Cofactors

The determinant of an $n \times n$ matrix A can be computed by multiplying the entries in any row (or column) by their cofactors and adding the resulting products; that is, for each $1 \leq i \leq n$ and $1 \leq j \leq n$,

$$\det(A) = a_{1j}C_{1j} + a_{2j}C_{2j} + \cdots + a_{nj}C_{nj}$$

(cofactor expansion along the jth column)

and

$$\det(A) = a_{i1}C_{i1} + a_{i2}C_{i2} + \cdots + a_{in}C_{in}$$

(cofactor expansion along the ith row)

Note that we may choose *any* row or *any* column.

EXAMPLE 3   Cofactor Expansion Along the First Column

Let A be the matrix in Example 2. Evaluate $\det(A)$ by cofactor expansion along the first column of A.

Solution

From (4)

$$\det(A) = \begin{vmatrix} 3 & 1 & 0 \\ -2 & -4 & 3 \\ 5 & 4 & -2 \end{vmatrix} = 3\begin{vmatrix} -4 & 3 \\ 4 & 2 \end{vmatrix} - (-2)\begin{vmatrix} 1 & 0 \\ 4 & -2 \end{vmatrix} + 5\begin{vmatrix} 1 & 0 \\ -4 & 3 \end{vmatrix}$$

$$= 3(-4) - (-2)(-2) + 5(3) = -1$$

This agrees with the result obtained in Example 2. ◆

REMARK   In this example we had to compute three cofactors, but in Example 2 we only had to compute two of them, since the third was multiplied by zero. In general, the best strategy for evaluating a determinant by cofactor expansion is to expand along a row or column having the largest number of zeros.

EXAMPLE 4   Smart Choice of Row or Column

If A is the 4×4 matrix

$$A = \begin{bmatrix} 1 & 0 & 0 & -1 \\ 3 & 1 & 2 & 2 \\ 1 & 0 & -2 & 1 \\ 2 & 0 & 0 & 1 \end{bmatrix}$$

then to find $\det(A)$ it will be easiest to use cofactor expansion along the second column, since it has the most zeros:

$$\det(A) = 1 \cdot \begin{vmatrix} 1 & 0 & -1 \\ 1 & -2 & 1 \\ 2 & 0 & 1 \end{vmatrix}$$

For the 3×3 determinant, it will be easiest to use cofactor expansion along its second column, since it has the most zeros:

$$\det(A) = 1 \cdot -2 \cdot \begin{vmatrix} 1 & -1 \\ 2 & 1 \end{vmatrix}$$
$$= -2(1+2)$$
$$= -6$$

We would have found the same answer if we had used any other row or column. ◆

Adjoint of a Matrix

In a cofactor expansion we compute $\det(A)$ by multiplying the entries in a row or column by their cofactors and adding the resulting products. It turns out that if one multiplies the entries in any row by the corresponding cofactors from a *different* row, the sum of these products is always zero. (This result also holds for columns.) Although we omit the general proof, the next example illustrates the idea of the proof in a special case.

EXAMPLE 5 Entries and Cofactors from Different Rows

Let

$$A = \begin{bmatrix} a_{11} & a_{12} & a_{13} \\ a_{21} & a_{22} & a_{23} \\ a_{31} & a_{32} & a_{33} \end{bmatrix}$$

Consider the quantity

$$a_{11}C_{31} + a_{12}C_{32} + a_{13}C_{33}$$

that is formed by multiplying the entries in the first row by the cofactors of the corresponding entries in the third row and adding the resulting products. We now show that this quantity is equal to zero by the following trick. Construct a new matrix A' by replacing the third row of A with another copy of the first row. Thus

$$A' = \begin{bmatrix} a_{11} & a_{12} & a_{13} \\ a_{21} & a_{22} & a_{23} \\ a_{11} & a_{12} & a_{13} \end{bmatrix}$$

Let $C'_{31}, C'_{32}, C'_{33}$ be the cofactors of the entries in the third row of A'. Since the first two rows of A and A' are the same, and since the computations of $C_{31}, C_{32}, C_{33}, C'_{31}$, C'_{32}, and C'_{33} involve only entries from the first two rows of A and A', it follows that

$$C_{31} = C'_{31}, \qquad C_{32} = C'_{32}, \qquad C_{33} = C'_{33}$$

Since A' has two identical rows, it follows from (3) that

$$\det(A') = 0 \tag{5}$$

On the other hand, evaluating $\det(A')$ by cofactor expansion along the third row gives

$$\det(A') = a_{11}C'_{31} + a_{12}C'_{32} + a_{13}C'_{33} = a_{11}C_{31} + a_{12}C_{32} + a_{13}C_{33} \tag{6}$$

From (5) and (6) we obtain

$$a_{11}C_{31} + a_{12}C_{32} + a_{13}C_{33} = 0 \quad \blacklozenge$$

Now we'll use this fact to get a formula for A^{-1}.

DEFINITION

If A is any $n \times n$ matrix and C_{ij} is the cofactor of a_{ij}, then the matrix

$$\begin{bmatrix} C_{11} & C_{12} & \cdots & C_{1n} \\ C_{21} & C_{22} & \cdots & C_{2n} \\ \vdots & \vdots & & \vdots \\ C_{n1} & C_{n2} & \cdots & C_{nn} \end{bmatrix}$$

is called the *matrix of cofactors from A*. The transpose of this matrix is called the *adjoint of A* and is denoted by adj(A).

EXAMPLE 6   Adjoint of a 3 × 3 Matrix

Let

$$A = \begin{bmatrix} 3 & 2 & -1 \\ 1 & 6 & 3 \\ 2 & -4 & 0 \end{bmatrix}$$

The cofactors of A are

$$C_{11} = 12 \qquad C_{12} = 6 \qquad C_{13} = -16$$
$$C_{21} = 4 \qquad C_{22} = 2 \qquad C_{23} = 16$$
$$C_{31} = 12 \qquad C_{32} = -10 \qquad C_{33} = 16$$

so the matrix of cofactors is

$$\begin{bmatrix} 12 & 6 & -16 \\ 4 & 2 & 16 \\ 12 & -10 & 16 \end{bmatrix}$$

and the adjoint of A is

$$\text{adj}(A) = \begin{bmatrix} 12 & 4 & 12 \\ 6 & 2 & -10 \\ -16 & 16 & 16 \end{bmatrix} \quad \blacklozenge$$

We are now in a position to derive a formula for the inverse of an invertible matrix. We need to use an important fact that will be proved in Section 2.3: The square matrix A is invertible if and only if $\det(A)$ is not zero.

THEOREM 2.1.2

Inverse of a Matrix Using Its Adjoint

If A is an invertible matrix, then

$$A^{-1} = \frac{1}{\det(A)} \text{adj}(A) \tag{7}$$

Proof   We show first that

$$A \, \text{adj}(A) = \det(A) I$$

Consider the product

$$A \, \text{adj}(A) = \begin{bmatrix} a_{11} & a_{12} & \dots & a_{1n} \\ a_{21} & a_{22} & \dots & a_{2n} \\ \vdots & \vdots & & \vdots \\ a_{i1} & a_{i2} & \dots & a_{in} \\ \vdots & \vdots & & \vdots \\ a_{n1} & a_{n2} & \dots & a_{nn} \end{bmatrix} \begin{bmatrix} C_{11} & C_{21} & \dots & C_{j1} & \dots & C_{n1} \\ C_{12} & C_{22} & \dots & C_{j2} & \dots & C_{n2} \\ \vdots & \vdots & & \vdots & & \vdots \\ C_{1n} & C_{2n} & \dots & C_{jn} & \dots & C_{nn} \end{bmatrix}$$

The entry in the ith row and jth column of the product $A \, \text{adj}(A)$ is

$$a_{i1}C_{j1} + a_{i2}C_{j2} + \dots + a_{in}C_{jn} \tag{8}$$

(see the shaded lines above).

If $i = j$, then (8) is the cofactor expansion of $\det(A)$ along the ith row of A (Theorem 2.1.1), and if $i \neq j$, then the a's and the cofactors come from different rows of A, so the value of (8) is zero. Therefore,

$$A \operatorname{adj}(A) = \begin{bmatrix} \det(A) & 0 & \cdots & 0 \\ 0 & \det(A) & \cdots & 0 \\ \vdots & \vdots & & \vdots \\ 0 & 0 & \cdots & \det(A) \end{bmatrix} = \det(A)I \qquad (9)$$

Since A is invertible, $\det(A) \neq 0$. Therefore, Equation (9) can be rewritten as

$$\frac{1}{\det(A)}[A \operatorname{adj}(A)] = I \quad \text{or} \quad A\left[\frac{1}{\det(A)}\operatorname{adj}(A)\right] = I$$

Multiplying both sides on the left by A^{-1} yields

$$A^{-1} = \frac{1}{\det(A)}\operatorname{adj}(A) \qquad \blacksquare$$

EXAMPLE 7   Using the Adjoint to Find an Inverse Matrix

Use (7) to find the inverse of the matrix A in Example 6.

Solution

The reader can check that $\det(A) = 64$. Thus

$$A^{-1} = \frac{1}{\det(A)}\operatorname{adj}(A) = \frac{1}{64}\begin{bmatrix} 12 & 4 & 12 \\ 6 & 2 & -10 \\ -16 & 16 & 16 \end{bmatrix} = \begin{bmatrix} \frac{12}{64} & \frac{4}{64} & \frac{12}{64} \\ \frac{6}{64} & \frac{2}{64} & -\frac{10}{64} \\ -\frac{16}{64} & \frac{16}{64} & \frac{16}{64} \end{bmatrix} \quad \blacklozenge$$

Applications of Formula (7)

Although the method in the preceding example is reasonable for inverting 3×3 matrices by hand, the inversion algorithm discussed in Section 1.5 is more efficient for larger matrices. It should be kept in mind, however, that the method of Section 1.5 is just a computational procedure, whereas Formula (7) is an actual formula for the inverse. As we shall now see, this formula is useful for deriving properties of the inverse.

In Section 1.7 we stated two results about inverses without proof.

- **Theorem 1.7.1c:** A triangular matrix is invertible if and only if its diagonal entries are all nonzero.

- **Theorem 1.7.1d:** The inverse of an invertible lower triangular matrix is lower triangular, and the inverse of an invertible upper triangular matrix is upper triangular.

We will now prove these results using the adjoint formula for the inverse. We need a preliminary result.

THEOREM 2.1.3

> *If A is an $n \times n$ triangular matrix (upper triangular, lower triangular, or diagonal), then $\det(A)$ is the product of the entries on the main diagonal of the matrix; that is, $\det(A) = a_{11}a_{22}\cdots a_{nn}$.*

For simplicity of notation, we will prove the result for a 4×4 lower triangular matrix

$$A = \begin{bmatrix} a_{11} & 0 & 0 & 0 \\ a_{21} & a_{22} & 0 & 0 \\ a_{31} & a_{32} & a_{33} & 0 \\ a_{41} & a_{42} & a_{43} & a_{44} \end{bmatrix}$$

The argument in the $n \times n$ case is similar, as is the case of upper triangular matrices.

Proof of Theorem 2.1.3 (4 × 4 lower triangular case)   By Theorem 2.1.1, the determinant of A may be found by cofactor expansion along the first row:

$$\det(A) = \begin{vmatrix} a_{11} & 0 & 0 & 0 \\ a_{21} & a_{22} & 0 & 0 \\ a_{31} & a_{32} & a_{33} & 0 \\ a_{41} & a_{42} & a_{43} & a_{44} \end{vmatrix}$$

$$= a_{11} \begin{vmatrix} a_{22} & 0 & 0 \\ a_{32} & a_{33} & 0 \\ a_{42} & a_{43} & a_{44} \end{vmatrix}$$

Once again, it's easy to expand along the first row:

$$\det(A) = a_{11}a_{22} \begin{vmatrix} a_{33} & 0 \\ a_{43} & a_{44} \end{vmatrix}$$

$$= a_{11}a_{22}a_{33}|a_{44}|$$

$$= a_{11}a_{22}a_{33}a_{44}$$

where we have used the convention that the determinant of a 1×1 matrix $[a]$ is a. ■

EXAMPLE 8   Determinant of an Upper Triangular Matrix

$$\begin{vmatrix} 2 & 7 & -3 & 8 & 3 \\ 0 & -3 & 7 & 5 & 1 \\ 0 & 0 & 6 & 7 & 6 \\ 0 & 0 & 0 & 9 & 8 \\ 0 & 0 & 0 & 0 & 4 \end{vmatrix} = (2)(-3)(6)(9)(4) = -1296 \quad \blacklozenge$$

Proof of Theorem 1.7.1c   Let $A = [a_{ij}]$ be a triangular matrix, so that its diagonal entries are

$$a_{11}, a_{22}, \ldots, a_{nn}$$

From Theorem 2.1.3, the matrix A is invertible if and only if

$$\det(A) = a_{11}a_{22} \cdots a_{nn}$$

is nonzero, which is true if and only if the diagonal entries are all nonzero. ■

We leave it as an exercise for the reader to use the adjoint formula for A^{-1} to show that if $A = [a_{ij}]$ is an invertible triangular matrix, then the successive diagonal entries of A^{-1} are

$$\frac{1}{a_{11}}, \frac{1}{a_{22}}, \ldots, \frac{1}{a_{nn}}$$

(See Example 3 of Section 1.7.)

Proof of Theorem 1.7.1d We will prove the result for upper triangular matrices and leave the lower triangular case as an exercise. Assume that A is upper triangular and invertible. Since

$$A^{-1} = \frac{1}{\det(A)} \text{adj}(A)$$

we can prove that A^{-1} is upper triangular by showing that $\text{adj}(A)$ is upper triangular, or, equivalently, that the matrix of cofactors is lower triangular. We can do this by showing that every cofactor C_{ij} with $i < j$ (i.e., above the main diagonal) is zero. Since

$$C_{ij} = (-1)^{i+j} M_{ij}$$

it suffices to show that each minor M_{ij} with $i < j$ is zero. For this purpose, let B_{ij} be the matrix that results when the ith row and jth column of A are deleted, so

$$M_{ij} = \det(B_{ij}) \tag{10}$$

From the assumption that $i < j$, it follows that B_{ij} is upper triangular (Exercise 32). Since A is upper triangular, its $(i + 1)$-st row begins with at least i zeros. But the ith row of B_{ij} is the $(i + 1)$-st row of A with the entry in the jth column removed. Since $i < j$, none of the first i zeros is removed by deleting the jth column; thus the ith row of B_{ij} starts with at least i zeros, which implies that this row has a zero on the main diagonal. It now follows from Theorem 2.1.3 that $\det(B_{ij}) = 0$ and from (10) that $M_{ij} = 0$. ∎

Cramer's Rule

The next theorem provides a formula for the solution of certain linear systems of n equations in n unknowns. This formula, known as ***Cramer's rule***, is of marginal interest for computational purposes, but it is useful for studying the mathematical properties of a solution without the need for solving the system.

THEOREM 2.1.4

Cramer's Rule

If $A\mathbf{x} = \mathbf{b}$ is a system of n linear equations in n unknowns such that $\det(A) \neq 0$, then the system has a unique solution. This solution is

$$x_1 = \frac{\det(A_1)}{\det(A)}, \quad x_2 = \frac{\det(A_2)}{\det(A)}, \ldots, \quad x_n = \frac{\det(A_n)}{\det(A)}$$

where A_j is the matrix obtained by replacing the entries in the jth column of A by the entries in the matrix

$$\mathbf{b} = \begin{bmatrix} b_1 \\ b_2 \\ \vdots \\ b_n \end{bmatrix}$$

Proof If $\det(A) \neq 0$, then A is invertible, and by Theorem 1.6.2, $\mathbf{x} = A^{-1}\mathbf{b}$ is the unique solution of $A\mathbf{x} = \mathbf{b}$. Therefore, by Theorem 2.1.2 we have

$$\mathbf{x} = A^{-1}\mathbf{b} = \frac{1}{\det(A)} \text{adj}(A)\mathbf{b} = \frac{1}{\det(A)} \begin{bmatrix} C_{11} & C_{21} & \cdots & C_{n1} \\ C_{12} & C_{22} & \cdots & C_{n2} \\ \vdots & \vdots & & \vdots \\ C_{1n} & C_{2n} & \cdots & C_{nn} \end{bmatrix} \begin{bmatrix} b_1 \\ b_2 \\ \vdots \\ b_n \end{bmatrix}$$

Multiplying the matrices out gives

$$\mathbf{x} = \frac{1}{\det(A)} \begin{bmatrix} b_1 C_{11} + b_2 C_{21} + \cdots + b_n C_{n1} \\ b_1 C_{12} + b_2 C_{22} + \cdots + b_n C_{n2} \\ \vdots \qquad \vdots \qquad\qquad \vdots \\ b_1 C_{1n} + b_2 C_{2n} + \cdots + b_n C_{nn} \end{bmatrix}$$

The entry in the jth row of $\mathbf{x}$ is therefore

$$x_j = \frac{b_1 C_{1j} + b_2 C_{2j} + \cdots + b_n C_{nj}}{\det(A)} \tag{11}$$

Now let

$$A_j = \begin{bmatrix} a_{11} & a_{12} & \cdots & a_{1j-1} & b_1 & a_{1j+1} & \cdots & a_{1n} \\ a_{21} & a_{22} & \cdots & a_{2j-1} & b_2 & a_{2j+1} & \cdots & a_{2n} \\ \vdots & \vdots & & \vdots & \vdots & \vdots & & \vdots \\ a_{n1} & a_{n2} & \cdots & a_{nj-1} & b_n & a_{nj+1} & \cdots & a_{nn} \end{bmatrix}$$

Since A_j differs from A only in the jth column, it follows that the cofactors of entries $b_1, b_2, \ldots, b_n$ in A_j are the same as the cofactors of the corresponding entries in the jth column of A. The cofactor expansion of $\det(A_j)$ along the jth column is therefore

$$\det(A_j) = b_1 C_{1j} + b_2 C_{2j} + \cdots + b_n C_{nj}$$

Substituting this result in (11) gives

$$x_j = \frac{\det(A_j)}{\det(A)} \qquad ■$$

Gabriel Cramer *(1704–1752)* was a Swiss mathematician. Although Cramer does not rank with the great mathematicians of his time, his contributions as a disseminator of mathematical ideas have earned him a well-deserved place in the history of mathematics. Cramer traveled extensively and met many of the leading mathematicians of his day.

Cramer's most widely known work, *Introduction à l'analyse des lignes courbes algébriques* (1750), was a study and classification of algebraic curves; Cramer's rule appeared in the appendix. Although the rule bears his name, variations of the idea were formulated earlier by various mathematicians. However, Cramer's superior notation helped clarify and popularize the technique.

Overwork combined with a fall from a carriage led to his death at the age of 48. Cramer was apparently a good-natured and pleasant person with broad interests. He wrote on philosophy of law and government and the history of mathematics. He served in public office, participated in artillery and fortifications activities for the government, instructed workers on techniques of cathedral repair, and undertook excavations of cathedral archives. Cramer received numerous honors for his activities.

EXAMPLE 9 Using Cramer's Rule to Solve a Linear System

Use Cramer's rule to solve

$$\begin{aligned} x_1 + \quad\;\; + 2x_3 &= 6 \\ -3x_1 + 4x_2 + 6x_3 &= 30 \\ -x_1 - 2x_2 + 3x_3 &= 8 \end{aligned}$$

Solution

$$A = \begin{bmatrix} 1 & 0 & 2 \\ -3 & 4 & 6 \\ -1 & -2 & 3 \end{bmatrix}, \quad A_1 = \begin{bmatrix} 6 & 0 & 2 \\ 30 & 4 & 6 \\ 8 & -2 & 3 \end{bmatrix},$$

$$A_2 = \begin{bmatrix} 1 & 6 & 2 \\ -3 & 30 & 6 \\ -1 & 8 & 3 \end{bmatrix}, \quad A_3 = \begin{bmatrix} 1 & 0 & 6 \\ -3 & 4 & 30 \\ -1 & -2 & 8 \end{bmatrix}$$

Therefore,

$$x_1 = \frac{\det(A_1)}{\det(A)} = \frac{-40}{44} = \frac{-10}{11}, \qquad x_2 = \frac{\det(A_2)}{\det(A)} = \frac{72}{44} = \frac{18}{11},$$

$$x_3 = \frac{\det(A_3)}{\det(A)} = \frac{152}{44} = \frac{38}{11} \qquad \blacklozenge$$

REMARK To solve a system of n equations in n unknowns by Cramer's rule, it is necessary to evaluate $n + 1$ determinants of $n \times n$ matrices. For systems with more than three equations, Gaussian elimination is far more efficient. However, Cramer's rule does give a formula for the solution if the determinant of the coefficient matrix is nonzero.

EXERCISE SET 2.1

1. Let

$$A = \begin{bmatrix} 1 & -2 & 3 \\ 6 & 7 & -1 \\ -3 & 1 & 4 \end{bmatrix}$$

(a) Find all the minors of A. (b) Find all the cofactors.

2. Let

$$A = \begin{bmatrix} 4 & -1 & 1 & 6 \\ 0 & 0 & -3 & 3 \\ 4 & 1 & 0 & 14 \\ 4 & 1 & 3 & 2 \end{bmatrix}$$

Find

(a) M_{13} and C_{13} (b) M_{23} and C_{23} (c) M_{22} and C_{22} (d) M_{21} and C_{21}

3. Evaluate the determinant of the matrix in Exercise 1 by a cofactor expansion along

(a) the first row (b) the first column (c) the second row

(d) the second column (e) the third row (f) the third column

4. For the matrix in Exercise 1, find

(a) adj(A) (b) A^{-1} using Theorem 2.1.2

In Exercises 5–10 evaluate det(A) by a cofactor expansion along a row or column of your choice.

5. $A = \begin{bmatrix} -3 & 0 & 7 \\ 2 & 5 & 1 \\ -1 & 0 & 5 \end{bmatrix}$ **6.** $A = \begin{bmatrix} 3 & 3 & 1 \\ 1 & 0 & -4 \\ 1 & -3 & 5 \end{bmatrix}$ **7.** $A = \begin{bmatrix} 1 & k & k^2 \\ 1 & k & k^2 \\ 1 & k & k^2 \end{bmatrix}$

8. $A = \begin{bmatrix} k+1 & k-1 & 7 \\ 2 & k-3 & 4 \\ 5 & k+1 & k \end{bmatrix}$ **9.** $A = \begin{bmatrix} 3 & 3 & 0 & 5 \\ 2 & 2 & 0 & -2 \\ 4 & 1 & -3 & 0 \\ 2 & 10 & 3 & 2 \end{bmatrix}$

10. $A = \begin{bmatrix} 4 & 0 & 0 & 1 & 0 \\ 3 & 3 & 3 & -1 & 0 \\ 1 & 2 & 4 & 2 & 3 \\ 9 & 4 & 6 & 2 & 3 \\ 2 & 2 & 4 & 2 & 3 \end{bmatrix}$

In Exercises 11–14 find A^{-1} using Theorem 2.1.2.

11. $A = \begin{bmatrix} 2 & 5 & 5 \\ -1 & -1 & 0 \\ 2 & 4 & 3 \end{bmatrix}$ **12.** $A = \begin{bmatrix} 2 & 0 & 3 \\ 0 & 3 & 2 \\ -2 & 0 & -4 \end{bmatrix}$

13. $A = \begin{bmatrix} 2 & -3 & 5 \\ 0 & 1 & -3 \\ 0 & 0 & 2 \end{bmatrix}$ **14.** $A = \begin{bmatrix} 2 & 0 & 0 \\ 8 & 1 & 0 \\ -5 & 3 & 6 \end{bmatrix}$

15. Let

$$A = \begin{bmatrix} 1 & 3 & 1 & 1 \\ 2 & 5 & 2 & 2 \\ 1 & 3 & 8 & 9 \\ 1 & 3 & 2 & 2 \end{bmatrix}$$

    (a) Evaluate A^{-1} using Theorem 2.1.2.

    (b) Evaluate A^{-1} using the method of Example 4 in Section 1.5.

    (c) Which method involves less computation?

In Exercises 16–21 solve by Cramer's rule, where it applies.

16. $7x_1 - 2x_2 = 3$
$\quad\;\, 3x_1 + x_2 = 5$

17. $4x + 5y \quad\;\;\, = 2$
$\quad\;\, 11x + y + 2z = 3$
$\quad\;\;\;\, x + 5y + 2z = 1$

18. $x - 4y + z = 6$
$\quad\;\, 4x - y + 2z = -1$
$\quad\;\, 2x + 2y - 3z = -20$

19. $x_1 - 3x_2 + x_3 = 4$
$\quad\;\, 2x_1 - x_2 \quad\;\;\, = -2$
$\quad\;\, 4x_1 \quad\;\;\; - 3x_3 = 0$

20. $-x_1 - 4x_2 + 2x_3 + x_4 = -32$
$\quad\;\, 2x_1 - x_2 + 7x_3 + 9x_4 = 14$
$\quad\;\, -x_1 + x_2 + 3x_3 + x_4 = 11$
$\quad\;\, x_1 - 2x_2 + x_3 - 4x_4 = -4$

21. $3x_1 - x_2 + x_3 = 4$
$\quad\;\, -x_1 + 7x_2 - 2x_3 = 1$
$\quad\;\, 2x_1 + 6x_2 - x_3 = 5$

22. Show that the matrix

$$A = \begin{bmatrix} \cos\theta & \sin\theta & 0 \\ -\sin\theta & \cos\theta & 0 \\ 0 & 0 & 1 \end{bmatrix}$$

is invertible for all values of θ; then find A^{-1} using Theorem 2.1.2.

23. Use Cramer's rule to solve for y without solving for x, z, and w.

$$4x + y + z + w = 6$$
$$3x + 7y - z + w = 1$$
$$7x + 3y - 5z + 8w = -3$$
$$x + y + z + 2w = 3$$

24. Let $A\mathbf{x} = \mathbf{b}$ be the system in Exercise 23.

    (a) Solve by Cramer's rule.   (b) Solve by Gauss–Jordan elimination.

    (c) Which method involves fewer computations?

25. Prove that if $\det(A) = 1$ and all the entries in A are integers, then all the entries in A^{-1} are integers.

26. Let $A\mathbf{x} = \mathbf{b}$ be a system of n linear equations in n unknowns with integer coefficients and integer constants. Prove that if $\det(A) = 1$, the solution $\mathbf{x}$ has integer entries.

27. Prove that if A is an invertible lower triangular matrix, then A^{-1} is lower triangular.

28. Derive the last cofactor expansion listed in Formula (4).

29. Prove: The equation of the line through the distinct points (a_1, b_1) and (a_2, b_2) can be written as

$$\begin{vmatrix} x & y & 1 \\ a_1 & b_1 & 1 \\ a_2 & b_2 & 1 \end{vmatrix} = 0$$

30. Prove: (x_1, y_1), (x_2, y_2), and (x_3, y_3) are collinear points if and only if

$$\begin{vmatrix} x_1 & y_1 & 1 \\ x_2 & y_2 & 1 \\ x_3 & y_3 & 1 \end{vmatrix} = 0$$

31. (a) If $A = \begin{bmatrix} A_{11} & A_{12} \\ \hline 0 & A_{22} \end{bmatrix}$ is an "upper triangular" block matrix, where A_{11} and A_{22} are

square matrices, then $\det(A) = \det(A_{11})\det(A_{22})$. Use this result to evaluate $\det(A)$ for

$$\begin{bmatrix} 2 & -1 & 2 & 5 & 6 \\ 4 & 3 & -1 & 3 & 4 \\ \hline 0 & 0 & 1 & 3 & 5 \\ 0 & 0 & -2 & 6 & 2 \\ 0 & 0 & 3 & 5 & 2 \end{bmatrix}$$

(b) Verify your answer in part (a) by using a cofactor expansion to evaluate $\det(A)$.

32. Prove that if A is upper triangular and B_{ij} is the matrix that results when the ith row and jth column of A are deleted, then B_{ij} is upper triangular if $i < j$.

Discussion & Discovery

33. What is the maximum number of zeros that a 4×4 matrix can have without having a zero determinant? Explain your reasoning.

34. Let A be a matrix of the form

$$A = \begin{bmatrix} * & * & 0 & 0 & 0 \\ * & * & 0 & 0 & 0 \\ * & * & 0 & 0 & 0 \\ * & * & * & * & * \\ * & * & * & * & * \end{bmatrix}$$

How many different values can you obtain for $\det(A)$ by substituting numerical values (not necessarily all the same) for the *'s? Explain your reasoning.

35. Indicate whether the statement is always true or sometimes false. Justify your answer by giving a logical argument or a counterexample.

(a) $A\,\mathrm{adj}(A)$ is a diagonal matrix for every square matrix A.

(b) In theory, Cramer's rule can be used to solve any system of linear equations, although the amount of computation may be enormous.

(c) If A is invertible, then $\mathrm{adj}(A)$ must also be invertible.

(d) If A has a row of zeros, then so does $\mathrm{adj}(A)$.

2.2
EVALUATING DETERMINANTS BY ROW REDUCTION

In this section we shall show that the determinant of a square matrix can be evaluated by reducing the matrix to row-echelon form. This method is important since it is the most computationally efficient way to find the determinant of a general matrix.

A Basic Theorem

We begin with a fundamental theorem that will lead us to an efficient procedure for evaluating the determinant of a matrix of any order n.

THEOREM 2.2.1

Let A be a square matrix. If A has a row of zeros or a column of zeros, then $\det(A) = 0$.

Proof By Theorem 2.1.1, the determinant of A found by cofactor expansion along the row or column of all zeros is

$$\det(A) = 0 \cdot C_1 + 0 \cdot C_2 + \cdots + 0 \cdot C_n$$

where $C_1, \ldots, C_n$ are the cofactors for that row or column. Hence $\det(A)$ is zero. ■

Here is another useful theorem:

THEOREM 2.2.2

Let A be a square matrix. Then $\det(A) = \det(A^T)$.

Proof By Theorem 2.1.1, the determinant of A found by cofactor expansion along its first row is the same as the determinant of A^T found by cofactor expansion along its first column. ■

REMARK Because of Theorem 2.2.2, nearly every theorem about determinants that contains the word *row* in its statement is also true when the word *column* is substituted for *row*. To prove a column statement, one need only transpose the matrix in question, to convert the column statement to a row statement, and then apply the corresponding known result for rows.

Elementary Row Operations

The next theorem shows how an elementary row operation on a matrix affects the value of its determinant.

THEOREM 2.2.3

Let A be an $n \times n$ matrix.

(a) If B is the matrix that results when a single row or single column of A is multiplied by a scalar k, then $\det(B) = k \det(A)$.

(b) If B is the matrix that results when two rows or two columns of A are interchanged, then $\det(B) = -\det(A)$.

(c) If B is the matrix that results when a multiple of one row of A is added to another row or when a multiple of one column is added to another column, then $\det(B) = \det(A)$.

We omit the proof but give the following example that illustrates the theorem for 3×3 determinants.

EXAMPLE 1   Theorem 2.2.3 Applied to 3 × 3 Determinants

We will verify the equation in the first row of Table 1 and leave the last two for the reader. By Theorem 2.1.1, the determinant of B may be found by cofactor expansion along the first row:

$$\det(B) = \begin{vmatrix} ka_{11} & ka_{12} & ka_{13} \\ a_{21} & a_{22} & a_{23} \\ a_{31} & a_{32} & a_{33} \end{vmatrix}$$

$$= ka_{11}C_{11} + ka_{12}C_{12} + ka_{33}C_{13}$$

$$= k(a_{11}C_{11} + a_{12}C_{12} + a_{33}C_{13})$$

$$= k \det(A)$$

Table 1

Relationship	Operation
$\begin{vmatrix} ka_{11} & ka_{12} & ka_{13} \\ a_{21} & a_{22} & a_{23} \\ a_{31} & a_{32} & a_{33} \end{vmatrix} = k \begin{vmatrix} a_{11} & a_{12} & a_{13} \\ a_{21} & a_{22} & a_{23} \\ a_{31} & a_{32} & a_{33} \end{vmatrix}$ $\det(B) = k \det(A)$	The first row of A is multiplied by k.
$\begin{vmatrix} a_{21} & a_{22} & a_{23} \\ a_{11} & a_{12} & a_{13} \\ a_{31} & a_{32} & a_{33} \end{vmatrix} = - \begin{vmatrix} a_{11} & a_{12} & a_{13} \\ a_{21} & a_{22} & a_{23} \\ a_{31} & a_{32} & a_{33} \end{vmatrix}$ $\det(B) = -\det(A)$	The first and second rows of A are interchanged.
$\begin{vmatrix} a_{11}+ka_{21} & a_{12}+ka_{22} & a_{13}+ka_{23} \\ a_{21} & a_{22} & a_{23} \\ a_{31} & a_{32} & a_{33} \end{vmatrix} = \begin{vmatrix} a_{11} & a_{12} & a_{13} \\ a_{21} & a_{22} & a_{23} \\ a_{31} & a_{32} & a_{33} \end{vmatrix}$ $\det(B) = \det(A)$	A multiple of the second row of A is added to the first row.

since C_{11}, C_{12}, and C_{13} do not depend on the first row of the matrix, and A and B differ only in their first rows. ◆

REMARK As illustrated by the first equation in Table 1, part (*a*) of Theorem 2.2.3 enables us to bring a "common factor" from any row (or column) through the determinant sign.

Elementary Matrices

Recall that an elementary matrix results from performing a single elementary row operation on an identity matrix; thus, if we let $A = I_n$ in Theorem 2.2.3 [so that we have $\det(A) = \det(I_n) = 1$], then the matrix B is an elementary matrix, and the theorem yields the following result about determinants of elementary matrices.

THEOREM 2.2.4

Let E be an n × n elementary matrix.

(*a*) *If E results from multiplying a row of I_n by k, then $\det(E) = k$.*
(*b*) *If E results from interchanging two rows of I_n, then $\det(E) = -1$.*
(*c*) *If E results from adding a multiple of one row of I_n to another, then $\det(E) = 1$.*

EXAMPLE 2 Determinants of Elementary Matrices

The following determinants of elementary matrices, which are evaluated by inspection, illustrate Theorem 2.2.4.

$$\begin{vmatrix} 1 & 0 & 0 & 0 \\ 0 & 3 & 0 & 0 \\ 0 & 0 & 1 & 0 \\ 0 & 0 & 0 & 1 \end{vmatrix} = 3, \qquad \begin{vmatrix} 0 & 0 & 0 & 1 \\ 0 & 1 & 0 & 0 \\ 0 & 0 & 1 & 0 \\ 1 & 0 & 0 & 0 \end{vmatrix} = -1, \qquad \begin{vmatrix} 1 & 0 & 0 & 7 \\ 0 & 1 & 0 & 0 \\ 0 & 0 & 1 & 0 \\ 0 & 0 & 0 & 1 \end{vmatrix} = 1 \quad ◆$$

The second row of I_4 was multiplied by 3.   The first and last rows of I_4 were interchanged.   7 times the last row of I_4 was added to the first row.

Matrices with Proportional Rows or Columns

If a square matrix A has two proportional rows, then a row of zeros can be introduced by adding a suitable multiple of one of the rows to the other. Similarly for columns. But adding a multiple of one row or column to another does not change the determinant, so from Theorem 2.2.1, we must have $\det(A) = 0$. This proves the following theorem.

THEOREM 2.2.5

> *If A is a square matrix with two proportional rows or two proportional columns, then* $\det(A) = 0$.

EXAMPLE 3   Introducing Zero Rows

The following computation illustrates the introduction of a row of zeros when there are two proportional rows:

$$\begin{vmatrix} 1 & 3 & -2 & 4 \\ 2 & 6 & -4 & 8 \\ 3 & 9 & 1 & 5 \\ 1 & 1 & 4 & 8 \end{vmatrix} = \begin{vmatrix} 1 & 3 & -2 & 4 \\ 0 & 0 & 0 & 0 \\ 3 & 9 & 1 & 5 \\ 1 & 1 & 4 & 8 \end{vmatrix} = 0$$

← The second row is 2 times the first, so we added -2 times the first row to the second to introduce a row of zeros.

Each of the following matrices has two proportional rows or columns; thus, each has a determinant of zero.

$$\begin{bmatrix} -1 & 4 \\ -2 & 8 \end{bmatrix}, \qquad \begin{bmatrix} 1 & -2 & 7 \\ -4 & 8 & 5 \\ 2 & -4 & 3 \end{bmatrix}, \qquad \begin{bmatrix} 3 & -1 & 4 & -5 \\ 6 & -2 & 5 & 2 \\ 5 & 8 & 1 & 4 \\ -9 & 3 & -12 & 15 \end{bmatrix} \quad \blacklozenge$$

Evaluating Determinants by Row Reduction

We shall now give a method for evaluating determinants that involves substantially less computation than the cofactor expansion method. The idea of the method is to reduce the given matrix to upper triangular form by elementary row operations, then compute the determinant of the upper triangular matrix (an easy computation), and then relate that determinant to that of the original matrix. Here is an example:

EXAMPLE 4   Using Row Reduction to Evaluate a Determinant

Evaluate $\det(A)$ where

$$A = \begin{bmatrix} 0 & 1 & 5 \\ 3 & -6 & 9 \\ 2 & 6 & 1 \end{bmatrix}$$

Solution

We will reduce A to row-echelon form (which is upper triangular) and apply Theorem 2.2.3:

$$\det(A) = \begin{vmatrix} 0 & 1 & 5 \\ 3 & -6 & 9 \\ 2 & 6 & 1 \end{vmatrix} = -\begin{vmatrix} 3 & -6 & 9 \\ 0 & 1 & 5 \\ 2 & 6 & 1 \end{vmatrix}$$

← The first and second rows of A were interchanged.

$$= -3\begin{vmatrix} 1 & -2 & 3 \\ 0 & 1 & 5 \\ 2 & 6 & 1 \end{vmatrix}$$

← A common factor of 3 from the first row was taken through the determinant sign.

$$= -3 \begin{vmatrix} 1 & -2 & 3 \\ 0 & 1 & 5 \\ 0 & 10 & -5 \end{vmatrix}$$ ◄——— -2 times the first row was added to the third row.

$$= -3 \begin{vmatrix} 1 & -2 & 3 \\ 0 & 1 & 5 \\ 0 & 0 & -55 \end{vmatrix}$$ ◄——— -10 times the second row was added to the third row.

$$= (-3)(-55) \begin{vmatrix} 1 & -2 & 3 \\ 0 & 1 & 5 \\ 0 & 0 & 1 \end{vmatrix}$$ ◄——— A common factor of -55 from the last row was taken through the determinant sign.

$$= (-3)(-55)(1) = 165 \quad \blacklozenge$$

REMARK The method of row reduction is well suited for computer evaluation of determinants because it is computationally efficient and easily programmed. However, cofactor expansion is often easier for hand computation.

EXAMPLE 5 Using Column Operations to Evaluate a Determinant

Compute the determinant of

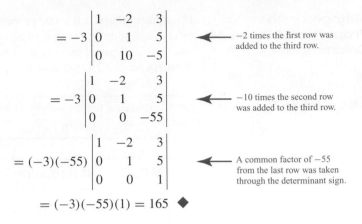

$$A = \begin{bmatrix} 1 & 0 & 0 & 3 \\ 2 & 7 & 0 & 6 \\ 0 & 6 & 3 & 0 \\ 7 & 3 & 1 & -5 \end{bmatrix}$$

Solution

This determinant could be computed as above by using elementary row operations to reduce A to row-echelon form, but we can put A in lower triangular form in one step by adding -3 times the first column to the fourth to obtain

$$\det(A) = \det \begin{bmatrix} 1 & 0 & 0 & 0 \\ 2 & 7 & 0 & 0 \\ 0 & 6 & 3 & 0 \\ 7 & 3 & 1 & -26 \end{bmatrix} = (1)(7)(3)(-26) = -546$$

This example points out the utility of keeping an eye open for column operations that can shorten computations. ◆

Cofactor expansion and row or column operations can sometimes be used in combination to provide an effective method for evaluating determinants. The following example illustrates this idea.

EXAMPLE 6 Row Operations and Cofactor Expansion

Evaluate $\det(A)$ where

$$A = \begin{bmatrix} 3 & 5 & -2 & 6 \\ 1 & 2 & -1 & 1 \\ 2 & 4 & 1 & 5 \\ 3 & 7 & 5 & 3 \end{bmatrix}$$

Solution

By adding suitable multiples of the second row to the remaining rows, we obtain

$$\det(A) = \begin{vmatrix} 0 & -1 & 1 & 3 \\ 1 & 2 & -1 & 1 \\ 0 & 0 & 3 & 3 \\ 0 & 1 & 8 & 0 \end{vmatrix}$$

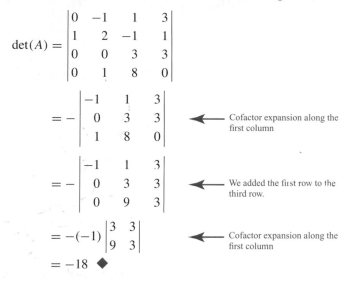

$$= -\begin{vmatrix} -1 & 1 & 3 \\ 0 & 3 & 3 \\ 1 & 8 & 0 \end{vmatrix} \qquad \longleftarrow \quad \text{Cofactor expansion along the first column}$$

$$= -\begin{vmatrix} -1 & 1 & 3 \\ 0 & 3 & 3 \\ 0 & 9 & 3 \end{vmatrix} \qquad \longleftarrow \quad \text{We added the first row to the third row.}$$

$$= -(-1)\begin{vmatrix} 3 & 3 \\ 9 & 3 \end{vmatrix} \qquad \longleftarrow \quad \text{Cofactor expansion along the first column}$$

$$= -18 \quad \blacklozenge$$

EXERCISE SET 2.2

1. Verify that $\det(A) = \det(A^T)$ for

   (a) $A = \begin{bmatrix} -2 & 3 \\ 1 & 4 \end{bmatrix}$ (b) $A = \begin{bmatrix} 2 & -1 & 3 \\ 1 & 2 & 4 \\ 5 & -3 & 6 \end{bmatrix}$

2. Evaluate the following determinants by inspection.

   (a) $\begin{vmatrix} 3 & -17 & 4 \\ 0 & 5 & 1 \\ 0 & 0 & -2 \end{vmatrix}$ (b) $\begin{vmatrix} \sqrt{2} & 0 & 0 & 0 \\ -8 & \sqrt{2} & 0 & 0 \\ 7 & 0 & -1 & 0 \\ 9 & 5 & 6 & 1 \end{vmatrix}$

   (c) $\begin{vmatrix} -2 & 1 & 3 \\ 1 & -7 & 4 \\ -2 & 1 & 3 \end{vmatrix}$ (d) $\begin{vmatrix} 1 & -2 & 3 \\ 2 & -4 & 6 \\ 5 & -8 & 1 \end{vmatrix}$

3. Find the determinants of the following elementary matrices by inspection.

   (a) $\begin{bmatrix} 1 & 0 & 0 & 0 \\ 0 & 1 & 0 & 0 \\ 0 & 0 & -5 & 0 \\ 0 & 0 & 0 & 1 \end{bmatrix}$ (b) $\begin{bmatrix} 1 & 0 & 0 & 0 \\ 0 & 0 & 1 & 0 \\ 0 & 1 & 0 & 0 \\ 0 & 0 & 0 & 1 \end{bmatrix}$ (c) $\begin{bmatrix} 1 & 0 & 0 & 0 \\ 0 & 1 & 0 & -9 \\ 0 & 0 & 1 & 0 \\ 0 & 0 & 0 & 1 \end{bmatrix}$

In Exercises 4–11 evaluate the determinant of the given matrix by reducing the matrix to row-echelon form.

4. $\begin{bmatrix} 3 & 6 & -9 \\ 0 & 0 & -2 \\ -2 & 1 & 5 \end{bmatrix}$ 5. $\begin{bmatrix} 0 & 3 & 1 \\ 1 & 1 & 2 \\ 3 & 2 & 4 \end{bmatrix}$ 6. $\begin{bmatrix} 1 & -3 & 0 \\ -2 & 4 & 1 \\ 5 & -2 & 2 \end{bmatrix}$

7. $\begin{bmatrix} 3 & -6 & 9 \\ -2 & 7 & -2 \\ 0 & 1 & 5 \end{bmatrix}$ 8. $\begin{bmatrix} 1 & -2 & 3 & 1 \\ 5 & -9 & 6 & 3 \\ -1 & 2 & -6 & -2 \\ 2 & 8 & 6 & 1 \end{bmatrix}$ 9. $\begin{bmatrix} 2 & 1 & 3 & 1 \\ 1 & 0 & 1 & 1 \\ 0 & 2 & 1 & 0 \\ 0 & 1 & 2 & 3 \end{bmatrix}$

10. $\begin{bmatrix} 0 & 1 & 1 & 1 \\ \frac{1}{2} & \frac{1}{2} & 1 & \frac{1}{2} \\ \frac{2}{3} & \frac{1}{3} & \frac{1}{3} & 0 \\ -\frac{1}{3} & \frac{2}{3} & 0 & 0 \end{bmatrix}$

11. $\begin{bmatrix} 1 & 3 & 1 & 5 & 3 \\ -2 & -7 & 0 & -4 & 2 \\ 0 & 0 & 1 & 0 & 1 \\ 0 & 0 & 2 & 1 & 1 \\ 0 & 0 & 0 & 1 & 1 \end{bmatrix}$

12. Given that $\begin{vmatrix} a & b & c \\ d & e & f \\ g & h & i \end{vmatrix} = -6$, find

(a) $\begin{vmatrix} d & e & f \\ g & h & i \\ a & b & c \end{vmatrix}$

(b) $\begin{vmatrix} 3a & 3b & 3c \\ -d & -e & -f \\ 4g & 4h & 4i \end{vmatrix}$

(c) $\begin{vmatrix} a+g & b+h & c+i \\ d & e & f \\ g & h & i \end{vmatrix}$

(d) $\begin{vmatrix} -3a & -3b & -3c \\ d & e & f \\ g-4d & h-4e & i-4f \end{vmatrix}$

13. Use row reduction to show that

$$\begin{vmatrix} 1 & 1 & 1 \\ a & b & c \\ a^2 & b^2 & c^2 \end{vmatrix} = (b-a)(c-a)(c-b)$$

14. Use an argument like that in the proof of Theorem 2.1.3 to show that

(a) $\det \begin{bmatrix} 0 & 0 & a_{13} \\ 0 & a_{22} & a_{23} \\ a_{31} & a_{32} & a_{33} \end{bmatrix} = -a_{13}a_{22}a_{31}$

(b) $\det \begin{bmatrix} 0 & 0 & 0 & a_{14} \\ 0 & 0 & a_{23} & a_{24} \\ 0 & a_{32} & a_{33} & a_{34} \\ a_{41} & a_{42} & a_{43} & a_{44} \end{bmatrix} = a_{14}a_{23}a_{32}a_{41}$

15. Prove the following special cases of Theorem 2.2.3.

(a) $\begin{vmatrix} a_{21} & a_{22} & a_{23} \\ a_{11} & a_{12} & a_{13} \\ a_{31} & a_{32} & a_{33} \end{vmatrix} = - \begin{vmatrix} a_{11} & a_{12} & a_{13} \\ a_{21} & a_{22} & a_{23} \\ a_{31} & a_{32} & a_{33} \end{vmatrix}$

(b) $\begin{vmatrix} a_{11}+ka_{21} & a_{12}+ka_{22} & a_{13}+ka_{23} \\ a_{21} & a_{22} & a_{23} \\ a_{31} & a_{32} & a_{33} \end{vmatrix} = \begin{vmatrix} a_{11} & a_{12} & a_{13} \\ a_{21} & a_{22} & a_{23} \\ a_{31} & a_{32} & a_{33} \end{vmatrix}$

16. Repeat Exercises 4–7 using a combination of row reduction and cofactor expansion, as in Example 6.

17. Repeat Exercises 8–11 using a combination of row reduction and cofactor expansion, as in Example 6.

Discussion & Discovery

18. In each part, find $\det(A)$ by inspection, and explain your reasoning.

(a) $A = \begin{bmatrix} 0 & 0 & 1 \\ 0 & 1 & 0 \\ 1 & 0 & 0 \end{bmatrix}$

(b) $A = \begin{bmatrix} 0 & 0 & 0 & 1 \\ 0 & 0 & 1 & 0 \\ 0 & 1 & 0 & 0 \\ 1 & 0 & 0 & 0 \end{bmatrix}$

19. By inspection, solve the equation

$$
\begin{vmatrix}
x & 5 & 7 \\
0 & x+1 & 6 \\
0 & 0 & 2x-1
\end{vmatrix} = 0
$$

Explain your reasoning.

20. (a) By inspection, find two solutions of the equation

$$
\begin{vmatrix}
1 & x & x^2 \\
1 & 1 & 1 \\
1 & -3 & 9
\end{vmatrix} = 0
$$

(b) Is it possible that there are other solutions? Justify your answer.

21. How many arithmetic operations are needed, in general, to find $\det(A)$ by row reduction? By cofactor expansion?

2.3
PROPERTIES OF THE DETERMINANT FUNCTION

In this section we shall develop some of the fundamental properties of the determinant function. Our work here will give us some further insight into the relationship between a square matrix and its determinant. One of the immediate consequences of this material will be the determinant test for the invertibility of a matrix.

Basic Properties of Determinants

Suppose that A and B are $n \times n$ matrices and k is any scalar. We begin by considering possible relationships between $\det(A)$, $\det(B)$, and

$$\det(kA), \quad \det(A+B), \quad \text{and} \quad \det(AB)$$

Since a common factor of any row of a matrix can be moved through the det sign, and since each of the n rows in kA has a common factor of k, we obtain

$$\det(kA) = k^n \det(A) \tag{1}$$

For example,

$$
\begin{vmatrix}
ka_{11} & ka_{12} & ka_{13} \\
ka_{21} & ka_{22} & ka_{23} \\
ka_{31} & ka_{32} & ka_{33}
\end{vmatrix}
= k^3
\begin{vmatrix}
a_{11} & a_{12} & a_{13} \\
a_{21} & a_{22} & a_{23} \\
a_{31} & a_{32} & a_{33}
\end{vmatrix}
$$

Unfortunately, no simple relationship exists among $\det(A)$, $\det(B)$, and $\det(A+B)$. In particular, we emphasize that $\det(A+B)$ will usually *not* be equal to $\det(A) + \det(B)$. The following example illustrates this fact.

EXAMPLE 1 det(A + B) ≠ det(A) + det(B)

Consider

$$
A = \begin{bmatrix} 1 & 2 \\ 2 & 5 \end{bmatrix}, \qquad
B = \begin{bmatrix} 3 & 1 \\ 1 & 3 \end{bmatrix}, \qquad
A + B = \begin{bmatrix} 4 & 3 \\ 3 & 8 \end{bmatrix}
$$

We have $\det(A) = 1$, $\det(B) = 8$, and $\det(A + B) = 23$; thus

$$\det(A + B) \neq \det(A) + \det(B) \quad \blacklozenge$$

In spite of the negative tone of the preceding example, there is one important relationship concerning sums of determinants that is often useful. To obtain it, consider two 2×2 matrices that differ only in the second row:

$$A = \begin{bmatrix} a_{11} & a_{12} \\ a_{21} & a_{22} \end{bmatrix} \quad \text{and} \quad B = \begin{bmatrix} a_{11} & a_{12} \\ b_{21} & b_{22} \end{bmatrix}$$

We have

$$\begin{aligned} \det(A) + \det(B) &= (a_{11}a_{22} - a_{12}a_{21}) + (a_{11}b_{22} - a_{12}b_{21}) \\ &= a_{11}(a_{22} + b_{22}) - a_{12}(a_{21} + b_{21}) \\ &= \det \begin{bmatrix} a_{11} & a_{12} \\ a_{21} + b_{21} & a_{22} + b_{22} \end{bmatrix} \end{aligned}$$

Thus

$$\det \begin{bmatrix} a_{11} & a_{12} \\ a_{21} & a_{22} \end{bmatrix} + \det \begin{bmatrix} a_{11} & a_{12} \\ b_{21} & b_{22} \end{bmatrix} = \det \begin{bmatrix} a_{11} & a_{12} \\ a_{21} + b_{21} & a_{22} + b_{22} \end{bmatrix}$$

This is a special case of the following general result.

THEOREM 2.3.1

Let A, B, and C be $n \times n$ matrices that differ only in a single row, say the rth, and assume that the rth row of C can be obtained by adding corresponding entries in the rth rows of A and B. Then

$$\det(C) = \det(A) + \det(B)$$

The same result holds for columns.

EXAMPLE 2 Using Theorem 2.3.1

By evaluating the determinants, the reader can check that

$$\det \begin{bmatrix} 1 & 7 & 5 \\ 2 & 0 & 3 \\ 1+0 & 4+1 & 7+(-1) \end{bmatrix} = \det \begin{bmatrix} 1 & 7 & 5 \\ 2 & 0 & 3 \\ 1 & 4 & 7 \end{bmatrix} + \det \begin{bmatrix} 1 & 7 & 5 \\ 2 & 0 & 3 \\ 0 & 1 & -1 \end{bmatrix} \quad \blacklozenge$$

Determinant of a Matrix Product

When one considers the complexity of the definitions of matrix multiplication and determinants, it would seem unlikely that any simple relationship should exist between them. This is what makes the elegant simplicity of the following result so surprising: We will show that if A and B are square matrices of the same size, then

$$\det(AB) = \det(A)\det(B) \tag{2}$$

The proof of this theorem is fairly intricate, so we will have to develop some preliminary results first. We begin with the special case of (2) in which A is an elementary matrix. Because this special case is only a prelude to (2), we call it a lemma.

LEMMA 2.3.2

If B is an $n \times n$ matrix and E is an $n \times n$ elementary matrix, then

$$\det(EB) = \det(E)\det(B)$$

Proof We shall consider three cases, each depending on the row operation that produces matrix E.

Case 1. If E results from multiplying a row of I_n by k, then by Theorem 1.5.1, EB results from B by multiplying a row by k; so from Theorem 2.2.3a we have

$$\det(EB) = k\det(B)$$

But from Theorem 2.2.4a we have $\det(E) = k$, so

$$\det(EB) = \det(E)\det(B)$$

Cases 2 and 3. The proofs of the cases where E results from interchanging two rows of I_n or from adding a multiple of one row to another follow the same pattern as Case 1 and are left as exercises. ∎

REMARK It follows by repeated applications of Lemma 2.3.2 that if B is an $n \times n$ matrix and $E_1, E_2, \ldots, E_r$ are $n \times n$ elementary matrices, then

$$\det(E_1 E_2 \cdots E_r B) = \det(E_1)\det(E_2)\cdots\det(E_r)\det(B) \tag{3}$$

For example,

$$\det(E_1 E_2 B) = \det(E_1)\det(E_2 B) = \det(E_1)\det(E_2)\det(B)$$

Determinant Test for Invertibility

The next theorem provides an important criterion for invertibility in terms of determinants, and it will be used in proving (2).

THEOREM 2.3.3

> *A square matrix A is invertible if and only if* $\det(A) \neq 0$.

Proof Let R be the reduced row-echelon form of A. As a preliminary step, we will show that $\det(A)$ and $\det(R)$ are both zero or both nonzero: Let $E_1, E_2, \ldots, E_r$ be the elementary matrices that correspond to the elementary row operations that produce R from A. Thus

$$R = E_r \cdots E_2 E_1 A$$

and from (3),

$$\det(R) = \det(E_r)\cdots\det(E_2)\det(E_1)\det(A) \tag{4}$$

But from Theorem 2.2.4 the determinants of the elementary matrices are all nonzero. (Keep in mind that multiplying a row by zero is *not* an allowable elementary row operation, so $k \neq 0$ in this application of Theorem 2.2.4.) Thus, it follows from (4) that $\det(A)$ and $\det(R)$ are both zero or both nonzero. Now to the main body of the proof.

If A is invertible, then by Theorem 1.6.4 we have $R = I$, so $\det(R) = 1 \neq 0$ and consequently $\det(A) \neq 0$. Conversely, if $\det(A) \neq 0$, then $\det(R) \neq 0$, so R cannot have a row of zeros. It follows from Theorem 1.4.3 that $R = I$, so A is invertible by Theorem 1.6.4. ∎

It follows from Theorems 2.3.3 and 2.2.5 that a square matrix with two proportional rows or columns is not invertible.

EXAMPLE 3   Determinant Test for Invertibility

Since the first and third rows of

$$A = \begin{bmatrix} 1 & 2 & 3 \\ 1 & 0 & 1 \\ 2 & 4 & 6 \end{bmatrix}$$

are proportional, $\det(A) = 0$. Thus A is not invertible. ◆

We are now ready for the result concerning products of matrices.

THEOREM 2.3.4

If A and B are square matrices of the same size, then

$$\det(AB) = \det(A)\det(B)$$

Proof   We divide the proof into two cases that depend on whether or not A is invertible. If the matrix A is not invertible, then by Theorem 1.6.5 neither is the product AB. Thus, from Theorem 2.3.3, we have $\det(AB) = 0$ and $\det(A) = 0$, so it follows that $\det(AB) = \det(A)\det(B)$.

Now assume that A is invertible. By Theorem 1.6.4, the matrix A is expressible as a product of elementary matrices, say

$$A = E_1 E_2 \cdots E_r \tag{5}$$

so

$$AB = E_1 E_2 \cdots E_r B$$

Applying (3) to this equation yields

$$\det(AB) = \det(E_1)\det(E_2)\cdots\det(E_r)\det(B)$$

and applying (3) again yields

$$\det(AB) = \det(E_1 E_2 \cdots E_r)\det(B)$$

which, from (5), can be written as $\det(AB) = \det(A)\det(B)$. ∎

EXAMPLE 4   Verifying That det(*AB*) = det(*A*) det(*B*)

Consider the matrices

$$A = \begin{bmatrix} 3 & 1 \\ 2 & 1 \end{bmatrix}, \qquad B = \begin{bmatrix} -1 & 3 \\ 5 & 8 \end{bmatrix}, \qquad AB = \begin{bmatrix} 2 & 17 \\ 3 & 14 \end{bmatrix}$$

We leave it for the reader to verify that

$$\det(A) = 1, \quad \det(B) = -23, \quad \text{and} \quad \det(AB) = -23$$

Thus $\det(AB) = \det(A)\det(B)$, as guaranteed by Theorem 2.3.4. ◆

The following theorem gives a useful relationship between the determinant of an invertible matrix and the determinant of its inverse.

THEOREM 2.3.5

> *If A is invertible, then*
>
> $$\det(A^{-1}) = \frac{1}{\det(A)}$$

Proof   Since $A^{-1}A = I$, it follows that $\det(A^{-1}A) = \det(I)$. Therefore, we must have $\det(A^{-1})\det(A) = 1$. Since $\det(A) \neq 0$, the proof can be completed by dividing through by $\det(A)$.   ∎

Linear Systems of the Form $A\mathbf{x} = \lambda\mathbf{x}$

Many applications of linear algebra are concerned with systems of n linear equations in n unknowns that are expressed in the form

$$A\mathbf{x} = \lambda\mathbf{x} \tag{6}$$

where λ is a scalar. Such systems are really homogeneous linear systems in disguise, since (6) can be rewritten as $\lambda\mathbf{x} - A\mathbf{x} = \mathbf{0}$ or, by inserting an identity matrix and factoring, as

$$(\lambda I - A)\mathbf{x} = \mathbf{0} \tag{7}$$

Here is an example:

EXAMPLE 5   Finding $\lambda I - A$

The linear system

$$x_1 + 3x_2 = \lambda x_1$$
$$4x_1 + 2x_2 = \lambda x_2$$

can be written in matrix form as

$$\begin{bmatrix} 1 & 3 \\ 4 & 2 \end{bmatrix}\begin{bmatrix} x_1 \\ x_2 \end{bmatrix} = \lambda\begin{bmatrix} x_1 \\ x_2 \end{bmatrix}$$

which is of form (6) with

$$A = \begin{bmatrix} 1 & 3 \\ 4 & 2 \end{bmatrix} \quad \text{and} \quad \mathbf{x} = \begin{bmatrix} x_1 \\ x_2 \end{bmatrix}$$

This system can be rewritten as

$$\lambda\begin{bmatrix} x_1 \\ x_2 \end{bmatrix} - \begin{bmatrix} 1 & 3 \\ 4 & 2 \end{bmatrix}\begin{bmatrix} x_1 \\ x_2 \end{bmatrix} = \begin{bmatrix} 0 \\ 0 \end{bmatrix}$$

or

$$\lambda\begin{bmatrix} 1 & 0 \\ 0 & 1 \end{bmatrix}\begin{bmatrix} x_1 \\ x_2 \end{bmatrix} - \begin{bmatrix} 1 & 3 \\ 4 & 2 \end{bmatrix}\begin{bmatrix} x_1 \\ x_2 \end{bmatrix} = \begin{bmatrix} 0 \\ 0 \end{bmatrix}$$

or

$$\begin{bmatrix} \lambda - 1 & -3 \\ -4 & \lambda - 2 \end{bmatrix}\begin{bmatrix} x_1 \\ x_2 \end{bmatrix} = \begin{bmatrix} 0 \\ 0 \end{bmatrix}$$

which is of form (7) with

$$\lambda I - A = \begin{bmatrix} \lambda - 1 & -3 \\ -4 & \lambda - 2 \end{bmatrix} \quad ◆$$

The primary problem of interest for linear systems of the form (7) is to determine those values of λ for which the system has a nontrivial solution; such a value of λ is called a **characteristic value** or an **eigenvalue**[†] of A. If λ is an eigenvalue of A, then the nontrivial solutions of (7) are called the **eigenvectors** of A corresponding to λ.

It follows from Theorem 2.3.3 that the system $(\lambda I - A)\mathbf{x} = \mathbf{0}$ has a nontrivial solution if and only if

$$\det(\lambda I - A) = 0 \tag{8}$$

This is called the **characteristic equation** of A; the eigenvalues of A can be found by solving this equation for λ.

Eigenvalues and eigenvectors will be studied again in subsequent chapters, where we will discuss their geometric interpretation and develop their properties in more depth.

EXAMPLE 6 Eigenvalues and Eigenvectors

Find the eigenvalues and corresponding eigenvectors of the matrix A in Example 5.

Solution

The characteristic equation of A is

$$\det(\lambda I - A) = \begin{vmatrix} \lambda - 1 & -3 \\ -4 & \lambda - 2 \end{vmatrix} = 0 \quad \text{or} \quad \lambda^2 - 3\lambda - 10 = 0$$

The factored form of this equation is $(\lambda + 2)(\lambda - 5) = 0$, so the eigenvalues of A are $\lambda = -2$ and $\lambda = 5$.

By definition,

$$\mathbf{x} = \begin{bmatrix} x_1 \\ x_2 \end{bmatrix}$$

is an eigenvector of A if and only if $\mathbf{x}$ is a nontrivial solution of $(\lambda I - A)\mathbf{x} = \mathbf{0}$; that is,

$$\begin{bmatrix} \lambda - 1 & -3 \\ -4 & \lambda - 2 \end{bmatrix} \begin{bmatrix} x_1 \\ x_2 \end{bmatrix} = \begin{bmatrix} 0 \\ 0 \end{bmatrix} \tag{9}$$

If $\lambda = -2$, then (9) becomes

$$\begin{bmatrix} -3 & -3 \\ -4 & -4 \end{bmatrix} \begin{bmatrix} x_1 \\ x_2 \end{bmatrix} = \begin{bmatrix} 0 \\ 0 \end{bmatrix}$$

Solving this system yields (verify) $x_1 = -t$, $x_2 = t$, so the eigenvectors corresponding to $\lambda = -2$ are the nonzero solutions of the form

$$\mathbf{x} = \begin{bmatrix} x_1 \\ x_2 \end{bmatrix} = \begin{bmatrix} -t \\ t \end{bmatrix}$$

Again from (9), the eigenvectors of A corresponding to $\lambda = 5$ are the nontrivial solutions of

$$\begin{bmatrix} 4 & -3 \\ -4 & 3 \end{bmatrix} \begin{bmatrix} x_1 \\ x_2 \end{bmatrix} = \begin{bmatrix} 0 \\ 0 \end{bmatrix}$$

[†]The word *eigenvalue* is a mixture of German and English. The German prefix *eigen* can be translated as "proper," which stems from the older literature where eigenvalues were known as *proper values*; they were also called *latent roots*.

We leave it for the reader to solve this system and show that the eigenvectors of A corresponding to $\lambda = 5$ are the nonzero solutions of the form

$$\mathbf{x} = \begin{bmatrix} \tfrac{3}{4}t \\ t \end{bmatrix} \blacklozenge$$

Summary

In Theorem 1.6.4 we listed five results that are equivalent to the invertibility of a matrix A. We conclude this section by merging Theorem 2.3.3 with that list to produce the following theorem that relates all of the major topics we have studied thus far.

THEOREM 2.3.6

Equivalent Statements

If A is an $n \times n$ matrix, then the following statements are equivalent.

(*a*) *A is invertible.*
(*b*) *$A\mathbf{x} = \mathbf{0}$ has only the trivial solution.*
(*c*) *The reduced row-echelon form of A is I_n.*
(*d*) *A can be expressed as a product of elementary matrices.*
(*e*) *$A\mathbf{x} = \mathbf{b}$ is consistent for every $n \times 1$ matrix $\mathbf{b}$.*
(*f*) *$A\mathbf{x} = \mathbf{b}$ has exactly one solution for every $n \times 1$ matrix $\mathbf{b}$.*
(*g*) *$\det(A) \neq 0$.*

EXERCISE SET 2.3

1. Verify that $\det(kA) = k^n \det(A)$ for

   (a) $A = \begin{bmatrix} -1 & 2 \\ 3 & 4 \end{bmatrix}$; $k = 2$ (b) $A = \begin{bmatrix} 2 & -1 & 3 \\ 3 & 2 & 1 \\ 1 & 4 & 5 \end{bmatrix}$; $k = -2$

2. Verify that $\det(AB) = \det(A)\det(B)$ for

   $$A = \begin{bmatrix} 2 & 1 & 0 \\ 3 & 4 & 0 \\ 0 & 0 & 2 \end{bmatrix} \quad \text{and} \quad B = \begin{bmatrix} 1 & -1 & 3 \\ 7 & 1 & 2 \\ 5 & 0 & 1 \end{bmatrix}$$

   Is $\det(A + B) = \det(A) + \det(B)$?

3. By inspection, explain why $\det(A) = 0$.

   $$A = \begin{bmatrix} -2 & 8 & 1 & 4 \\ 3 & 2 & 5 & 1 \\ 1 & 10 & 6 & 5 \\ 4 & -6 & 4 & -3 \end{bmatrix}$$

4. Use Theorem 2.3.3 to determine which of the following matrices are invertible.

   (a) $\begin{bmatrix} 1 & 0 & -1 \\ 9 & -1 & 4 \\ 8 & 9 & -1 \end{bmatrix}$ (b) $\begin{bmatrix} 4 & 2 & 8 \\ -2 & 1 & -4 \\ 3 & 1 & 6 \end{bmatrix}$

   (c) $\begin{bmatrix} \sqrt{2} & -\sqrt{7} & 0 \\ 3\sqrt{2} & -3\sqrt{7} & 0 \\ 5 & -9 & 0 \end{bmatrix}$ (d) $\begin{bmatrix} -3 & 0 & 1 \\ 5 & 0 & 6 \\ 8 & 0 & 3 \end{bmatrix}$

5. Let

$$A = \begin{bmatrix} a & b & c \\ d & e & f \\ g & h & i \end{bmatrix}$$

Assuming that $\det(A) = -7$, find

(a) $\det(3A)$ (b) $\det(A^{-1})$ (c) $\det(2A^{-1})$

(d) $\det((2A)^{-1})$ (e) $\det \begin{bmatrix} a & g & d \\ b & h & e \\ c & i & f \end{bmatrix}$

6. Without directly evaluating, show that $x = 0$ and $x = 2$ satisfy

$$\begin{vmatrix} x^2 & x & 2 \\ 2 & 1 & 1 \\ 0 & 0 & -5 \end{vmatrix} = 0$$

7. Without directly evaluating, show that

$$\det \begin{bmatrix} b+c & c+a & b+a \\ a & b & c \\ 1 & 1 & 1 \end{bmatrix} = 0$$

In Exercises 8–11 prove the identity without evaluating the determinants.

8. $\begin{vmatrix} a_1 & b_1 & a_1 + b_1 + c_1 \\ a_2 & b_2 & a_2 + b_2 + c_2 \\ a_3 & b_3 & a_3 + b_3 + c_3 \end{vmatrix} = \begin{vmatrix} a_1 & b_1 & c_1 \\ a_2 & b_2 & c_2 \\ a_3 & b_3 & c_3 \end{vmatrix}$

9. $\begin{vmatrix} a_1 + b_1 & a_1 - b_1 & c_1 \\ a_2 + b_2 & a_2 - b_2 & c_2 \\ a_3 + b_3 & a_3 - b_3 & c_3 \end{vmatrix} = -2 \begin{vmatrix} a_1 & b_1 & c_1 \\ a_2 & b_2 & c_2 \\ a_3 & b_3 & c_3 \end{vmatrix}$

10. $\begin{vmatrix} a_1 + b_1 t & a_2 + b_2 t & a_3 + b_3 t \\ a_1 t + b_1 & a_2 t + b_2 & a_3 t + b_3 \\ c_1 & c_2 & c_3 \end{vmatrix} = (1 - t^2) \begin{vmatrix} a_1 & a_2 & a_3 \\ b_1 & b_2 & b_3 \\ c_1 & c_2 & c_3 \end{vmatrix}$

11. $\begin{vmatrix} a_1 & b_1 + t a_1 & c_1 + r b_1 + s a_1 \\ a_2 & b_2 + t a_2 & c_2 + r b_2 + s a_2 \\ a_3 & b_3 + t a_3 & c_3 + r b_3 + s a_3 \end{vmatrix} = \begin{vmatrix} a_1 & a_2 & a_3 \\ b_1 & b_2 & b_3 \\ c_1 & c_2 & c_3 \end{vmatrix}$

12. For which value(s) of k does A fail to be invertible?

(a) $A = \begin{bmatrix} k - 3 & -2 \\ -2 & k - 2 \end{bmatrix}$ (b) $A = \begin{bmatrix} 1 & 2 & 4 \\ 3 & 1 & 6 \\ k & 3 & 2 \end{bmatrix}$

13. Use Theorem 2.3.3 to show that

$$\begin{bmatrix} \sin^2 \alpha & \sin^2 \beta & \sin^2 \gamma \\ \cos^2 \alpha & \cos^2 \beta & \cos^2 \gamma \\ 1 & 1 & 1 \end{bmatrix}$$

is not invertible for any values of α, β, and γ.

14. Express the following linear systems in the form $(\lambda I - A)\mathbf{x} = \mathbf{0}$.

(a) $\begin{aligned} x_1 + 2x_2 &= \lambda x_1 \\ 2x_1 + x_2 &= \lambda x_2 \end{aligned}$ (b) $\begin{aligned} 2x_1 + 3x_2 &= \lambda x_1 \\ 4x_1 + 3x_2 &= \lambda x_2 \end{aligned}$ (c) $\begin{aligned} 3x_1 + x_2 &= \lambda x_1 \\ -5x_1 - 3x_2 &= \lambda x_2 \end{aligned}$

15. For each of the systems in Exercise 14, find
    (i) the characteristic equation;
    (ii) the eigenvalues;
    (iii) the eigenvectors corresponding to each of the eigenvalues.

16. Let A and B be $n \times n$ matrices. Show that if A is invertible, then $\det(B) = \det(A^{-1}BA)$.

17. (a) Express
$$\begin{vmatrix} a_1 + b_1 & c_1 + d_1 \\ a_2 + b_2 & c_2 + d_2 \end{vmatrix}$$

as a sum of four determinants whose entries contain no sums.

   (b) Express
$$\begin{vmatrix} a_1 + b_1 & c_1 + d_1 & e_1 + f_1 \\ a_2 + b_2 & c_2 + d_2 & e_2 + f_2 \\ a_3 + b_3 & c_3 + d_3 & e_3 + f_3 \end{vmatrix}$$

as a sum of eight determinants whose entries contain no sums.

18. Prove that a square matrix A is invertible if and only if $A^T A$ is invertible.

19. Prove Cases 2 and 3 of Lemma 2.3.2.

Discussion & Discovery

20. Let A and B be $n \times n$ matrices. You know from earlier work that AB and BA need not be equal. Is the same true for $\det(AB)$ and $\det(BA)$? Explain your reasoning.

21. Let A and B be $n \times n$ matrices. You know from earlier work that AB is invertible if A and B are invertible. What can you say about the invertibility of AB if one or both of the factors are singular? Explain your reasoning.

22. Indicate whether the statement is always true or sometimes false. Justify each answer by giving a logical argument or a counterexample.
    (a) $\det(2A) = 2\det(A)$
    (b) $|A^2| = |A|^2$
    (c) $\det(I + A) = 1 + \det(A)$
    (d) If $\det(A) = 0$, then the homogeneous system $A\mathbf{x} = \mathbf{0}$ has infinitely many solutions.

23. Indicate whether the statement is always true or sometimes false. Justify your answer by giving a logical argument or a counterexample.
    (a) If $\det(A) = 0$, then A is not expressible as a product of elementary matrices.
    (b) If the reduced row-echelon form of A has a row of zeros, then $\det(A) = 0$.
    (c) The determinant of a matrix is unchanged if the columns are written in reverse order.
    (d) There is no square matrix A such that $\det(AA^T) = -1$.

2.4
A COMBINATORIAL APPROACH TO DETERMINANTS

There is a combinatorial view of determinants that actually predates matrices. In this section we explore this connection.

There is another way to approach determinants that complements the cofactor expansion approach. It is based on permutations.

> **DEFINITION**
>
> A *permutation* of the set of integers $\{1, 2, \ldots, n\}$ is an arrangement of these integers in some order without omissions or repetitions.

EXAMPLE 1 Permutations of Three Integers

There are six different permutations of the set of integers $\{1, 2, 3\}$. These are

$$(1, 2, 3) \qquad (2, 1, 3) \qquad (3, 1, 2)$$
$$(1, 3, 2) \qquad (2, 3, 1) \qquad (3, 2, 1) \; \blacklozenge$$

One convenient method of systematically listing permutations is to use a *permutation tree*. This method is illustrated in our next example.

EXAMPLE 2 Permutations of Four Integers

List all permutations of the set of integers $\{1, 2, 3, 4\}$.

Solution

Consider Figure 2.4.1. The four dots labeled 1, 2, 3, 4 at the top of the figure represent the possible choices for the first number in the permutation. The three branches emanating from these dots represent the possible choices for the second position in the permutation. Thus, if the permutation begins $(2, -, -, -)$, the three possibilities for the second position are 1, 3, and 4. The two branches emanating from each dot in the second position represent the possible choices for the third position. Thus, if the permutation begins $(2, 3, -, -)$, the two possible choices for the third position are 1 and 4. Finally, the single branch emanating from each dot in the third position represents the only possible choice for the fourth position. Thus, if the permutation begins with $(2, 3, 4, -)$, the only choice for the fourth position is 1. The different permutations can now be listed by tracing out all the possible paths through the "tree" from the first position to the last position. We obtain the following list by this process.

$$
\begin{array}{llll}
(1, 2, 3, 4) & (2, 1, 3, 4) & (3, 1, 2, 4) & (4, 1, 2, 3) \\
(1, 2, 4, 3) & (2, 1, 4, 3) & (3, 1, 4, 2) & (4, 1, 3, 2) \\
(1, 3, 2, 4) & (2, 3, 1, 4) & (3, 2, 1, 4) & (4, 2, 1, 3) \\
(1, 3, 4, 2) & (2, 3, 4, 1) & (3, 2, 4, 1) & (4, 2, 3, 1) \\
(1, 4, 2, 3) & (2, 4, 1, 3) & (3, 4, 1, 2) & (4, 3, 1, 2) \\
(1, 4, 3, 2) & (2, 4, 3, 1) & (3, 4, 2, 1) & (4, 3, 2, 1) \; \blacklozenge
\end{array}
$$

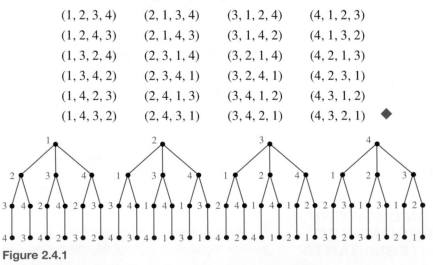

Figure 2.4.1

From this example we see that there are 24 permutations of $\{1, 2, 3, 4\}$. This result could have been anticipated without actually listing the permutations by arguing as follows. Since the first position can be filled in four ways and then the second position in three ways, there are $4 \cdot 3$ ways of filling the first two positions. Since the third position can then be filled in two ways, there are $4 \cdot 3 \cdot 2$ ways of filling the first three positions. Finally, since the last position can then be filled in only one way, there are $4 \cdot 3 \cdot 2 \cdot 1 = 24$ ways of filling all four positions. In general, the set $\{1, 2, \ldots, n\}$ will have $n(n-1)(n-2) \cdots 2 \cdot 1 = n!$ different permutations.

We will denote a general permutation of the set $\{1, 2, \ldots, n\}$ by $(j_1, j_2, \ldots, j_n)$. Here, j_1 is the first integer in the permutation, j_2 is the second, and so on. An ***inversion*** is said to occur in a permutation $(j_1, j_2, \ldots, j_n)$ whenever a larger integer precedes a smaller one. The total number of inversions occurring in a permutation can be obtained as follows: (1) find the number of integers that are less than j_1 and that follow j_1 in the permutation; (2) find the number of integers that are less than j_2 and that follow j_2 in the permutation. Continue this counting process for $j_3, \ldots, j_{n-1}$. The sum of these numbers will be the total number of inversions in the permutation.

EXAMPLE 3 Counting Inversions

Determine the number of inversions in the following permutations:

$$\text{(a)} \ (6, 1, 3, 4, 5, 2) \qquad \text{(b)} \ (2, 4, 1, 3) \qquad \text{(c)} \ (1, 2, 3, 4)$$

Solution

(a) The number of inversions is $5 + 0 + 1 + 1 + 1 = 8$.

(b) The number of inversions is $1 + 2 + 0 = 3$.

(c) There are zero inversions in this permutation. ◆

> **DEFINITION**
>
> A permutation is called ***even*** if the total number of inversions is an even integer and is called ***odd*** if the total number of inversions is an odd integer.

EXAMPLE 4 Classifying Permutations

The following table classifies the various permutations of $\{1, 2, 3\}$ as even or odd.

Permutation	Number of Inversions	Classification
$(1, 2, 3)$	0	even
$(1, 3, 2)$	1	odd
$(2, 1, 3)$	1	odd
$(2, 3, 1)$	2	even
$(3, 1, 2)$	2	even
$(3, 2, 1)$	3	odd

◆

Combinatorial Definition of the Determinant

By an *elementary product* from an $n \times n$ matrix A we shall mean any product of n entries from A, no two of which come from the same row or the same column.

EXAMPLE 5 Elementary Products

List all elementary products from the matrices

$$\text{(a)} \begin{bmatrix} a_{11} & a_{12} \\ a_{21} & a_{22} \end{bmatrix} \qquad \text{(b)} \begin{bmatrix} a_{11} & a_{12} & a_{13} \\ a_{21} & a_{22} & a_{23} \\ a_{31} & a_{32} & a_{33} \end{bmatrix}$$

Solution (a)

Since each elementary product has two factors, and since each factor comes from a different row, an elementary product can be written in the form

$$a_{1\_}a_{2\_}$$

where the blanks designate column numbers. Since no two factors in the product come from the same column, the column numbers must be $\underline{1\ 2}$ or $\underline{2\ 1}$. Thus the only elementary products are $a_{11}a_{22}$ and $a_{12}a_{21}$.

Solution (b)

Since each elementary product has three factors, each of which comes from a different row, an elementary product can be written in the form

$$a_{1\_}a_{2\_}a_{3\_}$$

Since no two factors in the product come from the same column, the column numbers have no repetitions; consequently, they must form a permutation of the set $\{1, 2, 3\}$. These $3! = 6$ permutations yield the following list of elementary products.

$$a_{11}a_{22}a_{33} \qquad a_{12}a_{21}a_{33} \qquad a_{13}a_{21}a_{32}$$
$$a_{11}a_{23}a_{32} \qquad a_{12}a_{23}a_{31} \qquad a_{13}a_{22}a_{31} \ \blacklozenge$$

As this example points out, an $n \times n$ matrix A has $n!$ elementary products. They are the products of the form $a_{1j_1}a_{2j_2}\cdots a_{nj_n}$, where $(j_1, j_2, \ldots, j_n)$ is a permutation of the set $\{1, 2, \ldots, n\}$. By a *signed elementary product from A* we shall mean an elementary product $a_{1j_1}a_{2j_2}\cdots a_{nj_n}$ multiplied by $+1$ or -1. We use the $+$ if $(j_1, j_2, \ldots, j_n)$ is an even permutation and the $-$ if $(j_1, j_2, \ldots, j_n)$ is an odd permutation.

EXAMPLE 6 Signed Elementary Products

List all signed elementary products from the matrices

$$\text{(a)} \begin{bmatrix} a_{11} & a_{12} \\ a_{21} & a_{22} \end{bmatrix} \qquad \text{(b)} \begin{bmatrix} a_{11} & a_{12} & a_{13} \\ a_{21} & a_{22} & a_{23} \\ a_{31} & a_{32} & a_{33} \end{bmatrix}$$

Solution

(a)

Elementary Product	Associated Permutation	Even or Odd	Signed Elementary Product
$a_{11}a_{22}$	$(1, 2)$	even	$a_{11}a_{22}$
$a_{12}a_{21}$	$(2, 1)$	odd	$-a_{12}a_{21}$

(b)

Elementary Product	Associated Permutation	Even or Odd	Signed Elementary Product
$a_{11}a_{22}a_{33}$	$(1, 2, 3)$	even	$a_{11}a_{22}a_{33}$
$a_{11}a_{23}a_{32}$	$(1, 3, 2)$	odd	$-a_{11}a_{23}a_{32}$
$a_{12}a_{21}a_{33}$	$(2, 1, 3)$	odd	$-a_{12}a_{21}a_{33}$
$a_{12}a_{23}a_{31}$	$(2, 3, 1)$	even	$a_{12}a_{23}a_{31}$
$a_{13}a_{21}a_{32}$	$(3, 1, 2)$	even	$a_{13}a_{21}a_{32}$
$a_{13}a_{22}a_{31}$	$(3, 2, 1)$	odd	$-a_{13}a_{22}a_{31}$

◆

We are now in a position to give the combinatorial definition of the determinant function.

DEFINITION

Let A be a square matrix. We define $\det(A)$ to be the sum of all signed elementary products from A.

EXAMPLE 7 Determinants of 2 × 2 and 3 × 3 Matrices

Referring to Example 6, we obtain

(a) $\det \begin{bmatrix} a_{11} & a_{12} \\ a_{21} & a_{22} \end{bmatrix} = a_{11}a_{22} - a_{12}a_{21}$

(b) $\det \begin{bmatrix} a_{11} & a_{12} & a_{13} \\ a_{21} & a_{22} & a_{23} \\ a_{31} & a_{32} & a_{33} \end{bmatrix} = a_{11}a_{22}a_{33} + a_{12}a_{23}a_{31} + a_{13}a_{21}a_{32}$
$$- a_{13}a_{22}a_{31} - a_{12}a_{21}a_{33} - a_{11}a_{23}a_{32} \quad ◆$$

Of course, this definition of $\det(A)$ agrees with the definition in Section 2.1, although we will not prove this.

These expressions suggest the mnemonic devices given in Figure 2.4.2. The formula in part (a) of Example 7 is obtained from Figure 2.4.2*a* by multiplying the entries on the rightward arrow and subtracting the product of the entries on the leftward arrow. The formula in part (b) of Example 7 is obtained by recopying the first and second columns

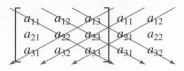

(a) Determinant of a 2 × 2 matrix   (b) Determinant of a 3 × 3 matrix

Figure 2.4.2

as shown in Figure 2.4.2b. The determinant is then computed by summing the products on the rightward arrows and subtracting the products on the leftward arrows.

WARNING   We emphasize that the methods shown in Figure 2.4.2 do not work for determinants of 4 × 4 matrices or higher.

EXAMPLE 8   Evaluating Determinants

Evaluate the determinants of

$$A = \begin{bmatrix} 3 & 1 \\ 4 & -2 \end{bmatrix} \quad \text{and} \quad B = \begin{bmatrix} 1 & 2 & 3 \\ -4 & 5 & 6 \\ 7 & -8 & 9 \end{bmatrix}$$

Solution

Using the method of Figure 2.4.2a gives

$$\det(A) = (3)(-2) - (1)(4) = -10$$

Using the method of Figure 2.4.2b gives

$$\det(B) = (45) + (84) + (96) - (105) - (-48) - (-72) = 240$$

$$\begin{bmatrix} 1 & 2 & 3 \\ -4 & 5 & 6 \\ 7 & -8 & 9 \end{bmatrix} \begin{matrix} 1 & 2 \\ -4 & 5 \\ 7 & -8 \end{matrix} \quad \blacklozenge$$

The determinant of A may be written as

$$\det(A) = \sum \pm a_{1j_1} a_{2j_2} \cdots a_{nj_n} \tag{1}$$

where $\sum$ indicates that the terms are to be summed over all permutations $(j_1, j_2, \ldots, j_n)$ and the $+$ or $-$ is selected in each term according to whether the permutation is even or odd. This notation is useful when the combinatorial definition of a determinant needs to be emphasized.

REMARK   Evaluating determinants directly from this definition leads to computational difficulties. Indeed, evaluating a 4 × 4 determinant directly would involve computing $4! = 24$ signed elementary products, and a 10 × 10 determinant would require the computation of $10! = 3,628,800$ signed elementary products. Even the fastest of digital computers cannot handle the computation of a 25×25 determinant by this method in a practical amount of time.

EXERCISE SET 2.4

1. Find the number of inversions in each of the following permutations of $\{1, 2, 3, 4, 5\}$.
   - (a) $(4\,1\,3\,5\,2)$
   - (b) $(5\,3\,4\,2\,1)$
   - (c) $(3\,2\,5\,4\,1)$
   - (d) $(5\,4\,3\,2\,1)$
   - (e) $(1\,2\,3\,4\,5)$
   - (f) $(1\,4\,2\,3\,5)$

2. Classify each of the permutations in Exercise 1 as even or odd.

In Exercises 3–12 evaluate the determinant using the method of this section.

3. $\begin{vmatrix} 3 & 5 \\ -2 & 4 \end{vmatrix}$
4. $\begin{vmatrix} 4 & 1 \\ 8 & 2 \end{vmatrix}$
5. $\begin{vmatrix} -5 & 6 \\ -7 & -2 \end{vmatrix}$

6. $\begin{vmatrix} \sqrt{2} & \sqrt{6} \\ 4 & \sqrt{3} \end{vmatrix}$
7. $\begin{vmatrix} a-3 & 5 \\ -3 & a-2 \end{vmatrix}$
8. $\begin{vmatrix} -2 & 7 & 6 \\ 5 & 1 & -2 \\ 3 & 8 & 4 \end{vmatrix}$

9. $\begin{vmatrix} -2 & 1 & 4 \\ 3 & 5 & -7 \\ 1 & 6 & 2 \end{vmatrix}$
10. $\begin{vmatrix} -1 & 1 & 2 \\ 3 & 0 & -5 \\ 1 & 7 & 2 \end{vmatrix}$

11. $\begin{vmatrix} 3 & 0 & 0 \\ 2 & -1 & 5 \\ 1 & 9 & -4 \end{vmatrix}$
12. $\begin{vmatrix} c & -4 & 3 \\ 2 & 1 & c^2 \\ 4 & c-1 & 2 \end{vmatrix}$

13. Find all values of λ for which $\det(A) = 0$, using the method of this section.
   - (a) $\begin{bmatrix} \lambda-2 & 1 \\ -5 & \lambda+4 \end{bmatrix}$
   - (b) $\begin{bmatrix} \lambda-4 & 0 & 0 \\ 0 & \lambda & 2 \\ 0 & 3 & \lambda-1 \end{bmatrix}$

14. Classify each permutation of $\{1, 2, 3, 4\}$ as even or odd.

15. (a) Use the results in Exercise 14 to construct a formula for the determinant of a 4×4 matrix.
    (b) Why do the mnemonics of Figure 2.4.2 fail for a 4×4 matrix?

16. Use the formula obtained in Exercise 15 to evaluate
$$\begin{vmatrix} 4 & -9 & 9 & 2 \\ -2 & 5 & 6 & 4 \\ 1 & 2 & -5 & -3 \\ 1 & -2 & 0 & -2 \end{vmatrix}$$

17. Use the combinatorial definition of the determinant to evaluate
   - (a) $\begin{vmatrix} 0 & 0 & 0 & 0 & -3 \\ 0 & 0 & 0 & -4 & 0 \\ 0 & 0 & -1 & 0 & 0 \\ 0 & 2 & 0 & 0 & 0 \\ 5 & 0 & 0 & 0 & 0 \end{vmatrix}$
   - (b) $\begin{vmatrix} 5 & 0 & 0 & 0 & 0 \\ 0 & 0 & 0 & 0 & -4 \\ 0 & 0 & 3 & 0 & 0 \\ 0 & 0 & 0 & 1 & 0 \\ 0 & -2 & 0 & 0 & 0 \end{vmatrix}$

18. Solve for x.
$$\begin{vmatrix} x & -1 \\ 3 & 1-x \end{vmatrix} = \begin{vmatrix} 1 & 0 & -3 \\ 2 & x & -6 \\ 1 & 3 & x-5 \end{vmatrix}$$

19. Show that the value of the determinant

$$\begin{vmatrix} \sin\theta & \cos\theta & 0 \\ -\cos\theta & \sin\theta & 0 \\ \sin\theta - \cos\theta & \sin\theta + \cos\theta & 1 \end{vmatrix}$$

does not depend on θ, using the method of this section.

20. Prove that the matrices

$$A = \begin{bmatrix} a & b \\ 0 & c \end{bmatrix} \quad \text{and} \quad B = \begin{bmatrix} d & e \\ 0 & f \end{bmatrix}$$

commute if and only if

$$\begin{vmatrix} b & a-c \\ e & d-f \end{vmatrix} = 0$$

Discussion & Discovery

21. Explain why the determinant of an $n \times n$ matrix with integer entries must be an integer, using the method of this section.

22. What can you say about the determinant of an $n \times n$ matrix all of whose entries are 1? Explain your reasoning, using the method of this section.

23. (a) Explain why the determinant of an $n \times n$ matrix with a row of zeros must have a zero determinant, using the method of this section.

(b) Explain why the determinant of an $n \times n$ matrix with a column of zeros must have a zero determinant.

24. Use Formula (1) to discover a formula for the determinant of an $n \times n$ diagonal matrix. Express the formula in words.

25. Use Formula (1) to discover a formula for the determinant of an $n \times n$ upper triangular matrix. Express the formula in words. Do the same for a lower triangular matrix.

CHAPTER 2
Supplementary Exercises

1. Use Cramer's rule to solve for x' and y' in terms of x and y.

$$x = \tfrac{3}{5}x' - \tfrac{4}{5}y'$$
$$y = \tfrac{4}{5}x' + \tfrac{3}{5}y'$$

2. Use Cramer's rule to solve for x' and y' in terms of x and y.

$$x = x'\cos\theta - y'\sin\theta$$
$$y = x'\sin\theta + y'\cos\theta$$

3. By examining the determinant of the coefficient matrix, show that the following system has a nontrivial solution if and only if $\alpha = \beta$.

$$x + y + \alpha z = 0$$
$$x + y + \beta z = 0$$
$$\alpha x + \beta y + z = 0$$

4. Let A be a 3×3 matrix, each of whose entries is 1 or 0. What is the largest possible value for $\det(A)$?

5. (a) For the triangle in the accompanying figure, use trigonometry to show that

$$b\cos\gamma + c\cos\beta = a$$
$$c\cos\alpha + a\cos\gamma = b$$
$$a\cos\beta + b\cos\alpha = c$$

Figure Ex-5

and then apply Cramer's rule to show that

$$\cos \alpha = \frac{b^2 + c^2 - a^2}{2bc}$$

(b)   Use Cramer's rule to obtain similar formulas for $\cos \beta$ and $\cos \gamma$.

6.   Use determinants to show that for all real values of λ, the only solution of

$$x - 2y = \lambda x$$
$$x - \ \ y = \lambda y$$

is $x = 0$, $y = 0$.

7.   Prove: If A is invertible, then adj(A) is invertible and

$$[\text{adj}(A)]^{-1} = \frac{1}{\det(A)} A = \text{adj}(A^{-1})$$

8.   Prove: If A is an $n \times n$ matrix, then $\det[\text{adj}(A)] = [\det(A)]^{n-1}$.

9.   **(For Readers Who Have Studied Calculus)** Show that if $f_1(x)$, $f_2(x)$, $g_1(x)$, and $g_2(x)$ are differentiable functions, and if

$$W = \begin{vmatrix} f_1(x) & f_2(x) \\ g_1(x) & g_2(x) \end{vmatrix}, \quad \text{then} \quad \frac{dW}{dx} = \begin{vmatrix} f_1'(x) & f_2'(x) \\ g_1(x) & g_2(x) \end{vmatrix} + \begin{vmatrix} f_1(x) & f_2(x) \\ g_1'(x) & g_2'(x) \end{vmatrix}$$

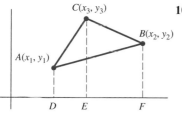

Figure Ex-10

10.   (a)   In the accompanying figure, the area of the triangle ABC can be expressed as

$$\text{area } ABC = \text{area } ADEC + \text{area } CEFB - \text{area } ADFB$$

Use this and the fact that the area of a trapezoid equals $\frac{1}{2}$ the altitude times the sum of the parallel sides to show that

$$\text{area } ABC = \frac{1}{2} \begin{vmatrix} x_1 & y_1 & 1 \\ x_2 & y_2 & 1 \\ x_3 & y_3 & 1 \end{vmatrix}$$

Note   In the derivation of this formula, the vertices are labeled such that the triangle is traced counterclockwise proceeding from (x_1, y_1) to (x_2, y_2) to (x_3, y_3). For a clockwise orientation, the determinant above yields the *negative* of the area.

(b)   Use the result in (a) to find the area of the triangle with vertices $(3, 3)$, $(4, 0)$, $(-2, -1)$.

11.   Prove: If the entries in each row of an $n \times n$ matrix A add up to zero, then the determinant of A is zero.

Hint   Consider the product AX, where X is the $n \times 1$ matrix, each of whose entries is one.

12.   Let A be an $n \times n$ matrix and B the matrix that results when the rows of A are written in reverse order (last row becomes the first, and so forth). How are $\det(A)$ and $\det(B)$ related?

13.   Indicate how A^{-1} will be affected if

(a)   the ith and jth rows of A are interchanged.

(b)   the ith row of A is multiplied by a nonzero scalar, c.

(c) c times the ith row of A is added to the jth row.

14.   Let A be an $n \times n$ matrix. Suppose that B_1 is obtained by adding the same number t to each entry in the ith row of A and that B_2 is obtained by subtracting t from each entry in the ith row of A. Show that $\det(A) = \frac{1}{2}[\det(B_1) + \det(B_2)]$.

15.   Let

$$A = \begin{bmatrix} a_{11} & a_{12} & a_{13} \\ a_{21} & a_{22} & a_{23} \\ a_{31} & a_{32} & a_{33} \end{bmatrix}$$

(a)   Express $\det(\lambda I - A)$ as a polynomial $p(\lambda) = \lambda^3 + b\lambda^2 + c\lambda + d$.

(b)   Express the coefficients b and d in terms of determinants and traces.

16. Without directly evaluating the determinant, show that

$$\begin{vmatrix} \sin\alpha & \cos\alpha & \sin(\alpha+\delta) \\ \sin\beta & \cos\beta & \sin(\beta+\delta) \\ \sin\gamma & \cos\gamma & \sin(\gamma+\delta) \end{vmatrix} = 0$$

17. Use the fact that 21,375, 38,798, 34,162, 40,223, and 79,154 are all divisible by 19 to show that

$$\begin{vmatrix} 2 & 1 & 3 & 7 & 5 \\ 3 & 8 & 7 & 9 & 8 \\ 3 & 4 & 1 & 6 & 2 \\ 4 & 0 & 2 & 2 & 3 \\ 7 & 9 & 1 & 5 & 4 \end{vmatrix}$$

is divisible by 19 without directly evaluating the determinant.

18. Find the eigenvalues and corresponding eigenvectors for each of the following systems.

(a)
$$\begin{aligned} x_2 + 9x_3 &= \lambda x_1 \\ x_1 + 4x_2 - 7x_3 &= \lambda x_2 \\ x_1 \qquad\quad - 3x_3 &= \lambda x_3 \end{aligned}$$

(b)
$$\begin{aligned} x_2 + x_3 &= \lambda x_1 \\ x_1 \qquad - x_3 &= \lambda x_2 \\ x_1 + 5x_2 + 3x_3 &= \lambda x_3 \end{aligned}$$

CHAPTER 2

Technology Exercises

The following exercises are designed to be solved using a technology utility. Typically, this will be MATLAB, *Mathematica*, Maple, Derive, or Mathcad, but it may also be some other type of linear algebra software or a scientific calculator with some linear algebra capabilities. For each exercise you will need to read the relevant documentation for the particular utility you are using. The goal of these exercises is to provide you with a basic proficiency with your technology utility. Once you have mastered the techniques in these exercises, you will be able to use your technology utility to solve many of the problems in the regular exercise sets.

Section 2.1 **T1.** **(Determinants)** Read your documentation on how to compute determinants, and then compute several determinants.

T2. **(Minors, Cofactors, and Adjoints)** Technology utilities vary widely in their treatment of minors, cofactors, and adjoints. For example, some utilities have commands for computing minors but not cofactors, and some provide direct commands for finding adjoints, whereas others do not. Thus, depending on your utility, you may have to piece together commands or do some sign adjustment by hand to find cofactors and adjoints. Read your documentation, and then find the adjoint of the matrix A in Example 6.

T3. Use Cramer's rule to find a polynomial of degree 3 that passes through the points $(0, 1)$, $(1, -1)$, $(2, -1)$, and $(3, 7)$. Verify your results by plotting the points and the curve on one graph.

Section 2.2 **T1.** **(Determinant of a Transpose)** Confirm part (*b*) of Theorem 2.2.1 using some matrices of your choice.

Section 2.3 **T1.** **(Determinant of a Product)** Confirm Theorem 2.3.4 for some matrices of your choice.

T2. **(Determinant of an Inverse)** Confirm Theorem 2.3.5 for some matrices of your choice.

T3. **(Characteristic Equation)** If you are working with a CAS, use it to find the characteristic equation of the matrix A in Example 6. Also, read your documentation on how to solve equations, and then solve the equation $\det(\lambda I - A) = 0$ for the eigenvalues of A.

Section 2.4 **T1.** **(Determinant Formulas)** If you are working with a CAS, use it to confirm the formulas in Example 7. Also, use it to obtain the formula requested in Exercise 15 of Section 2.4.

T2. **(Simplification)** If you are working with a CAS, read the documentation on simplifying algebraic expressions, and then use the determinant and simplification commands in combination to show that

$$\begin{vmatrix} a & b & c & d \\ -b & a & d & -c \\ -c & -d & a & b \\ -d & c & -b & a \end{vmatrix} = (a^2 + b^2 + c^2 + d^2)^2$$

T3. Use the method of Exercise T2 to find a simple formula for the determinant

$$\begin{vmatrix} (a+b)^2 & c^2 & c^2 \\ a^2 & (b+c)^2 & a^2 \\ b^2 & b^2 & (c+a)^2 \end{vmatrix}$$

Vectors in 2-Space and 3-Space

CHAPTER CONTENTS

INTRODUCTION: Many physical quantities, such as area, length, mass, and temperature, are completely described once the magnitude of the quantity is given. Such quantities are called **scalars**. Other physical quantities are not completely determined until both a magnitude and a direction are specified. These quantities are called **vectors**. For example, wind movement is usually described by giving the speed and direction, say 20 mph northeast. The wind speed and wind direction form a vector called the wind **velocity**. Other examples of vectors are **force** and **displacement**. In this chapter our goal is to review some of the basic theory of vectors in two and three dimensions.

Note. *Readers already familiar with the contents of this chapter can go to Chapter 4 with no loss of continuity.*

3.1

INTRODUCTION TO VECTORS (GEOMETRIC)

In this section, vectors in 2-space and 3-space will be introduced geometrically, arithmetic operations on vectors will be defined, and some basic properties of these arithmetic operations will be established.

Geometric Vectors

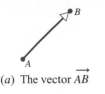

(*a*) The vector $\overrightarrow{AB}$

(*b*) Equivalent vectors

Figure 3.1.1

Vectors can be represented geometrically as directed line segments or arrows in 2-space or 3-space. The direction of the arrow specifies the direction of the vector, and the length of the arrow describes its magnitude. The tail of the arrow is called the ***initial point*** of the vector, and the tip of the arrow the ***terminal point***. Symbolically, we shall denote vectors in lowercase boldface type (for instance, **a**, **k**, **v**, **w**, and **x**). When discussing vectors, we shall refer to numbers as ***scalars***. For now, all our scalars will be real numbers and will be denoted in lowercase italic type (for instance, a, k, v, w, and x).

If, as in Figure 3.1.1*a*, the initial point of a vector **v** is A and the terminal point is B, we write

$$\mathbf{v} = \overrightarrow{AB}$$

Vectors with the same length and same direction, such as those in Figure 3.1.1*b*, are called ***equivalent***. Since we want a vector to be determined solely by its length and direction, equivalent vectors are regarded as ***equal*** even though they may be located in different positions. If **v** and **w** are equivalent, we write

$$\mathbf{v} = \mathbf{w}$$

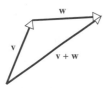

(*a*) The sum **v** + **w**

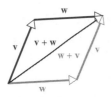

(*b*) **v** + **w** = **w** + **v**

Figure 3.1.2

DEFINITION

If **v** and **w** are any two vectors, then the ***sum* v + w** is the vector determined as follows: Position the vector **w** so that its initial point coincides with the terminal point of **v**. The vector **v** + **w** is represented by the arrow from the initial point of **v** to the terminal point of **w** (Figure 3.1.2*a*).

In Figure 3.1.2*b* we have constructed two sums, **v** + **w** (color arrows) and **w** + **v** (gray arrows). It is evident that

$$\mathbf{v} + \mathbf{w} = \mathbf{w} + \mathbf{v}$$

and that the sum coincides with the diagonal of the parallelogram determined by **v** and **w** when these vectors are positioned so that they have the same initial point.

The vector of length zero is called the ***zero vector*** and is denoted by **0**. We define

$$\mathbf{0} + \mathbf{v} = \mathbf{v} + \mathbf{0} = \mathbf{v}$$

for every vector **v**. Since there is no natural direction for the zero vector, we shall agree that it can be assigned any direction that is convenient for the problem being considered. If **v** is any nonzero vector, then $-\mathbf{v}$, the ***negative*** of **v**, is defined to be the vector that has the same magnitude as **v** but is oppositely directed (Figure 3.1.3). This vector has the property

$$\mathbf{v} + (-\mathbf{v}) = \mathbf{0}$$

(Why?) In addition, we define $-\mathbf{0} = \mathbf{0}$. Subtraction of vectors is defined as follows:

Figure 3.1.3 The negative of **v** has the same length as **v** but is oppositely directed.

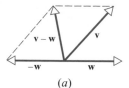

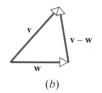

(a)

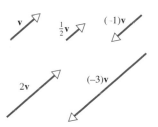

(b)

Figure 3.1.4

DEFINITION

If **v** and **w** are any two vectors, then the ***difference*** of **w** from **v** is defined by

$$\mathbf{v} - \mathbf{w} = \mathbf{v} + (-\mathbf{w})$$

(Figure 3.1.4*a*).

To obtain the difference $\mathbf{v} - \mathbf{w}$ without constructing $-\mathbf{w}$, position **v** and **w** so that their initial points coincide; the vector from the terminal point of **w** to the terminal point of **v** is then the vector $\mathbf{v} - \mathbf{w}$ (Figure 3.1.4*b*).

DEFINITION

If **v** is a nonzero vector and k is a nonzero real number (scalar), then the ***product*** $k\mathbf{v}$ is defined to be the vector whose length is $|k|$ times the length of **v** and whose direction is the same as that of **v** if $k > 0$ and opposite to that of **v** if $k < 0$. We define $k\mathbf{v} = \mathbf{0}$ if $k = 0$ or $\mathbf{v} = \mathbf{0}$.

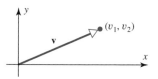

Figure 3.1.5

Figure 3.1.5 illustrates the relation between a vector **v** and the vectors $\frac{1}{2}\mathbf{v}$, $(-1)\mathbf{v}$, $2\mathbf{v}$, and $(-3)\mathbf{v}$. Note that the vector $(-1)\mathbf{v}$ has the same length as **v** but is oppositely directed. Thus $(-1)\mathbf{v}$ is just the negative of **v**; that is,

$$(-1)\mathbf{v} = -\mathbf{v}$$

A vector of the form $k\mathbf{v}$ is called a ***scalar multiple*** of **v**. As evidenced by Figure 3.1.5, vectors that are scalar multiples of each other are parallel. Conversely, it can be shown that nonzero parallel vectors are scalar multiples of each other. We omit the proof.

Vectors in Coordinate Systems

Problems involving vectors can often be simplified by introducing a rectangular coordinate system. For the moment we shall restrict the discussion to vectors in 2-space (the plane). Let **v** be any vector in the plane, and assume, as in Figure 3.1.6, that **v** has been positioned so that its initial point is at the origin of a rectangular coordinate system. The coordinates (v_1, v_2) of the terminal point of **v** are called the ***components of*** **v**, and we write

$$\mathbf{v} = (v_1, v_2)$$

Figure 3.1.6 v_1 and v_2 are the components of **v**.

If equivalent vectors, **v** and **w**, are located so that their initial points fall at the origin, then it is obvious that their terminal points must coincide (since the vectors have the same length and direction); thus the vectors have the same components. Conversely, vectors with the same components are equivalent since they have the same length and the same direction. In summary, two vectors

$$\mathbf{v} = (v_1, v_2) \quad \text{and} \quad \mathbf{w} = (w_1, w_2)$$

are equivalent if and only if

$$v_1 = w_1 \quad \text{and} \quad v_2 = w_2$$

The operations of vector addition and multiplication by scalars are easy to carry out in terms of components. As illustrated in Figure 3.1.7, if

$$\mathbf{v} = (v_1, v_2) \quad \text{and} \quad \mathbf{w} = (w_1, w_2)$$

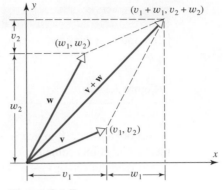

Figure 3.1.7

then

$$\mathbf{v} + \mathbf{w} = (v_1 + w_1, v_2 + w_2) \tag{1}$$

If $\mathbf{v} = (v_1, v_2)$ and k is any scalar, then by using a geometric argument involving similar triangles, it can be shown (Exercise 16) that

$$k\mathbf{v} = (kv_1, kv_2) \tag{2}$$

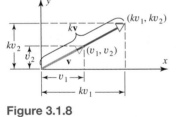

Figure 3.1.8

(Figure 3.1.8). Thus, for example, if $\mathbf{v} = (1, -2)$ and $\mathbf{w} = (7, 6)$, then

$$\mathbf{v} + \mathbf{w} = (1, -2) + (7, 6) = (1 + 7, -2 + 6) = (8, 4)$$

and

$$4\mathbf{v} = 4(1, -2) = (4(1), 4(-2)) = (4, -8)$$

Since $\mathbf{v} - \mathbf{w} = \mathbf{v} + (-1)\mathbf{w}$, it follows from Formulas (1) and (2) that

$$\mathbf{v} - \mathbf{w} = (v_1 - w_1, v_2 - w_2)$$

(Verify.)

Vectors in 3-Space

Just as vectors in the plane can be described by pairs of real numbers, vectors in 3-space can be described by triples of real numbers by introducing a ***rectangular coordinate*** system. To construct such a coordinate system, select a point O, called the ***origin***, and choose three mutually perpendicular lines, called ***coordinate axes***, passing through the origin. Label these axes x, y, and z, and select a positive direction for each coordinate

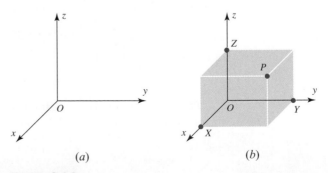

(a)   (b)

Figure 3.1.9

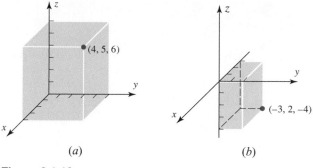

(a)   (b)

Figure 3.1.10

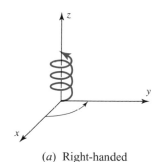

(a) Right-handed

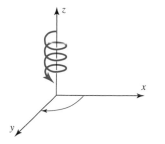

(b) Left-handed

Figure 3.1.11

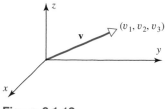

Figure 3.1.12

axis as well as a unit of length for measuring distances (Figure 3.1.9*a*). Each pair of coordinate axes determines a plane called a *coordinate plane*. These are referred to as the *xy-plane*, the *xz-plane*, and the *yz-plane*. To each point P in 3-space we assign a triple of numbers (x, y, z), called the *coordinates of P*, as follows: Pass three planes through P parallel to the coordinate planes, and denote the points of intersection of these planes with the three coordinate axes by X, Y, and Z (Figure 3.1.9*b*). The coordinates of P are defined to be the signed lengths

$$x = OX, \qquad y = OY, \qquad z = OZ$$

In Figure 3.1.10*a* we have constructed the point whose coordinates are $(4, 5, 6)$ and in Figure 3.1.10*b* the point whose coordinates are $(-3, 2, -4)$.

Rectangular coordinate systems in 3-space fall into two categories, *left-handed* and *right-handed*. A right-handed system has the property that an ordinary screw pointed in the positive direction on the z-axis would be advanced if the positive x-axis were rotated 90° toward the positive y-axis (Figure 3.1.11*a*); the system is left-handed if the screw would be retracted (Figure 3.1.11*b*).

REMARK   In this book we shall use only right-handed coordinate systems.

If, as in Figure 3.1.12, a vector **v** in 3-space is positioned so its initial point is at the origin of a rectangular coordinate system, then the coordinates of the terminal point are called the *components* of **v**, and we write

$$\mathbf{v} = (v_1, v_2, v_3)$$

If $\mathbf{v} = (v_1, v_2, v_3)$ and $\mathbf{w} = (w_1, w_2, w_3)$ are two vectors in 3-space, then arguments similar to those used for vectors in a plane can be used to establish the following results.

> **v** and **w** are equivalent if and only if $v_1 = w_1$, $v_2 = w_2$, and $v_3 = w_3$
>
> $\mathbf{v} + \mathbf{w} = (v_1 + w_1, v_2 + w_2, v_3 + w_3)$
>
> $k\mathbf{v} = (kv_1, kv_2, kv_3)$, where k is any scalar

EXAMPLE 1   Vector Computations with Components

If $\mathbf{v} = (1, -3, 2)$ and $\mathbf{w} = (4, 2, 1)$, then

$$\mathbf{v} + \mathbf{w} = (5, -1, 3), \qquad 2\mathbf{v} = (2, -6, 4), \qquad -\mathbf{w} = (-4, -2, -1),$$
$$\mathbf{v} - \mathbf{w} = \mathbf{v} + (-\mathbf{w}) = (-3, -5, 1) \quad \blacklozenge$$

Application to Computer Color Models

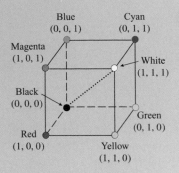

Blue (0, 0, 1)
Cyan (0, 1, 1)
Magenta (1, 0, 1)
White (1, 1, 1)
Black (0, 0, 0)
Green (0, 1, 0)
Red (1, 0, 0)
Yellow (1, 1, 0)

Colors on computer monitors are commonly based on what is called the **RGB** *color model*. Colors in this system are created by adding together percentages of the primary colors red (R), green (G), and blue (B). One way to do this is to identify the primary colors with the vectors

$$\mathbf{r} = (1, 0, 0) \quad \text{(pure red)},$$
$$\mathbf{g} = (0, 1, 0) \quad \text{(pure green)},$$
$$\mathbf{b} = (0, 0, 1) \quad \text{(pure blue)}$$

in R^3 and to create all other colors by forming linear combinations of $\mathbf{r}$, $\mathbf{g}$, and $\mathbf{b}$ using coefficients between 0 and 1, inclusive; these coefficients represent the percentage of each pure color in the mix. The set of all such color vectors is called **RGB** *space* or the **RGB** *color cube*. Thus, each color vector $\mathbf{c}$ in this cube is expressible as a linear combination of the form

$$\mathbf{c} = c_1\mathbf{r} + c_2\mathbf{g} + c_3\mathbf{b}$$
$$= c_1(1, 0, 0) + c_2(0, 1, 0) + c_3(0, 0, 1)$$
$$= (c_1, c_2, c_3)$$

where $0 \le c_i \le 1$. As indicated in the figure, the corners of the cube represent the pure primary colors together with the colors, black, white, magenta, cyan, and yellow. The vectors along the diagonal running from black to white correspond to shades of gray.

Sometimes a vector is positioned so that its initial point is not at the origin. If the vector $\overrightarrow{P_1P_2}$ has initial point $P_1(x_1, y_1, z_1)$ and terminal point $P_2(x_2, y_2, z_2)$, then

$$\overrightarrow{P_1P_2} = (x_2 - x_1, y_2 - y_1, z_2 - z_1)$$

That is, the components of $\overrightarrow{P_1P_2}$ are obtained by subtracting the coordinates of the initial point from the coordinates of the terminal point. This may be seen using Figure 3.1.13: The vector $\overrightarrow{P_1P_2}$ is the difference of vectors $\overrightarrow{OP_2}$ and $\overrightarrow{OP_1}$, so

$$\overrightarrow{P_1P_2} = \overrightarrow{OP_2} - \overrightarrow{OP_1} = (x_2, y_2, z_2) - (x_1, y_1, z_1) = (x_2 - x_1, y_2 - y_1, z_2 - z_1)$$

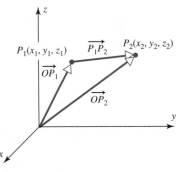

Figure 3.1.13

EXAMPLE 2 Finding the Components of a Vector

The components of the vector $\mathbf{v} = \overrightarrow{P_1P_2}$ with initial point $P_1(2, -1, 4)$ and terminal point $P_2(7, 5, -8)$ are

$$\mathbf{v} = (7 - 2, 5 - (-1), (-8) - 4) = (5, 6, -12) \quad \blacklozenge$$

In 2-space the vector with initial point $P_1(x_1, y_1)$ and terminal point $P_2(x_2, y_2)$ is

$$\overrightarrow{P_1 P_2} = (x_2 - x_1, y_2 - y_1)$$

Translation of Axes

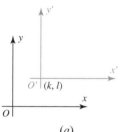

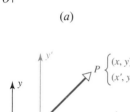

(a)

The solutions to many problems can be simplified by translating the coordinate axes to obtain new axes parallel to the original ones.

In Figure 3.1.14a we have translated the axes of an xy-coordinate system to obtain an $x'y'$-coordinate system whose origin O' is at the point $(x, y) = (k, l)$. A point P in 2-space now has both (x, y) coordinates and (x', y') coordinates. To see how the two are related, consider the vector $\overrightarrow{O'P}$ (Figure 3.1.14b). In the xy-system its initial point is at (k, l) and its terminal point is at (x, y), so $\overrightarrow{O'P} = (x - k, y - l)$. In the $x'y'$-system its initial point is at $(0, 0)$ and its terminal point is at (x', y'), so $\overrightarrow{O'P} = (x', y')$. Therefore,

$$x' = x - k, \qquad y' = y - l$$

These formulas are called the **translation equations**.

(b)

Figure 3.1.14

EXAMPLE 3 Using the Translation Equations

Suppose that an xy-coordinate system is translated to obtain an $x'y'$-coordinate system whose origin has xy-coordinates $(k, l) = (4, 1)$.

(a) Find the $x'y'$-coordinates of the point with the xy-coordinates $P(2, 0)$.
(b) Find the xy-coordinates of the point with $x'y'$-coordinates $Q(-1, 5)$.

Solution (a)

The translation equations are

$$x' = x - 4, \qquad y' = y - 1$$

so the $x'y'$-coordinates of $P(2, 0)$ are $x' = 2 - 4 = -2$ and $y' = 0 - 1 = -1$.

Solution (b)

The translation equations in (a) can be rewritten as

$$x = x' + 4, \qquad y = y' + 1$$

so the xy-coordinates of Q are $x = -1 + 4 = 3$ and $y = 5 + 1 = 6$. ◆

In 3-space the translation equations are

$$x' = x - k, \qquad y' = y - l, \qquad z' = z - m$$

where (k, l, m) are the xyz-coordinates of the $x'y'z'$-origin.

EXERCISE SET
3.1

1. Draw a right-handed coordinate system and locate the points whose coordinates are
   (a) $(3, 4, 5)$ (b) $(-3, 4, 5)$ (c) $(3, -4, 5)$ (d) $(3, 4, -5)$
   (e) $(-3, -4, 5)$ (f) $(-3, 4, -5)$ (g) $(3, -4, -5)$ (h) $(-3, -4, -5)$
   (i) $(-3, 0, 0)$ (j) $(3, 0, 3)$ (k) $(0, 0, -3)$ (l) $(0, 3, 0)$

2. Sketch the following vectors with the initial points located at the origin:
   (a) $\mathbf{v}_1 = (3, 6)$ (b) $\mathbf{v}_2 = (-4, -8)$ (c) $\mathbf{v}_3 = (-4, -3)$
   (d) $\mathbf{v}_4 = (5, -4)$ (e) $\mathbf{v}_5 = (3, 0)$ (f) $\mathbf{v}_6 = (0, -7)$
   (g) $\mathbf{v}_7 = (3, 4, 5)$ (h) $\mathbf{v}_8 = (3, 3, 0)$ (i) $\mathbf{v}_9 = (0, 0, -3)$

3. Find the components of the vector having initial point P_1 and terminal point P_2.
   (a) $P_1(4, 8), P_2(3, 7)$ (b) $P_1(3, -5), P_2(-4, -7)$
   (c) $P_1(-5, 0), P_2(-3, 1)$ (d) $P_1(0, 0), P_2(a, b)$
   (e) $P_1(3, -7, 2), P_2(-2, 5, -4)$ (f) $P_1(-1, 0, 2), P_2(0, -1, 0)$
   (g) $P_1(a, b, c), P_2(0, 0, 0)$ (h) $P_1(0, 0, 0), P_2(a, b, c)$

4. Find a nonzero vector $\mathbf{u}$ with initial point $P(-1, 3, -5)$ such that
   (a) $\mathbf{u}$ has the same direction as $\mathbf{v} = (6, 7, -3)$
   (b) $\mathbf{u}$ is oppositely directed to $\mathbf{v} = (6, 7, -3)$

5. Find a nonzero vector $\mathbf{u}$ with terminal point $Q(3, 0, -5)$ such that
   (a) $\mathbf{u}$ has the same direction as $\mathbf{v} = (4, -2, -1)$
   (b) $\mathbf{u}$ is oppositely directed to $\mathbf{v} = (4, -2, -1)$

6. Let $\mathbf{u} = (-3, 1, 2)$, $\mathbf{v} = (4, 0, -8)$, and $\mathbf{w} = (6, -1, -4)$. Find the components of
   (a) $\mathbf{v} - \mathbf{w}$ (b) $6\mathbf{u} + 2\mathbf{v}$ (c) $-\mathbf{v} + \mathbf{u}$
   (d) $5(\mathbf{v} - 4\mathbf{u})$ (e) $-3(\mathbf{v} - 8\mathbf{w})$ (f) $(2\mathbf{u} - 7\mathbf{w}) - (8\mathbf{v} + \mathbf{u})$

7. Let $\mathbf{u}$, $\mathbf{v}$, and $\mathbf{w}$ be the vectors in Exercise 6. Find the components of the vector $\mathbf{x}$ that satisfies
   $2\mathbf{u} - \mathbf{v} + \mathbf{x} = 7\mathbf{x} + \mathbf{w}$.

8. Let $\mathbf{u}$, $\mathbf{v}$, and $\mathbf{w}$ be the vectors in Exercise 6. Find scalars c_1, c_2, and c_3 such that
   $$c_1\mathbf{u} + c_2\mathbf{v} + c_3\mathbf{w} = (2, 0, 4)$$

9. Show that there do not exist scalars c_1, c_2, and c_3 such that
   $$c_1(-2, 9, 6) + c_2(-3, 2, 1) + c_3(1, 7, 5) = (0, 5, 4)$$

10. Find all scalars c_1, c_2, and c_3 such that
    $$c_1(1, 2, 0) + c_2(2, 1, 1) + c_3(0, 3, 1) = (0, 0, 0)$$

11. Let P be the point $(2, 3, -2)$ and Q the point $(7, -4, 1)$.
    (a) Find the midpoint of the line segment connecting P and Q.
    (b) Find the point on the line segment connecting P and Q that is $\frac{3}{4}$ of the way from P to Q.

12. Suppose an xy-coordinate system is translated to obtain an $x'y'$-coordinate system whose origin O' has xy-coordinates $(2, -3)$.
    (a) Find the $x'y'$-coordinates of the point P whose xy-coordinates are $(7, 5)$.
    (b) Find the xy-coordinates of the point Q whose $x'y'$-coordinates are $(-3, 6)$.
    (c) Draw the xy and $x'y'$-coordinate axes and locate the points P and Q.
    (d) If $\mathbf{v} = (3, 7)$ is a vector in the xy-coordinate system, what are the components of $\mathbf{v}$ in the $x'y'$-coordinate system?
    (e) If $\mathbf{v} = (v_1, v_2)$ is a vector in the xy-coordinate system, what are the components of $\mathbf{v}$ in the $x'y'$-coordinate system?

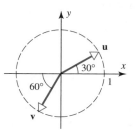

Figure Ex-15

13. Let P be the point $(1, 3, 7)$. If the point $(4, 0, -6)$ is the midpoint of the line segment connecting P and Q, what is Q?

14. Suppose that an xyz-coordinate system is translated to obtain an $x'y'z'$-coordinate system. Let $\mathbf{v}$ be a vector whose components are $\mathbf{v} = (v_1, v_2, v_3)$ in the xyz-system. Show that $\mathbf{v}$ has the same components in the $x'y'z'$-system.

15. Find the components of $\mathbf{u}$, $\mathbf{v}$, $\mathbf{u} + \mathbf{v}$, and $\mathbf{u} - \mathbf{v}$ for the vectors shown in the accompanying figure.

16. Prove geometrically that if $\mathbf{v} = (v_1, v_2)$, then $k\mathbf{v} = (kv_1, kv_2)$. (Restrict the proof to the case $k > 0$ illustrated in Figure 3.1.8. The complete proof would involve various cases that depend on the sign of k and the quadrant in which the vector falls.)

Discussion & Discovery

17. Consider Figure 3.1.13. Discuss a geometric interpretation of the vector

$$\mathbf{u} = \overrightarrow{OP_1} + \tfrac{1}{2}(\overrightarrow{OP_2} - \overrightarrow{OP_1})$$

18. Draw a picture that shows four nonzero vectors whose sum is zero.

19. If you were given four nonzero vectors, how would you construct geometrically a fifth vector that is equal to the sum of the first four? Draw a picture to illustrate your method.

20. Consider a clock with vectors drawn from the center to each hour as shown in the accompanying figure.

    (a) What is the sum of the 12 vectors that result if the vector terminating at 12 is doubled in length and the other vectors are left alone?

    (b) What is the sum of the 12 vectors that result if the vectors terminating at 3 and 9 are each tripled and the others are left alone?

    (c) What is the sum of the 9 vectors that remain if the vectors terminating at 5, 11, and 8 are removed?

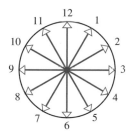

Figure Ex-20

21. Indicate whether the statement is true (T) or false (F). Justify your answer.

    (a) If $\mathbf{x} + \mathbf{y} = \mathbf{x} + \mathbf{z}$, then $\mathbf{y} = \mathbf{x}$.

    (b) If $\mathbf{u} + \mathbf{v} = \mathbf{0}$, then $a\mathbf{u} + b\mathbf{v} = \mathbf{0}$ for all a and b.

    (c) Parallel vectors with the same length are equal.

    (d) If $a\mathbf{x} = \mathbf{0}$, then either $a = 0$ or $\mathbf{x} = \mathbf{0}$.

    (e) If $a\mathbf{u} + b\mathbf{v} = \mathbf{0}$, then $\mathbf{u}$ and $\mathbf{v}$ are parallel vectors.

    (f) The vectors $\mathbf{u} = (\sqrt{2}, \sqrt{3})$ and $\mathbf{v} = \left(\tfrac{1}{\sqrt{2}}, \tfrac{1}{2}\sqrt{3}\right)$ are equivalent.

3.2
NORM OF A VECTOR; VECTOR ARITHMETIC

In this section we shall establish the basic rules of vector arithmetic.

Properties of Vector Operations

The following theorem lists the most important properties of vectors in 2-space and 3-space.

THEOREM 3.2.1

> **Properties of Vector Arithmetic**
>
> *If* $\mathbf{u}$, $\mathbf{v}$, *and* $\mathbf{w}$ *are vectors in 2- or 3-space and k and l are scalars, then the following relationships hold.*
>
> (a) $\mathbf{u} + \mathbf{v} = \mathbf{v} + \mathbf{u}$
>
> (b) $(\mathbf{u} + \mathbf{v}) + \mathbf{w} = \mathbf{u} + (\mathbf{v} + \mathbf{w})$
>
> (c) $\mathbf{u} + \mathbf{0} = \mathbf{0} + \mathbf{u} = \mathbf{u}$
>
> (d) $\mathbf{u} + (-\mathbf{u}) = \mathbf{0}$

$$(e)\quad k(l\mathbf{u}) = (kl)\mathbf{u} \qquad\qquad (f)\quad k(\mathbf{u} + \mathbf{v}) = k\mathbf{u} + k\mathbf{v}$$
$$(g)\quad (k+l)\mathbf{u} = k\mathbf{u} + l\mathbf{u} \qquad (h)\quad 1\mathbf{u} = \mathbf{u}$$

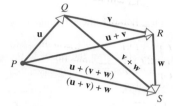

Figure 3.2.1 The vectors $\mathbf{u} + (\mathbf{v} + \mathbf{w})$ and $(\mathbf{u} + \mathbf{v}) + \mathbf{w}$ are equal.

Before discussing the proof, we note that we have developed two approaches to vectors: *geometric*, in which vectors are represented by arrows or directed line segments, and *analytic*, in which vectors are represented by pairs or triples of numbers called components. As a consequence, the equations in Theorem 1 can be proved either geometrically or analytically. To illustrate, we shall prove part (*b*) both ways. The remaining proofs are left as exercises.

Proof of part (b) (analytic) We shall give the proof for vectors in 3-space; the proof for 2-space is similar. If $\mathbf{u} = (u_1, u_2, u_3)$, $\mathbf{v} = (v_1, v_2, v_3)$, and $\mathbf{w} = (w_1, w_2, w_3)$, then

$$
\begin{aligned}
(\mathbf{u} + \mathbf{v}) + \mathbf{w} &= [(u_1, u_2, u_3) + (v_1, v_2, v_3)] + (w_1, w_2, w_3) \\
&= (u_1 + v_1, u_2 + v_2, u_3 + v_3) + (w_1, w_2, w_3) \\
&= ([u_1 + v_1] + w_1, [u_2 + v_2] + w_2, [u_3 + v_3] + w_3) \\
&= (u_1 + [v_1 + w_1], u_2 + [v_2 + w_2], u_3 + [v_3 + w_3]) \\
&= (u_1, u_2, u_3) + (v_1 + w_1, v_2 + w_2, v_3 + w_3) \\
&= \mathbf{u} + (\mathbf{v} + \mathbf{w})
\end{aligned}
$$

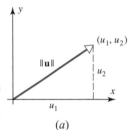

(a)

Proof of part (b) (geometric) Let $\mathbf{u}$, $\mathbf{v}$, and $\mathbf{w}$ be represented by $\overrightarrow{PQ}$, $\overrightarrow{QR}$, and $\overrightarrow{RS}$ as shown in Figure 3.2.1. Then

$$\mathbf{v} + \mathbf{w} = \overrightarrow{QS} \quad\text{and}\quad \mathbf{u} + (\mathbf{v} + \mathbf{w}) = \overrightarrow{PS}$$

Also,

$$\mathbf{u} + \mathbf{v} = \overrightarrow{PR} \quad\text{and}\quad (\mathbf{u} + \mathbf{v}) + \mathbf{w} = \overrightarrow{PS}$$

Therefore,

$$\mathbf{u} + (\mathbf{v} + \mathbf{w}) = (\mathbf{u} + \mathbf{v}) + \mathbf{w} \qquad\blacksquare$$

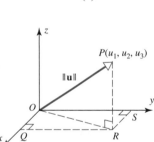

(b)

Figure 3.2.2

REMARK In light of part (*b*) of this theorem, the symbol $\mathbf{u} + \mathbf{v} + \mathbf{w}$ is unambiguous since the same sum is obtained no matter where parentheses are inserted. Moreover, if the vectors $\mathbf{u}$, $\mathbf{v}$, and $\mathbf{w}$ are placed "tip to tail," then the sum $\mathbf{u} + \mathbf{v} + \mathbf{w}$ is the vector from the initial point of $\mathbf{u}$ to the terminal point of $\mathbf{w}$ (Figure 3.2.1).

Norm of a Vector

The ***length*** of a vector $\mathbf{u}$ is often called the ***norm*** of $\mathbf{u}$ and is denoted by $\|\mathbf{u}\|$. It follows from the Theorem of Pythagoras that the norm of a vector $\mathbf{u} = (u_1, u_2)$ in 2-space is

$$\|\mathbf{u}\| = \sqrt{u_1^2 + u_2^2} \tag{1}$$

(Figure 3.2.2*a*). Let $\mathbf{u} = (u_1, u_2, u_3)$ be a vector in 3-space. Using Figure 3.2.2*b* and two applications of the Theorem of Pythagoras, we obtain

$$\|\mathbf{u}\|^2 = (OR)^2 + (RP)^2 = (OQ)^2 + (OS)^2 + (RP)^2 = u_1^2 + u_2^2 + u_3^2$$

Thus

$$\|\mathbf{u}\| = \sqrt{u_1^2 + u_2^2 + u_3^2} \tag{2}$$

A vector of norm 1 is called a ***unit vector***.

Global Positioning

GPS (*Global Positioning System*) is the system used by the military, ships, airplane pilots, surveyors, utility companies, automobiles, and hikers to locate current positions by communicating with a system of satellites. The system, which is operated by the U.S. Department of Defense, nominally uses 24 satellites that orbit the Earth every 12 hours at a height of about 11,000 miles. These satellites move in six orbital planes that have been chosen to make between five and eight satellites visible from any point on Earth.

To explain how the system works, assume that the Earth is a sphere, and suppose that there is an xyz-coordinate system with its origin at the Earth's center and its z-axis through the North Pole. Let us assume that relative to this coordinate system a ship is at an unknown point (x, y, z) at some time t. For simplicity, assume that distances are measured in units equal to the Earth's radius, so that the coordinates of the ship always satisfy the equation

$$x^2 + y^2 + z^2 = 1$$

The GPS identifies the ship's coordinates (x, y, z) at a time t using a triangulation system and computed distances from four satellites. These distances are computed using the speed of light (approximately 0.469 Earth radii per hundredth of a second) and the time it takes for the signal to travel from the satellite to the ship. For example, if the ship receives the signal at time t and the satellite indicates that it transmitted the signal at time t_0, then the distance d traveled by the signal will be

$$d = 0.469(t - t_0)$$

In theory, knowing three ship-to-satellite distances would suffice to determine the three unknown coordinates of the ship. However, the problem is that the ships (or other GPS users) do not generally have clocks that can compute t with sufficient accuracy for global positioning. Thus, the variable t must be regarded as a fourth unknown, and hence the need for the distance to a fourth satellite. Suppose that in addition to transmitting the time t_0, each satellite also transmits its coordinates (x_0, y_0, z_0) at that time, thereby allowing d to be computed as

$$d = \sqrt{(x - x_0)^2 + (y - y_0)^2 + (z - z_0)^2}$$

If we now equate the squares of d from both equations and round off to three decimal places, then we obtain the second-degree equation

$$(x - x_0)^2 + (y - y_0)^2 + (z - z_0)^2 = 0.22(t - t_0)^2$$

Since there are four different satellites, and we can get an equation like this for each one, we can produce four equations in the unknowns x, y, z, and t_0. Although these are second-degree equations, it is possible to use these equations and some algebra to produce a system of linear equations that can be solved for the unknowns.

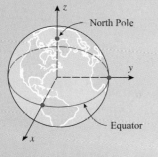

North Pole

Equator

z

y

x

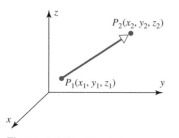

Figure 3.2.3 The distance between P_1 and P_2 is the norm of the vector $\overrightarrow{P_1 P_2}$.

z

$P_2(x_2, y_2, z_2)$

$P_1(x_1, y_1, z_1)$

y

x

If $P_1(x_1, y_1, z_1)$ and $P_2(x_2, y_2, z_2)$ are two points in 3-space, then the ***distance*** d between them is the norm of the vector $\overrightarrow{P_1 P_2}$ (Figure 3.2.3). Since

$$\overrightarrow{P_1 P_2} = (x_2 - x_1, y_2 - y_1, z_2 - z_1)$$

it follows from (2) that

$$d = \sqrt{(x_2 - x_1)^2 + (y_2 - y_1)^2 + (z_2 - z_1)^2} \tag{3}$$

Similarly, if $P_1(x_1, y_1)$ and $P_2(x_2, y_2)$ are points in 2-space, then the distance between them is given by

$$d = \sqrt{(x_2 - x_1)^2 + (y_2 - y_1)^2} \tag{4}$$

EXAMPLE 1 Finding Norm and Distance

The norm of the vector $\mathbf{u} = (-3, 2, 1)$ is

$$\|\mathbf{u}\| = \sqrt{(-3)^2 + (2)^2 + (1)^2} = \sqrt{14}$$

The distance d between the points $P_1(2, -1, -5)$ and $P_2(4, -3, 1)$ is

$$d = \sqrt{(4-2)^2 + (-3+1)^2 + (1+5)^2} = \sqrt{44} = 2\sqrt{11} \quad \blacklozenge$$

From the definition of the product $k\mathbf{u}$, the length of the vector $k\mathbf{u}$ is $|k|$ times the length of $\mathbf{u}$. Expressed as an equation, this statement says that

$$\|k\mathbf{u}\| = |k|\,\|\mathbf{u}\| \tag{5}$$

This useful formula is applicable in both 2-space and 3-space.

EXERCISE SET 3.2

1. Find the norm of $\mathbf{v}$.
   (a) $\mathbf{v} = (4, -3)$ (b) $\mathbf{v} = (2, 3)$ (c) $\mathbf{v} = (-5, 0)$
   (d) $\mathbf{v} = (2, 2, 2)$ (e) $\mathbf{v} = (-7, 2, -1)$ (f) $\mathbf{v} = (0, 6, 0)$

2. Find the distance between P_1 and P_2.
   (a) $P_1(3, 4)$, $P_2(5, 7)$ (b) $P_1(-3, 6)$, $P_2(-1, -4)$
   (c) $P_1(7, -5, 1)$, $P_2(-7, -2, -1)$ (d) $P_1(3, 3, 3)$, $P_2(6, 0, 3)$

3. Let $\mathbf{u} = (2, -2, 3)$, $\mathbf{v} = (1, -3, 4)$, $\mathbf{w} = (3, 6, -4)$. In each part, evaluate the expression.
   (a) $\|\mathbf{u} + \mathbf{v}\|$ (b) $\|\mathbf{u}\| + \|\mathbf{v}\|$ (c) $\|-2\mathbf{u}\| + 2\|\mathbf{u}\|$

   (d) $\|3\mathbf{u} - 5\mathbf{v} + \mathbf{w}\|$ (e) $\dfrac{1}{\|\mathbf{w}\|}\mathbf{w}$ (f) $\left\|\dfrac{1}{\|\mathbf{w}\|}\mathbf{w}\right\|$

4. If $\|\mathbf{v}\| = 2$ and $\|\mathbf{w}\| = 3$, what are the largest and smallest values possible for $\|\mathbf{v} - \mathbf{w}\|$? Give a geometric explanation of your results.

5. Let $\mathbf{u} = (2, 0, 4)$ and $\mathbf{v} = (1, 3, -6)$. In each of the following, determine, if possible, scalars k, l such that
   (a) $k\mathbf{u} + l\mathbf{v} = (5, 9, -14)$ (b) $k\mathbf{u} + l\mathbf{v} = (9, 15, -21)$

6. Let $\mathbf{u} = (2, 6, -7)$, $\mathbf{v} = (-1, -1, 8)$, and $k = 3$. If $(2, 14, 11) = k\mathbf{u} + l\mathbf{v}$, what is the value of l?

7. Let $\mathbf{v} = (-1, 2, 5)$. Find all scalars k such that $\|k\mathbf{v}\| = 4$.

8. Let $\mathbf{u} = (7, -3, 1)$, $\mathbf{v} = (9, 6, 6)$, $\mathbf{w} = (2, 1, -8)$, $k = -2$, and $l = 5$. Verify that these vectors and scalars satisfy the stated equalities from Theorem 1.
   (a) part (b) (b) part (e) (c) part (f) (d) part (g)

9. (a) Show that if $\mathbf{v}$ is any nonzero vector, then $\dfrac{1}{\|\mathbf{v}\|}\mathbf{v}$ is a unit vector.

   (b) Use the result in part (a) to find a unit vector that has the same direction as the vector $\mathbf{v} = (3, 4)$.

   (c) Use the result in part (a) to find a unit vector that is oppositely directed to the vector $\mathbf{v} = (-2, 3, -6)$.

10. (a) Show that the components of the vector $\mathbf{v} = (v_1, v_2)$ in Figure Ex-10a are $v_1 = \|\mathbf{v}\| \cos\theta$ and $v_2 = \|\mathbf{v}\| \sin\theta$.

(b) Let $\mathbf{u}$ and $\mathbf{v}$ be the vectors in Figure Ex-10b. Use the result in part (a) to find the components of $4\mathbf{u} - 5\mathbf{v}$.

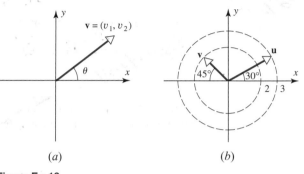

(a)   *(b)*

Figure Ex-10

11. Let $\mathbf{p}_0 = (x_0, y_0, z_0)$ and $\mathbf{p} = (x, y, z)$. Describe the set of all points (x, y, z) for which $\|\mathbf{p} - \mathbf{p}_0\| = 1$.

12. Prove geometrically that if $\mathbf{u}$ and $\mathbf{v}$ are vectors in 2- or 3-space, then $\|\mathbf{u} + \mathbf{v}\| \le \|\mathbf{u}\| + \|\mathbf{v}\|$.

13. Prove parts (a), (c), and (e) of Theorem 1 analytically.

14. Prove parts (d), (g), and (h) of Theorem 1 analytically.

Discussion & Discovery

15. For the inequality stated in Exercise 9, is it possible to have $\|\mathbf{u} + \mathbf{v}\| = \|\mathbf{u}\| + \|\mathbf{v}\|$? Explain your reasoning.

16. (a) What relationship must hold for the point $\mathbf{p} = (a, b, c)$ to be equidistant from the origin and the xz-plane? Make sure that the relationship you state is valid for positive and negative values of a, b, and c.

(b) What relationship must hold for the point $\mathbf{p} = (a, b, c)$ to be farther from the origin than from the xz-plane? Make sure that the relationship you state is valid for positive and negative values of a, b, and c.

17. (a) What does the inequality $\|\mathbf{x}\| < 1$ tell you about the location of the point $\mathbf{x}$ in the plane?

(b) Write down an inequality that describes the set of points that lie outside the circle of radius 1, centered at the point $\mathbf{x}_0$.

18. The triangles in the accompanying figure should suggest a geometric proof of Theorem 3.2.1(f) for the case where $k > 0$. Give the proof.

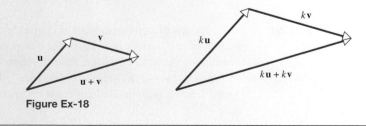

Figure Ex-18

3.3
DOT PRODUCT; PROJECTIONS

In this section we shall discuss an important way of multiplying vectors in 2-space or 3-space. We shall then give some applications of this multiplication to geometry.

Dot Product of Vectors

Let **u** and **v** be two nonzero vectors in 2-space or 3-space, and assume these vectors have been positioned so that their initial points coincide. By the ***angle between* u *and* v**, we shall mean the angle θ determined by **u** and **v** that satisfies $0 \le \theta \le \pi$ (Figure 3.3.1).

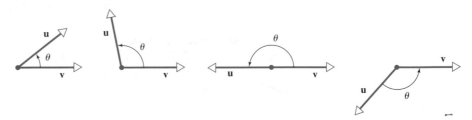

Figure 3.3.1 The angle θ between **u** and **v** satisfies $0 \le \theta \le \pi$.

> **DEFINITION**
>
> If **u** and **v** are vectors in 2-space or 3-space and θ is the angle between **u** and **v**, then the ***dot product*** or ***Euclidean inner product*** **u · v** is defined by
>
> $$\mathbf{u} \cdot \mathbf{v} = \begin{cases} \|\mathbf{u}\|\|\mathbf{v}\| \cos\theta & \text{if } \mathbf{u} \ne \mathbf{0} \text{ and } \mathbf{v} \ne \mathbf{0} \\ 0 & \text{if } \mathbf{u} = \mathbf{0} \text{ or } \mathbf{v} = \mathbf{0} \end{cases} \qquad (1)$$

EXAMPLE 1 Dot Product

As shown in Figure 3.3.2, the angle between the vectors **u** = (0, 0, 1) and **v** = (0, 2, 2) is 45°. Thus

$$\mathbf{u} \cdot \mathbf{v} = \|\mathbf{u}\|\|\mathbf{v}\| \cos\theta = (\sqrt{0^2 + 0^2 + 1^2})(\sqrt{0^2 + 2^2 + 2^2})\left(\frac{1}{\sqrt{2}}\right) = 2 \quad \blacklozenge$$

Figure 3.3.2

Component Form of the Dot Product

For purposes of computation, it is desirable to have a formula that expresses the dot product of two vectors in terms of the components of the vectors. We will derive such a formula for vectors in 3-space; the derivation for vectors in 2-space is similar.

Let **u** = (u_1, u_2, u_3) and **v** = (v_1, v_2, v_3) be two nonzero vectors. If, as shown in Figure 3.3.3, θ is the angle between **u** and **v**, then the law of cosines yields

$$\|\overrightarrow{PQ}\|^2 = \|\mathbf{u}\|^2 + \|\mathbf{v}\|^2 - 2\|\mathbf{u}\|\|\mathbf{v}\| \cos\theta \qquad (2)$$

Since $\overrightarrow{PQ} = \mathbf{v} - \mathbf{u}$, we can rewrite (2) as

$$\|\mathbf{u}\|\|\mathbf{v}\| \cos\theta = \tfrac{1}{2}(\|\mathbf{u}\|^2 + \|\mathbf{v}\|^2 - \|\mathbf{v} - \mathbf{u}\|^2)$$

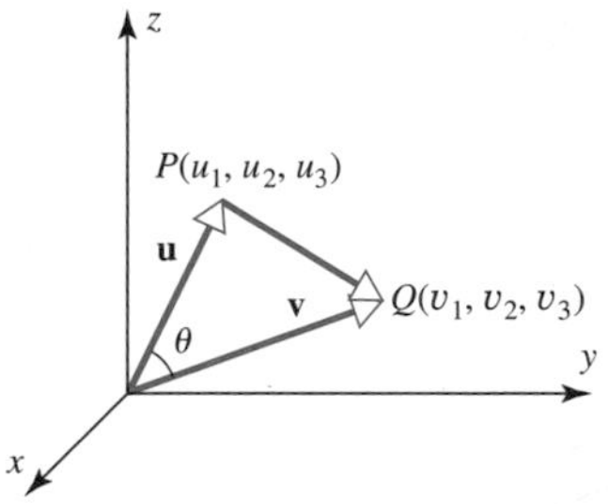

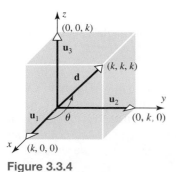

Figure 3.3.3

or

$$\mathbf{u} \cdot \mathbf{v} = \tfrac{1}{2}(\|\mathbf{u}\|^2 + \|\mathbf{v}\|^2 - \|\mathbf{v} - \mathbf{u}\|^2)$$

Substituting

$$\|\mathbf{u}\|^2 = u_1^2 + u_2^2 + u_3^2, \qquad \|\mathbf{v}\|^2 = v_1^2 + v_2^2 + v_3^2,$$

and

$$\|\mathbf{v} - \mathbf{u}\|^2 = (v_1 - u_1)^2 + (v_2 - u_2)^2 + (v_3 - u_3)^2$$

we obtain, after simplifying,

$$\mathbf{u} \cdot \mathbf{v} = u_1 v_1 + u_2 v_2 + u_3 v_3 \tag{3}$$

Although we derived this formula under the assumption that $\mathbf{u}$ and $\mathbf{v}$ are nonzero, the formula is also valid if $\mathbf{u} = \mathbf{0}$ or $\mathbf{v} = \mathbf{0}$ (verify).

If $\mathbf{u} = (u_1, u_2)$ and $\mathbf{v} = (v_1, v_2)$ are two vectors in 2-space, then the formula corresponding to (3) is

$$\mathbf{u} \cdot \mathbf{v} = u_1 v_1 + u_2 v_2 \tag{4}$$

Finding the Angle Between Vectors

If $\mathbf{u}$ and $\mathbf{v}$ are nonzero vectors, then Formula (1) can be written as

$$\cos\theta = \frac{\mathbf{u} \cdot \mathbf{v}}{\|\mathbf{u}\|\|\mathbf{v}\|} \tag{5}$$

EXAMPLE 2 Dot Product Using (3)

Consider the vectors $\mathbf{u} = (2, -1, 1)$ and $\mathbf{v} = (1, 1, 2)$. Find $\mathbf{u} \cdot \mathbf{v}$ and determine the angle θ between $\mathbf{u}$ and $\mathbf{v}$.

Solution

$$\mathbf{u} \cdot \mathbf{v} = u_1 v_1 + u_2 v_2 + u_3 v_3 = (2)(1) + (-1)(1) + (1)(2) = 3$$

For the given vectors we have $\|\mathbf{u}\| = \|\mathbf{v}\| = \sqrt{6}$, so from (5),

$$\cos\theta = \frac{\mathbf{u} \cdot \mathbf{v}}{\|\mathbf{u}\|\|\mathbf{v}\|} = \frac{3}{\sqrt{6}\sqrt{6}} = \frac{1}{2}$$

Thus, $\theta = 60°$. ◆

EXAMPLE 3 A Geometric Problem

Find the angle between a diagonal of a cube and one of its edges.

Solution

Let k be the length of an edge and introduce a coordinate system as shown in Figure 3.3.4. If we let $\mathbf{u}_1 = (k, 0, 0)$, $\mathbf{u}_2 = (0, k, 0)$, and $\mathbf{u}_3 = (0, 0, k)$, then the vector

$$\mathbf{d} = (k, k, k) = \mathbf{u}_1 + \mathbf{u}_2 + \mathbf{u}_3$$

Figure 3.3.4

is a diagonal of the cube. The angle θ between $\mathbf{d}$ and the edge $\mathbf{u}_1$ satisfies

$$\cos \theta = \frac{\mathbf{u}_1 \cdot \mathbf{d}}{\|\mathbf{u}_1\| \|\mathbf{d}\|} = \frac{k^2}{(k)(\sqrt{3k^2})} = \frac{1}{\sqrt{3}}$$

Thus

$$\theta = \cos^{-1}\left(\frac{1}{\sqrt{3}}\right) \approx 54.74°$$

Note that this is independent of k, as expected. ◆

The following theorem shows how the dot product can be used to obtain information about the angle between two vectors; it also establishes an important relationship between the norm and the dot product.

THEOREM 3.3.1

> *Let $\mathbf{u}$ and $\mathbf{v}$ be vectors in 2- or 3-space.*
>
> (a) $\mathbf{v} \cdot \mathbf{v} = \|\mathbf{v}\|^2$; *that is,* $\|\mathbf{v}\| = (\mathbf{v} \cdot \mathbf{v})^{1/2}$
>
> (b) *If the vectors $\mathbf{u}$ and $\mathbf{v}$ are nonzero and θ is the angle between them, then*
>
> | | | |
> |---|---|---|
> | θ *is acute* | *if and only if* | $\mathbf{u} \cdot \mathbf{v} > 0$ |
> | θ *is obtuse* | *if and only if* | $\mathbf{u} \cdot \mathbf{v} < 0$ |
> | $\theta = \pi/2$ | *if and only if* | $\mathbf{u} \cdot \mathbf{v} = 0$ |

Proof (a)   Since the angle θ between $\mathbf{v}$ and $\mathbf{v}$ is 0, we have

$$\mathbf{v} \cdot \mathbf{v} = \|\mathbf{v}\| \|\mathbf{v}\| \cos \theta = \|\mathbf{v}\|^2 \cos 0 = \|\mathbf{v}\|^2$$

Proof (b)   Since θ satisfies $0 \leq \theta \leq \pi$, it follows that θ is acute if and only if $\cos \theta > 0$, that θ is obtuse if and only if $\cos \theta < 0$, and that $\theta = \pi/2$ if and only if $\cos \theta = 0$. But $\cos \theta$ has the same sign as $\mathbf{u} \cdot \mathbf{v}$ since $\mathbf{u} \cdot \mathbf{v} = \|\mathbf{u}\| \|\mathbf{v}\| \cos \theta$, $\|\mathbf{u}\| > 0$, and $\|\mathbf{v}\| > 0$. Thus, the result follows. ∎

EXAMPLE 4   Finding Dot Products from Components

If $\mathbf{u} = (1, -2, 3)$, $\mathbf{v} = (-3, 4, 2)$, and $\mathbf{w} = (3, 6, 3)$, then

$$\mathbf{u} \cdot \mathbf{v} = (1)(-3) + (-2)(4) + (3)(2) = -5$$
$$\mathbf{v} \cdot \mathbf{w} = (-3)(3) + (4)(6) + (2)(3) = 21$$
$$\mathbf{u} \cdot \mathbf{w} = (1)(3) + (-2)(6) + (3)(3) = 0$$

Therefore, $\mathbf{u}$ and $\mathbf{v}$ make an obtuse angle, $\mathbf{v}$ and $\mathbf{w}$ make an acute angle, and $\mathbf{u}$ and $\mathbf{w}$ are perpendicular. ◆

Orthogonal Vectors

Perpendicular vectors are also called **orthogonal** vectors. In light of Theorem 3.3.1*b*, two *nonzero* vectors are orthogonal if and only if their dot product is zero. If we agree to consider $\mathbf{u}$ and $\mathbf{v}$ to be perpendicular when either or both of these vectors is $\mathbf{0}$, then we can state without exception that *two vectors $\mathbf{u}$ and $\mathbf{v}$ are orthogonal (perpendicular) if and only if* $\mathbf{u} \cdot \mathbf{v} = 0$. To indicate that $\mathbf{u}$ and $\mathbf{v}$ are orthogonal vectors, we write $\mathbf{u} \perp \mathbf{v}$.

EXAMPLE 5 A Vector Perpendicular to a Line

Show that in 2-space the nonzero vector $\mathbf{n} = (a, b)$ is perpendicular to the line $ax + by + c = 0$.

Solution

Let $P_1(x_1, y_1)$ and $P_2(x_2, y_2)$ be distinct points on the line, so that

$$ax_1 + by_1 + c = 0 \qquad\qquad (6)$$
$$ax_2 + by_2 + c = 0$$

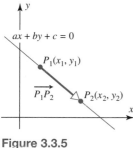

Figure 3.3.5

Since the vector $\overrightarrow{P_1 P_2} = (x_2 - x_1, y_2 - y_1)$ runs along the line (Figure 3.3.5), we need only show that $\mathbf{n}$ and $\overrightarrow{P_1 P_2}$ are perpendicular. But on subtracting the equations in (6), we obtain

$$a(x_2 - x_1) + b(y_2 - y_1) = 0$$

which can be expressed in the form

$$(a, b) \cdot (x_2 - x_1, y_2 - y_1) = 0 \quad \text{or} \quad \mathbf{n} \cdot \overrightarrow{P_1 P_2} = 0$$

Thus $\mathbf{n}$ and $\overrightarrow{P_1 P_2}$ are perpendicular. ◆

The following theorem lists the most important properties of the dot product. They are useful in calculations involving vectors.

THEOREM 3.3.2

> ### Properties of the Dot Product
>
> *If $\mathbf{u}$, $\mathbf{v}$, and $\mathbf{w}$ are vectors in 2- or 3-space and k is a scalar, then*
>
> (a) $\mathbf{u} \cdot \mathbf{v} = \mathbf{v} \cdot \mathbf{u}$
> (b) $\mathbf{u} \cdot (\mathbf{v} + \mathbf{w}) = \mathbf{u} \cdot \mathbf{v} + \mathbf{u} \cdot \mathbf{w}$
> (c) $k(\mathbf{u} \cdot \mathbf{v}) = (k\mathbf{u}) \cdot \mathbf{v} = \mathbf{u} \cdot (k\mathbf{v})$
> (d) $\mathbf{v} \cdot \mathbf{v} > 0$ if $\mathbf{v} \neq \mathbf{0}$, and $\mathbf{v} \cdot \mathbf{v} = 0$ if $\mathbf{v} = \mathbf{0}$

Proof We shall prove (c) for vectors in 3-space and leave the remaining proofs as exercises. Let $\mathbf{u} = (u_1, u_2, u_3)$ and $\mathbf{v} = (v_1, v_2, v_3)$; then

$$\begin{aligned}
k(\mathbf{u} \cdot \mathbf{v}) &= k(u_1 v_1 + u_2 v_2 + u_3 v_3) \\
&= (k u_1) v_1 + (k u_2) v_2 + (k u_3) v_3 \\
&= (k\mathbf{u}) \cdot \mathbf{v}
\end{aligned}$$

Similarly,

$$k(\mathbf{u} \cdot \mathbf{v}) = \mathbf{u} \cdot (k\mathbf{v}) \qquad\qquad \blacksquare$$

An Orthogonal Projection

In many applications it is of interest to "decompose" a vector $\mathbf{u}$ into a sum of two terms, one parallel to a specified nonzero vector $\mathbf{a}$ and the other perpendicular to $\mathbf{a}$. If $\mathbf{u}$ and $\mathbf{a}$ are positioned so that their initial points coincide at a point Q, we can decompose the vector $\mathbf{u}$ as follows (Figure 3.3.6): Drop a perpendicular from the tip of $\mathbf{u}$ to the line through $\mathbf{a}$, and construct the vector $\mathbf{w}_1$ from Q to the foot of this perpendicular. Next form the difference

$$\mathbf{w}_2 = \mathbf{u} - \mathbf{w}_1$$

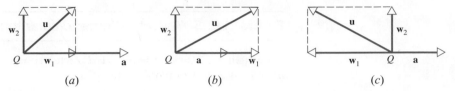

Figure 3.3.6 The vector $\mathbf{u}$ is the sum of $\mathbf{w}_1$ and $\mathbf{w}_2$, where $\mathbf{w}_1$ is parallel to $\mathbf{a}$ and $\mathbf{w}_2$ is perpendicular to $\mathbf{a}$.

As indicated in Figure 3.3.6, the vector $\mathbf{w}_1$ is parallel to $\mathbf{a}$, the vector $\mathbf{w}_2$ is perpendicular to $\mathbf{a}$, and

$$\mathbf{w}_1 + \mathbf{w}_2 = \mathbf{w}_1 + (\mathbf{u} - \mathbf{w}_1) = \mathbf{u}$$

The vector $\mathbf{w}_1$ is called the ***orthogonal projection of*** $\mathbf{u}$ ***on*** $\mathbf{a}$ or sometimes the ***vector component of*** $\mathbf{u}$ ***along*** $\mathbf{a}$. It is denoted by

$$\text{proj}_{\mathbf{a}}\mathbf{u} \tag{7}$$

The vector $\mathbf{w}_2$ is called the ***vector component of*** $\mathbf{u}$ ***orthogonal to*** $\mathbf{a}$. Since we have $\mathbf{w}_2 = \mathbf{u} - \mathbf{w}_1$, this vector can be written in notation (7) as

$$\mathbf{w}_2 = \mathbf{u} - \text{proj}_{\mathbf{a}}\mathbf{u}$$

The following theorem gives formulas for calculating $\text{proj}_{\mathbf{a}}\mathbf{u}$ and $\mathbf{u} - \text{proj}_{\mathbf{a}}\mathbf{u}$.

THEOREM 3.3.3

> *If* $\mathbf{u}$ *and* $\mathbf{a}$ *are vectors in 2-space or 3-space and if* $\mathbf{a} \neq \mathbf{0}$, *then*
>
> $$\text{proj}_{\mathbf{a}}\mathbf{u} = \frac{\mathbf{u} \cdot \mathbf{a}}{\|\mathbf{a}\|^2}\mathbf{a} \quad \text{(\textit{vector component of} \textbf{u} \textit{along} \textbf{a})}$$
>
> $$\mathbf{u} - \text{proj}_{\mathbf{a}}\mathbf{u} = \mathbf{u} - \frac{\mathbf{u} \cdot \mathbf{a}}{\|\mathbf{a}\|^2}\mathbf{a} \quad \text{(\textit{vector component of} \textbf{u} \textit{orthogonal to} \textbf{a})}$$

Proof Let $\mathbf{w}_1 = \text{proj}_{\mathbf{a}}\mathbf{u}$ and $\mathbf{w}_2 = \mathbf{u} - \text{proj}_{\mathbf{a}}\mathbf{u}$. Since $\mathbf{w}_1$ is parallel to $\mathbf{a}$, it must be a scalar multiple of $\mathbf{a}$, so it can be written in the form $\mathbf{w}_1 = k\mathbf{a}$. Thus

$$\mathbf{u} = \mathbf{w}_1 + \mathbf{w}_2 = k\mathbf{a} + \mathbf{w}_2 \tag{8}$$

Taking the dot product of both sides of (8) with $\mathbf{a}$ and using Theorems 3.3.1a and 3.3.2 yields

$$\mathbf{u} \cdot \mathbf{a} = (k\mathbf{a} + \mathbf{w}_2) \cdot \mathbf{a} = k\|\mathbf{a}\|^2 + \mathbf{w}_2 \cdot \mathbf{a} \tag{9}$$

But $\mathbf{w}_2 \cdot \mathbf{a} = 0$ since $\mathbf{w}_2$ is perpendicular to $\mathbf{a}$; so (9) yields

$$k = \frac{\mathbf{u} \cdot \mathbf{a}}{\|\mathbf{a}\|^2}$$

Since $\text{proj}_{\mathbf{a}}\mathbf{u} = \mathbf{w}_1 = k\mathbf{a}$, we obtain

$$\text{proj}_{\mathbf{a}}\mathbf{u} = \frac{\mathbf{u} \cdot \mathbf{a}}{\|\mathbf{a}\|^2}\mathbf{a} \qquad \blacksquare$$

EXAMPLE 6 Vector Component of u Along a

Let $\mathbf{u} = (2, -1, 3)$ and $\mathbf{a} = (4, -1, 2)$. Find the vector component of $\mathbf{u}$ along $\mathbf{a}$ and the vector component of $\mathbf{u}$ orthogonal to $\mathbf{a}$.

Solution

$$\mathbf{u} \cdot \mathbf{a} = (2)(4) + (-1)(-1) + (3)(2) = 15$$
$$\|\mathbf{a}\|^2 = 4^2 + (-1)^2 + 2^2 = 21$$

Thus the vector component of $\mathbf{u}$ along $\mathbf{a}$ is

$$\text{proj}_{\mathbf{a}}\mathbf{u} = \frac{\mathbf{u} \cdot \mathbf{a}}{\|\mathbf{a}\|^2}\mathbf{a} = \tfrac{15}{21}(4, -1, 2) = \left(\tfrac{20}{7}, -\tfrac{5}{7}, \tfrac{10}{7}\right)$$

and the vector component of $\mathbf{u}$ orthogonal to $\mathbf{a}$ is

$$\mathbf{u} - \text{proj}_{\mathbf{a}}\mathbf{u} = (2, -1, 3) - \left(\tfrac{20}{7}, -\tfrac{5}{7}, \tfrac{10}{7}\right) = \left(-\tfrac{6}{7}, -\tfrac{2}{7}, \tfrac{11}{7}\right)$$

As a check, the reader may wish to verify that the vectors $\mathbf{u} - \text{proj}_{\mathbf{a}}\mathbf{u}$ and $\mathbf{a}$ are perpendicular by showing that their dot product is zero. ◆

A formula for the length of the vector component of $\mathbf{u}$ along $\mathbf{a}$ can be obtained by writing

$$\|\text{proj}_{\mathbf{a}}\mathbf{u}\| = \left\|\frac{\mathbf{u} \cdot \mathbf{a}}{\|\mathbf{a}\|^2}\mathbf{a}\right\|$$

$$= \left|\frac{\mathbf{u} \cdot \mathbf{a}}{\|\mathbf{a}\|^2}\right| \|\mathbf{a}\| \qquad \longleftarrow \text{Formula (5) of Section 3.2}$$

$$= \frac{|\mathbf{u} \cdot \mathbf{a}|}{\|\mathbf{a}\|^2} \|\mathbf{a}\| \qquad \longleftarrow \text{Since } \|\mathbf{a}\|^2 > 0$$

which yields

$$\|\text{proj}_{\mathbf{a}}\mathbf{u}\| = \frac{|\mathbf{u} \cdot \mathbf{a}|}{\|\mathbf{a}\|} \qquad (10)$$

If θ denotes the angle between $\mathbf{u}$ and $\mathbf{a}$, then $\mathbf{u} \cdot \mathbf{a} = \|\mathbf{u}\|\|\mathbf{a}\| \cos\theta$, so (10) can also be written as

$$\|\text{proj}_{\mathbf{a}}\mathbf{u}\| = \|\mathbf{u}\| |\cos\theta| \qquad (11)$$

(Verify.) A geometric interpretation of this result is given in Figure 3.3.7.

As an example, we will use vector methods to derive a formula for the distance from a point in the plane to a line.

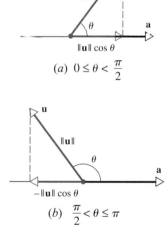

(a) $0 \leq \theta < \dfrac{\pi}{2}$

(b) $\dfrac{\pi}{2} < \theta \leq \pi$

Figure 3.3.7

EXAMPLE 7 Distance Between a Point and a Line

Find a formula for the distance D between point $P_0(x_0, y_0)$ and the line $ax + by + c = 0$.

Solution

Let $Q(x_1, y_1)$ be any point on the line, and position the vector $\mathbf{n} = (a, b)$ so that its initial point is at Q.

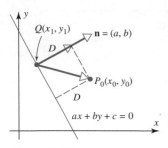

Figure 3.3.8

By virtue of Example 5, the vector $\mathbf{n}$ is perpendicular to the line (Figure 3.3.8). As indicated in the figure, the distance D is equal to the length of the orthogonal projection of $\overrightarrow{QP_0}$ on $\mathbf{n}$; thus, from (10),

$$D = \|\text{proj}_{\mathbf{n}}\overrightarrow{QP_0}\| = \frac{|\overrightarrow{QP_0} \cdot \mathbf{n}|}{\|\mathbf{n}\|}$$

But

$$\overrightarrow{QP_0} = (x_0 - x_1, y_0 - y_1)$$

$$\overrightarrow{QP_0} \cdot \mathbf{n} = a(x_0 - x_1) + b(y_0 - y_1)$$

$$\|\mathbf{n}\| = \sqrt{a^2 + b^2}$$

so

$$D = \frac{|a(x_0 - x_1) + b(y_0 - y_1)|}{\sqrt{a^2 + b^2}} \qquad (12)$$

Since the point $Q(x_1, y_1)$ lies on the line, its coordinates satisfy the equation of the line, so

$$ax_1 + by_1 + c = 0 \quad \text{or} \quad c = -ax_1 - by_1$$

Substituting this expression in (12) yields the formula

$$D = \frac{|ax_0 + by_0 + c|}{\sqrt{a^2 + b^2}} \qquad \blacklozenge \qquad (13)$$

EXAMPLE 8 Using the Distance Formula

It follows from Formula (13) that the distance D from the point $(1, -2)$ to the line $3x + 4y - 6 = 0$ is

$$D = \frac{|(3)(1) + 4(-2) - 6|}{\sqrt{3^2 + 4^2}} = \frac{|-11|}{\sqrt{25}} = \frac{11}{5} \qquad \blacklozenge$$

EXERCISE SET 3.3

1. Find $\mathbf{u} \cdot \mathbf{v}$.
   (a) $\mathbf{u} = (2, 3)$, $\mathbf{v} = (5, -7)$ (b) $\mathbf{u} = (-6, -2)$, $\mathbf{v} = (4, 0)$
   (c) $\mathbf{u} = (1, -5, 4)$, $\mathbf{v} = (3, 3, 3)$ (d) $\mathbf{u} = (-2, 2, 3)$, $\mathbf{v} = (1, 7, -4)$

2. In each part of Exercise 1, find the cosine of the angle θ between $\mathbf{u}$ and $\mathbf{v}$.

3. Determine whether $\mathbf{u}$ and $\mathbf{v}$ make an acute angle, make an obtuse angle, or are orthogonal.
   (a) $\mathbf{u} = (6, 1, 4)$, $\mathbf{v} = (2, 0, -3)$ (b) $\mathbf{u} = (0, 0, -1)$, $\mathbf{v} = (1, 1, 1)$
   (c) $\mathbf{u} = (-6, 0, 4)$, $\mathbf{v} = (3, 1, 6)$ (d) $\mathbf{u} = (2, 4, -8)$, $\mathbf{v} = (5, 3, 7)$

4. Find the orthogonal projection of $\mathbf{u}$ on $\mathbf{a}$.
   (a) $\mathbf{u} = (6, 2)$, $\mathbf{a} = (3, -9)$ (b) $\mathbf{u} = (-1, -2)$, $\mathbf{a} = (-2, 3)$
   (c) $\mathbf{u} = (3, 1, -7)$, $\mathbf{a} = (1, 0, 5)$ (d) $\mathbf{u} = (1, 0, 0)$, $\mathbf{a} = (4, 3, 8)$

5. In each part of Exercise 4, find the vector component of $\mathbf{u}$ orthogonal to $\mathbf{a}$.

6. In each part, find $\|\text{proj}_{\mathbf{a}}\mathbf{u}\|$.
   (a) $\mathbf{u} = (1, -2)$, $\mathbf{a} = (-4, -3)$ (b) $\mathbf{u} = (5, 6)$, $\mathbf{a} = (2, -1)$
   (c) $\mathbf{u} = (3, 0, 4)$, $\mathbf{a} = (2, 3, 3)$ (d) $\mathbf{u} = (3, -2, 6)$, $\mathbf{a} = (1, 2, -7)$

7. Let $\mathbf{u} = (5, -2, 1)$, $\mathbf{v} = (1, 6, 3)$, and $k = -4$. Verify Theorem 3.3.2 for these quantities.

8. (a) Show that $\mathbf{v} = (a, b)$ and $\mathbf{w} = (-b, a)$ are orthogonal vectors.

   (b) Use the result in part (a) to find two vectors that are orthogonal to $\mathbf{v} = (2, -3)$.

   (c) Find two unit vectors that are orthogonal to $(-3, 4)$.

9. Let $\mathbf{u} = (3, 4)$, $\mathbf{v} = (5, -1)$, and $\mathbf{w} = (7, 1)$. Evaluate the expressions.

   (a) $\mathbf{u} \cdot (7\mathbf{v} + \mathbf{w})$ (b) $\|(\mathbf{u} \cdot \mathbf{w})\mathbf{w}\|$ (c) $\|\mathbf{u}\|(\mathbf{v} \cdot \mathbf{w})$ (d) $(\|\mathbf{u}\|\mathbf{v}) \cdot \mathbf{w}$

10. Find five different nonzero vectors that are orthogonal to $\mathbf{u} = (5, -2, 3)$.

11. Use vectors to find the cosines of the interior angles of the triangle with vertices $(0, -1)$, $(1, -2)$, and $(4, 1)$.

12. Show that $A(3, 0, 2)$, $B(4, 3, 0)$, and $C(8, 1, -1)$ are vertices of a right triangle. At which vertex is the right angle?

13. Find a unit vector that is orthogonal to both $\mathbf{u} = (1, 0, 1)$ and $\mathbf{v} = (0, 1, 1)$.

14. A vector $\mathbf{a}$ in the xy-plane has a length of 9 units and points in a direction that is $120°$ counterclockwise from the positive x-axis, and a vector $\mathbf{b}$ in that plane has a length of 5 units and points in the positive y-direction. Find $\mathbf{a} \cdot \mathbf{b}$.

15. A vector $\mathbf{a}$ in the xy-plane points in a direction that is $47°$ counterclockwise from the positive x-axis, and a vector $\mathbf{b}$ in that plane points in a direction that is $43°$ clockwise from the positive x-axis. What can you say about the value of $\mathbf{a} \cdot \mathbf{b}$?

16. Let $\mathbf{p} = (2, k)$ and $\mathbf{q} = (3, 5)$. Find k such that

   (a) $\mathbf{p}$ and $\mathbf{q}$ are parallel   (b) $\mathbf{p}$ and $\mathbf{q}$ are orthogonal

   (c) the angle between $\mathbf{p}$ and $\mathbf{q}$ is $\pi/3$ (d) the angle between $\mathbf{p}$ and $\mathbf{q}$ is $\pi/4$

17. Use Formula (13) to calculate the distance between the point and the line.

   (a) $4x + 3y + 4 = 0$; $(-3, 1)$ (b) $y = -4x + 2$; $(2, -5)$ (c) $3x + y = 5$; $(1, 8)$

18. Establish the identity $\|\mathbf{u} + \mathbf{v}\|^2 + \|\mathbf{u} - \mathbf{v}\|^2 = 2\|\mathbf{u}\|^2 + 2\|\mathbf{v}\|^2$.

19. Establish the identity $\mathbf{u} \cdot \mathbf{v} = \frac{1}{4}\|\mathbf{u} + \mathbf{v}\|^2 - \frac{1}{4}\|\mathbf{u} - \mathbf{v}\|^2$.

20. Find the angle between a diagonal of a cube and one of its faces.

21. Let $\mathbf{i}$, $\mathbf{j}$, and $\mathbf{k}$ be unit vectors along the positive x, y, and z axes of a rectangular coordinate system in 3-space. If $\mathbf{v} = (a, b, c)$ is a nonzero vector, then the angles α, β, and γ between $\mathbf{v}$ and the vectors $\mathbf{i}$, $\mathbf{j}$, and $\mathbf{k}$, respectively, are called the ***direction angles*** of $\mathbf{v}$ (see accompanying figure), and the numbers $\cos\alpha$, $\cos\beta$, and $\cos\gamma$ are called the ***direction cosines*** of $\mathbf{v}$.

   (a) Show that $\cos\alpha = a/\|\mathbf{v}\|$.

   (b) Find $\cos\beta$ and $\cos\gamma$.

   (c) Show that $\mathbf{v}/\|\mathbf{v}\| = (\cos\alpha, \cos\beta, \cos\gamma)$.

   (d) Show that $\cos^2\alpha + \cos^2\beta + \cos^2\gamma = 1$.

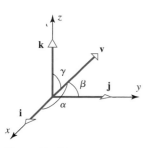

Figure Ex-21

22. Use the result in Exercise 21 to estimate, to the nearest degree, the angles that a diagonal of a box with dimensions 10 cm $\times$ 15 cm $\times$ 25 cm makes with the edges of the box.

   Note   A calculator is needed.

23. Referring to Exercise 21, show that two nonzero vectors, $\mathbf{v}_1$ and $\mathbf{v}_2$, in 3-space are perpendicular if and only if their direction cosines satisfy

$$\cos\alpha_1 \cos\alpha_2 + \cos\beta_1 \cos\beta_2 + \cos\gamma_1 \cos\gamma_2 = 0$$

24. (a) Find the area of the triangle with vertices $A(2, 3)$, $C(4, 7)$, and $D(-5, 8)$.

   (b) Find the coordinates of the point B such that the quadrilateral $ABCD$ is a parallelogram. What is the area of this parallelogram?

25. Show that if $\mathbf{v}$ is orthogonal to both $\mathbf{w}_1$ and $\mathbf{w}_2$, then $\mathbf{v}$ is orthogonal to $k_1\mathbf{w}_1 + k_2\mathbf{w}_2$ for all scalars k_1 and k_2.

26. Let **u** and **v** be nonzero vectors in 2- or 3-space, and let $k = \|\mathbf{u}\|$ and $l = \|\mathbf{v}\|$. Show that the vector $\mathbf{w} = l\mathbf{u} + k\mathbf{v}$ bisects the angle between **u** and **v**.

Discussion & Discovery

27. In each part, something is wrong with the expression. What?

    (a) $\mathbf{u} \cdot (\mathbf{v} \cdot \mathbf{w})$ (b) $(\mathbf{u} \cdot \mathbf{v}) + \mathbf{w}$ (c) $\|\mathbf{u} \cdot \mathbf{v}\|$ (d) $k \cdot (\mathbf{u} + \mathbf{v})$

28. Is it possible to have $\text{proj}_{\mathbf{a}}\mathbf{u} = \text{proj}_{\mathbf{u}}\mathbf{a}$? Explain your reasoning.

29. If $\mathbf{u} \neq \mathbf{0}$, is it valid to cancel **u** from both sides of the equation $\mathbf{u} \cdot \mathbf{v} = \mathbf{u} \cdot \mathbf{w}$ and conclude that $\mathbf{v} = \mathbf{w}$? Explain your reasoning.

30. Suppose that **u**, **v**, and **w** are mutually orthogonal nonzero vectors in 3-space, and suppose that you know the dot products of these vectors with a vector **r** in 3-space. Find an expression for **r** in terms of **u**, **v**, **w**, and the dot products.

    Hint Look for an expression of the form $\mathbf{r} = c_1\mathbf{u} + c_2\mathbf{v} + c_3\mathbf{w}$.

31. Suppose that **u** and **v** are orthogonal vectors in 2-space or 3-space. What famous theorem is described by the equation $\|\mathbf{u} + \mathbf{v}\|^2 = \|\mathbf{u}\|^2 + \|\mathbf{v}\|^2$? Draw a picture to support your answer.

3.4
CROSS PRODUCT

In many applications of vectors to problems in geometry, physics, and engineering, it is of interest to construct a vector in 3-space that is perpendicular to two given vectors. In this section we shall show how to do this.

Cross Product of Vectors

Recall from Section 3.3 that the dot product of two vectors in 2-space or 3-space produces a scalar. We will now define a type of vector multiplication that produces a vector as the product but that is applicable only in 3-space.

> **DEFINITION**
>
> If $\mathbf{u} = (u_1, u_2, u_3)$ and $\mathbf{v} = (v_1, v_2, v_3)$ are vectors in 3-space, then the **cross product** $\mathbf{u} \times \mathbf{v}$ is the vector defined by
>
> $$\mathbf{u} \times \mathbf{v} = (u_2v_3 - u_3v_2, \, u_3v_1 - u_1v_3, \, u_1v_2 - u_2v_1)$$
>
> or, in determinant notation,
>
> $$\mathbf{u} \times \mathbf{v} = \left(\begin{vmatrix} u_2 & u_3 \\ v_2 & v_3 \end{vmatrix}, \, -\begin{vmatrix} u_1 & u_3 \\ v_1 & v_3 \end{vmatrix}, \, \begin{vmatrix} u_1 & u_2 \\ v_1 & v_2 \end{vmatrix} \right) \tag{1}$$

REMARK   Instead of memorizing (1), you can obtain the components of $\mathbf{u} \times \mathbf{v}$ as follows:

- Form the 2×3 matrix $\begin{bmatrix} u_1 & u_2 & u_3 \\ v_1 & v_2 & v_3 \end{bmatrix}$ whose first row contains the components of **u** and whose second row contains the components of **v**.

- To find the first component of $\mathbf{u} \times \mathbf{v}$, delete the first column and take the determinant; to find the second component, delete the second column and take the negative of the determinant; and to find the third component, delete the third column and take the determinant.

EXAMPLE 1 Calculating a Cross Product

Find $\mathbf{u} \times \mathbf{v}$, where $\mathbf{u} = (1, 2, -2)$ and $\mathbf{v} = (3, 0, 1)$.

Solution

From either (1) or the mnemonic in the preceding remark, we have

$$\mathbf{u} \times \mathbf{v} = \left(\begin{vmatrix} 2 & -2 \\ 0 & 1 \end{vmatrix}, -\begin{vmatrix} 1 & -2 \\ 3 & 1 \end{vmatrix}, \begin{vmatrix} 1 & 2 \\ 3 & 0 \end{vmatrix} \right)$$

$$= (2, -7, -6) \ \blacklozenge$$

There is an important difference between the dot product and cross product of two vectors—the dot product is a scalar and the cross product is a vector. The following theorem gives some important relationships between the dot product and cross product and also shows that $\mathbf{u} \times \mathbf{v}$ is orthogonal to both $\mathbf{u}$ and $\mathbf{v}$.

THEOREM 3.4.1

> ### Relationships Involving Cross Product and Dot Product
>
> *If* $\mathbf{u}$, $\mathbf{v}$, *and* $\mathbf{w}$ *are vectors in 3-space, then*
>
> (a) $\mathbf{u} \cdot (\mathbf{u} \times \mathbf{v}) = 0$ ($\mathbf{u} \times \mathbf{v}$ *is orthogonal to* $\mathbf{u}$)
> (b) $\mathbf{v} \cdot (\mathbf{u} \times \mathbf{v}) = 0$ ($\mathbf{u} \times \mathbf{v}$ *is orthogonal to* $\mathbf{v}$)
> (c) $\|\mathbf{u} \times \mathbf{v}\|^2 = \|\mathbf{u}\|^2 \|\mathbf{v}\|^2 - (\mathbf{u} \cdot \mathbf{v})^2$ (*Lagrange's identity*)
> (d) $\mathbf{u} \times (\mathbf{v} \times \mathbf{w}) = (\mathbf{u} \cdot \mathbf{w})\mathbf{v} - (\mathbf{u} \cdot \mathbf{v})\mathbf{w}$ (*relationship between cross and dot products*)
> (e) $(\mathbf{u} \times \mathbf{v}) \times \mathbf{w} = (\mathbf{u} \cdot \mathbf{w})\mathbf{v} - (\mathbf{v} \cdot \mathbf{w})\mathbf{u}$ (*relationship between cross and dot products*)

Proof (a) Let $\mathbf{u} = (u_1, u_2, u_3)$ and $\mathbf{v} = (v_1, v_2, v_3)$. Then

$$\mathbf{u} \cdot (\mathbf{u} \times \mathbf{v}) = (u_1, u_2, u_3) \cdot (u_2 v_3 - u_3 v_2, u_3 v_1 - u_1 v_3, u_1 v_2 - u_2 v_1)$$

$$= u_1(u_2 v_3 - u_3 v_2) + u_2(u_3 v_1 - u_1 v_3) + u_3(u_1 v_2 - u_2 v_1) = 0$$

Proof (b) Similar to (a).

Proof (c) Since

$$\|\mathbf{u} \times \mathbf{v}\|^2 = (u_2 v_3 - u_3 v_2)^2 + (u_3 v_1 - u_1 v_3)^2 + (u_1 v_2 - u_2 v_1)^2 \tag{2}$$

and

$$\|\mathbf{u}\|^2 \|\mathbf{v}\|^2 - (\mathbf{u} \cdot \mathbf{v})^2 = (u_1^2 + u_2^2 + u_3^2)(v_1^2 + v_2^2 + v_3^2) - (u_1 v_1 + u_2 v_2 + u_3 v_3)^2 \tag{3}$$

the proof can be completed by "multiplying out" the right sides of (2) and (3) and verifying their equality.

Proof (d) and (e) See Exercises 26 and 27.   ■

EXAMPLE 2 $\mathbf{u} \times \mathbf{v}$ Is Perpendicular to $\mathbf{u}$ and to $\mathbf{v}$

Consider the vectors

$$\mathbf{u} = (1, 2, -2) \quad \text{and} \quad \mathbf{v} = (3, 0, 1)$$

Joseph Louis Lagrange *(1736–1813)* was a French-Italian mathematician and astronomer. Although his father wanted him to become a lawyer, Lagrange was attracted to mathematics and astronomy after reading a memoir by the astronomer Halley. At age 16 he began to study mathematics on his own and by age 19 was appointed to a professorship at the Royal Artillery School in Turin. The following year he solved some famous problems using new methods that eventually blossomed into a branch of mathematics called the *calculus of variations*. These methods and Lagrange's applications of them to problems in celestial mechanics were so monumental that by age 25 he was regarded by many of his contemporaries as the greatest living mathematician. One of Lagrange's most famous works is a memoir, *Mécanique Analytique*, in which he reduced the theory of mechanics to a few general formulas from which all other necessary equations could be derived.

Napoleon was a great admirer of Lagrange and showered him with many honors. In spite of his fame, Lagrange was a shy and modest man. On his death, he was buried with honor in the Pantheon.

In Example 1 we showed that

$$\mathbf{u} \times \mathbf{v} = (2, -7, -6)$$

Since

$$\mathbf{u} \cdot (\mathbf{u} \times \mathbf{v}) = (1)(2) + (2)(-7) + (-2)(-6) = 0$$

and

$$\mathbf{v} \cdot (\mathbf{u} \times \mathbf{v}) = (3)(2) + (0)(-7) + (1)(-6) = 0$$

$\mathbf{u} \times \mathbf{v}$ is orthogonal to both $\mathbf{u}$ and $\mathbf{v}$, as guaranteed by Theorem 3.4.1. ◆

The main arithmetic properties of the cross product are listed in the next theorem.

THEOREM 3.4.2

Properties of Cross Product

If $\mathbf{u}$, $\mathbf{v}$, and $\mathbf{w}$ are any vectors in 3-space and k is any scalar, then

(a) $\mathbf{u} \times \mathbf{v} = -(\mathbf{v} \times \mathbf{u})$
(b) $\mathbf{u} \times (\mathbf{v} + \mathbf{w}) = (\mathbf{u} \times \mathbf{v}) + (\mathbf{u} \times \mathbf{w})$
(c) $(\mathbf{u} + \mathbf{v}) \times \mathbf{w} = (\mathbf{u} \times \mathbf{w}) + (\mathbf{v} \times \mathbf{w})$
(d) $k(\mathbf{u} \times \mathbf{v}) = (k\mathbf{u}) \times \mathbf{v} = \mathbf{u} \times (k\mathbf{v})$
(e) $\mathbf{u} \times \mathbf{0} = \mathbf{0} \times \mathbf{u} = \mathbf{0}$
(f) $\mathbf{u} \times \mathbf{u} = \mathbf{0}$

The proofs follow immediately from Formula (1) and properties of determinants; for example, (a) can be proved as follows:

Proof (a) Interchanging $\mathbf{u}$ and $\mathbf{v}$ in (1) interchanges the rows of the three determinants on the right side of (1) and hence changes the sign of each component in the cross product. Thus $\mathbf{u} \times \mathbf{v} = -(\mathbf{v} \times \mathbf{u})$. ■

The proofs of the remaining parts are left as exercises.

EXAMPLE 3   Standard Unit Vectors

Consider the vectors

$$\mathbf{i} = (1, 0, 0), \qquad \mathbf{j} = (0, 1, 0), \qquad \mathbf{k} = (0, 0, 1)$$

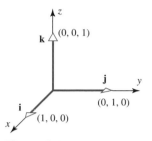

Figure 3.4.1   The standard unit vectors.

These vectors each have length 1 and lie along the coordinate axes (Figure 3.4.1). They are called the ***standard unit vectors*** in 3-space. Every vector $\mathbf{v} = (v_1, v_2, v_3)$ in 3-space is expressible in terms of $\mathbf{i}$, $\mathbf{j}$, and $\mathbf{k}$ since we can write

$$\mathbf{v} = (v_1, v_2, v_3) = v_1(1, 0, 0) + v_2(0, 1, 0) + v_3(0, 0, 1) = v_1\mathbf{i} + v_2\mathbf{j} + v_3\mathbf{k}$$

For example,

$$(2, -3, 4) = 2\mathbf{i} - 3\mathbf{j} + 4\mathbf{k}$$

From (1) we obtain

$$\mathbf{i} \times \mathbf{j} = \left(\begin{vmatrix} 0 & 0 \\ 1 & 0 \end{vmatrix}, -\begin{vmatrix} 1 & 0 \\ 0 & 0 \end{vmatrix}, \begin{vmatrix} 1 & 0 \\ 0 & 1 \end{vmatrix} \right) = (0, 0, 1) = \mathbf{k} \quad \blacklozenge$$

The reader should have no trouble obtaining the following results:

Figure 3.4.2

$$\mathbf{i} \times \mathbf{i} = \mathbf{0} \qquad \mathbf{j} \times \mathbf{j} = \mathbf{0} \qquad \mathbf{k} \times \mathbf{k} = \mathbf{0}$$
$$\mathbf{i} \times \mathbf{j} = \mathbf{k} \qquad \mathbf{j} \times \mathbf{k} = \mathbf{i} \qquad \mathbf{k} \times \mathbf{i} = \mathbf{j}$$
$$\mathbf{j} \times \mathbf{i} = -\mathbf{k} \qquad \mathbf{k} \times \mathbf{j} = -\mathbf{i} \qquad \mathbf{i} \times \mathbf{k} = -\mathbf{j}$$

Figure 3.4.2 is helpful for remembering these results. Referring to this diagram, the cross product of two consecutive vectors going clockwise is the next vector around, and the cross product of two consecutive vectors going counterclockwise is the negative of the next vector around.

Determinant Form of Cross Product

It is also worth noting that a cross product can be represented symbolically in the form of a formal 3×3 determinant:

$$\mathbf{u} \times \mathbf{v} = \begin{vmatrix} \mathbf{i} & \mathbf{j} & \mathbf{k} \\ u_1 & u_2 & u_3 \\ v_1 & v_2 & v_3 \end{vmatrix} = \begin{vmatrix} u_2 & u_3 \\ v_2 & v_3 \end{vmatrix} \mathbf{i} - \begin{vmatrix} u_1 & u_3 \\ v_1 & v_3 \end{vmatrix} \mathbf{j} + \begin{vmatrix} u_1 & u_2 \\ v_1 & v_2 \end{vmatrix} \mathbf{k} \qquad (4)$$

For example, if $\mathbf{u} = (1, 2, -2)$ and $\mathbf{v} = (3, 0, 1)$, then

$$\mathbf{u} \times \mathbf{v} = \begin{vmatrix} \mathbf{i} & \mathbf{j} & \mathbf{k} \\ 1 & 2 & -2 \\ 3 & 0 & 1 \end{vmatrix} = 2\mathbf{i} - 7\mathbf{j} - 6\mathbf{k}$$

which agrees with the result obtained in Example 1.

WARNING   It is not true in general that $\mathbf{u} \times (\mathbf{v} \times \mathbf{w}) = (\mathbf{u} \times \mathbf{v}) \times \mathbf{w}$. For example,

$$\mathbf{i} \times (\mathbf{j} \times \mathbf{j}) = \mathbf{i} \times \mathbf{0} = \mathbf{0}$$

and

$$(\mathbf{i} \times \mathbf{j}) \times \mathbf{j} = \mathbf{k} \times \mathbf{j} = -\mathbf{i}$$

so

$$\mathbf{i} \times (\mathbf{j} \times \mathbf{j}) \neq (\mathbf{i} \times \mathbf{j}) \times \mathbf{j}$$

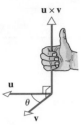

Figure 3.4.3

Geometric Interpretation of Cross Product

We know from Theorem 3.4.1 that $\mathbf{u} \times \mathbf{v}$ is orthogonal to both $\mathbf{u}$ and $\mathbf{v}$. If $\mathbf{u}$ and $\mathbf{v}$ are nonzero vectors, it can be shown that the direction of $\mathbf{u} \times \mathbf{v}$ can be determined using the following "right-hand rule"[†] (Figure 3.4.3): Let θ be the angle between $\mathbf{u}$ and $\mathbf{v}$, and suppose $\mathbf{u}$ is rotated through the angle θ until it coincides with $\mathbf{v}$. If the fingers of the right hand are cupped so that they point in the direction of rotation, then the thumb indicates (roughly) the direction of $\mathbf{u} \times \mathbf{v}$.

The reader may find it instructive to practice this rule with the products

$$\mathbf{i} \times \mathbf{j} = \mathbf{k}, \qquad \mathbf{j} \times \mathbf{k} = \mathbf{i}, \qquad \mathbf{k} \times \mathbf{i} = \mathbf{j}$$

If $\mathbf{u}$ and $\mathbf{v}$ are vectors in 3-space, then the norm of $\mathbf{u} \times \mathbf{v}$ has a useful geometric interpretation. Lagrange's identity, given in Theorem 3.4.1, states that

$$\|\mathbf{u} \times \mathbf{v}\|^2 = \|\mathbf{u}\|^2\|\mathbf{v}\|^2 - (\mathbf{u} \cdot \mathbf{v})^2 \tag{5}$$

If θ denotes the angle between $\mathbf{u}$ and $\mathbf{v}$, then $\mathbf{u} \cdot \mathbf{v} = \|\mathbf{u}\|\|\mathbf{v}\| \cos\theta$, so (5) can be rewritten as

$$\begin{aligned} \|\mathbf{u} \times \mathbf{v}\|^2 &= \|\mathbf{u}\|^2\|\mathbf{v}\|^2 - \|\mathbf{u}\|^2\|\mathbf{v}\|^2 \cos^2\theta \\ &= \|\mathbf{u}\|^2\|\mathbf{v}\|^2(1 - \cos^2\theta) \\ &= \|\mathbf{u}\|^2\|\mathbf{v}\|^2 \sin^2\theta \end{aligned}$$

Since $0 \le \theta \le \pi$, it follows that $\sin\theta \ge 0$, so this can be rewritten as

$$\|\mathbf{u} \times \mathbf{v}\| = \|\mathbf{u}\|\|\mathbf{v}\| \sin\theta \tag{6}$$

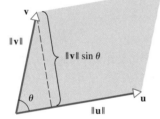

Figure 3.4.4

But $\|\mathbf{v}\| \sin\theta$ is the altitude of the parallelogram determined by $\mathbf{u}$ and $\mathbf{v}$ (Figure 3.4.4). Thus, from (6), the area A of this parallelogram is given by

$$A = (\text{base})(\text{altitude}) = \|\mathbf{u}\|\|\mathbf{v}\| \sin\theta = \|\mathbf{u} \times \mathbf{v}\|$$

This result is even correct if $\mathbf{u}$ and $\mathbf{v}$ are collinear, since the parallelogram determined by $\mathbf{u}$ and $\mathbf{v}$ has zero area and from (6) we have $\mathbf{u} \times \mathbf{v} = \mathbf{0}$ because $\theta = 0$ in this case. Thus we have the following theorem.

THEOREM 3.4.3

> ### Area of a Parallelogram
>
> *If $\mathbf{u}$ and $\mathbf{v}$ are vectors in 3-space, then $\|\mathbf{u} \times \mathbf{v}\|$ is equal to the area of the parallelogram determined by $\mathbf{u}$ and $\mathbf{v}$.*

EXAMPLE 4 Area of a Triangle

Find the area of the triangle determined by the points $P_1(2, 2, 0)$, $P_2(-1, 0, 2)$, and $P_3(0, 4, 3)$.

Solution

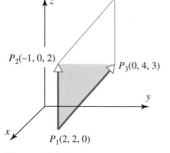

Figure 3.4.5

The area A of the triangle is $\frac{1}{2}$ the area of the parallelogram determined by the vectors $\overrightarrow{P_1P_2}$ and $\overrightarrow{P_1P_3}$ (Figure 3.4.5). Using the method discussed in Example 2 of Section 3.1,

---

[†]Recall that we agreed to consider only right-handed coordinate systems in this text. Had we used left-handed systems instead, a "left-hand rule" would apply here.

$\overrightarrow{P_1 P_2} = (-3, -2, 2)$ and $\overrightarrow{P_1 P_3} = (-2, 2, 3)$. It follows that

$$\overrightarrow{P_1 P_2} \times \overrightarrow{P_1 P_3} = (-10, 5, -10)$$

and consequently,

$$A = \tfrac{1}{2} \| \overrightarrow{P_1 P_2} \times \overrightarrow{P_1 P_3} \| = \tfrac{1}{2}(15) = \tfrac{15}{2} \quad \blacklozenge$$

DEFINITION

If $\mathbf{u}$, $\mathbf{v}$, and $\mathbf{w}$ are vectors in 3-space, then

$$\mathbf{u} \cdot (\mathbf{v} \times \mathbf{w})$$

is called the *scalar triple product* of $\mathbf{u}$, $\mathbf{v}$, and $\mathbf{w}$.

The scalar triple product of $\mathbf{u} = (u_1, u_2, u_3)$, $\mathbf{v} = (v_1, v_2, v_3)$, and $\mathbf{w} = (w_1, w_2, w_3)$ can be calculated from the formula

$$\mathbf{u} \cdot (\mathbf{v} \times \mathbf{w}) = \begin{vmatrix} u_1 & u_2 & u_3 \\ v_1 & v_2 & v_3 \\ w_1 & w_2 & w_3 \end{vmatrix} \tag{7}$$

This follows from Formula (4) since

$$\mathbf{u} \cdot (\mathbf{v} \times \mathbf{w}) = \mathbf{u} \cdot \left(\begin{vmatrix} v_2 & v_3 \\ w_2 & w_3 \end{vmatrix} \mathbf{i} - \begin{vmatrix} v_1 & v_3 \\ w_1 & w_3 \end{vmatrix} \mathbf{j} + \begin{vmatrix} v_1 & v_2 \\ w_1 & w_2 \end{vmatrix} \mathbf{k} \right)$$

$$= \begin{vmatrix} v_2 & v_3 \\ w_2 & w_3 \end{vmatrix} u_1 - \begin{vmatrix} v_1 & v_3 \\ w_1 & w_3 \end{vmatrix} u_2 + \begin{vmatrix} v_1 & v_2 \\ w_1 & w_2 \end{vmatrix} u_3$$

$$= \begin{vmatrix} u_1 & u_2 & u_3 \\ v_1 & v_2 & v_3 \\ w_1 & w_2 & w_3 \end{vmatrix}$$

EXAMPLE 5 Calculating a Scalar Triple Product

Calculate the scalar triple product $\mathbf{u} \cdot (\mathbf{v} \times \mathbf{w})$ of the vectors

$$\mathbf{u} = 3\mathbf{i} - 2\mathbf{j} - 5\mathbf{k}, \qquad \mathbf{v} = \mathbf{i} + 4\mathbf{j} - 4\mathbf{k}, \qquad \mathbf{w} = 3\mathbf{j} + 2\mathbf{k}$$

Solution

From (7),

$$\mathbf{u} \cdot (\mathbf{v} \times \mathbf{w}) = \begin{vmatrix} 3 & -2 & -5 \\ 1 & 4 & -4 \\ 0 & 3 & 2 \end{vmatrix}$$

$$= 3 \begin{vmatrix} 4 & -4 \\ 3 & 2 \end{vmatrix} - (-2) \begin{vmatrix} 1 & -4 \\ 0 & 2 \end{vmatrix} + (-5) \begin{vmatrix} 1 & 4 \\ 0 & 3 \end{vmatrix}$$

$$= 60 + 4 - 15 = 49 \quad \blacklozenge$$

REMARK The symbol $(\mathbf{u} \cdot \mathbf{v}) \times \mathbf{w}$ makes no sense because we cannot form the cross product of a scalar and a vector. Thus no ambiguity arises if we write $\mathbf{u} \cdot \mathbf{v} \times \mathbf{w}$ rather than $\mathbf{u} \cdot (\mathbf{v} \times \mathbf{w})$. However, for clarity we shall usually keep the parentheses.

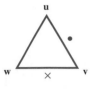

It follows from (7) that

$$\mathbf{u} \cdot (\mathbf{v} \times \mathbf{w}) = \mathbf{w} \cdot (\mathbf{u} \times \mathbf{v}) = \mathbf{v} \cdot (\mathbf{w} \times \mathbf{u})$$

Figure 3.4.6

since the 3×3 determinants that represent these products can be obtained from one another by *two* row interchanges. (Verify.) These relationships can be remembered by moving the vectors $\mathbf{u}$, $\mathbf{v}$, and $\mathbf{w}$ clockwise around the vertices of the triangle in Figure 3.4.6.

Geometric Interpretation of Determinants

The next theorem provides a useful geometric interpretation of 2×2 and 3×3 determinants.

THEOREM 3.4.4

(a) *The absolute value of the determinant*

$$\det \begin{bmatrix} u_1 & u_2 \\ v_1 & v_2 \end{bmatrix}$$

is equal to the area of the parallelogram in 2-space determined by the vectors $\mathbf{u} = (u_1, u_2)$ *and* $\mathbf{v} = (v_1, v_2)$. (*See Figure 3.4.7a.*)

(b) *The absolute value of the determinant*

$$\det \begin{bmatrix} u_1 & u_2 & u_3 \\ v_1 & v_2 & v_3 \\ w_1 & w_2 & w_3 \end{bmatrix}$$

is equal to the volume of the parallelepiped in 3-space determined by the vectors $\mathbf{u} = (u_1, u_2, u_3)$, $\mathbf{v} = (v_1, v_2, v_3)$, *and* $\mathbf{w} = (w_1, w_2, w_3)$. (*See Figure 3.4.7b.*)

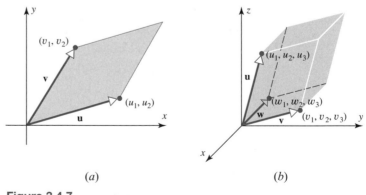

(a)   (b)

Figure 3.4.7

Proof (a) The key to the proof is to use Theorem 3.4.3. However, that theorem applies to vectors in 3-space, whereas $\mathbf{u} = (u_1, u_2)$ and $\mathbf{v} = (v_1, v_2)$ are vectors in 2-space. To circumvent this "dimension problem," we shall view $\mathbf{u}$ and $\mathbf{v}$ as vectors in the xy-plane of an xyz-coordinate system (Figure 3.4.8a), in which case these vectors are expressed

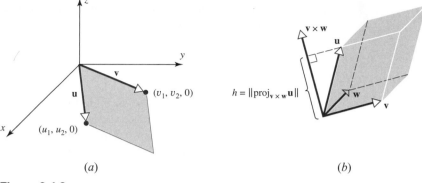

(a) (b)

Figure 3.4.8

as $\mathbf{u} = (u_1, u_2, 0)$ and $\mathbf{v} = (v_1, v_2, 0)$. Thus

$$\mathbf{u} \times \mathbf{v} = \begin{vmatrix} \mathbf{i} & \mathbf{j} & \mathbf{k} \\ u_1 & u_2 & 0 \\ v_1 & v_2 & 0 \end{vmatrix} = \begin{vmatrix} u_1 & u_2 \\ v_1 & v_2 \end{vmatrix} \mathbf{k} = \det \begin{bmatrix} u_1 & u_2 \\ v_1 & v_2 \end{bmatrix} \mathbf{k}$$

It now follows from Theorem 3.4.3 and the fact that $\|\mathbf{k}\| = 1$ that the area A of the parallelogram determined by $\mathbf{u}$ and $\mathbf{v}$ is

$$A = \|\mathbf{u} \times \mathbf{v}\| = \left\| \det \begin{bmatrix} u_1 & u_2 \\ v_1 & v_2 \end{bmatrix} \mathbf{k} \right\| = \left| \det \begin{bmatrix} u_1 & u_2 \\ v_1 & v_2 \end{bmatrix} \right| \|\mathbf{k}\| = \left| \det \begin{bmatrix} u_1 & u_2 \\ v_1 & v_2 \end{bmatrix} \right|$$

which completes the proof.

Proof (b)   As shown in Figure 3.4.8*b*, take the base of the parallelepiped determined by $\mathbf{u}$, $\mathbf{v}$, and $\mathbf{w}$ to be the parallelogram determined by $\mathbf{v}$ and $\mathbf{w}$. It follows from Theorem 3.4.3 that the area of the base is $\|\mathbf{v} \times \mathbf{w}\|$ and, as illustrated in Figure 3.4.8*b*, the height h of the parallelepiped is the length of the orthogonal projection of $\mathbf{u}$ on $\mathbf{v} \times \mathbf{w}$. Therefore, by Formula (10) of Section 3.3,

$$h = \|\mathrm{proj}_{\mathbf{v} \times \mathbf{w}} \mathbf{u}\| = \frac{|\mathbf{u} \cdot (\mathbf{v} \times \mathbf{w})|}{\|\mathbf{v} \times \mathbf{w}\|}$$

It follows that the volume V of the parallelepiped is

$$V = (\text{area of base}) \cdot \text{height} = \|\mathbf{v} \times \mathbf{w}\| \frac{|\mathbf{u} \cdot (\mathbf{v} \times \mathbf{w})|}{\|\mathbf{v} \times \mathbf{w}\|} = |\mathbf{u} \cdot (\mathbf{v} \times \mathbf{w})|$$

so from (7),

$$V = \left| \det \begin{bmatrix} u_1 & u_2 & u_3 \\ v_1 & v_2 & v_3 \\ w_1 & w_2 & w_3 \end{bmatrix} \right|$$

which completes the proof.   ∎

Remark   If V denotes the volume of the parallelepiped determined by vectors $\mathbf{u}$, $\mathbf{v}$, and $\mathbf{w}$, then it follows from Theorem 3.3 and Formula (7) that

$$V = \begin{bmatrix} \text{volume of parallelepiped} \\ \text{determined by } \mathbf{u}, \mathbf{v}, \text{ and } \mathbf{w} \end{bmatrix} = |\mathbf{u} \cdot (\mathbf{v} \times \mathbf{w})| \qquad (8)$$

From this and Theorem 3.3.1b, we can conclude that

$$\mathbf{u} \cdot (\mathbf{v} \times \mathbf{w}) = \pm V$$

where the $+$ or $-$ results depending on whether $\mathbf{u}$ makes an acute or an obtuse angle with $\mathbf{v} \times \mathbf{w}$.

Formula (8) leads to a useful test for ascertaining whether three given vectors lie in the same plane. Since three vectors not in the same plane determine a parallelepiped of positive volume, it follows from (8) that $|\mathbf{u} \cdot (\mathbf{v} \times \mathbf{w})| = 0$ if and only if the vectors $\mathbf{u}$, $\mathbf{v}$, and $\mathbf{w}$ lie in the same plane. Thus we have the following result.

THEOREM 3.4.5

If the vectors $\mathbf{u} = (u_1, u_2, u_3)$, $\mathbf{v} = (v_1, v_2, v_3)$, *and* $\mathbf{w} = (w_1, w_2, w_3)$ *have the same initial point, then they lie in the same plane if and only if*

$$\mathbf{u} \cdot (\mathbf{v} \times \mathbf{w}) = \begin{vmatrix} u_1 & u_2 & u_3 \\ v_1 & v_2 & v_3 \\ w_1 & w_2 & w_3 \end{vmatrix} = 0$$

Independence of Cross Product and Coordinates

Initially, we defined a vector to be a directed line segment or arrow in 2-space or 3-space; coordinate systems and components were introduced later in order to simplify computations with vectors. Thus, a vector has a "mathematical existence" regardless of whether a coordinate system has been introduced. Further, the components of a vector are not determined by the vector alone; they depend as well on the coordinate system chosen. For example, in Figure 3.4.9 we have indicated a fixed vector $\mathbf{v}$ in the plane and two different coordinate systems. In the xy-coordinate system the components of $\mathbf{v}$ are $(1, 1)$, and in the $x'y'$-system they are $(\sqrt{2}, 0)$.

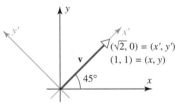

Figure 3.4.9

This raises an important question about our definition of cross product. Since we defined the cross product $\mathbf{u} \times \mathbf{v}$ in terms of the components of $\mathbf{u}$ and $\mathbf{v}$, and since these components depend on the coordinate system chosen, it seems possible that two *fixed* vectors $\mathbf{u}$ and $\mathbf{v}$ might have different cross products in different coordinate systems. Fortunately, this is not the case. To see that this is so, we need only recall that

* $\mathbf{u} \times \mathbf{v}$ is perpendicular to both $\mathbf{u}$ and $\mathbf{v}$.
* The orientation of $\mathbf{u} \times \mathbf{v}$ is determined by the right-hand rule.
* $\|\mathbf{u} \times \mathbf{v}\| = \|\mathbf{u}\| \|\mathbf{v}\| \sin \theta$.

These three properties completely determine the vector $\mathbf{u} \times \mathbf{v}$: the first and second properties determine the direction, and the third property determines the length. Since these properties of $\mathbf{u} \times \mathbf{v}$ depend only on the lengths and relative positions of $\mathbf{u}$ and $\mathbf{v}$ and not on the particular right-hand coordinate system being used, the vector $\mathbf{u} \times \mathbf{v}$ will remain unchanged if a different right-hand coordinate system is introduced. We say that the definition of $\mathbf{u} \times \mathbf{v}$ is ***coordinate free***. This result is of importance to physicists and engineers who often work with many coordinate systems in the same problem.

EXAMPLE 6   u × v Is Independent of the Coordinate System

Consider two perpendicular vectors $\mathbf{u}$ and $\mathbf{v}$, each of length 1 (Figure 3.4.10*a*). If we introduce an xyz-coordinate system as shown in Figure 3.4.10*b*, then

$$\mathbf{u} = (1, 0, 0) = \mathbf{i} \quad \text{and} \quad \mathbf{v} = (0, 1, 0) = \mathbf{j}$$

so that

$$\mathbf{u} \times \mathbf{v} = \mathbf{i} \times \mathbf{j} = \mathbf{k} = (0, 0, 1)$$

However, if we introduce an $x'y'z'$-coordinate system as shown in Figure 3.4.10*c*, then

$$\mathbf{u} = (0, 0, 1) = \mathbf{k} \quad \text{and} \quad \mathbf{v} = (1, 0, 0) = \mathbf{i}$$

so that

$$\mathbf{u} \times \mathbf{v} = \mathbf{k} \times \mathbf{i} = \mathbf{j} = (0, 1, 0)$$

But it is clear from Figures 3.4.10*b* and 3.4.10*c* that the vector $(0, 0, 1)$ in the xyz-system is the same as the vector $(0, 1, 0)$ in the $x'y'z'$-system. Thus we obtain the same vector $\mathbf{u} \times \mathbf{v}$ whether we compute with coordinates from the xyz-system or with coordinates from the $x'y'z'$-system. ◆

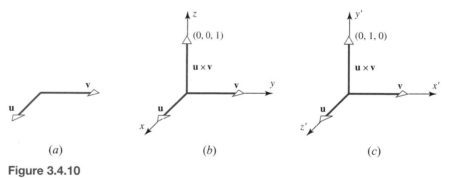

Figure 3.4.10

EXERCISE SET
3.4

1. Let $\mathbf{u} = (3, 2, -1)$, $\mathbf{v} = (0, 2, -3)$, and $\mathbf{w} = (2, 6, 7)$. Compute
   (a) $\mathbf{v} \times \mathbf{w}$ (b) $\mathbf{u} \times (\mathbf{v} \times \mathbf{w})$ (c) $(\mathbf{u} \times \mathbf{v}) \times \mathbf{w}$
   (d) $(\mathbf{u} \times \mathbf{v}) \times (\mathbf{v} \times \mathbf{w})$ (e) $\mathbf{u} \times (\mathbf{v} - 2\mathbf{w})$ (f) $(\mathbf{u} \times \mathbf{v}) - 2\mathbf{w}$

2. Find a vector that is orthogonal to both $\mathbf{u}$ and $\mathbf{v}$.
   (a) $\mathbf{u} = (-6, 4, 2)$, $\mathbf{v} = (3, 1, 5)$ (b) $\mathbf{u} = (-2, 1, 5)$, $\mathbf{v} = (3, 0, -3)$

3. Find the area of the parallelogram determined by $\mathbf{u}$ and $\mathbf{v}$.
   (a) $\mathbf{u} = (1, -1, 2)$, $\mathbf{v} = (0, 3, 1)$ (b) $\mathbf{u} = (2, 3, 0)$, $\mathbf{v} = (-1, 2, -2)$
   (c) $\mathbf{u} = (3, -1, 4)$, $\mathbf{v} = (6, -2, 8)$

4. Find the area of the triangle having vertices P, Q, and R.
   (a) $P(2, 6, -1)$, $Q(1, 1, 1)$, $R(4, 6, 2)$ (b) $P(1, -1, 2)$, $Q(0, 3, 4)$, $R(6, 1, 8)$

5. Verify parts (a), (b), and (c) of Theorem 3.4.1 for the vectors $\mathbf{u} = (4, 2, 1)$ and $\mathbf{v} = (-3, 2, 7)$.

6. Verify parts (a), (b), and (c) of Theorem 3.4.2 for $\mathbf{u} = (5, -1, 2)$, $\mathbf{v} = (6, 0, -2)$, and $\mathbf{w} = (1, 2, -1)$.

7. Find a vector $\mathbf{v}$ that is orthogonal to the vector $\mathbf{u} = (2, -3, 5)$.

8. Find the scalar triple product $\mathbf{u} \cdot (\mathbf{v} \times \mathbf{w})$.

   (a) $\mathbf{u} = (-1, 2, 4)$, $\mathbf{v} = (3, 4, -2)$, $\mathbf{w} = (-1, 2, 5)$

   (b) $\mathbf{u} = (3, -1, 6)$, $\mathbf{v} = (2, 4, 3)$, $\mathbf{w} = (5, -1, 2)$

9. Suppose that $\mathbf{u} \cdot (\mathbf{v} \times \mathbf{w}) = 3$. Find

   (a) $\mathbf{u} \cdot (\mathbf{w} \times \mathbf{v})$ (b) $(\mathbf{v} \times \mathbf{w}) \cdot \mathbf{u}$ (c) $\mathbf{w} \cdot (\mathbf{u} \times \mathbf{v})$

   (d) $\mathbf{v} \cdot (\mathbf{u} \times \mathbf{w})$ (e) $(\mathbf{u} \times \mathbf{w}) \cdot \mathbf{v}$ (f) $\mathbf{v} \cdot (\mathbf{w} \times \mathbf{w})$

10. Find the volume of the parallelepiped with sides $\mathbf{u}$, $\mathbf{v}$, and $\mathbf{w}$.

    (a) $\mathbf{u} = (2, -6, 2)$, $\mathbf{v} = (0, 4, -2)$, $\mathbf{w} = (2, 2, -4)$

    (b) $\mathbf{u} = (3, 1, 2)$, $\mathbf{v} = (4, 5, 1)$, $\mathbf{w} = (1, 2, 4)$

11. Determine whether $\mathbf{u}$, $\mathbf{v}$, and $\mathbf{w}$ lie in the same plane when positioned so that their initial points coincide.

    (a) $\mathbf{u} = (-1, -2, 1)$, $\mathbf{v} = (3, 0, -2)$, $\mathbf{w} = (5, -4, 0)$

    (b) $\mathbf{u} = (5, -2, 1)$, $\mathbf{v} = (4, -1, 1)$, $\mathbf{w} = (1, -1, 0)$

    (c) $\mathbf{u} = (4, -8, 1)$, $\mathbf{v} = (2, 1, -2)$, $\mathbf{w} = (3, -4, 12)$

12. Find all unit vectors parallel to the yz-plane that are perpendicular to the vector $(3, -1, 2)$.

13. Find all unit vectors in the plane determined by $\mathbf{u} = (3, 0, 1)$ and $\mathbf{v} = (1, -1, 1)$ that are perpendicular to the vector $\mathbf{w} = (1, 2, 0)$.

14. Let $\mathbf{a} = (a_1, a_2, a_3)$, $\mathbf{b} = (b_1, b_2, b_3)$, $\mathbf{c} = (c_1, c_2, c_3)$, and $\mathbf{d} = (d_1, d_2, d_3)$. Show that

$$(\mathbf{a} + \mathbf{d}) \cdot (\mathbf{b} \times \mathbf{c}) = \mathbf{a} \cdot (\mathbf{b} \times \mathbf{c}) + \mathbf{d} \cdot (\mathbf{b} \times \mathbf{c})$$

15. Simplify $(\mathbf{u} + \mathbf{v}) \times (\mathbf{u} - \mathbf{v})$.

16. Use the cross product to find the sine of the angle between the vectors $\mathbf{u} = (2, 3, -6)$ and $\mathbf{v} = (2, 3, 6)$.

17. (a) Find the area of the triangle having vertices $A(1, 0, 1)$, $B(0, 2, 3)$, and $C(2, 1, 0)$.

    (b) Use the result of part (a) to find the length of the altitude from vertex C to side AB.

18. Show that if $\mathbf{u}$ is a vector from any point on a line to a point P not on the line, and $\mathbf{v}$ is a vector parallel to the line, then the distance between P and the line is given by $\|\mathbf{u} \times \mathbf{v}\|/\|\mathbf{v}\|$.

19. Use the result of Exercise 18 to find the distance between the point P and the line through the points A and B.

    (a) $P(-3, 1, 2)$, $A(1, 1, 0)$, $B(-2, 3, -4)$ (b) $P(4, 3, 0)$, $A(2, 1, -3)$, $B(0, 2, -1)$

20. Prove: If θ is the angle between $\mathbf{u}$ and $\mathbf{v}$ and $\mathbf{u} \cdot \mathbf{v} \neq 0$, then $\tan\theta = \|\mathbf{u} \times \mathbf{v}\|/(\mathbf{u} \cdot \mathbf{v})$.

21. Consider the parallelepiped with sides $\mathbf{u} = (3, 2, 1)$, $\mathbf{v} = (1, 1, 2)$, and $\mathbf{w} = (1, 3, 3)$.

    (a) Find the area of the face determined by $\mathbf{u}$ and $\mathbf{w}$.

    (b) Find the angle between $\mathbf{u}$ and the plane containing the face determined by $\mathbf{v}$ and $\mathbf{w}$.

      Note   The ***angle between a vector and a plane*** is defined to be the complement of the angle θ between the vector and that normal to the plane for which $0 \leq \theta \leq \pi/2$.

22. Find a vector $\mathbf{n}$ that is perpendicular to the plane determined by the points $A(0, -2, 1)$, $B(1, -1, -2)$, and $C(-1, 1, 0)$. [See the note in Exercise 21.]

23. Let $\mathbf{m}$ and $\mathbf{n}$ be vectors whose components in the xyz-system of Figure 3.4.10 are $\mathbf{m} = (0, 0, 1)$ and $\mathbf{n} = (0, 1, 0)$.

(a)   Find the components of **m** and **n** in the $x'y'z'$-system of Figure 3.4.10.

(b)   Compute **m** $\times$ **n** using the components in the xyz-system.

(c)   Compute **m** $\times$ **n** using the components in the $x'y'z'$-system.

(d)   Show that the vectors obtained in (b) and (c) are the same.

24.   Prove the following identities.

(a) $(\mathbf{u} + k\mathbf{v}) \times \mathbf{v} = \mathbf{u} \times \mathbf{v}$ (b) $\mathbf{u} \cdot (\mathbf{v} \times \mathbf{z}) = -(\mathbf{u} \times \mathbf{z}) \cdot \mathbf{v}$

25.   Let **u**, **v**, and **w** be nonzero vectors in 3-space with the same initial point, but such that no two of them are collinear. Show that

(a) $\mathbf{u} \times (\mathbf{v} \times \mathbf{w})$ lies in the plane determined by **v** and **w**

(b) $(\mathbf{u} \times \mathbf{v}) \times \mathbf{w}$ lies in the plane determined by **u** and **v**

26.   Prove part (d) of Theorem 3.4.1.

Hint   First prove the result in the case where $\mathbf{w} = \mathbf{i} = (1, 0, 0)$, then when $\mathbf{w} = \mathbf{j} = (0, 1, 0)$, and then when $\mathbf{w} = \mathbf{k} = (0, 0, 1)$. Finally, prove it for an arbitrary vector $\mathbf{w} = (w_1, w_2, w_3)$ by writing $\mathbf{w} = w_1\mathbf{i} + w_2\mathbf{j} + w_3\mathbf{k}$.

27.   Prove part (e) of Theorem 3.4.1.

Hint   Apply part (a) of Theorem 3.4.2 to the result in part (d) of Theorem 3.4.1.

28.   Let $\mathbf{u} = (1, 3, -1)$, $\mathbf{v} = (1, 1, 2)$, and $\mathbf{w} = (3, -1, 2)$. Calculate $\mathbf{u} \times (\mathbf{v} \times \mathbf{w})$ using the technique of Exercise 26; then check your result by calculating directly.

29.   Prove: If **a**, **b**, **c**, and **d** lie in the same plane, then $(\mathbf{a} \times \mathbf{b}) \times (\mathbf{c} \times \mathbf{d}) = \mathbf{0}$.

30.   It is a theorem of solid geometry that the volume of a tetrahedron is $\frac{1}{3}$(area of base) · (height). Use this result to prove that the volume of a tetrahedron whose sides are the vectors **a**, **b**, and **c** is $\frac{1}{6}|\mathbf{a} \cdot (\mathbf{b} \times \mathbf{c})|$ (see the accompanying figure).

Figure Ex-30

31.   Use the result of Exercise 30 to find the volume of the tetrahedron with vertices P, Q, R, S.

(a) $P(-1, 2, 0)$, $Q(2, 1, -3)$, $R(1, 0, 1)$, $S(3, -2, 3)$

(b) $P(0, 0, 0)$, $Q(1, 2, -1)$, $R(3, 4, 0)$, $S(-1, -3, 4)$

32.   Prove part (b) of Theorem 3.4.2.

33.   Prove parts (c) and (d) of Theorem 3.4.2.

34.   Prove parts (e) and (f) of Theorem 3.4.2.

Discussion & Discovery

35.   (a)   Suppose that **u** and **v** are noncollinear vectors with their initial points at the origin in 3-space. Make a sketch that illustrates how $\mathbf{w} = \mathbf{v} \times (\mathbf{u} \times \mathbf{v})$ is oriented in relation to **u** and **v**.

(b)   For **w** as in part (a), what can you say about the values of $\mathbf{u} \cdot \mathbf{w}$ and $\mathbf{v} \cdot \mathbf{w}$? Explain your reasoning.

36.   If $\mathbf{u} \neq \mathbf{0}$, is it valid to cancel **u** from both sides of the equation $\mathbf{u} \times \mathbf{v} = \mathbf{u} \times \mathbf{w}$ and conclude that $\mathbf{v} = \mathbf{w}$? Explain your reasoning.

37.   Something is wrong with one of the following expressions. Which one is it and what is wrong?

$$\mathbf{u} \cdot (\mathbf{v} \times \mathbf{w}), \qquad \mathbf{u} \times \mathbf{v} \times \mathbf{w}, \qquad (\mathbf{u} \cdot \mathbf{v}) \times \mathbf{w}$$

38.   What can you say about the vectors **u** and **v** if $\mathbf{u} \times \mathbf{v} = \mathbf{0}$?

39.   Give some examples of algebraic rules that hold for multiplication of real numbers but not for the cross product of vectors.

3.5
LINES AND PLANES IN 3-SPACE

In this section we shall use vectors to derive equations of lines and planes in 3-space. We shall then use these equations to solve some basic geometric problems.

Planes in 3-Space

In analytic geometry a line in 2-space can be specified by giving its slope and one of its points. Similarly, one can specify a plane in 3-space by giving its inclination and specifying one of its points. A convenient method for describing the inclination of a plane is to specify a nonzero vector, called a ***normal***, that is perpendicular to the plane.

Suppose that we want to find the equation of the plane passing through the point $P_0(x_0, y_0, z_0)$ and having the nonzero vector $\mathbf{n} = (a, b, c)$ as a normal. It is evident from Figure 3.5.1 that the plane consists precisely of those points $P(x, y, z)$ for which the vector $\overrightarrow{P_0P}$ is orthogonal to $\mathbf{n}$; that is,

$$\mathbf{n} \cdot \overrightarrow{P_0P} = 0 \tag{1}$$

Since $\overrightarrow{P_0P} = (x - x_0, y - y_0, z - z_0)$, Equation (1) can be written as

$$a(x - x_0) + b(y - y_0) + c(z - z_0) = 0 \tag{2}$$

We call this the ***point-normal*** form of the equation of a plane.

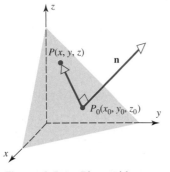

Figure 3.5.1 Plane with normal vector.

EXAMPLE 1   Finding the Point-Normal Equation of a Plane

Find an equation of the plane passing through the point $(3, -1, 7)$ and perpendicular to the vector $\mathbf{n} = (4, 2, -5)$.

Solution

From (2) a point-normal form is

$$4(x - 3) + 2(y + 1) - 5(z - 7) = 0 \quad \blacklozenge$$

By multiplying out and collecting terms, we can rewrite (2) in the form

$$ax + by + cz + d = 0$$

where a, b, c, and d are constants, and a, b, and c are not all zero. For example, the equation in Example 1 can be rewritten as

$$4x + 2y - 5z + 25 = 0$$

As the next theorem shows, planes in 3-space are represented by equations of the form $ax + by + cz + d = 0$.

THEOREM 3.5.1

If a, b, c, and d are constants and a, b, and c are not all zero, then the graph of the equation

$$ax + by + cz + d = 0 \tag{3}$$

is a plane having the vector $\mathbf{n} = (a, b, c)$ as a normal.

Equation (3) is a linear equation in x, y, and z; it is called the ***general form*** of the equation of a plane.

Proof By hypothesis, the coefficients a, b, and c are not all zero. Assume, for the moment, that $a \neq 0$. Then the equation $ax + by + cz + d = 0$ can be rewritten in the form $a(x + (d/a)) + by + cz = 0$. But this is a point-normal form of the plane passing through the point $(-d/a, 0, 0)$ and having $\mathbf{n} = (a, b, c)$ as a normal.

If $a = 0$, then either $b \neq 0$ or $c \neq 0$. A straightforward modification of the above argument will handle these other cases. ■

Just as the solutions of a system of linear equations

$$ax + by = k_1$$
$$cx + dy = k_2$$

correspond to points of intersection of the lines $ax + by = k_1$ and $cx + dy = k_2$ in the xy-plane, so the solutions of a system

$$ax + by + cz = k_1$$
$$dx + ey + fz = k_2 \tag{4}$$
$$gx + hy + iz = k_3$$

correspond to the points of intersection of the three planes $ax + by + cz = k_1$, $dx + ey + fz = k_2$, and $gx + hy + iz = k_3$.

In Figure 3.5.2 we have illustrated the geometric possibilities that occur when (4) has zero, one, or infinitely many solutions.

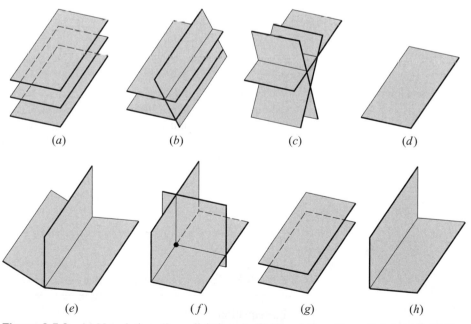

Figure 3.5.2 (*a*) No solutions (3 parallel planes). (*b*) No solutions (2 parallel planes). (*c*) No solutions (3 planes with no common intersection). (*d*) Infinitely many solutions (3 coincident planes). (*e*) Infinitely many solutions (3 planes intersecting in a line). (*f*) One solution (3 planes intersecting at a point). (*g*) No solutions (2 coincident planes parallel to a third plane). (*h*) Infinitely many solutions (2 coincident planes intersecting a third plane).

EXAMPLE 2 Equation of a Plane Through Three Points

Find the equation of the plane passing through the points $P_1(1, 2, -1)$, $P_2(2, 3, 1)$, and $P_3(3, -1, 2)$.

Solution

Since the three points lie in the plane, their coordinates must satisfy the general equation $ax + by + cz + d = 0$ of the plane. Thus

$$a + 2b - c + d = 0$$
$$2a + 3b + c + d = 0$$
$$3a - b + 2c + d = 0$$

Solving this system gives $a = -\frac{9}{16}t$, $b = -\frac{1}{16}t$, $c = \frac{5}{16}t$, $d = t$. Letting $t = -16$, for example, yields the desired equation

$$9x + y - 5z - 16 = 0$$

We note that any other choice of t gives a multiple of this equation, so that any value of $t \neq 0$ would also give a valid equation of the plane.

Alternative Solution

Since the points $P_1(1, 2, -1)$, $P_2(2, 3, 1)$, and $P_3(3, -1, 2)$ lie in the plane, the vectors $\overrightarrow{P_1 P_2} = (1, 1, 2)$ and $\overrightarrow{P_1 P_3} = (2, -3, 3)$ are parallel to the plane. Therefore, the equation $\overrightarrow{P_1 P_2} \times \overrightarrow{P_1 P_3} = (9, 1, -5)$ is normal to the plane, since it is perpendicular to both $\overrightarrow{P_1 P_2}$ and $\overrightarrow{P_1 P_3}$. From this and the fact that P_1 lies in the plane, a point-normal form for the equation of the plane is

$$9(x - 1) + (y - 2) - 5(z + 1) = 0 \quad \text{or} \quad 9x + y - 5z - 16 = 0 \quad \blacklozenge$$

Vector Form of Equation of a Plane

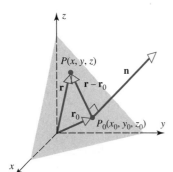

Figure 3.5.3

Vector notation provides a useful alternative way of writing the point-normal form of the equation of a plane: Referring to Figure 3.5.3, let $\mathbf{r} = (x, y, z)$ be the vector from the origin to the point $P(x, y, z)$, let $\mathbf{r}_0 = (x_0, y_0, z_0)$ be the vector from the origin to the point $P_0(x_0, y_0, z_0)$, and let $\mathbf{n} = (a, b, c)$ be a vector normal to the plane. Then $\overrightarrow{P_0 P} = \mathbf{r} - \mathbf{r}_0$, so Formula (1) can be rewritten as

$$\mathbf{n} \cdot (\mathbf{r} - \mathbf{r}_0) = 0 \tag{5}$$

This is called the **vector form of the equation of a plane**.

EXAMPLE 3 Vector Equation of a Plane Using (5)

The equation

$$(-1, 2, 5) \cdot (x - 6, y - 3, z + 4) = 0$$

is the vector equation of the plane that passes through the point $(6, 3, -4)$ and is perpendicular to the vector $\mathbf{n} = (-1, 2, 5)$. $\blacklozenge$

Lines in 3-Space

We shall now show how to obtain equations for lines in 3-space. Suppose that l is the line in 3-space through the point $P_0(x_0, y_0, z_0)$ and parallel to the nonzero vector

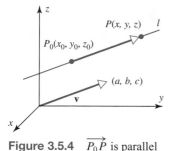

Figure 3.5.4 $\overrightarrow{P_0P}$ is parallel to $\mathbf{v}$.

$\mathbf{v} = (a, b, c)$. It is clear (Figure 3.5.4) that l consists precisely of those points $P(x, y, z)$ for which the vector $\overrightarrow{P_0P}$ is parallel to $\mathbf{v}$—that is, for which there is a scalar t such that

$$\overrightarrow{P_0P} = t\mathbf{v} \qquad (6)$$

In terms of components, (6) can be written as

$$(x - x_0, y - y_0, z - z_0) = (ta, tb, tc)$$

from which it follows that $x - x_0 = ta$, $y - y_0 = tb$, and $z - z_0 = tc$, so

$$x = x_0 + ta, \qquad y = y_0 + tb, \qquad z = z_0 + tc$$

As the parameter t varies from $-\infty$ to $+\infty$, the point $P(x, y, z)$ traces out the line l. The equations

$$x = x_0 + ta, \quad y = y_0 + tb, \quad z = z_0 + tc \qquad (-\infty < t < +\infty) \qquad (7)$$

are called ***parametric equations*** for l.

EXAMPLE 4 Parametric Equations of a Line

The line through the point $(1, 2, -3)$ and parallel to the vector $\mathbf{v} = (4, 5, -7)$ has parametric equations

$$x = 1 + 4t, \quad y = 2 + 5t, \quad z = -3 - 7t \qquad (-\infty < t < +\infty) \; \blacklozenge$$

EXAMPLE 5 Intersection of a Line and the *xy*-Plane

(a) Find parametric equations for the line l passing through the points $P_1(2, 4, -1)$ and $P_2(5, 0, 7)$.

(b) Where does the line intersect the xy-plane?

Solution (a)

Since the vector $\overrightarrow{P_1P_2} = (3, -4, 8)$ is parallel to l and $P_1(2, 4, -1)$ lies on l, the line l is given by

$$x = 2 + 3t, \quad y = 4 - 4t, \quad z = -1 + 8t \qquad (-\infty < t < +\infty)$$

Solution (b)

The line intersects the xy-plane at the point where $z = -1 + 8t = 0$, that is, where $t = \frac{1}{8}$. Substituting this value of t in the parametric equations for l yields, as the point of intersection,

$$(x, y, z) = \left(\tfrac{19}{8}, \tfrac{7}{2}, 0\right) \qquad \blacksquare$$

EXAMPLE 6 Line of Intersection of Two Planes

Find parametric equations for the line of intersection of the planes

$$3x + 2y - 4z - 6 = 0 \quad \text{and} \quad x - 3y - 2z - 4 = 0$$

Solution

The line of intersection consists of all points (x, y, z) that satisfy the two equations in the system

$$3x + 2y - 4z = 6$$
$$x - 3y - 2z = 4$$

Solving this system by Gaussian elimination gives $x = \frac{26}{11} + \frac{16}{11}t$, $y = -\frac{6}{11} - \frac{2}{11}t$, $z = t$. Therefore, the line of intersection can be represented by the parametric equations

$$x = \tfrac{26}{11} + \tfrac{16}{11}t, \quad y = -\tfrac{6}{11} - \tfrac{2}{11}t, \quad z = t \qquad (-\infty < t < +\infty) \qquad \blacksquare$$

Vector Form of Equation of a Line

Vector notation provides a useful alternative way of writing the parametric equations of a line: Referring to Figure 3.5.5, let $\mathbf{r} = (x, y, z)$ be the vector from the origin to the point $P(x, y, z)$, let $\mathbf{r}_0 = (x_0, y_0, z_0)$ be the vector from the origin to the point $P_0(x_0, y_0, z_0)$, and let $\mathbf{v} = (a, b, c)$ be a vector parallel to the line. Then $\overrightarrow{P_0 P} = \mathbf{r} - \mathbf{r}_0$, so Formula (6) can be rewritten as

$$\mathbf{r} - \mathbf{r}_0 = t\mathbf{v}$$

Taking into account the range of t-values, this can be rewritten as

$$\mathbf{r} = \mathbf{r}_0 + t\mathbf{v} \qquad (-\infty < t < +\infty) \tag{8}$$

This is called the *vector form of the equation of a line* in 3-space.

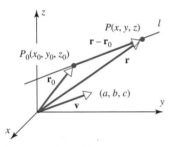

Figure 3.5.5 Vector interpretation of a line in 3-space.

EXAMPLE 7 A Line Parallel to a Given Vector

The equation

$$(x, y, z) = (-2, 0, 3) + t(4, -7, 1) \qquad (-\infty < t < +\infty)$$

is the vector equation of the line through the point $(-2, 0, 3)$ that is parallel to the vector $\mathbf{v} = (4, -7, 1)$. ◆

Problems Involving Distance

We conclude this section by discussing two basic "distance problems" in 3-space:

Problems:

(a) Find the distance between a point and a plane.

(b) Find the distance between two parallel planes.

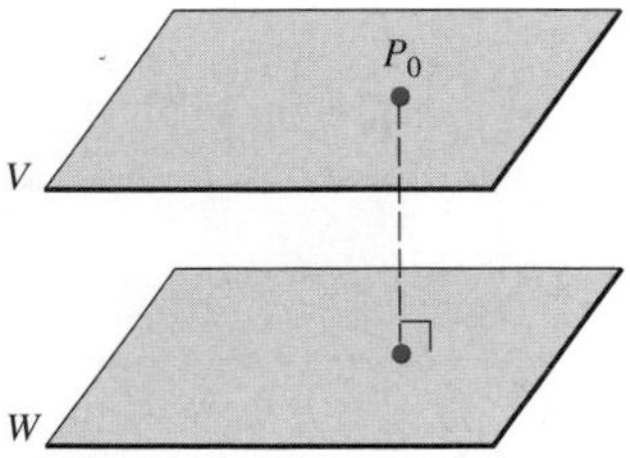

The two problems are related. If we can find the distance between a point and a plane, then we can find the distance between parallel planes by computing the distance between either one of the planes and an arbitrary point P_0 in the other (Figure 3.5.6).

Figure 3.5.6 The distance between the parallel planes V and W is equal to the distance between P_0 and W.

THEOREM 3.5.2

Distance Between a Point and a Plane

The distance D between a point $P_0(x_0, y_0, z_0)$ and the plane $ax + by + cz + d = 0$ is

$$D = \frac{|ax_0 + by_0 + cz_0 + d|}{\sqrt{a^2 + b^2 + c^2}} \qquad (9)$$

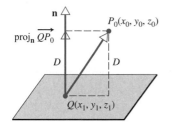

Figure 3.5.7 Distance from P_0 to plane.

Proof Let $Q(x_1, y_1, z_1)$ be any point in the plane. Position the normal $\mathbf{n} = (a, b, c)$ so that its initial point is at Q. As illustrated in Figure 3.5.7, the distance D is equal to the length of the orthogonal projection of $\overrightarrow{QP_0}$ on $\mathbf{n}$. Thus, from (10) of Section 3.3,

$$D = \|\text{proj}_{\mathbf{n}} \overrightarrow{QP_0}\| = \frac{|\overrightarrow{QP_0} \cdot \mathbf{n}|}{\|\mathbf{n}\|}$$

But

$$\overrightarrow{QP_0} = (x_0 - x_1, y_0 - y_1, z_0 - z_1)$$

$$\overrightarrow{QP_0} \cdot \mathbf{n} = a(x_0 - x_1) + b(y_0 - y_1) + c(z_0 - z_1)$$

$$\|\mathbf{n}\| = \sqrt{a^2 + b^2 + c^2}$$

Thus

$$D = \frac{|a(x_0 - x_1) + b(y_0 - y_1) + c(z_0 - z_1)|}{\sqrt{a^2 + b^2 + c^2}} \qquad (10)$$

Since the point $Q(x_1, y_1, z_1)$ lies in the plane, its coordinates satisfy the equation of the plane; thus

$$ax_1 + by_1 + cz_1 + d = 0$$

or

$$d = -ax_1 - by_1 - cz_1$$

Substituting this expression in (10) yields (9). ∎

REMARK Note the similarity between (9) and the formula for the distance between a point and a line in 2-space [(13) of Section 3.3].

EXAMPLE 8 Distance Between a Point and a Plane

Find the distance D between the point $(1, -4, -3)$ and the plane $2x - 3y + 6z = -1$.

Solution

To apply (9), we first rewrite the equation of the plane in the form

$$2x - 3y + 6z + 1 = 0$$

Then

$$D = \frac{|2(1) + (-3)(-4) + 6(-3) + 1|}{\sqrt{2^2 + (-3)^2 + 6^2}} = \frac{|-3|}{7} = \frac{3}{7} \quad \blacklozenge$$

Given two planes, either they intersect, in which case we can ask for their line of intersection, as in Example 6, or they are parallel, in which case we can ask for the distance between them. The following example illustrates the latter problem.

EXAMPLE 9 Distance Between Parallel Planes

The planes

$$x + 2y - 2z = 3 \quad \text{and} \quad 2x + 4y - 4z = 7$$

are parallel since their normals, $(1, 2, -2)$ and $(2, 4, -4)$, are parallel vectors. Find the distance between these planes.

Solution

To find the distance D between the planes, we may select an arbitrary point in one of the planes and compute its distance to the other plane. By setting $y = z = 0$ in the equation $x + 2y - 2z = 3$, we obtain the point $P_0(3, 0, 0)$ in this plane. From (9), the distance between P_0 and the plane $2x + 4y - 4z = 7$ is

$$D = \frac{|2(3) + 4(0) + (-4)(0) - 7|}{\sqrt{2^2 + 4^2 + (-4)^2}} = \frac{1}{6} \qquad \blacksquare$$

EXERCISE SET 3.5

1. Find a point-normal form of the equation of the plane passing through P and having $\mathbf{n}$ as a normal.
   - (a) $P(-1, 3, -2)$; $\mathbf{n} = (-2, 1, -1)$
   - (b) $P(1, 1, 4)$; $\mathbf{n} = (1, 9, 8)$
   - (c) $P(2, 0, 0)$; $\mathbf{n} = (0, 0, 2)$
   - (d) $P(0, 0, 0)$; $\mathbf{n} = (1, 2, 3)$

2. Write the equations of the planes in Exercise 1 in general form.

3. Find a point-normal form of the equations of the following planes.
   - (a) $-3x + 7y + 2z = 10$
   - (b) $x - 4z = 0$

4. Find an equation for the plane passing through the given points.
   - (a) $P(-4, -1, -1)$, $Q(-2, 0, 1)$, $R(-1, -2, -3)$
   - (b) $P(5, 4, 3)$, $Q(4, 3, 1)$, $R(1, 5, 4)$

5. Determine whether the planes are parallel.
   - (a) $4x - y + 2z = 5$ and $7x - 3y + 4z = 8$
   - (b) $x - 4y - 3z - 2 = 0$ and $3x - 12y - 9z - 7 = 0$
   - (c) $2y = 8x - 4z + 5$ and $x = \frac{1}{2}z + \frac{1}{4}y$

6. Determine whether the line and plane are parallel.
   - (a) $x = -5 - 4t$, $y = 1 - t$, $z = 3 + 2t$; $x + 2y + 3z - 9 = 0$
   - (b) $x = 3t$, $y = 1 + 2t$, $z = 2 - t$; $4x - y + 2z = 1$

7. Determine whether the planes are perpendicular.
   - (a) $3x - y + z - 4 = 0$, $x + 2z = -1$
   - (b) $x - 2y + 3z = 4$, $-2x + 5y + 4z = -1$

8. Determine whether the line and plane are perpendicular.
   - (a) $x = -2 - 4t$, $y = 3 - 2t$, $z = 1 + 2t$; $2x + y - z = 5$
   - (b) $x = 2 + t$, $y = 1 - t$, $z = 5 + 3t$; $6x + 6y - 7 = 0$

9. Find parametric equations for the line passing through P and parallel to $\mathbf{n}$.
   - (a) $P(3, -1, 2)$; $\mathbf{n} = (2, 1, 3)$
   - (b) $P(-2, 3, -3)$; $\mathbf{n} = (6, -6, -2)$
   - (c) $P(2, 2, 6)$; $\mathbf{n} = (0, 1, 0)$
   - (d) $P(0, 0, 0)$; $\mathbf{n} = (1, -2, 3)$

10. Find parametric equations for the line passing through the given points.

  (a) $(5, -2, 4)$, $(7, 2, -4)$ (b) $(0, 0, 0)$, $(2, -1, -3)$

11. Find parametric equations for the line of intersection of the given planes.

  (a) $7x - 2y + 3z = -2$ and $-3x + y + 2z + 5 = 0$

  (b) $2x + 3y - 5z = 0$ and $y = 0$

12. Find the vector form of the equation of the plane that passes through P_0 and has normal **n**.

  (a) $P_0(-1, 2, 4)$; $\mathbf{n} = (-2, 4, 1)$ (b) $P_0(2, 0, -5)$; $\mathbf{n} = (-1, 4, 3)$

  (c) $P_0(5, -2, 1)$; $\mathbf{n} = (-1, 0, 0)$.   (d) $P_0(0, 0, 0)$; $\mathbf{n} = (a, b, c)$

13. Determine whether the planes are parallel.

  (a) $(-1, 2, 4) \cdot (x - 5, y + 3, z - 7) = 0$; $(2, -4, -8) \cdot (x + 3, y + 5, z - 9) = 0$

  (b) $(3, 0, -1) \cdot (x + 1, y - 2, z - 3) = 0$; $(-1, 0, 3) \cdot (x + 1, y - z, z - 3) = 0$

14. Determine whether the planes are perpendicular.

  (a) $(-2, 1, 4) \cdot (x - 1, y, z + 3) = 0$; $(1, -2, 1) \cdot (x + 3, y - 5, z) = 0$

  (b) $(3, 0, -2) \cdot (x + 4, y - 7, z + 1) = 0$; $(1, 1, 1) \cdot (x, y, z) = 0$

15. Find the vector form of the equation of the line through P_0 and parallel to **v**.

  (a) $P_0(-1, 2, 3)$; $\mathbf{v} = (7, -1, 5)$ (b) $P_0(2, 0, -1)$; $\mathbf{v} = (1, 1, 1)$

  (c) $P_0(2, -4, 1)$; $\mathbf{v} = (0, 0, -2)$ (d) $P_0(0, 0, 0)$; $\mathbf{v} = (a, b, c)$

16. Show that the line

$$x = 0, \quad y = t, \quad z = t \qquad (-\infty < t < +\infty)$$

  (a) lies in the plane $6x + 4y - 4z = 0$

  (b) is parallel to and below the plane $5x - 3y + 3z = 1$

  (c) is parallel to and above the plane $6x + 2y - 2z = 3$

17. Find an equation for the plane through $(-2, 1, 7)$ that is perpendicular to the line $x - 4 = 2t$, $y + 2 = 3t, z = -5t$.

18. Find an equation of

  (a) the xy-plane   (b) the xz-plane   (c) the yz-plane

19. Find an equation of the plane that contains the point (x_0, y_0, z_0) and is

  (a) parallel to the xy-plane

  (b) parallel to the yz-plane

  (c) parallel to the xz-plane

20. Find an equation for the plane that passes through the origin and is parallel to the plane $7x + 4y - 2z + 3 = 0$.

21. Find an equation for the plane that passes through the point $(3, -6, 7)$ and is parallel to the plane $5x - 2y + z - 5 = 0$.

22. Find the point of intersection of the line

$$x - 9 = -5t, \quad y + 1 = -t, \quad z - 3 = t \qquad (-\infty < t < +\infty)$$

  and the plane $2x - 3y + 4z + 7 = 0$.

23. Find an equation for the plane that contains the line $x = -1 + 3t, y = 5 + 2t, z = 2 - t$ and is perpendicular to the plane $2x - 4y + 2z = 9$.

24. Find an equation for the plane that passes through $(2, 4, -1)$ and contains the line of intersection of the planes $x - y - 4z = 2$ and $-2x + y + 2z = 3$.

25. Show that the points $(-1, -2, -3)$, $(-2, 0, 1)$, $(-4, -1, -1)$, and $(2, 0, 1)$ lie in the same plane.

26. Find parametric equations for the line through $(-2, 5, 0)$ that is parallel to the planes $2x + y - 4z = 0$ and $-x + 2y + 3z + 1 = 0$.

27. Find an equation for the plane through $(-2, 1, 5)$ that is perpendicular to the planes $4x - 2y + 2z = -1$ and $3x + 3y - 6z = 5$.

28. Find an equation for the plane through $(2, -1, 4)$ that is perpendicular to the line of intersection of the planes $4x + 2y + 2z = -1$ and $3x + 6y + 3z = 7$.

29. Find an equation for the plane that is perpendicular to the plane $8x - 2y + 6z = 1$ and passes through the points $P_1(-1, 2, 5)$ and $P_2(2, 1, 4)$.

30. Show that the lines

$$x = 3 - 2t, \quad y = 4 + t, \quad z = 1 - t \qquad (-\infty < t < +\infty)$$

and

$$x = 5 + 2t, \quad y = 1 - t, \quad z = 7 + t \qquad (-\infty < t < +\infty)$$

are parallel, and find an equation for the plane they determine.

31. Find an equation for the plane that contains the point $(1, -1, 2)$ and the line $x = t$, $y = t + 1$, $z = -3 + 2t$.

32. Find an equation for the plane that contains the line $x = 1 + t$, $y = 3t$, $z = 2t$ and is parallel to the line of intersection of the planes $-x + 2y + z = 0$ and $x + z + 1 = 0$.

33. Find an equation for the plane, each of whose points is equidistant from $(-1, -4, -2)$ and $(0, -2, 2)$.

34. Show that the line

$$x - 5 = -t, \quad y + 3 = 2t, \quad z + 1 = -5t \qquad (-\infty < t < +\infty)$$

is parallel to the plane $-3x + y + z - 9 = 0$.

35. Show that the lines

$$x - 3 = 4t, \quad y - 4 = t, \quad z - 1 = 0 \qquad (-\infty < t < +\infty)$$

and

$$x + 1 = 12t, \quad y - 7 = 6t, \quad z - 5 = 3t \qquad (-\infty < t < +\infty)$$

intersect, and find the point of intersection.

36. Find an equation for the plane containing the lines in Exercise 35.

37. Find parametric equations for the line of intersection of the planes
    (a) $-3x + 2y + z = -5$ and $7x + 3y - 2z = -2$
    (b) $5x - 7y + 2z = 0$ and $y = 0$

38. Show that the plane whose intercepts with the coordinate axes are $x = a$, $y = b$, and $z = c$ has equation

$$\frac{x}{a} + \frac{y}{b} + \frac{z}{c} = 1$$

provided that a, b, and c are nonzero.

39. Find the distance between the point and the plane.
    (a) $(3, 1, -2)$; $x + 2y - 2z = 4$
    (b) $(-1, 2, 1)$; $2x + 3y - 4z = 1$
    (c) $(0, 3, -2)$; $x - y - z = 3$

40. Find the distance between the given parallel planes.
    (a) $3x - 4y + z = 1$ and $6x - 8y + 2z = 3$
    (b) $-4x + y - 3z = 0$ and $8x - 2y + 6z = 0$
    (c) $2x - y + z = 1$ and $2x - y + z = -1$

41. Find the distance between the line $x = 3t - 1$, $y = 2 - t$, $z = t$ and each of the following points.

(a) $(0, 0, 0)$ (b) $(2, 0, -5)$ (c) $(2, 1, 1)$

42. Show that if a, b, and c are nonzero, then the line

$$x = x_0 + at, \quad y = y_0 + bt, \quad z = z_0 + ct \qquad (-\infty < t < +\infty)$$

consists of all points (x, y, z) that satisfy

$$\frac{x - x_0}{a} = \frac{y - y_0}{b} = \frac{z - z_0}{c}$$

These are called **symmetric equations** for the line.

43. Find symmetric equations for the lines in parts (a) and (b) of Exercise 9.

Note See Exercise 42 for terminology.

44. In each part, find equations for two planes whose intersection is the given line.

(a) $x = 7 - 4t, \quad y = -5 - 2t, \quad z = 5 + t \qquad (-\infty < t < +\infty)$

(b) $x = 4t, \quad y = 2t, \quad z = 7t \qquad (-\infty < t < +\infty)$

Hint Each equality in the symmetric equations of a line represents a plane containing the line. See Exercise 42 for terminology.

45. Two intersecting planes in 3-space determine two angles of intersection: an acute angle $(0 \le \theta \le 90°)$ and its supplement $180° - \theta$ (see the accompanying figure). If $\mathbf{n}_1$ and $\mathbf{n}_2$ are nonzero normals to the planes, then the angle between $\mathbf{n}_1$ and $\mathbf{n}_2$ is θ or $180° - \theta$, depending on the directions of the normals (see the accompanying figure). In each part, find the acute angle of intersection of the planes to the nearest degree.

(a) $x = 0$ and $2x - y + z - 4 = 0$

(b) $x + 2y - 2z = 5$ and $6x - 3y + 2z = 8$

Note A calculator is needed.

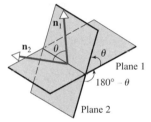

Figure Ex-45

46. Find the acute angle between the plane $x - y - 3z = 5$ and the line $x = 2 - t$, $y = 2t$, $z = 3t - 1$ to the nearest degree.

Hint See Exercise 45.

Discussion & Discovery

47. What do the lines $\mathbf{r} = \mathbf{r}_0 + t\mathbf{v}$ and $\mathbf{r} = \mathbf{r}_0 - t\mathbf{v}$ have in common? Explain.

48. What is the relationship between the line $x = x_0 + at$, $y = y_0 + bt$, $z = z_0 + ct$ and the plane $ax + by + cz = 0$? Explain your reasoning.

49. Let $\mathbf{r}_1$ and $\mathbf{r}_2$ be vectors from the origin to the points $P_1(x_1, y_1, z_1)$ and $P_2(x_2, y_2, z_2)$, respectively. What does the equation

$$\mathbf{r} = (1 - t)\mathbf{r}_1 + t\mathbf{r}_2 \qquad (0 \le t \le 1)$$

represent geometrically? Explain your reasoning.

50. Write parametric equations for two perpendicular lines through the point (x_0, y_0, z_0).

51. How can you tell whether the line $\mathbf{x} = \mathbf{x}_0 + t\mathbf{v}$ in 3-space is parallel to the plane $\mathbf{x} = \mathbf{x}_0 + t_1\mathbf{v}_1 + t_2\mathbf{v}_2$?

52. Indicate whether the statement is true (T) or false (F). Justify your answer.

(a) If a, b, and c are not all zero, then the line $x = at$, $y = bt$, $z = ct$ is perpendicular to the plane $ax + by + cz = 0$.

(b) Two nonparallel lines in 3-space must intersect in at least one point.

(c) If $\mathbf{u}$, $\mathbf{v}$, and $\mathbf{w}$ are vectors in 3-space such that $\mathbf{u} + \mathbf{v} + \mathbf{w} = \mathbf{0}$, then the three vectors lie in some plane.

(d) The equation $\mathbf{x} = t\mathbf{v}$ represents a line for every vector $\mathbf{v}$ in 2-space.

CHAPTER 3

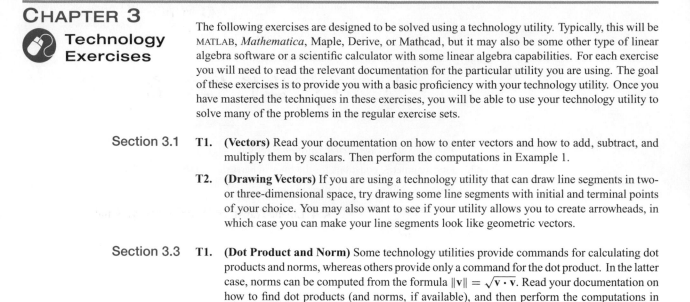

Technology Exercises

The following exercises are designed to be solved using a technology utility. Typically, this will be MATLAB, *Mathematica*, Maple, Derive, or Mathcad, but it may also be some other type of linear algebra software or a scientific calculator with some linear algebra capabilities. For each exercise you will need to read the relevant documentation for the particular utility you are using. The goal of these exercises is to provide you with a basic proficiency with your technology utility. Once you have mastered the techniques in these exercises, you will be able to use your technology utility to solve many of the problems in the regular exercise sets.

Section 3.1

T1. **(Vectors)** Read your documentation on how to enter vectors and how to add, subtract, and multiply them by scalars. Then perform the computations in Example 1.

T2. **(Drawing Vectors)** If you are using a technology utility that can draw line segments in two- or three-dimensional space, try drawing some line segments with initial and terminal points of your choice. You may also want to see if your utility allows you to create arrowheads, in which case you can make your line segments look like geometric vectors.

Section 3.3

T1. **(Dot Product and Norm)** Some technology utilities provide commands for calculating dot products and norms, whereas others provide only a command for the dot product. In the latter case, norms can be computed from the formula $\|\mathbf{v}\| = \sqrt{\mathbf{v} \cdot \mathbf{v}}$. Read your documentation on how to find dot products (and norms, if available), and then perform the computations in Example 2.

T2. **(Projections)** See if you can program your utility to calculate $\text{proj}_{\mathbf{a}}\mathbf{u}$ when the user enters the vectors **a** and **u**. Check your work by having your program perform the computations in Example 6.

Section 3.4

T1. **(Cross Product)** Read your documentation on how to find cross products, and then perform the computation in Example 1.

T2. **(Cross Product Formula)** If you are working with a CAS, use it to confirm Formula (1a).

T3. **(Cross Product Properties)** If you are working with a CAS, use it to prove the results in Theorem 3.4.1.

T4. **(Area of a Triangle)** See if you can program your technology utility to find the area of the triangle in 3-space determined by three points when the user enters their coordinates. Check your work by calculating the area of the triangle in Example 4.

T5. **(Triple Scalar Product Formula)** If you are working with a CAS, use it to prove Formula (7) by showing that the difference between the two sides is zero.

T6. **(Volume of a Parallelepiped)** See if you can program your technology utility to find the volume of the parallelepiped in 3-space determined by vectors **u**, **v**, and **w** when the user enters the vectors. Check your work by solving Exercise 10 in Exercise Set 3.4.

Euclidean Vector Spaces

CHAPTER CONTENTS

I NTRODUCTION: The idea of using pairs of numbers to locate points in the plane and triples of numbers to locate points in 3-space was first clearly spelled out in the mid-seventeenth century. By the latter part of the eighteenth century, mathematicians and physicists began to realize that there was no need to stop with triples. It was recognized that quadruples of numbers (a_1, a_2, a_3, a_4) could be regarded as points in "four-dimensional" space, quintuples $(a_1, a_2, a_3, a_4, a_5)$ as points in "five-dimensional" space, and so on, an n-tuple of numbers being a point in "n-dimensional" space. Our goal in this chapter is to study the properties of operations on vectors in this kind of space.

4.1
EUCLIDEAN n-SPACE

Although our geometric visualization does not extend beyond 3-space, it is nevertheless possible to extend many familiar ideas beyond 3-space by working with analytic or numerical properties of points and vectors rather than the geometric properties. In this section we shall make these ideas more precise.

Vectors in n-Space

We begin with a definition.

> **DEFINITION**
>
> If n is a positive integer, then an ***ordered n-tuple*** is a sequence of n real numbers $(a_1, a_2, \ldots, a_n)$. The set of all ordered n-tuples is called ***n-space*** and is denoted by R^n.

When $n = 2$ or 3, it is customary to use the terms ***ordered pair*** and ***ordered triple***, respectively, rather than *ordered 2-tuple* and *ordered 3-tuple*. When $n = 1$, each ordered n-tuple consists of one real number, so R^1 may be viewed as the set of real numbers. It is usual to write R rather than R^1 for this set.

It might have occurred to you in the study of 3-space that the symbol (a_1, a_2, a_3) has two different geometric interpretations: it can be interpreted as a point, in which case a_1, a_2, and a_3 are the coordinates (Figure 4.1.1a), or it can be interpreted as a vector, in which case a_1, a_2, and a_3 are the components (Figure 4.1.1b). It follows, therefore, that an ordered n-tuple $(a_1, a_2, \ldots, a_n)$ can be viewed either as a "generalized point" or as a "generalized vector"—the distinction is mathematically unimportant. Thus we can describe the 5-tuple $(-2, 4, 0, 1, 6)$ either as a point in R^5 or as a vector in R^5.

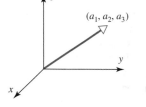

(a)

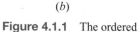

(b)

Figure 4.1.1 The ordered triple (a_1, a_2, a_3) can be interpreted geometrically as a point or as a vector.

> **DEFINITION**
>
> Two vectors $\mathbf{u} = (u_1, u_2, \ldots, u_n)$ and $\mathbf{v} = (v_1, v_2, \ldots, v_n)$ in R^n are called ***equal*** if
>
> $$u_1 = v_1, \quad u_2 = v_2, \ldots, \quad u_n = v_n$$
>
> The ***sum $\mathbf{u} + \mathbf{v}$*** is defined by
>
> $$\mathbf{u} + \mathbf{v} = (u_1 + v_1, u_2 + v_2, \ldots, u_n + v_n)$$
>
> and if k is any scalar, the ***scalar multiple $k\mathbf{u}$*** is defined by
>
> $$k\mathbf{u} = (ku_1, ku_2, \ldots, ku_n)$$

The operations of addition and scalar multiplication in this definition are called the ***standard operations*** on R^n.

The ***zero vector*** in R^n is denoted by $\mathbf{0}$ and is defined to be the vector

$$\mathbf{0} = (0, 0, \ldots, 0)$$

If $\mathbf{u} = (u_1, u_2, \ldots, u_n)$ is any vector in R^n, then the ***negative*** (or ***additive inverse***) of $\mathbf{u}$ is denoted by $-\mathbf{u}$ and is defined by

$$-\mathbf{u} = (-u_1, -u_2, \ldots, -u_n)$$

The ***difference*** of vectors in R^n is defined by

$$\mathbf{v} - \mathbf{u} = \mathbf{v} + (-\mathbf{u})$$

Some Examples of Vectors in Higher-Dimensional Spaces

- **Experimental Data**—A scientist performs an experiment and makes n numerical measurements each time the experiment is performed. The result of each experiment can be regarded as a vector $\mathbf{y} = (y_1, y_2, \ldots, y_n)$ in R^n in which $y_1, y_2, \ldots, y_n$ are the measured values.

- **Storage and Warehousing**—A national trucking company has 15 depots for storing and servicing its trucks. At each point in time the distribution of trucks in the service depots can be described by a 15-tuple $\mathbf{x} = (x_1, x_2, \ldots, x_{15})$ in which x_1 is the number of trucks in the first depot, x_2 is the number in the second depot, and so forth.

- **Electrical Circuits**—A certain kind of processing chip is designed to receive four input voltages and produces three output voltages in response. The input voltages can be regarded as vectors in R^4 and the output voltages as vectors in R^3. Thus, the chip can be viewed as a device that transforms each input vector $\mathbf{v} = (v_1, v_2, v_3, v_4)$ in R^4 into some output vector $\mathbf{w} = (w_1, w_2, w_3)$ in R^3.

- **Graphical Images**—One way in which color images are created on computer screens is by assigning each pixel (an addressable point on the screen) three numbers that describe the *hue*, *saturation*, and *brightness* of the pixel. Thus, a complete color image can be viewed as a set of 5-tuples of the form $\mathbf{v} = (x, y, h, s, b)$ in which x and y are the screen coordinates of a pixel and h, s, and b are its hue, saturation, and brightness.

- **Economics**—Our approach to economic analysis is to divide an economy into sectors (manufacturing, services, utilities, and so forth) and to measure the output of each sector by a dollar value. Thus, in an economy with 10 sectors the economic output of the entire economy can be represented by a 10-tuple $\mathbf{s} = (s_1, s_2, \ldots, s_{10})$ in which the numbers $s_1, s_2, \ldots, s_{10}$ are the outputs of the individual sectors.

- **Mechanical Systems**—Suppose that six particles move along the same coordinate line so that at time t their coordinates are $x_1, x_2, \ldots, x_6$ and their velocities are $v_1, v_2, \ldots, v_6$, respectively. This information can be represented by the vector

$$\mathbf{v} = (x_1, x_2, x_3, x_4, x_5, x_6,$$
$$v_1, v_2, v_3, v_4, v_5, v_6, t)$$

in R^{13}. This vector is called the *state* of the particle system at time t.

- **Physics**—In string theory the smallest, indivisible components of the Universe are not particles but loops that behave like vibrating strings. Whereas Einstein's space-time universe was four-dimensional, strings reside in an 11-dimensional world.

or, in terms of components,

$$\mathbf{v} - \mathbf{u} = (v_1 - u_1, v_2 - u_2, \ldots, v_n - u_n)$$

Properties of Vector Operations in n-Space

The most important arithmetic properties of addition and scalar multiplication of vectors in R^n are listed in the following theorem. The proofs are all easy and are left as exercises.

THEOREM 4.1.1

Properties of Vectors in R^n

If $\mathbf{u} = (u_1, u_2, \ldots, u_n)$, $\mathbf{v} = (v_1, v_2, \ldots, v_n)$, *and* $\mathbf{w} = (w_1, w_2, \ldots, w_n)$ *are vectors in R^n and k and m are scalars, then:*

(a) $\mathbf{u} + \mathbf{v} = \mathbf{v} + \mathbf{u}$ (b) $\mathbf{u} + (\mathbf{v} + \mathbf{w}) = (\mathbf{u} + \mathbf{v}) + \mathbf{w}$

(c) $\mathbf{u} + \mathbf{0} = \mathbf{0} + \mathbf{u} = \mathbf{u}$ (d) $\mathbf{u} + (-\mathbf{u}) = \mathbf{0}$; *that is,* $\mathbf{u} - \mathbf{u} = \mathbf{0}$

(e) $k(m\mathbf{u}) = (km)\mathbf{u}$ (f) $k(\mathbf{u} + \mathbf{v}) = k\mathbf{u} + k\mathbf{v}$

(g) $(k + m)\mathbf{u} = k\mathbf{u} + m\mathbf{u}$ (h) $1\mathbf{u} = \mathbf{u}$

This theorem enables us to manipulate vectors in R^n without expressing the vectors in terms of components. For example, to solve the vector equation $\mathbf{x} + \mathbf{u} = \mathbf{v}$ for $\mathbf{x}$, we can add $-\mathbf{u}$ to both sides and proceed as follows:

$$(\mathbf{x} + \mathbf{u}) + (-\mathbf{u}) = \mathbf{v} + (-\mathbf{u})$$
$$\mathbf{x} + (\mathbf{u} - \mathbf{u}) = \mathbf{v} - \mathbf{u}$$
$$\mathbf{x} + \mathbf{0} = \mathbf{v} - \mathbf{u}$$
$$\mathbf{x} = \mathbf{v} - \mathbf{u}$$

The reader will find it instructive to name the parts of Theorem 4.1.1 that justify the last three steps in this computation.

Euclidean n-Space

To extend the notions of distance, norm, and angle to R^n, we begin with the following generalization of the dot product on R^2 and R^3 [Formulas (3) and (4) of Section 3.3].

DEFINITION

If $\mathbf{u} = (u_1, u_2, \ldots, u_n)$ and $\mathbf{v} = (v_1, v_2, \ldots, v_n)$ are any vectors in R^n, then the *Euclidean inner product* $\mathbf{u} \cdot \mathbf{v}$ is defined by

$$\mathbf{u} \cdot \mathbf{v} = u_1 v_1 + u_2 v_2 + \cdots + u_n v_n$$

Observe that when $n = 2$ or 3, the Euclidean inner product is the ordinary dot product.

EXAMPLE 1 Inner Product of Vectors in R^4

The Euclidean inner product of the vectors

$$\mathbf{u} = (-1, 3, 5, 7) \quad \text{and} \quad \mathbf{v} = (5, -4, 7, 0)$$

in R^4 is

$$\mathbf{u} \cdot \mathbf{v} = (-1)(5) + (3)(-4) + (5)(7) + (7)(0) = 18 \quad \blacklozenge$$

Since so many of the familiar ideas from 2-space and 3-space carry over to n-space, it is common to refer to R^n, with the operations of addition, scalar multiplication, and the Euclidean inner product, as *Euclidean n-space*.

The four main arithmetic properties of the Euclidean inner product are listed in the next theorem.

THEOREM 4.1.2

Properties of Euclidean Inner Product

If $\mathbf{u}$, $\mathbf{v}$, *and* $\mathbf{w}$ *are vectors in R^n and k is any scalar, then:*

(a) $\mathbf{u} \cdot \mathbf{v} = \mathbf{v} \cdot \mathbf{u}$ (b) $(\mathbf{u} + \mathbf{v}) \cdot \mathbf{w} = \mathbf{u} \cdot \mathbf{w} + \mathbf{v} \cdot \mathbf{w}$

(c) $(k\mathbf{u}) \cdot \mathbf{v} = k(\mathbf{u} \cdot \mathbf{v})$ (d) $\mathbf{v} \cdot \mathbf{v} \geq 0$. *Further,* $\mathbf{v} \cdot \mathbf{v} = 0$ *if and only if* $\mathbf{v} = \mathbf{0}$.

We shall prove parts (b) and (d) and leave proofs of the rest as exercises.

Application of Dot Products to ISBNs

Most books published in the last 25 years have been assigned a unique 10-digit number called an *International Standard Book Number* or ISBN. The first nine digits of this number are split into three groups—the first group representing the country or group of countries in which the book originates, the second identifying the publisher, and the third assigned to the book title itself. The tenth and final digit, called a *check digit*, is computed from the first nine digits and is used to ensure that an electronic transmission of the ISBN, say over the Internet, occurs without error.

To explain how this is done, regard the first nine digits of the ISBN as a vector **b** in R^9, and let **a** be the vector

$$\mathbf{a} = (1, 2, 3, 4, 5, 6, 7, 8, 9)$$

Then the check digit c is computed using the following procedure:

1. Form the dot product $\mathbf{a} \cdot \mathbf{b}$.

2. Divide $\mathbf{a} \cdot \mathbf{b}$ by 11, thereby producing a remainder c that is an integer between 0 and 10, inclusive. The check digit is taken to be c, with the proviso that $c = 10$ is written as X to avoid double digits.

For example, the ISBN of the brief edition of *Calculus*, sixth edition, by Howard Anton is

$$0\text{-}471\text{-}15307\text{-}9$$

which has a check digit of 9. This is consistent with the first nine digits of the ISBN, since

$$\mathbf{a} \cdot \mathbf{b} = (1, 2, 3, 4, 5, 6, 7, 8, 9)$$
$$\cdot (0, 4, 7, 1, 1, 5, 3, 0, 7) = 152$$

Dividing 152 by 11 produces a quotient of 13 and a remainder of 9, so the check digit is $c = 9$. If an electronic order is placed for a book with a certain ISBN, then the warehouse can use the above procedure to verify that the check digit is consistent with the first nine digits, thereby reducing the possibility of a costly shipping error.

Proof (b) Let $\mathbf{u} = (u_1, u_2, \ldots, u_n)$, $\mathbf{v} = (v_1, v_2, \ldots, v_n)$, and $\mathbf{w} = (w_1, w_2, \ldots, w_n)$. Then

$$(\mathbf{u} + \mathbf{v}) \cdot \mathbf{w} = (u_1 + v_1, u_2 + v_2, \ldots, u_n + v_n) \cdot (w_1, w_2, \ldots, w_n)$$
$$= (u_1 + v_1)w_1 + (u_2 + v_2)w_2 + \cdots + (u_n + v_n)w_n$$
$$= (u_1 w_1 + u_2 w_2 + \cdots + u_n w_n) + (v_1 w_1 + v_2 w_2 + \cdots + v_n w_n)$$
$$= \mathbf{u} \cdot \mathbf{w} + \mathbf{v} \cdot \mathbf{w}$$

Proof (d) We have $\mathbf{v} \cdot \mathbf{v} = v_1^2 + v_2^2 + \cdots + v_n^2 \geq 0$. Further, equality holds if and only if $v_1 = v_2 = \cdots = v_n = 0$—that is, if and only if $\mathbf{v} = \mathbf{0}$. ∎

EXAMPLE 2 Length and Distance in R^4

Theorem 4.1.2 allows us to perform computations with Euclidean inner products in much the same way as we perform them with ordinary arithmetic products. For example,

$$(3\mathbf{u} + 2\mathbf{v}) \cdot (4\mathbf{u} + \mathbf{v}) = (3\mathbf{u}) \cdot (4\mathbf{u} + \mathbf{v}) + (2\mathbf{v}) \cdot (4\mathbf{u} + \mathbf{v})$$
$$= (3\mathbf{u}) \cdot (4\mathbf{u}) + (3\mathbf{u}) \cdot \mathbf{v} + (2\mathbf{v}) \cdot (4\mathbf{u}) + (2\mathbf{v}) \cdot \mathbf{v}$$
$$= 12(\mathbf{u} \cdot \mathbf{u}) + 3(\mathbf{u} \cdot \mathbf{v}) + 8(\mathbf{v} \cdot \mathbf{u}) + 2(\mathbf{v} \cdot \mathbf{v})$$
$$= 12(\mathbf{u} \cdot \mathbf{u}) + 11(\mathbf{u} \cdot \mathbf{v}) + 2(\mathbf{v} \cdot \mathbf{v})$$

The reader should determine which parts of Theorem 4.1.2 were used in each step. ◆

Norm and Distance in Euclidean _n_-Space

By analogy with the familiar formulas in R^2 and R^3, we define the **_Euclidean norm_** (or **_Euclidean length_**) of a vector $\mathbf{u} = (u_1, u_2, \ldots, u_n)$ in R^n by

$$\|\mathbf{u}\| = (\mathbf{u} \cdot \mathbf{u})^{1/2} = \sqrt{u_1^2 + u_2^2 + \cdots + u_n^2} \tag{1}$$

[Compare this formula to Formulas (1) and (2) in Section 3.2.]

Similarly, the **_Euclidean distance_** between the points $\mathbf{u} = (u_1, u_2, \ldots, u_n)$ and $\mathbf{v} = (v_1, v_2, \ldots, v_n)$ in R^n is defined by

$$d(\mathbf{u}, \mathbf{v}) = \|\mathbf{u} - \mathbf{v}\| = \sqrt{(u_1 - v_1)^2 + (u_2 - v_2)^2 + \cdots + (u_n - v_n)^2} \tag{2}$$

[See Formulas (3) and (4) of Section 3.2.]

EXAMPLE 3   Finding Norm and Distance

If $\mathbf{u} = (1, 3, -2, 7)$ and $\mathbf{v} = (0, 7, 2, 2)$, then in the Euclidean space R^4,

$$\|\mathbf{u}\| = \sqrt{(1)^2 + (3)^2 + (-2)^2 + (7)^2} = \sqrt{63} = 3\sqrt{7}$$

and

$$d(\mathbf{u}, \mathbf{v}) = \sqrt{(1 - 0)^2 + (3 - 7)^2 + (-2 - 2)^2 + (7 - 2)^2} = \sqrt{58} \quad \blacklozenge$$

The following theorem provides one of the most important inequalities in linear algebra: the **_Cauchy–Schwarz_** inequality.

THEOREM 4.1.3

Cauchy–Schwarz Inequality in R^n

_If_ $\mathbf{u} = (u_1, u_2, \ldots, u_n)$ _and_ $\mathbf{v} = (v_1, v_2, \ldots, v_n)$ _are vectors in_ R^n, _then_

$$|\mathbf{u} \cdot \mathbf{v}| \leq \|\mathbf{u}\| \|\mathbf{v}\| \tag{3}$$

In terms of components, (3) is the same as

$$|u_1 v_1 + u_2 v_2 + \cdots + u_n v_n| \leq (u_1^2 + u_2^2 + \cdots + u_n^2)^{1/2} (v_1^2 + v_2^2 + \cdots + v_n^2)^{1/2} \tag{4}$$

We omit the proof at this time, since a more general version of this theorem will be proved later in the text. However, for vectors in R^2 and R^3, this result is a simple consequence of Formula (1) of Section 3.3: If $\mathbf{u}$ and $\mathbf{v}$ are nonzero vectors in R^2 or R^3, then

$$|\mathbf{u} \cdot \mathbf{v}| = |\|\mathbf{u}\| \|\mathbf{v}\| \cos\theta| = \|\mathbf{u}\| \|\mathbf{v}\| |\cos\theta| \leq \|\mathbf{u}\| \|\mathbf{v}\| \tag{5}$$

and if either $\mathbf{u} = \mathbf{0}$ or $\mathbf{v} = \mathbf{0}$, then both sides of (3) are zero, so the inequality holds in this case as well.

The next two theorems list the basic properties of length and distance in Euclidean n-space.

Augustin Louis (Baron de) Cauchy
(1789–1857), French mathematician. Cauchy's early education was acquired from his father, a barrister and master of the classics. Cauchy entered L'Ecole Polytechnique in 1805 to study engineering, but because of poor health, he was advised to concentrate on mathematics. His major mathematical work began in 1811 with a series of brilliant solutions to some difficult outstanding problems.

Cauchy's mathematical contributions for the next 35 years were brilliant and staggering in quantity: over 700 papers filling 26 modern volumes. Cauchy's work initiated the era of modern analysis; he brought to mathematics standards of precision and rigor undreamed of by earlier mathematicians.

Cauchy's life was inextricably tied to the political upheavals of the time. A strong partisan of the Bourbons, he left his wife and children in 1830 to follow the Bourbon king Charles X into exile. For his loyalty he was made a baron by the ex-king. Cauchy eventually returned to France but refused to accept a university position until the government waived its requirement that he take a loyalty oath.

It is difficult to get a clear picture of the man. Devoutly Catholic, he sponsored charitable work for unwed mothers and criminals and relief for Ireland. Yet other aspects of his life cast him in an unfavorable light. The Norwegian mathematician Abel described him as "mad, infinitely Catholic, and bigoted." Some writers praise his teaching, yet others say he rambled incoherently and, according to a report of the day, he once devoted an entire lecture to extracting the square root of seventeen to ten decimal places by a method well known to his students. In any event, Cauchy is undeniably one of the greatest minds in the history of science.

Herman Amandus Schwarz *(1843–1921)*, German mathematician. Schwarz was the leading mathematician in Berlin in the first part of the twentieth century. Because of a devotion to his teaching duties at the University of Berlin and a propensity for treating both important and trivial facts with equal thoroughness, he did not publish in great volume. He tended to focus on narrow concrete problems, but his techniques were often extremely clever and influenced the work of other mathematicians. A version of the inequality that bears his name appeared in a paper about surfaces of minimal area published in 1885.

Augustin Louis (Baron de) Cauchy

Herman Amandus Schwarz

THEOREM 4.1.4	**Properties of Length in R^n**

If $\mathbf{u}$ and $\mathbf{v}$ are vectors in R^n and k is any scalar, then:

(a) $\|\mathbf{u}\| \geq 0$ (b) $\|\mathbf{u}\| = 0$ *if and only if* $\mathbf{u} = \mathbf{0}$

(c) $\|k\mathbf{u}\| = |k|\|\mathbf{u}\|$ (d) $\|\mathbf{u} + \mathbf{v}\| \leq \|\mathbf{u}\| + \|\mathbf{v}\|$ (*Triangle inequality*)

We shall prove (c) and (d) and leave (a) and (b) as exercises.

Proof (c) If $\mathbf{u} = (u_1, u_2, \ldots, u_n)$, then $k\mathbf{u} = (ku_1, ku_2, \ldots, ku_n)$, so

$$\|k\mathbf{u}\| = \sqrt{(ku_1)^2 + (ku_2)^2 + \cdots + (ku_n)^2}$$

$$= |k|\sqrt{u_1^2 + u_2^2 + \cdots + u_n^2}$$

$$= |k|\|\mathbf{u}\|$$

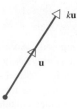

(a) $\|k\mathbf{u}\| = |k|\,\|\mathbf{u}\|$

Proof (d)

$$\|\mathbf{u} + \mathbf{v}\|^2 = (\mathbf{u} + \mathbf{v}) \cdot (\mathbf{u} + \mathbf{v}) = (\mathbf{u} \cdot \mathbf{u}) + 2(\mathbf{u} \cdot \mathbf{v}) + (\mathbf{v} \cdot \mathbf{v})$$

$$= \|\mathbf{u}\|^2 + 2(\mathbf{u} \cdot \mathbf{v}) + \|\mathbf{v}\|^2$$

$$\leq \|\mathbf{u}\|^2 + 2|\mathbf{u} \cdot \mathbf{v}| + \|\mathbf{v}\|^2 \qquad \longleftarrow \text{Property of absolute value}$$

$$\leq \|\mathbf{u}\|^2 + 2\|\mathbf{u}\|\,\|\mathbf{v}\| + \|\mathbf{v}\|^2 \qquad \longleftarrow \text{Cauchy–Schwarz inequality}$$

$$= (\|\mathbf{u}\| + \|\mathbf{v}\|)^2$$

The result now follows on taking square roots of both sides. ∎

Part (c) of this theorem states that multiplying a vector by a scalar k multiplies the length of that vector by a factor of $|k|$ (Figure 4.1.2a). Part (d) of this theorem is known as the ***triangle inequality*** because it generalizes the familiar result from Euclidean geometry that states that the sum of the lengths of any two sides of a triangle is at least as large as the length of the third side (Figure 4.1.2b).

The results in the next theorem are immediate consequences of those in Theorem 4.1.4, as applied to the distance function $d(\mathbf{u}, \mathbf{v})$ on R^n. They generalize the familiar results for R^2 and R^3.

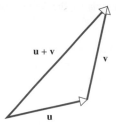

(b) $\|\mathbf{u} + \mathbf{v}\| \leq \|\mathbf{u}\| + \|\mathbf{v}\|$

Figure 4.1.2

THEOREM 4.1.5

> **Properties of Distance in R^n**
>
> *If* $\mathbf{u}$, $\mathbf{v}$, *and* $\mathbf{w}$ *are vectors in* R^n *and* k *is any scalar, then:*
> (a) $d(\mathbf{u}, \mathbf{v}) \geq 0$ $\qquad$ (b) $d(\mathbf{u}, \mathbf{v}) = 0$ *if and only if* $\mathbf{u} = \mathbf{v}$
> (c) $d(\mathbf{u}, \mathbf{v}) = d(\mathbf{v}, \mathbf{u})$ $\qquad$ (d) $d(\mathbf{u}, \mathbf{v}) \leq d(\mathbf{u}, \mathbf{w}) + d(\mathbf{w}, \mathbf{v})$ *(Triangle inequality)*

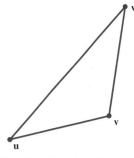

$d(\mathbf{u}, \mathbf{w}) \leq d(\mathbf{u}, \mathbf{v}) + d(\mathbf{v}, \mathbf{w})$

Figure 4.1.3

We shall prove part (d) and leave the remaining parts as exercises.

Proof (d) From (2) and part (d) of Theorem 4.1.4, we have

$$d(\mathbf{u}, \mathbf{v}) = \|\mathbf{u} - \mathbf{v}\| = \|(\mathbf{u} - \mathbf{w}) + (\mathbf{w} - \mathbf{v})\|$$

$$\leq \|\mathbf{u} - \mathbf{w}\| + \|\mathbf{w} - \mathbf{v}\| = d(\mathbf{u}, \mathbf{w}) + d(\mathbf{w}, \mathbf{v}) \quad ∎$$

Part (d) of this theorem, which is also called the *triangle inequality*, generalizes the familiar result from Euclidean geometry that states that the shortest distance between two points is along a straight line (Figure 4.1.3).

Formula (1) expresses the norm of a vector in terms of a dot product. The following useful theorem expresses the dot product in terms of norms.

THEOREM 4.1.6

> *If* $\mathbf{u}$ *and* $\mathbf{v}$ *are vectors in* R^n *with the Euclidean inner product, then*
> $$\mathbf{u} \cdot \mathbf{v} = \tfrac{1}{4}\|\mathbf{u} + \mathbf{v}\|^2 - \tfrac{1}{4}\|\mathbf{u} - \mathbf{v}\|^2 \qquad (6)$$

Proof

$$\|\mathbf{u} + \mathbf{v}\|^2 = (\mathbf{u} + \mathbf{v}) \cdot (\mathbf{u} + \mathbf{v}) = \|\mathbf{u}\|^2 + 2(\mathbf{u} \cdot \mathbf{v}) + \|\mathbf{v}\|^2$$

$$\|\mathbf{u} - \mathbf{v}\|^2 = (\mathbf{u} - \mathbf{v}) \cdot (\mathbf{u} - \mathbf{v}) = \|\mathbf{u}\|^2 - 2(\mathbf{u} \cdot \mathbf{v}) + \|\mathbf{v}\|^2$$

from which (6) follows by simple algebra. ∎

Some problems that use this theorem are given in the exercises.

Orthogonality

Recall that in the Euclidean spaces R^2 and R^3, two vectors **u** and **v** are defined to be *orthogonal* (perpendicular) if $\mathbf{u} \cdot \mathbf{v} = 0$ (Section 3.3). Motivated by this, we make the following definition.

> **DEFINITION**
>
> Two vectors **u** and **v** in R^n are called *orthogonal* if $\mathbf{u} \cdot \mathbf{v} = 0$.

EXAMPLE 4 Orthogonal Vectors in R^4

In the Euclidean space R^4 the vectors

$$\mathbf{u} = (-2, 3, 1, 4) \quad \text{and} \quad \mathbf{v} = (1, 2, 0, -1)$$

are orthogonal, since

$$\mathbf{u} \cdot \mathbf{v} = (-2)(1) + (3)(2) + (1)(0) + (4)(-1) = 0 \ \blacklozenge$$

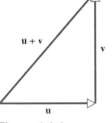
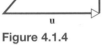

Figure 4.1.4

Properties of orthogonal vectors will be discussed in more detail later in the text, but we note at this point that many of the familiar properties of orthogonal vectors in the Euclidean spaces R^2 and R^3 continue to hold in the Euclidean space R^n. For example, if **u** and **v** are orthogonal vectors in R^2 or R^3, then **u**, **v**, and $\mathbf{u} + \mathbf{v}$ form the sides of a right triangle (Figure 4.1.4); thus, by the Theorem of Pythagoras,

$$\|\mathbf{u} + \mathbf{v}\|^2 = \|\mathbf{u}\|^2 + \|\mathbf{v}\|^2$$

The following theorem shows that this result extends to R^n.

THEOREM 4.1.7

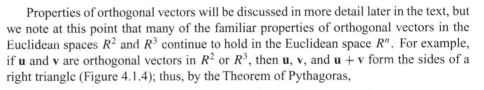

> **Pythagorean Theorem in R^n**
>
> *If **u** and **v** are orthogonal vectors in R^n with the Euclidean inner product, then*
>
> $$\|\mathbf{u} + \mathbf{v}\|^2 = \|\mathbf{u}\|^2 + \|\mathbf{v}\|^2$$

Proof

$$\|\mathbf{u} + \mathbf{v}\|^2 = (\mathbf{u} + \mathbf{v}) \cdot (\mathbf{u} + \mathbf{v}) = \|\mathbf{u}\|^2 + 2(\mathbf{u} \cdot \mathbf{v}) + \|\mathbf{v}\|^2 = \|\mathbf{u}\|^2 + \|\mathbf{v}\|^2 \quad \blacksquare$$

Alternative Notations for Vectors in R^n

It is often useful to write a vector $\mathbf{u} = (u_1, u_2, \ldots, u_n)$ in R^n in matrix notation as a row matrix or a column matrix:

$$\mathbf{u} = \begin{bmatrix} u_1 \\ u_2 \\ \vdots \\ u_n \end{bmatrix} \quad \text{or} \quad \mathbf{u} = [u_1 \quad u_2 \quad \cdots \quad u_n]$$

This is justified because the matrix operations

$$\mathbf{u} + \mathbf{v} = \begin{bmatrix} u_1 \\ u_2 \\ \vdots \\ u_n \end{bmatrix} + \begin{bmatrix} v_1 \\ v_2 \\ \vdots \\ v_n \end{bmatrix} = \begin{bmatrix} u_1 + v_1 \\ u_2 + v_2 \\ \vdots \\ u_n + v_n \end{bmatrix}, \quad k\mathbf{u} = k\begin{bmatrix} u_1 \\ u_2 \\ \vdots \\ u_n \end{bmatrix} = \begin{bmatrix} ku_1 \\ ku_2 \\ \vdots \\ ku_n \end{bmatrix}$$

or

$$\mathbf{u} + \mathbf{v} = [u_1 \quad u_2 \quad \cdots \quad u_n] + [v_1 \quad v_2 \quad \cdots \quad v_n]$$
$$= [u_1 + v_1 \quad u_2 + v_2 \quad \cdots \quad u_n + v_n]$$
$$k\mathbf{u} = k[u_1 \quad u_2 \quad \cdots \quad u_n] = [ku_1 \quad ku_2 \quad \cdots \quad ku_n]$$

produce the same results as the vector operations

$$\mathbf{u} + \mathbf{v} = (u_1, u_2, \ldots, u_n) + (v_1, v_2, \ldots, v_n) = (u_1 + v_1, u_2 + v_2, \ldots, u_n + v_n)$$
$$k\mathbf{u} = k(u_1, u_2, \ldots, u_n) = (ku_1, ku_2, \ldots, ku_n)$$

The only difference is the form in which the vectors are written.

A Matrix Formula for the Dot Product

If we use column matrix notation for the vectors

$$\mathbf{u} = \begin{bmatrix} u_1 \\ u_2 \\ \vdots \\ u_n \end{bmatrix} \quad \text{and} \quad \mathbf{v} = \begin{bmatrix} v_1 \\ v_2 \\ \vdots \\ v_n \end{bmatrix}$$

and omit the brackets on 1×1 matrices, then it follows that

$$\mathbf{v}^T \mathbf{u} = [v_1 \quad v_2 \quad \cdots \quad v_n] \begin{bmatrix} u_1 \\ u_2 \\ \vdots \\ u_n \end{bmatrix} = [u_1 v_1 + u_2 v_2 + \cdots + u_n v_n] = [\mathbf{u} \cdot \mathbf{v}] = \mathbf{u} \cdot \mathbf{v}$$

Thus, for vectors in column matrix notation, we have the following formula for the Euclidean inner product:

$$\mathbf{u} \cdot \mathbf{v} = \mathbf{v}^T \mathbf{u} \tag{7}$$

For example, if

$$\mathbf{u} = \begin{bmatrix} -1 \\ 3 \\ 5 \\ 7 \end{bmatrix} \quad \text{and} \quad \mathbf{v} = \begin{bmatrix} 5 \\ -4 \\ 7 \\ 0 \end{bmatrix}$$

then

$$\mathbf{u} \cdot \mathbf{v} = \mathbf{v}^T \mathbf{u} = [5 \quad -4 \quad 7 \quad 0] \begin{bmatrix} -1 \\ 3 \\ 5 \\ 7 \end{bmatrix} = [18] = 18$$

If A is an $n \times n$ matrix, then it follows from Formula (7) and properties of the transpose that

$$A\mathbf{u} \cdot \mathbf{v} = \mathbf{v}^T (A\mathbf{u}) = (\mathbf{v}^T A)\mathbf{u} = (A^T \mathbf{v})^T \mathbf{u} = \mathbf{u} \cdot A^T \mathbf{v}$$
$$\mathbf{u} \cdot A\mathbf{v} = (A\mathbf{v})^T \mathbf{u} = (\mathbf{v}^T A^T)\mathbf{u} = \mathbf{v}^T (A^T \mathbf{u}) = A^T \mathbf{u} \cdot \mathbf{v}$$

The resulting formulas

$$A\mathbf{u} \cdot \mathbf{v} = \mathbf{u} \cdot A^T \mathbf{v} \tag{8}$$

$$\mathbf{u} \cdot A\mathbf{v} = A^T \mathbf{u} \cdot \mathbf{v} \tag{9}$$

provide an important link between multiplication by an $n \times n$ matrix A and multiplication by A^T.

EXAMPLE 5 Verifying That $A\mathbf{u} \cdot \mathbf{v} = \mathbf{u} \cdot A^T\mathbf{v}$

Suppose that

$$A = \begin{bmatrix} 1 & -2 & 3 \\ 2 & 4 & 1 \\ -1 & 0 & 1 \end{bmatrix}, \quad \mathbf{u} = \begin{bmatrix} -1 \\ 2 \\ 4 \end{bmatrix}, \quad \mathbf{v} = \begin{bmatrix} -2 \\ 0 \\ 5 \end{bmatrix}$$

Then

$$A\mathbf{u} = \begin{bmatrix} 1 & -2 & 3 \\ 2 & 4 & 1 \\ -1 & 0 & 1 \end{bmatrix}\begin{bmatrix} -1 \\ 2 \\ 4 \end{bmatrix} = \begin{bmatrix} 7 \\ 10 \\ 5 \end{bmatrix}$$

$$A^T\mathbf{v} = \begin{bmatrix} 1 & 2 & -1 \\ -2 & 4 & 0 \\ 3 & 1 & 1 \end{bmatrix}\begin{bmatrix} -2 \\ 0 \\ 5 \end{bmatrix} = \begin{bmatrix} -7 \\ 4 \\ -1 \end{bmatrix}$$

from which we obtain

$$A\mathbf{u} \cdot \mathbf{v} = 7(-2) + 10(0) + 5(5) = 11$$
$$\mathbf{u} \cdot A^T\mathbf{v} = (-1)(-7) + 2(4) + 4(-1) = 11$$

Thus $A\mathbf{u} \cdot \mathbf{v} = \mathbf{u} \cdot A^T\mathbf{v}$ as guaranteed by Formula (8). We leave it for the reader to verify that (9) also holds. ◆

A Dot Product View of Matrix Multiplication

Dot products provide another way of thinking about matrix multiplication. Recall that if $A = [a_{ij}]$ is an $m \times r$ matrix and $B = [b_{ij}]$ is an $r \times n$ matrix, then the ijth entry of AB is

$$a_{i1}b_{1j} + a_{i2}b_{2j} + \cdots + a_{ir}b_{rj}$$

which is the dot product of the ith row vector of A

$$[a_{i1} \quad a_{i2} \quad \cdots \quad a_{ir}]$$

and the jth column vector of B

$$\begin{bmatrix} b_{1j} \\ b_{2j} \\ \vdots \\ b_{rj} \end{bmatrix}$$

Thus, if the row vectors of A are $\mathbf{r}_1, \mathbf{r}_2, \ldots, \mathbf{r}_m$ and the column vectors of B are $\mathbf{c}_1, \mathbf{c}_2, \ldots, \mathbf{c}_n$, then the matrix product AB can be expressed as

$$AB = \begin{bmatrix} \mathbf{r}_1 \cdot \mathbf{c}_1 & \mathbf{r}_1 \cdot \mathbf{c}_2 & \cdots & \mathbf{r}_1 \cdot \mathbf{c}_n \\ \mathbf{r}_2 \cdot \mathbf{c}_1 & \mathbf{r}_2 \cdot \mathbf{c}_2 & \cdots & \mathbf{r}_2 \cdot \mathbf{c}_n \\ \vdots & \vdots & & \vdots \\ \mathbf{r}_m \cdot \mathbf{c}_1 & \mathbf{r}_m \cdot \mathbf{c}_2 & \cdots & \mathbf{r}_m \cdot \mathbf{c}_n \end{bmatrix} \quad (10)$$

In particular, a linear system $A\mathbf{x} = \mathbf{b}$ can be expressed in dot product form as

$$\begin{bmatrix} \mathbf{r}_1 \cdot \mathbf{x} \\ \mathbf{r}_2 \cdot \mathbf{x} \\ \vdots \\ \mathbf{r}_m \cdot \mathbf{x} \end{bmatrix} = \begin{bmatrix} b_1 \\ b_2 \\ \vdots \\ b_m \end{bmatrix} \quad (11)$$

where $\mathbf{r}_1, \mathbf{r}_2, \ldots, \mathbf{r}_m$ are the row vectors of A, and $b_1, b_2, \ldots, b_m$ are the entries of $\mathbf{b}$.

EXAMPLE 6   A Linear System Written in Dot Product Form

The following is an example of a linear system expressed in dot product form (11).

$$
\begin{array}{cc}
\textbf{System} & \textbf{Dot Product Form} \\[4pt]
\begin{aligned}
3x_1 - 4x_2 + \ x_3 &= 1 \\
2x_1 - 7x_2 - 4x_3 &= 5 \\
x_1 + 5x_2 - 8x_3 &= 0
\end{aligned}
&
\begin{bmatrix}
(3, -4, 1) \cdot (x_1, x_2, x_3) \\
(2, -7, -4) \cdot (x_1, x_2, x_3) \\
(1, 5, -8) \cdot (x_1, x_2, x_3)
\end{bmatrix}
=
\begin{bmatrix}
1 \\
5 \\
0
\end{bmatrix}
\ \blacklozenge
\end{array}
$$

EXERCISE SET 4.1

1. Let $\mathbf{u} = (-3, 2, 1, 0)$, $\mathbf{v} = (4, 7, -3, 2)$, and $\mathbf{w} = (5, -2, 8, 1)$. Find

   (a) $\mathbf{v} - \mathbf{w}$ (b) $2\mathbf{u} + 7\mathbf{v}$ (c) $-\mathbf{u} + (\mathbf{v} - 4\mathbf{w})$

   (d) $6(\mathbf{u} - 3\mathbf{v})$ (e) $-\mathbf{v} - \mathbf{w}$ (f) $(6\mathbf{v} - \mathbf{w}) - (4\mathbf{u} + \mathbf{v})$

2. Let $\mathbf{u}$, $\mathbf{v}$, and $\mathbf{w}$ be the vectors in Exercise 1. Find the vector $\mathbf{x}$ that satisfies $5\mathbf{x} - 2\mathbf{v} = 2(\mathbf{w} - 5\mathbf{x})$.

3. Let $\mathbf{u}_1 = (-1, 3, 2, 0)$, $\mathbf{u}_2 = (2, 0, 4, -1)$, $\mathbf{u}_3 = (7, 1, 1, 4)$, and $\mathbf{u}_4 = (6, 3, 1, 2)$. Find scalars c_1, c_2, c_3, and c_4 such that $c_1\mathbf{u}_1 + c_2\mathbf{u}_2 + c_3\mathbf{u}_3 + c_4\mathbf{u}_4 = (0, 5, 6, -3)$.

4. Show that there do not exist scalars c_1, c_2, and c_3 such that

   $$c_1(1, 0, 1, 0) + c_2(1, 0, -2, 1) + c_3(2, 0, 1, 2) = (1, -2, 2, 3)$$

5. In each part, compute the Euclidean norm of the vector.

   (a) $(-2, 5)$ (b) $(1, 2, -2)$ (c) $(3, 4, 0, -12)$ (d) $(-2, 1, 1, -3, 4)$

6. Let $\mathbf{u} = (4, 1, 2, 3)$, $\mathbf{v} = (0, 3, 8, -2)$, and $\mathbf{w} = (3, 1, 2, 2)$. Evaluate each expression.

   (a) $\|\mathbf{u} + \mathbf{v}\|$ (b) $\|\mathbf{u}\| + \|\mathbf{v}\|$ (c) $\|-2\mathbf{u}\| + 2\|\mathbf{u}\|$

   (d) $\|3\mathbf{u} - 5\mathbf{v} + \mathbf{w}\|$ (e) $\dfrac{1}{\|\mathbf{w}\|}\mathbf{w}$ (f) $\left\|\dfrac{1}{\|\mathbf{w}\|}\mathbf{w}\right\|$

7. Show that if $\mathbf{v}$ is a nonzero vector in R^n, then $(1/\|\mathbf{v}\|)\mathbf{v}$ has Euclidean norm 1.

8. Let $\mathbf{v} = (-2, 3, 0, 6)$. Find all scalars k such that $\|k\mathbf{v}\| = 5$.

9. Find the Euclidean inner product $\mathbf{u} \cdot \mathbf{v}$.

   (a) $\mathbf{u} = (2, 5)$, $\mathbf{v} = (-4, 3)$

   (b) $\mathbf{u} = (4, 8, 2)$, $\mathbf{v} = (0, 1, 3)$

   (c) $\mathbf{u} = (3, 1, 4, -5)$, $\mathbf{v} = (2, 2, -4, -3)$

   (d) $\mathbf{u} = (-1, 1, 0, 4, -3)$, $\mathbf{v} = (-2, -2, 0, 2, -1)$

10. (a) Find two vectors in R^2 with Euclidean norm 1 whose Euclidean inner product with $(3, -1)$ is zero.

    (b) Show that there are infinitely many vectors in R^3 with Euclidean norm 1 whose Euclidean inner product with $(1, -3, 5)$ is zero.

11. Find the Euclidean distance between $\mathbf{u}$ and $\mathbf{v}$.

    (a) $\mathbf{u} = (1, -2)$, $\mathbf{v} = (2, 1)$

    (b) $\mathbf{u} = (2, -2, 2)$, $\mathbf{v} = (0, 4, -2)$

    (c) $\mathbf{u} = (0, -2, -1, 1)$, $\mathbf{v} = (-3, 2, 4, 4)$

    (d) $\mathbf{u} = (3, -3, -2, 0, -3)$, $\mathbf{v} = (-4, 1, -1, 5, 0)$

12. Verify parts (b), (e), (f), and (g) of Theorem 4.1.1 for $\mathbf{u} = (2, 0, -3, 1)$, $\mathbf{v} = (4, 0, 3, 5)$, $\mathbf{w} = (1, 6, 2, -1)$, $k = 5$, and $l = -3$.

13. Verify parts (b) and (c) of Theorem 4.1.2 for the values of **u**, **v**, **w**, and k in Exercise 12.

14. In each part, determine whether the given vectors are orthogonal.
    (a) $\mathbf{u} = (-1, 3, 2)$, $\mathbf{v} = (4, 2, -1)$ (b) $\mathbf{u} = (-2, -2, -2)$, $\mathbf{v} = (1, 1, 1)$
    (c) $\mathbf{u} = (u_1, u_2, u_3)$, $\mathbf{v} = (0, 0, 0)$ (d) $\mathbf{u} = (-4, 6, -10, 1)$, $\mathbf{v} = (2, 1, -2, 9)$
    (e) $\mathbf{u} = (0, 3, -2, 1)$, $\mathbf{v} = (5, 2, -1, 0)$ (f) $\mathbf{u} = (a, b)$, $\mathbf{v} = (-b, a)$

15. For which values of k are **u** and **v** orthogonal?
    (a) $\mathbf{u} = (2, 1, 3)$, $\mathbf{v} = (1, 7, k)$ (b) $\mathbf{u} = (k, k, 1)$, $\mathbf{v} = (k, 5, 6)$

16. Find two vectors of norm 1 that are orthogonal to the three vectors $\mathbf{u} = (2, 1, -4, 0)$, $\mathbf{v} = (-1, -1, 2, 2)$, and $\mathbf{w} = (3, 2, 5, 4)$.

17. In each part, verify that the Cauchy–Schwarz inequality holds.
    (a) $\mathbf{u} = (3, 2)$, $\mathbf{v} = (4, -1)$ (b) $\mathbf{u} = (-3, 1, 0)$, $\mathbf{v} = (2, -1, 3)$
    (c) $\mathbf{u} = (-4, 2, 1)$, $\mathbf{v} = (8, -4, -2)$ (d) $\mathbf{u} = (0, -2, 2, 1)$, $\mathbf{v} = (-1, -1, 1, 1)$

18. In each part, verify that Formulas (8) and (9) hold.

    (a) $A = \begin{bmatrix} 2 & -1 \\ 3 & 4 \end{bmatrix}$, $\mathbf{u} = \begin{bmatrix} 3 \\ 1 \end{bmatrix}$, $\mathbf{v} = \begin{bmatrix} -2 \\ 6 \end{bmatrix}$

    (b) $A = \begin{bmatrix} -1 & 2 & 4 \\ 3 & 1 & 0 \\ 5 & -2 & 3 \end{bmatrix}$, $\mathbf{u} = \begin{bmatrix} -1 \\ 2 \\ 5 \end{bmatrix}$, $\mathbf{v} = \begin{bmatrix} 0 \\ 2 \\ -4 \end{bmatrix}$

19. Solve the following linear system for x_1, x_2, and x_3.

    $$(1, -1, 4) \cdot (x_1, x_2, x_3) = 10$$
    $$(3, 2, 0) \cdot (x_1, x_2, x_3) = 1$$
    $$(4, -5, -1) \cdot (x_1, x_2, x_3) = 7$$

20. Find $\mathbf{u} \cdot \mathbf{v}$ given that $\|\mathbf{u} + \mathbf{v}\| = 1$ and $\|\mathbf{u} - \mathbf{v}\| = 5$.

21. Use Theorem 4.1.6 to show that **u** and **v** are orthogonal vectors in R^n if $\|\mathbf{u} + \mathbf{v}\| = \|\mathbf{u} - \mathbf{v}\|$. Interpret this result geometrically in R^2.

22. The formulas for the vector components in Theorem 3.3.3 hold in R^n as well. Given that $\mathbf{a} = (-1, 1, 2, 3)$ and $\mathbf{u} = (2, 1, 4, -1)$, find the vector component of **u** along **a** and the vector component of **u** orthogonal to **a**.

23. Determine whether the two lines

    $$\mathbf{r} = (3, 2, 3, -1) + t(4, 6, 4, -2) \quad \text{and} \quad \mathbf{r} = (0, 3, 5, 4) + s(1, -3, -4, -2)$$

    intersect in R^4.

24. Prove the following generalization of Theorem 4.1.7. If $\mathbf{v}_1, \mathbf{v}_2, \dots, \mathbf{v}_r$ are pairwise orthogonal vectors in R^n, then

    $$\|\mathbf{v}_1 + \mathbf{v}_2 + \cdots + \mathbf{v}_r\|^2 = \|\mathbf{v}_1\|^2 + \|\mathbf{v}_2\|^2 + \cdots + \|\mathbf{v}_r\|^2$$

25. Prove: If **u** and **v** are $n \times 1$ matrices and A is an $n \times n$ matrix, then

    $$(\mathbf{v}^T A^T A \mathbf{u})^2 \leq (\mathbf{u}^T A^T A \mathbf{u})(\mathbf{v}^T A^T A \mathbf{v})$$

26. Use the Cauchy–Schwarz inequality to prove that for all real values of a, b, and θ,

    $$(a \cos \theta + b \sin \theta)^2 \leq a^2 + b^2$$

27. Prove: If **u**, **v**, and **w** are vectors in R^n and k is any scalar, then
    (a) $\mathbf{u} \cdot (k\mathbf{v}) = k(\mathbf{u} \cdot \mathbf{v})$ (b) $\mathbf{u} \cdot (\mathbf{v} + \mathbf{w}) = \mathbf{u} \cdot \mathbf{v} + \mathbf{u} \cdot \mathbf{w}$

28. Prove parts (a) through (d) of Theorem 4.1.1.

29. Prove parts (e) through (h) of Theorem 4.1.1.

30. Prove parts (a) and (c) of Theorem 4.1.2.

31. Prove parts (a) and (b) of Theorem 4.1.4.

32. Prove parts (a), (b), and (c) of Theorem 4.1.5.

33. Suppose that $a_1, a_2, \ldots, a_n$ are positive real numbers. In R^2, the vectors $\mathbf{v}_1 = (a_1, 0)$ and $\mathbf{v}_2 = (0, a_2)$ determine a rectangle of area $A = a_1 a_2$ (see the accompanying figure), and in R^3, the vectors $\mathbf{v}_1 = (a_1, 0, 0)$, $\mathbf{v}_2 = (0, a_2, 0)$, and $\mathbf{v}_3 = (0, 0, a_3)$ determine a box of volume $V = a_1 a_2 a_3$ (see the accompanying figure). The area A and the volume V are sometimes called the **Euclidean measure** of the rectangle and box, respectively.

(a) How would you define the Euclidean measure of the "box" in R^n that is determined by the vectors

$$\mathbf{v}_1 = (a_1, 0, 0, \ldots, 0), \quad \mathbf{v}_2 = (0, a_2, 0, \ldots, 0), \ldots, \quad \mathbf{v}_n = (0, 0, 0, \ldots, a_n)?$$

(b) How would you define the Euclidean length of the "diagonal" of the box in part (a)?

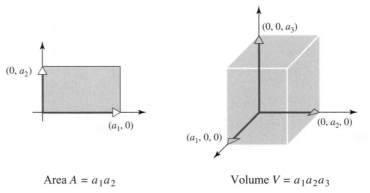

Area $A = a_1 a_2$ Volume $V = a_1 a_2 a_3$

Figure Ex-33

Discussion & Discovery

34. (a) Suppose that $\mathbf{u}$ and $\mathbf{v}$ are vectors in R^n. Show that

$$\|\mathbf{u} + \mathbf{v}\|^2 + \|\mathbf{u} - \mathbf{v}\|^2 = 2(\|\mathbf{u}\|^2 + \|\mathbf{v}\|^2)$$

(b) The result in part (a) states a theorem about parallelograms in R^2. What is the theorem?

35. (a) If $\mathbf{u}$ and $\mathbf{v}$ are orthogonal vectors in R^n such that $\|\mathbf{u}\| = 1$ and $\|\mathbf{v}\| = 1$, then $d(\mathbf{u}, \mathbf{v}) =$____.

(b) Draw a picture to illustrate this result.

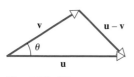

Figure Ex-36

36. In the accompanying figure the vectors $\mathbf{u}$, $\mathbf{v}$, and $\mathbf{u} - \mathbf{v}$ form a triangle in R^2, and θ denotes the angle between $\mathbf{u}$ and $\mathbf{v}$. It follows from the law of cosines in trigonometry that

$$\|\mathbf{u} - \mathbf{v}\|^2 = \|\mathbf{u}\|^2 + \|\mathbf{v}\|^2 - 2\|\mathbf{u}\|\|\mathbf{v}\|\cos\theta$$

Do you think that this formula still holds if $\mathbf{u}$ and $\mathbf{v}$ are vectors in R^n? Justify your answer.

37. Indicate whether each statement is always true or sometimes false. Justify your answer by giving a logical argument or a counterexample.

(a) If $\|\mathbf{u} + \mathbf{v}\|^2 = \|\mathbf{u}\|^2 + \|\mathbf{v}\|^2$, then $\mathbf{u}$ and $\mathbf{v}$ are orthogonal.

(b) If $\mathbf{u}$ is orthogonal to $\mathbf{v}$ and $\mathbf{w}$, then $\mathbf{u}$ is orthogonal to $\mathbf{v} + \mathbf{w}$.

(c) If $\mathbf{u}$ is orthogonal to $\mathbf{v} + \mathbf{w}$, then $\mathbf{u}$ is orthogonal to $\mathbf{v}$ and $\mathbf{w}$.

(d) If $\|\mathbf{u} - \mathbf{v}\| = 0$, then $\mathbf{u} = \mathbf{v}$.

(e) If $\|k\mathbf{u}\| = k\|\mathbf{u}\|$, then $k \geq 0$.

4.2

LINEAR TRANS-
FORMATIONS
FROM R^n TO R^m

In this section we shall begin the study of functions of the form $\mathbf{w} = F(\mathbf{x})$, where the independent variable $\mathbf{x}$ is a vector in R^n and the dependent variable $\mathbf{w}$ is a vector in R^m. We shall concentrate on a special class of such functions called "linear transformations." Linear transformations are fundamental in the study of linear algebra and have many important applications in physics, engineering, social sciences, and various branches of mathematics.

Functions from R^n to R

Recall that a ***function*** is a rule f that associates with each element in a set A one and only one element in a set B. If f associates the element b with the element a, then we write $b = f(a)$ and say that b is the ***image*** of a under f or that $f(a)$ is the ***value*** of f at a. The set A is called the ***domain*** of f and the set B is called the ***codomain*** of f. The subset of B consisting of all possible values for f as a varies over A is called the ***range*** of f. For the most common functions, A and B are sets of real numbers, in which case f is called a ***real-valued function of a real variable***. Other common functions occur when B is a set of real numbers and A is a set of vectors in R^2, R^3, or, more generally, R^n. Some examples are shown in Table 1. Two functions f_1 and f_2 are regarded as ***equal***, written $f_1 = f_2$, if they have the same domain and $f_1(a) = f_2(a)$ for all a in the domain.

Table 1

Formula	Example	Classification	Description
$f(x)$	$f(x) = x^2$	Real-valued function of a real variable	Function from R to R
$f(x, y)$	$f(x, y) = x^2 + y^2$	Real-valued function of two real variables	Function from R^2 to R
$f(x, y, z)$	$f(x, y, z) = x^2 + y^2 + z^2$	Real-valued function of three real variables	Function from R^3 to R
$f(x_1, x_2, \ldots, x_n)$	$f(x_1, x_2, \ldots, x_n) = x_1^2 + x_2^2 + \cdots + x_n^2$	Real-valued function of n real variables	Function from R^n to R

Functions from R^n to R^m

If the domain of a function f is R^n and the codomain is R^m (m and n possibly the same), then f is called a ***map*** or ***transformation*** from R^n to R^m, and we say that the function f ***maps*** R^n into R^m. We denote this by writing $f: R^n \to R^m$. The functions in Table 1 are transformations for which $m = 1$. In the case where $m = n$, the transformation $f: R^n \to R^n$ is called an ***operator*** on R^n. The first entry in Table 1 is an operator on R.

To illustrate one important way in which transformations can arise, suppose that $f_1, f_2, \ldots, f_m$ are real-valued functions of n real variables, say

$$
\begin{aligned}
w_1 &= f_1(x_1, x_2, \ldots, x_n) \\
w_2 &= f_2(x_1, x_2, \ldots, x_n) \\
&\vdots \qquad\qquad \vdots \\
w_m &= f_m(x_1, x_2, \ldots, x_n)
\end{aligned}
\tag{1}
$$

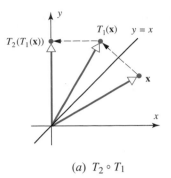

(a) $T_2 \circ T_1$

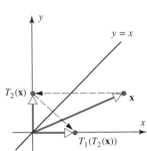

(b) $T_1 \circ T_2$

Figure 4.2.8

and $T_2 \circ T_1$ have different effects on a vector $\mathbf{x}$. This same conclusion can be reached by showing that the standard matrices for T_1 and T_2 do not commute:

$$[T_1 \circ T_2] = [T_1][T_2] = \begin{bmatrix} 0 & 1 \\ 1 & 0 \end{bmatrix}\begin{bmatrix} 0 & 0 \\ 0 & 1 \end{bmatrix} = \begin{bmatrix} 0 & 1 \\ 0 & 0 \end{bmatrix}$$

$$[T_2 \circ T_1] = [T_2][T_1] = \begin{bmatrix} 0 & 0 \\ 0 & 1 \end{bmatrix}\begin{bmatrix} 0 & 1 \\ 1 & 0 \end{bmatrix} = \begin{bmatrix} 0 & 0 \\ 1 & 0 \end{bmatrix}$$

so $[T_2 \circ T_1] \neq [T_1 \circ T_2]$. ◆

EXAMPLE 8 Composition of Two Reflections

Let $T_1: R^2 \to R^2$ be the reflection about the y-axis, and let $T_2: R^2 \to R^2$ be the reflection about the x-axis. In this case $T_1 \circ T_2$ and $T_2 \circ T_1$ are the same; both map each vector $\mathbf{x} = (x, y)$ into its negative $-\mathbf{x} = (-x, -y)$ (Figure 4.2.9):

$$(T_1 \circ T_2)(x, y) = T_1(x, -y) = (-x, -y)$$
$$(T_2 \circ T_1)(x, y) = T_2(-x, y) = (-x, -y)$$

The equality of $T_1 \circ T_2$ and $T_2 \circ T_1$ can also be deduced by showing that the standard matrices for T_1 and T_2 commute:

$$[T_1 \circ T_2] = [T_1][T_2] = \begin{bmatrix} -1 & 0 \\ 0 & 1 \end{bmatrix}\begin{bmatrix} 1 & 0 \\ 0 & -1 \end{bmatrix} = \begin{bmatrix} -1 & 0 \\ 0 & -1 \end{bmatrix}$$

$$[T_2 \circ T_1] = [T_2][T_1] = \begin{bmatrix} 1 & 0 \\ 0 & -1 \end{bmatrix}\begin{bmatrix} -1 & 0 \\ 0 & 1 \end{bmatrix} = \begin{bmatrix} -1 & 0 \\ 0 & -1 \end{bmatrix}$$

The operator $T(\mathbf{x}) = -\mathbf{x}$ on R^2 or R^3 is called the ***reflection about the origin***. As the computations above show, the standard matrix for this operator on R^2 is

$$[T] = \begin{bmatrix} -1 & 0 \\ 0 & -1 \end{bmatrix} ◆$$

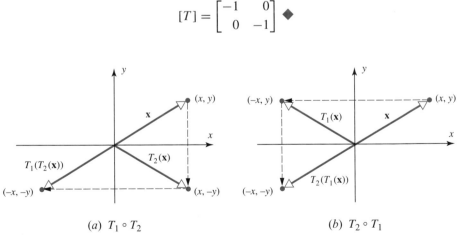

(a) $T_1 \circ T_2$

(b) $T_2 \circ T_1$

Figure 4.2.9

Compositions of Three or More Linear Transformations

Compositions can be defined for three or more linear transformations. For example, consider the linear transformations

$$T_1: R^n \to R^k, \qquad T_2: R^k \to R^l, \qquad T_3: R^l \to R^m$$

We define the composition $(T_3 \circ T_2 \circ T_1): R^n \to R^m$ by

$$(T_3 \circ T_2 \circ T_1)(\mathbf{x}) = T_3(T_2(T_1(\mathbf{x})))$$

It can be shown that this composition is a linear transformation and that the standard matrix for $T_3 \circ T_2 \circ T_1$ is related to the standard matrices for T_1, T_2, and T_3 by

$$[T_3 \circ T_2 \circ T_1] = [T_3][T_2][T_1] \tag{22}$$

which is a generalization of (21). If the standard matrices for T_1, T_2, and T_3 are denoted by A, B, and C, respectively, then we also have the following generalization of (20):

$$T_C \circ T_B \circ T_A = T_{CBA} \tag{23}$$

EXAMPLE 9 Composition of Three Transformations

Find the standard matrix for the linear operator $T: R^3 \to R^3$ that first rotates a vector counterclockwise about the z-axis through an angle θ, then reflects the resulting vector about the yz-plane, and then projects that vector orthogonally onto the xy-plane.

Solution

The linear transformation T can be expressed as the composition

$$T = T_3 \circ T_2 \circ T_1$$

where T_1 is the rotation about the z-axis, T_2 is the reflection about the yz-plane, and T_3 is the orthogonal projection on the xy-plane. From Tables 3, 5, and 7, the standard matrices for these linear transformations are

$$[T_1] = \begin{bmatrix} \cos\theta & -\sin\theta & 0 \\ \sin\theta & \cos\theta & 0 \\ 0 & 0 & 1 \end{bmatrix}, \qquad [T_2] = \begin{bmatrix} -1 & 0 & 0 \\ 0 & 1 & 0 \\ 0 & 0 & 1 \end{bmatrix}, \qquad [T_3] = \begin{bmatrix} 1 & 0 & 0 \\ 0 & 1 & 0 \\ 0 & 0 & 0 \end{bmatrix}$$

Thus, from (22) the standard matrix for T is $[T] = [T_3][T_2][T_1]$; that is,

$$[T] = \begin{bmatrix} 1 & 0 & 0 \\ 0 & 1 & 0 \\ 0 & 0 & 0 \end{bmatrix} \begin{bmatrix} -1 & 0 & 0 \\ 0 & 1 & 0 \\ 0 & 0 & 1 \end{bmatrix} \begin{bmatrix} \cos\theta & -\sin\theta & 0 \\ \sin\theta & \cos\theta & 0 \\ 0 & 0 & 1 \end{bmatrix}$$

$$= \begin{bmatrix} -\cos\theta & \sin\theta & 0 \\ \sin\theta & \cos\theta & 0 \\ 0 & 0 & 0 \end{bmatrix} \blacklozenge$$

EXERCISE SET
4.2

1. Find the domain and codomain of the transformation defined by the equations, and determine whether the transformation is linear.

(a) $w_1 = 3x_1 - 2x_2 + 4x_3$
$w_2 = 5x_1 - 8x_2 + x_3$

(b) $w_1 = 2x_1 x_2 - x_2$
$w_2 = x_1 + 3x_1 x_2$
$w_3 = x_1 + x_2$

(c) $w_1 = 5x_1 - x_2 + x_3$
$w_2 = -x_1 + x_2 + 7x_3$
$w_3 = 2x_1 - 4x_2 - x_3$

(d) $w_1 = x_1^2 - 3x_2 + x_3 - 2x_4$
$w_2 = 3x_1 - 4x_2 - x_3^2 + x_4$

2. Find the standard matrix for the linear transformation defined by the equations.

   (a) $w_1 = 2x_1 - 3x_2 + x_4$
   $w_2 = 3x_1 + 5x_2 - x_4$

   (b) $w_1 = 7x_1 + 2x_2 - 8x_3$
   $w_2 = - x_2 + 5x_3$
   $w_3 = 4x_1 + 7x_2 - x_3$

   (c) $w_1 = -x_1 + x_2$
   $w_2 = 3x_1 - 2x_2$
   $w_3 = 5x_1 - 7x_2$

   (d) $w_1 = x_1$
   $w_2 = x_1 + x_2$
   $w_3 = x_1 + x_2 + x_3$
   $w_4 = x_1 + x_2 + x_3 + x_4$

3. Find the standard matrix for the linear operator $T: R^3 \to R^3$ given by

$$w_1 = 3x_1 + 5x_2 - x_3$$
$$w_2 = 4x_1 - x_2 + x_3$$
$$w_3 = 3x_1 + 2x_2 - x_3$$

   and then calculate $T(-1, 2, 4)$ by directly substituting in the equations and also by matrix multiplication.

4. Find the standard matrix for the linear operator T defined by the formula.

   (a) $T(x_1, x_2) = (2x_1 - x_2, x_1 + x_2)$

   (b) $T(x_1, x_2) = (x_1, x_2)$

   (c) $T(x_1, x_2, x_3) = (x_1 + 2x_2 + x_3, x_1 + 5x_2, x_3)$

   (d) $T(x_1, x_2, x_3) = (4x_1, 7x_2, -8x_3)$

5. Find the standard matrix for the linear transformation T defined by the formula.

   (a) $T(x_1, x_2) = (x_2, -x_1, x_1 + 3x_2, x_1 - x_2)$

   (b) $T(x_1, x_2, x_3, x_4) = (7x_1 + 2x_2 - x_3 + x_4, x_2 + x_3, -x_1)$

   (c) $T(x_1, x_2, x_3) = (0, 0, 0, 0, 0)$

   (d) $T(x_1, x_2, x_3, x_4) = (x_4, x_1, x_3, x_2, x_1 - x_3)$

6. In each part, the standard matrix $[T]$ of a linear transformation T is given. Use it to find $T(x)$. [Express the answers in matrix form.]

   (a) $[T] = \begin{bmatrix} 1 & 2 \\ 3 & 4 \end{bmatrix}$; $\mathbf{x} = \begin{bmatrix} 3 \\ -2 \end{bmatrix}$

   (b) $[T] = \begin{bmatrix} -1 & 2 & 0 \\ 3 & 1 & 5 \end{bmatrix}$; $\mathbf{x} = \begin{bmatrix} -1 \\ 1 \\ 3 \end{bmatrix}$

   (c) $[T] = \begin{bmatrix} -2 & 1 & 4 \\ 3 & 5 & 7 \\ 6 & 0 & -1 \end{bmatrix}$; $\mathbf{x} = \begin{bmatrix} x_1 \\ x_2 \\ x_3 \end{bmatrix}$

   (d) $[T] = \begin{bmatrix} -1 & 1 \\ 2 & 4 \\ 7 & 8 \end{bmatrix}$; $\mathbf{x} = \begin{bmatrix} x_1 \\ x_2 \end{bmatrix}$

7. In each part, use the standard matrix for T to find $T(\mathbf{x})$; then check the result by calculating $T(\mathbf{x})$ directly.

   (a) $T(x_1, x_2) = (-x_1 + x_2, x_2)$; $\mathbf{x} = (-1, 4)$

   (b) $T(x_1, x_2, x_3) = (2x_1 - x_2 + x_3, x_2 + x_3, 0)$; $\mathbf{x} = (2, 1, -3)$

8. Use matrix multiplication to find the reflection of $(-1, 2)$ about

   (a) the x-axis (b) the y-axis (c) the line $y = x$

9. Use matrix multiplication to find the reflection of $(2, -5, 3)$ about

   (a) the xy-plane (b) the xz-plane (c) the yz-plane

10. Use matrix multiplication to find the orthogonal projection of $(2, -5)$ on

    (a) the x-axis (b) the y-axis

11. Use matrix multiplication to find the orthogonal projection of $(-2, 1, 3)$ on

    (a) the xy-plane (b) the xz-plane (c) the yz-plane

12. Use matrix multiplication to find the image of the vector $(3, -4)$ when it is rotated through an angle of

 (a) $\theta = 30°$ (b) $\theta = -60°$ (c) $\theta = 45°$ (d) $\theta = 90°$

13. Use matrix multiplication to find the image of the vector $(-2, 1, 2)$ if it is rotated

 (a) $30°$ about the x-axis   (b) $45°$ about the y-axis   (c) $90°$ about the z-axis

14. Find the standard matrix for the linear operator that rotates a vector in R^3 through an angle of $-60°$ about

 (a) the x-axis   (b) the y-axis   (c) the z-axis

15. Use matrix multiplication to find the image of the vector $(-2, 1, 2)$ if it is rotated

 (a) $-30°$ about the x-axis   (b) $-45°$ about the y-axis   (c) $-90°$ about the z-axis

16. Find the standard matrix for the stated composition of linear operators on R^2.

 (a) A rotation of $90°$, followed by a reflection about the line $y = x$.

 (b) An orthogonal projection on the y-axis, followed by a contraction with factor $k = \frac{1}{2}$.

 (c) A reflection about the x-axis, followed by a dilation with factor $k = 3$.

17. Find the standard matrix for the stated composition of linear operators on R^2.

 (a) A rotation of $60°$, followed by an orthogonal projection on the x-axis, followed by a reflection about the line $y = x$.

 (b) A dilation with factor $k = 2$, followed by a rotation of $45°$, followed by a reflection about the y-axis.

 (c) A rotation of $15°$, followed by a rotation of $105°$, followed by a rotation of $60°$.

18. Find the standard matrix for the stated composition of linear operators on R^3.

 (a) A reflection about the yz-plane, followed by an orthogonal projection on the xz-plane.

 (b) A rotation of $45°$ about the y-axis, followed by a dilation with factor $k = \sqrt{2}$.

 (c) An orthogonal projection on the xy-plane, followed by a reflection about the yz-plane.

19. Find the standard matrix for the stated composition of linear operators on R^3.

 (a) A rotation of $30°$ about the x-axis, followed by a rotation of $30°$ about the z-axis, followed by a contraction with factor $k = \frac{1}{4}$.

 (b) A reflection about the xy-plane, followed by a reflection about the xz-plane, followed by an orthogonal projection on the yz-plane.

 (c) A rotation of $270°$ about the x-axis, followed by a rotation of $90°$ about the y-axis, followed by a rotation of $180°$ about the z-axis.

20. Determine whether $T_1 \circ T_2 = T_2 \circ T_1$.

 (a) $T_1 \colon R^2 \to R^2$ is the orthogonal projection on the x-axis, and $T_2 \colon R^2 \to R^2$ is the orthogonal projection on the y-axis.

 (b) $T_1 \colon R^2 \to R^2$ is the rotation through an angle θ_1, and $T_2 \colon R^2 \to R^2$ is the rotation through an angle θ_2.

 (c) $T_1 \colon R^2 \to R^2$ is the orthogonal projection on the x-axis, and $T_2 \colon R^2 \to R^2$ is the rotation through an angle θ.

21. Determine whether $T_1 \circ T_2 = T_2 \circ T_1$.

 (a) $T_1 \colon R^3 \to R^3$ is a dilation by a factor k, and $T_2 \colon R^3 \to R^3$ is the rotation about the z-axis through an angle θ.

 (b) $T_1 \colon R^3 \to R^3$ is the rotation about the x-axis through an angle θ_1, and $T_2 \colon R^3 \to R^3$ is the rotation about the z-axis through an angle θ_2.

22. In R^3 the **orthogonal projections** on the x-axis, y-axis, and z-axis are defined by

$$T_1(x, y, z) = (x, 0, 0), \qquad T_2(x, y, z) = (0, y, 0), \qquad T_3(x, y, z) = (0, 0, z)$$

respectively.

    (a) Show that the orthogonal projections on the coordinate axes are linear operators, and find their standard matrices.

    (b) Show that if $T: R^3 \to R^3$ is an orthogonal projection on one of the coordinate axes, then for every vector $\mathbf{x}$ in R^3, the vectors $T(\mathbf{x})$ and $\mathbf{x} - T(\mathbf{x})$ are orthogonal vectors.

    (c) Make a sketch showing $\mathbf{x}$ and $\mathbf{x} - T(\mathbf{x})$ in the case where T is the orthogonal projection on the x-axis.

23. Derive the standard matrices for the rotations about the x-axis, y-axis, and z-axis in R^3 from Formula (17).

24. Use Formula (17) to find the standard matrix for a rotation of $\pi/2$ radians about the axis determined by the vector $\mathbf{v} = (1, 1, 1)$.

    Note Formula (17) requires that the vector defining the axis of rotation have length 1.

25. Verify Formula (21) for the given linear transformations.

    (a) $T_1(x_1, x_2) = (x_1 + x_2, x_1 - x_2)$ and $T_2(x_1, x_2) = (3x_1, 2x_1 + 4x_2)$

    (b) $T_1(x_1, x_2) = (4x_1, -2x_1 + x_2, -x_1 - 3x_2)$ and
$T_2(x_1, x_2, x_3) = (x_1 + 2x_2 - x_3, 4x_1 - x_3)$

    (c) $T_1(x_1, x_2, x_3) = (-x_1 + x_2, -x_2 + x_3, -x_3 + x_1)$ and
$T_2(x_1, x_2, x_3) = (-2x_1, 3x_3, -4x_2)$

26. It can be proved that if A is a 2×2 matrix with $\det(A) = 1$ and such that the column vectors of A are orthogonal and have length 1, then multiplication by A is a rotation through some angle θ. Verify that

$$A = \begin{bmatrix} -1/\sqrt{2} & -1/\sqrt{2} \\ 1/\sqrt{2} & -1/\sqrt{2} \end{bmatrix}$$

satisfies the stated conditions and find the angle of rotation.

27. The result stated in Exercise 26 is also true in R^3: It can be proved that if A is a 3×3 matrix with $\det(A) = 1$ and such that the column vectors of A are pairwise orthogonal and have length 1, then multiplication by A is a rotation about some axis of rotation through some angle θ. Use Formula (17) to show that if A satisfies the stated conditions, then the angle of rotation satisfies the equation

$$\cos \theta = \frac{\text{tr}(A) - 1}{2}$$

28. Let A be a 3×3 matrix (other than the identity matrix) satisfying the conditions stated in Exercise 27. It can be shown that if $\mathbf{x}$ is any nonzero vector in R^3, then the vector $\mathbf{u} = A\mathbf{x} + A^T\mathbf{x} + [1 - \text{tr}(A)]\mathbf{x}$ determines an axis of rotation when $\mathbf{u}$ is positioned with its initial point at the origin. [See "The Axis of Rotation: Analysis, Algebra, Geometry," by Dan Kalman, *Mathematics Magazine*, Vol. 62, No. 4, October 1989.]

    (a) Show that multiplication by

$$A = \begin{bmatrix} \frac{1}{9} & -\frac{4}{9} & \frac{8}{9} \\ \frac{8}{9} & \frac{4}{9} & \frac{1}{9} \\ -\frac{4}{9} & \frac{7}{9} & \frac{4}{9} \end{bmatrix}$$

    is a rotation.

    (b) Find a vector of length 1 that defines an axis for the rotation.

    (c) Use the result in Exercise 27 to find the angle of rotation about the axis obtained in part (b).

Discussion & Discovery

29. In words, describe the geometric effect of multiplying a vector $\mathbf{x}$ by the matrix A.

    (a) $A = \begin{bmatrix} 2 & 0 \\ 0 & 0 \end{bmatrix}$ (b) $A = \begin{bmatrix} 2 & 0 \\ 0 & -2 \end{bmatrix}$

30. In words, describe the geometric effect of multiplying a vector **x** by the matrix A.

(a) $A = \begin{bmatrix} 2 & 0 \\ 0 & 3 \end{bmatrix}$ (b) $A = \begin{bmatrix} \sqrt{3}/2 & -1/2 \\ 1/2 & \sqrt{3}/2 \end{bmatrix}$

31. In words, describe the geometric effect of multiplying a vector **x** by the matrix

$$A = \begin{bmatrix} \cos^2\theta - \sin^2\theta & -2\sin\theta\cos\theta \\ 2\sin\theta\cos\theta & \cos^2\theta - \sin^2\theta \end{bmatrix}$$

32. If multiplication by A rotates a vector **x** in the xy-plane through an angle θ, what is the effect of multiplying **x** by A^T? Explain your reasoning.

33. Let $\mathbf{x}_0$ be a nonzero column vector in R^2, and suppose that $T: R^2 \to R^2$ is the transformation defined by $T(\mathbf{x}) = \mathbf{x}_0 + R_\theta \mathbf{x}$, where R_θ is the standard matrix of the rotation of R^2 about the origin through the angle θ. Give a geometric description of this transformation. Is it a linear transformation? Explain.

34. A function of the form $f(x) = mx + b$ is commonly called a "linear function" because the graph of $y = mx + b$ is a line. Is f a linear transformation on R?

35. Let $\mathbf{x} = \mathbf{x}_0 + t\mathbf{v}$ be a line in R^n, and let $T: R^n \to R^n$ be a linear operator on R^n. What kind of geometric object is the image of this line under the operator T? Explain your reasoning.

4.3

PROPERTIES OF LINEAR TRANS- FORMATIONS FROM R^n TO R^m

In this section we shall investigate the relationship between the invertibility of a matrix and properties of the corresponding matrix transformation. We shall also obtain a characterization of linear transformations from R^n to R^m that will form the basis for more general linear transformations to be discussed in subsequent sections, and we shall discuss some geometric properties of eigenvectors.

One-to-One Linear Transformations

Linear transformations that map distinct vectors (or points) into distinct vectors (or points) are of special importance. One example of such a transformation is the linear operator $T: R^2 \to R^2$ that rotates each vector through an angle θ. It is obvious geometrically that if **u** and **v** are distinct vectors in R^2, then so are the rotated vectors $T(\mathbf{u})$ and $T(\mathbf{v})$ (Figure 4.3.1).

In contrast, if $T: R^3 \to R^3$ is the orthogonal projection of R^3 on the xy-plane, then distinct points on the same vertical line are mapped into the same point in the xy-plane (Figure 4.3.2).

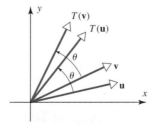

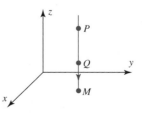

Figure 4.3.1 Distinct vectors **u** and **v** are rotated into distinct vectors $T(\mathbf{u})$ and $T(\mathbf{v})$.

Figure 4.3.2 The distinct points P and Q are mapped into the same point M.

> **DEFINITION**
>
> A linear transformation $T\colon R^n \to R^m$ is said to be ***one-to-one*** if T maps distinct vectors (points) in R^n into distinct vectors (points) in R^m.

REMARK   It follows from this definition that for each vector $\mathbf{w}$ in the range of a one-to-one linear transformation T, there is exactly one vector $\mathbf{x}$ such that $T(\mathbf{x}) = \mathbf{w}$.

EXAMPLE 1   One-to-One Linear Transformations

In the terminology of the preceding definition, the rotation operator of Figure 4.3.1 is one-to-one, but the orthogonal projection operator of Figure 4.3.2 is not. ◆

Let A be an $n \times n$ matrix, and let $T_A\colon R^n \to R^n$ be multiplication by A. We shall now investigate relationships between the invertibility of A and properties of T_A.

Recall from Theorem 2.3.6 (with $\mathbf{w}$ in place of $\mathbf{b}$) that the following are equivalent:

- A is invertible.
- $A\mathbf{x} = \mathbf{w}$ is consistent for every $n \times 1$ matrix $\mathbf{w}$.
- $A\mathbf{x} = \mathbf{w}$ has exactly one solution for every $n \times 1$ matrix $\mathbf{w}$.

However, the last of these statements is actually stronger than necessary. One can show that the following are equivalent (Exercise 24):

- A is invertible.
- $A\mathbf{x} = \mathbf{w}$ is consistent for every $n \times 1$ matrix $\mathbf{w}$.
- $A\mathbf{x} = \mathbf{w}$ has exactly one solution when the system is consistent.

Translating these into the corresponding statements about the linear operator T_A, we deduce that the following are equivalent:

- A is invertible.
- For every vector $\mathbf{w}$ in R^n, there is some vector $\mathbf{x}$ in R^n such that $T_A(\mathbf{x}) = \mathbf{w}$. Stated another way, the range of T_A is all of R^n.
- For every vector $\mathbf{w}$ in the range of T_A, there is exactly one vector $\mathbf{x}$ in R^n such that $T_A(\mathbf{x}) = \mathbf{w}$. Stated another way, T_A is one-to-one.

In summary, we have established the following theorem about linear operators on R^n.

THEOREM 4.3.1

> **Equivalent Statements**
>
> *If A is an $n \times n$ matrix and $T_A\colon R^n \to R^n$ is multiplication by A, then the following statements are equivalent.*
>
> *(a) A is invertible.*   *(b) The range of T_A is R^n.*   *(c) T_A is one-to-one.*

EXAMPLE 2   Applying Theorem 4.3.1

In Example 1 we observed that the rotation operator $T\colon R^2 \to R^2$ illustrated in Figure 4.3.1 is one-to-one. It follows from Theorem 4.3.1 that the range of T must be all of

R^2 and that the standard matrix for T must be invertible. To show that the range of T is all of R^2, we must show that every vector $\mathbf{w}$ in R^2 is the image of some vector $\mathbf{x}$ under T. But this is clearly so, since the vector $\mathbf{x}$ obtained by rotating $\mathbf{w}$ through the angle $-\theta$ maps into $\mathbf{w}$ when rotated through the angle θ. Moreover, from Table 6 of Section 4.2, the standard matrix for T is

$$[T] = \begin{bmatrix} \cos\theta & -\sin\theta \\ \sin\theta & \cos\theta \end{bmatrix}$$

which is invertible, since

$$\det[T] = \begin{vmatrix} \cos\theta & -\sin\theta \\ \sin\theta & \cos\theta \end{vmatrix} = \cos^2\theta + \sin^2\theta = 1 \neq 0 \quad \blacklozenge$$

EXAMPLE 3 Applying Theorem 4.3.1

In Example 1 we observed that the projection operator $T: R^3 \to R^3$ illustrated in Figure 4.3.2 is not one-to-one. It follows from Theorem 4.3.1 that the range of T is *not* all of R^3 and that the standard matrix for T is not invertible. To show directly that the range of T is not all of R^3, we must find a vector $\mathbf{w}$ in R^3 that is not the image of any vector $\mathbf{x}$ under T. But any vector $\mathbf{w}$ outside of the xy-plane has this property, since all images under T lie in the xy-plane. Moreover, from Table 5 of Section 4.2, the standard matrix for T is

$$[T] = \begin{bmatrix} 1 & 0 & 0 \\ 0 & 1 & 0 \\ 0 & 0 & 0 \end{bmatrix}$$

which is not invertible, since $\det[T] = 0$. $\quad \blacklozenge$

Inverse of a One-to-One Linear Operator

If $T_A: R^n \to R^n$ is a one-to-one linear operator, then from Theorem 4.3.1 the matrix A is invertible. Thus, $T_{A^{-1}}: R^n \to R^n$ is itself a linear operator; it is called the ***inverse of*** T_A. The linear operators T_A and $T_{A^{-1}}$ cancel the effect of one another in the sense that for all $\mathbf{x}$ in R^n,

$$T_A(T_{A^{-1}}(\mathbf{x})) = AA^{-1}\mathbf{x} = I\mathbf{x} = \mathbf{x}$$
$$T_{A^{-1}}(T_A(\mathbf{x})) = A^{-1}A\mathbf{x} = I\mathbf{x} = \mathbf{x}$$

or, equivalently,

$$T_A \circ T_{A^{-1}} = T_{AA^{-1}} = T_I$$
$$T_{A^{-1}} \circ T_A = T_{A^{-1}A} = T_I$$

From a more geometric viewpoint, if $\mathbf{w}$ is the image of $\mathbf{x}$ under T_A, then $T_{A^{-1}}$ maps $\mathbf{w}$ back into $\mathbf{x}$, since

$$T_{A^{-1}}(\mathbf{w}) = T_{A^{-1}}(T_A(\mathbf{x})) = \mathbf{x}$$

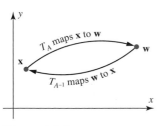

Figure 4.3.3

(Figure 4.3.3).

Before turning to an example, it will be helpful to touch on a notational matter. When a one-to-one linear operator on R^n is written as $T: R^n \to R^n$ (rather than $T_A: R^n \to R^n$), then the inverse of the operator T is denoted by T^{-1} (rather than $T_{A^{-1}}$). Since the standard matrix for T^{-1} is the inverse of the standard matrix for T, we have

$$[T^{-1}] = [T]^{-1} \tag{1}$$

EXAMPLE 4   Standard Matrix for T^{-1}

Let $T: R^2 \rightarrow R^2$ be the operator that rotates each vector in R^2 through the angle θ, so from Table 6 of Section 4.2,

$$[T] = \begin{bmatrix} \cos\theta & -\sin\theta \\ \sin\theta & \cos\theta \end{bmatrix} \tag{2}$$

It is evident geometrically that to undo the effect of T, one must rotate each vector in R^2 through the angle $-\theta$. But this is exactly what the operator T^{-1} does, since the standard matrix for T^{-1} is

$$[T^{-1}] = [T]^{-1} = \begin{bmatrix} \cos\theta & \sin\theta \\ -\sin\theta & \cos\theta \end{bmatrix} = \begin{bmatrix} \cos(-\theta) & -\sin(-\theta) \\ \sin(-\theta) & \cos(-\theta) \end{bmatrix}$$

(verify), which is identical to (2) except that θ is replaced by $-\theta$. ◆

EXAMPLE 5   Finding T^{-1}

Show that the linear operator $T: R^2 \rightarrow R^2$ defined by the equations

$$\begin{aligned} w_1 &= 2x_1 + x_2 \\ w_2 &= 3x_1 + 4x_2 \end{aligned}$$

is one-to-one, and find $T^{-1}(w_1, w_2)$.

Solution

The matrix form of these equations is

$$\begin{bmatrix} w_1 \\ w_2 \end{bmatrix} = \begin{bmatrix} 2 & 1 \\ 3 & 4 \end{bmatrix} \begin{bmatrix} x_1 \\ x_2 \end{bmatrix}$$

so the standard matrix for T is

$$[T] = \begin{bmatrix} 2 & 1 \\ 3 & 4 \end{bmatrix}$$

This matrix is invertible (so T is one-to-one) and the standard matrix for T^{-1} is

$$[T^{-1}] = [T]^{-1} = \begin{bmatrix} \frac{4}{5} & -\frac{1}{5} \\ -\frac{3}{5} & \frac{2}{5} \end{bmatrix}$$

Thus

$$[T^{-1}] \begin{bmatrix} w_1 \\ w_2 \end{bmatrix} = \begin{bmatrix} \frac{4}{5} & -\frac{1}{5} \\ -\frac{3}{5} & \frac{2}{5} \end{bmatrix} \begin{bmatrix} w_1 \\ w_2 \end{bmatrix} = \begin{bmatrix} \frac{4}{5}w_1 - \frac{1}{5}w_2 \\ -\frac{3}{5}w_1 + \frac{2}{5}w_2 \end{bmatrix}$$

from which we conclude that

$$T^{-1}(w_1, w_2) = (\tfrac{4}{5}w_1 - \tfrac{1}{5}w_2, -\tfrac{3}{5}w_1 + \tfrac{2}{5}w_2)$$ ◆

Linearity Properties

In the preceding section we defined a transformation $T: R^n \rightarrow R^m$ to be linear if the equations relating $\mathbf{x}$ and $\mathbf{w} = T(\mathbf{x})$ are linear equations. The following theorem provides an alternative characterization of linearity. This theorem is fundamental and will be the basis for extending the concept of a linear transformation to more general settings later in this text.

THEOREM 4.3.2

Properties of Linear Transformations

A transformation $T: R^n \to R^m$ is linear if and only if the following relationships hold for all vectors $\mathbf{u}$ and $\mathbf{v}$ in R^n and for every scalar c.

(a) $T(\mathbf{u} + \mathbf{v}) = T(\mathbf{u}) + T(\mathbf{v})$ (b) $T(c\mathbf{u}) = cT(\mathbf{u})$

Proof Assume first that T is a linear transformation, and let A be the standard matrix for T. It follows from the basic arithmetic properties of matrices that

$$T(\mathbf{u} + \mathbf{v}) = A(\mathbf{u} + \mathbf{v}) = A\mathbf{u} + A\mathbf{v} = T(\mathbf{u}) + T(\mathbf{v})$$

and

$$T(c\mathbf{u}) = A(c\mathbf{u}) = c(A\mathbf{u}) = cT(\mathbf{u})$$

Conversely, assume that properties (a) and (b) hold for the transformation T. We can prove that T is linear by finding a matrix A with the property that

$$T(\mathbf{x}) = A\mathbf{x} \tag{3}$$

for all vectors $\mathbf{x}$ in R^n. This will show that T is multiplication by A and therefore linear. But before we can produce this matrix, we need to observe that property (a) can be extended to three or more terms; for example, if $\mathbf{u}$, $\mathbf{v}$, and $\mathbf{w}$ are any vectors in R^n, then by first grouping $\mathbf{v}$ and $\mathbf{w}$ and applying property (a), we obtain

$$T(\mathbf{u} + \mathbf{v} + \mathbf{w}) = T(\mathbf{u} + (\mathbf{v} + \mathbf{w})) = T(\mathbf{u}) + T(\mathbf{v} + \mathbf{w}) = T(\mathbf{u}) + T(\mathbf{v}) + T(\mathbf{w})$$

More generally, for any vectors $\mathbf{v}_1, \mathbf{v}_2, \ldots, \mathbf{v}_k$ in R^n, we have

$$T(\mathbf{v}_1 + \mathbf{v}_2 + \cdots + \mathbf{v}_k) = T(\mathbf{v}_1) + T(\mathbf{v}_2) + \cdots + T(\mathbf{v}_k)$$

Now, to find the matrix A, let $\mathbf{e}_1, \mathbf{e}_2, \ldots, \mathbf{e}_n$ be the vectors

$$\mathbf{e}_1 = \begin{bmatrix} 1 \\ 0 \\ 0 \\ \vdots \\ 0 \end{bmatrix}, \quad \mathbf{e}_2 = \begin{bmatrix} 0 \\ 1 \\ 0 \\ \vdots \\ 0 \end{bmatrix}, \ldots, \quad \mathbf{e}_n = \begin{bmatrix} 0 \\ 0 \\ 0 \\ \vdots \\ 1 \end{bmatrix} \tag{4}$$

and let A be the matrix whose successive column vectors are $T(\mathbf{e}_1), T(\mathbf{e}_2), \ldots, T(\mathbf{e}_n)$; that is,

$$A = [T(\mathbf{e}_1) \mid T(\mathbf{e}_2) \mid \cdots \mid T(\mathbf{e}_n)] \tag{5}$$

If

$$\mathbf{x} = \begin{bmatrix} x_1 \\ x_2 \\ \vdots \\ x_n \end{bmatrix}$$

is any vector in R^n, then as discussed in Section 1.3, the product $A\mathbf{x}$ is a linear combination of the column vectors of A with coefficients from $\mathbf{x}$, so

$$\begin{aligned} A\mathbf{x} &= x_1 T(\mathbf{e}_1) + x_2 T(\mathbf{e}_2) + \cdots + x_n T(\mathbf{e}_n) \\ &= T(x_1 \mathbf{e}_1) + T(x_2 \mathbf{e}_2) + \cdots + T(x_n \mathbf{e}_n) \quad \longleftarrow \text{Property } (b) \\ &= T(x_1 \mathbf{e}_1 + x_2 \mathbf{e}_2 + \cdots + x_n \mathbf{e}_n) \quad \longleftarrow \text{Property } (a) \text{ for } n \text{ terms} \\ &= T(\mathbf{x}) \end{aligned}$$

which completes the proof. ∎

Expression (5) is important in its own right, since it provides an explicit formula for the standard matrix of a linear operator $T: R^n \rightarrow R^m$ in terms of the images of the vectors $\mathbf{e}_1, \mathbf{e}_2, \ldots, \mathbf{e}_n$ under T. For reasons that will be discussed later, the vectors $\mathbf{e}_1, \mathbf{e}_2, \ldots, \mathbf{e}_n$ in (4) are called the **standard basis** vectors for R^n. In R^2 and R^3 these are the vectors of length 1 along the coordinate axes (Figure 4.3.4).

Because of its importance, we shall state (5) as a theorem for future reference.

THEOREM 4.3.3

If $T: R^n \rightarrow R^m$ is a linear transformation, and $\mathbf{e}_1, \mathbf{e}_2, \ldots, \mathbf{e}_n$ are the standard basis vectors for R^n, then the standard matrix for T is

$$[T] = [T(\mathbf{e}_1) \mid T(\mathbf{e}_2) \mid \cdots \mid T(\mathbf{e}_n)] \tag{6}$$

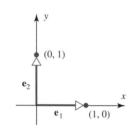

(a) Standard basis for R^2

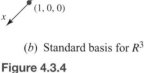

(b) Standard basis for R^3

Figure 4.3.4

Formula (6) is a powerful tool for finding standard matrices and analyzing the geometric effect of a linear transformation. For example, suppose that $T: R^3 \rightarrow R^3$ is the orthogonal projection on the xy-plane. Referring to Figure 4.3.4, it is evident geometrically that

$$T(\mathbf{e}_1) = \mathbf{e}_1 = \begin{bmatrix} 1 \\ 0 \\ 0 \end{bmatrix}, \qquad T(\mathbf{e}_2) = \mathbf{e}_2 = \begin{bmatrix} 0 \\ 1 \\ 0 \end{bmatrix}, \qquad T(\mathbf{e}_3) = \mathbf{0} = \begin{bmatrix} 0 \\ 0 \\ 0 \end{bmatrix}$$

so by (6),

$$[T] = \begin{bmatrix} 1 & 0 & 0 \\ 0 & 1 & 0 \\ 0 & 0 & 0 \end{bmatrix}$$

which agrees with the result in Table 5 of Section 4.2.

Using (6) another way, suppose that $T_A: R^3 \rightarrow R^2$ is multiplication by

$$A = \begin{bmatrix} -1 & 2 & 1 \\ 3 & 0 & 6 \end{bmatrix}$$

The images of the standard basis vectors can be read directly from the columns of the matrix A:

$$T_A\left(\begin{bmatrix} 1 \\ 0 \\ 0 \end{bmatrix} \right) = \begin{bmatrix} -1 \\ 3 \end{bmatrix}, \qquad T_A\left(\begin{bmatrix} 0 \\ 1 \\ 0 \end{bmatrix} \right) = \begin{bmatrix} 2 \\ 0 \end{bmatrix}, \qquad T_A\left(\begin{bmatrix} 0 \\ 0 \\ 1 \end{bmatrix} \right) = \begin{bmatrix} 1 \\ 6 \end{bmatrix}$$

EXAMPLE 6 **Standard Matrix for a Projection Operator**

Let l be the line in the xy-plane that passes through the origin and makes an angle θ with the positive x-axis, where $0 \leq \theta < \pi$. As illustrated in Figure 4.3.5a, let $T: R^2 \rightarrow R^2$ be a linear operator that maps each vector into its orthogonal projection on l.

(a) Find the standard matrix for T.

(b) Find the orthogonal projection of the vector $\mathbf{x} = (1, 5)$ onto the line through the origin that makes an angle of $\theta = \pi/6$ with the positive x-axis.

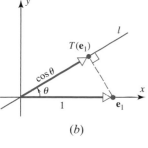

(a)

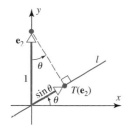

(b)

(c)

Figure 4.3.5

Solution (a)

From (6),

$$[T] = [T(\mathbf{e}_1) \mid T(\mathbf{e}_2)]$$

where $\mathbf{e}_1$ and $\mathbf{e}_2$ are the standard basis vectors for R^2. We consider the case where $0 \leq \theta \leq \pi/2$; the case where $\pi/2 < \theta < \pi$ is similar. Referring to Figure 4.3.5b, we have $\|T(\mathbf{e}_1)\| = \cos\theta$, so

$$T(\mathbf{e}_1) = \begin{bmatrix} \|T(\mathbf{e}_1)\| \cos\theta \\ \|T(\mathbf{e}_1)\| \sin\theta \end{bmatrix} = \begin{bmatrix} \cos^2\theta \\ \sin\theta\cos\theta \end{bmatrix}$$

and referring to Figure 4.3.5c, we have $\|T(\mathbf{e}_2)\| = \sin\theta$, so

$$T(\mathbf{e}_2) = \begin{bmatrix} \|T(\mathbf{e}_2)\| \cos\theta \\ \|T(\mathbf{e}_2)\| \sin\theta \end{bmatrix} = \begin{bmatrix} \sin\theta\cos\theta \\ \sin^2\theta \end{bmatrix}$$

Thus the standard matrix for T is

$$[T] = \begin{bmatrix} \cos^2\theta & \sin\theta\cos\theta \\ \sin\theta\cos\theta & \sin^2\theta \end{bmatrix}$$

Solution (b)

Since $\sin \pi/6 = 1/2$ and $\cos \pi/6 = \sqrt{3}/2$, it follows from part (a) that the standard matrix for this projection operator is

$$[T] = \begin{bmatrix} 3/4 & \sqrt{3}/4 \\ \sqrt{3}/4 & 1/4 \end{bmatrix}$$

Thus

$$T\left(\begin{bmatrix} 1 \\ 5 \end{bmatrix}\right) = \begin{bmatrix} 3/4 & \sqrt{3}/4 \\ \sqrt{3}/4 & 1/4 \end{bmatrix}\begin{bmatrix} 1 \\ 5 \end{bmatrix} = \begin{bmatrix} \dfrac{3+5\sqrt{3}}{4} \\ \dfrac{\sqrt{3}+5}{4} \end{bmatrix}$$

or, in point notation,

$$T(1,5) = \left(\frac{3+5\sqrt{3}}{4}, \frac{\sqrt{3}+5}{4}\right) \quad \blacklozenge$$

Geometric Interpretation of Eigenvectors

Recall from Section 2.3 that if A is an $n \times n$ matrix, then λ is called an *eigenvalue* of A if there is a nonzero vector $\mathbf{x}$ such that

$$A\mathbf{x} = \lambda\mathbf{x} \quad \text{or, equivalently,} \quad (\lambda I - A)\mathbf{x} = \mathbf{0}$$

The nonzero vectors $\mathbf{x}$ satisfying this equation are called the *eigenvectors* of A corresponding to λ.

Eigenvalues and eigenvectors can also be defined for linear operators on R^n; the definitions parallel those for matrices.

DEFINITION

If $T: R^n \to R^n$ is a linear operator, then a scalar λ is called an ***eigenvalue of T*** if there is a nonzero $\mathbf{x}$ in R^n such that

$$T(\mathbf{x}) = \lambda\mathbf{x} \tag{7}$$

Those nonzero vectors $\mathbf{x}$ that satisfy this equation are called the ***eigenvectors of T corresponding to*** λ.

(a) $\lambda \geq 0$

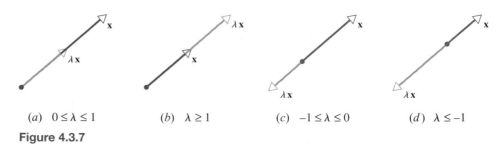

(b) $\lambda \leq 0$

Figure 4.3.6

Observe that if A is the standard matrix for T, then (7) can be written as

$$A\mathbf{x} = \lambda\mathbf{x}$$

from which it follows that

• The eigenvalues of T are precisely the eigenvalues of its standard matrix A.

• $\mathbf{x}$ is an eigenvector of T corresponding to λ if and only if $\mathbf{x}$ is an eigenvector of A corresponding to λ.

If λ is an eigenvalue of A and $\mathbf{x}$ is a corresponding eigenvector, then $A\mathbf{x} = \lambda\mathbf{x}$, so multiplication by A maps $\mathbf{x}$ into a scalar multiple of itself. In R^2 and R^3, this means that *multiplication by A maps each eigenvector $\mathbf{x}$ into a vector that lies on the same line as* $\mathbf{x}$ (Figure 4.3.6).

Recall from Section 4.2 that if $\lambda \geq 0$, then the linear operator $A\mathbf{x} = \lambda\mathbf{x}$ compresses $\mathbf{x}$ by a factor of λ if $0 \leq \lambda \leq 1$ or stretches $\mathbf{x}$ by a factor of λ if $\lambda \geq 1$. If $\lambda < 0$, then $A\mathbf{x} = \lambda\mathbf{x}$ reverses the direction of $\mathbf{x}$ and compresses the reversed vector by a factor of $|\lambda|$ if $0 \leq |\lambda| \leq 1$ or stretches the reversed vector by a factor of $|\lambda|$ if $|\lambda| \geq 1$ (Figure 4.3.7).

(a) $0 \leq \lambda \leq 1$ (b) $\lambda \geq 1$ (c) $-1 \leq \lambda \leq 0$ (d) $\lambda \leq -1$

Figure 4.3.7

EXAMPLE 7   Eigenvalues of a Linear Operator

Let $T: R^2 \rightarrow R^2$ be the linear operator that rotates each vector through an angle θ. It is evident geometrically that unless θ is a multiple of π, T does not map any nonzero vector $\mathbf{x}$ onto the same line as $\mathbf{x}$; consequently, T has no real eigenvalues. But if θ *is* a multiple of π, then every nonzero vector $\mathbf{x}$ is mapped onto the same line as $\mathbf{x}$, so *every* nonzero vector is an eigenvector of T. Let us verify these geometric observations algebraically. The standard matrix for T is

$$A = \begin{bmatrix} \cos\theta & -\sin\theta \\ \sin\theta & \cos\theta \end{bmatrix}$$

As discussed in Section 2.3, the eigenvalues of this matrix are the solutions of the characteristic equation

$$\det(\lambda I - A) = \begin{vmatrix} \lambda - \cos\theta & \sin\theta \\ -\sin\theta & \lambda - \cos\theta \end{vmatrix} = 0$$

that is,

$$(\lambda - \cos\theta)^2 + \sin^2\theta = 0 \tag{8}$$

But if θ is not a multiple of π, then $\sin^2\theta > 0$, so this equation has no real solution for λ, and consequently A has no real eigenvalues.[†] If θ *is* a multiple of π, then $\sin\theta = 0$ and

---

[†]There are applications that require complex scalars and vectors with complex components. In such cases, complex eigenvalues and eigenvectors with complex components are allowed. However, these have no direct geometric significance here. In later chapters we will discuss such eigenvalues and eigenvectors, but until explicitly stated otherwise, it will be assumed that only real eigenvalues and eigenvectors with real components are to be considered.

either $\cos\theta = 1$ or $\cos\theta = -1$, depending on the particular multiple of π. In the case where $\sin\theta = 0$ and $\cos\theta = 1$, the characteristic equation (8) becomes $(\lambda - 1)^2 = 0$, so $\lambda = 1$ is the only eigenvalue of A. In this case the matrix A is

$$A = \begin{bmatrix} 1 & 0 \\ 0 & 1 \end{bmatrix} = I$$

Thus, for all $\mathbf{x}$ in R^2,

$$T(\mathbf{x}) = A\mathbf{x} = I\mathbf{x} = \mathbf{x}$$

so T maps every vector to itself, and hence to the same line. In the case where $\sin\theta = 0$ and $\cos\theta = -1$, the characteristic equation (8) becomes $(\lambda + 1)^2 = 0$, so $\lambda = -1$ is the only eigenvalue of A. In this case the matrix A is

$$A = \begin{bmatrix} -1 & 0 \\ 0 & -1 \end{bmatrix} = -I$$

Thus, for all $\mathbf{x}$ in R^2,

$$T(\mathbf{x}) = A\mathbf{x} = -I\mathbf{x} = -\mathbf{x}$$

so T maps every vector to its negative, and hence to the same line as $\mathbf{x}$. ◆

EXAMPLE 8 Eigenvalues of a Linear Operator

Let $T: R^3 \to R^3$ be the orthogonal projection on the xy-plane. Vectors in the xy-plane are mapped into themselves under T, so each nonzero vector in the xy-plane is an eigenvector corresponding to the eigenvalue $\lambda = 1$. Every vector $\mathbf{x}$ along the z-axis is mapped into $\mathbf{0}$ under T, which is on the same line as $\mathbf{x}$, so every nonzero vector on the z-axis is an eigenvector corresponding to the eigenvalue $\lambda = 0$. Vectors that are not in the xy-plane or along the z-axis are not mapped into scalar multiples of themselves, so there are no other eigenvectors or eigenvalues.

To verify these geometric observations algebraically, recall from Table 5 of Section 4.2 that the standard matrix for T is

$$A = \begin{bmatrix} 1 & 0 & 0 \\ 0 & 1 & 0 \\ 0 & 0 & 0 \end{bmatrix}$$

The characteristic equation of A is

$$\det(\lambda I - A) = \begin{vmatrix} \lambda - 1 & 0 & 0 \\ 0 & \lambda - 1 & 0 \\ 0 & 0 & \lambda \end{vmatrix} = 0 \quad \text{or} \quad (\lambda - 1)^2 \lambda = 0$$

which has the solutions $\lambda = 0$ and $\lambda = 1$ anticipated above.

As discussed in Section 2.3, the eigenvectors of the matrix A corresponding to an eigenvalue λ are the nonzero solutions of

$$\begin{bmatrix} \lambda - 1 & 0 & 0 \\ 0 & \lambda - 1 & 0 \\ 0 & 0 & \lambda \end{bmatrix} \begin{bmatrix} x_1 \\ x_2 \\ x_3 \end{bmatrix} = \begin{bmatrix} 0 \\ 0 \\ 0 \end{bmatrix} \tag{9}$$

If $\lambda = 0$, this system is

$$\begin{bmatrix} -1 & 0 & 0 \\ 0 & -1 & 0 \\ 0 & 0 & 0 \end{bmatrix} \begin{bmatrix} x_1 \\ x_2 \\ x_3 \end{bmatrix} = \begin{bmatrix} 0 \\ 0 \\ 0 \end{bmatrix}$$

which has the solutions $x_1 = 0$, $x_2 = 0$, $x_3 = t$ (verify), or, in matrix form,

$$\begin{bmatrix} x_1 \\ x_2 \\ x_3 \end{bmatrix} = \begin{bmatrix} 0 \\ 0 \\ t \end{bmatrix}$$

As anticipated, these are the vectors along the z-axis. If $\lambda = 1$, then system (9) is

$$\begin{bmatrix} 0 & 0 & 0 \\ 0 & 0 & 0 \\ 0 & 0 & 1 \end{bmatrix} \begin{bmatrix} x_1 \\ x_2 \\ x_3 \end{bmatrix} = \begin{bmatrix} 0 \\ 0 \\ 0 \end{bmatrix}$$

which has the solutions $x_1 = s$, $x_2 = t$, $x_3 = 0$ (verify), or, in matrix form,

$$\begin{bmatrix} x_1 \\ x_2 \\ x_3 \end{bmatrix} = \begin{bmatrix} s \\ t \\ 0 \end{bmatrix}$$

As anticipated, these are the vectors in the xy-plane. ◆

Summary

In Theorem 2.3.6 we listed six results that are equivalent to the invertibility of a matrix A. We conclude this section by merging Theorem 4.3.1 with that list to produce the following theorem that relates all of the major topics we have studied thus far.

THEOREM 4.3.4

Equivalent Statements

If A is an $n \times n$ matrix, and if $T_A : R^n \to R^n$ is multiplication by A, then the following are equivalent.

(a) *A is invertible.*
(b) *$A\mathbf{x} = \mathbf{0}$ has only the trivial solution.*
(c) *The reduced row-echelon form of A is I_n.*
(d) *A is expressible as a product of elementary matrices.*
(e) *$A\mathbf{x} = \mathbf{b}$ is consistent for every $n \times 1$ matrix $\mathbf{b}$.*
(f) *$A\mathbf{x} = \mathbf{b}$ has exactly one solution for every $n \times 1$ matrix $\mathbf{b}$.*
(g) *$\det(A) \neq 0$.*
(h) *The range of T_A is R^n.*
(i) *T_A is one-to-one.*

EXERCISE SET 4.3

1. By inspection, determine whether the linear operator is one-to-one.
   (a) the orthogonal projection on the x-axis in R^2
   (b) the reflection about the y-axis in R^2
   (c) the reflection about the line $y = x$ in R^2
   (d) a contraction with factor $k > 0$ in R^2
   (e) a rotation about the z-axis in R^3
   (f) a reflection about the xy-plane in R^3
   (g) a dilation with factor $k > 0$ in R^3

2. Find the standard matrix for the linear operator defined by the equations, and use Theorem 4.3.4 to determine whether the operator is one-to-one.

   (a) $w_1 = 8x_1 + 4x_2$
       $w_2 = 2x_1 + \ x_2$

   (b) $w_1 = 2x_1 - 3x_2$
       $w_2 = 5x_1 + \ x_2$

   (c) $w_1 = -x_1 + 3x_2 + 2x_3$
       $w_2 = \ 2x_1 \qquad + 4x_3$
       $w_3 = \ \ x_1 + 3x_2 + 6x_3$

   (d) $w_1 = \ \ x_1 + 2x_2 + 3x_3$
       $w_2 = 2x_1 + 5x_2 + 3x_3$
       $w_3 = \ \ x_1 \qquad + 8x_3$

3. Show that the range of the linear operator defined by the equations

$$w_1 = 4x_1 - 2x_2$$
$$w_2 = 2x_1 - \ x_2$$

   is not all of R^2, and find a vector that is not in the range.

4. Show that the range of the linear operator defined by the equations

$$w_1 = \ \ x_1 - 2x_2 + \ x_3$$
$$w_2 = 5x_1 - \ x_2 + 3x_3$$
$$w_3 = 4x_1 + \ x_2 + 2x_3$$

   is not all of R^3, and find a vector that is not in the range.

5. Determine whether the linear operator $T: R^2 \to R^2$ defined by the equations is one-to-one; if so, find the standard matrix for the inverse operator, and find $T^{-1}(w_1, w_2)$.

   (a) $w_1 = \ \ x_1 + 2x_2$
       $w_2 = -x_1 + \ x_2$

   (b) $w_1 = \ \ 4x_1 - 6x_2$
       $w_2 = -2x_1 + 3x_2$

   (c) $w_1 = -x_2$
       $w_2 = -x_1$

   (d) $w_1 = \ \ 3x_1$
       $w_2 = -5x_1$

6. Determine whether the linear operator $T: R^3 \to R^3$ defined by the equations is one-to-one; if so, find the standard matrix for the inverse operator, and find $T^{-1}(w_1, w_2, w_3)$.

   (a) $w_1 = \ \ x_1 - 2x_2 + 2x_3$
       $w_2 = 2x_1 + \ x_2 + \ x_3$
       $w_3 = \ \ x_1 + \ x_2$

   (b) $w_1 = \ \ x_1 - 3x_2 + 4x_3$
       $w_2 = -x_1 + \ x_2 + \ x_3$
       $w_3 = \qquad - 2x_2 + 5x_3$

   (c) $w_1 = \ \ x_1 + 4x_2 - x_3$
       $w_2 = 2x_1 + 7x_2 + x_3$
       $w_3 = \ \ x_1 + 3x_2$

   (d) $w_1 = \ \ x_1 + 2x_2 + \ x_3$
       $w_2 = -2x_1 + \ x_2 + 4x_3$
       $w_3 = \ \ 7x_1 + 4x_2 - 5x_3$

7. By inspection, determine the inverse of the given one-to-one linear operator.

   (a) the reflection about the x-axis in R^2

   (b) the rotation through an angle of $\pi/4$ in R^2

   (c) the dilation by a factor of 3 in R^2

   (d) the reflection about the yz-plane in R^3

   (e) the contraction by a factor of $\frac{1}{5}$ in R^3

In Exercises 8 and 9 use Theorem 4.3.2 to determine whether $T: R^2 \to R^2$ is a linear operator.

8. (a) $T(x, y) = (2x, y)$
   (b) $T(x, y) = (x^2, y)$
   (c) $T(x, y) = (-y, x)$
   (d) $T(x, y) = (x, 0)$

9. (a) $T(x, y) = (2x + y, x - y)$
   (b) $T(x, y) = (x + 1, y)$
   (c) $T(x, y) = (y, y)$
   (d) $T(x, y) = (\sqrt[3]{x}, \sqrt[3]{y})$

In Exercises 10 and 11 use Theorem 4.3.2 to determine whether $T: R^3 \to R^2$ is a linear transformation.

10. (a) $T(x, y, z) = (x, x + y + z)$
    (b) $T(x, y, z) = (1, 1)$

11. (a) $T(x, y, z) = (0, 0)$
    (b) $T(x, y, z) = (3x - 4y, 2x - 5z)$

12. In each part, use Theorem 4.3.3 to find the standard matrix for the linear operator from the images of the standard basis vectors.

(a) the reflection operators on R^2 in Table 2 of Section 4.2

(b) the reflection operators on R^3 in Table 3 of Section 4.2

(c) the projection operators on R^2 in Table 4 of Section 4.2

(d) the projection operators on R^3 in Table 5 of Section 4.2

(e) the rotation operators on R^2 in Table 6 of Section 4.2

(f) the dilation and contraction operators on R^3 in Table 9 of Section 4.2

13. Use Theorem 4.3.3 to find the standard matrix for $T: R^2 \to R^2$ from the images of the standard basis vectors.

(a) $T: R^2 \to R^2$ projects a vector orthogonally onto the x-axis and then reflects that vector about the y-axis.

(b) $T: R^2 \to R^2$ reflects a vector about the line $y = x$ and then reflects that vector about the x-axis.

(c) $T: R^2 \to R^2$ dilates a vector by a factor of 3, then reflects that vector about the line $y = x$, and then projects that vector orthogonally onto the y-axis.

14. Use Theorem 4.3.3 to find the standard matrix for $T: R^3 \to R^3$ from the images of the standard basis vectors.

(a) $T: R^3 \to R^3$ reflects a vector about the xz-plane and then contracts that vector by a factor of $\frac{1}{5}$.

(b) $T: R^3 \to R^3$ projects a vector orthogonally onto the xz-plane and then projects that vector orthogonally onto the xy-plane.

(c) $T: R^3 \to R^3$ reflects a vector about the xy-plane, then reflects that vector about the xz-plane, and then reflects that vector about the yz-plane.

15. Let $T_A: R^3 \to R^3$ be multiplication by

$$A = \begin{bmatrix} -1 & 3 & 0 \\ 2 & 1 & 2 \\ 4 & 5 & -3 \end{bmatrix}$$

and let e_1, e_2, and e_3 be the standard basis vectors for R^3. Find the following vectors by inspection.

(a) $T_A(e_1)$, $T_A(e_2)$, and $T_A(e_3)$ (b) $T_A(e_1 + e_2 + e_3)$ (c) $T_A(7e_3)$

16. Determine whether multiplication by A is a one-to-one linear transformation.

(a) $A = \begin{bmatrix} 1 & -1 \\ 2 & 0 \\ 3 & -4 \end{bmatrix}$ (b) $A = \begin{bmatrix} 1 & 2 & 3 \\ -1 & 0 & -4 \end{bmatrix}$ (c) $\begin{bmatrix} 1 & 2 & 1 \\ 0 & 1 & 1 \\ 1 & 1 & 0 \\ 1 & 0 & -1 \end{bmatrix}$

17. Use the result in Example 6 to find the orthogonal projection of x onto the line through the origin that makes an angle θ with the positive x-axis.

(a) $x = (-1, 2)$; $\theta = 45°$ (b) $x = (1, 0)$; $\theta = 30°$ (c) $x = (1, 5)$; $\theta = 120°$

18. Use the type of argument given in Example 8 to find the eigenvalues and corresponding eigenvectors of T. Check your conclusions by calculating the eigenvalues and corresponding eigenvectors from the standard matrix for T.

(a) $T: R^2 \to R^2$ is the reflection about the x-axis.

(b) $T: R^2 \to R^2$ is the reflection about the line $y = x$.

(c) $T: R^2 \to R^2$ is the orthogonal projection on the x-axis.

(d) $T: R^2 \to R^2$ is the contraction by a factor of $\frac{1}{2}$.

19. Follow the directions of Exercise 18.

(a) $T: R^3 \to R^3$ is the reflection about the yz-plane.

(b) $T: R^3 \to R^3$ is the orthogonal projection on the xz-plane.

(c) $T: R^3 \to R^3$ is the dilation by a factor of 2.

(d) $T: R^3 \to R^3$ is a rotation of $\pi/4$ about the z-axis.

20. (a) Is a composition of one-to-one linear transformations one-to-one? Justify your conclusion.

(b) Can the composition of a one-to-one linear transformation and a linear transformation that is not one-to-one be one-to-one? Account for both possible orders of composition and justify your conclusion.

21. Show that $T(x, y) = (0, 0)$ defines a linear operator on R^2 but $T(x, y) = (1, 1)$ does not.

22. (a) Prove that if $T: R^n \to R^m$ is a linear transformation, then $T(\mathbf{0}) = \mathbf{0}$—that is, T maps the zero vector in R^n into the zero vector in R^m.

(b) The converse of this is not true. Find an example of a function that satisfies $T(0) = 0$ but is not a linear transformation.

23. Let l be the line in the xy-plane that passes through the origin and makes an angle θ with the positive x-axis, where $0 \leq \theta < \pi$. Let $T: R^2 \to R^2$ be the linear operator that reflects each vector about l (see the accompanying figure).

(a) Use the method of Example 6 to find the standard matrix for T.

(b) Find the reflection of the vector $\mathbf{x} = (1, 5)$ about the line l through the origin that makes an angle of $\theta = 30°$ with the positive x-axis.

24. Prove: An $n \times n$ matrix A is invertible if and only if the linear system $A\mathbf{x} = \mathbf{w}$ has exactly one solution for every vector $\mathbf{w}$ in R^n for which the system is consistent.

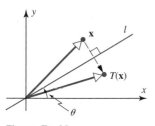

Figure Ex-23

Discussion & Discovery

25. Indicate whether each statement is always true or sometimes false. Justify your answer by giving a logical argument or a counterexample.

(a) If T maps R^n into R^m, and $T(\mathbf{0}) = \mathbf{0}$, then T is linear.

(b) If $T: R^n \to R^m$ is a one-to-one linear transformation, then there are no distinct vectors $\mathbf{u}$ and $\mathbf{v}$ in R^n such that $T(\mathbf{u} - \mathbf{v}) = \mathbf{0}$.

(c) If $T: R^n \to R^n$ is a linear operator, and if $T(\mathbf{x}) = 2\mathbf{x}$ for some vector $\mathbf{x}$, then $\lambda = 2$ is an eigenvalue of T.

(d) If T maps R^n into R^m, and if $T(c_1\mathbf{u} + c_2\mathbf{v}) = c_1 T(\mathbf{u}) + c_2 T(\mathbf{v})$ for all scalars c_1 and c_2 and for all vectors $\mathbf{u}$ and $\mathbf{v}$ in R^n, then T is linear.

26. Indicate whether each statement is always true, sometimes true, or always false.

(a) If $T: R^n \to R^m$ is a linear transformation and $m > n$, then T is one-to-one.

(b) If $T: R^n \to R^m$ is a linear transformation and $m < n$, then T is one-to-one.

(c) If $T: R^n \to R^m$ is a linear transformation and $m = n$, then T is one-to-one.

27. Let A be an $n \times n$ matrix such that $\det(A) = \mathbf{0}$, and let $T: R^n \to R^n$ be multiplication by A.

(a) What can you say about the range of the linear operator T? Give an example that illustrates your conclusion.

(b) What can you say about the number of vectors that T maps into $\mathbf{0}$?

28. In each part, make a conjecture about the eigenvectors and eigenvalues of the matrix A corresponding to the given transformation by considering the geometric properties of multiplication by A. Confirm each of your conjectures with computations.

(a) Reflection about the line $y = c$. (b) Contraction by a factor of $\frac{1}{2}$.

4.4

LINEAR TRANS-
FORMATIONS
AND
POLYNOMIALS

In this section we shall apply our new knowledge of linear transformations to polynomials. This is the beginning of a general strategy of using our ideas about R^n to solve problems that are in different, yet somehow analogous, settings.

Polynomials and Vectors

Suppose that we have a polynomial function, say

$$p(x) = ax^2 + bx + c$$

where x is a real-valued variable. To form the related function $2p(x)$, we multiply each of its coefficients by 2:

$$2p(x) = 2ax^2 + 2bx + 2c$$

That is, if the coefficients of the polynomial $p(x)$ are a, b, c in descending order of the power of x with which they are associated, then $2p(x)$ is also a polynomial, and its coefficients are $2a$, $2b$, $2c$ in the same order.

Similarly, if $q(x) = dx^2 + ex + f$ is another polynomial function, then $p(x) + q(x)$ is also a polynomial, and its coefficients are $a + d$, $b + e$, $c + f$. We add polynomials by adding corresponding coefficients.

This suggests that associating a polynomial with the vector consisting of its coefficients may be useful.

EXAMPLE 1 Correspondence between Polynomials and Vectors

Consider the quadratic function $p(x) = ax^2 + bx + c$. Define the vector

$$\mathbf{z} = \begin{bmatrix} a \\ b \\ c \end{bmatrix}$$

consisting of the coefficients of this polynomial in descending order of the corresponding power of x. Then multiplication of $p(x)$ by a scalar s gives $sp(x) = sax^2 + sbx + sc$, and this corresponds exactly to the scalar multiple

$$s\mathbf{z} = \begin{bmatrix} sa \\ sb \\ sc \end{bmatrix}$$

of $\mathbf{z}$. Similarly, $p(x) + p(x)$ is $2ax^2 + 2bx + 2c$, and this corresponds exactly to the vector sum $\mathbf{z} + \mathbf{z}$:

$$\mathbf{z} + \mathbf{z} = \begin{bmatrix} a \\ b \\ c \end{bmatrix} + \begin{bmatrix} a \\ b \\ c \end{bmatrix}$$

$$= \begin{bmatrix} 2a \\ 2b \\ 2c \end{bmatrix} \quad \blacklozenge$$

In general, given a polynomial $p(x) = a_n x^n + a_{n-1} x^{n-1} + \cdots + a_1 x + a_0$, we associate with it the vector

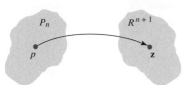

Figure 4.4.1 The vector z is associated with the polynomial p.

$$\mathbf{z} = \begin{bmatrix} a_n \\ a_{n-1} \\ \vdots \\ a_1 \\ a_0 \end{bmatrix}$$

in R^{n+1} (Figure 4.4.1). It is then possible to view operations like $p(x) \to 2p(x)$ as being equivalent to a linear transformation on R^{n+1}, namely $T(\mathbf{z}) = 2\mathbf{z}$. We can perform the desired operations in R^{n+1} rather than on the polynomials themselves.

EXAMPLE 2 Addition of Polynomials by Adding Vectors

Let $p(x) = 4x^3 - 2x + 1$ and $q(x) = 3x^3 - 3x^2 + x$. Then to compute $r(x) = 4p(x) - 2q(x)$, we could define

$$\mathbf{u} = \begin{bmatrix} 4 \\ 0 \\ -2 \\ 1 \end{bmatrix}, \qquad \mathbf{v} = \begin{bmatrix} 3 \\ -3 \\ 1 \\ 0 \end{bmatrix}$$

and perform the corresponding operation on these vectors:

$$4\mathbf{u} - 2\mathbf{v} = 4\begin{bmatrix} 4 \\ 0 \\ -2 \\ 1 \end{bmatrix} - 2\begin{bmatrix} 3 \\ -3 \\ 1 \\ 0 \end{bmatrix}$$

$$ = \begin{bmatrix} 10 \\ 6 \\ -10 \\ 4 \end{bmatrix}$$

Hence $r(x) = 10x^3 + 6x^2 - 10x + 4$. ◆

This association between polynomials of degree n and vectors in R^{n+1} would be useful for someone writing a computer program to perform polynomial computations, as in a computer algebra system. The coefficients of polynomial functions could be stored as vectors, and computations could be performed on these vectors.

For convenience, we define P_n to be the set of all polynomials of degree at most n (including the zero polynomial, all the coefficients of which are zero). This is also called the *space* of polynomials of degree at most n. The use of the word *space* indicates that this set has some sort of structure to it. The structure of P_n will be explored in Chapter 8.

Calculus Required   **EXAMPLE 3** Differentiation of Polynomials

---

Differentiation takes polynomials of degree n to polynomials of degree $n - 1$, so the corresponding transformation on vectors must take vectors in R^{n+1} to vectors in R^n. Hence, if differentiation corresponds to a linear transformation, it must be represented by a $n \times (n + 1)$ matrix. For example, if p is an element of P_2—that is,

$$p(x) = ax^2 + bx + c$$

for some real numbers a, b, and c—then

$$\frac{d}{dx}p(x) = 2ax + b$$

Evidently, if $p(x)$ in P_2 corresponds to the vector (a, b, c) in R^3, then its derivative is in P_1 and corresponds to the vector $(2a, b)$ in R^2. Note that

$$\begin{bmatrix} 2a \\ b \end{bmatrix} = \begin{bmatrix} 2 & 0 & 0 \\ 0 & 1 & 0 \end{bmatrix} \begin{bmatrix} a \\ b \\ c \end{bmatrix}$$

The operation differentiation, $D: P_2 \to P_1$, corresponds to a linear transformation $T_A: R^3 \to R^2$, where

$$A = \begin{bmatrix} 2 & 0 & 0 \\ 0 & 1 & 0 \end{bmatrix} \quad \blacklozenge$$

Some transformations from P_n to P_m do not correspond to linear transformations from R^{n+1} to R^{m+1}. For example, if we consider the transformation of $ax^2 + bx + c$ in P_2 to $|a|$ in P_0, the space of all constants (viewed as polynomials of degree zero, plus the zero polynomial), then we find that there is no matrix that maps (a, b, c) in R^3 to $|a|$ in R. Other transformations may correspond to transformations that are not *quite* linear, in the following sense.

DEFINITION

An *affine transformation* from R^n to R^m is a mapping of the form $S(\mathbf{u}) = T(\mathbf{u}) + \mathbf{f}$, where T is a linear transformation from R^n to R^m and $\mathbf{f}$ is a (constant) vector in R^m.

The affine transformation S is a linear transformation if $\mathbf{f}$ is the zero vector. Otherwise, it isn't linear, because it doesn't satisfy Theorem 4.3.2. This may seem surprising because the form of S looks like a natural generalization of an equation describing a line, but linear transformations satisfy the *Principle of Superposition*

$$T(c_1\mathbf{u} + c_2\mathbf{v}) = c_1 T(\mathbf{u}) + c_2 T(\mathbf{v})$$

for any scalars c_1, c_2 and any vectors $\mathbf{u}$, $\mathbf{v}$ in their domain. (This is just a restatement of Theorem 4.3.2.) Affine transformations with $\mathbf{f}$ nonzero don't have this property.

EXAMPLE 4 Affine Transformations

The mapping

$$S(\mathbf{u}) = \begin{bmatrix} 0 & 1 \\ -1 & 0 \end{bmatrix} \mathbf{u} + \begin{bmatrix} 1 \\ 1 \end{bmatrix}$$

is an affine transformation on R^2. If $\mathbf{u} = (a, b)$, then

$$S(\mathbf{u}) = \begin{bmatrix} 0 & 1 \\ -1 & 0 \end{bmatrix} \begin{bmatrix} a \\ b \end{bmatrix} + \begin{bmatrix} 1 \\ 1 \end{bmatrix}$$

$$= \begin{bmatrix} b + 1 \\ -a + 1 \end{bmatrix}$$

The corresponding operation from P_1 to P_1 takes $ax + b$ to $(b + 1)x - a + 1$. $\blacklozenge$

The relationship between an action on P_n and its corresponding action on the vector of coefficients in R^{n+1}, and the similarities between P_n and R^{n+1}, will be explored in more detail later in this text.

Interpolating Polynomials

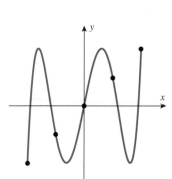

Figure 4.4.2   Interpolation

Consider the problem of interpolating a polynomial to a set of $n + 1$ points $(x_0, y_0), \ldots,$ (x_n, y_n). That is, we seek to find a curve $p(x) = a_m x^m + a_{m-1} x^{m-1} + \cdots + a_1 x + a_0$ of minimum degree that goes through each of these data points (Figure 4.4.2). Such a curve must satisfy

$$
\begin{aligned}
y_0 &= a_m x_0^m + a_{m-1} x_0^{m-1} + \cdots + a_1 x_0 + a_0 \\
y_1 &= a_m x_1^m + a_{m-1} x_1^{m-1} + \cdots + a_1 x_1 + a_0 \\
&\ \ \vdots \qquad \vdots \qquad \vdots \qquad\qquad \vdots \qquad \vdots \\
y_n &= a_m x_n^m + a_{m-1} x_n^{m-1} + \cdots + a_1 x_n + a_0
\end{aligned}
$$

Because the x_i are known, this leads to the following matrix system:

$$
\begin{bmatrix}
1 & x_0 & x_0^2 & \cdots & x_0^m \\
1 & x_1 & x_1^2 & \cdots & x_1^m \\
\vdots & \vdots & \vdots & \cdots & \vdots \\
1 & x_{n-1} & x_{n-1}^2 & \cdots & x_{n-1}^m \\
1 & x_n & x_n^2 & \cdots & x_n^m
\end{bmatrix}
\begin{bmatrix}
a_0 \\
a_1 \\
\vdots \\
a_{m-1} \\
a_m
\end{bmatrix}
=
\begin{bmatrix}
y_0 \\
y_1 \\
\vdots \\
y_{n-1} \\
y_n
\end{bmatrix}
$$

Note that this is a square system when $n = m$. Taking $n = m$ gives the following system for the coefficients of the interpolating polynomial $p(x)$:

$$
\begin{bmatrix}
1 & x_0 & x_0^2 & \cdots & x_0^n \\
1 & x_1 & x_1^2 & \cdots & x_1^n \\
\vdots & \vdots & \vdots & \cdots & \vdots \\
1 & x_{n-1} & x_{n-1}^2 & \cdots & x_{n-1}^n \\
1 & x_n & x_n^2 & \cdots & x_n^n
\end{bmatrix}
\begin{bmatrix}
a_0 \\
a_1 \\
\vdots \\
a_{n-1} \\
a_n
\end{bmatrix}
=
\begin{bmatrix}
y_0 \\
y_1 \\
\vdots \\
y_{n-1} \\
y_n
\end{bmatrix}
\qquad (1)
$$

The matrix in (1) is known as a ***Vandermonde matrix***; column j is the second column raised elementwise to the $j - 1$ power. The linear system in (1) is said to be a ***Vandermonde system***.

EXAMPLE 5   Interpolating a Cubic

To interpolate a polynomial to the data $(-2, 11), (-1, 2), (1, 2), (2, -1)$, we form the Vandermonde system (1):

$$
\begin{bmatrix}
1 & x_0 & x_0^2 & x_0^3 \\
1 & x_1 & x_1^2 & x_1^3 \\
1 & x_2 & x_2^2 & x_2^3 \\
1 & x_3 & x_3^2 & x_3^3
\end{bmatrix}
\begin{bmatrix}
a_0 \\
a_1 \\
a_2 \\
a_3
\end{bmatrix}
=
\begin{bmatrix}
y_0 \\
y_1 \\
y_2 \\
y_3
\end{bmatrix}
$$

For this data, we have

$$
\begin{bmatrix}
1 & -2 & 4 & -8 \\
1 & -1 & 1 & -1 \\
1 & 1 & 1 & 1 \\
1 & 2 & 4 & 8
\end{bmatrix}
\begin{bmatrix}
a_0 \\
a_1 \\
a_2 \\
a_3
\end{bmatrix}
=
\begin{bmatrix}
11 \\
2 \\
2 \\
-1
\end{bmatrix}
$$

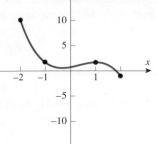

The solution, found by Gaussian elimination, is

$$\begin{bmatrix} a_0 \\ a_1 \\ a_2 \\ a_3 \end{bmatrix} = \begin{bmatrix} 1 \\ 1 \\ 1 \\ -1 \end{bmatrix}$$

and so the interpolant is $p(x) = -x^3 + x^2 + x + 1$. This is plotted in Figure 4.4.3, together with the data points, and we see that $p(x)$ does indeed interpolate the data, as required. ◆

Figure 4.4.3 The interpolant of Example 4

Newton Form

The interpolating polynomial $p(x) = a_n x^n + a_{n-1} x^{n-1} + \cdots + a_1 x + a_0$ is said to be written in its natural, or standard, form. But there is convenience in using other forms. For example, suppose we seek a cubic interpolant to the data (x_0, y_0), (x_1, y_1), (x_2, y_2), (x_3, y_3). If we write

$$p(x) = a_3 x^3 + a_2 x^2 + a_1 x + a_0 \tag{2}$$

in the equivalent form

$$p(x) = a_3(x - x_0)^3 + a_2(x - x_0)^2 + a_1(x - x_0) + a_0$$

then the interpolation condition $p(x_0) = y_0$ immediately gives $a_0 = y_0$. This reduces the size of the system that must be solved from $(n + 1) \times (n + 1)$ to $n \times n$. That is not much of a savings, but if we take this idea further, we may write (2) in the equivalent form

$$p(x) = b_3(x - x_0)(x - x_1)(x - x_2) + b_2(x - x_0)(x - x_1) + b_1(x - x_0) + b_0 \tag{3}$$

which is called the ***Newton form*** of the interpolant. Set $h_i = x_i - x_{i-1}$ for $i = 1, 2, 3$. The interpolation conditions give

$$p(x_0) = b_0$$
$$p(x_1) = b_1 h_1 + b_0$$
$$p(x_2) = b_2(h_1 + h_2)h_2 + b_1(h_1 + h_2) + b_0$$
$$p(x_3) = b_3(h_1 + h_2 + h_3)(h_2 + h_3)h_3 + b_2(h_1 + h_2 + h_3)(h_2 + h_3) + b_1(h_1 + h_2 + h_3) + b_0$$

that is,

$$\begin{bmatrix} 1 & 0 & 0 & 0 \\ 1 & h_1 & 0 & 0 \\ 1 & h_1 + h_2 & (h_1 + h_2)h_2 & 0 \\ 1 & h_1 + h_2 + h_3 & (h_1 + h_2 + h_3)(h_2 + h_3) & (h_1 + h_2 + h_3)(h_2 + h_3)h_3 \end{bmatrix} \begin{bmatrix} b_0 \\ b_1 \\ b_2 \\ b_3 \end{bmatrix} = \begin{bmatrix} y_0 \\ y_1 \\ y_2 \\ y_3 \end{bmatrix} \tag{4}$$

Unlike the Vandermonde system (1), this system has a lower triangular coefficient matrix. This is a much simpler system. We may solve for the coefficients very easily and efficiently by ***forward-substitution***, in analogy with back-substitution. In the case of equally spaced points arranged in increasing order, we have $h_i = h > 0$, so (4) becomes

$$\begin{bmatrix} 1 & 0 & 0 & 0 \\ 1 & h & 0 & 0 \\ 1 & 2h & 2h^2 & 0 \\ 1 & 3h & 6h^2 & 6h^3 \end{bmatrix} \begin{bmatrix} b_0 \\ b_1 \\ b_2 \\ b_3 \end{bmatrix} = \begin{bmatrix} y_0 \\ y_1 \\ y_2 \\ y_3 \end{bmatrix}$$

Note that the determinant of (4) is nonzero exactly when h_i is nonzero for each i, so there exists a unique interpolant whenever the x_i are distinct. Because the Vandermonde system computes a different form of the same interpolant, it too must have a unique solution exactly when the x_i are distinct.

EXAMPLE 6 Interpolating a Cubic in Newton Form

To interpolate a polynomial in Newton form to the data $(-2, 11), (-1, 2), (1, 2), (2, -1)$ of Example 5, we form the system (4):

$$\begin{bmatrix} 1 & 0 & 0 & 0 \\ 1 & 1 & 0 & 0 \\ 1 & 3 & 6 & 0 \\ 1 & 4 & 12 & 12 \end{bmatrix} \begin{bmatrix} b_0 \\ b_1 \\ b_2 \\ b_3 \end{bmatrix} = \begin{bmatrix} 11 \\ 2 \\ 2 \\ -1 \end{bmatrix}$$

The solution, found by forward-substitution, is

$$\begin{aligned} b_0 &= 11 \\ b_0 + b_1 &= 2 & b_1 &= -9 \\ b_0 + 3b_1 + 6b_2 &= 2 & b_2 &= 3 \\ b_0 + 4b_1 + 12b_2 + 12b_3 &= -1 & b_3 &= -1 \end{aligned}$$

and so, from (3), the interpolant is

$$\begin{aligned} p(x) &= -1 \cdot (x+2)(x+1)(x-1) + 3 \cdot (x+2)(x+1) + (-9) \cdot (x+2) + 11 \\ &= -(x+2)(x+1)(x-1) + 3(x+2)(x+1) - 9(x+2) + 11 \quad \blacklozenge \end{aligned}$$

Converting between Forms

The Newton form offers other advantages, but now we turn to the following question: If we have the coefficients of the interpolating polynomial in Newton form, what are the coefficients in the standard form? For example, if we know the coefficients in

$$p(x) = b_3(x - x_0)(x - x_1)(x - x_2) + b_2(x - x_0)(x - x_1) + b_1(x - x_0) + b_0$$

because we have solved (4) in order to avoid having to solve the more complicated Vandermonde system (1), how can we get the coefficients in (2),

$$p(x) = a_3 x^3 + a_2 x^2 + a_1 x + a_0$$

from b_0, b_1, b_2, b_3? Expanding the products in (3) gives

$$\begin{aligned} p(x) &= b_3(x - x_0)(x - x_1)(x - x_2) + b_2(x - x_0)(x - x_1) + b_1(x - x_0) + b_0 \\ &= b_3 x^3 + (b_2 - b_3(x_0 + x_1 + x_2))x^2 \\ &\quad + (b_1 - b_2(x_0 + x_1) + b_3(x_0 x_1 + x_0 x_2 + x_1 x_2))x \\ &\quad + b_0 - x_0 b_1 + x_0 x_1 b_2 - x_0 x_1 x_2 b_3 \end{aligned}$$

so

$$\begin{aligned} a_0 &= b_0 - x_0 b_1 + x_0 x_1 b_2 - x_0 x_1 x_2 b_3 \\ a_1 &= b_1 - b_2(x_0 + x_1) + b_3(x_0 x_1 + x_0 x_2 + x_1 x_2) \\ a_2 &= b_2 - b_3(x_0 + x_1 + x_2) \\ a_3 &= b_3 \end{aligned}$$

This can be expressed as

$$
\begin{bmatrix} a_0 \\ a_1 \\ a_2 \\ a_3 \end{bmatrix} = \begin{bmatrix} 1 & -x_0 & x_0x_1 & -x_0x_1x_2 \\ 0 & 1 & -(x_0+x_1) & x_0x_1+x_0x_2+x_1x_2 \\ 0 & 0 & 1 & -(x_0+x_1+x_2) \\ 0 & 0 & 0 & 1 \end{bmatrix} \begin{bmatrix} b_0 \\ b_1 \\ b_2 \\ b_3 \end{bmatrix} \tag{5}
$$

This is an important result! Solving the Vandermonde system (1) by Gaussian elimination would require us to form an $n \times n$ matrix that might have no nonzero entries and then to solve it using a number of arithmetic operations that grows in proportion to n^3 for large n. But solving the lower triangular system (4) requires an amount of work that grows in proportion to n^2 for large n, and using (5) to compute the coefficients a_0, a_1, a_2, a_3 also requires an amount of work that grows in proportion to n^2 for large n. Hence, for large n, the latter approach is an order of magnitude more efficient. The two-step procedure of solving (4) and then using the linear transformation (5) is a superior approach to solving (1) when n is large (Figure 4.4.4).

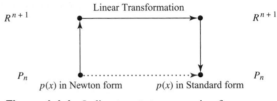

Figure 4.4.4 Indirect route to conversion from Newton form to standard form

---

EXAMPLE 7   Changing Forms

In Example 4 we found that $a_0 = 1$, $a_1 = 1$, $a_2 = 1$, $a_3 = -1$, whereas in Example 5 we found that $b_0 = 11$, $b_1 = -9$, $b_2 = 3$, $b_3 = -1$ for the same data. From (5), with $x_0 = -2$, $x_1 = -1$, $x_1 = 1$, we expect that

$$
\begin{bmatrix} 1 \\ 1 \\ 1 \\ -1 \end{bmatrix} = \begin{bmatrix} 1 & 2 & 2 & -2 \\ 0 & 1 & 3 & -1 \\ 0 & 0 & 1 & 2 \\ 0 & 0 & 0 & 1 \end{bmatrix} \begin{bmatrix} 11 \\ -9 \\ 3 \\ -1 \end{bmatrix}
$$

which checks. ◆

There is another approach to solving (1), based on the Fast Fourier Transform, that also requires an amount of work proportional to n^2. The point for now is to see that the use of linear transformations on R^{n+1} can help us perform computations involving polynomials. The original problem—to fit a polynomial of minimum degree to a set of data points—was not couched in the language of linear algebra at all. But rephrasing it in those terms and using matrices and the notation of linear transformations on R^{n+1} has allowed us to see when a unique solution must exist, how to compute it efficiently, and how to transform it among various forms.

EXERCISE SET 4.4

1. Identify the operations on polynomials that correspond to the following operations on vectors. Give the resulting polynomial.

   (a) $\begin{bmatrix} 1 \\ 2 \\ -1 \end{bmatrix} - 2 \begin{bmatrix} 3 \\ 0 \\ 2 \end{bmatrix}$ (b) $5 \begin{bmatrix} 4 \\ 3 \\ 0 \end{bmatrix} + 6 \begin{bmatrix} 1 \\ 2 \\ 1 \end{bmatrix}$

   (c) $\begin{bmatrix} 1 \\ 2 \\ 1 \\ -2 \\ 1 \end{bmatrix} - \begin{bmatrix} 0 \\ 2 \\ 0 \\ -2 \\ 0 \end{bmatrix}$ (d) $\pi \begin{bmatrix} 4 \\ -3 \\ 7 \\ 1 \end{bmatrix}$

2. (a) Consider the operation on P_2 that takes $ax^2 + bx + c$ to $cx^2 + bx + a$. Does it correspond to a linear transformation from R^3 to R^3? If so, what is its matrix?

   (b) Consider the operation on P_3 that takes $ax^3 + bx^2 + cx + d$ to $cx^3 - bx^2 - ax + d$. Does it correspond to a linear transformation from R^3 to R^3? If so, what is its matrix?

3. (a) Consider the transformation of $ax^2 + bx + c$ in P_2 to $|a|$ in P_0. Show that it does not correspond to a linear transformation by showing that there is no matrix that maps (a, b, c) in R^3 to $|a|$ in R.

   (b) Does the transformation of $ax^2 + bx + c$ in P_2 to a in P_0 correspond to a linear transformation from R^3 to R?

4. (a) Consider the operation $M: P_2 \rightarrow P_3$ that takes $p(x)$ in P_2 to $xp(x)$ in P_3. Does this correspond to a linear transformation from R^3 to R^4? If so, what is its matrix?

   (b) Consider the operation $N: P_2 \rightarrow P_3$ that takes $p(x)$ in P_n to $(x - 1)p(x)$ in P_{n+1}. Does this correspond to a linear transformation from R^3 to R^4? If so, what is its matrix?

   (c) Consider the operation $W: P_2 \rightarrow P_3$ that takes $p(x)$ in P_n to $xp(x) + 1$ in P_{n+1}. Does this correspond to a linear transformation from R^3 to R^4? If so, what is its matrix?

5. **(For Readers Who Have Studied Calculus)** What matrix corresponds to differentiation in each case?

   (a) $D: P_3 \rightarrow P_2$ (b) $D: P_4 \rightarrow P_3$ (c) $D: P_5 \rightarrow P_4$

6. **(For Readers Who Have Studied Calculus)** What matrix corresponds to differentiation in each case, assuming we represent $p(x) = a_n x^n + a_{n-1} x^{n-1} + \cdots + a_1 x + a_0$ as the vector $(a_0, a_1, \ldots, a_{n-1}, a_n)$?

   Note This is the opposite of the ordering of coefficients we have been using.

   (a) $D: P_3 \rightarrow P_2$ (b) $D: P_4 \rightarrow P_3$ (c) $D: P_5 \rightarrow P_4$

7. Consider the following matrices. What is the corresponding transformation on polynomials? Indicate the domain P_i and the codomain P_j.

   (a) $\begin{bmatrix} 1 & 1 \\ 1 & -1 \end{bmatrix}$ (b) $\begin{bmatrix} 1 & 0 \\ 1 & 1 \\ 2 & -1 \end{bmatrix}$ (c) $\begin{bmatrix} 1 & 0 & 2 & -1 \\ 2 & 1 & 1 & 3 \end{bmatrix}$

   (d) $\begin{bmatrix} 0 & 0 & 0 \\ 0 & 1 & 0 \\ 0 & 0 & 0 \end{bmatrix}$ (e) $\begin{bmatrix} 0 & 1 & 0 \end{bmatrix}$

8. Consider the space of all functions of the form $a + b\cos(x) + c\sin(x)$ where a, b, c are scalars.

   (a) What matrix, if any, corresponds to the change of variables $x \rightarrow x - \pi/2$, assuming that we represent a function in this space as the vector (a, b, c)?

   (b) What matrix corresponds to differentiation of functions on this space?

9. Consider the space of all functions of the form $a + bt + ce^t + de^{-t}$, where a, b, c, d are scalars.

    (a) What function in the space corresponds to the sum of $(1, 2, 3, 4)$ and $(-1, -2, 0, -1)$, assuming that we represent a function in this space as the vector (a, b, c, d)?

    (b) Is $\cosh(t)$ in this space? That is, does $\cosh(t)$ correspond to some choice of a, b, c, d?

    (c) What matrix corresponds to differentiation of functions on this space?

10. Show that the Principle of Superposition is equivalent to Theorem 4.3.2.

11. Show that an affine transformation with **f** nonzero is not a linear transformation.

12. Find a quadratic interpolant to the data $(-1, 2)$, $(0, 0)$, $(1, 2)$ using the Vandermonde system approach.

13. (a) Find a quadratic interpolant to the data $(-2, 1)$, $(0, 1)$, $(1, 4)$ using the Vandermonde system approach from (1).

    (b) Repeat using the Newton approach from (4).

14. (a) Find a polynomial interpolant to the data $(-1, 0)$, $(0, 0)$, $(1, 0)$, $(2, 6)$ using the Vandermonde system approach from (1).

    (b) Repeat using the Newton approach from (4).

    (c) Use (5) to get your answer in part (a) from your answer in part (b).

    (d) Use (5) to get your answer in part (b) from your answer in part (a) by finding the inverse of the matrix.

    (e) What happens if you change the data to $(-1, 0)$, $(0, 0)$, $(1, 0)$, $(2, 0)$?

15. (a) Find a polynomial interpolant to the data $(-2, -10)$, $(-1, 2)$, $(1, 2)$, $(2, 14)$ using the Vandermonde system approach from (1).

    (b) Repeat using the Newton approach from (4).

    (c) Use (5) to get your answer in part (a) from your answer in part (b).

    (d) Use (5) to get your answer in part (b) from your answer in part (a) by finding the inverse of the matrix.

16. Show that the determinant of the 2×2 Vandermonde matrix

$$\begin{bmatrix} 1 & a \\ 1 & b \end{bmatrix}$$

can be written as $(b - a)$ and that the determinant of the 3×3 Vandermonde matrix

$$\det \begin{bmatrix} 1 & a & a^2 \\ 1 & b & b^2 \\ 1 & c & c^2 \end{bmatrix}$$

can be written as $(b - a)(c - a)(c - b)$. Conclude that a unique straight line can be fit through any two points (x_0, y_0), (x_1, y_1) with x_0 and x_1 distinct, and that a unique parabola (which may be degenerate, such as a line) can be fit through any three points (x_0, y_0), (x_1, y_1), (x_2, y_2) with x_0, x_1, and x_2 distinct.

17. (a) What form does (5) take for lines?

    (b) What form does (5) take for quadratics?

    (c) What form does (5) take for quartics?

Discussion & Discovery

18. **(For Readers Who Have Studied Calculus)**

    (a) Does indefinite integration of functions in P_n correspond to some linear transformation from R^{n+1} to R^{n+2}?

(b) Does definite integration (from $x = 0$ to $x = 1$) of functions in P_n correspond to some linear transformation from R^{n+1} to R?

19. **(For Readers Who Have Studied Calculus)**

   (a) What matrix corresponds to second differentiation of functions from P_2 (giving functions in P_0)?

   (b) What matrix corresponds to second differentiation of functions from P_3 (giving functions in P_1)?

   (c) Is the matrix for second differentiation the square of the matrix for (first) differentiation?

20. Consider the transformation from P_2 to P_2 associated with the matrix

$$\begin{bmatrix} 0 & 0 & 0 \\ 0 & 1 & 0 \\ 0 & 0 & 0 \end{bmatrix}$$

and the transformation from P_2 to P_0 associated with the matrix

$$\begin{bmatrix} 0 & 1 & 0 \end{bmatrix}$$

These differ only in their codomains. Comment on this difference. In what ways (if any) is it important?

21. The third major technique for polynomial interpolation is interpolation using **_Lagrange interpolating polynomials_**. Given a set of distinct x-values $x_0, x_1, \ldots, x_n$, define the $n + 1$ Lagrange interpolating polynomials for these values by (for $i = 0, 1, \ldots, n$)

$$L_i(x) = \frac{(x - x_0)(x - x_1) \cdots (x - x_{i-1})(x - x_{i+1}) \cdots (x - x_n)}{(x_i - x_0)(x_i - x_1) \cdots (x_i - x_{i-1})(x_i - x_{i+1}) \cdots (x_i - x_n)}$$

Note that $L_i(x)$ is a polynomial of exact degree n and that $L_i(x_j) = 0$ if $i \neq j$, and $L_i(x_i) = 1$. It follows that we can write the polynomial interpolant to $(x_0, y_0), \ldots, (x_n, y_n)$ in the form

$$p(x) = c_0 L_0(x) + c_1 L_1(x) + \cdots + c_n L_n(x)$$

where $c_i = y_i$, $i = 0, 1, \ldots, n$.

   (a) Verify that $p(x) = y_0 L_0(x) + y_1 L_1(x) + \cdots + y_n L_n(x)$ is the unique interpolating polynomial for this data.

   (b) What is the linear system for the coefficients $c_0, c_1, \ldots, c_n$, corresponding to (1) for the Vandermonde approach and to (4) for the Newton approach?

   (c) Compare the three approaches to polynomial interpolation that we have seen. Which is most efficient with respect to finding the coefficients? Which is most efficient with respect to evaluating the interpolant somewhere between data points?

22. Generalize the result in Problem 16 by finding a formula for the determinant of an $n \times n$ Vandermonde matrix for arbitrary n.

23. The **_norm_** of a linear transformation $T_A: R^n \to R^n$ can be defined by

$$\|T\|_E = \max \frac{\|T(\mathbf{x})\|}{\|\mathbf{x}\|}$$

where the maximum is taken over all nonzero $\mathbf{x}$ in R^n. (The subscript indicates that the norm of the linear transformation on the left is found using the Euclidean vector norm on the right.) It is a fact that the largest value is always achieved—that is, there is always some $\mathbf{x}_0$ in R^n such that $\|T\|_E = \max(\|T(\mathbf{x}_0)\|/\|\mathbf{x}_0\|)$. What are the norms of the linear transformations T_A with the following matrices?

(a) $\begin{bmatrix} 2 & 0 \\ 0 & 1 \end{bmatrix}$ (b) $\begin{bmatrix} 1 & 0 \\ 0 & -1 \end{bmatrix}$ (c) $\begin{bmatrix} 2 & 0 \\ 0 & -3 \end{bmatrix}$ (d) $\begin{bmatrix} 1/\sqrt{2} & 1/\sqrt{2} \\ 1/\sqrt{2} & -1/\sqrt{2} \end{bmatrix}$

CHAPTER 4

Technology Exercises

The following exercises are designed to be solved using a technology utility. Typically, this will be MATLAB, *Mathematica*, Maple, Derive, or Mathcad, but it may also be some other type of linear algebra software or a scientific calculator with some linear algebra capabilities. For each exercise you will need to read the relevant documentation for the particular utility you are using. The goal of these exercises is to provide you with a basic proficiency with your technology utility. Once you have mastered the techniques in these exercises, you will be able to use your technology utility to solve many of the problems in the regular exercise sets.

Section 4.1 **T1.** **(Vector Operations in R^n)** With most technology utilities, the commands for operating on vectors in R^n are the same as those for operating on vectors in R^2 and R^3, and the command for computing a dot product produces the Euclidean inner product in R^n. Use your utility to perform computations in Exercises 1, 3, and 9 of Section 4.1.

Section 4.2 **T1.** **(Rotations)** Find the standard matrix for the linear operator on R^3 that performs a counterclockwise rotation of $45°$ about the x-axis, followed by a counterclockwise rotation of $60°$ about the y-axis, followed by a counterclockwise rotation of $30°$ about the z-axis. Then find the image of the point $(1, 1, 1)$ under this operator.

Section 4.3 **T1.** **(Projections)** Use your utility to perform the computations for $\theta = \pi/6$ in Example 6. Then project the vectors $(1, 1)$ and $(1, -5)$. Repeat for $\theta = \pi/4, \pi/3, \pi/2, \pi$.

Section 4.4 **T1.** **(Interpolation)** Most technology utilities have a command that performs polynomial interpolation. Read your documentation, and find the command or commands for fitting a polynomial interpolant to given data. Then use it (or them) to confirm the result of Example 5.

General Vector Spaces

CHAPTER CONTENTS

INTRODUCTION: In the last chapter we generalized vectors from 2- and 3-space to vectors in n-space. In this chapter we shall generalize the concept of vector still further. We shall state a set of axioms that, if satisfied by a class of objects, will entitle those objects to be called "vectors." These generalized vectors will include, among other things, various kinds of matrices and functions. Our work in this chapter is not an idle exercise in theoretical mathematics; it will provide a powerful tool for extending our geometric visualization to a wide variety of important mathematical problems where geometric intuition would not otherwise be available. We can visualize vectors in R^2 and R^3 as arrows, which enables us to draw or form mental pictures to help solve problems. Because the axioms we give to define our new kinds of vectors will be based on properties of vectors in R^2 and R^3, the new vectors will have many familiar properties. Consequently, when we want to solve a problem involving our new kinds of vectors, say matrices or functions, we may be able to get a foothold on the problem by visualizing what the corresponding problem would be like in R^2 and R^3.

5.1
REAL VECTOR SPACES

In this section we shall extend the concept of a vector by extracting the most important properties of familiar vectors and turning them into axioms. Thus, when a set of objects satisfies these axioms, they will automatically have the most important properties of familiar vectors, thereby making it reasonable to regard these objects as new kinds of vectors.

Vector Space Axioms

The following definition consists of ten axioms. As you read each axiom, keep in mind that you have already seen each of them as parts of various definitions and theorems in the preceding two chapters (for instance, see Theorem 4.1.1). Remember, too, that you do not prove axioms; they are simply the "rules of the game."

DEFINITION

Let V be an arbitrary nonempty set of objects on which two operations are defined: addition, and multiplication by scalars (numbers). By *addition* we mean a rule for associating with each pair of objects $\mathbf{u}$ and $\mathbf{v}$ in V an object $\mathbf{u} + \mathbf{v}$, called the *sum* of $\mathbf{u}$ and $\mathbf{v}$; by *scalar multiplication* we mean a rule for associating with each scalar k and each object $\mathbf{u}$ in V an object $k\mathbf{u}$, called the *scalar multiple* of $\mathbf{u}$ by k. If the following axioms are satisfied by all objects $\mathbf{u}$, $\mathbf{v}$, $\mathbf{w}$ in V and all scalars k and m, then we call V a *vector space* and we call the objects in V *vectors*.

1. If $\mathbf{u}$ and $\mathbf{v}$ are objects in V, then $\mathbf{u} + \mathbf{v}$ is in V.
2. $\mathbf{u} + \mathbf{v} = \mathbf{v} + \mathbf{u}$
3. $\mathbf{u} + (\mathbf{v} + \mathbf{w}) = (\mathbf{u} + \mathbf{v}) + \mathbf{w}$
4. There is an object $\mathbf{0}$ in V, called a *zero vector* for V, such that $\mathbf{0} + \mathbf{u} = \mathbf{u} + \mathbf{0} = \mathbf{u}$ for all $\mathbf{u}$ in V.
5. For each $\mathbf{u}$ in V, there is an object $-\mathbf{u}$ in V, called a *negative* of $\mathbf{u}$, such that $\mathbf{u} + (-\mathbf{u}) = (-\mathbf{u}) + \mathbf{u} = \mathbf{0}$.
6. If k is any scalar and $\mathbf{u}$ is any object in V, then $k\mathbf{u}$ is in V.
7. $k(\mathbf{u} + \mathbf{v}) = k\mathbf{u} + k\mathbf{v}$
8. $(k + m)\mathbf{u} = k\mathbf{u} + m\mathbf{u}$
9. $k(m\mathbf{u}) = (km)(\mathbf{u})$
10. $1\mathbf{u} = \mathbf{u}$

REMARK Depending on the application, scalars may be real numbers or complex numbers. Vector spaces in which the scalars are complex numbers are called *complex vector spaces*, and those in which the scalars must be real are called *real vector spaces*. In Chapter 10 we shall discuss complex vector spaces; until then, *all of our scalars will be real numbers*.

The reader should keep in mind that the definition of a vector space specifies neither the nature of the vectors nor the operations. Any kind of object can be a vector, and the operations of addition and scalar multiplication may not have any relationship or similarity to the standard vector operations on R^n. The only requirement is that the ten vector space axioms be satisfied. Some authors use the notations $\oplus$ and $\odot$ for vector addition and scalar multiplication to distinguish these operations from addition and multiplication of real numbers; we will not use this convention, however.

Examples of Vector Spaces

The following examples will illustrate the variety of possible vector spaces. In each example we will specify a nonempty set V and two operations, addition and scalar multiplication; then we shall verify that the ten vector space axioms are satisfied, thereby entitling V, with the specified operations, to be called a vector space.

EXAMPLE 1 R^n Is a Vector Space

The set $V = R^n$ with the standard operations of addition and scalar multiplication defined in Section 4.1 is a vector space. Axioms 1 and 6 follow from the definitions of the standard operations on R^n; the remaining axioms follow from Theorem 4.1.1. ◆

The three most important special cases of R^n are R (the real numbers), R^2 (the vectors in the plane), and R^3 (the vectors in 3-space).

EXAMPLE 2   A Vector Space of 2 × 2 Matrices

Show that the set V of all 2×2 matrices with real entries is a vector space if addition is defined to be matrix addition and scalar multiplication is defined to be matrix scalar multiplication.

Solution

In this example we will find it convenient to verify the axioms in the following order: 1, 6, 2, 3, 7, 8, 9, 4, 5, and 10. Let

$$\mathbf{u} = \begin{bmatrix} u_{11} & u_{12} \\ u_{21} & u_{22} \end{bmatrix} \quad \text{and} \quad \mathbf{v} = \begin{bmatrix} v_{11} & v_{12} \\ v_{21} & v_{22} \end{bmatrix}$$

To prove Axiom 1, we must show that $\mathbf{u} + \mathbf{v}$ is an object in V; that is, we must show that $\mathbf{u} + \mathbf{v}$ is a 2×2 matrix. But this follows from the definition of matrix addition, since

$$\mathbf{u} + \mathbf{v} = \begin{bmatrix} u_{11} & u_{12} \\ u_{21} & u_{22} \end{bmatrix} + \begin{bmatrix} v_{11} & v_{12} \\ v_{21} & v_{22} \end{bmatrix} = \begin{bmatrix} u_{11} + v_{11} & u_{12} + v_{12} \\ u_{21} + v_{21} & u_{22} + v_{22} \end{bmatrix}$$

Similarly, Axiom 6 holds because for any real number k, we have

$$k\mathbf{u} = k \begin{bmatrix} u_{11} & u_{12} \\ u_{21} & u_{22} \end{bmatrix} = \begin{bmatrix} ku_{11} & ku_{12} \\ ku_{21} & ku_{22} \end{bmatrix}$$

so $k\mathbf{u}$ is a 2×2 matrix and consequently is an object in V.

Axiom 2 follows from Theorem 1.4.1a since

$$\mathbf{u} + \mathbf{v} = \begin{bmatrix} u_{11} & u_{12} \\ u_{21} & u_{22} \end{bmatrix} + \begin{bmatrix} v_{11} & v_{12} \\ v_{21} & v_{22} \end{bmatrix} = \begin{bmatrix} v_{11} & v_{12} \\ v_{21} & v_{22} \end{bmatrix} + \begin{bmatrix} u_{11} & u_{12} \\ u_{21} & u_{22} \end{bmatrix} = \mathbf{v} + \mathbf{u}$$

Similarly, Axiom 3 follows from part (b) of that theorem; and Axioms 7, 8, and 9 follow from parts (h), (j), and (l), respectively.

To prove Axiom 4, we must find an object $\mathbf{0}$ in V such that $\mathbf{0} + \mathbf{u} = \mathbf{u} + \mathbf{0} = \mathbf{u}$ for all $\mathbf{u}$ in V. This can be done by defining $\mathbf{0}$ to be

$$\mathbf{0} = \begin{bmatrix} 0 & 0 \\ 0 & 0 \end{bmatrix}$$

With this definition,

$$\mathbf{0} + \mathbf{u} = \begin{bmatrix} 0 & 0 \\ 0 & 0 \end{bmatrix} + \begin{bmatrix} u_{11} & u_{12} \\ u_{21} & u_{22} \end{bmatrix} = \begin{bmatrix} u_{11} & u_{12} \\ u_{21} & u_{22} \end{bmatrix} = \mathbf{u}$$

and similarly $\mathbf{u} + \mathbf{0} = \mathbf{u}$. To prove Axiom 5, we must show that each object $\mathbf{u}$ in V has a negative $-\mathbf{u}$ such that $\mathbf{u} + (-\mathbf{u}) = \mathbf{0}$ and $(-\mathbf{u}) + \mathbf{u} = \mathbf{0}$. This can be done by defining the negative of $\mathbf{u}$ to be

$$-\mathbf{u} = \begin{bmatrix} -u_{11} & -u_{12} \\ -u_{21} & -u_{22} \end{bmatrix}$$

With this definition,

$$\mathbf{u} + (-\mathbf{u}) = \begin{bmatrix} u_{11} & u_{12} \\ u_{21} & u_{22} \end{bmatrix} + \begin{bmatrix} -u_{11} & -u_{12} \\ -u_{21} & -u_{22} \end{bmatrix} = \begin{bmatrix} 0 & 0 \\ 0 & 0 \end{bmatrix} = \mathbf{0}$$

and similarly $(-\mathbf{u}) + \mathbf{u} = \mathbf{0}$. Finally, Axiom 10 is a simple computation:

$$1\mathbf{u} = 1\begin{bmatrix} u_{11} & u_{12} \\ u_{21} & u_{22} \end{bmatrix} = \begin{bmatrix} u_{11} & u_{12} \\ u_{21} & u_{22} \end{bmatrix} = \mathbf{u} \quad \blacklozenge$$

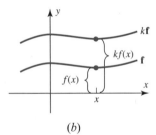

(a)

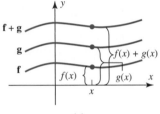

(b)

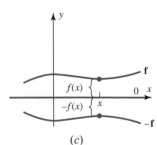

(c)

Figure 5.1.1

EXAMPLE 3 A Vector Space of $m \times n$ Matrices

Example 2 is a special case of a more general class of vector spaces. The arguments in that example can be adapted to show that the set V of all $m \times n$ matrices with real entries, together with the operations of matrix addition and scalar multiplication, is a vector space. The $m \times n$ zero matrix is the zero vector $\mathbf{0}$, and if $\mathbf{u}$ is the $m \times n$ matrix U, then the matrix $-U$ is the negative $-\mathbf{u}$ of the vector $\mathbf{u}$. We shall denote this vector space by the symbol M_{mn}. $\blacklozenge$

EXAMPLE 4 A Vector Space of Real-Valued Functions

Let V be the set of real-valued functions defined on the entire real line $(-\infty, \infty)$. If $\mathbf{f} = f(x)$ and $\mathbf{g} = g(x)$ are two such functions and k is any real number, define the sum function $\mathbf{f} + \mathbf{g}$ and the scalar multiple $k\mathbf{f}$, respectively, by

$$(\mathbf{f} + \mathbf{g})(x) = f(x) + g(x) \quad \text{and} \quad (k\mathbf{f})(x) = kf(x)$$

In other words, the value of the function $\mathbf{f} + \mathbf{g}$ at x is obtained by adding together the values of $\mathbf{f}$ and $\mathbf{g}$ at x (Figure 5.1.1a). Similarly, the value of $k\mathbf{f}$ at x is k times the value of $\mathbf{f}$ at x (Figure 5.1.1b). In the exercises we shall ask you to show that V is a vector space with respect to these operations. This vector space is denoted by $F(-\infty, \infty)$. If $\mathbf{f}$ and $\mathbf{g}$ are vectors in this space, then to say that $\mathbf{f} = \mathbf{g}$ is equivalent to saying that $f(x) = g(x)$ for all x in the interval $(-\infty, \infty)$.

The vector $\mathbf{0}$ in $F(-\infty, \infty)$ is the constant function that is identically zero for all values of x. The graph of this function is the line that coincides with the x-axis. The negative of a vector $\mathbf{f}$ is the function $-\mathbf{f} = -f(x)$. Geometrically, the graph of $-\mathbf{f}$ is the reflection of the graph of $\mathbf{f}$ across the x-axis (Figure 5.1.1c). $\blacklozenge$

REMARK In the preceding example we focused on the interval $(-\infty, \infty)$. Had we restricted our attention to some closed interval $[a, b]$ or some open interval (a, b), the functions defined on those intervals with the operations stated in the example would also have produced vector spaces. Those vector spaces are denoted by $F[a, b]$ and $F(a, b)$, respectively.

EXAMPLE 5   A Set That Is Not a Vector Space

Let $V = R^2$ and define addition and scalar multiplication operations as follows: If $\mathbf{u} = (u_1, u_2)$ and $\mathbf{v} = (v_1, v_2)$, then define

$$\mathbf{u} + \mathbf{v} = (u_1 + v_1, u_2 + v_2)$$

and if k is any real number, then define

$$k\mathbf{u} = (ku_1, 0)$$

For example, if $\mathbf{u} = (2, 4)$, $\mathbf{v} = (-3, 5)$, and $k = 7$, then

$$\mathbf{u} + \mathbf{v} = (2 + (-3), 4 + 5) = (-1, 9)$$
$$k\mathbf{u} = 7\mathbf{u} = (7 \cdot 2, 0) = (14, 0)$$

The addition operation is the standard addition operation on R^2, but the scalar multiplication operation is not the standard scalar multiplication. In the exercises we will ask you to show that the first nine vector space axioms are satisfied; however, there are values of $\mathbf{u}$ for which Axiom 10 fails to hold. For example, if $\mathbf{u} = (u_1, u_2)$ is such that $u_2 \neq 0$, then

$$1\mathbf{u} = 1(u_1, u_2) = (1 \cdot u_1, 0) = (u_1, 0) \neq \mathbf{u}$$

Thus V is not a vector space with the stated operations. ◆

EXAMPLE 6   Every Plane through the Origin Is a Vector Space

Let V be any plane through the origin in R^3. We shall show that the points in V form a vector space under the standard addition and scalar multiplication operations for vectors in R^3. From Example 1, we know that R^3 itself is a vector space under these operations. Thus Axioms 2, 3, 7, 8, 9, and 10 hold for all points in R^3 and consequently for all points in the plane V. We therefore need only show that Axioms 1, 4, 5, and 6 are satisfied.

Since the plane V passes through the origin, it has an equation of the form

$$ax + by + cz = 0 \tag{1}$$

(Theorem 3.5.1). Thus, if $\mathbf{u} = (u_1, u_2, u_3)$ and $\mathbf{v} = (v_1, v_2, v_3)$ are points in V, then $au_1 + bu_2 + cu_3 = 0$ and $av_1 + bv_2 + cv_3 = 0$. Adding these equations gives

$$a(u_1 + v_1) + b(u_2 + v_2) + c(u_3 + v_3) = 0$$

This equality tells us that the coordinates of the point

$$\mathbf{u} + \mathbf{v} = (u_1 + v_1, u_2 + v_2, u_3 + v_3)$$

satisfy (1); thus $\mathbf{u} + \mathbf{v}$ lies in the plane V. This proves that Axiom 1 is satisfied. The verifications of Axioms 4 and 6 are left as exercises; however, we shall prove that Axiom 5 is satisfied. Multiplying $au_1 + bu_2 + cu_3 = 0$ through by -1 gives

$$a(-u_1) + b(-u_2) + c(-u_3) = 0$$

Thus $-\mathbf{u} = (-u_1, -u_2, -u_3)$ lies in V. This establishes Axiom 5. ◆

EXAMPLE 7 The Zero Vector Space

Let V consist of a single object, which we denote by $\mathbf{0}$, and define

$$\mathbf{0} + \mathbf{0} = \mathbf{0} \quad \text{and} \quad k\mathbf{0} = \mathbf{0}$$

for all scalars k. It is easy to check that all the vector space axioms are satisfied. We call this the *zero vector space*. ◆

Some Properties of Vectors

As we progress, we shall add more examples of vector spaces to our list. We conclude this section with a theorem that gives a useful list of vector properties.

THEOREM 5.1.1

> *Let V be a vector space, $\mathbf{u}$ a vector in V, and k a scalar; then:*
>
> (*a*) $0\mathbf{u} = \mathbf{0}$
> (*b*) $k\mathbf{0} = \mathbf{0}$
> (*c*) $(-1)\mathbf{u} = -\mathbf{u}$
> (*d*) *If* $k\mathbf{u} = \mathbf{0}$, *then* $k = 0$ *or* $\mathbf{u} = \mathbf{0}$.

We shall prove parts (*a*) and (*c*) and leave proofs of the remaining parts as exercises.

Proof (a)   We can write

$$0\mathbf{u} + 0\mathbf{u} = (0 + 0)\mathbf{u} \quad \text{[Axiom 8]}$$
$$= 0\mathbf{u} \qquad \text{[Property of the number 0]}$$

By Axiom 5 the vector $0\mathbf{u}$ has a negative, $-0\mathbf{u}$. Adding this negative to both sides above yields

$$[0\mathbf{u} + 0\mathbf{u}] + (-0\mathbf{u}) = 0\mathbf{u} + (-0\mathbf{u})$$

or

$$0\mathbf{u} + [0\mathbf{u} + (-0\mathbf{u})] = 0\mathbf{u} + (-0\mathbf{u}) \quad \text{[Axiom 3]}$$
$$0\mathbf{u} + \mathbf{0} = \mathbf{0} \qquad \text{[Axiom 5]}$$
$$0\mathbf{u} = \mathbf{0} \qquad \text{[Axiom 4]}$$

Proof (c)   To show that $(-1)\mathbf{u} = -\mathbf{u}$, we must demonstrate that $\mathbf{u} + (-1)\mathbf{u} = \mathbf{0}$. To see this, observe that

$$\mathbf{u} + (-1)\mathbf{u} = 1\mathbf{u} + (-1)\mathbf{u} \quad \text{[Axiom 10]}$$
$$= (1 + (-1))\mathbf{u} \quad \text{[Axiom 8]}$$
$$= 0\mathbf{u} \qquad \text{[Property of numbers]}$$
$$= \mathbf{0} \qquad \text{[Part (\textit{a}) above]} \quad ■$$

EXERCISE SET 5.1

In Exercises 1–16 a set of objects is given, together with operations of addition and scalar multiplication. Determine which sets are vector spaces under the given operations. For those that are not vector spaces, list all axioms that fail to hold.

1.   The set of all triples of real numbers (x, y, z) with the operations

$$(x, y, z) + (x', y', z') = (x + x', y + y', z + z') \quad \text{and} \quad k(x, y, z) = (kx, y, z)$$

2. The set of all triples of real numbers (x, y, z) with the operations
$$(x, y, z) + (x', y', z') = (x + x', y + y', z + z') \quad \text{and} \quad k(x, y, z) = (0, 0, 0)$$

3. The set of all pairs of real numbers (x, y) with the operations
$$(x, y) + (x', y') = (x + x', y + y') \quad \text{and} \quad k(x, y) = (2kx, 2ky)$$

4. The set of all real numbers x with the standard operations of addition and multiplication.

5. The set of all pairs of real numbers of the form $(x, 0)$ with the standard operations on R^2.

6. The set of all pairs of real numbers of the form (x, y), where $x \geq 0$, with the standard operations on R^2.

7. The set of all n-tuples of real numbers of the form $(x, x, \ldots, x)$ with the standard operations on R^n.

8. The set of all pairs of real numbers (x, y) with the operations
$$(x, y) + (x', y') = (x + x' + 1, y + y' + 1) \quad \text{and} \quad k(x, y) = (kx, ky)$$

9. The set of all 2×2 matrices of the form
$$\begin{bmatrix} a & 1 \\ 1 & b \end{bmatrix}$$
with the standard matrix addition and scalar multiplication.

10. The set of all 2×2 matrices of the form
$$\begin{bmatrix} a & 0 \\ 0 & b \end{bmatrix}$$
with the standard matrix addition and scalar multiplication.

11. The set of all real-valued functions f defined everywhere on the real line and such that $f(1) = 0$, with the operations defined in Example 4.

12. The set of all 2×2 matrices of the form
$$\begin{bmatrix} a & a+b \\ a+b & b \end{bmatrix}$$
with matrix addition and scalar multiplication.

13. The set of all pairs of real numbers of the form $(1, x)$ with the operations
$$(1, y) + (1, y') = (1, y + y') \quad \text{and} \quad k(1, y) = (1, ky)$$

14. The set of polynomials of the form $a + bx$ with the operations
$$(a_0 + a_1 x) + (b_0 + b_1 x) = (a_0 + b_0) + (a_1 + b_1)x \quad \text{and} \quad k(a_0 + a_1 x) = (ka_0) + (ka_1)x$$

15. The set of all positive real numbers with the operations
$$x + y = xy \quad \text{and} \quad kx = x^k$$

16. The set of all pairs of real numbers (x, y) with the operations
$$(x, y) + (x', y') = (xx', yy') \quad \text{and} \quad k(x, y) = (kx, ky)$$

17. Show that the following sets with the given operations fail to be vector spaces by identifying all axioms that fail to hold.

    (a) The set of all triples of real numbers with the standard vector addition but with scalar multiplication defined by $k(x, y, z) = (k^2 x, k^2 y, k^2 z)$.

    (b) The set of all triples of real numbers with addition defined by $(x, y, z) + (u, v, w) = (z + w, y + v, x + u)$ and standard scalar multiplication.

    (c) The set of all 2×2 invertible matrices with the standard matrix addition and scalar multiplication.

18. Show that the set of all 2×2 matrices of the form $\begin{bmatrix} a & 1 \\ 1 & b \end{bmatrix}$ with addition defined by

$$\begin{bmatrix} a & 1 \\ 1 & b \end{bmatrix} + \begin{bmatrix} c & 1 \\ 1 & d \end{bmatrix} = \begin{bmatrix} a+c & 1 \\ 1 & b+d \end{bmatrix}$$ and scalar multiplication defined by $k\begin{bmatrix} a & 1 \\ 1 & b \end{bmatrix} =$

$\begin{bmatrix} ka & 1 \\ 1 & kb \end{bmatrix}$ is a vector space. What is the zero vector in this space?

19. (a) Show that the set of all points in R^2 lying on a line is a vector space, with respect to the standard operations of vector addition and scalar multiplication, exactly when the line passes through the origin.

    (b) Show that the set of all points in R^3 lying on a plane is a vector space, with respect to the standard operations of vector addition and scalar multiplication, exactly when the plane passes through the origin.

20. Consider the set of all 2×2 invertible matrices with vector addition defined to be matrix *multiplication* and the standard scalar multiplication. Is this a vector space?

21. Show that the first nine vector space axioms are satisfied if $V = R^2$ has the addition and scalar multiplication operations defined in Example 5.

22. Prove that a line passing through the origin in R^3 is a vector space under the standard operations on R^3.

23. Complete the unfinished details of Example 4.

24. Complete the unfinished details of Example 6.

Discussion & Discovery

25. We showed in Example 6 that every plane in R^3 that passes through the origin is a vector space under the standard operations on R^3. Is the same true for planes that do not pass through the origin? Explain your reasoning.

26. It was shown in Exercise 14 above that the set of polynomials of degree 1 or less is a vector space under the operations stated in that exercise. Is the set of polynomials whose degree is exactly 1 a vector space under those operations? Explain your reasoning.

27. Consider the set whose only element is the moon. Is this set a vector space under the operations moon + moon = moon and $k(\text{moon}) = \text{moon}$ for every real number k? Explain your reasoning.

28. Do you think that it is possible to have a vector space with exactly two distinct vectors in it? Explain your reasoning.

29. The following is a proof of part (b) of Theorem 5.1.1. Justify each step by filling in the blank line with the word *hypothesis* or by specifying the number of one of the vector space axioms given in this section.

    Hypothesis: Let $\mathbf{u}$ be any vector in a vector space V, $\mathbf{0}$ the zero vector in V, and k a scalar.

    Conclusion: Then $k\mathbf{0} = \mathbf{0}$.

    Proof: (1) First, $k\mathbf{0} + k\mathbf{u} = k(\mathbf{0} + \mathbf{u})$. _____

        (2) $= k\mathbf{u}$ _____

        (3) Since $k\mathbf{u}$ is in V, $-k\mathbf{u}$ is in V. _____

        (4) Therefore, $(k\mathbf{0} + k\mathbf{u}) + (-k\mathbf{u}) = k\mathbf{u} + (-k\mathbf{u})$. _____

        (5) $k\mathbf{0} + (k\mathbf{u} + (-k\mathbf{u})) = k\mathbf{u} + (-k\mathbf{u})$ _____

        (6) $k\mathbf{0} + \mathbf{0} = \mathbf{0}$ _____

        (7) Finally, $k\mathbf{0} = \mathbf{0}$. _____

30. Prove part (d) of Theorem 5.1.1.

31. The following is a proof that the cancellation law for addition holds in a vector space. Justify each step by filling in the blank line with the word *hypothesis* or by specifying the number of one of the vector space axioms given in this section.

Hypothesis: Let $\mathbf{u}$, $\mathbf{v}$, and $\mathbf{w}$ be vectors in a vector space V and suppose that $\mathbf{u} + \mathbf{w} = \mathbf{v} + \mathbf{w}$.

Conclusion: Then $\mathbf{u} = \mathbf{v}$.

Proof: (1) First, $(\mathbf{u} + \mathbf{w}) + (-\mathbf{w})$ and $(\mathbf{v} + \mathbf{w}) + (-\mathbf{w})$ are vectors in V. _____

(2) Then $(\mathbf{u} + \mathbf{w}) + (-\mathbf{w}) = (\mathbf{v} + \mathbf{w}) + (-\mathbf{w})$. _____

(3) The left side of the equality in step (2) is

$$(\mathbf{u} + \mathbf{w}) + (-\mathbf{w}) = \mathbf{u} + (\mathbf{w} + (-\mathbf{w}))$$ _____

$$= \mathbf{u}$$ _____

(4) The right side of the equality in step (2) is

$$(\mathbf{v} + \mathbf{w}) + (-\mathbf{w}) = \mathbf{v} + (\mathbf{w} + (-\mathbf{w}))$$ _____

$$= \mathbf{v}$$ _____

From the equality in step (2), it follows from steps (3) and (4) that $\mathbf{u} = \mathbf{v}$.

32. Do you think it is possible for a vector space to have two different zero vectors? That is, is it possible to have two *different* vectors $\mathbf{0}_1$ and $\mathbf{0}_2$ such that these vectors both satisfy Axiom 4? Explain your reasoning.

33. Do you think that it is possible for a vector $\mathbf{u}$ in a vector space to have two different negatives? That is, is it possible to have two *different* vectors $(-\mathbf{u})_1$ and $(-\mathbf{u})_2$, both of which satisfy Axiom 5? Explain your reasoning.

34. The set of ten axioms of a vector space is not an independent set because Axiom 2 can be deduced from other axioms in the set. Using the expression

$$(\mathbf{u} + \mathbf{v}) - (\mathbf{v} + \mathbf{u})$$

and Axiom 7 as a starting point, prove that $\mathbf{u} + \mathbf{v} = \mathbf{v} + \mathbf{u}$.

Hint You can use Theorem 5.1.1 since the proof of each part of that theorem does not use Axiom 2.

5.2
SUBSPACES

It is possible for one vector space to be contained within another vector space. For example, we showed in the preceding section that planes through the origin are vector spaces that are contained in the vector space R^3. In this section we shall study this important concept in detail.

A subset of a vector space V that is itself a vector space with respect to the operations of vector addition and scalar multiplication defined on V is given a special name.

> **DEFINITION**
>
> A subset W of a vector space V is called a ***subspace*** of V if W is itself a vector space under the addition and scalar multiplication defined on V.

In general, one must verify the ten vector space axioms to show that a set W with addition and scalar multiplication forms a vector space. However, if W is part of a larger set V that is already known to be a vector space, then certain axioms need not be verified for W because they are "inherited" from V. For example, there is no need to check that $\mathbf{u} + \mathbf{v} = \mathbf{v} + \mathbf{u}$ (Axiom 2) for W because this holds for all vectors in V and consequently for all vectors in W. Other axioms inherited by W from V are 3, 7, 8, 9, and 10. Thus, to show that a set W is a subspace of a vector space V, we need only verify Axioms 1, 4, 5, and 6. The following theorem shows that even Axioms 4 and 5 can be omitted.

THEOREM 5.2.1

> *If W is a set of one or more vectors from a vector space V, then W is a subspace of V if and only if the following conditions hold.*
>
> (a) *If $\mathbf{u}$ and $\mathbf{v}$ are vectors in W, then $\mathbf{u} + \mathbf{v}$ is in W.*
> (b) *If k is any scalar and $\mathbf{u}$ is any vector in W, then $k\mathbf{u}$ is in W.*

Proof If W is a subspace of V, then all the vector space axioms are satisfied; in particular, Axioms 1 and 6 hold. But these are precisely conditions (a) and (b).

Conversely, assume conditions (a) and (b) hold. Since these conditions are vector space Axioms 1 and 6, we need only show that W satisfies the remaining eight axioms. Axioms 2, 3, 7, 8, 9, and 10 are automatically satisfied by the vectors in W since they are satisfied by all vectors in V. Therefore, to complete the proof, we need only verify that Axioms 4 and 5 are satisfied by vectors in W.

Let $\mathbf{u}$ be any vector in W. By condition (b), $k\mathbf{u}$ is in W for every scalar k. Setting $k = 0$, it follows from Theorem 5.1.1 that $0\mathbf{u} = \mathbf{0}$ is in W, and setting $k = -1$, it follows that $(-1)\mathbf{u} = -\mathbf{u}$ is in W. ∎

REMARK A set W of one or more vectors from a vector space V is said to be ***closed under addition*** if condition (a) in Theorem 5.2.1 holds and ***closed under scalar multiplication*** if condition (b) holds. Thus Theorem 5.2.1 states that W is a subspace of V if and only if W is closed under addition and closed under scalar multiplication.

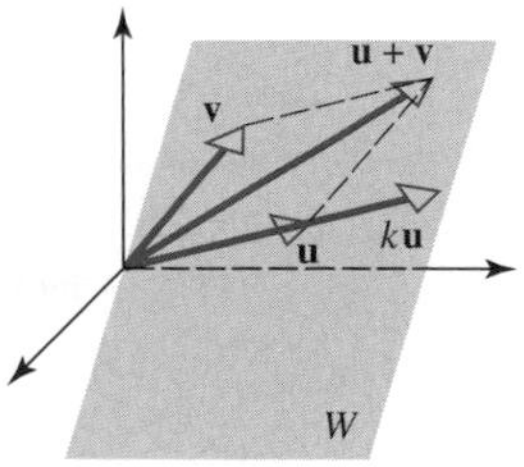

Figure 5.2.1 The vectors $\mathbf{u} + \mathbf{v}$ and $k\mathbf{u}$ both lie in the same plane as $\mathbf{u}$ and $\mathbf{v}$.

EXAMPLE 1 Testing for a Subspace

In Example 6 of Section 5.1 we verified the ten vector space axioms to show that the points in a plane through the origin of R^3 form a subspace of R^3. In light of Theorem 5.2.1 we can see that much of that work was unnecessary; it would have been sufficient to verify that the plane is closed under addition and scalar multiplication (Axioms 1 and 6). In Section 5.1 we verified those two axioms algebraically; however, they can also be proved geometrically as follows: Let W be any plane through the origin, and let $\mathbf{u}$ and $\mathbf{v}$ be any vectors in W. Then $\mathbf{u} + \mathbf{v}$ must lie in W because it is the diagonal of the parallelogram determined by $\mathbf{u}$ and $\mathbf{v}$ (Figure 5.2.1), and $k\mathbf{u}$ must lie in W for any scalar k because $k\mathbf{u}$ lies on a line through $\mathbf{u}$. Thus W is closed under addition and scalar multiplication, so it is a subspace of R^3. ◆

EXAMPLE 2 Lines through the Origin Are Subspaces

Show that a line through the origin of R^3 is a subspace of R^3.

Solution

Let W be a line through the origin of R^3. It is evident geometrically that the sum of two vectors on this line also lies on the line and that a scalar multiple of a vector on the

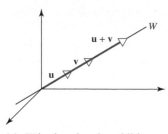

(a) W is closed under addition.

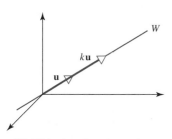

(b) W is closed under scalar multiplication.

Figure 5.2.2

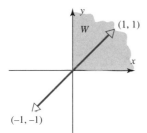

Figure 5.2.3 *W is not closed under scalar multiplication.*

line is on the line as well (Figure 5.2.2). Thus W is closed under addition and scalar multiplication, so it is a subspace of R^3. In the exercises we will ask you to prove this result algebraically using parametric equations for the line. ◆

EXAMPLE 3 A Subset of R^2 That Is Not a Subspace

Let W be the set of all points (x, y) in R^2 such that $x \geq 0$ and $y \geq 0$. These are the points in the first quadrant. The set W is *not* a subspace of R^2 since it is not closed under scalar multiplication. For example, $\mathbf{v} = (1, 1)$ lies in W, but its negative $(-1)\mathbf{v} = -\mathbf{v} = (-1, -1)$ does not (Figure 5.2.3). ◆

Every nonzero vector space V has at least two subspaces: V itself is a subspace, and the set $\{\mathbf{0}\}$ consisting of just the zero vector in V is a subspace called the ***zero subspace***. Combining this with Examples 1 and 2, we obtain the following list of subspaces of R^2 and R^3:

Subspaces of R^2	Subspaces of R^3
• $\{\mathbf{0}\}$	• $\{\mathbf{0}\}$
• Lines through the origin	• Lines through the origin
• R^2	• Planes through the origin
	• R^3

Later, we will show that these are the only subspaces of R^2 and R^3.

EXAMPLE 4 Subspaces of M_{nn}

From Theorem 1.7.2, the sum of two symmetric matrices is symmetric, and a scalar multiple of a symmetric matrix is symmetric. Thus the set of $n \times n$ symmetric matrices is a subspace of the vector space M_{nn} of all $n \times n$ matrices. Similarly, the set of $n \times n$ upper triangular matrices, the set of $n \times n$ lower triangular matrices, and the set of $n \times n$ diagonal matrices all form subspaces of M_{nn}, since each of these sets is closed under addition and scalar multiplication. ◆

EXAMPLE 5 A Subspace of Polynomials of Degree $\leq n$

Let n be a nonnegative integer, and let W consist of all functions expressible in the form

$$p(x) = a_0 + a_1 x + \cdots + a_n x^n \tag{1}$$

where $a_0, \ldots, a_n$ are real numbers. Thus W consists of all real polynomials of degree n or less. The set W is a subspace of the vector space of all real-valued functions discussed in Example 4 of the preceding section. To see this, let $\mathbf{p}$ and $\mathbf{q}$ be the polynomials

$$p(x) = a_0 + a_1 x + \cdots + a_n x^n \quad \text{and} \quad q(x) = b_0 + b_1 x + \cdots + b_n x^n$$

Then

$$(\mathbf{p} + \mathbf{q})(x) = p(x) + q(x) = (a_0 + b_0) + (a_1 + b_1)x + \cdots + (a_n + b_n)x^n$$

The CMYK Color Model

Color magazines and books are printed using what is called a **CMYK** *color model*. Colors in this model are created using four colored inks: cyan (C), magenta (M), yellow (Y), and black (K). The colors can be created either by mixing inks of the four types and printing with the mixed inks (the *spot color method*) or by printing dot patterns (called *rosettes*) with the four colors and allowing the reader's eye and perception process to create the desired color combination (the *process color method*). There is a numbering system for commercial inks, called the *Pantone Matching System*, that assigns every commercial ink color a number in accordance with its percentages of cyan, magenta, yellows, and black. One way to represent a Pantone color is by associating the four base colors with the vectors

$$\mathbf{c} = (1, 0, 0, 0) \quad \text{(pure cyan)}$$

$$\mathbf{m} = (0, 1, 0, 0) \quad \text{(pure magenta)}$$

$$\mathbf{y} = (0, 0, 1, 0) \quad \text{(pure yellow)}$$

$$\mathbf{k} = (0, 0, 0, 1) \quad \text{(pure black)}$$

in R^4 and describing the ink color as a linear combination of these using coefficients between 0 and 1, inclusive. Thus, an ink color $\mathbf{p}$ is represented as a linear combination of the form

$$\mathbf{p} = c_1\mathbf{c} + c_2\mathbf{m} + c_3\mathbf{y} + c_4\mathbf{k} = (c_1, c_2, c_3, c_4)$$

where $0 \leq c_i \leq 1$. The set of all such linear combinations is called **CMYK** *space*, although it is not a subspace of R^4. (Why?) For example, Pantone color 876CVC is a mixture of 38% cyan, 59% magenta, 73% yellow, and 7% black; Pantone color 216CVC is a mixture of 0% cyan, 83% magenta, 34% yellow, and 47% black; and Pantone color 328CVC is a mixture of 100% cyan, 0% magenta, 47% yellow, and 30% black. We can denote these colors by $\mathbf{p}_{876} = (0.38, 0.59, 0.73, 0.07)$, $\mathbf{p}_{216} = (0, 0.83, 0.34, 0.47)$, and $\mathbf{p}_{328} = (1, 0, 0.47, 0.30)$, respectively.

and

$$(k\mathbf{p})(x) = kp(x) = (ka_0) + (ka_1)x + \cdots + (ka_n)x^n$$

These functions have the form given in (1), so $\mathbf{p} + \mathbf{q}$ and $k\mathbf{p}$ lie in W. As in Section 4.4, we shall denote the vector space W in this example by the symbol P_n. ◆

Calculus Required

EXAMPLE 6 Subspaces of Functions Continuous on $(-\infty, \infty)$

Recall from calculus that if $\mathbf{f}$ and $\mathbf{g}$ are continuous functions on the interval $(-\infty, \infty)$ and k is a constant, then $\mathbf{f} + \mathbf{g}$ and $k\mathbf{f}$ are also continuous. Thus the continuous functions on the interval $(-\infty, \infty)$ form a subspace of $F(-\infty, \infty)$, since they are closed under addition and scalar multiplication. We denote this subspace by $C(-\infty, \infty)$. Similarly, if $\mathbf{f}$ and $\mathbf{g}$ have continuous first derivatives on $(-\infty, \infty)$, then so do $\mathbf{f} + \mathbf{g}$ and $k\mathbf{f}$. Thus the functions with continuous first derivatives on $(-\infty, \infty)$ form a subspace of $F(-\infty, \infty)$. We denote this subspace by $C^1(-\infty, \infty)$, where the superscript 1 is used to emphasize the *first* derivative. However, it is a theorem of calculus that every differentiable function is continuous, so $C^1(-\infty, \infty)$ is actually a subspace of $C(-\infty, \infty)$.

To take this a step further, for each positive integer m, the functions with continuous mth derivatives on $(-\infty, \infty)$ form a subspace of $C^1(-\infty, \infty)$ as do the functions that have continuous derivatives of all orders. We denote the subspace of functions with continuous mth derivatives on $(-\infty, \infty)$ by $C^m(-\infty, \infty)$, and we denote the subspace of functions that have continuous derivatives of all orders on $(-\infty, \infty)$ by $C^\infty(-\infty, \infty)$. Finally, it is a theorem of calculus that polynomials have continuous derivatives of all orders, so P_n is a subspace of $C^\infty(-\infty, \infty)$. The hierarchy of subspaces discussed in this example is illustrated in Figure 5.2.4. ◆

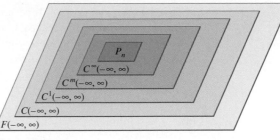

Figure 5.2.4

REMARK　In the preceding example we focused on the interval $(-\infty, \infty)$. Had we focused on a closed interval $[a, b]$, then the subspaces corresponding to those defined in the example would be denoted by $C[a, b]$, $C^m[a, b]$, and $C^\infty[a, b]$. Similarly, on an open interval (a, b) they would be denoted by $C(a, b)$, $C^m(a, b)$, and $C^\infty(a, b)$.

Solution Spaces of Homogeneous Systems

If $A\mathbf{x} = \mathbf{b}$ is a system of linear equations, then each vector $\mathbf{x}$ that satisfies this equation is called a ***solution vector*** of the system. The following theorem shows that the solution vectors of a *homogeneous* linear system form a vector space, which we shall call the ***solution space*** of the system.

THEOREM 5.2.2

> *If $A\mathbf{x} = \mathbf{0}$ is a homogeneous linear system of m equations in n unknowns, then the set of solution vectors is a subspace of R^n.*

Proof　Let W be the set of solution vectors. There is at least one vector in W, namely $\mathbf{0}$. To show that W is closed under addition and scalar multiplication, we must show that if $\mathbf{x}$ and $\mathbf{x}'$ are any solution vectors and k is any scalar, then $\mathbf{x} + \mathbf{x}'$ and $k\mathbf{x}$ are also solution vectors. But if $\mathbf{x}$ and $\mathbf{x}'$ are solution vectors, then

$$A\mathbf{x} = \mathbf{0} \quad \text{and} \quad A\mathbf{x}' = \mathbf{0}$$

from which it follows that

$$A(\mathbf{x} + \mathbf{x}') = A\mathbf{x} + A\mathbf{x}' = \mathbf{0} + \mathbf{0} = \mathbf{0}$$

and

$$A(k\mathbf{x}) = kA\mathbf{x} = k\mathbf{0} = \mathbf{0}$$

which proves that $\mathbf{x} + \mathbf{x}'$ and $k\mathbf{x}$ are solution vectors.　∎

EXAMPLE 7　Solution Spaces That Are Subspaces of R^3

Consider the linear systems

(a) $\begin{bmatrix} 1 & -2 & 3 \\ 2 & -4 & 6 \\ 3 & -6 & 9 \end{bmatrix} \begin{bmatrix} x \\ y \\ z \end{bmatrix} = \begin{bmatrix} 0 \\ 0 \\ 0 \end{bmatrix}$
(b) $\begin{bmatrix} 1 & -2 & 3 \\ -3 & 7 & -8 \\ -2 & 4 & -6 \end{bmatrix} \begin{bmatrix} x \\ y \\ z \end{bmatrix} = \begin{bmatrix} 0 \\ 0 \\ 0 \end{bmatrix}$

(c) $\begin{bmatrix} 1 & -2 & 3 \\ -3 & 7 & -8 \\ 4 & 1 & 2 \end{bmatrix} \begin{bmatrix} x \\ y \\ z \end{bmatrix} = \begin{bmatrix} 0 \\ 0 \\ 0 \end{bmatrix}$
(d) $\begin{bmatrix} 0 & 0 & 0 \\ 0 & 0 & 0 \\ 0 & 0 & 0 \end{bmatrix} \begin{bmatrix} x \\ y \\ z \end{bmatrix} = \begin{bmatrix} 0 \\ 0 \\ 0 \end{bmatrix}$

Each of these systems has three unknowns, so the solutions form subspaces of R^3. Geometrically, this means that each solution space must be the origin only, a line through the origin, a plane through the origin, or all of R^3. We shall now verify that this is so (leaving it to the reader to solve the systems).

Solution

(a)   The solutions are

$$x = 2s - 3t, \qquad y = s, \qquad z = t$$

from which it follows that

$$x = 2y - 3z \quad \text{or} \quad x - 2y + 3z = 0$$

This is the equation of the plane through the origin with $\mathbf{n} = (1, -2, 3)$ as a normal vector.

(b)   The solutions are

$$x = -5t, \qquad y = -t, \qquad z = t$$

which are parametric equations for the line through the origin parallel to the vector $\mathbf{v} = (-5, -1, 1)$.

(c)   The solution is $x = 0$, $y = 0$, $z = 0$, so the solution space is the origin only—that is, $\{\mathbf{0}\}$.

(d)   The solutions are

$$x = r, \qquad y = s, \qquad z = t$$

where r, s, and t have arbitrary values, so the solution space is all of R^3. ◆

In Section 1.3 we introduced the concept of a linear combination of column vectors. The following definition extends this idea to more general vectors.

DEFINITION

A vector $\mathbf{w}$ is called a ***linear combination*** of the vectors $\mathbf{v}_1, \mathbf{v}_2, \ldots, \mathbf{v}_r$ if it can be expressed in the form

$$\mathbf{w} = k_1\mathbf{v}_1 + k_2\mathbf{v}_2 + \cdots + k_r\mathbf{v}_r$$

where $k_1, k_2, \ldots, k_r$ are scalars.

REMARK   If $r = 1$, then the equation in the preceding definition reduces to $\mathbf{w} = k_1\mathbf{v}_1$; that is, $\mathbf{w}$ is a linear combination of a single vector $\mathbf{v}_1$ if it is a scalar multiple of $\mathbf{v}_1$.

EXAMPLE 8   Vectors in R^3 Are Linear Combinations of i, j, and k

---

Every vector $\mathbf{v} = (a, b, c)$ in R^3 is expressible as a linear combination of the standard basis vectors

$$\mathbf{i} = (1, 0, 0), \qquad \mathbf{j} = (0, 1, 0), \qquad \mathbf{k} = (0, 0, 1)$$

since

$$\mathbf{v} = (a, b, c) = a(1, 0, 0) + b(0, 1, 0) + c(0, 0, 1) = a\mathbf{i} + b\mathbf{j} + c\mathbf{k} \quad ◆$$

EXAMPLE 9 Checking a Linear Combination

Consider the vectors $\mathbf{u} = (1, 2, -1)$ and $\mathbf{v} = (6, 4, 2)$ in R^3. Show that $\mathbf{w} = (9, 2, 7)$ is a linear combination of $\mathbf{u}$ and $\mathbf{v}$ and that $\mathbf{w}' = (4, -1, 8)$ is *not* a linear combination of $\mathbf{u}$ and $\mathbf{v}$.

Solution

In order for $\mathbf{w}$ to be a linear combination of $\mathbf{u}$ and $\mathbf{v}$, there must be scalars k_1 and k_2 such that $\mathbf{w} = k_1\mathbf{u} + k_2\mathbf{v}$; that is,

$$(9, 2, 7) = k_1(1, 2, -1) + k_2(6, 4, 2)$$

or

$$(9, 2, 7) = (k_1 + 6k_2, 2k_1 + 4k_2, -k_1 + 2k_2)$$

Equating corresponding components gives

$$k_1 + 6k_2 = 9$$
$$2k_1 + 4k_2 = 2$$
$$-k_1 + 2k_2 = 7$$

Solving this system using Gaussian elimination yields $k_1 = -3$, $k_2 = 2$, so

$$\mathbf{w} = -3\mathbf{u} + 2\mathbf{v}$$

Similarly, for $\mathbf{w}'$ to be a linear combination of $\mathbf{u}$ and $\mathbf{v}$, there must be scalars k_1 and k_2 such that $\mathbf{w}' = k_1\mathbf{u} + k_2\mathbf{v}$; that is,

$$(4, -1, 8) = k_1(1, 2, -1) + k_2(6, 4, 2)$$

or

$$(4, -1, 8) = (k_1 + 6k_2, 2k_1 + 4k_2, -k_1 + 2k_2)$$

Equating corresponding components gives

$$k_1 + 6k_2 = 4$$
$$2k_1 + 4k_2 = -1$$
$$-k_1 + 2k_2 = 8$$

This system of equations is inconsistent (verify), so no such scalars k_1 and k_2 exist. Consequently, $\mathbf{w}'$ is not a linear combination of $\mathbf{u}$ and $\mathbf{v}$. ◆

Spanning

If $\mathbf{v}_1, \mathbf{v}_2, \ldots, \mathbf{v}_r$ are vectors in a vector space V, then generally some vectors in V may be linear combinations of $\mathbf{v}_1, \mathbf{v}_2, \ldots, \mathbf{v}_r$ and others may not. The following theorem shows that if we construct a set W consisting of all those vectors that are expressible as linear combinations of $\mathbf{v}_1, \mathbf{v}_2, \ldots, \mathbf{v}_r$, then W forms a subspace of V.

THEOREM 5.2.3

If $\mathbf{v}_1, \mathbf{v}_2, \ldots, \mathbf{v}_r$ are vectors in a vector space V, then

(a) The set W of all linear combinations of $\mathbf{v}_1, \mathbf{v}_2, \ldots, \mathbf{v}_r$ is a subspace of V.
(b) W is the smallest subspace of V that contains $\mathbf{v}_1, \mathbf{v}_2, \ldots, \mathbf{v}_r$ in the sense that every other subspace of V that contains $\mathbf{v}_1, \mathbf{v}_2, \ldots, \mathbf{v}_r$ must contain W.

Proof (a) To show that W is a subspace of V, we must prove that it is closed under addition and scalar multiplication. There is at least one vector in W—namely $\mathbf{0}$, since $\mathbf{0} = 0\mathbf{v}_1 + 0\mathbf{v}_2 + \cdots + 0\mathbf{v}_r$. If $\mathbf{u}$ and $\mathbf{v}$ are vectors in W, then

$$\mathbf{u} = c_1\mathbf{v}_1 + c_2\mathbf{v}_2 + \cdots + c_r\mathbf{v}_r$$

and

$$\mathbf{v} = k_1\mathbf{v}_1 + k_2\mathbf{v}_2 + \cdots + k_r\mathbf{v}_r$$

where $c_1, c_2, \ldots, c_r, k_1, k_2, \ldots, k_r$ are scalars. Therefore,

$$\mathbf{u} + \mathbf{v} = (c_1 + k_1)\mathbf{v}_1 + (c_2 + k_2)\mathbf{v}_2 + \cdots + (c_r + k_r)\mathbf{v}_r$$

and, for any scalar k,

$$k\mathbf{u} = (kc_1)\mathbf{v}_1 + (kc_2)\mathbf{v}_2 + \cdots + (kc_r)\mathbf{v}_r$$

Thus $\mathbf{u} + \mathbf{v}$ and $k\mathbf{u}$ are linear combinations of $\mathbf{v}_1, \mathbf{v}_2, \ldots, \mathbf{v}_r$ and consequently lie in W. Therefore, W is closed under addition and scalar multiplication.

Proof (b) Each vector $\mathbf{v}_i$ is a linear combination of $\mathbf{v}_1, \mathbf{v}_2, \ldots, \mathbf{v}_r$ since we can write

$$\mathbf{v}_i = 0\mathbf{v}_1 + 0\mathbf{v}_2 + \cdots + 1\mathbf{v}_i + \cdots + 0\mathbf{v}_r$$

Therefore, the subspace W contains each of the vectors $\mathbf{v}_1, \mathbf{v}_2, \ldots, \mathbf{v}_r$. Let W' be any other subspace that contains $\mathbf{v}_1, \mathbf{v}_2, \ldots, \mathbf{v}_r$. Since W' is closed under addition and scalar multiplication, it must contain all linear combinations of $\mathbf{v}_1, \mathbf{v}_2, \ldots, \mathbf{v}_r$. Thus, W' contains each vector of W. ∎

We make the following definition.

DEFINITION

If $S = \{\mathbf{v}_1, \mathbf{v}_2, \ldots, \mathbf{v}_r\}$ is a set of vectors in a vector space V, then the subspace W of V consisting of all linear combinations of the vectors in S is called the ***space spanned*** by $\mathbf{v}_1, \mathbf{v}_2, \ldots, \mathbf{v}_r$, and we say that the vectors $\mathbf{v}_1, \mathbf{v}_2, \ldots, \mathbf{v}_r$ ***span*** W. To indicate that W is the space spanned by the vectors in the set $S = \{\mathbf{v}_1, \mathbf{v}_2, \ldots, \mathbf{v}_r\}$, we write

$$W = \operatorname{span}(S) \quad \text{or} \quad W = \operatorname{span}\{\mathbf{v}_1, \mathbf{v}_2, \ldots, \mathbf{v}_r\}$$

EXAMPLE 10 Spaces Spanned by One or Two Vectors

If $\mathbf{v}_1$ and $\mathbf{v}_2$ are noncollinear vectors in R^3 with their initial points at the origin, then $\operatorname{span}\{\mathbf{v}_1, \mathbf{v}_2\}$, which consists of all linear combinations $k_1\mathbf{v}_1 + k_2\mathbf{v}_2$, is the plane determined by $\mathbf{v}_1$ and $\mathbf{v}_2$ (see Figure 5.2.5a). Similarly, if $\mathbf{v}$ is a nonzero vector in R^2 or R^3, then $\operatorname{span}\{\mathbf{v}\}$, which is the set of all scalar multiples $k\mathbf{v}$, is the line determined by $\mathbf{v}$ (see Figure 5.2.5b). ◆

EXAMPLE 11 Spanning Set for P_n

The polynomials $1, x, x^2, \ldots, x^n$ span the vector space P_n defined in Example 5 since each polynomial $\mathbf{p}$ in P_n can be written as

$$\mathbf{p} = a_0 + a_1x + \cdots + a_nx^n$$

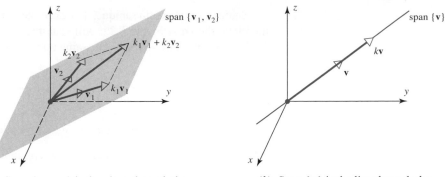

(*a*) Span $\{\mathbf{v}_1, \mathbf{v}_2\}$ is the plane through the origin determined by $\mathbf{v}_1$ and $\mathbf{v}_2$.

(*b*) Span $\{\mathbf{v}\}$ is the line through the origin determined by $\mathbf{v}$.

Figure 5.2.5

which is a linear combination of $1, x, x^2, \ldots, x^n$. We can denote this by writing

$$P_n = \text{span}\{1, x, x^2, \ldots, x^n\} \quad \blacklozenge$$

EXAMPLE 12   Three Vectors That Do Not Span R^3

Determine whether $\mathbf{v}_1 = (1, 1, 2)$, $\mathbf{v}_2 = (1, 0, 1)$, and $\mathbf{v}_3 = (2, 1, 3)$ span the vector space R^3.

Solution

We must determine whether an arbitrary vector $\mathbf{b} = (b_1, b_2, b_3)$ in R^3 can be expressed as a linear combination

$$\mathbf{b} = k_1\mathbf{v}_1 + k_2\mathbf{v}_2 + k_3\mathbf{v}_3$$

of the vectors $\mathbf{v}_1$, $\mathbf{v}_2$, and $\mathbf{v}_3$. Expressing this equation in terms of components gives

$$(b_1, b_2, b_3) = k_1(1, 1, 2) + k_2(1, 0, 1) + k_3(2, 1, 3)$$

or

$$(b_1, b_2, b_3) = (k_1 + k_2 + 2k_3, k_1 + k_3, 2k_1 + k_2 + 3k_3)$$

or

$$
\begin{aligned}
k_1 + k_2 + 2k_3 &= b_1 \\
k_1 \qquad + k_3 &= b_2 \\
2k_1 + k_2 + 3k_3 &= b_3
\end{aligned}
$$

The problem thus reduces to determining whether this system is consistent for all values of b_1, b_2, and b_3. By parts (*e*) and (*g*) of Theorem 4.3.4, this system is consistent for all b_1, b_2, and b_3 if and only if the coefficient matrix

$$A = \begin{bmatrix} 1 & 1 & 2 \\ 1 & 0 & 1 \\ 2 & 1 & 3 \end{bmatrix}$$

has a nonzero determinant. However, $\det(A) = 0$ (verify), so $\mathbf{v}_1$, $\mathbf{v}_2$, and $\mathbf{v}_3$ do not span R^3. $\blacklozenge$

Spanning sets are not unique. For example, any two noncollinear vectors that lie in the plane shown in Figure 5.2.5 will span that same plane, and any nonzero vector on the line in that figure will span the same line. We leave the proof of the following useful theorem as an exercise.

THEOREM 5.2.4

If $S = \{\mathbf{v}_1, \mathbf{v}_2, \ldots, \mathbf{v}_r\}$ and $S' = \{\mathbf{w}_1, \mathbf{w}_2, \ldots, \mathbf{w}_k\}$ are two sets of vectors in a vector space V, then

$$span\{\mathbf{v}_1, \mathbf{v}_2, \ldots, \mathbf{v}_r\} = span\{\mathbf{w}_1, \mathbf{w}_2, \ldots, \mathbf{w}_k\}$$

if and only if each vector in S is a linear combination of those in S' and each vector in S' is a linear combination of those in S.

EXERCISE SET 5.2

1. Use Theorem 5.2.1 to determine which of the following are subspaces of R^3.

   (a) all vectors of the form $(a, 0, 0)$

   (b) all vectors of the form $(a, 1, 1)$

   (c) all vectors of the form (a, b, c), where $b = a + c$

   (d) all vectors of the form (a, b, c), where $b = a + c + 1$

   (e) all vectors of the form $(a, b, 0)$

2. Use Theorem 5.2.1 to determine which of the following are subspaces of M_{22}.

   (a) all 2×2 matrices with integer entries

   (b) all matrices

   $$\begin{bmatrix} a & b \\ c & d \end{bmatrix}$$

   where $a + b + c + d = 0$

   (c) all 2×2 matrices A such that $\det(A) = 0$

   (d) all matrices of the form

   $$\begin{bmatrix} a & b \\ 0 & c \end{bmatrix}$$

   (e) all matrices of the form

   $$\begin{bmatrix} a & a \\ -a & -a \end{bmatrix}$$

3. Use Theorem 5.2.1 to determine which of the following are subspaces of P_3.

   (a) all polynomials $a_0 + a_1 x + a_2 x^2 + a_3 x^3$ for which $a_0 = 0$

   (b) all polynomials $a_0 + a_1 x + a_2 x^2 + a_3 x^3$ for which $a_0 + a_1 + a_2 + a_3 = 0$

   (c) all polynomials $a_0 + a_1 x + a_2 x^2 + a_3 x^3$ for which a_0, a_1, a_2, and a_3 are integers

   (d) all polynomials of the form $a_0 + a_1 x$, where a_0 and a_1 are real numbers

4. Use Theorem 5.2.1 to determine which of the following are subspaces of the space $F(-\infty, \infty)$.

   (a) all f such that $f(x) \leq 0$ for all x (b) all f such that $f(0) = 0$

   (c) all f such that $f(0) = 2$ (d) all constant functions

   (e) all f of the form $k_1 + k_2 \sin x$, where k_1 and k_2 are real numbers

5. Use Theorem 5.2.1 to determine which of the following are subspaces of M_{nn}.

   (a) all $n \times n$ matrices A such that $\text{tr}(A) = 0$

   (b) all $n \times n$ matrices A such that $A^T = -A$

(c)   all $n \times n$ matrices A such that the linear system $A\mathbf{x} = \mathbf{0}$ has only the trivial solution

(d)   all $n \times n$ matrices A such that $AB = BA$ for a fixed $n \times n$ matrix B

6. Determine whether the solution space of the system $A\mathbf{x} = \mathbf{0}$ is a line through the origin, a plane through the origin, or the origin only. If it is a plane, find an equation for it; if it is a line, find parametric equations for it.

(a) $A = \begin{bmatrix} -1 & 1 & 1 \\ 3 & -1 & 0 \\ 2 & -4 & -5 \end{bmatrix}$
(b) $A = \begin{bmatrix} 1 & -2 & 3 \\ -3 & 6 & 9 \\ -2 & 4 & -6 \end{bmatrix}$

(c) $A = \begin{bmatrix} 1 & 2 & 3 \\ 2 & 5 & 3 \\ 1 & 0 & 8 \end{bmatrix}$
(d) $A = \begin{bmatrix} 1 & 2 & -6 \\ 1 & 4 & 4 \\ 3 & 10 & 6 \end{bmatrix}$

(e) $A = \begin{bmatrix} 1 & -1 & 1 \\ 2 & -1 & 4 \\ 3 & 1 & 11 \end{bmatrix}$
(f) $A = \begin{bmatrix} 1 & -3 & 1 \\ 2 & -6 & 2 \\ 3 & -9 & 3 \end{bmatrix}$

7. Which of the following are linear combinations of $\mathbf{u} = (0, -2, 2)$ and $\mathbf{v} = (1, 3, -1)$?

(a) $(2, 2, 2)$ (b) $(3, 1, 5)$ (c) $(0, 4, 5)$ (d) $(0, 0, 0)$

8. Express the following as linear combinations of $\mathbf{u} = (2, 1, 4)$, $\mathbf{v} = (1, -1, 3)$, and $\mathbf{w} = (3, 2, 5)$.

(a) $(-9, -7, -15)$ (b) $(6, 11, 6)$ (c) $(0, 0, 0)$ (d) $(7, 8, 9)$

9. Express the following as linear combinations of $\mathbf{p}_1 = 2 + x + 4x^2$, $\mathbf{p}_2 = 1 - x + 3x^2$, and $\mathbf{p}_3 = 3 + 2x + 5x^2$.

(a) $-9 - 7x - 15x^2$ (b) $6 + 11x + 6x^2$ (c) 0 (d) $7 + 8x + 9x^2$

10. Which of the following are linear combinations of

$$A = \begin{bmatrix} 4 & 0 \\ -2 & -2 \end{bmatrix}, \quad B = \begin{bmatrix} 1 & -1 \\ 2 & 3 \end{bmatrix}, \quad C = \begin{bmatrix} 0 & 2 \\ 1 & 4 \end{bmatrix}?$$

(a) $\begin{bmatrix} 6 & -8 \\ -1 & -8 \end{bmatrix}$
(b) $\begin{bmatrix} 0 & 0 \\ 0 & 0 \end{bmatrix}$
(c) $\begin{bmatrix} 6 & 0 \\ 3 & 8 \end{bmatrix}$
(d) $\begin{bmatrix} -1 & 5 \\ 7 & 1 \end{bmatrix}$

11. In each part, determine whether the given vectors span R^3.

(a) $\mathbf{v}_1 = (2, 2, 2)$, $\mathbf{v}_2 = (0, 0, 3)$, $\mathbf{v}_3 = (0, 1, 1)$

(b) $\mathbf{v}_1 = (2, -1, 3)$, $\mathbf{v}_2 = (4, 1, 2)$, $\mathbf{v}_3 = (8, -1, 8)$

(c) $\mathbf{v}_1 = (3, 1, 4)$, $\mathbf{v}_2 = (2, -3, 5)$, $\mathbf{v}_3 = (5, -2, 9)$, $\mathbf{v}_4 = (1, 4, -1)$

(d) $\mathbf{v}_1 = (1, 2, 6)$, $\mathbf{v}_2 = (3, 4, 1)$, $\mathbf{v}_3 = (4, 3, 1)$, $\mathbf{v}_4 = (3, 3, 1)$

12. Let $\mathbf{f} = \cos^2 x$ and $\mathbf{g} = \sin^2 x$. Which of the following lie in the space spanned by $\mathbf{f}$ and $\mathbf{g}$?

(a) $\cos 2x$ (b) $3 + x^2$ (c) 1 (d) $\sin x$ (e) 0

13. Determine whether the following polynomials span P_2.

$$\mathbf{p}_1 = 1 - x + 2x^2, \qquad \mathbf{p}_2 = 3 + x, \qquad \mathbf{p}_3 = 5 - x + 4x^2, \qquad \mathbf{p}_4 = -2 - 2x + 2x^2$$

14. Let $\mathbf{v}_1 = (2, 1, 0, 3)$, $\mathbf{v}_2 = (3, -1, 5, 2)$, and $\mathbf{v}_3 = (-1, 0, 2, 1)$. Which of the following vectors are in $\text{span}\{\mathbf{v}_1, \mathbf{v}_2, \mathbf{v}_3\}$?

(a) $(2, 3, -7, 3)$ (b) $(0, 0, 0, 0)$ (c) $(1, 1, 1, 1)$ (d) $(-4, 6, -13, 4)$

15. Find an equation for the plane spanned by the vectors $\mathbf{u} = (-1, 1, 1)$ and $\mathbf{v} = (3, 4, 4)$.

16. Find parametric equations for the line spanned by the vector $\mathbf{u} = (3, -2, 5)$.

17. Show that the solution vectors of a consistent nonhomogeneous system of m linear equations in n unknowns do not form a subspace of R^n.

18. Prove Theorem 5.2.4.

19. Use Theorem 5.2.4 to show that $\mathbf{v}_1 = (1, 6, 4)$, $\mathbf{v}_2 = (2, 4, -1)$, $\mathbf{v}_3 = (-1, 2, 5)$, and $\mathbf{w}_1 = (1, -2, -5)$, $\mathbf{w}_2 = (0, 8, 9)$ span the same subspace of R^3.

20. A line L through the origin in R^3 can be represented by parametric equations of the form $x = at$, $y = bt$, and $z = ct$. Use these equations to show that L is a subspace of R^3; that is, show that if $\mathbf{v}_1 = (x_1, y_1, z_1)$ and $\mathbf{v}_2 = (x_2, y_2, z_2)$ are points on L and k is any real number, then $k\mathbf{v}_1$ and $\mathbf{v}_1 + \mathbf{v}_2$ are also points on L.

21. **(For Readers Who Have Studied Calculus)** Show that the following sets of functions are subspaces of $F(-\infty, \infty)$.

(a) all everywhere continuous functions

(b) all everywhere differentiable functions

(c) all everywhere differentiable functions that satisfy $\mathbf{f}' + 2\mathbf{f} = \mathbf{0}$

22. **(For Readers Who Have Studied Calculus)** Show that the set of continuous functions $\mathbf{f} = f(x)$ on $[a, b]$ such that

$$\int_a^b f(x)\, dx = 0$$

is a subspace of $C[a, b]$.

Discussion & Discovery

23. Indicate whether each statement is always true or sometimes false. Justify your answer by giving a logical argument or a counterexample.

(a) If $A\mathbf{x} = \mathbf{b}$ is any consistent linear system of m equations in n unknowns, then the solution set is a subspace of R^n.

(b) If W is a set of one or more vectors from a vector space V, and if $k\mathbf{u} + \mathbf{v}$ is a vector in W for all vectors $\mathbf{u}$ and $\mathbf{v}$ in W and for all scalars k, then W is a subspace of V.

(c) If S is a finite set of vectors in a vector space V, then span(S) must be closed under addition and scalar multiplication.

(d) The intersection of two subspaces of a vector space V is also a subspace of V.

(e) If span$(S_1) =$ span(S_2), then $S_1 = S_2$.

24. (a) Under what conditions will two vectors in R^3 span a plane? A line?

(b) Under what conditions will it be true that span$\{\mathbf{u}\} =$ span$\{\mathbf{v}\}$? Explain.

(c) If $A\mathbf{x} = \mathbf{b}$ is a consistent system of m equations in n unknowns, under what conditions will it be true that the solution set is a subspace of R^n? Explain.

25. Recall that lines through the origin are subspaces of R^2. If W_1 is the line $y = x$ and W_2 is the line $y = -x$, is the union $W_1 \cup W_2$ a subspace of R^2? Explain your reasoning.

26. (a) Let M_{22} be the vector space of 2×2 matrices. Find four matrices that span M_{22}.

(b) In words, describe a set of matrices that spans M_{nn}.

27. We showed in Example 8 that the vectors $\mathbf{i}$, $\mathbf{j}$, $\mathbf{k}$ span R^3. However, spanning sets are not unique. What geometric property must a set of three vectors in R^3 have if they are to span R^3?

5.3

LINEAR INDEPENDENCE

In the preceding section we learned that a set of vectors $S = \{\mathbf{v}_1, \mathbf{v}_2, \ldots, \mathbf{v}_r\}$ spans a given vector space V if every vector in V is expressible as a linear combination of the vectors in S. In general, there may be more than one way to express a vector in V as a linear combination of vectors in a spanning set. In this section we shall study conditions under which each vector in V is expressible as a linear combination of the spanning vectors in exactly one way. Spanning sets with this property play a fundamental role in the study of vector spaces.

> **DEFINITION**
>
> If $S = \{\mathbf{v}_1, \mathbf{v}_2, \ldots, \mathbf{v}_r\}$ is a nonempty set of vectors, then the vector equation
>
> $$k_1\mathbf{v}_1 + k_2\mathbf{v}_2 + \cdots + k_r\mathbf{v}_r = \mathbf{0}$$
>
> has at least one solution, namely
>
> $$k_1 = 0, \quad k_2 = 0, \ldots, \quad k_r = 0$$
>
> If this is the only solution, then S is called a ***linearly independent*** set. If there are other solutions, then S is called a ***linearly dependent*** set.

EXAMPLE 1 A Linearly Dependent Set

If $\mathbf{v}_1 = (2, -1, 0, 3)$, $\mathbf{v}_2 = (1, 2, 5, -1)$, and $\mathbf{v}_3 = (7, -1, 5, 8)$, then the set of vectors $S = \{\mathbf{v}_1, \mathbf{v}_2, \mathbf{v}_3\}$ is linearly dependent, since $3\mathbf{v}_1 + \mathbf{v}_2 - \mathbf{v}_3 = \mathbf{0}$. ◆

EXAMPLE 2 A Linearly Dependent Set

The polynomials

$$\mathbf{p}_1 = 1 - x, \quad \mathbf{p}_2 = 5 + 3x - 2x^2, \quad \text{and} \quad \mathbf{p}_3 = 1 + 3x - x^2$$

form a linearly dependent set in P_2 since $3\mathbf{p}_1 - \mathbf{p}_2 + 2\mathbf{p}_3 = \mathbf{0}$. ◆

EXAMPLE 3 Linearly Independent Sets

Consider the vectors $\mathbf{i} = (1, 0, 0)$, $\mathbf{j} = (0, 1, 0)$, and $\mathbf{k} = (0, 0, 1)$ in R^3. In terms of components, the vector equation

$$k_1\mathbf{i} + k_2\mathbf{j} + k_3\mathbf{k} = \mathbf{0}$$

becomes

$$k_1(1, 0, 0) + k_2(0, 1, 0) + k_3(0, 0, 1) = (0, 0, 0)$$

or, equivalently,

$$(k_1, k_2, k_3) = (0, 0, 0)$$

This implies that $k_1 = 0$, $k_2 = 0$, and $k_3 = 0$, so the set $S = \{\mathbf{i}, \mathbf{j}, \mathbf{k}\}$ is linearly independent. A similar argument can be used to show that the vectors

$$\mathbf{e}_1 = (1, 0, 0, \ldots, 0), \quad \mathbf{e}_2 = (0, 1, 0, \ldots, 0), \ldots, \quad \mathbf{e}_n = (0, 0, 0, \ldots, 1)$$

form a linearly independent set in R^n. ◆

EXAMPLE 4 Determining Linear Independence/Dependence

Determine whether the vectors

$$\mathbf{v}_1 = (1, -2, 3), \qquad \mathbf{v}_2 = (5, 6, -1), \qquad \mathbf{v}_3 = (3, 2, 1)$$

form a linearly dependent set or a linearly independent set.

Proof (b) Let $S' = \{\mathbf{w}_1, \mathbf{w}_2, \ldots, \mathbf{w}_m\}$ be any set of m vectors in V, where $m < n$. We want to show that S' does not span V. The proof will be by contradiction: We will show that assuming S' spans V leads to a contradiction of the linear independence of $\{\mathbf{v}_1, \mathbf{v}_2, \ldots, \mathbf{v}_n\}$.

If S' spans V, then every vector in V is a linear combination of the vectors in S'. In particular, each basis vector $\mathbf{v}_i$ is a linear combination of the vectors in S', say

$$
\begin{aligned}
\mathbf{v}_1 &= a_{11}\mathbf{w}_1 + a_{21}\mathbf{w}_2 + \cdots + a_{m1}\mathbf{w}_m \\
\mathbf{v}_2 &= a_{12}\mathbf{w}_1 + a_{22}\mathbf{w}_2 + \cdots + a_{m2}\mathbf{w}_m \\
&\ \vdots \qquad\quad \vdots \qquad\quad \vdots \qquad\qquad\quad \vdots \\
\mathbf{v}_n &= a_{1n}\mathbf{w}_1 + a_{2n}\mathbf{w}_2 + \cdots + a_{mn}\mathbf{w}_m
\end{aligned}
\tag{9}
$$

To obtain our contradiction, we will show that there are scalars $k_1, k_2, \ldots, k_n$, not all zero, such that

$$
k_1\mathbf{v}_1 + k_2\mathbf{v}_2 + \cdots + k_n\mathbf{v}_n = \mathbf{0}
\tag{10}
$$

But observe that (9) and (10) have the same form as (6) and (7) except that m and n are interchanged and the $\mathbf{w}$'s and $\mathbf{v}$'s are interchanged. Thus the computations that led to (8) now yield

$$
\begin{aligned}
a_{11}k_1 + a_{12}k_2 + \cdots + a_{1n}k_n &= 0 \\
a_{21}k_1 + a_{22}k_2 + \cdots + a_{2n}k_n &= 0 \\
\ \vdots \qquad\quad \vdots \qquad\qquad \vdots \qquad\quad \vdots \\
a_{m1}k_1 + a_{m2}k_2 + \cdots + a_{mn}k_n &= 0
\end{aligned}
$$

This linear system has more unknowns than equations and hence has nontrivial solutions by Theorem 1.2.1. ■

It follows from the preceding theorem that if $S = \{\mathbf{v}_1, \mathbf{v}_2, \ldots, \mathbf{v}_n\}$ is any basis for a vector space V, then all sets in V that simultaneously span V and are linearly independent must have precisely n vectors. Thus, all bases for V must have the same number of vectors as the arbitrary basis S. This yields the following result, which is one of the most important in linear algebra.

THEOREM 5.4.3

> *All bases for a finite-dimensional vector space have the same number of vectors.*

To see how this theorem is related to the concept of "dimension," recall that the standard basis for R^n has n vectors (Example 2). Thus Theorem 5.4.3 implies that all bases for R^n have n vectors. In particular, every basis for R^3 has three vectors, every basis for R^2 has two vectors, and every basis for $R^1 (=R)$ has one vector. Intuitively, R^3 is three-dimensional, R^2 (a plane) is two-dimensional, and R (a line) is one-dimensional. Thus, for familiar vector spaces, the number of vectors in a basis is the same as the dimension. This suggests the following definition.

DEFINITION

The ***dimension*** of a finite-dimensional vector space V, denoted by $\dim(V)$, is defined to be the number of vectors in a basis for V. In addition, we define the zero vector space to have dimension zero.

REMARK From here on we shall follow a common convention of regarding the empty set to be a basis for the zero vector space. This is consistent with the preceding definition, since the empty set has no vectors and the zero vector space has dimension zero.

EXAMPLE 9   Dimensions of Some Vector Spaces

$\dim(R^n) = n$ **[The standard basis has n vectors (Example 2).]**

$\dim(P_n) = n + 1$ **[The standard basis has $n + 1$ vectors (Example 5).]**

$\dim(M_{mn}) = mn$ **[The standard basis has mn vectors (Example 6).]** ◆

EXAMPLE 10   Dimension of a Solution Space

Determine a basis for and the dimension of the solution space of the homogeneous system

$$
\begin{aligned}
2x_1 + 2x_2 - x_3 \qquad + x_5 &= 0 \\
-x_1 - x_2 + 2x_3 - 3x_4 + x_5 &= 0 \\
x_1 + x_2 - 2x_3 \qquad - x_5 &= 0 \\
x_3 + x_4 + x_5 &= 0
\end{aligned}
$$

Solution

In Example 7 of Section 1.2 it was shown that the general solution of the given system is

$$
x_1 = -s - t, \qquad x_2 = s, \qquad x_3 = -t, \qquad x_4 = 0, \qquad x_5 = t
$$

Therefore, the solution vectors can be written as

$$
\begin{bmatrix} x_1 \\ x_2 \\ x_3 \\ x_4 \\ x_5 \end{bmatrix}
=
\begin{bmatrix} -s - t \\ s \\ -t \\ 0 \\ t \end{bmatrix}
=
\begin{bmatrix} -s \\ s \\ 0 \\ 0 \\ 0 \end{bmatrix}
+
\begin{bmatrix} -t \\ 0 \\ -t \\ 0 \\ t \end{bmatrix}
= s
\begin{bmatrix} -1 \\ 1 \\ 0 \\ 0 \\ 0 \end{bmatrix}
+ t
\begin{bmatrix} -1 \\ 0 \\ -1 \\ 0 \\ 1 \end{bmatrix}
$$

which shows that the vectors

$$
\mathbf{v}_1 = \begin{bmatrix} -1 \\ 1 \\ 0 \\ 0 \\ 0 \end{bmatrix}
\quad \text{and} \quad
\mathbf{v}_2 = \begin{bmatrix} -1 \\ 0 \\ -1 \\ 0 \\ 1 \end{bmatrix}
$$

span the solution space. Since they are also linearly independent (verify), $\{\mathbf{v}_1, \mathbf{v}_2\}$ is a basis, and the solution space is two-dimensional. ◆

Some Fundamental Theorems

We shall devote the remainder of this section to a series of theorems that reveal the subtle interrelationships among the concepts of spanning, linear independence, basis, and dimension. These theorems are not idle exercises in mathematical theory—they are essential to the understanding of vector spaces, and many practical applications of linear algebra build on them.

The following theorem, which we call the *Plus/Minus Theorem* (our own name), establishes two basic principles on which most of the theorems to follow will rely.

THEOREM 5.4.4

> **Plus/Minus Theorem**
>
> *Let S be a nonempty set of vectors in a vector space V.*
>
> *(a) If S is a linearly independent set, and if **v** is a vector in V that is outside of span(S), then the set $S \cup \{\mathbf{v}\}$ that results by inserting **v** into S is still linearly independent.*
>
> *(b) If **v** is a vector in S that is expressible as a linear combination of other vectors in S, and if $S - \{\mathbf{v}\}$ denotes the set obtained by removing **v** from S, then S and $S - \{\mathbf{v}\}$ span the same space; that is,*
>
> $$span(S) = span(S - \{\mathbf{v}\})$$

We shall defer the proof to the end of the section, so that we may move more immediately to the consequences of the theorem. However, the theorem can be visualized in R^3 as follows:

(a) A set S of two linearly independent vectors in R^3 spans a plane through the origin. If we enlarge S by inserting any vector **v** outside of this plane (Figure 5.4.5*a*), then the resulting set of three vectors is still linearly independent since none of the three vectors lies in the same plane as the other two.

(b) If S is a set of three noncollinear vectors in R^3 that lie in a common plane through the origin (Figure 5.4.5*b*, *c*), then the three vectors span the plane. However, if we remove from S any vector **v** that is a linear combination of the other two, then the remaining set of two vectors still spans the plane.

In general, to show that a set of vectors $\{\mathbf{v}_1, \mathbf{v}_2, \ldots, \mathbf{v}_n\}$ is a basis for a vector space V, we must show that the vectors are linearly independent and span V. However, if we happen to know that V has dimension n (so that $\{\mathbf{v}_1, \mathbf{v}_2, \ldots, \mathbf{v}_n\}$ contains the right number of vectors for a basis), then it suffices to check *either* linear independence *or* spanning—the remaining condition will hold automatically. This is the content of the following theorem.

THEOREM 5.4.5

> *If V is an n-dimensional vector space, and if S is a set in V with exactly n vectors, then S is a basis for V if either S spans V or S is linearly independent.*

Proof Assume that S has exactly n vectors and spans V. To prove that S is a basis, we must show that S is a linearly independent set. But if this is not so, then some vector **v** in S is a linear combination of the remaining vectors. If we remove this vector from S, then it follows from the Plus/Minus Theorem (5.4.4*b*) that the remaining set of $n - 1$ vectors still spans V. But this is impossible, since it follows from Theorem 5.4.2*b* that no

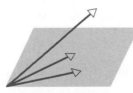 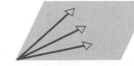 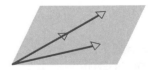

(*a*) None of the three vectors lies in the same plane as the other two.

(*b*) Any of the vectors can be removed, and the remaining two will still span the plane.

(*c*) Either of the collinear vectors can be removed, and the remaining two will still span the plane.

Figure 5.4.5

set with fewer than n vectors can span an n-dimensional vector space. Thus S is linearly independent.

Assume that S has exactly n vectors and is a linearly independent set. To prove that S is a basis, we must show that S spans V. But if this is not so, then there is some vector $\mathbf{v}$ in V that is not in span(S). If we insert this vector into S, then it follows from the Plus/Minus Theorem (5.4.4*a*) that this set of $n + 1$ vectors is still linearly independent. But this is impossible, since it follows from Theorem 5.4.2*a* that no set with more than n vectors in an n-dimensional vector space can be linearly independent. Thus S spans V. ∎

EXAMPLE 11 Checking for a Basis

(a) Show that $\mathbf{v}_1 = (-3, 7)$ and $\mathbf{v}_2 = (5, 5)$ form a basis for R^2 by inspection.

(b) Show that $\mathbf{v}_1 = (2, 0, -1)$, $\mathbf{v}_2 = (4, 0, 7)$, and $\mathbf{v}_3 = (-1, 1, 4)$ form a basis for R^3 by inspection.

Solution (a)

Since neither vector is a scalar multiple of the other, the two vectors form a linearly independent set in the two-dimensional space R^2, and hence they form a basis by Theorem 5.4.5.

Solution (b)

The vectors $\mathbf{v}_1$ and $\mathbf{v}_2$ form a linearly independent set in the xz-plane (why?). The vector $\mathbf{v}_3$ is outside of the xz-plane, so the set $\{\mathbf{v}_1, \mathbf{v}_2, \mathbf{v}_3\}$ is also linearly independent. Since R^3 is three-dimensional, Theorem 5.4.5 implies that $\{\mathbf{v}_1, \mathbf{v}_2, \mathbf{v}_3\}$ is a basis for R^3. ◆

The following theorem shows that for a finite-dimensional vector space V, every set that spans V contains a basis for V within it, and every linearly independent set in V is part of some basis for V.

THEOREM 5.4.6

> *Let S be a finite set of vectors in a finite-dimensional vector space V.*
>
> (*a*) *If S spans V but is not a basis for V, then S can be reduced to a basis for V by removing appropriate vectors from S.*
> (*b*) *If S is a linearly independent set that is not already a basis for V, then S can be enlarged to a basis for V by inserting appropriate vectors into S.*

Proof (a) If S is a set of vectors that spans V but is not a basis for V, then S is a linearly dependent set. Thus some vector $\mathbf{v}$ in S is expressible as a linear combination of the other vectors in S. By the Plus/Minus Theorem (5.4.4*b*), we can remove $\mathbf{v}$ from S, and the resulting set S' will still span V. If S' is linearly independent, then S' is a basis for V, and we are done. If S' is linearly dependent, then we can remove some appropriate vector from S' to produce a set S'' that still spans V. We can continue removing vectors in this way until we finally arrive at a set of vectors in S that is linearly independent and spans V. This subset of S is a basis for V.

Proof (b) Suppose that $\dim(V) = n$. If S is a linearly independent set that is not already a basis for V, then S fails to span V, and there is some vector $\mathbf{v}$ in V that is not in span(S). By the Plus/Minus Theorem (5.4.4*a*), we can insert $\mathbf{v}$ into S, and the resulting set S' will still be linearly independent. If S' spans V, then S' is a basis for V, and we are finished. If S' does not span V, then we can insert an appropriate vector into S' to produce a set S'' that is still linearly independent. We can continue inserting vectors in this way until we reach a set with n linearly independent vectors in V. This set will be a basis for V by Theorem 5.4.5. ∎

It can be proved (Exercise 30) that any subspace of a finite-dimensional vector space is finite-dimensional. We conclude this section with a theorem showing that the dimension of a subspace of a finite-dimensional vector space V cannot exceed the dimension of V itself and that the only way a subspace can have the same dimension as V is if the subspace is the entire vector space V. Figure 5.4.6 illustrates this idea in R^3. In that figure, observe that successively larger subspaces increase in dimension.

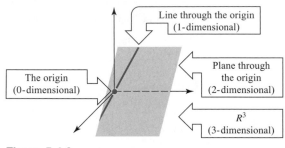

Line through the origin
(1-dimensional)

Plane through
the origin
(2-dimensional)

The origin
(0-dimensional)

R^3
(3-dimensional)

Figure 5.4.6

THEOREM 5.4.7

If W is a subspace of a finite-dimensional vector space V, then $\dim(W) \leq \dim(V)$; moreover, if $\dim(W) = \dim(V)$, then $W = V$.

Proof Since V is finite-dimensional, so is W by Exercise 30. Accordingly, suppose that $S = \{\mathbf{w}_1, \mathbf{w}_2, \ldots, \mathbf{w}_m\}$ is a basis for W. Either S is also a basis for V or it is not. If it is, then $\dim(W) = \dim(V) = m$. If it is not, then by Theorem 5.4.6b, vectors can be added to the linearly independent set S to make it into a basis for V, so $\dim(W) < \dim(V)$. Thus $\dim(W) \leq \dim(V)$ in all cases. If $\dim(W) = \dim(V)$, then S is a set of m linearly independent vectors in the m-dimensional vector space V; hence S is a basis for V by Theorem 5.4.5. This implies that $W = V$ (why?). ∎

Additional Proofs

Proof of Theorem 5.4.4a Assume that $S = \{\mathbf{v}_1, \mathbf{v}_2, \ldots, \mathbf{v}_r\}$ is a linearly independent set of vectors in V, and $\mathbf{v}$ is a vector in V outside of span(S). To show that $S' = \{\mathbf{v}_1, \mathbf{v}_2, \ldots, \mathbf{v}_r, \mathbf{v}\}$ is a linearly independent set, we must show that the only scalars that satisfy

$$k_1\mathbf{v}_1 + k_2\mathbf{v}_2 + \cdots + k_r\mathbf{v}_r + k_{r+1}\mathbf{v} = \mathbf{0} \tag{11}$$

are $k_1 = k_2 = \cdots = k_r = k_{r+1} = 0$. But we must have $k_{r+1} = 0$; otherwise, we could solve (11) for $\mathbf{v}$ as a linear combination of $\mathbf{v}_1, \mathbf{v}_2, \ldots, \mathbf{v}_r$, contradicting the assumption that $\mathbf{v}$ is outside of span(S). Thus (11) simplifies to

$$k_1\mathbf{v}_1 + k_2\mathbf{v}_2 + \cdots + k_r\mathbf{v}_r = \mathbf{0} \tag{12}$$

which, by the linear independence of $\{\mathbf{v}_1, \mathbf{v}_2, \ldots, \mathbf{v}_r\}$, implies that

$$k_1 = k_2 = \cdots = k_r = 0$$

Proof of Theorem 5.4.4b Assume that $S = \{\mathbf{v}_1, \mathbf{v}_2, \ldots, \mathbf{v}_r\}$ is a set of vectors in V, and to be specific, suppose that $\mathbf{v}_r$ is a linear combination of $\mathbf{v}_1, \mathbf{v}_2, \ldots, \mathbf{v}_{r-1}$, say

$$\mathbf{v}_r = c_1\mathbf{v}_1 + c_2\mathbf{v}_2 + \cdots + c_{r-1}\mathbf{v}_{r-1} \tag{13}$$

We want to show that if $\mathbf{v}_r$ is removed from S, then the remaining set of vectors $\{\mathbf{v}_1, \mathbf{v}_2, \ldots, \mathbf{v}_{r-1}\}$ still spans span(S); that is, we must show that every vector $\mathbf{w}$ in span(S) is expressible as a linear combination of $\{\mathbf{v}_1, \mathbf{v}_2, \ldots, \mathbf{v}_{r-1}\}$. But if $\mathbf{w}$ is in span(S), then $\mathbf{w}$ is expressible in the form

$$\mathbf{w} = k_1\mathbf{v}_1 + k_2\mathbf{v}_2 + \cdots + k_{r-1}\mathbf{v}_{r-1} + k_r\mathbf{v}_r$$

or, on substituting (13),

$$\mathbf{w} = k_1\mathbf{v}_1 + k_2\mathbf{v}_2 + \cdots + k_{r-1}\mathbf{v}_{r-1} + k_r(c_1\mathbf{v}_1 + c_2\mathbf{v}_2 + \cdots + c_{r-1}\mathbf{v}_{r-1})$$

which expresses $\mathbf{w}$ as a linear combination of $\mathbf{v}_1, \mathbf{v}_2, \ldots, \mathbf{v}_{r-1}$. ∎

EXERCISE SET 5.4

1. Explain why the following sets of vectors are *not* bases for the indicated vector spaces. (Solve this problem by inspection.)

   (a) $\mathbf{u}_1 = (1, 2)$, $\mathbf{u}_2 = (0, 3)$, $\mathbf{u}_3 = (2, 7)$ for R^2

   (b) $\mathbf{u}_1 = (-1, 3, 2)$, $\mathbf{u}_2 = (6, 1, 1)$ for R^3

   (c) $\mathbf{p}_1 = 1 + x + x^2$, $\mathbf{p}_2 = x - 1$ for P_2

   (d) $A = \begin{bmatrix} 1 & 1 \\ 2 & 3 \end{bmatrix}$, $B = \begin{bmatrix} 6 & 0 \\ -1 & 4 \end{bmatrix}$, $C = \begin{bmatrix} 3 & 0 \\ 1 & 7 \end{bmatrix}$, $D = \begin{bmatrix} 5 & 1 \\ 4 & 2 \end{bmatrix}$, $E = \begin{bmatrix} 7 & 1 \\ 2 & 9 \end{bmatrix}$

   for M_{22}

2. Which of the following sets of vectors are bases for R^2?

   (a) $(2, 1)$, $(3, 0)$ (b) $(4, 1)$, $(-7, -8)$

   (c) $(0, 0)$, $(1, 3)$ (d) $(3, 9)$, $(-4, \ 12)$

3. Which of the following sets of vectors are bases for R^3?

   (a) $(1, 0, 0)$, $(2, 2, 0)$, $(3, 3, 3)$ (b) $(3, 1, -4)$, $(2, 5, 6)$, $(1, 4, 8)$

   (c) $(2, -3, 1)$, $(4, 1, 1)$, $(0, -7, 1)$ (d) $(1, 6, 4)$, $(2, 4, -1)$, $(-1, 2, 5)$

4. Which of the following sets of vectors are bases for P_2?

   (a) $1 - 3x + 2x^2$, $1 + x + 4x^2$, $1 - 7x$

   (b) $4 + 6x + x^2$, $-1 + 4x + 2x^2$, $5 + 2x - x^2$

   (c) $1 + x + x^2$, $x + x^2$, x^2

   (d) $-4 + x + 3x^2$, $6 + 5x + 2x^2$, $8 + 4x + x^2$

5. Show that the following set of vectors is a basis for M_{22}.

   $$\begin{bmatrix} 3 & 6 \\ 3 & -6 \end{bmatrix}, \quad \begin{bmatrix} 0 & -1 \\ -1 & 0 \end{bmatrix}, \quad \begin{bmatrix} 0 & -8 \\ -12 & -4 \end{bmatrix}, \quad \begin{bmatrix} 1 & 0 \\ -1 & 2 \end{bmatrix}$$

6. Let V be the space spanned by $\mathbf{v}_1 = \cos^2 x$, $\mathbf{v}_2 = \sin^2 x$, $\mathbf{v}_3 = \cos 2x$.

   (a) Show that $S = \{\mathbf{v}_1, \mathbf{v}_2, \mathbf{v}_3\}$ is not a basis for V.   (b) Find a basis for V.

7. Find the coordinate vector of $\mathbf{w}$ relative to the basis $S = \{\mathbf{u}_1, \mathbf{u}_2\}$ for R^2.

   (a) $\mathbf{u}_1 = (1, 0)$, $\mathbf{u}_2 = (0, 1)$; $\mathbf{w} = (3, -7)$

   (b) $\mathbf{u}_1 = (2, -4)$, $\mathbf{u}_2 = (3, 8)$; $\mathbf{w} = (1, 1)$

   (c) $\mathbf{u}_1 = (1, 1)$, $\mathbf{u}_2 = (0, 2)$; $\mathbf{w} = (a, b)$

8. Find the coordinate vector of $\mathbf{w}$ relative to the basis $S = \{\mathbf{u}_1, \mathbf{u}_2\}$ of R^2.

(a) $\mathbf{u}_1 = (1, -1), \ \mathbf{u}_2 = (1, 1); \ \mathbf{w} = (1, 0)$

(b) $\mathbf{u}_1 = (1, -1), \ \mathbf{u}_2 = (1, 1); \ \mathbf{w} = (0, 1)$

(c) $\mathbf{u}_1 = (1, -1), \ \mathbf{u}_2 = (1, 1); \ \mathbf{w} = (1, 1)$

9. Find the coordinate vector of $\mathbf{v}$ relative to the basis $S = \{\mathbf{v}_1, \mathbf{v}_2, \mathbf{v}_3\}$.

(a) $\mathbf{v} = (2, -1, 3); \ \mathbf{v}_1 = (1, 0, 0), \ \mathbf{v}_2 = (2, 2, 0), \ \mathbf{v}_3 = (3, 3, 3)$

(b) $\mathbf{v} = (5, -12, 3); \ \mathbf{v}_1 = (1, 2, 3), \ \mathbf{v}_2 = (-4, 5, 6), \ \mathbf{v}_3 = (7, -8, 9)$

10. Find the coordinate vector of $\mathbf{p}$ relative to the basis $S = \{\mathbf{p}_1, \mathbf{p}_2, \mathbf{p}_3\}$.

(a) $\mathbf{p} = 4 - 3x + x^2; \ \mathbf{p}_1 = 1, \ \mathbf{p}_2 = x, \ \mathbf{p}_3 = x^2$

(b) $\mathbf{p} = 2 - x + x^2; \ \mathbf{p}_1 = 1 + x, \ \mathbf{p}_2 = 1 + x^2, \ \mathbf{p}_3 = x + x^2$

11. Find the coordinate vector of A relative to the basis $S = \{A_1, A_2, A_3, A_4\}$.

$$A = \begin{bmatrix} 2 & 0 \\ -1 & 3 \end{bmatrix}; \quad A_1 = \begin{bmatrix} -1 & 1 \\ 0 & 0 \end{bmatrix}, \quad A_2 = \begin{bmatrix} 1 & 1 \\ 0 & 0 \end{bmatrix}$$

$$A_3 = \begin{bmatrix} 0 & 0 \\ 1 & 0 \end{bmatrix}, \quad A_4 = \begin{bmatrix} 0 & 0 \\ 0 & 1 \end{bmatrix}$$

In Exercises 12–17 determine the dimension of and a basis for the solution space of the system.

12. $\begin{aligned} x_1 + x_2 - \ x_3 &= 0 \\ -2x_1 - x_2 + 2x_3 &= 0 \\ -x_1 \quad\ \ + \ x_3 &= 0 \end{aligned}$ **13.** $\begin{aligned} 3x_1 + x_2 + x_3 + x_4 &= 0 \\ 5x_1 - x_2 + x_3 - x_4 &= 0 \end{aligned}$

14. $\begin{aligned} x_1 - 4x_2 + 3x_3 - \ x_4 &= 0 \\ 2x_1 - 8x_2 + 6x_3 - 2x_4 &= 0 \end{aligned}$ **15.** $\begin{aligned} x_1 - 3x_2 + \ x_3 &= 0 \\ 2x_1 - 6x_2 + 2x_3 &= 0 \\ 3x_1 - 9x_2 + 3x_3 &= 0 \end{aligned}$

16. $\begin{aligned} 2x_1 + x_2 + 3x_3 &= 0 \\ x_1 \quad\ \ + 5x_3 &= 0 \\ x_2 + \ x_3 &= 0 \end{aligned}$ **17.** $\begin{aligned} x + \ y + \ z &= 0 \\ 3x + 2y - 2z &= 0 \\ 4x + 3y - \ z &= 0 \\ 6x + 5y + \ z &= 0 \end{aligned}$

18. Determine bases for the following subspaces of R^3.

(a) the plane $3x - 2y + 5z = 0$

(b) the plane $x - y = 0$

(c) the line $x = 2t, \ y = -t, z = 4t$

(d) all vectors of the form (a, b, c), where $b = a + c$

19. Determine the dimensions of the following subspaces of R^4.

(a) all vectors of the form $(a, b, c, 0)$

(b) all vectors of the form (a, b, c, d), where $d = a + b$ and $c = a - b$

(c) all vectors of the form (a, b, c, d), where $a = b = c = d$

20. Determine the dimension of the subspace of P_3 consisting of all polynomials $a_0 + a_1 x + a_2 x^2 + a_3 x^3$ for which $a_0 = 0$.

21. Find a standard basis vector that can be added to the set $\{\mathbf{v}_1, \mathbf{v}_2\}$ to produce a basis for R^3.

(a) $\mathbf{v}_1 = (-1, 2, 3), \ \mathbf{v}_2 = (1, -2, -2)$ (b) $\mathbf{v}_1 = (1, -1, 0), \ \mathbf{v}_2 = (3, 1, -2)$

22. Find standard basis vectors that can be added to the set $\{\mathbf{v}_1, \mathbf{v}_2\}$ to produce a basis for R^4.

$$\mathbf{v}_1 = (1, -4, 2, -3), \qquad \mathbf{v}_2 = (-3, 8, -4, 6)$$

23. Let $\{\mathbf{v}_1, \mathbf{v}_2, \mathbf{v}_3\}$ be a basis for a vector space V. Show that $\{\mathbf{u}_1, \mathbf{u}_2, \mathbf{u}_3\}$ is also a basis, where $\mathbf{u}_1 = \mathbf{v}_1, \mathbf{u}_2 = \mathbf{v}_1 + \mathbf{v}_2$, and $\mathbf{u}_3 = \mathbf{v}_1 + \mathbf{v}_2 + \mathbf{v}_3$.

24. (a) Show that for every positive integer n, one can find $n + 1$ linearly independent vectors in $F(-\infty, \infty)$.

   Hint Look for polynomials.

   (b) Use the result in part (a) to prove that $F(-\infty, \infty)$ is infinite-dimensional.

   (c) Prove that $C(-\infty, \infty)$, $C^m(-\infty, \infty)$, and $C^\infty(-\infty, \infty)$ are infinite-dimensional vector spaces.

25. Let S be a basis for an n-dimensional vector space V. Show that if $\mathbf{v}_1, \mathbf{v}_2, \ldots, \mathbf{v}_r$ form a linearly independent set of vectors in V, then the coordinate vectors $(\mathbf{v}_1)_S, (\mathbf{v}_2)_S, \ldots, (\mathbf{v}_r)_S$ form a linearly independent set in R^n, and conversely.

26. Using the notation from Exercise 25, show that if $\mathbf{v}_1, \mathbf{v}_2, \ldots, \mathbf{v}_r$ span V, then the coordinate vectors $(\mathbf{v}_1)_S, (\mathbf{v}_2)_S, \ldots, (\mathbf{v}_r)_S$ span R^n, and conversely.

27. Find a basis for the subspace of P_2 spanned by the given vectors.

   (a) $-1 + x - 2x^2$, $3 + 3x + 6x^2$, 9

   (b) $1 + x$, x^2, $-2 + 2x^2$, $-3x$

   (c) $1 + x - 3x^2$, $2 + 2x - 6x^2$, $3 + 3x - 9x^2$

   Hint Let S be the standard basis for P_2 and work with the coordinate vectors relative to S; note Exercises 25 and 26.

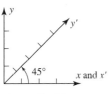

Figure Ex-28

28. The accompanying figure shows a rectangular xy-coordinate system and an $x'y'$-coordinate system with skewed axes. Assuming that 1-unit scales are used on all the axes, find the $x'y'$-coordinates of the points whose xy-coordinates are given.

   (a) $(1, 1)$ (b) $(1, 0)$ (c) $(0, 1)$ (d) (a, b)

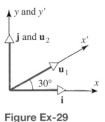

Figure Ex-29

29. The accompanying figure shows a rectangular xy-coordinate system determined by the unit basis vectors $\mathbf{i}$ and $\mathbf{j}$ and an $x'y'$-coordinate system determined by unit basis vectors $\mathbf{u}_1$ and $\mathbf{u}_2$. Find the $x'y'$-coordinates of the points whose xy-coordinates are given.

   (a) $(\sqrt{3}, 1)$ (b) $(1, 0)$ (c) $(0, 1)$ (d) (a, b)

30. Prove: Any subspace of a finite-dimensional vector space is finite-dimensional.

Discussion & Discovery

31. The basis that we gave for M_{22} in Example 6 consisted of noninvertible matrices. Do you think that there is a basis for M_{22} consisting of invertible matrices? Justify your answer.

32. (a) The vector space of all diagonal $n \times n$ matrices has dimension _____.

   (b) The vector space of all symmetric $n \times n$ matrices has dimension _____.

   (c) The vector space of all upper triangular $n \times n$ matrices has dimension _____.

33. (a) For a 3×3 matrix A, explain in words why the set $I_3, A, A^2, \ldots, A^9$ must be linearly dependent if the ten matrices are distinct.

   (b) State a corresponding result for an $n \times n$ matrix A.

34. State the two parts of Theorem 5.4.2 in contrapositive form. [See Exercise 34 of Section 1.4.]

35. (a) The equation $x_1 + x_2 + \cdots + x_n = 0$ can be viewed as a linear system of one equation in n unknowns. Make a conjecture about the dimension of its solution space.

   (b) Confirm your conjecture by finding a basis.

36. (a) Show that the set W of polynomials in P_2 such that $p(1) = 0$ is a subspace of P_2.

   (b) Make a conjecture about the dimension of W.

   (c) Confirm your conjecture by finding a basis for W.

5.5
ROW SPACE, COLUMN SPACE, AND NULLSPACE

In this section we shall study three important vector spaces that are associated with matrices. Our work here will provide us with a deeper understanding of the relationships between the solutions of a linear system of equations and properties of its coefficient matrix.

We begin with some definitions.

DEFINITION

For an $m \times n$ matrix

$$A = \begin{bmatrix} a_{11} & a_{12} & \cdots & a_{1n} \\ a_{21} & a_{22} & \cdots & a_{2n} \\ \vdots & \vdots & & \vdots \\ a_{m1} & a_{m2} & \cdots & a_{mn} \end{bmatrix}$$

the vectors

$$\mathbf{r}_1 = [a_{11} \quad a_{12} \quad \cdots \quad a_{1n}]$$
$$\mathbf{r}_2 = [a_{21} \quad a_{22} \quad \cdots \quad a_{2n}]$$
$$\vdots \qquad\qquad \vdots$$
$$\mathbf{r}_m = [a_{m1} \quad a_{m2} \quad \cdots \quad a_{mn}]$$

in R^n formed from the rows of A are called the ***row vectors*** of A, and the vectors

$$\mathbf{c}_1 = \begin{bmatrix} a_{11} \\ a_{21} \\ \vdots \\ a_{m1} \end{bmatrix}, \quad \mathbf{c}_2 = \begin{bmatrix} a_{12} \\ a_{22} \\ \vdots \\ a_{m2} \end{bmatrix}, \dots, \quad \mathbf{c}_n = \begin{bmatrix} a_{1n} \\ a_{2n} \\ \vdots \\ a_{mn} \end{bmatrix}$$

in R^m formed from the columns of A are called the ***column vectors*** of A.

EXAMPLE 1   Row and Column Vectors in a 2 x 3 Matrix

Let

$$A = \begin{bmatrix} 2 & 1 & 0 \\ 3 & -1 & 4 \end{bmatrix}$$

The row vectors of A are

$$\mathbf{r}_1 = [2 \quad 1 \quad 0] \quad \text{and} \quad \mathbf{r}_2 = [3 \quad -1 \quad 4]$$

and the column vectors of A are

$$\mathbf{c}_1 = \begin{bmatrix} 2 \\ 3 \end{bmatrix}, \quad \mathbf{c}_2 = \begin{bmatrix} 1 \\ -1 \end{bmatrix}, \quad \text{and} \quad \mathbf{c}_3 = \begin{bmatrix} 0 \\ 4 \end{bmatrix} \quad \blacklozenge$$

The following definition defines three important vector spaces associated with a matrix.

DEFINITION

If A is an $m \times n$ matrix, then the subspace of R^n spanned by the row vectors of A is called the ***row space*** of A, and the subspace of R^m spanned by the column vectors of A is called the ***column space*** of A. The solution space of the homogeneous system of equations $A\mathbf{x} = \mathbf{0}$, which is a subspace of R^n, is called the ***nullspace*** of A.

In this section and the next we shall be concerned with the following two general questions:

- What relationships exist between the solutions of a linear system $A\mathbf{x} = \mathbf{b}$ and the row space, column space, and nullspace of the coefficient matrix A?
- What relationships exist among the row space, column space, and nullspace of a matrix?

To investigate the first of these questions, suppose that

$$A = \begin{bmatrix} a_{11} & a_{12} & \cdots & a_{1n} \\ a_{21} & a_{22} & \cdots & a_{2n} \\ \vdots & \vdots & & \vdots \\ a_{m1} & a_{m2} & \cdots & a_{mn} \end{bmatrix} \quad \text{and} \quad \mathbf{x} = \begin{bmatrix} x_1 \\ x_2 \\ \vdots \\ x_n \end{bmatrix}$$

It follows from Formula (10) of Section 1.3 that if $\mathbf{c}_1, \mathbf{c}_2, \ldots, \mathbf{c}_n$ denote the column vectors of A, then the product $A\mathbf{x}$ can be expressed as a linear combination of these column vectors with coefficients from $\mathbf{x}$; that is,

$$A\mathbf{x} = x_1\mathbf{c}_1 + x_2\mathbf{c}_2 + \cdots + x_n\mathbf{c}_n \tag{1}$$

Thus a linear system, $A\mathbf{x} = \mathbf{b}$, of m equations in n unknowns can be written as

$$x_1\mathbf{c}_1 + x_2\mathbf{c}_2 + \cdots + x_n\mathbf{c}_n = \mathbf{b} \tag{2}$$

from which we conclude that $A\mathbf{x} = \mathbf{b}$ is consistent if and only if $\mathbf{b}$ is expressible as a linear combination of the column vectors of A or, equivalently, if and only if $\mathbf{b}$ is in the column space of A. This yields the following theorem.

THEOREM 5.5.1

> *A system of linear equations $A\mathbf{x} = \mathbf{b}$ is consistent if and only if $\mathbf{b}$ is in the column space of A.*

EXAMPLE 2   A Vector b in the Column Space of A

Let $A\mathbf{x} = \mathbf{b}$ be the linear system

$$\begin{bmatrix} -1 & 3 & 2 \\ 1 & 2 & -3 \\ 2 & 1 & -2 \end{bmatrix} \begin{bmatrix} x_1 \\ x_2 \\ x_3 \end{bmatrix} = \begin{bmatrix} 1 \\ -9 \\ -3 \end{bmatrix}$$

Show that $\mathbf{b}$ is in the column space of A, and express $\mathbf{b}$ as a linear combination of the column vectors of A.

Solution

Solving the system by Gaussian elimination yields (verify)

$$x_1 = 2, \qquad x_2 = -1, \qquad x_3 = 3$$

Since the system is consistent, $\mathbf{b}$ is in the column space of A. Moreover, from (2) and the solution obtained, it follows that

$$2\begin{bmatrix} -1 \\ 1 \\ 2 \end{bmatrix} - \begin{bmatrix} 3 \\ 2 \\ 1 \end{bmatrix} + 3\begin{bmatrix} 2 \\ -3 \\ -2 \end{bmatrix} = \begin{bmatrix} 1 \\ -9 \\ -3 \end{bmatrix} \quad \blacklozenge$$

The next theorem establishes a fundamental relationship between the solutions of a nonhomogeneous linear system $A\mathbf{x} = \mathbf{b}$ and those of the corresponding homogeneous linear system $A\mathbf{x} = \mathbf{0}$ with the same coefficient matrix.

THEOREM 5.5.2

> *If $\mathbf{x}_0$ denotes any single solution of a consistent linear system $A\mathbf{x} = \mathbf{b}$, and if $\mathbf{v}_1, \mathbf{v}_2, \ldots, \mathbf{v}_k$ form a basis for the nullspace of A—that is, the solution space of the homogeneous system $A\mathbf{x} = \mathbf{0}$—then every solution of $A\mathbf{x} = \mathbf{b}$ can be expressed in the form*
>
> $$\mathbf{x} = \mathbf{x}_0 + c_1\mathbf{v}_1 + c_2\mathbf{v}_2 + \cdots + c_k\mathbf{v}_k \qquad (3)$$
>
> *and, conversely, for all choices of scalars $c_1, c_2, \ldots, c_k$, the vector $\mathbf{x}$ in this formula is a solution of $A\mathbf{x} = \mathbf{b}$.*

Proof Assume that $\mathbf{x}_0$ is any fixed solution of $A\mathbf{x} = \mathbf{b}$ and that $\mathbf{x}$ is an arbitrary solution. Then

$$A\mathbf{x}_0 = \mathbf{b} \quad \text{and} \quad A\mathbf{x} = \mathbf{b}$$

Subtracting these equations yields

$$A\mathbf{x} - A\mathbf{x}_0 = \mathbf{0} \quad \text{or} \quad A(\mathbf{x} - \mathbf{x}_0) = \mathbf{0}$$

which shows that $\mathbf{x} - \mathbf{x}_0$ is a solution of the homogeneous system $A\mathbf{x} = \mathbf{0}$. Since $\mathbf{v}_1, \mathbf{v}_2, \ldots, \mathbf{v}_k$ is a basis for the solution space of this system, we can express $\mathbf{x} - \mathbf{x}_0$ as a linear combination of these vectors, say

$$\mathbf{x} - \mathbf{x}_0 = c_1\mathbf{v}_1 + c_2\mathbf{v}_2 + \cdots + c_k\mathbf{v}_k$$

Thus,

$$\mathbf{x} = \mathbf{x}_0 + c_1\mathbf{v}_1 + c_2\mathbf{v}_2 + \cdots + c_k\mathbf{v}_k$$

which proves the first part of the theorem. Conversely, for all choices of the scalars $c_1, c_2, \ldots, c_k$ in (3), we have

$$A\mathbf{x} = A(\mathbf{x}_0 + c_1\mathbf{v}_1 + c_2\mathbf{v}_2 + \cdots + c_k\mathbf{v}_k)$$

or

$$A\mathbf{x} = A\mathbf{x}_0 + c_1(A\mathbf{v}_1) + c_2(A\mathbf{v}_2) + \cdots + c_k(A\mathbf{v}_k)$$

But $\mathbf{x}_0$ is a solution of the nonhomogeneous system, and $\mathbf{v}_1, \mathbf{v}_2, \ldots, \mathbf{v}_k$ are solutions of the homogeneous system, so the last equation implies that

$$A\mathbf{x} = \mathbf{b} + \mathbf{0} + \mathbf{0} + \cdots + \mathbf{0} = \mathbf{b}$$

which shows that $\mathbf{x}$ is a solution of $A\mathbf{x} = \mathbf{b}$. ∎

General and Particular Solutions

There is some terminology associated with Formula (3). The vector $\mathbf{x}_0$ is called a ***particular solution*** of $A\mathbf{x} = \mathbf{b}$. The expression $\mathbf{x}_0 + c_1\mathbf{v}_1 + c_2\mathbf{v}_2 + \cdots + c_k\mathbf{v}_k$ is called the ***general solution*** of $A\mathbf{x} = \mathbf{b}$, and the expression $c_1\mathbf{v}_1 + c_2\mathbf{v}_2 + \cdots + c_k\mathbf{v}_k$ is called the ***general solution*** of $A\mathbf{x} = \mathbf{0}$. With this terminology, Formula (3) states that *the general solution of $A\mathbf{x} = \mathbf{b}$ is the sum of any particular solution of $A\mathbf{x} = \mathbf{b}$ and the general solution of $A\mathbf{x} = \mathbf{0}$.*

For linear systems with two or three unknowns, Theorem 5.5.2 has a nice geometric interpretation in R^2 and R^3. For example, consider the case where $A\mathbf{x} = \mathbf{0}$ and $A\mathbf{x} = \mathbf{b}$ are linear systems with two unknowns. The solutions of $A\mathbf{x} = \mathbf{0}$ form a subspace of R^2 and hence constitute a line through the origin, the origin only, or all of R^2. From Theorem 5.5.2, the solutions of $A\mathbf{x} = \mathbf{b}$ can be obtained by adding any particular solution of $A\mathbf{x} = \mathbf{b}$, say $\mathbf{x}_0$, to the solutions of $A\mathbf{x} = \mathbf{0}$. Assuming that $\mathbf{x}_0$ is positioned with its initial point at the origin, this has the geometric effect of translating the solution space of $A\mathbf{x} = \mathbf{0}$ so that the point at the origin is moved to the tip of $\mathbf{x}_0$ (Figure 5.5.1). This means that the solution vectors of $A\mathbf{x} = \mathbf{b}$ form a line through the tip of $\mathbf{x}_0$, the point at the tip of $\mathbf{x}_0$, or all of R^2. (Can you visualize the last case?) Similarly, for linear systems with three unknowns, the solutions of $A\mathbf{x} = \mathbf{b}$ constitute a plane through the tip of any particular solution $\mathbf{x}_0$, a line through the tip of $\mathbf{x}_0$, the point at the tip of $\mathbf{x}_0$, or all of R^3.

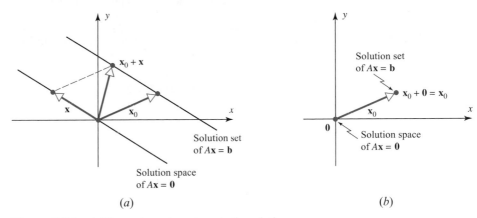

(a)   (b)

Figure 5.5.1   Adding $\mathbf{x}_0$ to each vector $\mathbf{x}$ in the solution space of $A\mathbf{x} = \mathbf{0}$ translates the solution space.

EXAMPLE 3   General Solution of a Linear System $A\mathbf{x} = \mathbf{b}$

In Example 4 of Section 1.2 we solved the nonhomogeneous linear system

$$\begin{array}{rcr}
x_1 + 3x_2 - 2x_3 + 2x_5 &=& 0 \\
2x_1 + 6x_2 - 5x_3 - 2x_4 + 4x_5 - 3x_6 &=& -1 \\
5x_3 + 10x_4 + 15x_6 &=& 5 \\
2x_1 + 6x_2 + 8x_4 + 4x_5 + 18x_6 &=& 6
\end{array} \qquad (4)$$

and obtained

$$x_1 = -3r - 4s - 2t, \qquad x_2 = r, \qquad x_3 = -2s, \qquad x_4 = s, \qquad x_5 = t, \qquad x_6 = \tfrac{1}{3}$$

This result can be written in vector form as

$$\begin{bmatrix} x_1 \\ x_2 \\ x_3 \\ x_4 \\ x_5 \\ x_6 \end{bmatrix} = \begin{bmatrix} -3r - 4s - 2t \\ r \\ -2s \\ s \\ t \\ \tfrac{1}{3} \end{bmatrix} = \underbrace{\begin{bmatrix} 0 \\ 0 \\ 0 \\ 0 \\ 0 \\ \tfrac{1}{3} \end{bmatrix}}_{\mathbf{x}_0} + \underbrace{r\begin{bmatrix} -3 \\ 1 \\ 0 \\ 0 \\ 0 \\ 0 \end{bmatrix} + s\begin{bmatrix} -4 \\ 0 \\ -2 \\ 1 \\ 0 \\ 0 \end{bmatrix} + t\begin{bmatrix} -2 \\ 0 \\ 0 \\ 0 \\ 1 \\ 0 \end{bmatrix}}_{\mathbf{x}} \qquad (5)$$

which is the general solution of (4). The vector $\mathbf{x}_0$ in (5) is a particular solution of (4); the linear combination $\mathbf{x}$ in (5) is the general solution of the homogeneous system

$$\begin{array}{rcl}
x_1 + 3x_2 - 2x_3 + 2x_5 &=& 0 \\
2x_1 + 6x_2 - 5x_3 - 2x_4 + 4x_5 - 3x_6 &=& 0 \\
5x_3 + 10x_4 + 15x_6 &=& 0 \\
2x_1 + 6x_2 + 8x_4 + 4x_5 + 18x_6 &=& 0
\end{array}$$

(verify). ◆

Bases for Row Spaces, Column Spaces, and Nullspaces

We first developed elementary row operations for the purpose of solving linear systems, and we know from that work that performing an elementary row operation on an augmented matrix does not change the solution set of the corresponding linear system. It follows that applying an elementary row operation to a matrix A does not change the solution set of the corresponding linear system $A\mathbf{x} = \mathbf{0}$, or, stated another way, it does not change the nullspace of A. Thus we have the following theorem.

THEOREM 5.5.3

Elementary row operations do not change the nullspace of a matrix.

EXAMPLE 4 Basis for Nullspace

Find a basis for the nullspace of

$$A = \begin{bmatrix} 2 & 2 & -1 & 0 & 1 \\ -1 & -1 & 2 & -3 & 1 \\ 1 & 1 & -2 & 0 & -1 \\ 0 & 0 & 1 & 1 & 1 \end{bmatrix}$$

Solution

The nullspace of A is the solution space of the homogeneous system

$$\begin{aligned} 2x_1 + 2x_2 - x_3 \quad + x_5 &= 0 \\ -x_1 - x_2 + 2x_3 - 3x_4 + x_5 &= 0 \\ x_1 + x_2 - 2x_3 \quad - x_5 &= 0 \\ x_3 + x_4 + x_5 &= 0 \end{aligned}$$

In Example 10 of Section 5.4 we showed that the vectors

$$\mathbf{v}_1 = \begin{bmatrix} -1 \\ 1 \\ 0 \\ 0 \\ 0 \end{bmatrix} \quad \text{and} \quad \mathbf{v}_2 = \begin{bmatrix} -1 \\ 0 \\ -1 \\ 0 \\ 1 \end{bmatrix}$$

form a basis for this space. ◆

The following theorem is a companion to Theorem 5.5.3.

THEOREM 5.5.4

Elementary row operations do not change the row space of a matrix.

Proof Suppose that the row vectors of a matrix A are $\mathbf{r}_1, \mathbf{r}_2, \ldots, \mathbf{r}_m$, and let B be obtained from A by performing an elementary row operation. We shall show that every vector in the row space of B is also in the row space of A and that, conversely, every vector in the row space of A is in the row space of B. We can then conclude that A and B have the same row space.

Consider the possibilities: If the row operation is a row interchange, then B and A have the same row vectors and consequently have the same row space. If the row operation is multiplication of a row by a nonzero scalar or the addition of a multiple of one row to another, then the row vectors $\mathbf{r}'_1, \mathbf{r}'_2, \ldots, \mathbf{r}'_m$ of B are linear combinations of

$\mathbf{r}_1, \mathbf{r}_2, \ldots, \mathbf{r}_m$; thus they lie in the row space of A. Since a vector space is closed under addition and scalar multiplication, all linear combinations of $\mathbf{r}'_1, \mathbf{r}'_2, \ldots, \mathbf{r}'_m$ will also lie in the row space of A. Therefore, each vector in the row space of B is in the row space of A.

Since B is obtained from A by performing a row operation, A can be obtained from B by performing the inverse operation (Section 1.5). Thus the argument above shows that the row space of A is contained in the row space of B. ∎

In light of Theorems 5.5.3 and 5.5.4, one might anticipate that elementary row operations should not change the column space of a matrix. However, this is *not* so— elementary row operations can change the column space. For example, consider the matrix

$$A = \begin{bmatrix} 1 & 3 \\ 2 & 6 \end{bmatrix}$$

The second column is a scalar multiple of the first, so the column space of A consists of all scalar multiples of the first column vector. However, if we add -2 times the first row of A to the second row, we obtain

$$B = \begin{bmatrix} 1 & 3 \\ 0 & 0 \end{bmatrix}$$

Here again the second column is a scalar multiple of the first, so the column space of B consists of all scalar multiples of the first column vector. This is not the same as the column space of A.

Although elementary row operations can change the column space of a matrix, we shall show that whatever relationships of linear independence or linear dependence exist among the column vectors prior to a row operation will also hold for the corresponding columns of the matrix that results from that operation. To make this more precise, suppose a matrix B results from performing an elementary row operation on an $m \times n$ matrix A. By Theorem 5.5.3, the two homogeneous linear systems

$$A\mathbf{x} = \mathbf{0} \quad \text{and} \quad B\mathbf{x} = \mathbf{0}$$

have the same solution set. Thus the first system has a nontrivial solution if and only if the same is true of the second. But if the column vectors of A and B, respectively, are

$$\mathbf{c}_1, \mathbf{c}_2, \ldots, \mathbf{c}_n \quad \text{and} \quad \mathbf{c}'_1, \mathbf{c}'_2, \ldots, \mathbf{c}'_n$$

then from (2) the two systems can be rewritten as

$$x_1 \mathbf{c}_1 + x_2 \mathbf{c}_2 + \cdots + x_n \mathbf{c}_n = \mathbf{0} \qquad (6)$$

and

$$x_1 \mathbf{c}'_1 + x_2 \mathbf{c}'_2 + \cdots + x_n \mathbf{c}'_n = \mathbf{0} \qquad (7)$$

Thus (6) has a nontrivial solution for $x_1, x_2, \ldots, x_n$ if and only if the same is true of (7). This implies that the column vectors of A are linearly independent if and only if the same is true of B. Although we shall omit the proof, this conclusion also applies to any subset of the column vectors. Thus we have the following result.

THEOREM 5.5.5

If A and B are row equivalent matrices, then

(a) *A given set of column vectors of A is linearly independent if and only if the corresponding column vectors of B are linearly independent.*

(b) *A given set of column vectors of A forms a basis for the column space of A if and only if the corresponding column vectors of B form a basis for the column space of B.*

The following theorem makes it possible to find bases for the row and column spaces of a matrix in row-echelon form by inspection.

THEOREM 5.5.6

> *If a matrix R is in row-echelon form, then the row vectors with the leading 1's (the nonzero row vectors) form a basis for the row space of R, and the column vectors with the leading 1's of the row vectors form a basis for the column space of R.*

Since this result is virtually self-evident when one looks at numerical examples, we shall omit the proof; the proof involves little more than an analysis of the positions of the 0's and 1's of R.

EXAMPLE 5 Bases for Row and Column Spaces

The matrix

$$R = \begin{bmatrix} 1 & -2 & 5 & 0 & 3 \\ 0 & 1 & 3 & 0 & 0 \\ 0 & 0 & 0 & 1 & 0 \\ 0 & 0 & 0 & 0 & 0 \end{bmatrix}$$

is in row-echelon form. From Theorem 5.5.6, the vectors

$$\mathbf{r}_1 = \begin{bmatrix} 1 & -2 & 5 & 0 & 3 \end{bmatrix}$$
$$\mathbf{r}_2 = \begin{bmatrix} 0 & 1 & 3 & 0 & 0 \end{bmatrix}$$
$$\mathbf{r}_3 = \begin{bmatrix} 0 & 0 & 0 & 1 & 0 \end{bmatrix}$$

form a basis for the row space of R, and the vectors

$$\mathbf{c}_1 = \begin{bmatrix} 1 \\ 0 \\ 0 \\ 0 \end{bmatrix}, \qquad \mathbf{c}_2 = \begin{bmatrix} -2 \\ 1 \\ 0 \\ 0 \end{bmatrix}, \qquad \mathbf{c}_4 = \begin{bmatrix} 0 \\ 0 \\ 1 \\ 0 \end{bmatrix}$$

form a basis for the column space of R. ◆

EXAMPLE 6 Bases for Row and Column Spaces

Find bases for the row and column spaces of

$$A = \begin{bmatrix} 1 & -3 & 4 & -2 & 5 & 4 \\ 2 & -6 & 9 & -1 & 8 & 2 \\ 2 & -6 & 9 & -1 & 9 & 7 \\ -1 & 3 & -4 & 2 & -5 & -4 \end{bmatrix}$$

Solution

Since elementary row operations do not change the row space of a matrix, we can find a basis for the row space of A by finding a basis for the row space of any row-echelon form of A. Reducing A to row-echelon form, we obtain (verify)

$$R = \begin{bmatrix} 1 & -3 & 4 & -2 & 5 & 4 \\ 0 & 0 & 1 & 3 & -2 & -6 \\ 0 & 0 & 0 & 0 & 1 & 5 \\ 0 & 0 & 0 & 0 & 0 & 0 \end{bmatrix}$$

By Theorem 5.5.6, the nonzero row vectors of R form a basis for the row space of R and hence form a basis for the row space of A. These basis vectors are

$$\mathbf{r}_1 = [1 \quad -3 \quad 4 \quad -2 \quad 5 \quad 4]$$
$$\mathbf{r}_2 = [0 \quad 0 \quad 1 \quad 3 \quad -2 \quad -6]$$
$$\mathbf{r}_3 = [0 \quad 0 \quad 0 \quad 0 \quad 1 \quad 5]$$

Keeping in mind that A and R may have different column spaces, we cannot find a basis for the column space of A *directly* from the column vectors of R. However, it follows from Theorem 5.5.5b that if we can find a set of column vectors of R that forms a basis for the column space of R, then the *corresponding* column vectors of A will form a basis for the column space of A.

The first, third, and fifth columns of R contain the leading 1's of the row vectors, so

$$\mathbf{c}_1' = \begin{bmatrix} 1 \\ 0 \\ 0 \\ 0 \end{bmatrix}, \quad \mathbf{c}_3' = \begin{bmatrix} 4 \\ 1 \\ 0 \\ 0 \end{bmatrix}, \quad \mathbf{c}_5' = \begin{bmatrix} 5 \\ -2 \\ 1 \\ 0 \end{bmatrix}$$

form a basis for the column space of R; thus the corresponding column vectors of A—namely,

$$\mathbf{c}_1 = \begin{bmatrix} 1 \\ 2 \\ 2 \\ -1 \end{bmatrix}, \quad \mathbf{c}_3 = \begin{bmatrix} 4 \\ 9 \\ 9 \\ -4 \end{bmatrix}, \quad \mathbf{c}_5 = \begin{bmatrix} 5 \\ 8 \\ 9 \\ -5 \end{bmatrix}$$

form a basis for the column space of A. ◆

EXAMPLE 7 Basis for a Vector Space Using Row Operations

Find a basis for the space spanned by the vectors

$$\mathbf{v}_1 = (1, -2, 0, 0, 3), \qquad \mathbf{v}_2 = (2, -5, -3, -2, 6),$$
$$\mathbf{v}_3 = (0, 5, 15, 10, 0), \qquad \mathbf{v}_4 = (2, 6, 18, 8, 6)$$

Solution

Except for a variation in notation, the space spanned by these vectors is the row space of the matrix

$$\begin{bmatrix} 1 & -2 & 0 & 0 & 3 \\ 2 & -5 & -3 & -2 & 6 \\ 0 & 5 & 15 & 10 & 0 \\ 2 & 6 & 18 & 8 & 6 \end{bmatrix}$$

Reducing this matrix to row-echelon form, we obtain

$$\begin{bmatrix} 1 & -2 & 0 & 0 & 3 \\ 0 & 1 & 3 & 2 & 0 \\ 0 & 0 & 1 & 1 & 0 \\ 0 & 0 & 0 & 0 & 0 \end{bmatrix}$$

The nonzero row vectors in this matrix are

$$\mathbf{w}_1 = (1, -2, 0, 0, 3), \qquad \mathbf{w}_2 = (0, 1, 3, 2, 0), \qquad \mathbf{w}_3 = (0, 0, 1, 1, 0)$$

These vectors form a basis for the row space and consequently form a basis for the subspace of R^5 spanned by $\mathbf{v}_1, \mathbf{v}_2, \mathbf{v}_3$, and $\mathbf{v}_4$. ◆

Observe that in Example 6 the basis vectors obtained for the column space of A consisted of column vectors of A, but the basis vectors obtained for the row space of A were not all row vectors of A. The following example illustrates a procedure for finding a basis for the row space of a matrix A that consists entirely of row vectors of A.

EXAMPLE 8   Basis for the Row Space of a Matrix

Find a basis for the row space of

$$A = \begin{bmatrix} 1 & -2 & 0 & 0 & 3 \\ 2 & -5 & -3 & -2 & 6 \\ 0 & 5 & 15 & 10 & 0 \\ 2 & 6 & 18 & 8 & 6 \end{bmatrix}$$

consisting entirely of row vectors from A.

Solution

We will transpose A, thereby converting the row space of A into the column space of A^T; then we will use the method of Example 6 to find a basis for the column space of A^T; and then we will transpose again to convert column vectors back to row vectors. Transposing A yields

$$A^T = \begin{bmatrix} 1 & 2 & 0 & 2 \\ -2 & -5 & 5 & 6 \\ 0 & -3 & 15 & 18 \\ 0 & -2 & 10 & 8 \\ 3 & 6 & 0 & 6 \end{bmatrix}$$

Reducing this matrix to row-echelon form yields

$$\begin{bmatrix} 1 & 2 & 0 & 2 \\ 0 & 1 & -5 & -10 \\ 0 & 0 & 0 & 1 \\ 0 & 0 & 0 & 0 \\ 0 & 0 & 0 & 0 \end{bmatrix}$$

The first, second, and fourth columns contain the leading 1's, so the corresponding column vectors in A^T form a basis for the column space of A^T; these are

$$\mathbf{c}_1 = \begin{bmatrix} 1 \\ -2 \\ 0 \\ 0 \\ 3 \end{bmatrix}, \quad \mathbf{c}_2 = \begin{bmatrix} 2 \\ -5 \\ -3 \\ -2 \\ 6 \end{bmatrix}, \quad \text{and} \quad \mathbf{c}_4 = \begin{bmatrix} 2 \\ 6 \\ 18 \\ 8 \\ 6 \end{bmatrix}$$

Transposing again and adjusting the notation appropriately yields the basis vectors

$$\mathbf{r}_1 = [1 \quad -2 \quad 0 \quad 0 \quad 3], \qquad \mathbf{r}_2 = [2 \quad -5 \quad -3 \quad -2 \quad 6],$$

and

$$\mathbf{r}_4 = [2 \quad 6 \quad 18 \quad 8 \quad 6]$$

for the row space of A. ◆

We know from Theorem 5.5.5 that elementary row operations do not alter relationships of linear independence and linear dependence among the column vectors; however, Formulas (6) and (7) imply an even deeper result. Because these formulas actually have *the same scalar coefficients* $x_1, x_2, \ldots, x_n$, it follows that elementary row operations do not alter the *formulas* (linear combinations) that relate linearly dependent column vectors. We omit the formal proof.

EXAMPLE 9 Basis and Linear Combinations

(a) Find a subset of the vectors

$$\mathbf{v}_1 = (1, -2, 0, 3), \qquad \mathbf{v}_2 = (2, -5, -3, 6),$$
$$\mathbf{v}_3 = (0, 1, 3, 0), \qquad \mathbf{v}_4 = (2, -1, 4, -7), \qquad \mathbf{v}_5 = (5, -8, 1, 2)$$

that forms a basis for the space spanned by these vectors.

(b) Express each vector not in the basis as a linear combination of the basis vectors.

Solution (a)

We begin by constructing a matrix that has $\mathbf{v}_1, \mathbf{v}_2, \ldots, \mathbf{v}_5$ as its column vectors:

$$\begin{bmatrix} 1 & 2 & 0 & 2 & 5 \\ -2 & -5 & 1 & -1 & -8 \\ 0 & -3 & 3 & 4 & 1 \\ 3 & 6 & 0 & -7 & 2 \end{bmatrix} \tag{8}$$
$$\begin{array}{ccccc} \uparrow & \uparrow & \uparrow & \uparrow & \uparrow \\ \mathbf{v}_1 & \mathbf{v}_2 & \mathbf{v}_3 & \mathbf{v}_4 & \mathbf{v}_5 \end{array}$$

The first part of our problem can be solved by finding a basis for the column space of this matrix. Reducing the matrix to *reduced* row-echelon form and denoting the column vectors of the resulting matrix by $\mathbf{w}_1, \mathbf{w}_2, \mathbf{w}_3, \mathbf{w}_4,$ and $\mathbf{w}_5$ yields

$$\begin{bmatrix} 1 & 0 & 2 & 0 & 1 \\ 0 & 1 & -1 & 0 & 1 \\ 0 & 0 & 0 & 1 & 1 \\ 0 & 0 & 0 & 0 & 0 \end{bmatrix} \tag{9}$$
$$\begin{array}{ccccc} \uparrow & \uparrow & \uparrow & \uparrow & \uparrow \\ \mathbf{w}_1 & \mathbf{w}_2 & \mathbf{w}_3 & \mathbf{w}_4 & \mathbf{w}_5 \end{array}$$

The leading 1's occur in columns 1, 2, and 4, so by Theorem 5.5.6,

$$\{\mathbf{w}_1, \ \mathbf{w}_2, \ \mathbf{w}_4\}$$

is a basis for the column space of (9), and consequently,

$$\{\mathbf{v}_1, \ \mathbf{v}_2, \ \mathbf{v}_4\}$$

is a basis for the column space of (9).

(d)   In general, if the column space of a 3×3 matrix is a plane through the origin in 3-space, what can you say about the geometric properties of the nullspace and row space? Explain your reasoning.

17.   Indicate whether each statement is always true or sometimes false. Justify your answer by giving a logical argument or a counterexample.

(a)   If A is not square, then the row vectors of A must be linearly dependent.

(b)   If A is square, then either the row vectors or the column vectors of A must be linearly independent.

(c)   If the row vectors and the column vectors of A are linearly independent, then A must be square.

(d)   Adding one additional column to a matrix A increases its rank by one.

18.   (a)   If A is a 3×5 matrix, then the number of leading 1's in the reduced row-echelon form of A is at most _____. Why?

(b)   If A is a 3×5 matrix, then the number of parameters in the general solution of $A\mathbf{x} = \mathbf{0}$ is at most _____. Why?

(c)   If A is a 5×3 matrix, then the number of leading 1's in the reduced row-echelon form of A is at most _____. Why?

(d)   If A is a 5×3 matrix, then the number of parameters in the general solution of $A\mathbf{x} = \mathbf{0}$ is at most _____. Why?

19.   (a)   If A is a 3×5 matrix, then the rank of A is at most _____. Why?

(b)   If A is a 3×5 matrix, then the nullity of A is at most _____. Why?

(c)   If A is a 3×5 matrix, then the rank of A^T is at most _____. Why?

(d)   If A is a 3×5 matrix, then the nullity of A^T is at most _____. Why?

CHAPTER 5

Supplementary Exercises

1.   In each part, the solution space is a subspace of R^3 and so must be a line through the origin, a plane through the origin, all of R^3, or the origin only. For each system, determine which is the case. If the subspace is a plane, find an equation for it, and if it is a line, find parametric equations.

(a) $0x + 0y + 0z = 0$

(b) $\begin{aligned} 2x - 3y + z &= 0 \\ 6x - 9y + 3z &= 0 \\ -4x + 6y - 2z &= 0 \end{aligned}$

(c) $\begin{aligned} x - 2y + 7z &= 0 \\ -4x + 8y + 5z &= 0 \\ 2x - 4y + 3z &= 0 \end{aligned}$

(d) $\begin{aligned} x + 4y + 8z &= 0 \\ 2x + 5y + 6z &= 0 \\ 3x + y - 4z &= 0 \end{aligned}$

2.   For what values of s is the solution space of

$$\begin{aligned} x_1 + x_2 + sx_3 &= 0 \\ x_1 + sx_2 + x_3 &= 0 \\ sx_1 + x_2 + x_3 &= 0 \end{aligned}$$

the origin only, a line through the origin, a plane through the origin, or all of R^3?

3.   (a)   Express $(4a, a - b, a + 2b)$ as a linear combination of $(4, 1, 1)$ and $(0, -1, 2)$.

(b)   Express $(3a + b + 3c, -a + 4b - c, 2a + b + 2c)$ as a linear combination of $(3, -1, 2)$ and $(1, 4, 1)$.

(c)   Express $(2a - b + 4c, 3a - c, 4b + c)$ as a linear combination of three nonzero vectors.

4. Let W be the space spanned by $\mathbf{f} = \sin x$ and $\mathbf{g} = \cos x$.

    (a) Show that for any value of θ, $\mathbf{f}_1 = \sin(x + \theta)$ and $\mathbf{g}_1 = \cos(x + \theta)$ are vectors in W.

    (b) Show that $\mathbf{f}_1$ and $\mathbf{g}_1$ form a basis for W.

5. (a) Express $\mathbf{v} = (1, 1)$ as a linear combination of $\mathbf{v}_1 = (1, -1)$, $\mathbf{v}_2 = (3, 0)$, and $\mathbf{v}_3 = (2, 1)$ in two different ways.

    (b) Explain why this does not violate Theorem 5.4.1.

6. Let A be an $n \times n$ matrix, and let $\mathbf{v}_1, \mathbf{v}_2, \ldots, \mathbf{v}_n$ be linearly independent vectors in R^n expressed as $n \times 1$ matrices. What must be true about A for $A\mathbf{v}_1, A\mathbf{v}_2, \ldots, A\mathbf{v}_n$ to be linearly independent?

7. Must a basis for P_n contain a polynomial of degree k for each $k = 0, 1, 2, \ldots, n$? Justify your answer.

8. For purposes of this problem, let us define a "checkerboard matrix" to be a square matrix $A = [a_{ij}]$ such that

    $$a_{ij} = \begin{cases} 1 & \text{if } i + j \text{ is even} \\ 0 & \text{if } i + j \text{ is odd} \end{cases}$$

    Find the rank and nullity of the following checkerboard matrices:

    (a) the 3×3 checkerboard matrix

    (b) the 4×4 checkerboard matrix

    (c) the $n \times n$ checkerboard matrix

9. For purposes of this exercise, let us define an "X-matrix" to be a square matrix with an odd number of rows and columns that has 0's everywhere except on the two diagonals, where it has 1's. Find the rank and nullity of the following X-matrices:

    (a) $\begin{bmatrix} 1 & 0 & 1 \\ 0 & 1 & 0 \\ 1 & 0 & 1 \end{bmatrix}$ (b) $\begin{bmatrix} 1 & 0 & 0 & 0 & 1 \\ 0 & 1 & 0 & 1 & 0 \\ 0 & 0 & 1 & 0 & 0 \\ 0 & 1 & 0 & 1 & 0 \\ 1 & 0 & 0 & 0 & 1 \end{bmatrix}$

    (c) the X-matrix of size $(2n + 1) \times (2n + 1)$

10. In each part, show that the set of polynomials is a subspace of P_n and find a basis for it.

    (a) all polynomials in P_n such that $p(-x) = p(x)$

    (b) all polynomials in P_n such that $p(0) = 0$

11. **(For Readers Who Have Studied Calculus)** Show that the set of all polynomials in P_n that have a horizontal tangent at $x = 0$ is a subspace of P_n. Find a basis for this subspace.

12. (a) Find a basis for the vector space of all 3×3 symmetric matrices.

    (b) Find a basis for the vector space of all 3×3 skew-symmetric matrices.

13. In advanced linear algebra, one proves the following determinant criterion for rank: *The rank of a matrix A is r if and only if A has some $r \times r$ submatrix with a nonzero determinant, and all square submatrices of larger size have determinant zero.* (A submatrix of A is any matrix obtained by deleting rows or columns of A. The matrix A itself is also considered to be a submatrix of A.) In each part, use this criterion to find the rank of the matrix.

    (a) $\begin{bmatrix} 1 & 2 & 0 \\ 2 & 4 & -1 \end{bmatrix}$ (b) $\begin{bmatrix} 1 & 2 & 3 \\ 2 & 4 & 6 \end{bmatrix}$

    (c) $\begin{bmatrix} 1 & 0 & 1 \\ 2 & -1 & 3 \\ 3 & -1 & 4 \end{bmatrix}$ (d) $\begin{bmatrix} 1 & -1 & 2 & 0 \\ 3 & 1 & 0 & 0 \\ -1 & 2 & 4 & 0 \end{bmatrix}$

14. Use the result in Exercise 13 to find the possible ranks for matrices of the form

$$
\begin{bmatrix}
0 & 0 & 0 & 0 & 0 & a_{16} \\
0 & 0 & 0 & 0 & 0 & a_{26} \\
0 & 0 & 0 & 0 & 0 & a_{36} \\
0 & 0 & 0 & 0 & 0 & a_{46} \\
a_{51} & a_{52} & a_{53} & a_{54} & a_{55} & a_{56}
\end{bmatrix}
$$

15. Prove: If S is a basis for a vector space V, then for any vectors $\mathbf{u}$ and $\mathbf{v}$ in V and any scalar k, the following relationships hold:

   (a) $(\mathbf{u} + \mathbf{v})_S = (\mathbf{u})_S + (\mathbf{v})_S$ (b) $(k\mathbf{u})_S = k(\mathbf{u})_S$

CHAPTER 5
Technology Exercises

The following exercises are designed to be solved using a technology utility. Typically, this will be MATLAB, *Mathematica*, Maple, Derive, or Mathcad, but it may also be some other type of linear algebra software or a scientific calculator with some linear algebra capabilities. For each exercise you will need to read the relevant documentation for the particular utility you are using. The goal of these exercises is to provide you with a basic proficiency with your technology utility. Once you have mastered the techniques in these exercises, you will be able to use your technology utility to solve many of the problems in the regular exercise sets.

Section 5.2 **T1.** (a) Some technology utilities do not have direct commands for finding linear combinations of vectors in R^n. However, you can use matrix multiplication to calculate a linear combination by creating a matrix A with the vectors as columns and a column vector $\mathbf{x}$ with the coefficients as entries. Use this method to compute the vector

$$\mathbf{v} = 6(8, -2, 1, -4) + 17(-3, 9, 11, 6) - 9(0, -1, 2, 4)$$

Check your work by hand.

   (b) Use your technology utility to determine whether the vector $(9, 1, 0)$ is a linear combination of the vectors $(1, 2, 3)$, $(1, 4, 6)$, and $(2, -3, -5)$.

Section 5.3 **T1.** Use your technology utility to perform the Wronskian test of linear independence on the sets in Exercise 20.

Section 5.4 **T1.** **(Linear Independence)** Devise three different procedures for using your technology utility to determine whether a set of n vectors in R^n is linearly independent, and use all of your procedures to determine whether the vectors

$$\mathbf{v}_1 = (4, -5, 2, 6), \quad \mathbf{v}_2 = (2, -2, 1, 3), \quad \mathbf{v}_3 = (6, -3, 3, 9), \quad \mathbf{v}_4 = (4, -1, 5, 6)$$

are linearly independent.

T2. **(Dimension)** Devise three different procedures for using your technology utility to determine the dimension of the subspace spanned by a set of vectors in R^n, and use all of your procedures to determine the dimension of the subspace of R^5 spanned by the vectors

$$\mathbf{v}_1 = (2, 2, -1, 0, 1), \qquad \mathbf{v}_2 = (-1, -1, 2, -3, 1),$$
$$\mathbf{v}_3 = (1, 1, -2, 0, -1), \qquad \mathbf{v}_4 = (0, 0, 1, 1, 1)$$

Section 5.5 **T1.** **(Basis for Row Space)** Some technology utilities provide a command for finding a basis for the row space of a matrix. If your utility has this capability, read the documentation and then use your utility to find a basis for the row space of the matrix in Example 6.

T2. **(Basis for Column Space)** Some technology utilities provide a command for finding a basis for the column space of a matrix. If your utility has this capability, read the documentation and then use your utility to find a basis for the column space of the matrix in Example 6.

T3. **(Nullspace)** Some technology utilities provide a command for finding a basis for the nullspace of a matrix. If your utility has this capability, read the documentation and then check your understanding of the procedure by finding a basis for the nullspace of the matrix A in Example 4. Use this result to find the general solution of the homogeneous system $A\mathbf{x} = \mathbf{0}$.

Section 5.6 **T1.** **(Rank and Nullity)** Read your documentation on finding the rank of a matrix, and then use your utility to find the rank of the matrix A in Example 1. Find the nullity of the matrix using Theorem 5.6.3 and the rank.

T2. There is a result, called ***Sylvester's inequality***, which states that if A and B are $n \times n$ matrices with rank r_A and r_B, respectively, then the rank r_{AB} of AB satisfies the inequality $r_A + r_B - n \le r_{AB} \le \min(r_A, r_B)$, where $\min(r_A, r_B)$ denotes the smaller of r_A and r_B or their common value if the two ranks are the same. Use your technology utility to confirm this result for some matrices of your choice.

CHAPTER

Inner Product Spaces

CHAPTER CONTENTS

INTRODUCTION: In Section 3.3 we defined the Euclidean inner product on the spaces R^2 and R^3. Then, in Section 4.1, we extended that concept to R^n and used it to define notions of length, distance, and angle in R^n. In this section we shall extend the concept of an inner product still further by extracting the most important properties of the Euclidean inner product on R^n and turning them into axioms that are applicable in general vector spaces. Thus, when these axioms are satisfied, they will produce generalized inner products that automatically have the most important properties of Euclidean inner products. It will then be reasonable to use these generalized inner products to define notions of length, distance, and angle in general vector spaces.

6.1
INNER PRODUCTS

In this section we shall use the most important properties of the Euclidean inner product as axioms for defining the general concept of an inner product. We will then show how an inner product can be used to define notions of length and distance in vector spaces other than R^n.

General Inner Products

In Section 4.1 we denoted the Euclidean inner product of two vectors in R^n by the notation $\mathbf{u} \cdot \mathbf{v}$. It will be convenient in this section to introduce the alternative notation $\langle \mathbf{u}, \mathbf{v} \rangle$ for the general inner product. With this new notation, the fundamental properties of the Euclidean inner product that were listed in Theorem 4.1.2 are precisely the axioms in the following definition.

DEFINITION

An ***inner product*** on a real vector space V is a function that associates a real number $\langle \mathbf{u}, \mathbf{v} \rangle$ with each pair of vectors $\mathbf{u}$ and $\mathbf{v}$ in V in such a way that the following axioms are satisfied for all vectors $\mathbf{u}$, $\mathbf{v}$, and $\mathbf{z}$ in V and all scalars k.

1. $\langle \mathbf{u}, \mathbf{v} \rangle = \langle \mathbf{v}, \mathbf{u} \rangle$ **[Symmetry axiom]**
2. $\langle \mathbf{u} + \mathbf{v}, \mathbf{z} \rangle = \langle \mathbf{u}, \mathbf{z} \rangle + \langle \mathbf{v}, \mathbf{z} \rangle$ **[Additivity axiom]**
3. $\langle k\mathbf{u}, \mathbf{v} \rangle = k\langle \mathbf{u}, \mathbf{v} \rangle$ **[Homogeneity axiom]**
4. $\langle \mathbf{v}, \mathbf{v} \rangle \geq 0$ **[Positivity axiom]**
   and $\langle \mathbf{v}, \mathbf{v} \rangle = 0$
   if and only if $\mathbf{v} = \mathbf{0}$

A real vector space with an inner product is called a ***real inner product space***.

REMARK In Chapter 10 we shall study inner products over complex vector spaces. However, until that time we shall use the term *inner product space* to mean "real inner product space."

Because the inner product axioms are based on properties of the Euclidean inner product, the Euclidean inner product satisfies these axioms; this is the content of the following example.

EXAMPLE 1   Euclidean Inner Product on R^n

If $\mathbf{u} = (u_1, u_2, \ldots, u_n)$ and $\mathbf{v} = (v_1, v_2, \ldots, v_n)$ are vectors in R^n, then the formula

$$\langle \mathbf{u}, \mathbf{v} \rangle = \mathbf{u} \cdot \mathbf{v} = u_1 v_1 + u_2 v_2 + \cdots + u_n v_n$$

defines $\langle \mathbf{u}, \mathbf{v} \rangle$ to be the Euclidean inner product on R^n. The four inner product axioms hold by Theorem 4.1.2. ◆

The Euclidean inner product is the most important inner product on R^n. However, there are various applications in which it is desirable to modify the Euclidean inner product by *weighting* its terms differently. More precisely, if

$$w_1, w_2, \ldots, w_n$$

are *positive* real numbers, which we shall call ***weights***, and if $\mathbf{u} = (u_1, u_2, \ldots, u_n)$ and $\mathbf{v} = (v_1, v_2, \ldots, v_n)$ are vectors in R^n, then it can be shown (Exercise 26) that the formula

$$\langle \mathbf{u}, \mathbf{v} \rangle = w_1 u_1 v_1 + w_2 u_2 v_2 + \cdots + w_n u_n v_n \tag{1}$$

defines an inner product on R^n; it is called the **weighted Euclidean inner product with weights** $w_1, w_2, \ldots, w_n$.

To illustrate one way in which a weighted Euclidean inner product can arise, suppose that some physical experiment can produce any of n possible numerical values

$$x_1, x_2, \ldots, x_n$$

and that a series of m repetitions of the experiment yields these values with various frequencies; that is, x_1 occurs f_1 times, x_2 occurs f_2 times, and so forth. Since there are a total of m repetitions of the experiment,

$$f_1 + f_2 + \cdots + f_n = m$$

Thus the **arithmetic average**, or **mean**, of the observed numerical values (denoted by $\bar{x}$) is

$$\bar{x} = \frac{f_1 x_1 + f_2 x_2 + \cdots + f_n x_n}{f_1 + f_2 + \cdots + f_n} = \frac{1}{m}(f_1 x_1 + f_2 x_2 + \cdots + f_n x_n) \tag{2}$$

If we let

$$\mathbf{f} = (f_1, f_2, \ldots, f_n)$$
$$\mathbf{x} = (x_1, x_2, \ldots, x_n)$$
$$w_1 = w_2 = \cdots = w_n = 1/m$$

then (2) can be expressed as the weighted inner product

$$\bar{x} = \langle \mathbf{f}, \mathbf{x} \rangle = w_1 f_1 x_1 + w_2 f_2 x_2 + \cdots + w_n f_n x_n$$

REMARK   It will always be assumed that R^n has the Euclidean inner product unless some other inner product is explicitly specified. As defined in Section 4.1, we refer to R^n with the Euclidean inner product as *Euclidean n-space*.

EXAMPLE 2   Weighted Euclidean Inner Product

Let $\mathbf{u} = (u_1, u_2)$ and $\mathbf{v} = (v_1, v_2)$ be vectors in R^2. Verify that the weighted Euclidean inner product

$$\langle \mathbf{u}, \mathbf{v} \rangle = 3u_1 v_1 + 2u_2 v_2$$

satisfies the four inner product axioms.

Solution

Note first that if $\mathbf{u}$ and $\mathbf{v}$ are interchanged in this equation, the right side remains the same. Therefore,

$$\langle \mathbf{u}, \mathbf{v} \rangle = \langle \mathbf{v}, \mathbf{u} \rangle$$

If $\mathbf{z} = (z_1, z_2)$, then

$$\begin{aligned}
\langle \mathbf{u} + \mathbf{v}, \mathbf{z} \rangle &= 3(u_1 + v_1)z_1 + 2(u_2 + v_2)z_2 \\
&= (3u_1 z_1 + 2u_2 z_2) + (3v_1 z_1 + 2v_2 z_2) \\
&= \langle \mathbf{u}, \mathbf{z} \rangle + \langle \mathbf{v}, \mathbf{z} \rangle
\end{aligned}$$

which establishes the second axiom.

Next,
$$\langle k\mathbf{u}, \mathbf{v}\rangle = 3(ku_1)v_1 + 2(ku_2)v_2 = k(3u_1v_1 + 2u_2v_2) = k\langle\mathbf{u}, \mathbf{v}\rangle$$
which establishes the third axiom.

Finally,
$$\langle\mathbf{v}, \mathbf{v}\rangle = 3v_1v_1 + 2v_2v_2 = 3v_1^2 + 2v_2^2$$

Obviously, $\langle\mathbf{v}, \mathbf{v}\rangle = 3v_1^2 + 2v_2^2 \geq 0$. Further, $\langle\mathbf{v}, \mathbf{v}\rangle = 3v_1^2 + 2v_2^2 = 0$ if and only if $v_1 = v_2 = 0$—that is, if and only if $\mathbf{v} = (v_1, v_2) = \mathbf{0}$. Thus the fourth axiom is satisfied. ◆

Length and Distance in Inner Product Spaces

Before discussing more examples of inner products, we shall pause to explain how inner products are used to introduce notions of length and distance in inner product spaces. Recall that in Euclidean n-space the Euclidean length of a vector $\mathbf{u} = (u_1, u_2, \ldots, u_n)$ can be expressed in terms of the Euclidean inner product as
$$\|\mathbf{u}\| = (\mathbf{u} \cdot \mathbf{u})^{1/2}$$
and the Euclidean distance between two arbitrary points $\mathbf{u} = (u_1, u_2, \ldots, u_n)$ and $\mathbf{v} = (v_1, v_2, \ldots, v_n)$ can be expressed as
$$d(\mathbf{u}, \mathbf{v}) = \|\mathbf{u} - \mathbf{v}\| = [(\mathbf{u} - \mathbf{v}) \cdot (\mathbf{u} - \mathbf{v})]^{1/2}$$
[see Formulas (1) and (2) of Section 4.1]. Motivated by these formulas, we make the following definition.

DEFINITION

If V is an inner product space, then the ***norm*** (or ***length***) of a vector $\mathbf{u}$ in V is denoted by $\|\mathbf{u}\|$ and is defined by
$$\|\mathbf{u}\| = \langle\mathbf{u}, \mathbf{u}\rangle^{1/2}$$
The ***distance*** between two points (vectors) $\mathbf{u}$ and $\mathbf{v}$ is denoted by $d(\mathbf{u}, \mathbf{v})$ and is defined by
$$d(\mathbf{u}, \mathbf{v}) = \|\mathbf{u} - \mathbf{v}\|$$

If a vector has norm 1, then we say that it is a ***unit vector***.

EXAMPLE 3 Norm and Distance in R^n

If $\mathbf{u} = (u_1, u_2, \ldots, u_n)$ and $\mathbf{v} = (v_1, v_2, \ldots, v_n)$ are vectors in R^n with the Euclidean inner product, then
$$\|\mathbf{u}\| = \langle\mathbf{u}, \mathbf{u}\rangle^{1/2} = (\mathbf{u} \cdot \mathbf{u})^{1/2} = \sqrt{u_1^2 + u_2^2 + \cdots + u_n^2}$$
and
$$d(\mathbf{u}, \mathbf{v}) = \|\mathbf{u} - \mathbf{v}\| = \langle\mathbf{u} - \mathbf{v}, \mathbf{u} - \mathbf{v}\rangle^{1/2} = [(\mathbf{u} - \mathbf{v}) \cdot (\mathbf{u} - \mathbf{v})]^{1/2}$$
$$= \sqrt{(u_1 - v_1)^2 + (u_2 - v_2)^2 + \cdots + (u_n - v_n)^2}$$
Observe that these are simply the standard formulas for the Euclidean norm and distance discussed in Section 4.1 [see Formulas (1) and (2) in that section]. ◆

EXAMPLE 4 Using a Weighted Euclidean Inner Product

It is important to keep in mind that norm and distance depend on the inner product being used. If the inner product is changed, then the norms and distances between vectors also change. For example, for the vectors $\mathbf{u} = (1, 0)$ and $\mathbf{v} = (0, 1)$ in R^2 with the Euclidean inner product, we have

$$\|\mathbf{u}\| = \sqrt{1^2 + 0^2} = 1$$

and

$$d(\mathbf{u}, \mathbf{v}) = \|\mathbf{u} - \mathbf{v}\| = \|(1, -1)\| = \sqrt{1^2 + (-1)^2} = \sqrt{2}$$

However, if we change to the weighted Euclidean inner product of Example 2,

$$\langle \mathbf{u}, \mathbf{v} \rangle = 3u_1v_1 + 2u_2v_2$$

then we obtain

$$\|\mathbf{u}\| = \langle \mathbf{u}, \mathbf{u} \rangle^{1/2} = [3(1)(1) + 2(0)(0)]^{1/2} = \sqrt{3}$$

and

$$d(\mathbf{u}, \mathbf{v}) = \|\mathbf{u} - \mathbf{v}\| = \langle (1, -1), (1, -1) \rangle^{1/2}$$
$$= [3(1)(1) + 2(-1)(-1)]^{1/2} = \sqrt{5} \quad \blacklozenge$$

Unit Circles and Spheres in Inner Product Spaces

If V is an inner product space, then the set of points in V that satisfy

$$\|\mathbf{u}\| = 1$$

is called the ***unit sphere*** or sometimes the ***unit circle*** in V. In R^2 and R^3 these are the points that lie 1 unit away from the origin.

(*a*) The unit circle with the Euclidian norm
$\|\mathbf{u}\| = \sqrt{x^2 + y^2}$

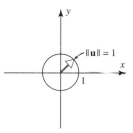

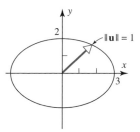

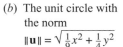

(*b*) The unit circle with the norm
$\|\mathbf{u}\| = \sqrt{\frac{1}{9}x^2 + \frac{1}{4}y^2}$

Figure 6.1.1

EXAMPLE 5 Unusual Unit Circles in R^2

(a) Sketch the unit circle in an xy-coordinate system in R^2 using the Euclidean inner product $\langle \mathbf{u}, \mathbf{v} \rangle = u_1v_1 + u_2v_2$.

(b) Sketch the unit circle in an xy-coordinate system in R^2 using the weighted Euclidean inner product $\langle \mathbf{u}, \mathbf{v} \rangle = \frac{1}{9}u_1v_1 + \frac{1}{4}u_2v_2$.

Solution (a)

If $\mathbf{u} = (x, y)$, then $\|\mathbf{u}\| = \langle \mathbf{u}, \mathbf{u} \rangle^{1/2} = \sqrt{x^2 + y^2}$, so the equation of the unit circle is $\sqrt{x^2 + y^2} = 1$, or, on squaring both sides,

$$x^2 + y^2 = 1$$

As expected, the graph of this equation is a circle of radius 1 centered at the origin (Figure 6.1.1*a*).

Solution (b)

If $\mathbf{u} = (x, y)$, then $\|\mathbf{u}\| = \langle \mathbf{u}, \mathbf{u} \rangle^{1/2} = \sqrt{\frac{1}{9}x^2 + \frac{1}{4}y^2}$, so the equation of the unit circle is $\sqrt{\frac{1}{9}x^2 + \frac{1}{4}y^2} = 1$, or, on squaring both sides,

$$\frac{x^2}{9} + \frac{y^2}{4} = 1$$

The graph of this equation is the ellipse shown in Figure 6.1.1*b*. $\blacklozenge$

It would be reasonable for you to feel uncomfortable with the results in the last example, because although our definitions of length and distance reduce to the standard definitions when applied to R^2 with the Euclidean inner product, it does require a stretch of the imagination to think of the unit "circle" as having an elliptical shape. However, even though nonstandard inner products distort familiar spaces and lead to strange values for lengths and distances, many of the basic theorems of Euclidean geometry continue to apply in these unusual spaces. For example, it is a basic fact in Euclidean geometry that the sum of the lengths of two sides of a triangle is at least as large as the length of the third side (Figure 6.1.2a). We shall see later that this familiar result holds in all inner product spaces, regardless of how unusual the inner product might be. As another example, recall the theorem from Euclidean geometry that states that the sum of the squares of the diagonals of a parallelogram is equal to the sum of the squares of the four sides (Figure 6.1.2b). This result also holds in all inner product spaces, regardless of the inner product (Exercise 20).

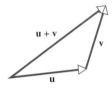

(a) $\|\mathbf{u} + \mathbf{v}\| \le \|\mathbf{u}\| + \|\mathbf{v}\|$ (b) $\|\mathbf{u} + \mathbf{v}\|^2 + \|\mathbf{u} - \mathbf{v}\|^2 = 2(\|\mathbf{u}\|^2 + \|\mathbf{v}\|^2)$

Figure 6.1.2

Inner Products Generated by Matrices

The Euclidean inner product and the weighted Euclidean inner products are special cases of a general class of inner products on R^n, which we shall now describe. Let

$$\mathbf{u} = \begin{bmatrix} u_1 \\ u_2 \\ \vdots \\ u_n \end{bmatrix} \quad \text{and} \quad \mathbf{v} = \begin{bmatrix} v_1 \\ v_2 \\ \vdots \\ v_n \end{bmatrix}$$

be vectors in R^n (expressed as $n \times 1$ matrices), and let A be an *invertible $n \times n$* matrix. It can be shown (Exercise 30) that if $\mathbf{u} \cdot \mathbf{v}$ is the Euclidean inner product on R^n, then the formula

$$\langle \mathbf{u}, \mathbf{v} \rangle = A\mathbf{u} \cdot A\mathbf{v} \tag{3}$$

defines an inner product; it is called the ***inner product on R^n generated by A.***

Recalling that the Euclidean inner product $\mathbf{u} \cdot \mathbf{v}$ can be written as the matrix product $\mathbf{v}^T \mathbf{u}$ [see (7) in Section 4.1], it follows that (3) can be written in the alternative form

$$\langle \mathbf{u}, \mathbf{v} \rangle = (A\mathbf{v})^T A\mathbf{u}$$

or, equivalently,

$$\langle \mathbf{u}, \mathbf{v} \rangle = \mathbf{v}^T A^T A\mathbf{u} \tag{4}$$

EXAMPLE 6   Inner Product Generated by the Identity Matrix

The inner product on R^n generated by the $n \times n$ identity matrix is the Euclidean inner product, since substituting $A = I$ in (3) yields

$$\langle \mathbf{u}, \mathbf{v} \rangle = I\mathbf{u} \cdot I\mathbf{v} = \mathbf{u} \cdot \mathbf{v}$$

The weighted Euclidean inner product $\langle \mathbf{u}, \mathbf{v} \rangle = 3u_1v_1 + 2u_2v_2$ discussed in Example 2 is the inner product on R^2 generated by

$$A = \begin{bmatrix} \sqrt{3} & 0 \\ 0 & \sqrt{2} \end{bmatrix}$$

because substituting this matrix in (4) yields

$$\langle \mathbf{u}, \mathbf{v} \rangle = \begin{bmatrix} v_1 & v_2 \end{bmatrix} \begin{bmatrix} \sqrt{3} & 0 \\ 0 & \sqrt{2} \end{bmatrix} \begin{bmatrix} \sqrt{3} & 0 \\ 0 & \sqrt{2} \end{bmatrix} \begin{bmatrix} u_1 \\ u_2 \end{bmatrix}$$

$$= \begin{bmatrix} v_1 & v_2 \end{bmatrix} \begin{bmatrix} 3 & 0 \\ 0 & 2 \end{bmatrix} \begin{bmatrix} u_1 \\ u_2 \end{bmatrix}$$

$$= 3u_1v_1 + 2u_2v_2$$

In general, the weighted Euclidean inner product

$$\langle \mathbf{u}, \mathbf{v} \rangle = w_1u_1v_1 + w_2u_2v_2 + \cdots + w_nu_nv_n$$

is the inner product on R^n generated by

$$A = \begin{bmatrix} \sqrt{w_1} & 0 & 0 & \cdots & 0 \\ 0 & \sqrt{w_2} & 0 & \cdots & 0 \\ \vdots & \vdots & \vdots & & \vdots \\ 0 & 0 & 0 & \cdots & \sqrt{w_n} \end{bmatrix} \tag{5}$$

(verify). ◆

In the following examples we shall describe some inner products on vector spaces other than R^n.

EXAMPLE 7   An Inner Product on M_{22}

If

$$U = \begin{bmatrix} u_1 & u_2 \\ u_3 & u_4 \end{bmatrix} \quad \text{and} \quad V = \begin{bmatrix} v_1 & v_2 \\ v_3 & v_4 \end{bmatrix}$$

are any two 2×2 matrices, then the following formula defines an inner product on M_{22} (verify):

$$\langle U, V \rangle = \text{tr}(U^T V) = \text{tr}(V^T U) = u_1v_1 + u_2v_2 + u_3v_3 + u_4v_4$$

(Refer to Section 1.3 for the definition of the trace.) For example, if

$$U = \begin{bmatrix} 1 & 2 \\ 3 & 4 \end{bmatrix} \quad \text{and} \quad V = \begin{bmatrix} -1 & 0 \\ 3 & 2 \end{bmatrix}$$

then

$$\langle U, V \rangle = 1(-1) + 2(0) + 3(3) + 4(2) = 16$$

The norm of a matrix U relative to this inner product is

$$\|U\| = \langle U, U \rangle^{1/2} = \sqrt{u_1^2 + u_2^2 + u_3^2 + u_4^2}$$

and the unit sphere in this space consists of all 2×2 matrices U whose entries satisfy the equation $\|U\| = 1$, which on squaring yields

$$u_1^2 + u_2^2 + u_3^2 + u_4^2 = 1 \quad \blacklozenge$$

EXAMPLE 8 An Inner Product on P_2

If

$$\mathbf{p} = a_0 + a_1 x + a_2 x^2 \quad \text{and} \quad \mathbf{q} = b_0 + b_1 x + b_2 x^2$$

are any two vectors in P_2, then the following formula defines an inner product on P_2 (verify):

$$\langle \mathbf{p}, \mathbf{q} \rangle = a_0 b_0 + a_1 b_1 + a_2 b_2$$

The norm of the polynomial $\mathbf{p}$ relative to this inner product is

$$\|\mathbf{p}\| = \langle \mathbf{p}, \mathbf{p} \rangle^{1/2} = \sqrt{a_0^2 + a_1^2 + a_2^2}$$

and the unit sphere in this space consists of all polynomials $\mathbf{p}$ in P_2 whose coefficients satisfy the equation $\|\mathbf{p}\| = 1$, which on squaring yields

$$a_0^2 + a_1^2 + a_2^2 = 1 \quad \blacklozenge$$

Calculus Required   ### EXAMPLE 9 An Inner Product on $C[a, b]$

Let $\mathbf{f} = f(x)$ and $\mathbf{g} = g(x)$ be two functions in $C[a, b]$ and define

$$\langle \mathbf{f}, \mathbf{g} \rangle = \int_a^b f(x)g(x)\,dx \tag{6}$$

This is well-defined since the functions in $C[a, b]$ are continuous. We shall show that this formula defines an inner product on $C[a, b]$ by verifying the four inner product axioms for functions $\mathbf{f} = f(x)$, $\mathbf{g} = g(x)$, and $\mathbf{s} = s(x)$ in $C[a, b]$:

1. $\langle \mathbf{f}, \mathbf{g} \rangle = \displaystyle\int_a^b f(x)g(x)\,dx = \int_a^b g(x)f(x)\,dx = \langle \mathbf{g}, \mathbf{f} \rangle$

    which proves that Axiom 1 holds.

2. $\langle \mathbf{f} + \mathbf{g}, \mathbf{s} \rangle = \displaystyle\int_a^b (f(x) + g(x))s(x)\,dx$

    $\qquad = \displaystyle\int_a^b f(x)s(x)\,dx + \int_a^b g(x)s(x)\,dx$

    $\qquad = \langle \mathbf{f}, \mathbf{s} \rangle + \langle \mathbf{g}, \mathbf{s} \rangle$

    which proves that Axiom 2 holds.

3. $\langle k\mathbf{f}, \mathbf{g} \rangle = \displaystyle\int_a^b kf(x)g(x)\,dx = k \int_a^b f(x)g(x)\,dx = k\langle \mathbf{f}, \mathbf{g} \rangle$

    which proves that Axiom 3 holds.

4. If $\mathbf{f} = f(x)$ is any function in $C[a, b]$, then $f^2(x) \geq 0$ for all x in $[a, b]$; therefore,

    $$\langle \mathbf{f}, \mathbf{f} \rangle = \int_a^b f^2(x)\,dx \geq 0$$

Further, because $f^2(x) \geq 0$ and $\mathbf{f} = f(x)$ is continuous on $[a, b]$, it follows that $\int_a^b f^2(x)\,dx = 0$ if and only if $f(x) = 0$ for all x in $[a, b]$. Therefore, we have $\langle \mathbf{f}, \mathbf{f} \rangle = \int_a^b f^2(x)\,dx = 0$ if and only if $\mathbf{f} = \mathbf{0}$. This proves that Axiom 4 holds. ◆

Calculus Required

EXAMPLE 10 Norm of a Vector in C[a, b]

If $C[a, b]$ has the inner product defined in the preceding example, then the norm of a function $\mathbf{f} = f(x)$ relative to this inner product is

$$\|\mathbf{f}\| = \langle \mathbf{f}, \mathbf{f} \rangle^{1/2} = \sqrt{\int_a^b f^2(x)\,dx} \tag{7}$$

and the unit sphere in this space consists of all functions $\mathbf{f}$ in $C[a, b]$ that satisfy the equation $\|\mathbf{f}\| = 1$, which on squaring yields

$$\int_a^b f^2(x)\,dx = 1 \quad ◆$$

Calculus Required

REMARK   Since polynomials are continuous functions on $(-\infty, \infty)$, they are continuous on any closed interval $[a, b]$. Thus, for all such intervals the vector space P_n is a subspace of $C[a, b]$, and Formula (6) defines an inner product on P_n.

Calculus Required

REMARK   Recall from calculus that the arc length of a curve $y = f(x)$ over an interval $[a, b]$ is given by the formula

$$L = \int_a^b \sqrt{1 + [f'(x)]^2}\,dx \tag{8}$$

Do not confuse this concept of arc length with $\|\mathbf{f}\|$, which is the length (norm) of $\mathbf{f}$ when $\mathbf{f}$ is viewed as a vector in $C[a, b]$. Formulas (7) and (8) are quite different.

The following theorem lists some basic algebraic properties of inner products.

THEOREM 6.1.1

> ## Properties of Inner Products
>
> *If* $\mathbf{u}$, $\mathbf{v}$, *and* $\mathbf{w}$ *are vectors in a real inner product space, and k is any scalar, then*
>
> (a) $\langle \mathbf{0}, \mathbf{v} \rangle = \langle \mathbf{v}, \mathbf{0} \rangle = 0$ $\qquad\qquad$ (b) $\langle \mathbf{u}, \mathbf{v} + \mathbf{w} \rangle = \langle \mathbf{u}, \mathbf{v} \rangle + \langle \mathbf{u}, \mathbf{w} \rangle$
> (c) $\langle \mathbf{u}, k\mathbf{v} \rangle = k\langle \mathbf{u}, \mathbf{v} \rangle$ $\qquad\qquad\quad$ (d) $\langle \mathbf{u} - \mathbf{v}, \mathbf{w} \rangle = \langle \mathbf{u}, \mathbf{w} \rangle - \langle \mathbf{v}, \mathbf{w} \rangle$
> (e) $\langle \mathbf{u}, \mathbf{v} - \mathbf{w} \rangle = \langle \mathbf{u}, \mathbf{v} \rangle - \langle \mathbf{u}, \mathbf{w} \rangle$

Proof   We shall prove part (b) and leave the proofs of the remaining parts as exercises.

$$
\begin{aligned}
\langle \mathbf{u}, \mathbf{v} + \mathbf{w} \rangle &= \langle \mathbf{v} + \mathbf{w}, \mathbf{u} \rangle && \text{[By symmetry]} \\
&= \langle \mathbf{v}, \mathbf{u} \rangle + \langle \mathbf{w}, \mathbf{u} \rangle && \text{[By additivity]} \\
&= \langle \mathbf{u}, \mathbf{v} \rangle + \langle \mathbf{u}, \mathbf{w} \rangle && \text{[By symmetry]}
\end{aligned}
$$
■

The following example illustrates how Theorem 6.1.1 and the defining properties of inner products can be used to perform algebraic computations with inner products. As you read through the example, you will find it instructive to justify the steps.

EXAMPLE 11 Calculating with Inner Products

$$\langle \mathbf{u} - 2\mathbf{v}, 3\mathbf{u} + 4\mathbf{v} \rangle = \langle \mathbf{u}, 3\mathbf{u} + 4\mathbf{v} \rangle - \langle 2\mathbf{v}, 3\mathbf{u} + 4\mathbf{v} \rangle$$
$$= \langle \mathbf{u}, 3\mathbf{u} \rangle + \langle \mathbf{u}, 4\mathbf{v} \rangle - \langle 2\mathbf{v}, 3\mathbf{u} \rangle - \langle 2\mathbf{v}, 4\mathbf{v} \rangle$$
$$= 3\langle \mathbf{u}, \mathbf{u} \rangle + 4\langle \mathbf{u}, \mathbf{v} \rangle - 6\langle \mathbf{v}, \mathbf{u} \rangle - 8\langle \mathbf{v}, \mathbf{v} \rangle$$
$$= 3\|\mathbf{u}\|^2 + 4\langle \mathbf{u}, \mathbf{v} \rangle - 6\langle \mathbf{u}, \mathbf{v} \rangle - 8\|\mathbf{v}\|^2$$
$$= 3\|\mathbf{u}\|^2 - 2\langle \mathbf{u}, \mathbf{v} \rangle - 8\|\mathbf{v}\|^2 \quad \blacklozenge$$

Since Theorem 6.1.1 is a general result, it is guaranteed to hold for *all* real inner product spaces. This is the real power of the axiomatic development of vector spaces and inner products—a single theorem proves a multitude of results at once. For example, we are guaranteed without any further proof that the five properties given in Theorem 6.1.1 are true for the inner product on R^n generated by any matrix A [Formula (3)]. For example, let us check part (*b*) of Theorem 6.1.1 for this inner product:

$$\langle \mathbf{u}, \mathbf{v} + \mathbf{w} \rangle = (\mathbf{v} + \mathbf{w})^T A^T A \mathbf{u}$$
$$= (\mathbf{v}^T + \mathbf{w}^T) A^T A \mathbf{u} \qquad \textbf{[Property of transpose]}$$
$$= (\mathbf{v}^T A^T A \mathbf{u}) + (\mathbf{w}^T A^T A \mathbf{u}) \qquad \textbf{[Property of matrix multiplication]}$$
$$= \langle \mathbf{u}, \mathbf{v} \rangle + \langle \mathbf{u}, \mathbf{w} \rangle$$

The reader will find it instructive to check the remaining parts of Theorem 6.1.1 for this inner product.

EXERCISE SET
6.1

1. Let $\langle \mathbf{u}, \mathbf{v} \rangle$ be the Euclidean inner product on R^2, and let $\mathbf{u} = (3, -2)$, $\mathbf{v} = (4, 5)$, $\mathbf{w} = (-1, 6)$, and $k = -4$. Verify that

   (a) $\langle \mathbf{u}, \mathbf{v} \rangle = \langle \mathbf{v}, \mathbf{u} \rangle$ (b) $\langle \mathbf{u} + \mathbf{v}, \mathbf{w} \rangle = \langle \mathbf{u}, \mathbf{w} \rangle + \langle \mathbf{v}, \mathbf{w} \rangle$

   (c) $\langle \mathbf{u}, \mathbf{v} + \mathbf{w} \rangle = \langle \mathbf{u}, \mathbf{v} \rangle + \langle \mathbf{u}, \mathbf{w} \rangle$ (d) $\langle k\mathbf{u}, \mathbf{v} \rangle = k\langle \mathbf{u}, \mathbf{v} \rangle = \langle \mathbf{u}, k\mathbf{v} \rangle$

   (e) $\langle \mathbf{0}, \mathbf{v} \rangle = \langle \mathbf{v}, \mathbf{0} \rangle = 0$

2. Repeat Exercise 1 for the weighted Euclidean inner product $\langle \mathbf{u}, \mathbf{v} \rangle = 4u_1 v_1 + 5u_2 v_2$.

3. Compute $\langle \mathbf{u}, \mathbf{v} \rangle$ using the inner product in Example 7.

   (a) $\mathbf{u} = \begin{bmatrix} 3 & -2 \\ 4 & 8 \end{bmatrix}$, $\mathbf{v} = \begin{bmatrix} -1 & 3 \\ 1 & 1 \end{bmatrix}$ (b) $\mathbf{u} = \begin{bmatrix} 1 & 2 \\ -3 & 5 \end{bmatrix}$, $\mathbf{v} = \begin{bmatrix} 4 & 6 \\ 0 & 8 \end{bmatrix}$

4. Compute $\langle \mathbf{p}, \mathbf{q} \rangle$ using the inner product in Example 8.

   (a) $\mathbf{p} = -2 + x + 3x^2$, $\mathbf{q} = 4 - 7x^2$ (b) $\mathbf{p} = -5 + 2x + x^2$, $\mathbf{q} = 3 + 2x - 4x^2$

5. (a) Use Formula (3) to show that $\langle \mathbf{u}, \mathbf{v} \rangle = 9u_1 v_1 + 4u_2 v_2$ is the inner product on R^2 generated by

   $$A = \begin{bmatrix} 3 & 0 \\ 0 & 2 \end{bmatrix}$$

   (b) Use the inner product in part (a) to compute $\langle \mathbf{u}, \mathbf{v} \rangle$ if $\mathbf{u} = (-3, 2)$ and $\mathbf{v} = (1, 7)$.

6. (a) Use Formula (3) to show that $\langle \mathbf{u}, \mathbf{v} \rangle = 5u_1 v_1 - u_1 v_2 - u_2 v_1 + 10u_2 v_2$ is the inner product on R^2 generated by

   $$A = \begin{bmatrix} 2 & 1 \\ -1 & 3 \end{bmatrix}$$

   (b) Use the inner product in part (a) to compute $\langle \mathbf{u}, \mathbf{v} \rangle$ if $\mathbf{u} = (0, -3)$ and $\mathbf{v} = (6, 2)$.

7. Let $\mathbf{u} = (u_1, u_2)$ and $\mathbf{v} = (v_1, v_2)$. In each part, the given expression is an inner product on R^2. Find a matrix that generates it.

   (a) $\langle \mathbf{u}, \mathbf{v} \rangle = 3u_1 v_1 + 5u_2 v_2$ (b) $\langle \mathbf{u}, \mathbf{v} \rangle = 4u_1 v_1 + 6u_2 v_2$

8. Let $\mathbf{u} = (u_1, u_2)$ and $\mathbf{v} = (v_1, v_2)$. Show that the following are inner products on R^2 by verifying that the inner product axioms hold.

   (a) $\langle \mathbf{u}, \mathbf{v} \rangle = 3u_1 v_1 + 5u_2 v_2$ (b) $\langle \mathbf{u}, \mathbf{v} \rangle = 4u_1 v_1 + u_2 v_1 + u_1 v_2 + 4u_2 v_2$

9. Let $\mathbf{u} = (u_1, u_2, u_3)$ and $\mathbf{v} = (v_1, v_2, v_3)$. Determine which of the following are inner products on R^3. For those that are not, list the axioms that do not hold.

   (a) $\langle \mathbf{u}, \mathbf{v} \rangle = u_1 v_1 + u_3 v_3$ (b) $\langle \mathbf{u}, \mathbf{v} \rangle = u_1^2 v_1^2 + u_2^2 v_2^2 + u_3^2 v_3^2$

   (c) $\langle \mathbf{u}, \mathbf{v} \rangle = 2u_1 v_1 + u_2 v_2 + 4u_3 v_3$ (d) $\langle \mathbf{u}, \mathbf{v} \rangle = u_1 v_1 - u_2 v_2 + u_3 v_3$

10. In each part, use the given inner product on R^2 to find $\|\mathbf{w}\|$, where $\mathbf{w} = (-1, 3)$.

    (a) the Euclidean inner product

    (b) the weighted Euclidean inner product $\langle \mathbf{u}, \mathbf{v} \rangle = 3u_1 v_1 + 2u_2 v_2$, where $\mathbf{u} = (u_1, u_2)$ and $\mathbf{v} = (v_1, v_2)$

    (c) the inner product generated by the matrix

$$A = \begin{bmatrix} 1 & 2 \\ -1 & 3 \end{bmatrix}$$

11. Use the inner products in Exercise 10 to find $d(\mathbf{u}, \mathbf{v})$ for $\mathbf{u} = (-1, 2)$ and $\mathbf{v} = (2, 5)$.

12. Let P_2 have the inner product in Example 8. In each part, find $\|\mathbf{p}\|$.

    (a) $\mathbf{p} = -2 + 3x + 2x^2$ (b) $\mathbf{p} = 4 - 3x^2$

13. Let M_{22} have the inner product in Example 7. In each part, find $\|A\|$.

    (a) $A = \begin{bmatrix} -2 & 5 \\ 3 & 6 \end{bmatrix}$ (b) $A = \begin{bmatrix} 0 & 0 \\ 0 & 0 \end{bmatrix}$

14. Let P_2 have the inner product in Example 8. Find $d(\mathbf{p}, \mathbf{q})$.

$$\mathbf{p} = 3 - x + x^2, \qquad \mathbf{q} = 2 + 5x^2$$

15. Let M_{22} have the inner product in Example 7. Find $d(A, B)$.

    (a) $A = \begin{bmatrix} 2 & 6 \\ 9 & 4 \end{bmatrix}, \quad B = \begin{bmatrix} -4 & 7 \\ 1 & 6 \end{bmatrix}$ (b) $A = \begin{bmatrix} -2 & 4 \\ 1 & 0 \end{bmatrix}, \quad B = \begin{bmatrix} -5 & 1 \\ 6 & 2 \end{bmatrix}$

16. Suppose that $\mathbf{u}$, $\mathbf{v}$, and $\mathbf{w}$ are vectors such that

$$\langle \mathbf{u}, \mathbf{v} \rangle = 2, \qquad \langle \mathbf{v}, \mathbf{w} \rangle = -3, \qquad \langle \mathbf{u}, \mathbf{w} \rangle = 5, \qquad \|\mathbf{u}\| = 1, \qquad \|\mathbf{v}\| = 2, \qquad \|\mathbf{w}\| = 7$$

Evaluate the given expression.

    (a) $\langle \mathbf{u} + \mathbf{v}, \mathbf{v} + \mathbf{w} \rangle$ (b) $\langle 2\mathbf{v} - \mathbf{w}, 3\mathbf{u} + 2\mathbf{w} \rangle$ (c) $\langle \mathbf{u} - \mathbf{v} - 2\mathbf{w}, 4\mathbf{u} + \mathbf{v} \rangle$

    (d) $\|\mathbf{u} + \mathbf{v}\|$ (e) $\|2\mathbf{w} - \mathbf{v}\|$ (f) $\|\mathbf{u} - 2\mathbf{v} + 4\mathbf{w}\|$

17. **(For Readers Who Have Studied Calculus)** Let the vector space P_2 have the inner product

$$\langle \mathbf{p}, \mathbf{q} \rangle = \int_{-1}^{1} p(x) q(x) \, dx$$

    (a) Find $\|\mathbf{p}\|$ for $\mathbf{p} = 1$, $\mathbf{p} = x$, and $\mathbf{p} = x^2$.   (b) Find $d(\mathbf{p}, \mathbf{q})$ if $\mathbf{p} = 1$ and $\mathbf{q} = x$.

18. Sketch the unit circle in R^2 using the given inner product.

    (a) $\langle \mathbf{u}, \mathbf{v} \rangle = \frac{1}{4} u_1 v_1 + \frac{1}{16} u_2 v_2$ (b) $\langle \mathbf{u}, \mathbf{v} \rangle = 2u_1 v_1 + u_2 v_2$

19. Find a weighted Euclidean inner product on R^2 for which the unit circle is the ellipse shown in the accompanying figure.

20. Show that the following identity holds for vectors in any inner product space.

$$\|\mathbf{u} + \mathbf{v}\|^2 + \|\mathbf{u} - \mathbf{v}\|^2 = 2\|\mathbf{u}\|^2 + 2\|\mathbf{v}\|^2$$

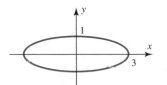

Figure Ex-19

21. Show that the following identity holds for vectors in any inner product space.

$$\langle \mathbf{u}, \mathbf{v} \rangle = \tfrac{1}{4} \|\mathbf{u} + \mathbf{v}\|^2 - \tfrac{1}{4} \|\mathbf{u} - \mathbf{v}\|^2$$

22. Let $U = \begin{bmatrix} u_1 & u_2 \\ u_3 & u_4 \end{bmatrix}$ and $V = \begin{bmatrix} v_1 & v_2 \\ v_3 & v_4 \end{bmatrix}$.

Show that $\langle U, V \rangle = u_1 v_1 + u_2 v_3 + u_3 v_2 + u_4 v_4$ is *not* an inner product on M_{22}.

23. Let $\mathbf{p} = p(x)$ and $\mathbf{q} = q(x)$ be polynomials in P_2. Show that

$$\langle \mathbf{p}, \mathbf{q} \rangle = p(0)q(0) + p\left(\tfrac{1}{2}\right) q\left(\tfrac{1}{2}\right) + p(1)q(1)$$

is an inner product on P_2. Is this an inner product on P_3? Explain.

24. Prove: If $\langle \mathbf{u}, \mathbf{v} \rangle$ is the Euclidean inner product on R^n, and if A is an $n \times n$ matrix, then

$$\langle \mathbf{u}, A\mathbf{v} \rangle = \langle A^T \mathbf{u}, \mathbf{v} \rangle$$

Hint Use the fact that $\langle \mathbf{u}, \mathbf{v} \rangle = \mathbf{u} \cdot \mathbf{v} = \mathbf{v}^T \mathbf{u}$.

25. Verify the result in Exercise 24 for the Euclidean inner product on R^3 and

$$\mathbf{u} = \begin{bmatrix} -1 \\ 2 \\ 4 \end{bmatrix}, \quad \mathbf{v} = \begin{bmatrix} 3 \\ 0 \\ -2 \end{bmatrix}, \quad A = \begin{bmatrix} 1 & -2 & 1 \\ 3 & 4 & 0 \\ 5 & -1 & 2 \end{bmatrix}$$

26. Let $\mathbf{u} = (u_1, u_2, \ldots, u_n)$ and $\mathbf{v} = (v_1, v_2, \ldots, v_n)$. Show that

$$\langle \mathbf{u}, \mathbf{v} \rangle = w_1 u_1 v_1 + w_2 u_2 v_2 + \cdots + w_n u_n v_n$$

is an inner product on R^n if $w_1, w_2, \ldots, w_n$ are positive real numbers.

27. **(For Readers Who Have Studied Calculus)** Use the inner product

$$\langle \mathbf{p}, \mathbf{q} \rangle = \int_{-1}^{1} p(x)q(x)\, dx$$

to compute $\langle \mathbf{p}, \mathbf{q} \rangle$ for the vectors $\mathbf{p} = p(x)$ and $\mathbf{q} = q(x)$ in P_3.

(a) $\mathbf{p} = 1 - x + x^2 + 5x^3, \quad \mathbf{q} = x - 3x^2$

(b) $\mathbf{p} = x - 5x^3, \quad \mathbf{q} = 2 + 8x^2$

28. **(For Readers Who Have Studied Calculus)** In each part, use the inner product

$$\langle \mathbf{f}, \mathbf{g} \rangle = \int_{0}^{1} f(x)g(x)\, dx$$

to compute $\langle \mathbf{f}, \mathbf{g} \rangle$ for the vectors $\mathbf{f} = f(x)$ and $\mathbf{g} = g(x)$ in $C[0, 1]$.

(a) $\mathbf{f} = \cos 2\pi x, \quad \mathbf{g} = \sin 2\pi x$ (b) $\mathbf{f} = x, \quad \mathbf{g} = e^x$ (c) $\mathbf{f} = \tan \dfrac{\pi}{4} x, \quad \mathbf{g} = 1$

29. Show that the inner product in Example 7 can be written as $\langle U, V \rangle = \operatorname{tr}(U^T V)$.

30. Prove that Formula (3) defines an inner product on R^n.

Hint Use the alternative version of Formula (3) given by (4).

31. Show that matrix (5) generates the weighted Euclidean inner product
$\langle \mathbf{u}, \mathbf{v} \rangle = w_1 u_1 v_1 + w_2 u_2 v_2 + \cdots + w_n u_n v_n$ on R^n.

Discussion & Discovery

32. The following is a proof of part (*c*) of Theorem 6.1.1. Fill in each blank line with the name of an inner product axiom that justifies the step.

Hypothesis: Let $\mathbf{u}$ and $\mathbf{v}$ be vectors in a real inner product space.

Conclusion: $\langle \mathbf{u}, k\mathbf{v} \rangle = k \langle \mathbf{u}, \mathbf{v} \rangle$.

Proof: (1) $\langle \mathbf{u}, k\mathbf{v} \rangle = \langle k\mathbf{v}, \mathbf{u} \rangle$ _____

       (2) $\qquad\qquad = k\langle \mathbf{v}, \mathbf{u} \rangle$ _____

       (3) $\qquad\qquad = k\langle \mathbf{u}, \mathbf{v} \rangle$ _____

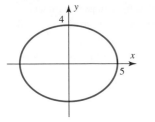

Figure Ex-34

33. Prove parts (a), (d), and (e) of Theorem 6.1.1, justifying each step with the name of a vector space axiom or by referring to previously established results.

34. Create a weighted Euclidean inner product $\langle \mathbf{u}, \mathbf{v} \rangle = au_1v_1 + bu_2v_2$ on R^2 for which the unit circle in an xy-coordinate system is the ellipse shown in the accompanying figure.

35. Generalize the result of Problem 34 for an ellipse with semimajor axis a and semiminor axis b, with a and b positive.

6.2
ANGLE AND ORTHOGONALITY IN INNER PRODUCT SPACES

In this section we shall define the notion of an angle between two vectors in an inner product space, and we shall use this concept to obtain some basic relations between vectors in an inner product, including a fundamental geometric relationship between the nullspace and column space of a matrix.

Cauchy–Schwarz Inequality

Recall from Formula (1) of Section 3.3 that if $\mathbf{u}$ and $\mathbf{v}$ are nonzero vectors in R^2 or R^3 and θ is the angle between them, then

$$\mathbf{u} \cdot \mathbf{v} = \|\mathbf{u}\| \|\mathbf{v}\| \cos \theta \tag{1}$$

or, alternatively,

$$\cos \theta = \frac{\mathbf{u} \cdot \mathbf{v}}{\|\mathbf{u}\| \|\mathbf{v}\|} \tag{2}$$

Our first goal in this section is to define the concept of an angle between two vectors in a general inner product space. For such a definition to be reasonable, we would want it to be consistent with Formula (2) when it is applied to the special case of R^2 and R^3 with the Euclidean inner product. Thus we will want our definition of the angle θ between two nonzero vectors in an inner product space to satisfy the relationship

$$\cos \theta = \frac{\langle \mathbf{u}, \mathbf{v} \rangle}{\|\mathbf{u}\| \|\mathbf{v}\|} \tag{3}$$

However, because $|\cos \theta| \leq 1$, there would be no hope of satisfying (3) unless we were assured that every pair of nonzero vectors in an inner product space satisfies the inequality

$$\left| \frac{\langle \mathbf{u}, \mathbf{v} \rangle}{\|\mathbf{u}\| \|\mathbf{v}\|} \right| \leq 1$$

Fortunately, we will be able to prove that this is the case by using the following generalization of the Cauchy–Schwarz inequality (see Theorem 4.1.3).

THEOREM 6.2.1

Cauchy–Schwarz Inequality

If $\mathbf{u}$ and $\mathbf{v}$ are vectors in a real inner product space, then

$$|\langle \mathbf{u}, \mathbf{v} \rangle| \leq \|\mathbf{u}\| \|\mathbf{v}\| \tag{4}$$

Proof We warn the reader in advance that the proof presented here depends on a clever trick that is not easy to motivate. If $\mathbf{u} = \mathbf{0}$, then $\langle \mathbf{u}, \mathbf{v} \rangle = \langle \mathbf{u}, \mathbf{u} \rangle = 0$, so the two sides of (4) are equal. Assume now that $\mathbf{u} \neq \mathbf{0}$. Let $a = \langle \mathbf{u}, \mathbf{u} \rangle$, $b = 2\langle \mathbf{u}, \mathbf{v} \rangle$, and $c = \langle \mathbf{v}, \mathbf{v} \rangle$, and let t be any real number. By the positivity axiom, the inner product of any vector with itself is always nonnegative. Therefore,

$$0 \leq \langle t\mathbf{u} + \mathbf{v}, t\mathbf{u} + \mathbf{v} \rangle = \langle \mathbf{u}, \mathbf{u} \rangle t^2 + 2\langle \mathbf{u}, \mathbf{v} \rangle t + \langle \mathbf{v}, \mathbf{v} \rangle$$
$$= at^2 + bt + c$$

This inequality implies that the quadratic polynomial $at^2 + bt + c$ has either no real roots or a repeated real root. Therefore, its discriminant must satisfy the inequality $b^2 - 4ac \leq 0$. Expressing the coefficients a, b, and c in terms of the vectors $\mathbf{u}$ and $\mathbf{v}$ gives $4\langle \mathbf{u}, \mathbf{v} \rangle^2 - 4\langle \mathbf{u}, \mathbf{u} \rangle\langle \mathbf{v}, \mathbf{v} \rangle \leq 0$, or, equivalently,

$$\langle \mathbf{u}, \mathbf{v} \rangle^2 \leq \langle \mathbf{u}, \mathbf{u} \rangle\langle \mathbf{v}, \mathbf{v} \rangle$$

Taking square roots of both sides and using the fact that $\langle \mathbf{u}, \mathbf{u} \rangle$ and $\langle \mathbf{v}, \mathbf{v} \rangle$ are nonnegative yields

$$|\langle \mathbf{u}, \mathbf{v} \rangle| \leq \langle \mathbf{u}, \mathbf{u} \rangle^{1/2}\langle \mathbf{v}, \mathbf{v} \rangle^{1/2} \quad \text{or, equivalently,} \quad |\langle \mathbf{u}, \mathbf{v} \rangle| \leq \|\mathbf{u}\|\|\mathbf{v}\|$$

which completes the proof. ∎

For reference, we note that the Cauchy–Schwarz inequality can be written in the following two alternative forms:

$$\langle \mathbf{u}, \mathbf{v} \rangle^2 \leq \langle \mathbf{u}, \mathbf{u} \rangle\langle \mathbf{v}, \mathbf{v} \rangle \tag{5}$$

$$\langle \mathbf{u}, \mathbf{v} \rangle^2 \leq \|\mathbf{u}\|^2\|\mathbf{v}\|^2 \tag{6}$$

The first of these formulas was obtained in the proof of Theorem 6.2.1, and the second is derived from the first using the fact that $\|\mathbf{u}\|^2 = \langle \mathbf{u}, \mathbf{u} \rangle$ and $\|\mathbf{v}\|^2 = \langle \mathbf{v}, \mathbf{v} \rangle$.

EXAMPLE 1 Cauchy–Schwarz Inequality in R^n

The Cauchy–Schwarz inequality for R^n (Theorem 4.1.3) follows as a special case of Theorem 6.2.1 by taking $\langle \mathbf{u}, \mathbf{v} \rangle$ to be the Euclidean inner product $\mathbf{u} \cdot \mathbf{v}$. ◆

The next two theorems show that the basic properties of length and distance that were established in Theorems 4.1.4 and 4.1.5 for vectors in Euclidean n-space continue to hold in general inner product spaces. This is strong evidence that our definitions of inner product, length, and distance are well chosen.

THEOREM 6.2.2

Properties of Length

If $\mathbf{u}$ and $\mathbf{v}$ are vectors in an inner product space V, and if k is any scalar, then

(a) $\|\mathbf{u}\| \geq 0$ (b) $\|\mathbf{u}\| = 0$ *if and only if* $\mathbf{u} = \mathbf{0}$
(c) $\|k\mathbf{u}\| = |k|\|\mathbf{u}\|$ (d) $\|\mathbf{u} + \mathbf{v}\| \leq \|\mathbf{u}\| + \|\mathbf{v}\|$ (*Triangle inequality*)

THEOREM 6.2.3

Properties of Distance

If **u**, **v**, *and* **w** *are vectors in an inner product space* V, *and if* k *is any scalar, then*

(a) $d(\mathbf{u}, \mathbf{v}) \geq 0$ (b) $d(\mathbf{u}, \mathbf{v}) = 0$ *if and only if* $\mathbf{u} = \mathbf{v}$

(c) $d(\mathbf{u}, \mathbf{v}) = d(\mathbf{v}, \mathbf{u})$ (d) $d(\mathbf{u}, \mathbf{v}) \leq d(\mathbf{u}, \mathbf{w}) + d(\mathbf{w}, \mathbf{v})$ (*Triangle inequality*)

We shall prove part (*d*) of Theorem 6.2.2 and leave the remaining parts of Theorems 6.2.2 and 6.2.3 as exercises.

Proof of Theorem 6.2.2d   By definition,

$$
\begin{aligned}
\|\mathbf{u} + \mathbf{v}\|^2 &= \langle \mathbf{u} + \mathbf{v}, \mathbf{u} + \mathbf{v} \rangle \\
&= \langle \mathbf{u}, \mathbf{u} \rangle + 2\langle \mathbf{u}, \mathbf{v} \rangle + \langle \mathbf{v}, \mathbf{v} \rangle \\
&\leq \langle \mathbf{u}, \mathbf{u} \rangle + 2|\langle \mathbf{u}, \mathbf{v} \rangle| + \langle \mathbf{v}, \mathbf{v} \rangle \quad \text{[Property of absolute value]} \\
&\leq \langle \mathbf{u}, \mathbf{u} \rangle + 2\|\mathbf{u}\|\|\mathbf{v}\| + \langle \mathbf{v}, \mathbf{v} \rangle \quad \text{[By (4)]} \\
&= \|\mathbf{u}\|^2 + 2\|\mathbf{u}\|\|\mathbf{v}\| + \|\mathbf{v}\|^2 \\
&= (\|\mathbf{u}\| + \|\mathbf{v}\|)^2
\end{aligned}
$$

Taking square roots gives $\|\mathbf{u} + \mathbf{v}\| \leq \|\mathbf{u}\| + \|\mathbf{v}\|$.   ■

Angle Between Vectors

We shall now show how the Cauchy–Schwarz inequality can be used to define angles in general inner product spaces. Suppose that **u** and **v** are nonzero vectors in an inner product space V. If we divide both sides of Formula (6) by $\|\mathbf{u}\|^2\|\mathbf{v}\|^2$, we obtain

$$
\left[\frac{\langle \mathbf{u}, \mathbf{v} \rangle}{\|\mathbf{u}\|\|\mathbf{v}\|} \right]^2 \leq 1
$$

or, equivalently,

$$
-1 \leq \frac{\langle \mathbf{u}, \mathbf{v} \rangle}{\|\mathbf{u}\|\|\mathbf{v}\|} \leq 1 \tag{7}
$$

Now if θ is an angle whose radian measure varies from 0 to π, then $\cos \theta$ assumes every value between -1 and 1 inclusive exactly once (Figure 6.2.1).

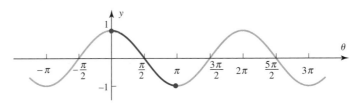

Figure 6.2.1

Thus, from (7), there is a unique angle θ such that

$$
\cos \theta = \frac{\langle \mathbf{u}, \mathbf{v} \rangle}{\|\mathbf{u}\|\|\mathbf{v}\|} \quad \text{and} \quad 0 \leq \theta \leq \pi \tag{8}
$$

We define θ to be the ***angle between*** **u** ***and*** **v**. Observe that in R^2 or R^3 with the Euclidean inner product, (8) agrees with the usual formula for the cosine of the angle between two nonzero vectors [Formula (2)].

EXAMPLE 2 Cosine of an Angle Between Two Vectors in R^4

Let R^4 have the Euclidean inner product. Find the cosine of the angle θ between the vectors $\mathbf{u} = (4, 3, 1, -2)$ and $\mathbf{v} = (-2, 1, 2, 3)$.

Solution

We leave it for the reader to verify that

$$\|\mathbf{u}\| = \sqrt{30}, \quad \|\mathbf{v}\| = \sqrt{18}, \quad \text{and} \quad \langle \mathbf{u}, \mathbf{v} \rangle = -9$$

so that

$$\cos\theta = \frac{\langle \mathbf{u}, \mathbf{v} \rangle}{\|\mathbf{u}\|\,\|\mathbf{v}\|} = -\frac{9}{\sqrt{30}\sqrt{18}} = -\frac{3}{2\sqrt{15}} \quad \blacklozenge$$

Orthogonality

Example 2 is primarily a mathematical exercise, for there is relatively little need to find angles between vectors, except in R^2 and R^3 with the Euclidean inner product. However, a problem of major importance in all inner product spaces is to determine whether two vectors are *orthogonal*—that is, whether the angle between them is $\theta = \pi/2$.

It follows from (8) that if $\mathbf{u}$ and $\mathbf{v}$ are *nonzero* vectors in an inner product space and θ is the angle between them, then $\cos\theta = 0$ if and only if $\langle \mathbf{u}, \mathbf{v} \rangle = 0$. Equivalently, for nonzero vectors we have $\theta = \pi/2$ if and only if $\langle \mathbf{u}, \mathbf{v} \rangle = 0$. If we agree to consider the angle between $\mathbf{u}$ and $\mathbf{v}$ to be $\pi/2$ when either or both of these vectors is $\mathbf{0}$, then we can state without exception that the angle between $\mathbf{u}$ and $\mathbf{v}$ is $\pi/2$ if and only if $\langle \mathbf{u}, \mathbf{v} \rangle = 0$. This suggests the following definition.

DEFINITION

Two vectors $\mathbf{u}$ and $\mathbf{v}$ in an inner product space are called **orthogonal** if $\langle \mathbf{u}, \mathbf{v} \rangle = 0$.

Observe that in the special case where $\langle \mathbf{u}, \mathbf{v} \rangle = \mathbf{u} \cdot \mathbf{v}$ is the Euclidean inner product on R^n, this definition reduces to the definition of orthogonality in Euclidean n-space given in Section 4.1. We also emphasize that orthogonality depends on the inner product; two vectors can be orthogonal with respect to one inner product but not another.

EXAMPLE 3 Orthogonal Vectors in M_{22}

If M_{22} has the inner product of Example 7 in the preceding section, then the matrices

$$U = \begin{bmatrix} 1 & 0 \\ 1 & 1 \end{bmatrix} \quad \text{and} \quad V = \begin{bmatrix} 0 & 2 \\ 0 & 0 \end{bmatrix}$$

are orthogonal, since

$$\langle U, V \rangle = 1(0) + 0(2) + 1(0) + 1(0) = 0 \quad \blacklozenge$$

Calculus Required

EXAMPLE 4 Orthogonal Vectors in P_2

Let P_2 have the inner product

$$\langle \mathbf{p}, \mathbf{q} \rangle = \int_{-1}^{1} p(x)q(x)\,dx$$

and let $\mathbf{p} = x$ and $\mathbf{q} = x^2$. Then

$$\|\mathbf{p}\| = \langle \mathbf{p}, \mathbf{p} \rangle^{1/2} = \left[\int_{-1}^{1} xx \, dx \right]^{1/2} = \left[\int_{-1}^{1} x^2 \, dx \right]^{1/2} = \sqrt{\frac{2}{3}}$$

$$\|\mathbf{q}\| = \langle \mathbf{q}, \mathbf{q} \rangle^{1/2} = \left[\int_{-1}^{1} x^2 x^2 \, dx \right]^{1/2} = \left[\int_{-1}^{1} x^4 \, dx \right]^{1/2} = \sqrt{\frac{2}{5}}$$

$$\langle \mathbf{p}, \mathbf{q} \rangle = \int_{-1}^{1} xx^2 \, dx = \int_{-1}^{1} x^3 \, dx = 0$$

Because $\langle \mathbf{p}, \mathbf{q} \rangle = 0$, the vectors $\mathbf{p} = x$ and $\mathbf{q} = x^2$ are orthogonal relative to the given inner product. ◆

In Section 4.1 we proved the Theorem of Pythagoras for vectors in Euclidean n-space. The following theorem extends this result to vectors in any inner product space.

THEOREM 6.2.4

> ### Generalized Theorem of Pythagoras
>
> *If* $\mathbf{u}$ *and* $\mathbf{v}$ *are orthogonal vectors in an inner product space, then*
>
> $$\|\mathbf{u} + \mathbf{v}\|^2 = \|\mathbf{u}\|^2 + \|\mathbf{v}\|^2$$

Proof The orthogonality of $\mathbf{u}$ and $\mathbf{v}$ implies that $\langle \mathbf{u}, \mathbf{v} \rangle = 0$, so

$$\|\mathbf{u} + \mathbf{v}\|^2 = \langle \mathbf{u} + \mathbf{v}, \mathbf{u} + \mathbf{v} \rangle = \|\mathbf{u}\|^2 + 2\langle \mathbf{u}, \mathbf{v} \rangle + \|\mathbf{v}\|^2$$
$$= \|\mathbf{u}\|^2 + \|\mathbf{v}\|^2 \qquad \blacksquare$$

Calculus Required

EXAMPLE 5   Theorem of Pythagoras in P_2

In Example 4 we showed that $\mathbf{p} = x$ and $\mathbf{q} = x^2$ are orthogonal relative to the inner product

$$\langle \mathbf{p}, \mathbf{q} \rangle = \int_{-1}^{1} p(x)q(x) \, dx$$

on P_2. It follows from the Theorem of Pythagoras that

$$\|\mathbf{p} + \mathbf{q}\|^2 = \|\mathbf{p}\|^2 + \|\mathbf{q}\|^2$$

Thus, from the computations in Example 4, we have

$$\|\mathbf{p} + \mathbf{q}\|^2 = \left(\sqrt{\frac{2}{3}} \right)^2 + \left(\sqrt{\frac{2}{5}} \right)^2 = \frac{2}{3} + \frac{2}{5} = \frac{16}{15}$$

We can check this result by direct integration:

$$\|\mathbf{p} + \mathbf{q}\|^2 = \langle \mathbf{p} + \mathbf{q}, \mathbf{p} + \mathbf{q} \rangle = \int_{-1}^{1} (x + x^2)(x + x^2) \, dx$$

$$= \int_{-1}^{1} x^2 \, dx + 2 \int_{-1}^{1} x^3 \, dx + \int_{-1}^{1} x^4 \, dx = \frac{2}{3} + 0 + \frac{2}{5} = \frac{16}{15} \quad ◆$$

Orthogonal Complements

If V is a plane through the origin of R^3 with the Euclidean inner product, then the set of all vectors that are orthogonal to every vector in V forms the line L through the origin that is perpendicular to V (Figure 6.2.2). In the language of linear algebra we say that the

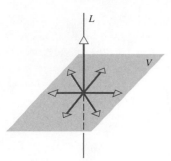

Figure 6.2.2 Every vector in *L* is orthogonal to every vector in *V*.

line and the plane are *orthogonal complements* of one another. The following definition extends this concept to general inner product spaces.

DEFINITION

Let *W* be a subspace of an inner product space *V*. A vector **u** in *V* is said to be *orthogonal to W* if it is orthogonal to every vector in *W*, and the set of all vectors in *V* that are orthogonal to *W* is called the *orthogonal complement of W*.

Recall from geometry that the symbol ⊥ is used to indicate perpendicularity. In linear algebra the orthogonal complement of a subspace *W* is denoted by $W^\perp$ (read "*W* perp"). The following theorem lists the basic properties of orthogonal complements.

THEOREM 6.2.5

Properties of Orthogonal Complements

If W is a subspace of a finite-dimensional inner product space V, then

(a) $W^\perp$ *is a subspace of V.*
(b) The only vector common to W and $W^\perp$ *is* **0**.
(c) The orthogonal complement of $W^\perp$ *is W; that is,* $(W^\perp)^\perp = W$.

We shall prove parts (*a*) and (*b*). The proof of (*c*) requires results covered later in this chapter, so its proof is left for the exercises at the end of the chapter.

Proof (a) Note first that $\langle \mathbf{0}, \mathbf{w} \rangle = 0$ for every vector **w** in *W*, so $W^\perp$ contains at least the zero vector. We want to show that $W^\perp$ is closed under addition and scalar multiplication; that is, we want to show that the sum of two vectors in $W^\perp$ is orthogonal to every vector in *W* and that any scalar multiple of a vector in $W^\perp$ is orthogonal to every vector in *W*. Let **u** and **v** be any vectors in $W^\perp$, let *k* be any scalar, and let **w** be any vector in *W*. Then, from the definition of $W^\perp$, we have $\langle \mathbf{u}, \mathbf{w} \rangle = 0$ and $\langle \mathbf{v}, \mathbf{w} \rangle = 0$. Using basic properties of the inner product, we have

$$\langle \mathbf{u} + \mathbf{v}, \mathbf{w} \rangle = \langle \mathbf{u}, \mathbf{w} \rangle + \langle \mathbf{v}, \mathbf{w} \rangle = 0 + 0 = 0$$
$$\langle k\mathbf{u}, \mathbf{w} \rangle = k\langle \mathbf{u}, \mathbf{w} \rangle = k(0) = 0$$

which proves that $\mathbf{u} + \mathbf{v}$ and $k\mathbf{u}$ are in $W^\perp$.

Proof (b) If **v** is common to *W* and $W^\perp$, then $\langle \mathbf{v}, \mathbf{v} \rangle = 0$, which implies that $\mathbf{v} = \mathbf{0}$ by Axiom 4 for inner products. ∎

REMARK Because *W* and $W^\perp$ are orthogonal complements of one another by part (*c*) of the preceding theorem, we shall say that *W* and $W^\perp$ are *orthogonal complements*.

A Geometric Link between Nullspace and Row Space

The following fundamental theorem provides a geometric link between the nullspace and row space of a matrix.

THEOREM 6.2.6

If A is an m × n matrix, then

(a) The nullspace of A and the row space of A are orthogonal complements in R^n with respect to the Euclidean inner product.
(b) The nullspace of A^T and the column space of A are orthogonal complements in R^m with respect to the Euclidean inner product.

Proof (a)   We want to show that the orthogonal complement of the row space of A is the nullspace of A. To do this, we must show that if a vector $\mathbf{v}$ is orthogonal to every vector in the row space, then $A\mathbf{v} = \mathbf{0}$, and conversely, that if $A\mathbf{v} = \mathbf{0}$, then $\mathbf{v}$ is orthogonal to every vector in the row space.

Assume first that $\mathbf{v}$ is orthogonal to every vector in the row space of A. Then in particular, $\mathbf{v}$ is orthogonal to the row vectors $\mathbf{r}_1, \mathbf{r}_2, \ldots, \mathbf{r}_m$ of A; that is,

$$\mathbf{r}_1 \cdot \mathbf{v} = \mathbf{r}_2 \cdot \mathbf{v} = \cdots = \mathbf{r}_m \cdot \mathbf{v} = 0 \tag{9}$$

But by Formula (11) of Section 4.1, the linear system $A\mathbf{x} = \mathbf{0}$ can be expressed in dot product notation as

$$\begin{bmatrix} \mathbf{r}_1 \cdot \mathbf{x} \\ \mathbf{r}_2 \cdot \mathbf{x} \\ \vdots \\ \mathbf{r}_m \cdot \mathbf{x} \end{bmatrix} = \begin{bmatrix} 0 \\ 0 \\ \vdots \\ 0 \end{bmatrix} \tag{10}$$

so it follows from (9) that $\mathbf{v}$ is a solution of this system and hence lies in the nullspace of A.

Conversely, assume that $\mathbf{v}$ is a vector in the nullspace of A, so $A\mathbf{v} = \mathbf{0}$. It follows from (10) that

$$\mathbf{r}_1 \cdot \mathbf{v} = \mathbf{r}_2 \cdot \mathbf{v} = \cdots = \mathbf{r}_m \cdot \mathbf{v} = 0$$

But if $\mathbf{r}$ is any vector in the row space of A, then $\mathbf{r}$ is expressible as a linear combination of the row vectors of A, say

$$\mathbf{r} = c_1\mathbf{r}_1 + c_2\mathbf{r}_2 + \cdots + c_m\mathbf{r}_m$$

Thus

$$\begin{aligned} \mathbf{r} \cdot \mathbf{v} &= (c_1\mathbf{r}_1 + c_2\mathbf{r}_2 + \cdots + c_m\mathbf{r}_m) \cdot \mathbf{v} \\ &= c_1(\mathbf{r}_1 \cdot \mathbf{v}) + c_2(\mathbf{r}_2 \cdot \mathbf{v}) + \cdots + c_m(\mathbf{r}_m \cdot \mathbf{v}) \\ &= 0 + 0 + \cdots + 0 = 0 \end{aligned}$$

which proves that $\mathbf{v}$ is orthogonal to every vector in the row space of A.

Proof (b)   Since the column space of A is the row space of A^T (except for a difference in notation), the proof follows by applying the result in part (a) to A^T. ∎

The following example shows how Theorem 6.2.6 can be used to find a basis for the orthogonal complement of a subspace of Euclidean n-space.

EXAMPLE 6   Basis for an Orthogonal Complement

Let W be the subspace of R^5 spanned by the vectors

$$\mathbf{w}_1 = (2, 2, -1, 0, 1), \qquad \mathbf{w}_2 = (-1, -1, 2, -3, 1),$$
$$\mathbf{w}_3 = (1, 1, -2, 0, -1), \qquad \mathbf{w}_4 = (0, 0, 1, 1, 1)$$

Find a basis for the orthogonal complement of W.

Solution

The space W spanned by $\mathbf{w}_1, \mathbf{w}_2, \mathbf{w}_3$, and $\mathbf{w}_4$ is the same as the row space of the matrix

$$A = \begin{bmatrix} 2 & 2 & -1 & 0 & 1 \\ -1 & -1 & 2 & -3 & 1 \\ 1 & 1 & -2 & 0 & -1 \\ 0 & 0 & 1 & 1 & 1 \end{bmatrix}$$

and by part (*a*) of Theorem 6.2.6, the nullspace of A is the orthogonal complement of W. In Example 4 of Section 5.5 we showed that

$$\mathbf{v}_1 = \begin{bmatrix} -1 \\ 1 \\ 0 \\ 0 \\ 0 \end{bmatrix} \quad \text{and} \quad \mathbf{v}_2 = \begin{bmatrix} -1 \\ 0 \\ -1 \\ 0 \\ 1 \end{bmatrix}$$

form a basis for this nullspace. Expressing these vectors in the same notation as $\mathbf{w}_1, \mathbf{w}_2, \mathbf{w}_3,$ and $\mathbf{w}_4$, we conclude that the vectors

$$\mathbf{v}_1 = (-1, 1, 0, 0, 0) \quad \text{and} \quad \mathbf{v}_2 = (-1, 0, -1, 0, 1)$$

form a basis for the orthogonal complement of W. As a check, the reader may want to verify that $\mathbf{v}_1$ and $\mathbf{v}_2$ are orthogonal to $\mathbf{w}_1, \mathbf{w}_2, \mathbf{w}_3,$ and $\mathbf{w}_4$ by calculating the necessary dot products. ◆

Summary

We leave it for the reader to show that in any inner product space V, the zero space $\{\mathbf{0}\}$ and the entire space V are orthogonal complements. Thus, if A is an $n \times n$ matrix, to say that $A\mathbf{x} = \mathbf{0}$ has only the trivial solution is equivalent to saying that the orthogonal complement of the nullspace of A is all of R^n, or, equivalently, that the rowspace of A is all of R^n. This enables us to add two new results to the seventeen listed in Theorem 5.6.9.

THEOREM 6.2.7

Equivalent Statements

If A is an $n \times n$ matrix, and if $T_A: R^n \to R^n$ is multiplication by A, then the following are equivalent.

(*a*) *A is invertible.*
(*b*) *$A\mathbf{x} = \mathbf{0}$ has only the trivial solution.*
(*c*) *The reduced row-echelon form of A is I_n.*
(*d*) *A is expressible as a product of elementary matrices.*
(*e*) *$A\mathbf{x} = \mathbf{b}$ is consistent for every $n \times 1$ matrix $\mathbf{b}$.*
(*f*) *$A\mathbf{x} = \mathbf{b}$ has exactly one solution for every $n \times 1$ matrix $\mathbf{b}$.*
(*g*) *$\det(A) \neq 0$.*
(*h*) *The range of T_A is R^n.*
(*i*) *T_A is one-to-one.*
(*j*) *The column vectors of A are linearly independent.*
(*k*) *The row vectors of A are linearly independent.*
(*l*) *The column vectors of A span R^n.*
(*m*) *The row vectors of A span R^n.*
(*n*) *The column vectors of A form a basis for R^n.*
(*o*) *The row vectors of A form a basis for R^n.*
(*p*) *A has rank n.*
(*q*) *A has nullity 0.*
(*r*) *The orthogonal complement of the nullspace of A is R^n.*
(*s*) *The orthogonal complement of the row space of A is $\{\mathbf{0}\}$.*

This theorem relates all of the major topics we have studied thus far.

EXERCISE SET
6.2

1. In each part, determine whether the given vectors are orthogonal with respect to the Euclidean inner product.
   (a) $\mathbf{u} = (-1, 3, 2)$, $\mathbf{v} = (4, 2, -1)$
   (b) $\mathbf{u} = (-2, -2, -2)$, $\mathbf{v} = (1, 1, 1)$
   (c) $\mathbf{u} = (u_1, u_2, u_3)$, $\mathbf{v} = (0, 0, 0)$
   (d) $\mathbf{u} = (-4, 6, -10, 1)$, $\mathbf{v} = (2, 1, -2, 9)$
   (e) $\mathbf{u} = (0, 3, -2, 1)$, $\mathbf{v} = (5, 2, -1, 0)$
   (f) $\mathbf{u} = (a, b)$, $\mathbf{v} = (-b, a)$

2. Do there exist scalars k, l such that the vectors $\mathbf{u} = (2, k, 6)$, $\mathbf{v} = (l, 5, 3)$, and $\mathbf{w} = (1, 2, 3)$ are mutually orthogonal with respect to the Euclidean inner product?

3. Let R^3 have the Euclidean inner product. Let $\mathbf{u} = (1, 1, -1)$ and $\mathbf{v} = (6, 7, -15)$. If $\|k\mathbf{u} + \mathbf{v}\| = 13$, what is k?

4. Let R^4 have the Euclidean inner product, and let $\mathbf{u} = (-1, 1, 0, 2)$. Determine whether the vector $\mathbf{u}$ is orthogonal to the subspace spanned by the vectors $\mathbf{w}_1 = (0, 0, 0, 0)$, $\mathbf{w}_2 = (1, -1, 3, 0)$, and $\mathbf{w}_3 = (4, 0, 9, 2)$.

5. Let R^2, R^3, and R^4 have the Euclidean inner product. In each part, find the cosine of the angle between $\mathbf{u}$ and $\mathbf{v}$.
   (a) $\mathbf{u} = (1, -3)$, $\mathbf{v} = (2, 4)$
   (b) $\mathbf{u} = (-1, 0)$, $\mathbf{v} = (3, 8)$
   (c) $\mathbf{u} = (-1, 5, 2)$, $\mathbf{v} = (2, 4, -9)$
   (d) $\mathbf{u} = (4, 1, 8)$, $\mathbf{v} = (1, 0, -3)$
   (e) $\mathbf{u} = (1, 0, 1, 0)$, $\mathbf{v} = (-3, -3, -3, -3)$
   (f) $\mathbf{u} = (2, 1, 7, -1)$, $\mathbf{v} = (4, 0, 0, 0)$

6. Let P_2 have the inner product in Example 8 of Section 6.1. Find the cosine of the angle between $\mathbf{p}$ and $\mathbf{q}$.
   (a) $\mathbf{p} = -1 + 5x + 2x^2$, $\mathbf{q} = 2 + 4x - 9x^2$
   (b) $\mathbf{p} = x - x^2$, $\mathbf{q} = 7 + 3x + 3x^2$

7. Show that $\mathbf{p} = 1 - x + 2x^2$ and $\mathbf{q} = 2x + x^2$ are orthogonal with respect to the inner product in Exercise 6.

8. Let M_{22} have the inner product in Example 7 of Section 6.1. Find the cosine of the angle between A and B.
   (a) $A = \begin{bmatrix} 2 & 6 \\ 1 & -3 \end{bmatrix}$, $B = \begin{bmatrix} 3 & 2 \\ 1 & 0 \end{bmatrix}$
   (b) $A = \begin{bmatrix} 2 & 4 \\ -1 & 3 \end{bmatrix}$, $B = \begin{bmatrix} -3 & 1 \\ 4 & 2 \end{bmatrix}$

9. Let
$$A = \begin{bmatrix} 2 & 1 \\ -1 & 3 \end{bmatrix}$$

   Which of the following matrices are orthogonal to A with respect to the inner product in Exercise 8?
   (a) $\begin{bmatrix} -3 & 0 \\ 0 & 2 \end{bmatrix}$
   (b) $\begin{bmatrix} 1 & 1 \\ 0 & -1 \end{bmatrix}$
   (c) $\begin{bmatrix} 0 & 0 \\ 0 & 0 \end{bmatrix}$
   (d) $\begin{bmatrix} 2 & 1 \\ 5 & 2 \end{bmatrix}$

10. Let R^3 have the Euclidean inner product. For which values of k are $\mathbf{u}$ and $\mathbf{v}$ orthogonal?
    (a) $\mathbf{u} = (2, 1, 3)$, $\mathbf{v} = (1, 7, k)$
    (b) $\mathbf{u} = (k, k, 1)$, $\mathbf{v} = (k, 5, 6)$

11. Let R^4 have the Euclidean inner product. Find two unit vectors that are orthogonal to the three vectors $\mathbf{u} = (2, 1, -4, 0)$, $\mathbf{v} = (-1, -1, 2, 2)$, and $\mathbf{w} = (3, 2, 5, 4)$.

12. In each part, verify that the Cauchy–Schwarz inequality holds for the given vectors using the Euclidean inner product.
    (a) $\mathbf{u} = (3, 2)$, $\mathbf{v} = (4, -1)$
    (b) $\mathbf{u} = (-3, 1, 0)$, $\mathbf{v} = (2, -1, 3)$
    (c) $\mathbf{u} = (-4, 2, 1)$, $\mathbf{v} = (8, -4, -2)$
    (d) $\mathbf{u} = (0, -2, 2, 1)$, $\mathbf{v} = (-1, -1, 1, 1)$

13. In each part, verify that the Cauchy–Schwarz inequality holds for the given vectors.
    (a) $\mathbf{u} = (-2, 1)$ and $\mathbf{v} = (1, 0)$ using the inner product of Example 2 of Section 6.1

(b) $U = \begin{bmatrix} -1 & 2 \\ 6 & 1 \end{bmatrix}$ and $V = \begin{bmatrix} 1 & 0 \\ 3 & 3 \end{bmatrix}$

using the inner product in Example 7 of Section 6.1

(c) $\mathbf{p} = -1 + 2x + x^2$ and $\mathbf{q} = 2 - 4x^2$ using the inner product given in Example 8 of Section 6.1

14. Let W be the line in R^2 with equation $y = 2x$. Find an equation for $W^\perp$.

15. (a) Let W be the plane in R^3 with equation $x - 2y - 3z = 0$. Find parametric equations for $W^\perp$.

   (b) Let W be the line in R^3 with parametric equations

   $$x = 2t, \quad y = -5t, \quad z = 4t \qquad (-\infty < t < \infty)$$

   Find an equation for $W^\perp$.

   (c) Let W be the intersection of the two planes

   $$x + y + z = 0 \quad \text{and} \quad x - y + z = 0$$

   in R^3. Find an equation for $W^\perp$.

16. Let

$$A = \begin{bmatrix} 1 & 2 & -1 & 2 \\ 3 & 5 & 0 & 4 \\ 1 & 1 & 2 & 0 \end{bmatrix}$$

   (a) Find bases for the row space and nullspace of A.

   (b) Verify that every vector in the row space is orthogonal to every vector in the nullspace (as guaranteed by Theorem 6.2.6a).

17. Let A be the matrix in Exercise 16.

   (a) Find bases for the column space of A and nullspace of A^T.

   (b) Verify that every vector in the column space of A is orthogonal to every vector in the nullspace of A^T (as guaranteed by Theorem 6.2.6b).

18. Find a basis for the orthogonal complement of the subspace of R^n spanned by the vectors.

   (a) $\mathbf{v}_1 = (1, -1, 3), \ \mathbf{v}_2 = (5, -4, -4), \ \mathbf{v}_3 = (7, -6, 2)$

   (b) $\mathbf{v}_1 = (2, 0, -1), \ \mathbf{v}_2 = (4, 0, -2)$

   (c) $\mathbf{v}_1 = (1, 4, 5, 2), \ \mathbf{v}_2 = (2, 1, 3, 0), \ \mathbf{v}_3 = (-1, 3, 2, 2)$

   (d) $\mathbf{v}_1 = (1, 4, 5, 6, 9), \ \mathbf{v}_2 = (3, -2, 1, 4, -1), \ \mathbf{v}_3 = (-1, 0, -1, -2, -1),$
   $\mathbf{v}_4 = (2, 3, 5, 7, 8)$

19. Let V be an inner product space. Show that if $\mathbf{u}$ and $\mathbf{v}$ are orthogonal unit vectors in V, then $\|\mathbf{u} - \mathbf{v}\| = \sqrt{2}$.

20. Let V be an inner product space. Show that if $\mathbf{w}$ is orthogonal to both $\mathbf{u}_1$ and $\mathbf{u}_2$, it is orthogonal to $k_1\mathbf{u}_1 + k_2\mathbf{u}_2$ for all scalars k_1 and k_2. Interpret this result geometrically in the case where V is R^3 with the Euclidean inner product.

21. Let V be an inner product space. Show that if $\mathbf{w}$ is orthogonal to each of the vectors $\mathbf{u}_1, \mathbf{u}_2, \ldots, \mathbf{u}_r$, then it is orthogonal to every vector in span$\{\mathbf{u}_1, \mathbf{u}_2, \ldots, \mathbf{u}_r\}$.

22. Let $\{\mathbf{v}_1, \mathbf{v}_2, \ldots, \mathbf{v}_r\}$ be a basis for an inner product space V. Show that the zero vector is the only vector in V that is orthogonal to all of the basis vectors.

23. Let $\{\mathbf{w}_1, \mathbf{w}_2, \ldots, \mathbf{w}_k\}$ be a basis for a subspace W of V. Show that $W^\perp$ consists of all vectors in V that are orthogonal to every basis vector.

24. Prove the following generalization of Theorem 6.2.4. If $\mathbf{v}_1, \mathbf{v}_2, \ldots, \mathbf{v}_r$ are pairwise orthogonal vectors in an inner product space V, then

$$\|\mathbf{v}_1 + \mathbf{v}_2 + \cdots + \mathbf{v}_r\|^2 = \|\mathbf{v}_1\|^2 + \|\mathbf{v}_2\|^2 + \cdots + \|\mathbf{v}_r\|^2$$

25. Prove the following parts of Theorem 6.2.2:

(a) part (*a*)   (b) part (*b*)   (c) part (*c*)

26. Prove the following parts of Theorem 6.2.3:

(a) part (*a*)   (b) part (*b*)   (c) part (*c*)   (d) part (*d*)

27. Prove: If **u** and **v** are $n \times 1$ matrices and A is an $n \times n$ matrix, then

$$(\mathbf{v}^T A^T A \mathbf{u})^2 \le (\mathbf{u}^T A^T A \mathbf{u})(\mathbf{v}^T A^T A \mathbf{v})$$

28. Use the Cauchy–Schwarz inequality to prove that for all real values of a, b, and θ,

$$(a \cos \theta + b \sin \theta)^2 \le a^2 + b^2$$

29. Prove: If $w_1, w_2, \ldots, w_n$ are positive real numbers and if $\mathbf{u} = (u_1, u_2, \ldots, u_n)$ and $\mathbf{v} = (v_1, v_2, \ldots, v_n)$ are any two vectors in R^n, then

$$|w_1 u_1 v_1 + w_2 u_2 v_2 + \cdots + w_n u_n v_n|$$
$$\le (w_1 u_1^2 + w_2 u_2^2 + \cdots + w_n u_n^2)^{1/2}(w_1 v_1^2 + w_2 v_2^2 + \cdots + w_n v_n^2)^{1/2}$$

30. Show that equality holds in the Cauchy–Schwarz inequality if and only if **u** and **v** are linearly dependent.

31. Use vector methods to prove that a triangle that is inscribed in a circle so that it has a diameter for a side must be a right triangle.

Hint Express the vectors $\overrightarrow{AB}$ and $\overrightarrow{BC}$ in the accompanying figure in terms of **u** and **v**.

32. With respect to the Euclidean inner product, the vectors $\mathbf{u} = (1, \sqrt{3})$ and $\mathbf{v} = (-1, \sqrt{3})$ have norm 2, and the angle between them is 60° (see the accompanying figure). Find a weighted Euclidean inner product with respect to which **u** and **v** are orthogonal unit vectors.

33. **(For Readers Who Have Studied Calculus)** Let $f(x)$ and $g(x)$ be continuous functions on $[0, 1]$. Prove:

(a)
$$\left[\int_0^1 f(x)g(x)\,dx \right]^2 \le \left[\int_0^1 f^2(x)\,dx \right]\left[\int_0^1 g^2(x)\,dx \right]$$

(b)
$$\left[\int_0^1 [f(x) + g(x)]^2\,dx \right]^{1/2} \le \left[\int_0^1 f^2(x)\,dx \right]^{1/2} + \left[\int_0^1 g^2(x)\,dx \right]^{1/2}$$

Hint Use the Cauchy–Schwarz inequality.

34. **(For Readers Who Have Studied Calculus)** Let $C[0, \pi]$ have the inner product

$$\langle \mathbf{f}, \mathbf{g} \rangle = \int_0^\pi f(x)g(x)\,dx$$

and let $\mathbf{f}_n = \cos nx$ $(n = 0, 1, 2, \ldots)$. Show that if $k \ne l$, then $\mathbf{f}_k$ and $\mathbf{f}_l$ are orthogonal with respect to the given inner product.

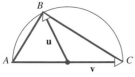

Figure Ex-31

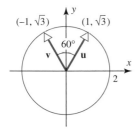

Figure Ex-32

Discussion & Discovery

35. (a) Let W be the line $y = x$ in an xy-coordinate system in R^2. Describe the subspace $W^\perp$.

(b) Let W be the y-axis in an xyz-coordinate system in R^3. Describe the subspace $W^\perp$.

(c) Let W be the yz-plane of an xyz-coordinate system in R^3. Describe the subspace $W^\perp$.

36. Let $A\mathbf{x} = \mathbf{0}$ be a homogeneous system of three equations in the unknowns x, y, and z.

(a) If the solution space is a line through the origin in R^3, what kind of geometric object is the row space of A? Explain your reasoning.

(b) If the column space of A is a line through the origin, what kind of geometric object is the solution space of the homogeneous system $A^T\mathbf{x} = \mathbf{0}$? Explain your reasoning.

(c) If the homogeneous system $A^T\mathbf{x} = \mathbf{0}$ has a unique solution, what can you say about the row space and column space of A? Explain your reasoning.

37. Indicate whether each statement is always true or sometimes false. Justify your answer by giving a logical argument or a counterexample.

(a) If V is a subspace of R^n and W is a subspace of V, then $W^\perp$ is a subspace of $V^\perp$.

(b) $\|\mathbf{u} + \mathbf{v} + \mathbf{w}\| \leq \|\mathbf{u}\| + \|\mathbf{v}\| + \|\mathbf{w}\|$ for all vectors $\mathbf{u}$, $\mathbf{v}$, and $\mathbf{w}$ in an inner product space.

(c) If $\mathbf{u}$ is in the row space and the nullspace of a square matrix A, then $\mathbf{u} = \mathbf{0}$.

(d) If $\mathbf{u}$ is in the row space and the column space of an $n \times n$ matrix A, then $\mathbf{u} = \mathbf{0}$.

38. Let M_{22} have the inner product $\langle U, V \rangle = \text{tr}(U^T V) = \text{tr}(V^T U)$ that was defined in Example 7 of Section 6.1. Describe the orthogonal complement of

(a) the subspace of all diagonal matrices

(b) the subspace of symmetric matrices

6.3
ORTHONORMAL BASES; GRAM–SCHMIDT PROCESS; *QR*-DECOMPOSITION

In many problems involving vector spaces, the problem solver is free to choose any basis for the vector space that seems appropriate. In inner product spaces, the solution of a problem is often greatly simplified by choosing a basis in which the vectors are orthogonal to one another. In this section we shall show how such bases can be obtained.

DEFINITION

A set of vectors in an inner product space is called an ***orthogonal set*** if all pairs of distinct vectors in the set are orthogonal. An orthogonal set in which each vector has norm 1 is called ***orthonormal***.

EXAMPLE 1   An Orthogonal Set in R^3

Let

$$\mathbf{u}_1 = (0, 1, 0), \qquad \mathbf{u}_2 = (1, 0, 1), \qquad \mathbf{u}_3 = (1, 0, -1)$$

and assume that R^3 has the Euclidean inner product. It follows that the set of vectors $S = \{\mathbf{u}_1, \mathbf{u}_2, \mathbf{u}_3\}$ is orthogonal since $\langle \mathbf{u}_1, \mathbf{u}_2 \rangle = \langle \mathbf{u}_1, \mathbf{u}_3 \rangle = \langle \mathbf{u}_2, \mathbf{u}_3 \rangle = 0$. ◆

If $\mathbf{v}$ is a nonzero vector in an inner product space, then by part (*c*) of Theorem 6.2.2, the vector

$$\frac{1}{\|\mathbf{v}\|}\mathbf{v}$$

has norm 1, since

$$\left\| \frac{1}{\|\mathbf{v}\|}\mathbf{v} \right\| = \left| \frac{1}{\|\mathbf{v}\|} \right| \|\mathbf{v}\| = \frac{1}{\|\mathbf{v}\|}\|\mathbf{v}\| = 1$$

The process of multiplying a nonzero vector $\mathbf{v}$ by the reciprocal of its length to obtain a unit vector is called ***normalizing*** $\mathbf{v}$. An orthogonal set of *nonzero* vectors can always be converted to an orthonormal set by normalizing each of its vectors.

EXAMPLE 2 Constructing an Orthonormal Set

The Euclidean norms of the vectors in Example 1 are

$$\|\mathbf{u}_1\| = 1, \qquad \|\mathbf{u}_2\| = \sqrt{2}, \qquad \|\mathbf{u}_3\| = \sqrt{2}$$

Consequently, normalizing $\mathbf{u}_1$, $\mathbf{u}_2$, and $\mathbf{u}_3$ yields

$$\mathbf{v}_1 = \frac{\mathbf{u}_1}{\|\mathbf{u}_1\|} = (0, 1, 0), \qquad \mathbf{v}_2 = \frac{\mathbf{u}_2}{\|\mathbf{u}_2\|} = \left(\frac{1}{\sqrt{2}}, 0, \frac{1}{\sqrt{2}}\right),$$

$$\mathbf{v}_3 = \frac{\mathbf{u}_3}{\|\mathbf{u}_3\|} = \left(\frac{1}{\sqrt{2}}, 0, -\frac{1}{\sqrt{2}}\right)$$

We leave it for you to verify that the set $S = \{\mathbf{v}_1, \mathbf{v}_2, \mathbf{v}_3\}$ is orthonormal by showing that

$$\langle \mathbf{v}_1, \mathbf{v}_2 \rangle = \langle \mathbf{v}_1, \mathbf{v}_3 \rangle = \langle \mathbf{v}_2, \mathbf{v}_3 \rangle = 0 \quad \text{and} \quad \|\mathbf{v}_1\| = \|\mathbf{v}_2\| = \|\mathbf{v}_3\| = 1 \quad \blacklozenge$$

In an inner product space, a basis consisting of orthonormal vectors is called an *orthonormal basis*, and a basis consisting of orthogonal vectors is called an *orthogonal basis*. A familiar example of an orthonormal basis is the standard basis for R^3 with the Euclidean inner product:

$$\mathbf{i} = (1, 0, 0), \qquad \mathbf{j} = (0, 1, 0), \qquad \mathbf{k} = (0, 0, 1)$$

This is the basis that is associated with rectangular coordinate systems (see Figure 5.4.4). More generally, in R^n with the Euclidean inner product, the standard basis

$$\mathbf{e}_1 = (1, 0, 0, \dots, 0), \quad \mathbf{e}_2 = (0, 1, 0, \dots, 0), \dots, \quad \mathbf{e}_n = (0, 0, 0, \dots, 1)$$

is orthonormal.

Coordinates Relative to Orthonormal Bases

The interest in finding orthonormal bases for inner product spaces is motivated in part by the following theorem, which shows that it is exceptionally simple to express a vector in terms of an orthonormal basis.

THEOREM 6.3.1

> *If $S = \{\mathbf{v}_1, \mathbf{v}_2, \dots, \mathbf{v}_n\}$ is an orthonormal basis for an inner product space V, and $\mathbf{u}$ is any vector in V, then*
>
> $$\mathbf{u} = \langle \mathbf{u}, \mathbf{v}_1 \rangle \mathbf{v}_1 + \langle \mathbf{u}, \mathbf{v}_2 \rangle \mathbf{v}_2 + \cdots + \langle \mathbf{u}, \mathbf{v}_n \rangle \mathbf{v}_n$$

Proof Since $S = \{\mathbf{v}_1, \mathbf{v}_2, \dots, \mathbf{v}_n\}$ is a basis, a vector $\mathbf{u}$ can be expressed in the form

$$\mathbf{u} = k_1 \mathbf{v}_1 + k_2 \mathbf{v}_2 + \cdots + k_n \mathbf{v}_n$$

We shall complete the proof by showing that $k_i = \langle \mathbf{u}, \mathbf{v}_i \rangle$ for $i = 1, 2, \dots, n$. For each vector $\mathbf{v}_i$ in S, we have

$$\langle \mathbf{u}, \mathbf{v}_i \rangle = \langle k_1 \mathbf{v}_1 + k_2 \mathbf{v}_2 + \cdots + k_n \mathbf{v}_n, \mathbf{v}_i \rangle$$
$$= k_1 \langle \mathbf{v}_1, \mathbf{v}_i \rangle + k_2 \langle \mathbf{v}_2, \mathbf{v}_i \rangle + \cdots + k_n \langle \mathbf{v}_n, \mathbf{v}_i \rangle$$

Since $S = \{\mathbf{v}_1, \mathbf{v}_2, \dots, \mathbf{v}_n\}$ is an orthonormal set, we have

$$\langle \mathbf{v}_i, \mathbf{v}_i \rangle = \|\mathbf{v}_i\|^2 = 1 \quad \text{and} \quad \langle \mathbf{v}_j, \mathbf{v}_i \rangle = 0 \quad \text{if } j \neq i$$

Therefore, the above expression for $\langle \mathbf{u}, \mathbf{v}_i \rangle$ simplifies to

$$\langle \mathbf{u}, \mathbf{v}_i \rangle = k_i \qquad \blacksquare$$

Using the terminology and notation introduced in Section 5.4, the scalars

$$\langle \mathbf{u}, \mathbf{v}_1 \rangle, \langle \mathbf{u}, \mathbf{v}_2 \rangle, \ldots, \langle \mathbf{u}, \mathbf{v}_n \rangle$$

in Theorem 6.3.1 are the coordinates of the vector $\mathbf{u}$ relative to the orthonormal basis $S = \{\mathbf{v}_1, \mathbf{v}_2, \ldots, \mathbf{v}_n\}$, and

$$(\mathbf{u})_S = (\langle \mathbf{u}, \mathbf{v}_1 \rangle, \langle \mathbf{u}, \mathbf{v}_2 \rangle, \ldots, \langle \mathbf{u}, \mathbf{v}_n \rangle)$$

is the coordinate vector of $\mathbf{u}$ relative to this basis.

EXAMPLE 3 Coordinate Vector Relative to an Orthonormal Basis

Let

$$\mathbf{v}_1 = (0, 1, 0), \qquad \mathbf{v}_2 = \left(-\tfrac{4}{5}, 0, \tfrac{3}{5}\right), \qquad \mathbf{v}_3 = \left(\tfrac{3}{5}, 0, \tfrac{4}{5}\right)$$

It is easy to check that $S = \{\mathbf{v}_1, \mathbf{v}_2, \mathbf{v}_3\}$ is an orthonormal basis for R^3 with the Euclidean inner product. Express the vector $\mathbf{u} = (1, 1, 1)$ as a linear combination of the vectors in S, and find the coordinate vector $(\mathbf{u})_S$.

Solution

$$\langle \mathbf{u}, \mathbf{v}_1 \rangle = 1, \quad \langle \mathbf{u}, \mathbf{v}_2 \rangle = -\tfrac{1}{5}, \quad \text{and} \quad \langle \mathbf{u}, \mathbf{v}_3 \rangle = \tfrac{7}{5}$$

Therefore, by Theorem 6.3.1 we have

$$\mathbf{u} = \mathbf{v}_1 - \tfrac{1}{5}\mathbf{v}_2 + \tfrac{7}{5}\mathbf{v}_3$$

that is,

$$(1, 1, 1) = (0, 1, 0) - \tfrac{1}{5}\left(-\tfrac{4}{5}, 0, \tfrac{3}{5}\right) + \tfrac{7}{5}\left(\tfrac{3}{5}, 0, \tfrac{4}{5}\right)$$

The coordinate vector of $\mathbf{u}$ relative to S is

$$(\mathbf{u})_S = (\langle \mathbf{u}, \mathbf{v}_1 \rangle, \langle \mathbf{u}, \mathbf{v}_2 \rangle, \langle \mathbf{u}, \mathbf{v}_3 \rangle) = \left(1, -\tfrac{1}{5}, \tfrac{7}{5}\right) \qquad \blacklozenge$$

REMARK The usefulness of Theorem 6.3.1 should be evident from this example if we remember that for nonorthonormal bases, it is usually necessary to solve a system of equations in order to express a vector in terms of the basis.

Orthonormal bases for inner product spaces are convenient because, as the following theorem shows, many familiar formulas hold for such bases.

THEOREM 6.3.2

If S is an orthonormal basis for an n-dimensional inner product space, and if

$$(\mathbf{u})_S = (u_1, u_2, \ldots, u_n) \quad and \quad (\mathbf{v})_S = (v_1, v_2, \ldots, v_n)$$

then

(a) $\|\mathbf{u}\| = \sqrt{u_1^2 + u_2^2 + \cdots + u_n^2}$

(b) $d(\mathbf{u}, \mathbf{v}) = \sqrt{(u_1 - v_1)^2 + (u_2 - v_2)^2 + \cdots + (u_n - v_n)^2}$

(c) $\langle \mathbf{u}, \mathbf{v} \rangle = u_1 v_1 + u_2 v_2 + \cdots + u_n v_n$

The proof is left for the exercises.

REMARK Observe that the right side of the equality in part (*a*) is the norm of the coordinate vector $(\mathbf{u})_S$ with respect to the Euclidean inner product on R^n, and the right side of the equality in part (*c*) is the Euclidean inner product of $(\mathbf{u})_S$ and $(\mathbf{v})_S$. Thus, by working with orthonormal bases, we can reduce the computation of general norms and inner products to the computation of Euclidean norms and inner products of the coordinate vectors.

EXAMPLE 4 Calculating Norms Using Orthonormal Bases

If R^3 has the Euclidean inner product, then the norm of the vector $\mathbf{u} = (1, 1, 1)$ is

$$\|\mathbf{u}\| = (\mathbf{u} \cdot \mathbf{u})^{1/2} = \sqrt{1^2 + 1^2 + 1^2} = \sqrt{3}$$

However, if we let R^3 have the orthonormal basis S in the last example, then we know from that example that the coordinate vector of $\mathbf{u}$ relative to S is

$$(\mathbf{u})_S = \left(1, -\tfrac{1}{5}, \tfrac{7}{5}\right)$$

The norm of $\mathbf{u}$ can also be calculated from this vector using part (*a*) of Theorem 6.3.2. This yields

$$\|\mathbf{u}\| = \sqrt{1^2 + \left(-\tfrac{1}{5}\right)^2 + \left(\tfrac{7}{5}\right)^2} = \sqrt{\tfrac{75}{25}} = \sqrt{3} \quad \blacklozenge$$

Coordinates Relative to Orthogonal Bases

If $S = \{\mathbf{v}_1, \mathbf{v}_2, \ldots, \mathbf{v}_n\}$ is an *orthogonal* basis for a vector space V, then normalizing each of these vectors yields the orthonormal basis

$$S' = \left\{ \frac{\mathbf{v}_1}{\|\mathbf{v}_1\|}, \frac{\mathbf{v}_2}{\|\mathbf{v}_2\|}, \ldots, \frac{\mathbf{v}_n}{\|\mathbf{v}_n\|} \right\}$$

Thus, if $\mathbf{u}$ is any vector in V, it follows from Theorem 6.3.1 that

$$\mathbf{u} = \left\langle \mathbf{u}, \frac{\mathbf{v}_1}{\|\mathbf{v}_1\|} \right\rangle \frac{\mathbf{v}_1}{\|\mathbf{v}_1\|} + \left\langle \mathbf{u}, \frac{\mathbf{v}_2}{\|\mathbf{v}_2\|} \right\rangle \frac{\mathbf{v}_2}{\|\mathbf{v}_2\|} + \cdots + \left\langle \mathbf{u}, \frac{\mathbf{v}_n}{\|\mathbf{v}_n\|} \right\rangle \frac{\mathbf{v}_n}{\|\mathbf{v}_n\|}$$

which, by part (*c*) of Theorem 6.1.1, can be rewritten as

$$\mathbf{u} = \frac{\langle \mathbf{u}, \mathbf{v}_1 \rangle}{\|\mathbf{v}_1\|^2} \mathbf{v}_1 + \frac{\langle \mathbf{u}, \mathbf{v}_2 \rangle}{\|\mathbf{v}_2\|^2} \mathbf{v}_2 + \cdots + \frac{\langle \mathbf{u}, \mathbf{v}_n \rangle}{\|\mathbf{v}_n\|^2} \mathbf{v}_n \tag{1}$$

This formula expresses $\mathbf{u}$ as a linear combination of the vectors in the orthogonal basis S. Some problems requiring the use of this formula are given in the exercises.

It is self-evident that if $\mathbf{v}_1$, $\mathbf{v}_2$, and $\mathbf{v}_3$ are three nonzero, mutually perpendicular vectors in R^3, then none of these vectors lies in the same plane as the other two; that is, the vectors are linearly independent. The following theorem generalizes this result.

THEOREM 6.3.3

If $S = \{\mathbf{v}_1, \mathbf{v}_2, \ldots, \mathbf{v}_n\}$ is an orthogonal set of nonzero vectors in an inner product space, then S is linearly independent.

Proof   Assume that

$$k_1\mathbf{v}_1 + k_2\mathbf{v}_2 + \cdots + k_n\mathbf{v}_n = \mathbf{0} \tag{2}$$

To demonstrate that $S = \{\mathbf{v}_1, \mathbf{v}_2, \ldots, \mathbf{v}_n\}$ is linearly independent, we must prove that $k_1 = k_2 = \cdots = k_n = 0$.

For each $\mathbf{v}_i$ in S, it follows from (2) that

$$\langle k_1\mathbf{v}_1 + k_2\mathbf{v}_2 + \cdots + k_n\mathbf{v}_n, \mathbf{v}_i \rangle = \langle \mathbf{0}, \mathbf{v}_i \rangle = 0$$

or, equivalently,

$$k_1\langle \mathbf{v}_1, \mathbf{v}_i \rangle + k_2\langle \mathbf{v}_2, \mathbf{v}_i \rangle + \cdots + k_n\langle \mathbf{v}_n, \mathbf{v}_i \rangle = 0$$

From the orthogonality of S it follows that $\langle \mathbf{v}_j, \mathbf{v}_i \rangle = 0$ when $j \neq i$, so this equation reduces to

$$k_i\langle \mathbf{v}_i, \mathbf{v}_i \rangle = 0$$

Since the vectors in S are assumed to be nonzero, $\langle \mathbf{v}_i, \mathbf{v}_i \rangle \neq 0$ by the positivity axiom for inner products. Therefore, $k_i = 0$. Since the subscript i is arbitrary, we have $k_1 = k_2 = \cdots = k_n = 0$; thus S is linearly independent. ∎

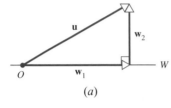

(a)

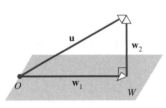

(b)

Figure 6.3.1

Orthogonal Projections

EXAMPLE 5   Using Theorem 6.3.3

In Example 2 we showed that the vectors

$$\mathbf{v}_1 = (0, 1, 0), \quad \mathbf{v}_2 = \left(\frac{1}{\sqrt{2}}, 0, \frac{1}{\sqrt{2}}\right), \quad \text{and} \quad \mathbf{v}_3 = \left(\frac{1}{\sqrt{2}}, 0, -\frac{1}{\sqrt{2}}\right)$$

form an orthonormal set with respect to the Euclidean inner product on R^3. By Theorem 6.3.3, these vectors form a linearly independent set, and since R^3 is three-dimensional, $S = \{\mathbf{v}_1, \mathbf{v}_2, \mathbf{v}_3\}$ is an orthonormal basis for R^3 by Theorem 5.4.5. ◆

We shall now develop some results that will help us to construct orthogonal and orthonormal bases for inner product spaces.

In R^2 or R^3 with the Euclidean inner product, it is evident geometrically that if W is a line or a plane through the origin, then each vector $\mathbf{u}$ in the space can be expressed as a sum

$$\mathbf{u} = \mathbf{w}_1 + \mathbf{w}_2$$

where $\mathbf{w}_1$ is in W and $\mathbf{w}_2$ is perpendicular to W (Figure 6.3.1). This result is a special case of the following general theorem whose proof is given at the end of this section.

THEOREM 6.3.4

Projection Theorem

If W is a finite-dimensional subspace of an inner product space V, then every vector $\mathbf{u}$ in V can be expressed in exactly one way as

$$\mathbf{u} = \mathbf{w}_1 + \mathbf{w}_2 \tag{3}$$

where $\mathbf{w}_1$ is in W and $\mathbf{w}_2$ is in $W^\perp$.

The vector $\mathbf{w}_1$ in the preceding theorem is called the ***orthogonal projection of* u *on*** W and is denoted by $\text{proj}_W \mathbf{u}$. The vector $\mathbf{w}_2$ is called the ***component of* u *orthogonal to W*** and is denoted by $\text{proj}_{W^\perp} \mathbf{u}$. Thus Formula (3) in the Projection Theorem can be expressed as

$$\mathbf{u} = \text{proj}_W \mathbf{u} + \text{proj}_{W^\perp} \mathbf{u} \tag{4}$$

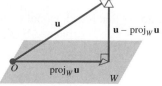

Figure 6.3.2

Since $\mathbf{w}_2 = \mathbf{u} - \mathbf{w}_1$, it follows that

$$\text{proj}_{W^\perp}\,\mathbf{u} = \mathbf{u} - \text{proj}_W\,\mathbf{u}$$

so Formula (4) can also be written as

$$\mathbf{u} = \text{proj}_W\,\mathbf{u} + (\mathbf{u} - \text{proj}_W\,\mathbf{u}) \tag{5}$$

(Figure 6.3.2).

The following theorem, whose proof is requested in the exercises, provides formulas for calculating orthogonal projections.

THEOREM 6.3.5

Let W be a finite-dimensional subspace of an inner product space V.

(a) If $\{\mathbf{v}_1, \mathbf{v}_2, \ldots, \mathbf{v}_r\}$ is an orthonormal basis for W, and $\mathbf{u}$ is any vector in V, then

$$\text{proj}_W\,\mathbf{u} = \langle \mathbf{u}, \mathbf{v}_1 \rangle \mathbf{v}_1 + \langle \mathbf{u}, \mathbf{v}_2 \rangle \mathbf{v}_2 + \cdots + \langle \mathbf{u}, \mathbf{v}_r \rangle \mathbf{v}_r \tag{6}$$

(b) If $\{\mathbf{v}_1, \mathbf{v}_2, \ldots, \mathbf{v}_r\}$ is an orthogonal basis for W, and $\mathbf{u}$ is any vector in V, then

$$\text{proj}_W\,\mathbf{u} = \frac{\langle \mathbf{u}, \mathbf{v}_1 \rangle}{\|\mathbf{v}_1\|^2}\mathbf{v}_1 + \frac{\langle \mathbf{u}, \mathbf{v}_2 \rangle}{\|\mathbf{v}_2\|^2}\mathbf{v}_2 + \cdots + \frac{\langle \mathbf{u}, \mathbf{v}_r \rangle}{\|\mathbf{v}_r\|^2}\mathbf{v}_r \tag{7}$$

EXAMPLE 6   Calculating Projections

Let R^3 have the Euclidean inner product, and let W be the subspace spanned by the orthonormal vectors $\mathbf{v}_1 = (0, 1, 0)$ and $\mathbf{v}_2 = \left(-\frac{4}{5}, 0, \frac{3}{5}\right)$. From (6) the orthogonal projection of $\mathbf{u} = (1, 1, 1)$ on W is

$$\begin{aligned}
\text{proj}_W\,\mathbf{u} &= \langle \mathbf{u}, \mathbf{v}_1 \rangle \mathbf{v}_1 + \langle \mathbf{u}, \mathbf{v}_2 \rangle \mathbf{v}_2 \\
&= (1)(0, 1, 0) + \left(-\tfrac{1}{5}\right)\left(-\tfrac{4}{5}, 0, \tfrac{3}{5}\right) \\
&= \left(\tfrac{4}{25}, 1, -\tfrac{3}{25}\right)
\end{aligned}$$

The component of $\mathbf{u}$ orthogonal to W is

$$\text{proj}_{W^\perp}\,\mathbf{u} = \mathbf{u} - \text{proj}_W\,\mathbf{u} = (1, 1, 1) - \left(\tfrac{4}{25}, 1, -\tfrac{3}{25}\right) = \left(\tfrac{21}{25}, 0, \tfrac{28}{25}\right)$$

Observe that $\text{proj}_{W^\perp}\,\mathbf{u}$ is orthogonal to both $\mathbf{v}_1$ and $\mathbf{v}_2$, so this vector is orthogonal to each vector in the space W spanned by $\mathbf{v}_1$ and $\mathbf{v}_2$, as it should be. ◆

Finding Orthogonal and Orthonormal Bases

We have seen that orthonormal bases exhibit a variety of useful properties. Our next theorem, which is the main result in this section, shows that every nonzero finite-dimensional vector space has an orthonormal basis. The proof of this result is extremely important, since it provides an algorithm, or method, for converting an arbitrary basis into an orthonormal basis.

THEOREM 6.3.6

Every nonzero finite-dimensional inner product space has an orthonormal basis.

Proof   Let V be any nonzero finite-dimensional inner product space, and suppose that $\{\mathbf{u}_1, \mathbf{u}_2, \ldots, \mathbf{u}_n\}$ is any basis for V. It suffices to show that V has an orthogonal basis, since the vectors in the orthogonal basis can be normalized to produce an orthonormal basis for V. The following sequence of steps will produce an orthogonal basis $\{\mathbf{v}_1, \mathbf{v}_2, \ldots, \mathbf{v}_n\}$ for V.

Jörgen Pederson Gram

(1850–1916) was a Danish actuary. Gram's early education was at village schools supplemented by private tutoring. After graduating from high school, he obtained a master's degree in mathematics with specialization in the newly developing modern algebra. Gram then took a position as an actuary for the Hafnia Life Insurance Company, where he developed mathematical foundations of accident insurance for the company Skjold. He served on the Board of Directors of Hafnia and directed Skjold until 1910, at which time he became director of the Danish Insurance Board. During his employ as an actuary, he earned a Ph.D. based on his dissertation "On Series Development Utilizing the Least Squares Method." It was in this thesis that his contributions to the Gram–Schmidt process were first formulated. Gram eventually became interested in abstract number theory and won a gold medal from the Royal Danish Society of Sciences and Letters for his contributions to that field. However, he also had a lifelong interest in the interplay between theoretical and applied mathematics that led to four treatises on Danish forest management. Gram was killed one evening in a bicycle collision on the way to a meeting of the Royal Danish Society.

Step 1.   Let $\mathbf{v}_1 = \mathbf{u}_1$.

Step 2.   As illustrated in Figure 6.3.3, we can obtain a vector $\mathbf{v}_2$ that is orthogonal to $\mathbf{v}_1$ by computing the component of $\mathbf{u}_2$ that is orthogonal to the space W_1 spanned by $\mathbf{v}_1$. We use Formula (7):

$$\mathbf{v}_2 = \mathbf{u}_2 - \operatorname{proj}_{W_1} \mathbf{u}_2 = \mathbf{u}_2 - \frac{\langle \mathbf{u}_2, \mathbf{v}_1 \rangle}{\|\mathbf{v}_1\|^2} \mathbf{v}_1$$

Of course, if $\mathbf{v}_2 = \mathbf{0}$, then $\mathbf{v}_2$ is not a basis vector. But this cannot happen, since it would then follow from the preceding formula for $\mathbf{v}_2$ that

$$\mathbf{u}_2 = \frac{\langle \mathbf{u}_2, \mathbf{v}_1 \rangle}{\|\mathbf{v}_1\|^2} \mathbf{v}_1 = \frac{\langle \mathbf{u}_2, \mathbf{v}_1 \rangle}{\|\mathbf{u}_1\|^2} \mathbf{u}_1$$

which says that $\mathbf{u}_2$ is a multiple of $\mathbf{u}_1$, contradicting the linear independence of the basis $S = \{\mathbf{u}_1, \mathbf{u}_2, \ldots, \mathbf{u}_n\}$.

Step 3.   To construct a vector $\mathbf{v}_3$ that is orthogonal to both $\mathbf{v}_1$ and $\mathbf{v}_2$, we compute the component of $\mathbf{u}_3$ orthogonal to the space W_2 spanned by $\mathbf{v}_1$ and $\mathbf{v}_2$ (Figure 6.3.4). From (7),

$$\mathbf{v}_3 = \mathbf{u}_3 - \operatorname{proj}_{W_2} \mathbf{u}_3 = \mathbf{u}_3 - \frac{\langle \mathbf{u}_3, \mathbf{v}_1 \rangle}{\|\mathbf{v}_1\|^2} \mathbf{v}_1 - \frac{\langle \mathbf{u}_3, \mathbf{v}_2 \rangle}{\|\mathbf{v}_2\|^2} \mathbf{v}_2$$

As in Step (2), the linear independence of $\{\mathbf{u}_1, \mathbf{u}_2, \ldots, \mathbf{u}_n\}$ ensures that $\mathbf{v}_3 \neq \mathbf{0}$. We leave the details as an exercise.

Step 4.   To determine a vector $\mathbf{v}_4$ that is orthogonal to $\mathbf{v}_1$, $\mathbf{v}_2$, and $\mathbf{v}_3$, we compute the component of $\mathbf{u}_4$ orthogonal to the space W_3 spanned by $\mathbf{v}_1$, $\mathbf{v}_2$, and $\mathbf{v}_3$. From (7),

$$\mathbf{v}_4 = \mathbf{u}_4 - \operatorname{proj}_{W_3} \mathbf{u}_4 = \mathbf{u}_4 - \frac{\langle \mathbf{u}_4, \mathbf{v}_1 \rangle}{\|\mathbf{v}_1\|^2} \mathbf{v}_1 - \frac{\langle \mathbf{u}_4, \mathbf{v}_2 \rangle}{\|\mathbf{v}_2\|^2} \mathbf{v}_2 - \frac{\langle \mathbf{u}_4, \mathbf{v}_3 \rangle}{\|\mathbf{v}_3\|^2} \mathbf{v}_3$$

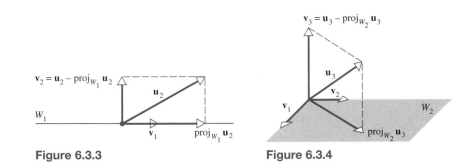

Figure 6.3.3   **Figure 6.3.4**

Continuing in this way, we will obtain, after n steps, an orthogonal set of vectors, $\{\mathbf{v}_1, \mathbf{v}_2, \ldots, \mathbf{v}_n\}$. Since V is n-dimensional and every orthogonal set is linearly independent, the set $\{\mathbf{v}_1, \mathbf{v}_2, \ldots, \mathbf{v}_n\}$ is an orthogonal basis for V. ■

The preceding step-by-step construction for converting an arbitrary basis into an orthogonal basis is called the ***Gram–Schmidt process***.

EXAMPLE 7   Using the Gram–Schmidt Process

Consider the vector space R^3 with the Euclidean inner product. Apply the Gram–Schmidt process to transform the basis vectors $\mathbf{u}_1 = (1, 1, 1)$, $\mathbf{u}_2 = (0, 1, 1)$, $\mathbf{u}_3 = (0, 0, 1)$ into an orthogonal basis $\{\mathbf{v}_1, \mathbf{v}_2, \mathbf{v}_3\}$; then normalize the orthogonal basis vectors to obtain an orthonormal basis $\{\mathbf{q}_1, \mathbf{q}_2, \mathbf{q}_3\}$.

Erhardt Schmidt
(1876–1959) was a German mathematician. Schmidt received his
doctoral degree from Göttingen
University in 1905, where he
studied under one of the giants
of mathematics, David Hilbert.
He eventually went to teach at
Berlin University in 1917, where
he stayed for the rest of his life.
Schmidt made important contributions to a variety of mathematical fields but is most note-
worthy for fashioning many of
Hilbert's diverse ideas into a general concept (called a Hilbert
space), which is fundamental in
the study of infinite-dimensional
vector spaces. Schmidt first described the process that bears his
name in a paper on integral equations published in 1907.

Solution

Step 1. $\mathbf{v}_1 = \mathbf{u}_1 = (1, 1, 1)$

Step 2. $\mathbf{v}_2 = \mathbf{u}_2 - \text{proj}_{W_1} \mathbf{u}_2 = \mathbf{u}_2 - \dfrac{\langle \mathbf{u}_2, \mathbf{v}_1 \rangle}{\|\mathbf{v}_1\|^2} \mathbf{v}_1$

$$= (0, 1, 1) - \frac{2}{3}(1, 1, 1) = \left(-\frac{2}{3}, \frac{1}{3}, \frac{1}{3}\right)$$

Step 3. $\mathbf{v}_3 = \mathbf{u}_3 - \text{proj}_{W_2} \mathbf{u}_3 = \mathbf{u}_3 - \dfrac{\langle \mathbf{u}_3, \mathbf{v}_1 \rangle}{\|\mathbf{v}_1\|^2} \mathbf{v}_1 - \dfrac{\langle \mathbf{u}_3, \mathbf{v}_2 \rangle}{\|\mathbf{v}_2\|^2} \mathbf{v}_2$

$$= (0, 0, 1) - \frac{1}{3}(1, 1, 1) - \frac{1/3}{2/3}\left(-\frac{2}{3}, \frac{1}{3}, \frac{1}{3}\right)$$

$$= \left(0, -\frac{1}{2}, \frac{1}{2}\right)$$

Thus

$$\mathbf{v}_1 = (1, 1, 1), \qquad \mathbf{v}_2 = \left(-\frac{2}{3}, \frac{1}{3}, \frac{1}{3}\right), \qquad \mathbf{v}_3 = \left(0, -\frac{1}{2}, \frac{1}{2}\right)$$

form an orthogonal basis for R^3. The norms of these vectors are

$$\|\mathbf{v}_1\| = \sqrt{3}, \qquad \|\mathbf{v}_2\| = \frac{\sqrt{6}}{3}, \qquad \|\mathbf{v}_3\| = \frac{1}{\sqrt{2}}$$

so an orthonormal basis for R^3 is

$$\mathbf{q}_1 = \frac{\mathbf{v}_1}{\|\mathbf{v}_1\|} = \left(\frac{1}{\sqrt{3}}, \frac{1}{\sqrt{3}}, \frac{1}{\sqrt{3}}\right), \qquad \mathbf{q}_2 = \frac{\mathbf{v}_2}{\|\mathbf{v}_2\|} = \left(-\frac{2}{\sqrt{6}}, \frac{1}{\sqrt{6}}, \frac{1}{\sqrt{6}}\right),$$

$$\mathbf{q}_3 = \frac{\mathbf{v}_3}{\|\mathbf{v}_3\|} = \left(0, -\frac{1}{\sqrt{2}}, \frac{1}{\sqrt{2}}\right) \quad \blacklozenge$$

REMARK In the preceding example we used the Gram–Schmidt process to produce an orthogonal basis; then, after the entire orthogonal basis was obtained, we normalized to obtain an orthonormal basis. Alternatively, one can normalize each orthogonal basis vector as soon as it is obtained, thereby generating the orthonormal basis step by step. However, this method has the slight disadvantage of producing more square roots to manipulate.

The Gram–Schmidt process with subsequent normalization not only converts an arbitrary basis $\{\mathbf{u}_1, \mathbf{u}_2, \ldots, \mathbf{u}_n\}$ into an orthonormal basis $\{\mathbf{q}_1, \mathbf{q}_2, \ldots, \mathbf{q}_n\}$ but does it in such a way that for $k \geq 2$ the following relationships hold:

- $\{\mathbf{q}_1, \mathbf{q}_2, \ldots, \mathbf{q}_k\}$ is an orthonormal basis for the space spanned by $\{\mathbf{u}_1, \mathbf{u}_2, \ldots, \mathbf{u}_k\}$.
- $\mathbf{q}_k$ is orthogonal to the space spanned by $\{\mathbf{u}_1, \mathbf{u}_2, \ldots, \mathbf{u}_{k-1}\}$.

We omit the proofs, but these facts should become evident after some thoughtful examination of the proof of Theorem 6.3.6.

QR-Decomposition

We pose the following problem.

Problem: If A is an $m \times n$ matrix with linearly independent column vectors, and if Q is the matrix with orthonormal column vectors that results from applying the

Gram–Schmidt process to the column vectors of A, what relationship, if any, exists between A and Q?

To solve this problem, suppose that the column vectors of A are $\mathbf{u}_1, \mathbf{u}_2, \ldots, \mathbf{u}_n$ and the orthonormal column vectors of Q are $\mathbf{q}_1, \mathbf{q}_2, \ldots, \mathbf{q}_n$; thus

$$A = [\mathbf{u}_1 \mid \mathbf{u}_2 \mid \cdots \mid \mathbf{u}_n] \quad \text{and} \quad Q = [\mathbf{q}_1 \mid \mathbf{q}_2 \mid \cdots \mid \mathbf{q}_n]$$

It follows from Theorem 6.3.1 that $\mathbf{u}_1, \mathbf{u}_2, \ldots, \mathbf{u}_n$ are expressible in terms of the vectors $\mathbf{q}_1, \mathbf{q}_2, \ldots, \mathbf{q}_n$ as

$$\mathbf{u}_1 = \langle \mathbf{u}_1, \mathbf{q}_1 \rangle \mathbf{q}_1 + \langle \mathbf{u}_1, \mathbf{q}_2 \rangle \mathbf{q}_2 + \cdots + \langle \mathbf{u}_1, \mathbf{q}_n \rangle \mathbf{q}_n$$
$$\mathbf{u}_2 = \langle \mathbf{u}_2, \mathbf{q}_1 \rangle \mathbf{q}_1 + \langle \mathbf{u}_2, \mathbf{q}_2 \rangle \mathbf{q}_2 + \cdots + \langle \mathbf{u}_2, \mathbf{q}_n \rangle \mathbf{q}_n$$
$$\vdots \qquad \vdots \qquad \vdots \qquad \vdots$$
$$\mathbf{u}_n = \langle \mathbf{u}_n, \mathbf{q}_1 \rangle \mathbf{q}_1 + \langle \mathbf{u}_n, \mathbf{q}_2 \rangle \mathbf{q}_2 + \cdots + \langle \mathbf{u}_n, \mathbf{q}_n \rangle \mathbf{q}_n$$

Recalling from Section 1.3 that the jth column vector of a matrix product is a linear combination of the column vectors of the first factor with coefficients coming from the jth column of the second factor, it follows that these relationships can be expressed in matrix form as

$$[\mathbf{u}_1 \mid \mathbf{u}_2 \mid \cdots \mid \mathbf{u}_n] = [\mathbf{q}_1 \mid \mathbf{q}_2 \mid \cdots \mid \mathbf{q}_n] \begin{bmatrix} \langle \mathbf{u}_1, \mathbf{q}_1 \rangle & \langle \mathbf{u}_2, \mathbf{q}_1 \rangle & \cdots & \langle \mathbf{u}_n, \mathbf{q}_1 \rangle \\ \langle \mathbf{u}_1, \mathbf{q}_2 \rangle & \langle \mathbf{u}_2, \mathbf{q}_2 \rangle & \cdots & \langle \mathbf{u}_n, \mathbf{q}_2 \rangle \\ \vdots & \vdots & & \vdots \\ \langle \mathbf{u}_1, \mathbf{q}_n \rangle & \langle \mathbf{u}_2, \mathbf{q}_n \rangle & \cdots & \langle \mathbf{u}_n, \mathbf{q}_n \rangle \end{bmatrix}$$

or more briefly as

$$A = QR \tag{8}$$

However, it is a property of the Gram–Schmidt process that for $j \geq 2$, the vector $\mathbf{q}_j$ is orthogonal to $\mathbf{u}_1, \mathbf{u}_2, \ldots, \mathbf{u}_{j-1}$; thus, all entries below the main diagonal of R are zero,

$$R = \begin{bmatrix} \langle \mathbf{u}_1, \mathbf{q}_1 \rangle & \langle \mathbf{u}_2, \mathbf{q}_1 \rangle & \cdots & \langle \mathbf{u}_n, \mathbf{q}_1 \rangle \\ 0 & \langle \mathbf{u}_2, \mathbf{q}_2 \rangle & \cdots & \langle \mathbf{u}_n, \mathbf{q}_2 \rangle \\ \vdots & \vdots & & \vdots \\ 0 & 0 & \cdots & \langle \mathbf{u}_n, \mathbf{q}_n \rangle \end{bmatrix} \tag{9}$$

We leave it as an exercise to show that the diagonal entries of R are nonzero, so R is invertible. Thus Equation (8) is a factorization of A into the product of a matrix Q with orthonormal column vectors and an invertible upper triangular matrix R. We call Equation (8) the ***QR-decomposition of A***. In summary, we have the following theorem.

THEOREM 6.3.7

QR-Decomposition

If A is an $m \times n$ matrix with linearly independent column vectors, then A can be factored as

$$A = QR$$

where Q is an $m \times n$ matrix with orthonormal column vectors, and R is an $n \times n$ invertible upper triangular matrix.

REMARK Recall from Theorem 6.2.7 that if A is an $n \times n$ matrix, then the invertibility of A is equivalent to linear independence of the column vectors; thus, every invertible matrix has a QR-decomposition.

EXAMPLE 8 *QR*-Decomposition of a 3 × 3 Matrix

Find the *QR*-decomposition of

$$A = \begin{bmatrix} 1 & 0 & 0 \\ 1 & 1 & 0 \\ 1 & 1 & 1 \end{bmatrix}$$

Solution

The column vectors of A are

$$\mathbf{u}_1 = \begin{bmatrix} 1 \\ 1 \\ 1 \end{bmatrix}, \qquad \mathbf{u}_2 = \begin{bmatrix} 0 \\ 1 \\ 1 \end{bmatrix}, \qquad \mathbf{u}_3 = \begin{bmatrix} 0 \\ 0 \\ 1 \end{bmatrix}$$

Applying the Gram–Schmidt process with subsequent normalization to these column vectors yields the orthonormal vectors (see Example 7)

$$\mathbf{q}_1 = \begin{bmatrix} 1/\sqrt{3} \\ 1/\sqrt{3} \\ 1/\sqrt{3} \end{bmatrix}, \qquad \mathbf{q}_2 = \begin{bmatrix} -2/\sqrt{6} \\ 1/\sqrt{6} \\ 1/\sqrt{6} \end{bmatrix}, \qquad \mathbf{q}_3 = \begin{bmatrix} 0 \\ -1/\sqrt{2} \\ 1/\sqrt{2} \end{bmatrix}$$

and from (9) the matrix R is

$$R = \begin{bmatrix} \langle \mathbf{u}_1, \mathbf{q}_1 \rangle & \langle \mathbf{u}_2, \mathbf{q}_1 \rangle & \langle \mathbf{u}_3, \mathbf{q}_1 \rangle \\ 0 & \langle \mathbf{u}_2, \mathbf{q}_2 \rangle & \langle \mathbf{u}_3, \mathbf{q}_2 \rangle \\ 0 & 0 & \langle \mathbf{u}_3, \mathbf{q}_3 \rangle \end{bmatrix} = \begin{bmatrix} 3/\sqrt{3} & 2/\sqrt{3} & 1/\sqrt{3} \\ 0 & 2/\sqrt{6} & 1/\sqrt{6} \\ 0 & 0 & 1/\sqrt{2} \end{bmatrix}$$

Thus the *QR*-decomposition of A is

$$\underbrace{\begin{bmatrix} 1 & 0 & 0 \\ 1 & 1 & 0 \\ 1 & 1 & 1 \end{bmatrix}}_{A} = \underbrace{\begin{bmatrix} 1/\sqrt{3} & -2/\sqrt{6} & 0 \\ 1/\sqrt{3} & 1/\sqrt{6} & -1/\sqrt{2} \\ 1/\sqrt{3} & 1/\sqrt{6} & 1/\sqrt{2} \end{bmatrix}}_{Q} \underbrace{\begin{bmatrix} 3/\sqrt{3} & 2/\sqrt{3} & 1/\sqrt{3} \\ 0 & 2/\sqrt{6} & 1/\sqrt{6} \\ 0 & 0 & 1/\sqrt{2} \end{bmatrix}}_{R} \quad \blacklozenge$$

The Role of the *QR*-Decomposition in Linear Algebra

In recent years the *QR*-decomposition has assumed growing importance as the mathematical foundation for a wide variety of practical numerical algorithms, including a widely used algorithm for computing eigenvalues of large matrices. Such algorithms are discussed in textbooks that deal with numerical linear algebra.

Additional Proof

Proof of Theorem 6.3.4   There are two parts to the proof. First we must find vectors $\mathbf{w}_1$ and $\mathbf{w}_2$ with the stated properties, and then we must show that these are the only such vectors.

By the Gram–Schmidt process, there is an orthonormal basis $\{\mathbf{v}_1, \mathbf{v}_2, \ldots, \mathbf{v}_n\}$ for W. Let

$$\mathbf{w}_1 = \langle \mathbf{u}, \mathbf{v}_1 \rangle \mathbf{v}_1 + \langle \mathbf{u}, \mathbf{v}_2 \rangle \mathbf{v}_2 + \cdots + \langle \mathbf{u}, \mathbf{v}_n \rangle \mathbf{v}_n \tag{10}$$

and

$$\mathbf{w}_2 = \mathbf{u} - \mathbf{w}_1 \tag{11}$$

It follows that $\mathbf{w}_1 + \mathbf{w}_2 = \mathbf{w}_1 + (\mathbf{u} - \mathbf{w}_1) = \mathbf{u}$, so it remains to show that $\mathbf{w}_1$ is in W and $\mathbf{w}_2$ is orthogonal to W. But $\mathbf{w}_1$ lies in W because it is a linear combination of the basis vectors for W. To show that $\mathbf{w}_2$ is orthogonal to W, we must show that $\langle \mathbf{w}_2, \mathbf{w} \rangle = 0$ for every vector $\mathbf{w}$ in W. But if $\mathbf{w}$ is any vector in W, it can be expressed as a linear combination

$$\mathbf{w} = k_1\mathbf{v}_1 + k_2\mathbf{v}_2 + \cdots + k_n\mathbf{v}_n$$

of the basis vectors $\mathbf{v}_1, \mathbf{v}_2, \ldots, \mathbf{v}_n$. Thus

$$\langle \mathbf{w}_2, \mathbf{w} \rangle = \langle \mathbf{u} - \mathbf{w}_1, \mathbf{w} \rangle = \langle \mathbf{u}, \mathbf{w} \rangle - \langle \mathbf{w}_1, \mathbf{w} \rangle \tag{12}$$

But

$$\langle \mathbf{u}, \mathbf{w} \rangle = \langle \mathbf{u}, k_1\mathbf{v}_1 + k_2\mathbf{v}_2 + \cdots + k_n\mathbf{v}_n \rangle$$
$$= k_1\langle \mathbf{u}, \mathbf{v}_1 \rangle + k_2\langle \mathbf{u}, \mathbf{v}_2 \rangle + \cdots + k_n\langle \mathbf{u}, \mathbf{v}_n \rangle$$

and by part (*c*) of Theorem 6.3.2,

$$\langle \mathbf{w}_1, \mathbf{w} \rangle = \langle \mathbf{u}, \mathbf{v}_1 \rangle k_1 + \langle \mathbf{u}, \mathbf{v}_2 \rangle k_2 + \cdots + \langle \mathbf{u}, \mathbf{v}_n \rangle k_n$$

Thus $\langle \mathbf{u}, \mathbf{w} \rangle$ and $\langle \mathbf{w}_1, \mathbf{w} \rangle$ are equal, so (12) yields $\langle \mathbf{w}_2, \mathbf{w} \rangle = 0$, which is what we want to show.

To see that (10) and (11) are the only vectors with the properties stated in the theorem, suppose that we can also write

$$\mathbf{u} = \mathbf{w}'_1 + \mathbf{w}'_2 \tag{13}$$

where $\mathbf{w}'_1$ is in W and $\mathbf{w}'_2$ is orthogonal to W. If we subtract from (13) the equation

$$\mathbf{u} = \mathbf{w}_1 + \mathbf{w}_2$$

we obtain

$$\mathbf{0} = (\mathbf{w}'_1 - \mathbf{w}_1) + (\mathbf{w}'_2 - \mathbf{w}_2)$$

or

$$\mathbf{w}_1 - \mathbf{w}'_1 = \mathbf{w}'_2 - \mathbf{w}_2 \tag{14}$$

Since $\mathbf{w}_2$ and $\mathbf{w}'_2$ are orthogonal to W, their difference is also orthogonal to W, since for any vector $\mathbf{w}$ in W, we can write

$$\langle \mathbf{w}, \mathbf{w}'_2 - \mathbf{w}_2 \rangle = \langle \mathbf{w}, \mathbf{w}'_2 \rangle - \langle \mathbf{w}, \mathbf{w}_2 \rangle = 0 - 0 = 0$$

But $\mathbf{w}'_2 - \mathbf{w}_2$ is itself a vector in W, since from (14) it is the difference of the two vectors $\mathbf{w}_1$ and $\mathbf{w}'_1$ that lie in the subspace W. Thus, $\mathbf{w}'_2 - \mathbf{w}_2$ must be orthogonal to itself; that is,

$$\langle \mathbf{w}'_2 - \mathbf{w}_2, \mathbf{w}'_2 - \mathbf{w}_2 \rangle = 0$$

But this implies that $\mathbf{w}'_2 - \mathbf{w}_2 = 0$ by Axiom 4 for inner products. Thus $\mathbf{w}'_2 = \mathbf{w}_2$, and by (14), $\mathbf{w}'_1 = \mathbf{w}_1$. ∎

EXERCISE SET
6.3

1. Which of the following sets of vectors are orthogonal with respect to the Euclidean inner product on R^2?

   (a) $(0, 1)$, $(2, 0)$ (b) $(-1/\sqrt{2}, 1/\sqrt{2})$, $(1/\sqrt{2}, 1/\sqrt{2})$

   (c) $(-1/\sqrt{2}, -1/\sqrt{2})$, $(1/\sqrt{2}, 1/\sqrt{2})$ (d) $(0, 0)$, $(0, 1)$

2. Which of the sets in Exercise 1 are orthonormal with respect to the Euclidean inner product on R^2?

3. Which of the following sets of vectors are orthogonal with respect to the Euclidean inner product on R^3?

(a) $\left(\frac{1}{\sqrt{2}}, 0, \frac{1}{\sqrt{2}}\right)$, $\left(\frac{1}{\sqrt{3}}, \frac{1}{\sqrt{3}}, -\frac{1}{\sqrt{3}}\right)$, $\left(-\frac{1}{\sqrt{2}}, 0, \frac{1}{\sqrt{2}}\right)$

(b) $\left(\frac{2}{3}, -\frac{2}{3}, \frac{1}{3}\right)$, $\left(\frac{2}{3}, \frac{1}{3}, -\frac{2}{3}\right)$, $\left(\frac{1}{3}, \frac{2}{3}, \frac{2}{3}\right)$

(c) $(1, 0, 0)$, $\left(0, \frac{1}{\sqrt{2}}, \frac{1}{\sqrt{2}}\right)$, $(0, 0, 1)$

(d) $\left(\frac{1}{\sqrt{6}}, \frac{1}{\sqrt{6}}, -\frac{2}{\sqrt{6}}\right)$, $\left(\frac{1}{\sqrt{2}}, -\frac{1}{\sqrt{2}}, 0\right)$

4. Which of the sets in Exercise 3 are orthonormal with respect to the Euclidean inner product on R^3?

5. Which of the following sets of polynomials are orthonormal with respect to the inner product on P_2 discussed in Example 8 of Section 6.1?

(a) $\frac{2}{3} - \frac{2}{3}x + \frac{1}{3}x^2$, $\frac{2}{3} + \frac{1}{3}x - \frac{2}{3}x^2$, $\frac{1}{3} + \frac{2}{3}x + \frac{2}{3}x^2$ (b) 1, $\frac{1}{\sqrt{2}}x + \frac{1}{\sqrt{2}}x^2$, x^2

6. Which of the following sets of matrices are orthonormal with respect to the inner product on M_{22} discussed in Example 7 of Section 6.1?

(a) $\begin{bmatrix} 1 & 0 \\ 0 & 0 \end{bmatrix}$, $\begin{bmatrix} 0 & \frac{2}{3} \\ \frac{1}{3} & -\frac{2}{3} \end{bmatrix}$, $\begin{bmatrix} 0 & \frac{2}{3} \\ -\frac{2}{3} & \frac{1}{3} \end{bmatrix}$, $\begin{bmatrix} 0 & \frac{1}{3} \\ \frac{2}{3} & \frac{2}{3} \end{bmatrix}$

(b) $\begin{bmatrix} 1 & 0 \\ 0 & 0 \end{bmatrix}$, $\begin{bmatrix} 0 & 1 \\ 0 & 0 \end{bmatrix}$, $\begin{bmatrix} 0 & 0 \\ 1 & 1 \end{bmatrix}$, $\begin{bmatrix} 0 & 0 \\ 1 & -1 \end{bmatrix}$

7. Verify that the given set of vectors is orthogonal with respect to the Euclidean inner product; then convert it to an orthonormal set by normalizing the vectors.

(a) $(-1, 2)$, $(6, 3)$

(b) $(1, 0, -1)$, $(2, 0, 2)$, $(0, 5, 0)$

(c) $\left(\frac{1}{5}, \frac{1}{5}, \frac{1}{5}\right)$, $\left(-\frac{1}{2}, \frac{1}{2}, 0\right)$, $\left(\frac{1}{3}, \frac{1}{3}, -\frac{2}{3}\right)$

8. Verify that the set of vectors $\{(1, 0), (0, 1)\}$ is orthogonal with respect to the inner product $\langle \mathbf{u}, \mathbf{v} \rangle = 4u_1 v_1 + u_2 v_2$ on R^2; then convert it to an orthonormal set by normalizing the vectors.

9. Verify that the vectors $\mathbf{v}_1 = \left(-\frac{3}{5}, \frac{4}{5}, 0\right)$, $\mathbf{v}_2 = \left(\frac{4}{5}, \frac{3}{5}, 0\right)$, $\mathbf{v}_3 = (0, 0, 1)$ form an orthonormal basis for R^3 with the Euclidean inner product; then use Theorem 6.3.1 to express each of the following as linear combinations of $\mathbf{v}_1$, $\mathbf{v}_2$, and $\mathbf{v}_3$.

(a) $(1, -1, 2)$ (b) $(3, -7, 4)$ (c) $\left(\frac{1}{7}, -\frac{3}{7}, \frac{5}{7}\right)$

10. Verify that the vectors

$$\mathbf{v}_1 = (1, -1, 2, -1), \quad \mathbf{v}_2 = (-2, 2, 3, 2), \quad \mathbf{v}_3 = (1, 2, 0, -1), \quad \mathbf{v}_4 = (1, 0, 0, 1)$$

form an orthogonal basis for R^4 with the Euclidean inner product; then use Formula (1) to express each of the following as linear combinations of $\mathbf{v}_1$, $\mathbf{v}_2$, $\mathbf{v}_3$, and $\mathbf{v}_4$.

(a) $(1, 1, 1, 1)$ (b) $(\sqrt{2}, -3\sqrt{2}, 5\sqrt{2}, -\sqrt{2})$ (c) $\left(-\frac{1}{3}, \frac{2}{3}, -\frac{1}{3}, \frac{4}{3}\right)$

11. In each part, an orthonormal basis relative to the Euclidean inner product is given. Use Theorem 6.3.1 to find the coordinate vector of $\mathbf{w}$ with respect to that basis.

(a) $\mathbf{w} = (3, 7)$; $\mathbf{u}_1 = \left(\frac{1}{\sqrt{2}}, -\frac{1}{\sqrt{2}}\right)$, $\mathbf{u}_2 = \left(\frac{1}{\sqrt{2}}, \frac{1}{\sqrt{2}}\right)$

(b) $\mathbf{w} = (-1, 0, 2)$; $\mathbf{u}_1 = \left(\frac{2}{3}, -\frac{2}{3}, \frac{1}{3}\right)$, $\mathbf{u}_2 = \left(\frac{2}{3}, \frac{1}{3}, -\frac{2}{3}\right)$, $\mathbf{u}_3 = \left(\frac{1}{3}, \frac{2}{3}, \frac{2}{3}\right)$

12. Let R^2 have the Euclidean inner product, and let $S = \{\mathbf{w}_1, \mathbf{w}_2\}$ be the orthonormal basis with $\mathbf{w}_1 = \left(\frac{3}{5}, -\frac{4}{5}\right)$, $\mathbf{w}_2 = \left(\frac{4}{5}, \frac{3}{5}\right)$.

(a) Find the vectors **u** and **v** that have coordinate vectors $(\mathbf{u})_S = (1, 1)$ and $(\mathbf{v})_S = (-1, 4)$.

(b) Compute $\|\mathbf{u}\|$, $d(\mathbf{u}, \mathbf{v})$, and $\langle \mathbf{u}, \mathbf{v} \rangle$ by applying Theorem 6.3.2 to the coordinate vectors $(\mathbf{u})_S$ and $(\mathbf{v})_S$; then check the results by performing the computations directly on **u** and **v**.

13. Let R^3 have the Euclidean inner product, and let $S = \{\mathbf{w}_1, \mathbf{w}_2, \mathbf{w}_3\}$ be the orthonormal basis with $\mathbf{w}_1 = \left(0, -\frac{3}{5}, \frac{4}{5}\right)$, $\mathbf{w}_2 = (1, 0, 0)$, and $\mathbf{w}_3 = \left(0, \frac{4}{5}, \frac{3}{5}\right)$.

(a) Find the vectors **u**, **v**, and **w** that have the coordinate vectors $(\mathbf{u})_S = (-2, 1, 2)$, $(\mathbf{v})_S = (3, 0, -2)$, and $(\mathbf{w})_S = (5, -4, 1)$.

(b) Compute $\|\mathbf{v}\|$, $d(\mathbf{u}, \mathbf{w})$, and $\langle \mathbf{w}, \mathbf{v} \rangle$ by applying Theorem 6.3.2 to the coordinate vectors $(\mathbf{u})_S$, $(\mathbf{v})_S$, and $(\mathbf{w})_S$; then check the results by performing the computations directly on **u**, **v**, and **w**.

14. In each part, S represents some orthonormal basis for a four-dimensional inner product space. Use the given information to find $\|\mathbf{u}\|$, $\|\mathbf{v} - \mathbf{w}\|$, $\|\mathbf{v} + \mathbf{w}\|$, and $\langle \mathbf{v}, \mathbf{w} \rangle$.

(a) $(\mathbf{u})_S = (-1, 2, 1, 3)$, $(\mathbf{v})_S = (0, -3, 1, 5)$, $(\mathbf{w})_S = (-2, -4, 3, 1)$

(b) $(\mathbf{u})_S = (0, 0, -1, -1)$, $(\mathbf{v})_S = (5, 5, -2, -2)$, $(\mathbf{w})_S = (3, 0, -3, 0)$

15. (a) Show that the vectors $\mathbf{v}_1 = (1, -2, 3, -4)$, $\mathbf{v}_2 = (2, 1, -4, -3)$, $\mathbf{v}_3 = (-3, 4, 1, -2)$, and $\mathbf{v}_4 = (4, 3, 2, 1)$ form an orthogonal basis for R^4 with the Euclidean inner product.

(b) Use (1) to express $\mathbf{u} = (-1, 2, 3, 7)$ as a linear combination of the vectors in part (a).

16. Let R^2 have the Euclidean inner product. Use the Gram–Schmidt process to transform the basis $\{\mathbf{u}_1, \mathbf{u}_2\}$ into an orthonormal basis. Draw both sets of basis vectors in the xy-plane.

(a) $\mathbf{u}_1 = (1, -3)$, $\mathbf{u}_2 = (2, 2)$ (b) $\mathbf{u}_1 = (1, 0)$, $\mathbf{u}_2 = (3, -5)$

17. Let R^3 have the Euclidean inner product. Use the Gram–Schmidt process to transform the basis $\{\mathbf{u}_1, \mathbf{u}_2, \mathbf{u}_3\}$ into an orthonormal basis.

(a) $\mathbf{u}_1 = (1, 1, 1)$, $\mathbf{u}_2 = (-1, 1, 0)$, $\mathbf{u}_3 = (1, 2, 1)$

(b) $\mathbf{u}_1 = (1, 0, 0)$, $\mathbf{u}_2 = (3, 7, -2)$, $\mathbf{u}_3 = (0, 4, 1)$

18. Let R^4 have the Euclidean inner product. Use the Gram–Schmidt process to transform the basis $\{\mathbf{u}_1, \mathbf{u}_2, \mathbf{u}_3, \mathbf{u}_4\}$ into an orthonormal basis.

$$\mathbf{u}_1 = (0, 2, 1, 0), \quad \mathbf{u}_2 = (1, -1, 0, 0), \quad \mathbf{u}_3 = (1, 2, 0, -1), \quad \mathbf{u}_4 = (1, 0, 0, 1)$$

19. Let R^3 have the Euclidean inner product. Find an orthonormal basis for the subspace spanned by $(0, 1, 2)$, $(-1, 0, 1)$, $(-1, 1, 3)$.

20. Let R^3 have the inner product $\langle \mathbf{u}, \mathbf{v} \rangle = u_1 v_1 + 2u_2 v_2 + 3u_3 v_3$. Use the Gram–Schmidt process to transform $\mathbf{u}_1 = (1, 1, 1)$, $\mathbf{u}_2 = (1, 1, 0)$, $\mathbf{u}_3 = (1, 0, 0)$ into an orthonormal basis.

21. The subspace of R^3 spanned by the vectors $\mathbf{u}_1 = \left(\frac{4}{5}, 0, -\frac{3}{5}\right)$ and $\mathbf{u}_2 = (0, 1, 0)$ is a plane passing through the origin. Express $\mathbf{w} = (1, 2, 3)$ in the form $\mathbf{w} = \mathbf{w}_1 + \mathbf{w}_2$, where $\mathbf{w}_1$ lies in the plane and $\mathbf{w}_2$ is perpendicular to the plane.

22. Repeat Exercise 21 with $\mathbf{u}_1 = (1, 1, 1)$ and $\mathbf{u}_2 = (2, 0, -1)$.

23. Let R^4 have the Euclidean inner product. Express $\mathbf{w} = (-1, 2, 6, 0)$ in the form $\mathbf{w} = \mathbf{w}_1 + \mathbf{w}_2$, where $\mathbf{w}_1$ is in the space W spanned by $\mathbf{u}_1 = (-1, 0, 1, 2)$ and $\mathbf{u}_2 = (0, 1, 0, 1)$, and $\mathbf{w}_2$ is orthogonal to W.

24. Find the QR-decomposition of the matrix, where possible.

(a) $\begin{bmatrix} 1 & -1 \\ 2 & 3 \end{bmatrix}$ (b) $\begin{bmatrix} 1 & 2 \\ 0 & 1 \\ 1 & 4 \end{bmatrix}$ (c) $\begin{bmatrix} 1 & 1 \\ -2 & 1 \\ 2 & 1 \end{bmatrix}$

(d) $\begin{bmatrix} 1 & 0 & 2 \\ 0 & 1 & 1 \\ 1 & 2 & 0 \end{bmatrix}$ (e) $\begin{bmatrix} 1 & 2 & 1 \\ 1 & 1 & 1 \\ 0 & 3 & 1 \end{bmatrix}$ (f) $\begin{bmatrix} 1 & 0 & 1 \\ -1 & 1 & 1 \\ 1 & 0 & 1 \\ -1 & 1 & 1 \end{bmatrix}$

25. Let $\{v_1, v_2, v_3\}$ be an orthonormal basis for an inner product space V. Show that if w is a vector in V, then $\|w\|^2 = \langle w, v_1 \rangle^2 + \langle w, v_2 \rangle^2 + \langle w, v_3 \rangle^2$.

26. Let $\{v_1, v_2, \ldots, v_n\}$ be an orthonormal basis for an inner product space V. Show that if w is a vector in V, then $\|w\|^2 = \langle w, v_1 \rangle^2 + \langle w, v_2 \rangle^2 + \cdots + \langle w, v_n \rangle^2$.

27. In Step 3 of the proof of Theorem 6.3.6, it was stated that "the linear independence of $\{u_1, u_2, \ldots, u_n\}$ ensures that $v_3 \neq 0$." Prove this statement.

28. Prove that the diagonal entries of R in Formula (9) are nonzero.

29. **(For Readers Who Have Studied Calculus)** Let the vector space P_2 have the inner product

$$\langle p, q \rangle = \int_{-1}^{1} p(x)q(x)\,dx$$

Apply the Gram–Schmidt process to transform the standard basis $S = \{1, x, x^2\}$ into an orthonormal basis. (The polynomials in the resulting basis are called the first three *normalized Legendre polynomials*.)

30. **(For Readers Who Have Studied Calculus)** Use Theorem 6.3.1 to express the following as linear combinations of the first three normalized Legendre polynomials (Exercise 29).
   (a) $1 + x + 4x^2$ (b) $2 - 7x^2$ (c) $4 + 3x$

31. **(For Readers Who Have Studied Calculus)** Let P_2 have the inner product

$$\langle p, q \rangle = \int_{0}^{1} p(x)q(x)\,dx$$

Apply the Gram–Schmidt process to transform the standard basis $S = \{1, x, x^2\}$ into an orthonormal basis.

32. Prove Theorem 6.3.2.

33. Prove Theorem 6.3.5.

Discussion & Discovery

34. (a) It follows from Theorem 6.3.6 that every plane through the origin in R^3 must have an orthonormal basis with respect to the Euclidean inner product. In words, explain how you would go about finding an orthonormal basis for a plane if you knew its equation.

   (b) Use your method to find an orthonormal basis for the plane $x + 2y - z = 0$.

35. Find vectors x and y in R^2 that are orthonormal with respect to the inner product $\langle u, v \rangle = 3u_1v_1 + 2u_2v_2$ but are not orthonormal with respect to the Euclidean inner product.

36. If W is a line through the origin of R^3 with the Euclidean inner product, and if u is a vector in R^3, then Theorem 6.3.4 implies that u can be expressed uniquely as $u = w_1 + w_2$, where w_1 is a vector in W and w_2 is a vector in $W^\perp$. Draw a picture that illustrates this.

37. Indicate whether each statement is always true or sometimes false. Justify your answer by giving a logical argument or a counterexample.
   (a) A linearly dependent set of vectors in an inner product space cannot be orthonormal.
   (b) Every finite-dimensional vector space has an orthonormal basis.
   (c) $\text{proj}_W u$ is orthogonal to $\text{proj}_{W^\perp} u$ in any inner product space.
   (d) Every matrix with a nonzero determinant has a QR-decomposition.

38. What happens if you apply the Gram–Schmidt process to a linearly dependent set of vectors?

6.4
BEST APPROXIMATION; LEAST SQUARES

In this section we shall show how orthogonal projections can be used to solve certain approximation problems. The results obtained in this section have a wide variety of applications in both mathematics and science.

Orthogonal Projections Viewed as Approximations

If P is a point in ordinary 3-space and W is a plane through the origin, then the point Q in W that is closest to P can be obtained by dropping a perpendicular from P to W (Figure 6.4.1a). Therefore, if we let $\mathbf{u} = \overrightarrow{OP}$, then the distance between P and W is given by

$$\| \mathbf{u} - \mathrm{proj}_W \mathbf{u} \|$$

In other words, among all vectors $\mathbf{w}$ in W, the vector $\mathbf{w} = \mathrm{proj}_W \mathbf{u}$ minimizes the distance $\| \mathbf{u} - \mathbf{w} \|$ (Figure 6.4.1b).

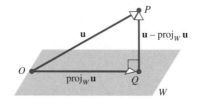 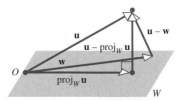

(a) Q is the point in W closest to P.   (b) $\| \mathbf{u} - \mathbf{w} \|$ is minimized by $\mathbf{w} = \mathrm{proj}_W \mathbf{u}$.

Figure 6.4.1

There is another way of thinking about this idea. View $\mathbf{u}$ as a fixed vector that we would like to approximate by a vector in W. Any such approximation $\mathbf{w}$ will result in an "error vector,"

$$\mathbf{u} - \mathbf{w}$$

that, unless $\mathbf{u}$ is in W, cannot be made equal to $\mathbf{0}$. However, by choosing

$$\mathbf{w} = \mathrm{proj}_W \mathbf{u}$$

we can make the length of the error vector

$$\| \mathbf{u} - \mathbf{w} \| = \| \mathbf{u} - \mathrm{proj}_W \mathbf{u} \|$$

as small as possible. Thus we can describe $\mathrm{proj}_W \mathbf{u}$ as the "best approximation" to $\mathbf{u}$ by vectors in W. The following theorem will make these intuitive ideas precise.

THEOREM 6.4.1

Best Approximation Theorem

*If W is a finite-dimensional subspace of an inner product space V, and if $\mathbf{u}$ is a vector in V, then $\mathrm{proj}_W \mathbf{u}$ is the **best approximation** to $\mathbf{u}$ from W in the sense that*

$$\| \mathbf{u} - \mathrm{proj}_W \mathbf{u} \| < \| \mathbf{u} - \mathbf{w} \|$$

for every vector $\mathbf{w}$ in W that is different from $\mathrm{proj}_W \mathbf{u}$.

Proof   For every vector $\mathbf{w}$ in W, we can write

$$\mathbf{u} - \mathbf{w} = (\mathbf{u} - \mathrm{proj}_W \mathbf{u}) + (\mathrm{proj}_W \mathbf{u} - \mathbf{w}) \tag{1}$$

But $\text{proj}_W \mathbf{u} - \mathbf{w}$, being a difference of vectors in W, is in W; and $\mathbf{u} - \text{proj}_W \mathbf{u}$ is orthogonal to W, so the two terms on the right side of (1) are orthogonal. Thus, by the Theorem of Pythagoras (Theorem 6.2.4),

$$\|\mathbf{u} - \mathbf{w}\|^2 = \|\mathbf{u} - \text{proj}_W \mathbf{u}\|^2 + \|\text{proj}_W \mathbf{u} - \mathbf{w}\|^2$$

If $\mathbf{w} \neq \text{proj}_W \mathbf{u}$, then the second term in this sum will be positive, so

$$\|\mathbf{u} - \mathbf{w}\|^2 > \|\mathbf{u} - \text{proj}_W \mathbf{u}\|^2$$

or, equivalently,

$$\|\mathbf{u} - \mathbf{w}\| > \|\mathbf{u} - \text{proj}_W \mathbf{u}\| \qquad \blacksquare$$

Applications of this theorem will be given later in the text.

Least Squares Solutions of Linear Systems

Up to now we have been concerned primarily with consistent systems of linear equations. However, inconsistent linear systems are also important in physical applications. It is a common situation that some physical problem leads to a linear system $A\mathbf{x} = \mathbf{b}$ that should be consistent on theoretical grounds but fails to be so because "measurement errors" in the entries of A and $\mathbf{b}$ perturb the system enough to cause inconsistency. In such situations one looks for a value of $\mathbf{x}$ that comes "as close as possible" to being a solution in the sense that it minimizes the value of $\|A\mathbf{x} - \mathbf{b}\|$ with respect to the Euclidean inner product. The quantity $\|A\mathbf{x} - \mathbf{b}\|$ can be viewed as a measure of the "error" that results from regarding $\mathbf{x}$ as an approximate solution of the linear system $A\mathbf{x} = \mathbf{b}$. If the system is consistent and $\mathbf{x}$ is an exact solution, then the error is zero, since $\|A\mathbf{x} - \mathbf{b}\| = \|\mathbf{0}\| = 0$. In general, the larger the value of $\|A\mathbf{x} - \mathbf{b}\|$, the more poorly $\mathbf{x}$ serves as an approximate solution of the system.

Least Squares Problem: Given a linear system $A\mathbf{x} = \mathbf{b}$ of m equations in n unknowns, find a vector $\mathbf{x}$, if possible, that minimizes $\|A\mathbf{x} - \mathbf{b}\|$ with respect to the Euclidean inner product on R^m. Such a vector is called a ***least squares solution*** of $A\mathbf{x} = \mathbf{b}$.

REMARK To understand the origin of the term *least squares*, let $\mathbf{e} = A\mathbf{x} - \mathbf{b}$, which we can view as the error vector that results from the approximation $\mathbf{x}$. If $\mathbf{e} = (e_1, e_2, \ldots, e_m)$, then a least squares solution minimizes $\|\mathbf{e}\| = (e_1^2 + e_2^2 + \cdots + e_m^2)^{1/2}$; hence it also minimizes $\|\mathbf{e}\|^2 = e_1^2 + e_2^2 + \cdots + e_m^2$. Hence the term *least squares*.

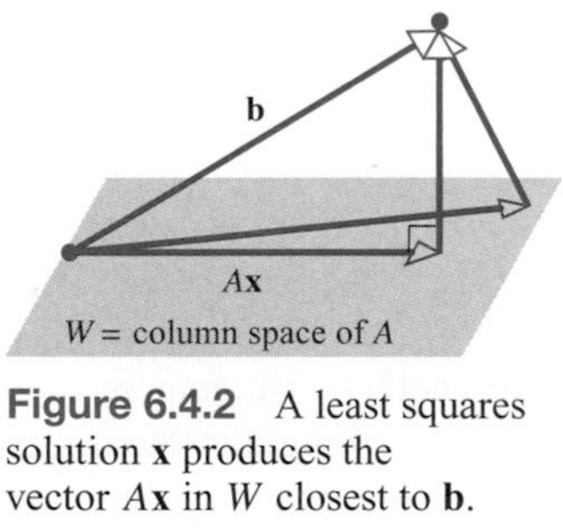

To solve the least squares problem, let W be the column space of A. For each $n \times 1$ matrix $\mathbf{x}$, the product $A\mathbf{x}$ is a linear combination of the column vectors of A. Thus, as $\mathbf{x}$ varies over R^n, the vector $A\mathbf{x}$ varies over all possible linear combinations of the column vectors of A; that is, $A\mathbf{x}$ varies over the entire column space W. Geometrically, solving the least squares problem amounts to finding a vector $\mathbf{x}$ in R^n such that $A\mathbf{x}$ is the closest vector in W to $\mathbf{b}$ (Figure 6.4.2).

It follows from the Best Approximation Theorem (6.4.1) that the closest vector in W to $\mathbf{b}$ is the orthogonal projection of $\mathbf{b}$ on W. Thus, for a vector $\mathbf{x}$ to be a least squares solution of $A\mathbf{x} = \mathbf{b}$, this vector must satisfy

$$A\mathbf{x} = \text{proj}_W \mathbf{b} \qquad (2)$$

Figure 6.4.2 A least squares solution $\mathbf{x}$ produces the vector $A\mathbf{x}$ in W closest to $\mathbf{b}$.

One could attempt to find least squares solutions of $A\mathbf{x} = \mathbf{b}$ by first calculating the vector $\text{proj}_W \mathbf{b}$ and then solving (2); however, there is a better approach. It follows from the Projection Theorem (6.3.4) and Formula (5) of Section 6.3 that

$$\mathbf{b} - A\mathbf{x} = \mathbf{b} - \text{proj}_W \mathbf{b}$$

is orthogonal to W. But W is the column space of A, so it follows from Theorem 6.2.6 that $\mathbf{b} - A\mathbf{x}$ lies in the nullspace of A^T. Therefore, a least squares solution of $A\mathbf{x} = \mathbf{b}$ must satisfy

$$A^T(\mathbf{b} - A\mathbf{x}) = 0$$

or, equivalently,

$$A^TA\mathbf{x} = A^T\mathbf{b} \tag{3}$$

This is called the ***normal system*** associated with $A\mathbf{x} = \mathbf{b}$, and the individual equations are called the ***normal equations*** associated with $A\mathbf{x} = \mathbf{b}$. Thus the problem of finding a least squares solution of $A\mathbf{x} = \mathbf{b}$ has been reduced to the problem of finding an exact solution of the associated normal system.

Note the following observations about the normal system:

- The normal system involves n equations in n unknowns (verify).
- The normal system is consistent, since it is satisfied by a least squares solution of $A\mathbf{x} = \mathbf{b}$.
- The normal system may have infinitely many solutions, in which case all of its solutions are least squares solutions of $A\mathbf{x} = \mathbf{b}$.

From these observations and Formula (2), we have the following theorem.

THEOREM 6.4.2

For any linear system $A\mathbf{x} = \mathbf{b}$, the associated normal system

$$A^TA\mathbf{x} = A^T\mathbf{b}$$

is consistent, and all solutions of the normal system are least squares solutions of $A\mathbf{x} = \mathbf{b}$. Moreover, if W is the column space of A, and $\mathbf{x}$ is any least squares solution of $A\mathbf{x} = \mathbf{b}$, then the orthogonal projection of $\mathbf{b}$ on W is

$$\operatorname{proj}_W \mathbf{b} = A\mathbf{x}$$

Uniqueness of Least Squares Solutions

Before we examine some numerical examples, we shall establish conditions under which a linear system is guaranteed to have a unique least squares solution. We shall need the following theorem.

THEOREM 6.4.3

If A is an $m \times n$ matrix, then the following are equivalent.

(a) A has linearly independent column vectors.   *(b) A^TA is invertible.*

Proof   We shall prove that $(a) \Rightarrow (b)$ and leave the proof that $(b) \Rightarrow (a)$ as an exercise.

(a) $\Rightarrow$ (b)   Assume that A has linearly independent column vectors. The matrix A^TA has size $n \times n$, so we can prove that this matrix is invertible by showing that the linear system $A^TA\mathbf{x} = \mathbf{0}$ has only the trivial solution. But if $\mathbf{x}$ is any solution of this system, then $A\mathbf{x}$ is in the nullspace of A^T and also in the column space of A. By Theorem 6.2.6 these spaces are orthogonal complements, so part (b) of Theorem 6.2.5 implies that $A\mathbf{x} = \mathbf{0}$. But A has linearly independent column vectors, so $\mathbf{x} = \mathbf{0}$ by Theorem 5.6.8.   ■

The next theorem is a direct consequence of Theorems 6.4.2 and 6.4.3. We omit the details.

THEOREM 6.4.4

> *If A is an $m \times n$ matrix with linearly independent column vectors, then for every $m \times 1$ matrix $\mathbf{b}$, the linear system $A\mathbf{x} = \mathbf{b}$ has a unique least squares solution. This solution is given by*
>
> $$\mathbf{x} = (A^TA)^{-1}A^T\mathbf{b} \qquad (4)$$
>
> *Moreover, if W is the column space of A, then the orthogonal projection of $\mathbf{b}$ on W is*
>
> $$\operatorname{proj}_W \mathbf{b} = A\mathbf{x} = A(A^TA)^{-1}A^T\mathbf{b} \qquad (5)$$

REMARK Formulas (4) and (5) have various theoretical applications, but they are very inefficient for numerical calculations. Least squares solutions of $A\mathbf{x} = \mathbf{b}$ are typically found by using Gaussian elimination to solve the normal equations, and the orthogonal projection of $\mathbf{b}$ on the column space of A, if needed, is best obtained by computing $A\mathbf{x}$, where $\mathbf{x}$ is the least squares solution of $A\mathbf{x} = \mathbf{b}$. The QR-decomposition of A is also used to find least squares solutions of $A\mathbf{x} = \mathbf{b}$.

EXAMPLE 1 Least Squares Solution

Find the least squares solution of the linear system $A\mathbf{x} = \mathbf{b}$ given by

$$
\begin{aligned}
x_1 - \ x_2 &= 4 \\
3x_1 + 2x_2 &= 1 \\
-2x_1 + 4x_2 &= 3
\end{aligned}
$$

and find the orthogonal projection of $\mathbf{b}$ on the column space of A.

Solution

Here

$$
A = \begin{bmatrix} 1 & -1 \\ 3 & 2 \\ -2 & 4 \end{bmatrix} \quad \text{and} \quad \mathbf{b} = \begin{bmatrix} 4 \\ 1 \\ 3 \end{bmatrix}
$$

Observe that A has linearly independent column vectors, so we know in advance that there is a unique least squares solution. We have

$$
A^TA = \begin{bmatrix} 1 & 3 & -2 \\ -1 & 2 & 4 \end{bmatrix} \begin{bmatrix} 1 & -1 \\ 3 & 2 \\ -2 & 4 \end{bmatrix} = \begin{bmatrix} 14 & -3 \\ -3 & 21 \end{bmatrix}
$$

$$
A^T\mathbf{b} = \begin{bmatrix} 1 & 3 & -2 \\ -1 & 2 & 4 \end{bmatrix} \begin{bmatrix} 4 \\ 1 \\ 3 \end{bmatrix} = \begin{bmatrix} 1 \\ 10 \end{bmatrix}
$$

so the normal system $A^TA\mathbf{x} = A^T\mathbf{b}$ in this case is

$$
\begin{bmatrix} 14 & -3 \\ -3 & 21 \end{bmatrix} \begin{bmatrix} x_1 \\ x_2 \end{bmatrix} = \begin{bmatrix} 1 \\ 10 \end{bmatrix}
$$

Solving this system yields the least squares solution

$$
x_1 = \tfrac{17}{95}, \qquad x_2 = \tfrac{143}{285}
$$

From Formula (5), the orthogonal projection of **b** on the column space of A is

$$A\mathbf{x} = \begin{bmatrix} 1 & -1 \\ 3 & 2 \\ -2 & 4 \end{bmatrix} \begin{bmatrix} \frac{17}{95} \\ \frac{143}{285} \end{bmatrix} = \begin{bmatrix} -\frac{92}{285} \\ \frac{439}{285} \\ \frac{94}{57} \end{bmatrix} \quad \blacklozenge$$

REMARK   The language used for least squares problems is somewhat misleading. A least squares solution of $A\mathbf{x} = \mathbf{b}$ is not in fact a solution of $A\mathbf{x} = \mathbf{b}$ unless $A\mathbf{x} = \mathbf{b}$ happens to be consistent; it is a solution of the related system $A^T A \mathbf{x} = A^T \mathbf{b}$ instead.

EXAMPLE 2 Orthogonal Projection on a Subspace

Find the orthogonal projection of the vector $\mathbf{u} = (-3, -3, 8, 9)$ on the subspace of R^4 spanned by the vectors

$$\mathbf{u}_1 = (3, 1, 0, 1), \qquad \mathbf{u}_2 = (1, 2, 1, 1), \qquad \mathbf{u}_3 = (-1, 0, 2, -1)$$

Solution

One could solve this problem by first using the Gram–Schmidt process to convert $\{\mathbf{u}_1, \mathbf{u}_2, \mathbf{u}_3\}$ into an orthonormal basis and then applying the method used in Example 6 of Section 6.3. However, the following method is more efficient.

The subspace W of R^4 spanned by $\mathbf{u}_1$, $\mathbf{u}_2$, and $\mathbf{u}_3$ is the column space of the matrix

$$A = \begin{bmatrix} 3 & 1 & -1 \\ 1 & 2 & 0 \\ 0 & 1 & 2 \\ 1 & 1 & -1 \end{bmatrix}$$

Thus, if **u** is expressed as a column vector, we can find the orthogonal projection of **u** on W by finding a least squares solution of the system $A\mathbf{x} = \mathbf{u}$ and then calculating $\text{proj}_W \mathbf{u} = A\mathbf{x}$ from the least squares solution. The computations are as follows: The system $A\mathbf{x} = \mathbf{u}$ is

$$\begin{bmatrix} 3 & 1 & -1 \\ 1 & 2 & 0 \\ 0 & 1 & 2 \\ 1 & 1 & -1 \end{bmatrix} \begin{bmatrix} x_1 \\ x_2 \\ x_3 \end{bmatrix} = \begin{bmatrix} -3 \\ -3 \\ 8 \\ 9 \end{bmatrix}$$

so

$$A^T A = \begin{bmatrix} 3 & 1 & 0 & 1 \\ 1 & 2 & 1 & 1 \\ -1 & 0 & 2 & -1 \end{bmatrix} \begin{bmatrix} 3 & 1 & -1 \\ 1 & 2 & 0 \\ 0 & 1 & 2 \\ 1 & 1 & -1 \end{bmatrix} = \begin{bmatrix} 11 & 6 & -4 \\ 6 & 7 & 0 \\ -4 & 0 & 6 \end{bmatrix}$$

$$A^T \mathbf{u} = \begin{bmatrix} 3 & 1 & 0 & 1 \\ 1 & 2 & 1 & 1 \\ -1 & 0 & 2 & -1 \end{bmatrix} \begin{bmatrix} -3 \\ -3 \\ 8 \\ 9 \end{bmatrix} = \begin{bmatrix} -3 \\ 8 \\ 10 \end{bmatrix}$$

The normal system $A^T A \mathbf{x} = A^T \mathbf{u}$ in this case is

$$\begin{bmatrix} 11 & 6 & -4 \\ 6 & 7 & 0 \\ -4 & 0 & 6 \end{bmatrix} \begin{bmatrix} x_1 \\ x_2 \\ x_3 \end{bmatrix} = \begin{bmatrix} -3 \\ 8 \\ 10 \end{bmatrix}$$

Solving this system yields

$$\mathbf{x} = \begin{bmatrix} x_1 \\ x_2 \\ x_3 \end{bmatrix} = \begin{bmatrix} -1 \\ 2 \\ 1 \end{bmatrix}$$

as the least squares solution of $A\mathbf{x} = \mathbf{u}$ (verify), so

$$\text{proj}_W \mathbf{u} = A\mathbf{x} = \begin{bmatrix} 3 & 1 & -1 \\ 1 & 2 & 0 \\ 0 & 1 & 2 \\ 1 & 1 & -1 \end{bmatrix} \begin{bmatrix} -1 \\ 2 \\ 1 \end{bmatrix} = \begin{bmatrix} -2 \\ 3 \\ 4 \\ 0 \end{bmatrix}$$

or, in horizontal notation (which is consistent with the original phrasing of the problem), $\text{proj}_W \mathbf{u} = (-2, 3, 4, 0)$. ◆

In Section 4.2 we discussed some basic orthogonal projection operators on R^2 and R^3 (Tables 4 and 5). The concept of an orthogonal projection operator can be extended to higher-dimensional Euclidean spaces as follows.

> **DEFINITION**
>
> If W is a subspace of R^m, then the transformation $P: R^m \to W$ that maps each vector $\mathbf{x}$ in R^m into its orthogonal projection $\text{proj}_W \mathbf{x}$ in W is called the ***orthogonal projection of R^m on W***.

We leave it as an exercise to show that orthogonal projections are linear operators. It follows from Formula (5) that the standard matrix for the orthogonal projection of R^m on W is

$$[P] = A(A^T A)^{-1} A^T \tag{6}$$

where A is constructed using any basis for W as its column vectors.

EXAMPLE 3 Verifying Formula (6)

In Table 5 of Section 4.2 we showed that the standard matrix for the orthogonal projection of R^3 on the xy-plane is

$$[P] = \begin{bmatrix} 1 & 0 & 0 \\ 0 & 1 & 0 \\ 0 & 0 & 0 \end{bmatrix} \tag{7}$$

To see that this is consistent with Formula (6), take the unit vectors along the positive x and y axes as a basis for the xy-plane, so that

$$A = \begin{bmatrix} 1 & 0 \\ 0 & 1 \\ 0 & 0 \end{bmatrix}$$

We leave it for the reader to verify that A^TA is the 2×2 identity matrix; thus Formula (6) simplifies to

$$[P] = AA^T = \begin{bmatrix} 1 & 0 \\ 0 & 1 \\ 0 & 0 \end{bmatrix} \begin{bmatrix} 1 & 0 & 0 \\ 0 & 1 & 0 \end{bmatrix} = \begin{bmatrix} 1 & 0 & 0 \\ 0 & 1 & 0 \\ 0 & 0 & 0 \end{bmatrix}$$

which agrees with (7). ◆

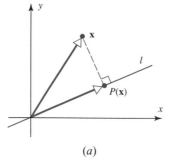

(a)

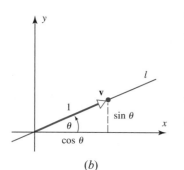

(b)

Figure 6.4.3

EXAMPLE 4 Standard Matrix for an Orthogonal Projection

Find the standard matrix for the orthogonal projection P of R^2 on the line l that passes through the origin and makes an angle θ with the positive x-axis.

Solution

The line l is a one-dimensional subspace of R^2. As illustrated in Figure 6.4.3, we can take $\mathbf{v} = (\cos\theta, \sin\theta)$ as a basis for this subspace, so

$$A = \begin{bmatrix} \cos\theta \\ \sin\theta \end{bmatrix}$$

We leave it for the reader to show that A^TA is the 1×1 identity matrix; thus Formula (6) simplifies to

$$[P] = AA^T = \begin{bmatrix} \cos\theta \\ \sin\theta \end{bmatrix} [\cos\theta \quad \sin\theta] = \begin{bmatrix} \cos^2\theta & \sin\theta\cos\theta \\ \sin\theta\cos\theta & \sin^2\theta \end{bmatrix}$$

Note that this agrees with Example 6 of Section 4.3. ◆

Summary

Theorem 6.4.3 enables us to add yet another result to Theorem 6.2.7.

THEOREM 6.4.5

Equivalent Statements

If A is an $n \times n$ matrix, and if $T_A: R^n \to R^n$ is multiplication by A, then the following are equivalent.

(a) *A is invertible.*
(b) *$A\mathbf{x} = \mathbf{0}$ has only the trivial solution.*
(c) *The reduced row-echelon form of A is I_n.*
(d) *A is expressible as a product of elementary matrices.*
(e) *$A\mathbf{x} = \mathbf{b}$ is consistent for every $n \times 1$ matrix $\mathbf{b}$.*
(f) *$A\mathbf{x} = \mathbf{b}$ has exactly one solution for every $n \times 1$ matrix $\mathbf{b}$.*
(g) *$\det(A) \neq 0$.*
(h) *The range of T_A is R^n.*
(i) *T_A is one-to-one.*
(j) *The column vectors of A are linearly independent.*
(k) *The row vectors of A are linearly independent.*
(l) *The column vectors of A span R^n.*
(m) *The row vectors of A span R^n.*
(n) *The column vectors of A form a basis for R^n.*
(o) *The row vectors of A form a basis for R^n.*

(*p*) *A has rank n.*
(*q*) *A has nullity* 0.
(*r*) *The orthogonal complement of the nullspace of A is R^n.*
(*s*) *The orthogonal complement of the row space of A is* {**0**}.
(*t*) *A^TA is invertible.*

This theorem relates all of the major topics we have studied thus far.

EXERCISE SET 6.4

1. Find the normal system associated with the given linear system.

    (a) $\begin{bmatrix} 1 & -1 \\ 2 & 3 \\ 4 & 5 \end{bmatrix} \begin{bmatrix} x_1 \\ x_2 \end{bmatrix} = \begin{bmatrix} 2 \\ -1 \\ 5 \end{bmatrix}$
    (b) $\begin{bmatrix} 2 & -1 & 0 \\ 3 & 1 & 2 \\ -1 & 4 & 5 \\ 1 & 2 & 4 \end{bmatrix} \begin{bmatrix} x_1 \\ x_2 \\ x_3 \end{bmatrix} = \begin{bmatrix} -1 \\ 0 \\ 1 \\ 2 \end{bmatrix}$

2. In each part, find $\det(A^TA)$, and apply Theorem 6.4.3 to determine whether A has linearly independent column vectors.

    (a) $A = \begin{bmatrix} -1 & 3 & 2 \\ 2 & 1 & 3 \\ 0 & 1 & 1 \end{bmatrix}$
    (b) $A = \begin{bmatrix} 2 & -1 & 3 \\ 0 & 1 & 1 \\ -1 & 0 & -2 \\ 4 & -5 & 3 \end{bmatrix}$

3. Find the least squares solution of the linear system $A\mathbf{x} = \mathbf{b}$, and find the orthogonal projection of $\mathbf{b}$ onto the column space of A.

    (a) $A = \begin{bmatrix} 1 & 1 \\ -1 & 1 \\ -1 & 2 \end{bmatrix}$, $\mathbf{b} = \begin{bmatrix} 7 \\ 0 \\ -7 \end{bmatrix}$
    (b) $A = \begin{bmatrix} 2 & -2 \\ 1 & 1 \\ 3 & 1 \end{bmatrix}$, $\mathbf{b} = \begin{bmatrix} 2 \\ -1 \\ 1 \end{bmatrix}$

    (c) $A = \begin{bmatrix} 1 & 0 & -1 \\ 2 & 1 & -2 \\ 1 & 1 & 0 \\ 1 & 1 & -1 \end{bmatrix}$, $\mathbf{b} = \begin{bmatrix} 6 \\ 0 \\ 9 \\ 3 \end{bmatrix}$
    (d) $A = \begin{bmatrix} 2 & 0 & -1 \\ 1 & -2 & 2 \\ 2 & -1 & 0 \\ 0 & 1 & -1 \end{bmatrix}$, $\mathbf{b} = \begin{bmatrix} 0 \\ 6 \\ 0 \\ 6 \end{bmatrix}$

4. Find the orthogonal projection of **u** onto the subspace of R^3 spanned by the vectors $\mathbf{v}_1$ and $\mathbf{v}_2$.

    (a) $\mathbf{u} = (2, 1, 3)$; $\mathbf{v}_1 = (1, 1, 0)$, $\mathbf{v}_2 = (1, 2, 1)$

    (b) $\mathbf{u} = (1, -6, 1)$; $\mathbf{v}_1 = (-1, 2, 1)$, $\mathbf{v}_2 = (2, 2, 4)$

5. Find the orthogonal projection of **u** onto the subspace of R^4 spanned by the vectors $\mathbf{v}_1$, $\mathbf{v}_2$, and $\mathbf{v}_3$.

    (a) $\mathbf{u} = (6, 3, 9, 6)$; $\mathbf{v}_1 = (2, 1, 1, 1)$, $\mathbf{v}_2 = (1, 0, 1, 1)$, $\mathbf{v}_3 = (-2, -1, 0, -1)$

    (b) $\mathbf{u} = (-2, 0, 2, 4)$; $\mathbf{v}_1 = (1, 1, 3, 0)$, $\mathbf{v}_2 = (-2, -1, -2, 1)$, $\mathbf{v}_3 = (-3, -1, 1, 3)$

6. Find the orthogonal projection of $\mathbf{u} = (5, 6, 7, 2)$ onto the solution space of the homogeneous linear system

    $$\begin{aligned} x_1 + x_2 + x_3 \qquad &= 0 \\ 2x_2 + x_3 + x_4 &= 0 \end{aligned}$$

7. Use Formula (6) and the method of Example 3 to find the standard matrix for the orthogonal projection $P: R^2 \to R^2$ onto

    (a) the x-axis   (b) the y-axis

    Note Compare your results to Table 4 of Section 4.2.

8. Use Formula (6) and the method of Example 3 to find the standard matrix for the orthogonal projection $P: R^3 \to R^3$ onto

   (a) the xz-plane   (b) the yz-plane

   Note Compare your results to Table 5 of Section 4.2.

9. Show that if $\mathbf{w} = (a, b, c)$ is a nonzero vector, then the standard matrix for the orthogonal projection of R^3 onto the line span$\{\mathbf{w}\}$ is

$$P = \frac{1}{a^2 + b^2 + c^2} \begin{bmatrix} a^2 & ab & ac \\ ab & b^2 & bc \\ ac & bc & c^2 \end{bmatrix}$$

10. Let W be the plane with equation $5x - 3y + z = 0$.

   (a) Find a basis for W.

   (b) Use Formula (6) to find the standard matrix for the orthogonal projection onto W.

   (c) Use the matrix obtained in (b) to find the orthogonal projection of a point $P_0(x_0, y_0, z_0)$ onto W.

   (d) Find the distance between the point $P_0(1, -2, 4)$ and the plane W, and check your result using Theorem 3.5.2.

11. Let W be the line with parametric equations

$$x = 2t, \quad y = -t, \quad z = 4t \quad (-\infty < t < \infty)$$

   (a) Find a basis for W.

   (b) Use Formula (6) to find the standard matrix for the orthogonal projection onto W.

   (c) Use the matrix obtained in (b) to find the orthogonal projection of a point $P_0(x_0, y_0, z_0)$ onto W.

   (d) Find the distance between the point $P_0(2, 1, -3)$ and the line W.

12. In R^3, consider the line l given by the equations $\{x = t, y = t, z = t\}$ and the line m given by the equations $\{x = s, y = 2s - 1, z = 1\}$. Let P be a point on l, and let Q be a point on m. Find the values of t and s that minimize the distance between the lines by minimizing the squared distance $\| P - Q \|^2$.

13. For the linear systems in Exercise 3, verify that the *error vector* $A\mathbf{x} - \mathbf{b}$ resulting from the least squares solution $\mathbf{x}$ is orthogonal to the column space of A.

14. Prove: If A has linearly independent column vectors, and if $A\mathbf{x} = \mathbf{b}$ is consistent, then the least squares solution of $A\mathbf{x} = \mathbf{b}$ and the exact solution of $A\mathbf{x} = \mathbf{b}$ are the same.

15. Prove: If A has linearly independent column vectors, and if $\mathbf{b}$ is orthogonal to the column space of A, then the least squares solution of $A\mathbf{x} = \mathbf{b}$ is $\mathbf{x} = \mathbf{0}$.

16. Let $P: R^m \to W$ be the orthogonal projection of R^m onto a subspace W.

   (a) Prove that $[P]^2 = [P]$.

   (b) What does the result in part (a) imply about the composition $P \circ P$?

   (c) Show that $[P]$ is symmetric.

   (d) Verify that the matrices in Tables 4 and 5 of Section 4.2 have the properties in parts (a) and (c).

17. Let A be an $m \times n$ matrix with linearly independent row vectors. Find a standard matrix for the orthogonal projection of R^n onto the row space of A.

   Hint Start with Formula (6).

18. The relationship between the current I through a resistor and the voltage drop V across it is given by Ohm's Law $V = IR$. Successive experiments are performed in which a known current (measured in amps) is passed through a resistor of unknown resistance R and the voltage drop (measured in volts) is measured. This results in the (I, V) data $(0.1, 1), (0.2, 2.1),$

(0.3, 2.9), (0.4, 4.2), (0.5, 5.1). The data is assumed to have measurement errors that prevent it from following Ohm's Law precisely.

(a) Set up a 5×1 linear system that represents the 5 equations $I_0 = RV_0, \ldots, I_4 = RV_4$.

(b) Is this system consistent?

(c) Find the least squares solution of this system and interpret your result.

19. Repeat Exercise 18 under the assumption that the relationship between the current I and the voltage drop V is best modeled by an equation of the form $V = IR + c$, where c is a constant offset value. This leads to a 5×2 linear system.

20. Use the techniques of Section 4.4 to fit a polynomial of degree 4 to the data of Exercise 18. Is there a physical interpretation of your result?

Discussion & Discovery

21. The following is the proof that $(b) \Rightarrow (a)$ in Theorem 6.4.3. Justify each line by filling in the blank appropriately.

Hypothesis: Suppose that A is an $m \times n$ matrix and $A^T A$ is invertible.

Conclusion: A has linearly independent column vectors.

Proof: (1) If $\mathbf{x}$ is a solution of $A\mathbf{x} = \mathbf{0}$, then $A^T A\mathbf{x} = \mathbf{0}$. _____

(2) Thus, $\mathbf{x} = \mathbf{0}$. _____

(3) Thus, the column vectors of A are linearly independent. _____

22. Let A be an $m \times n$ matrix with linearly independent column vectors, and let $\mathbf{b}$ be an $m \times 1$ matrix. Give a formula in terms of A and A^T for

(a) the vector in the column space of A that is closest to $\mathbf{b}$ relative to the Euclidean inner product;

(b) the least squares solution of $A\mathbf{x} = \mathbf{b}$ relative to the Euclidean inner product;

(c) the error in the least squares solution of $A\mathbf{x} = \mathbf{b}$ relative to the Euclidean inner product;

(d) the standard matrix for the orthogonal projection of R^m onto the column space of A relative to the Euclidean inner product.

23. Refer to Exercises 18–20. Contrast the techniques of polynomial interpolation and fitting a line by least squares. Give circumstances under which each is useful and appropriate.

6.5
CHANGE OF BASIS

A basis that is suitable for one problem may not be suitable for another, so it is a common process in the study of vector spaces to change from one basis to another. Because a basis is the vector space generalization of a coordinate system, changing bases is akin to changing coordinate axes in R^2 and R^3. In this section we shall study problems related to change of basis.

Coordinate Vectors

Recall from Theorem 5.4.1 that if $S = \{\mathbf{v}_1, \mathbf{v}_2, \ldots, \mathbf{v}_n\}$ is a basis for a vector space V, then each vector $\mathbf{v}$ in V can be expressed uniquely as a linear combination of the basis vectors, say

$$\mathbf{v} = k_1\mathbf{v}_1 + k_2\mathbf{v}_2 + \cdots + k_n\mathbf{v}_n$$

The scalars $k_1, k_2, \ldots, k_n$ are the coordinates of $\mathbf{v}$ relative to S, and the vector

$$(\mathbf{v})_S = (k_1, k_2, \ldots, k_n)$$

is the coordinate vector of **v** relative to S. In this section it will be convenient to list the coordinates as entries of an $n \times 1$ matrix. Thus we take

$$[\mathbf{v}]_S = \begin{bmatrix} k_1 \\ k_2 \\ \vdots \\ k_n \end{bmatrix}$$

to be the coordinate vector of **v** relative to S.

Change of Basis

In applications it is common to work with more than one coordinate system, and in such cases it is usually necessary to know the relationships between the coordinates of a fixed point or vector in the various coordinate systems. Since a basis is the vector space generalization of a coordinate system, we are led to consider the following problem.

Change-of-Basis Problem: If we change the basis for a vector space V from some old basis B to some new basis B', how is the old coordinate vector $[\mathbf{v}]_B$ of a vector **v** related to the new coordinate vector $[\mathbf{v}]_{B'}$?

For simplicity, we will solve this problem for two-dimensional spaces. The solution for n-dimensional spaces is similar and is left for the reader. Let

$$B = \{\mathbf{u}_1, \mathbf{u}_2\} \quad \text{and} \quad B' = \{\mathbf{u}_1', \mathbf{u}_2'\}$$

be the old and new bases, respectively. We will need the coordinate vectors for the new basis vectors relative to the old basis. Suppose they are

$$[\mathbf{u}_1']_B = \begin{bmatrix} a \\ b \end{bmatrix} \quad \text{and} \quad [\mathbf{u}_2']_B = \begin{bmatrix} c \\ d \end{bmatrix} \tag{1}$$

That is,

$$\begin{aligned} \mathbf{u}_1' &= a\mathbf{u}_1 + b\mathbf{u}_2 \\ \mathbf{u}_2' &= c\mathbf{u}_1 + d\mathbf{u}_2 \end{aligned} \tag{2}$$

Now let **v** be any vector in V, and let

$$[\mathbf{v}]_{B'} = \begin{bmatrix} k_1 \\ k_2 \end{bmatrix} \tag{3}$$

be the new coordinate vector, so that

$$\mathbf{v} = k_1\mathbf{u}_1' + k_2\mathbf{u}_2' \tag{4}$$

In order to find the old coordinates of **v**, we must express **v** in terms of the old basis B. To do this, we substitute (2) into (4). This yields

$$\mathbf{v} = k_1(a\mathbf{u}_1 + b\mathbf{u}_2) + k_2(c\mathbf{u}_1 + d\mathbf{u}_2)$$

or

$$\mathbf{v} = (k_1a + k_2c)\mathbf{u}_1 + (k_1b + k_2d)\mathbf{u}_2$$

Thus the old coordinate vector for **v** is

$$[\mathbf{v}]_B = \begin{bmatrix} k_1a + k_2c \\ k_1b + k_2d \end{bmatrix}$$

which can be written as

$$[\mathbf{v}]_B = \begin{bmatrix} a & c \\ b & d \end{bmatrix} \begin{bmatrix} k_1 \\ k_2 \end{bmatrix} \quad \text{or, from (3),} \quad [\mathbf{v}]_B = \begin{bmatrix} a & c \\ b & d \end{bmatrix} [\mathbf{v}]_{B'}$$

This equation states that the old coordinate vector $[\mathbf{v}]_B$ results when we multiply the new coordinate vector $[\mathbf{v}]_{B'}$ on the left by the matrix

$$P = \begin{bmatrix} a & c \\ b & d \end{bmatrix}$$

The columns of this matrix are the coordinates of the new basis vectors relative to the old basis [see (1)]. Thus we have the following solution of the change-of-basis problem.

Solution of the Change-of-Basis Problem: If we change the basis for a vector space V from the old basis $B = \{\mathbf{u}_1, \mathbf{u}_2, \ldots, \mathbf{u}_n\}$ to the new basis $B' = \{\mathbf{u}'_1, \mathbf{u}'_2, \ldots, \mathbf{u}'_n\}$, then the old coordinate vector $[\mathbf{v}]_B$ of a vector $\mathbf{v}$ is related to the new coordinate vector $[\mathbf{v}]_{B'}$ of the same vector $\mathbf{v}$ by the equation

$$[\mathbf{v}]_B = P[\mathbf{v}]_{B'} \tag{5}$$

where the columns of P are the coordinate vectors of the new basis vectors relative to the old basis; that is, the column vectors of P are

$$[\mathbf{u}'_1]_B, [\mathbf{u}'_2]_B, \ldots, [\mathbf{u}'_n]_B$$

Transition Matrices

The matrix P is called the ***transition matrix*** from B' to B; it can be expressed in terms of its column vectors as

$$P = \begin{bmatrix} [\mathbf{u}'_1]_B & | & [\mathbf{u}'_2]_B & | & \cdots & | & [\mathbf{u}'_n]_B \end{bmatrix} \tag{6}$$

EXAMPLE 1 Finding a Transition Matrix

Consider the bases $B = \{\mathbf{u}_1, \mathbf{u}_2\}$ and $B' = \{\mathbf{u}'_1, \mathbf{u}'_2\}$ for R^2, where

$$\mathbf{u}_1 = (1, 0); \quad \mathbf{u}_2 = (0, 1); \quad \mathbf{u}'_1 = (1, 1); \quad \mathbf{u}'_2 = (2, 1)$$

(a) Find the transition matrix from B' to B.

(b) Use (5) to find $[\mathbf{v}]_B$ if

$$[\mathbf{v}]_{B'} = \begin{bmatrix} -3 \\ 5 \end{bmatrix}$$

Solution (a)

First we must find the coordinate vectors for the new basis vectors $\mathbf{u}'_1$ and $\mathbf{u}'_2$ relative to the old basis B. By inspection,

$$\mathbf{u}'_1 = \mathbf{u}_1 + \mathbf{u}_2$$
$$\mathbf{u}'_2 = 2\mathbf{u}_1 + \mathbf{u}_2$$

so

$$[\mathbf{u}'_1]_B = \begin{bmatrix} 1 \\ 1 \end{bmatrix} \quad \text{and} \quad [\mathbf{u}'_2]_B = \begin{bmatrix} 2 \\ 1 \end{bmatrix}$$

Thus the transition matrix from B' to B is

$$P = \begin{bmatrix} 1 & 2 \\ 1 & 1 \end{bmatrix}$$

Solution (b)

Using (5) and the transition matrix in part (a) yields

$$[\mathbf{v}]_B = \begin{bmatrix} 1 & 2 \\ 1 & 1 \end{bmatrix} \begin{bmatrix} -3 \\ 5 \end{bmatrix} = \begin{bmatrix} 7 \\ 2 \end{bmatrix}$$

As a check, we should be able to recover the vector $\mathbf{v}$ either from $[\mathbf{v}]_B$ or $[\mathbf{v}]_{B'}$. We leave it for the reader to show that $-3\mathbf{u}_1' + 5\mathbf{u}_2' = 7\mathbf{u}_1 + 2\mathbf{u}_2 = \mathbf{v} = (7, 2)$. ◆

EXAMPLE 2   A Different Viewpoint on Example 1

Consider the vectors $\mathbf{u}_1 = (1, 0)$, $\mathbf{u}_2 = (0, 1)$, $\mathbf{u}_1' = (1, 1)$, $\mathbf{u}_2' = (2, 1)$. In Example 1 we found the transition matrix from the basis $B' = \{\mathbf{u}_1', \mathbf{u}_2'\}$ for R^2 to the basis $B = \{\mathbf{u}_1, \mathbf{u}_2\}$. However, we can just as well ask for the transition matrix from B to B'. To obtain this matrix, we simply change our point of view and regard B' as the old basis and B as the new basis. As usual, the columns of the transition matrix will be the coordinates of the new basis vectors relative to the old basis.

By equating corresponding components and solving the resulting linear system, the reader should be able to show that

$$\mathbf{u}_1 = -\mathbf{u}_1' + \mathbf{u}_2'$$
$$\mathbf{u}_2 = 2\mathbf{u}_1' - \mathbf{u}_2'$$

so

$$[\mathbf{u}_1]_{B'} = \begin{bmatrix} -1 \\ 1 \end{bmatrix} \quad \text{and} \quad [\mathbf{u}_2]_{B'} = \begin{bmatrix} 2 \\ -1 \end{bmatrix}$$

Thus the transition matrix from B to B' is

$$Q = \begin{bmatrix} -1 & 2 \\ 1 & -1 \end{bmatrix} \quad ◆$$

If we multiply the transition matrix from B' to B obtained in Example 1 and the transition matrix from B to B' obtained in Example 2, we find

$$PQ = \begin{bmatrix} 1 & 2 \\ 1 & 1 \end{bmatrix} \begin{bmatrix} -1 & 2 \\ 1 & -1 \end{bmatrix} = \begin{bmatrix} 1 & 0 \\ 0 & 1 \end{bmatrix} = I$$

which shows that $Q = P^{-1}$. The following theorem shows that this is not accidental.

THEOREM 6.5.1

> *If P is the transition matrix from a basis B' to a basis B for a finite-dimensional vector space V, then P is invertible, and P^{-1} is the transition matrix from B to B'.*

Proof   Let Q be the transition matrix from B to B'. We shall show that $PQ = I$ and thus conclude that $Q = P^{-1}$ to complete the proof.

Assume that $B = \{\mathbf{u}_1, \mathbf{u}_2, \ldots, \mathbf{u}_n\}$ and suppose that

$$PQ = \begin{bmatrix} c_{11} & c_{12} & \cdots & c_{1n} \\ c_{21} & c_{22} & \cdots & c_{2n} \\ \vdots & \vdots & & \vdots \\ c_{n1} & c_{n2} & \cdots & c_{nn} \end{bmatrix}$$

From (5),

$$[\mathbf{x}]_B = P[\mathbf{x}]_{B'} \quad \text{and} \quad [\mathbf{x}]_{B'} = Q[\mathbf{x}]_B$$

for all $\mathbf{x}$ in V. Multiplying the second equation through on the left by P and substituting the first gives

$$[\mathbf{x}]_B = PQ[\mathbf{x}]_B \tag{7}$$

for all $\mathbf{x}$ in V. Letting $\mathbf{x} = \mathbf{u}_1$ in (7) gives

$$\begin{bmatrix} 1 \\ 0 \\ \vdots \\ 0 \end{bmatrix} = \begin{bmatrix} c_{11} & c_{12} & \cdots & c_{1n} \\ c_{21} & c_{22} & \cdots & c_{2n} \\ \vdots & \vdots & & \vdots \\ c_{n1} & c_{n2} & \cdots & c_{nn} \end{bmatrix} \begin{bmatrix} 1 \\ 0 \\ \vdots \\ 0 \end{bmatrix} \quad \text{or} \quad \begin{bmatrix} 1 \\ 0 \\ \vdots \\ 0 \end{bmatrix} = \begin{bmatrix} c_{11} \\ c_{21} \\ \vdots \\ c_{n1} \end{bmatrix}$$

Similarly, successively substituting $\mathbf{x} = \mathbf{u}_2, \ldots, \mathbf{u}_n$ in (7) yields

$$\begin{bmatrix} c_{12} \\ c_{22} \\ \vdots \\ c_{n2} \end{bmatrix} = \begin{bmatrix} 0 \\ 1 \\ \vdots \\ 0 \end{bmatrix}, \ldots, \begin{bmatrix} c_{1n} \\ c_{2n} \\ \vdots \\ c_{nn} \end{bmatrix} = \begin{bmatrix} 0 \\ 0 \\ \vdots \\ 1 \end{bmatrix}$$

Therefore, $PQ = I$. ∎

To summarize, if P is the transition matrix from a basis B' to a basis B, then for every vector $\mathbf{v}$, the following relationships hold:

$$[\mathbf{v}]_B = P[\mathbf{v}]_{B'} \tag{8}$$

$$[\mathbf{v}]_{B'} = P^{-1}[\mathbf{v}]_B \tag{9}$$

EXERCISE SET 6.5

1. Find the coordinate vector for $\mathbf{w}$ relative to the basis $S = \{\mathbf{u}_1, \mathbf{u}_2\}$ for R^2.
   (a) $\mathbf{u}_1 = (1, 0)$, $\mathbf{u}_2 = (0, 1)$; $\mathbf{w} = (3, -7)$
   (b) $\mathbf{u}_1 = (2, -4)$, $\mathbf{u}_2 = (3, 8)$; $\mathbf{w} = (1, 1)$
   (c) $\mathbf{u}_1 = (1, 1)$, $\mathbf{u}_2 = (0, 2)$; $\mathbf{w} = (a, b)$

2. Find the coordinate vector for $\mathbf{v}$ relative to $S = \{\mathbf{v}_1, \mathbf{v}_2, \mathbf{v}_3\}$.
   (a) $\mathbf{v} = (2, -1, 3)$; $\mathbf{v}_1 = (1, 0, 0)$, $\mathbf{v}_2 = (2, 2, 0)$, $\mathbf{v}_3 = (3, 3, 3)$
   (b) $\mathbf{v} = (5, -12, 3)$; $\mathbf{v}_1 = (1, 2, 3)$, $\mathbf{v}_2 = (-4, 5, 6)$, $\mathbf{v}_3 = (7, -8, 9)$

3. Find the coordinate vector for $\mathbf{p}$ relative to $S = \{\mathbf{p}_1, \mathbf{p}_2, \mathbf{p}_3\}$.
   (a) $\mathbf{p} = 4 - 3x + x^2$; $\mathbf{p}_1 = 1$, $\mathbf{p}_2 = x$, $\mathbf{p}_3 = x^2$
   (b) $\mathbf{p} = 2 - x + x^2$; $\mathbf{p}_1 = 1 + x$, $\mathbf{p}_2 = 1 + x^2$, $\mathbf{p}_3 = x + x^2$

4. Find the coordinate vector for A relative to $S = \{A_1, A_2, A_3, A_4\}$.

$$A = \begin{bmatrix} 2 & 0 \\ -1 & 3 \end{bmatrix}, \quad A_1 = \begin{bmatrix} -1 & 1 \\ 0 & 0 \end{bmatrix}, \quad A_2 = \begin{bmatrix} 1 & 1 \\ 0 & 0 \end{bmatrix}, \quad A_3 = \begin{bmatrix} 0 & 0 \\ 1 & 0 \end{bmatrix}, \quad A_4 = \begin{bmatrix} 0 & 0 \\ 0 & 1 \end{bmatrix}$$

5. Consider the coordinate vectors

$$[\mathbf{w}]_S = \begin{bmatrix} 6 \\ -1 \\ 4 \end{bmatrix}, \quad [\mathbf{q}]_S = \begin{bmatrix} 3 \\ 0 \\ 4 \end{bmatrix}, \quad [B]_S = \begin{bmatrix} -8 \\ 7 \\ 6 \\ 3 \end{bmatrix}$$

(a) Find **w** if S is the basis in Exercise 2(a).

(b) Find **q** if S is the basis in Exercise 3(a).

(c) Find B if S is the basis in Exercise 4.

6. Consider the bases $B = \{\mathbf{u}_1, \mathbf{u}_2\}$ and $B' = \{\mathbf{v}_1, \mathbf{v}_2\}$ for R^2, where

$$\mathbf{u}_1 = \begin{bmatrix} 1 \\ 0 \end{bmatrix}, \quad \mathbf{u}_2 = \begin{bmatrix} 0 \\ 1 \end{bmatrix}, \quad \mathbf{v}_1 = \begin{bmatrix} 2 \\ 1 \end{bmatrix}, \quad \text{and} \quad \mathbf{v}_2 = \begin{bmatrix} -3 \\ 4 \end{bmatrix}$$

(a) Find the transition matrix from B' to B.

(b) Find the transition matrix from B to B'.

(c) Compute the coordinate vector $[\mathbf{w}]_B$, where

$$\mathbf{w} = \begin{bmatrix} 3 \\ -5 \end{bmatrix}$$

and use (9) to compute $[\mathbf{w}]_{B'}$.

(d) Check your work by computing $[\mathbf{w}]_{B'}$ directly.

7. Repeat the directions of Exercise 6 with the same vector **w** but with

$$\mathbf{u}_1 = \begin{bmatrix} 2 \\ 2 \end{bmatrix}, \quad \mathbf{u}_2 = \begin{bmatrix} 4 \\ -1 \end{bmatrix}, \quad \mathbf{v}_1 = \begin{bmatrix} 1 \\ 3 \end{bmatrix}, \quad \mathbf{v}_2 = \begin{bmatrix} -1 \\ -1 \end{bmatrix}$$

8. Consider the bases $B = \{\mathbf{u}_1, \mathbf{u}_2, \mathbf{u}_3\}$ and $B' = \{\mathbf{v}_1, \mathbf{v}_2, \mathbf{v}_3\}$ for R^3, where

$$\mathbf{u}_1 = \begin{bmatrix} -3 \\ 0 \\ -3 \end{bmatrix}, \quad \mathbf{u}_2 = \begin{bmatrix} -3 \\ 2 \\ -1 \end{bmatrix}, \quad \mathbf{u}_3 = \begin{bmatrix} 1 \\ 6 \\ -1 \end{bmatrix}, \quad \mathbf{v}_1 = \begin{bmatrix} -6 \\ -6 \\ 0 \end{bmatrix}, \quad \mathbf{v}_2 = \begin{bmatrix} -2 \\ -6 \\ 4 \end{bmatrix}, \quad \mathbf{v}_3 = \begin{bmatrix} -2 \\ -3 \\ 7 \end{bmatrix}$$

(a) Find the transition matrix from B to B'.

(b) Compute the coordinate vector $[\mathbf{w}]_B$, where

$$\mathbf{w} = \begin{bmatrix} -5 \\ 8 \\ -5 \end{bmatrix}$$

and use (9) to compute $[\mathbf{w}]_{B'}$.

(c) Check your work by computing $[\mathbf{w}]_{B'}$ directly.

9. Repeat the directions of Exercise 8 with the same vector **w**, but with

$$\mathbf{u}_1 = \begin{bmatrix} 2 \\ 1 \\ 1 \end{bmatrix}, \quad \mathbf{u}_2 = \begin{bmatrix} 2 \\ -1 \\ 1 \end{bmatrix}, \quad \mathbf{u}_3 = \begin{bmatrix} 1 \\ 2 \\ 1 \end{bmatrix}, \quad \mathbf{v}_1 = \begin{bmatrix} 3 \\ 1 \\ -5 \end{bmatrix}, \quad \mathbf{v}_2 = \begin{bmatrix} 1 \\ 1 \\ -3 \end{bmatrix}, \quad \mathbf{v}_3 = \begin{bmatrix} -1 \\ 0 \\ 2 \end{bmatrix}$$

10. Consider the bases $B = \{\mathbf{p}_1, \mathbf{p}_2\}$ and $B' = \{\mathbf{q}_1, \mathbf{q}_2\}$ for P_1, where

$$\mathbf{p}_1 = 6 + 3x, \quad \mathbf{p}_2 = 10 + 2x, \quad \mathbf{q}_1 = 2, \quad \mathbf{q}_2 = 3 + 2x$$

(a) Find the transition matrix from B' to B.

(b) Find the transition matrix from B to B'.

(c) Compute the coordinate vector $[\mathbf{p}]_B$, where $\mathbf{p} = -4 + x$, and use (9) to compute $[\mathbf{p}]_{B'}$.

(d) Check your work by computing $[\mathbf{p}]_{B'}$ directly.

11. Let V be the space spanned by $\mathbf{f}_1 = \sin x$ and $\mathbf{f}_2 = \cos x$.

(a) Show that $\mathbf{g}_1 = 2 \sin x + \cos x$ and $\mathbf{g}_2 = 3 \cos x$ form a basis for V.

(b) Find the transition matrix from $B' = \{\mathbf{g}_1, \mathbf{g}_2\}$ to $B = \{\mathbf{f}_1, \mathbf{f}_2\}$.

(c) Find the transition matrix from B to B'.

(d)   Compute the coordinate vector $[\mathbf{h}]_B$, where $\mathbf{h} = 2\sin x - 5\cos x$, and use (9) to obtain $[\mathbf{h}]_{B'}$.

(e)   Check your work by computing $[\mathbf{h}]_{B'}$ directly.

12.   If P is the transition matrix from a basis B' to a basis B, and Q is the transition matrix from B to a basis C, what is the transition matrix from B' to C? What is the transition matrix from C to B'?

13.   Refer to Section 4.4.

(a)   Identify the bases for P_3 used for interpolation in the standard form (found by using the Vandermonde system), the Newton form, and the Lagrange form, assuming $x_0 = -1$, $x_1 = 0$, and $x_2 = 1$.

(b)   What is the transition matrix from the Newton form basis to the standard basis?

14.   To write the coordinate vector for a vector, it is necessary to specify an order for the vectors in the basis. If P is the transition matrix from a basis B' to a basis B, what is the effect on P if we reverse the order of vectors in B from $\mathbf{v}_1, \ldots, \mathbf{v}_n$ to $\mathbf{v}_n, \ldots, \mathbf{v}_1$? What is the effect on P if we reverse the order of vectors in both B' and B?

Discussion & Discovery

15.   Consider the matrix

$$P = \begin{bmatrix} 1 & 1 & 0 \\ 1 & 0 & 2 \\ 0 & 2 & 1 \end{bmatrix}$$

(a) P is the transition matrix from what basis B to the standard basis $S = \{\mathbf{e}_1, \mathbf{e}_2, \mathbf{e}_3\}$ for R^3?

(b) P is the transition matrix from the standard basis $S = \{\mathbf{e}_1, \mathbf{e}_2, \mathbf{e}_3\}$ to what basis B for R^3?

16.   The matrix

$$P - \begin{bmatrix} 1 & 0 & 0 \\ 0 & 3 & 2 \\ 0 & 1 & 1 \end{bmatrix}$$

is the transition matrix from what basis B to the basis $\{(1, 1, 1), (1, 1, 0), (1, 0, 0)\}$ for R^3?

17.   If $[\mathbf{w}]_B = \mathbf{w}$ holds for all vectors $\mathbf{w}$ in R^n, what can you say about the basis B?

18.   Indicate whether each statement is always true or sometimes false. Justify your answer by giving a logical argument or a counterexample.

(a)   Given two bases for the same inner product space, there is always a transition matrix from one basis to the other basis.

(b)   The transition matrix from B to B is always the identify matrix.

(c)   Any invertible $n \times n$ matrix is the transition matrix for some pair of bases for R^n.

6.6 ORTHOGONAL MATRICES

In this section we shall develop properties of square matrices with orthonormal column vectors. Such matrices arise in many contexts, including problems involving a change from one orthonormal basis to another.

Matrices whose inverses can be obtained by transposition are sufficiently important that there is some terminology associated with them.

DEFINITION

A square matrix A with the property

$$A^{-1} = A^T$$

is said to be an **orthogonal matrix**.

It follows from this definition that *a square matrix A is orthogonal if and only if*

$$AA^T = A^TA = I \tag{1}$$

In fact, it follows from Theorem 1.6.3 that a square matrix A is orthogonal if *either* $AA^T = I$ or $A^TA = I$.

EXAMPLE 1   A 3 × 3 Orthogonal Matrix

The matrix

$$A = \begin{bmatrix} \frac{3}{7} & \frac{2}{7} & \frac{6}{7} \\ -\frac{6}{7} & \frac{3}{7} & \frac{2}{7} \\ \frac{2}{7} & \frac{6}{7} & -\frac{3}{7} \end{bmatrix}$$

is orthogonal, since

$$A^TA = \begin{bmatrix} \frac{3}{7} & -\frac{6}{7} & \frac{2}{7} \\ \frac{2}{7} & \frac{3}{7} & \frac{6}{7} \\ \frac{6}{7} & \frac{2}{7} & -\frac{3}{7} \end{bmatrix} \begin{bmatrix} \frac{3}{7} & \frac{2}{7} & \frac{6}{7} \\ -\frac{6}{7} & \frac{3}{7} & \frac{2}{7} \\ \frac{2}{7} & \frac{6}{7} & -\frac{3}{7} \end{bmatrix} = \begin{bmatrix} 1 & 0 & 0 \\ 0 & 1 & 0 \\ 0 & 0 & 1 \end{bmatrix} \quad \blacklozenge$$

EXAMPLE 2   A Rotation Matrix Is Orthogonal

Recall from Table 6 of Section 4.2 that the standard matrix for the counterclockwise rotation of R^2 through an angle θ is

$$A = \begin{bmatrix} \cos\theta & -\sin\theta \\ \sin\theta & \cos\theta \end{bmatrix}$$

This matrix is orthogonal for all choices of θ, since

$$A^TA = \begin{bmatrix} \cos\theta & \sin\theta \\ -\sin\theta & \cos\theta \end{bmatrix} \begin{bmatrix} \cos\theta & -\sin\theta \\ \sin\theta & \cos\theta \end{bmatrix} = \begin{bmatrix} 1 & 0 \\ 0 & 1 \end{bmatrix}$$

In fact, it is a simple matter to check that all of the "reflection matrices" in Tables 2 and 3 and all of the "rotation matrices" in Tables 6 and 7 of Section 4.2 are orthogonal matrices. $\blacklozenge$

Observe that for the orthogonal matrices in Examples 1 and 2, both the row vectors and the column vectors form orthonormal sets with respect to the Euclidean inner product (verify). This is not accidental; it is a consequence of the following theorem.

THEOREM 6.6.1

The following are equivalent for an n × n matrix A.

(a) *A is orthogonal.*

(b) *The row vectors of A form an orthonormal set in R^n with the Euclidean inner product.*

(c) *The column vectors of A form an orthonormal set in R^n with the Euclidean inner product.*

Proof We shall prove the equivalence of (*a*) and (*b*) and leave the equivalence of (*a*) and (*c*) as an exercise.

(a) ⇔ **(b)** The entry in the ith row and jth column of the matrix product AA^T is the dot product of the ith row vector of A and the jth column vector of A^T. But except for a difference in notation, the jth column vector of A^T is the jth row vector of A. Thus, if the row vectors of A are $\mathbf{r}_1, \mathbf{r}_2, \ldots, \mathbf{r}_n$, then the matrix product AA^T can be expressed as

$$AA^T = \begin{bmatrix} \mathbf{r}_1 \cdot \mathbf{r}_1 & \mathbf{r}_1 \cdot \mathbf{r}_2 & \cdots & \mathbf{r}_1 \cdot \mathbf{r}_n \\ \mathbf{r}_2 \cdot \mathbf{r}_1 & \mathbf{r}_2 \cdot \mathbf{r}_2 & \cdots & \mathbf{r}_2 \cdot \mathbf{r}_n \\ \vdots & \vdots & & \vdots \\ \mathbf{r}_n \cdot \mathbf{r}_1 & \mathbf{r}_n \cdot \mathbf{r}_2 & \cdots & \mathbf{r}_n \cdot \mathbf{r}_n \end{bmatrix}$$

Thus $AA^T = I$ if and only if

$$\mathbf{r}_1 \cdot \mathbf{r}_1 = \mathbf{r}_2 \cdot \mathbf{r}_2 = \cdots = \mathbf{r}_n \cdot \mathbf{r}_n = 1$$

and

$$\mathbf{r}_i \cdot \mathbf{r}_j = 0 \quad \text{when } i \neq j$$

which are true if and only if $\{\mathbf{r}_1, \mathbf{r}_2, \ldots, \mathbf{r}_n\}$ is an orthonormal set in R^n. ∎

REMARK In light of Theorem 6.6.1, it would seem more appropriate to call orthogonal matrices *orthonormal matrices*. However, we will not do so in deference to historical tradition.

The following theorem lists some additional fundamental properties of orthogonal matrices. The proofs are all straightforward and are left for the reader.

THEOREM 6.6.2

(a) *The inverse of an orthogonal matrix is orthogonal.*

(b) *A product of orthogonal matrices is orthogonal.*

(c) *If A is orthogonal, then* $\det(A) = 1$ *or* $\det(A) = -1$.

EXAMPLE 3 det(A) = ±1 for an Orthogonal Matrix A

The matrix

$$A = \begin{bmatrix} 1/\sqrt{2} & 1/\sqrt{2} \\ -1/\sqrt{2} & 1/\sqrt{2} \end{bmatrix}$$

is orthogonal since its row (and column) vectors form orthonormal sets in R^2. We leave it for the reader to check that $\det(A) = 1$. Interchanging the rows produces an orthogonal matrix for which $\det(A) = -1$. ◆

Orthogonal Matrices as Linear Operators

We observed in Example 2 that the standard matrices for the basic reflection and rotation operators on R^2 and R^3 are orthogonal. The next theorem will help explain why this is so.

THEOREM 6.6.3

> *If A is an $n \times n$ matrix, then the following are equivalent.*
>
> *(a)* *A is orthogonal*
> *(b)* $|A\mathbf{x}\| = \|\mathbf{x}\|$ *for all $\mathbf{x}$ in R^n.*
> *(c)* $A\mathbf{x} \cdot A\mathbf{y} = \mathbf{x} \cdot \mathbf{y}$ *for all $\mathbf{x}$ and $\mathbf{y}$ in R^n.*

Proof We shall prove the sequence of implications $(a) \Rightarrow (b) \Rightarrow (c) \Rightarrow (a)$.

$(a) \Rightarrow (b)$ Assume that A is orthogonal, so that $A^T A = I$. Then, from Formula (8) of Section 4.1,

$$\|A\mathbf{x}\| = (A\mathbf{x} \cdot A\mathbf{x})^{1/2} = (\mathbf{x} \cdot A^T A\mathbf{x})^{1/2} = (\mathbf{x} \cdot \mathbf{x})^{1/2} = \|\mathbf{x}\|$$

$(b) \Rightarrow (c)$ Assume that $\|A\mathbf{x}\| = \|\mathbf{x}\|$ for all $\mathbf{x}$ in R^n. From Theorem 4.1.6 we have

$$A\mathbf{x} \cdot A\mathbf{y} = \tfrac{1}{4}\|A\mathbf{x} + A\mathbf{y}\|^2 - \tfrac{1}{4}\|A\mathbf{x} - A\mathbf{y}\|^2 = \tfrac{1}{4}\|A(\mathbf{x} + \mathbf{y})\|^2 - \tfrac{1}{4}\|A(\mathbf{x} - \mathbf{y})\|^2$$
$$= \tfrac{1}{4}\|\mathbf{x} + \mathbf{y}\|^2 - \tfrac{1}{4}\|\mathbf{x} - \mathbf{y}\|^2 = \mathbf{x} \cdot \mathbf{y}$$

$(c) \Rightarrow (a)$ Assume that $A\mathbf{x} \cdot A\mathbf{y} = \mathbf{x} \cdot \mathbf{y}$ for all $\mathbf{x}$ and $\mathbf{y}$ in R^n. Then, from Formula (8) of Section 4.1, we have

$$\mathbf{x} \cdot \mathbf{y} = \mathbf{x} \cdot A^T A\mathbf{y}$$

which can be rewritten as

$$\mathbf{x} \cdot (A^T A\mathbf{y} - \mathbf{y}) = 0 \quad \text{or} \quad \mathbf{x} \cdot (A^T A - I)\mathbf{y} = 0$$

Since this holds for all $\mathbf{x}$ in R^n, it holds in particular if

$$\mathbf{x} = (A^T A - I)\mathbf{y}, \quad \text{so} \quad (A^T A - I)\mathbf{y} \cdot (A^T A - I)\mathbf{y} = 0$$

from which we can conclude that

$$(A^T A - I)\mathbf{y} = \mathbf{0} \tag{2}$$

(why?). Thus (2) is a homogeneous system of linear equations that is satisfied by every $\mathbf{y}$ in R^n. But this implies that the coefficient matrix must be zero (why?), so $A^T A = I$ and, consequently, A is orthogonal. ∎

If $T: R^n \to R^n$ is multiplication by an orthogonal matrix A, then T is called an **orthogonal operator** on R^n. It follows from parts (a) and (b) of the preceding theorem that the orthogonal operators on R^n are precisely those operators that leave the lengths of all vectors unchanged. Since reflections and rotations of R^2 and R^3 have this property, this explains our observation in Example 2 that the standard matrices for the basic reflections and rotations of R^2 and R^3 are orthogonal.

Change of Orthonormal Basis

The following theorem shows that in an inner product space, the transition matrix from one orthonormal basis to another is orthogonal.

THEOREM 6.6.4

> *If P is the transition matrix from one orthonormal basis to another orthonormal basis for an inner product space, then P is an orthogonal matrix; that is,*
>
> $$P^{-1} = P^T$$

Proof Assume that V is an n-dimensional inner product space and that P is the transition matrix from an orthonormal basis B' to an orthonormal basis B. To prove that P is orthogonal, we shall use Theorem 6.6.3 and show that $\| P\mathbf{x} \| = \| \mathbf{x} \|$ for every vector $\mathbf{x}$ in R^n.

Recall from Theorem 6.3.2*a* that for any orthonormal basis for V, the norm of any vector $\mathbf{u}$ in V is the same as the norm of its coordinate vector in R^n with respect to the Euclidean inner product. Thus for any vector $\mathbf{u}$ in V, we have

$$\| \mathbf{u} \| = \| [\mathbf{u}]_{B'} \| = \| [\mathbf{u}]_B \|$$

or

$$\| \mathbf{u} \| = \| [\mathbf{u}]_{B'} \| = \| P[\mathbf{u}]_{B'} \| \tag{3}$$

where the first norm is with respect to the inner product on V and the second and third are with respect to the Euclidean inner product on R^n.

Now let $\mathbf{x}$ be any vector in R^n, and let $\mathbf{u}$ be the vector in V whose coordinate vector with respect to the basis B' is $\mathbf{x}$; that is, $[\mathbf{u}]_{B'} = \mathbf{x}$. Thus, from (3),

$$\| \mathbf{u} \| = \| \mathbf{x} \| = \| P\mathbf{x} \|$$

which proves that P is orthogonal. ∎

EXAMPLE 4 Application to Rotation of Axes in 2-Space

In many problems a rectangular xy-coordinate system is given, and a new $x'y'$-coordinate system is obtained by rotating the xy-system counterclockwise about the origin through an angle θ. When this is done, each point Q in the plane has two sets of coordinates: coordinates (x, y) relative to the xy-system and coordinates (x', y') relative to the $x'y'$-system (Figure 6.6.1*a*).

By introducing unit vectors $\mathbf{u}_1$ and $\mathbf{u}_2$ along the positive x- and y-axes and unit vectors $\mathbf{u}_1'$ and $\mathbf{u}_2'$ along the positive x'- and y'-axes, we can regard this rotation as a change from an old basis $B = \{\mathbf{u}_1, \mathbf{u}_2\}$ to a new basis $B' = \{\mathbf{u}_1', \mathbf{u}_2'\}$ (Figure 6.6.1*b*). Thus, the new coordinates (x', y') and the old coordinates (x, y) of a point Q will be related by

$$\begin{bmatrix} x' \\ y' \end{bmatrix} = P^{-1} \begin{bmatrix} x \\ y \end{bmatrix} \tag{4}$$

where P is the transition from B' to B. To find P we must determine the coordinate matrices of the new basis vectors $\mathbf{u}_1'$ and $\mathbf{u}_2'$ relative to the old basis. As indicated in Figure 6.6.1*c*, the components of $\mathbf{u}_1'$ in the old basis are $\cos \theta$ and $\sin \theta$, so

$$[\mathbf{u}_1']_B = \begin{bmatrix} \cos \theta \\ \sin \theta \end{bmatrix}$$

Similarly, from Figure 6.6.1*d*, we see that the components of $\mathbf{u}_2'$ in the old basis are $\cos(\theta + \pi/2) = -\sin \theta$ and $\sin(\theta + \pi/2) = \cos \theta$, so

$$[\mathbf{u}_2']_B = \begin{bmatrix} -\sin \theta \\ \cos \theta \end{bmatrix}$$

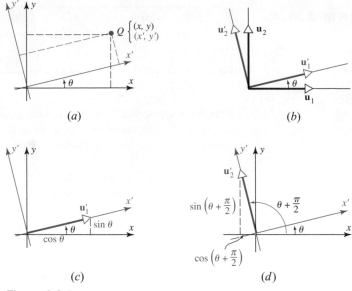

Figure 6.6.1

Thus the transition matrix from B' to B is

$$P = \begin{bmatrix} \cos\theta & -\sin\theta \\ \sin\theta & \cos\theta \end{bmatrix}$$

Observe that P is an orthogonal matrix, as expected, since B and B' are orthonormal bases. Thus

$$P^{-1} = P^T = \begin{bmatrix} \cos\theta & \sin\theta \\ -\sin\theta & \cos\theta \end{bmatrix}$$

so (4) yields

$$\begin{bmatrix} x' \\ y' \end{bmatrix} = \begin{bmatrix} \cos\theta & \sin\theta \\ -\sin\theta & \cos\theta \end{bmatrix} \begin{bmatrix} x \\ y \end{bmatrix} \tag{5}$$

or, equivalently,

$$\begin{aligned} x' &= x\cos\theta + y\sin\theta \\ y' &= -x\sin\theta + y\cos\theta \end{aligned}$$

For example, if the axes are rotated $\theta = \pi/4$, then since

$$\sin\frac{\pi}{4} = \cos\frac{\pi}{4} = \frac{1}{\sqrt{2}}$$

Equation (5) becomes

$$\begin{bmatrix} x' \\ y' \end{bmatrix} = \begin{bmatrix} \dfrac{1}{\sqrt{2}} & \dfrac{1}{\sqrt{2}} \\ -\dfrac{1}{\sqrt{2}} & \dfrac{1}{\sqrt{2}} \end{bmatrix} \begin{bmatrix} x \\ y \end{bmatrix}$$

Thus, if the old coordinates of a point Q are $(x, y) = (2, -1)$, then

$$\begin{bmatrix} x' \\ y' \end{bmatrix} = \begin{bmatrix} \dfrac{1}{\sqrt{2}} & \dfrac{1}{\sqrt{2}} \\ -\dfrac{1}{\sqrt{2}} & \dfrac{1}{\sqrt{2}} \end{bmatrix} \begin{bmatrix} 2 \\ -1 \end{bmatrix} = \begin{bmatrix} \dfrac{1}{\sqrt{2}} \\ -\dfrac{3}{\sqrt{2}} \end{bmatrix}$$

so the new coordinates of Q are $(x', y') = (1/\sqrt{2}, -3/\sqrt{2})$. ◆

REMARK Observe that the coefficient matrix in (5) is the same as the standard matrix for the linear operator that rotates the vectors of R^2 through the angle $-\theta$ (Table 6 of Section 4.2). This is to be expected since rotating the coordinate axes through the angle θ with the vectors of R^2 kept fixed has the same effect as rotating the vectors through the angle $-\theta$ with the axes kept fixed.

EXAMPLE 5 Application to Rotation of Axes in 3-Space

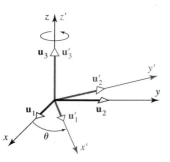

Figure 6.6.2

Suppose that a rectangular xyz-coordinate system is rotated around its z-axis counterclockwise (looking down the positive z-axis) through an angle θ (Figure 6.6.2). If we introduce unit vectors $\mathbf{u}_1$, $\mathbf{u}_2$, and $\mathbf{u}_3$ along the positive x-, y-, and z-axes and unit vectors $\mathbf{u}_1'$, $\mathbf{u}_2'$, and $\mathbf{u}_3'$ along the positive x', y', and z' axes, we can regard the rotation as a change from the old basis $B = \{\mathbf{u}_1, \mathbf{u}_2, \mathbf{u}_3\}$ to the new basis $B' = \{\mathbf{u}_1', \mathbf{u}_2', \mathbf{u}_3'\}$. In light of Example 4, it should be evident that

$$[\mathbf{u}_1']_B = \begin{bmatrix} \cos\theta \\ \sin\theta \\ 0 \end{bmatrix} \quad \text{and} \quad [\mathbf{u}_2']_B = \begin{bmatrix} -\sin\theta \\ \cos\theta \\ 0 \end{bmatrix}$$

Moreover, since $\mathbf{u}_3'$ extends 1 unit up the positive z'-axis,

$$[\mathbf{u}_3']_B = \begin{bmatrix} 0 \\ 0 \\ 1 \end{bmatrix}$$

Thus the transition matrix from B' to B is

$$P = \begin{bmatrix} \cos\theta & -\sin\theta & 0 \\ \sin\theta & \cos\theta & 0 \\ 0 & 0 & 1 \end{bmatrix}$$

and the transition matrix from B to B' is

$$P^{-1} = \begin{bmatrix} \cos\theta & \sin\theta & 0 \\ -\sin\theta & \cos\theta & 0 \\ 0 & 0 & 1 \end{bmatrix}$$

(verify). Thus the new coordinates (x', y', z') of a point Q can be computed from its old coordinates (x, y, z) by

$$\begin{bmatrix} x' \\ y' \\ z' \end{bmatrix} = \begin{bmatrix} \cos\theta & \sin\theta & 0 \\ -\sin\theta & \cos\theta & 0 \\ 0 & 0 & 1 \end{bmatrix} \begin{bmatrix} x \\ y \\ z \end{bmatrix} \qquad \blacklozenge$$

EXERCISE SET
6.6

1. (a) Show that the matrix

$$A = \begin{bmatrix} \frac{4}{5} & 0 & -\frac{3}{5} \\ -\frac{9}{25} & \frac{4}{5} & -\frac{12}{25} \\ \frac{12}{25} & \frac{3}{5} & \frac{16}{25} \end{bmatrix}$$

is orthogonal in three ways: by calculating $A^T A$, by using part (*b*) of Theorem 6.6.1, and by using part (*c*) of Theorem 6.6.1.

(b) Find the inverse of the matrix A in part (a).

2. (a) Show that the matrix

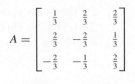

$$A = \begin{bmatrix} \frac{1}{3} & \frac{2}{3} & \frac{2}{3} \\ \frac{2}{3} & -\frac{2}{3} & \frac{1}{3} \\ -\frac{2}{3} & -\frac{1}{3} & \frac{2}{3} \end{bmatrix}$$

is orthogonal.

(b) Let $T: R^3 \to R^3$ be multiplication by the matrix A in part (a). Find $T(\mathbf{x})$ for the vector $\mathbf{x} = (-2, 3, 5)$. Using the Euclidean inner product on R^3, verify that $\|T(\mathbf{x})\| = \|\mathbf{x}\|$.

3. Determine which of the following matrices are orthogonal. For those that are orthogonal, find the inverse.

(a) $\begin{bmatrix} 1 & 0 \\ 0 & 1 \end{bmatrix}$

(b) $\begin{bmatrix} 1/\sqrt{2} & -1/\sqrt{2} \\ 1/\sqrt{2} & 1/\sqrt{2} \end{bmatrix}$

(c) $\begin{bmatrix} 0 & 1 & 1/\sqrt{2} \\ 1 & 0 & 0 \\ 0 & 0 & 1/\sqrt{2} \end{bmatrix}$

(d) $\begin{bmatrix} -1/\sqrt{2} & 1/\sqrt{6} & 1/\sqrt{3} \\ 0 & -2/\sqrt{6} & 1/\sqrt{3} \\ 1/\sqrt{2} & 1/\sqrt{6} & 1/\sqrt{3} \end{bmatrix}$

(e) $\begin{bmatrix} \frac{1}{2} & \frac{1}{2} & \frac{1}{2} & \frac{1}{2} \\ \frac{1}{2} & -\frac{5}{6} & \frac{1}{6} & \frac{1}{6} \\ \frac{1}{2} & \frac{1}{6} & \frac{1}{6} & -\frac{5}{6} \\ \frac{1}{2} & \frac{1}{6} & -\frac{5}{6} & \frac{1}{6} \end{bmatrix}$

(f) $\begin{bmatrix} 1 & 0 & 0 & 0 \\ 0 & 1/\sqrt{3} & -1/2 & 0 \\ 0 & 1/\sqrt{3} & 0 & 1 \\ 0 & 1/\sqrt{3} & 1/2 & 0 \end{bmatrix}$

4. (a) Show that if A is orthogonal, then A^T is orthogonal.

(b) What is the normal system for $A\mathbf{x} = \mathbf{b}$ when A is orthogonal?

5. Verify that the reflection matrices in Tables 2 and 3 of Section 4.2 are orthogonal.

6. Let a rectangular $x'y'$-coordinate system be obtained by rotating a rectangular xy-coordinate system counterclockwise through the angle $\theta = 3\pi/4$.

(a) Find the $x'y'$-coordinates of the point whose xy-coordinates are $(-2, 6)$.

(b) Find the xy-coordinates of the point whose $x'y'$-coordinates are $(5, 2)$.

7. Repeat Exercise 6 with $\theta = \pi/3$.

8. Let a rectangular $x'y'z'$-coordinate system be obtained by rotating a rectangular xyz-coordinate system counterclockwise about the z-axis (looking down the z-axis) through the angle $\theta = \pi/4$.

(a) Find the $x'y'z'$-coordinates of the point whose xyz-coordinates are $(-1, 2, 5)$.

(b) Find the xyz-coordinates of the point whose $x'y'z'$-coordinates are $(1, 6, -3)$.

9. Repeat Exercise 8 for a rotation of $\theta = \pi/3$ counterclockwise about the y-axis (looking along the positive y-axis toward the origin).

10. Repeat Exercise 8 for a rotation of $\theta = 3\pi/4$ counterclockwise about the x-axis (looking along the positive x-axis toward the origin).

11. (a) A rectangular $x'y'z'$-coordinate system is obtained by rotating an xyz-coordinate system counterclockwise about the y-axis through an angle θ (looking along the positive y-axis toward the origin). Find a matrix A such that

$$\begin{bmatrix} x' \\ y' \\ z' \end{bmatrix} = A \begin{bmatrix} x \\ y \\ z \end{bmatrix}$$

where (x, y, z) and (x', y', z') are the coordinates of the same point in the xyz- and $x'y'z'$-systems, respectively.

(b) Repeat part (a) for a rotation about the x-axis.

12. A rectangular $x''y''z''$-coordinate system is obtained by first rotating a rectangular xyz-coordinate system 60° counterclockwise about the z-axis (looking down the positive z-axis) to obtain an $x'y'z'$-coordinate system, and then rotating the $x'y'z'$-coordinate system 45° counterclockwise about the y'-axis (looking along the positive y'-axis toward the origin). Find a matrix A such that

$$\begin{bmatrix} x'' \\ y'' \\ z'' \end{bmatrix} = A \begin{bmatrix} x \\ y \\ z \end{bmatrix}$$

where (x, y, z) and (x'', y'', z'') are the xyz- and $x''y''z''$-coordinates of the same point.

13. What conditions must a and b satisfy for the matrix

$$\begin{bmatrix} a+b & b-a \\ a-b & b+a \end{bmatrix}$$

to be orthogonal?

14. Prove that a 2×2 orthogonal matrix A has one of two possible forms:

$$A = \begin{bmatrix} \cos\theta & -\sin\theta \\ \sin\theta & \cos\theta \end{bmatrix} \quad \text{or} \quad A = \begin{bmatrix} \cos\theta & \sin\theta \\ \sin\theta & -\cos\theta \end{bmatrix}$$

where $0 \le \theta < 2\pi$.

Hint Start with a general 2×2 matrix $A = (a_{ij})$, and use the fact that the column vectors form an orthonormal set in R^2.

15. (a) Use the result in Exercise 14 to prove that multiplication by a 2×2 orthogonal matrix is either a rotation or a rotation followed by a reflection about the x-axis.

   (b) Show that multiplication by A is a rotation if $\det(A) = 1$ and that a rotation followed by a reflection if $\det(A) = -1$.

16. Use the result in Exercise 15 to determine whether multiplication by A is a rotation or a rotation followed by a reflection about the x-axis. Find the angle of rotation in either case.

   (a) $A = \begin{bmatrix} -1/\sqrt{2} & 1/\sqrt{2} \\ -1/\sqrt{2} & -1/\sqrt{2} \end{bmatrix}$ (b) $A = \begin{bmatrix} -1/2 & \sqrt{3}/2 \\ \sqrt{3}/2 & 1/2 \end{bmatrix}$

17. The result in Exercise 15 has an analog for 3×3 orthogonal matrices: It can be proved that multiplication by a 3×3 orthogonal matrix A is a rotation about some axis if $\det(A) = 1$ and is a rotation about some axis followed by a reflection about some coordinate plane if $\det(A) = -1$. Determine whether multiplication by A is a rotation or a rotation followed by a reflection.

   (a) $A = \begin{bmatrix} \frac{3}{7} & \frac{2}{7} & \frac{6}{7} \\ -\frac{6}{7} & \frac{3}{7} & \frac{2}{7} \\ \frac{2}{7} & \frac{6}{7} & -\frac{3}{7} \end{bmatrix}$ (b) $A = \begin{bmatrix} \frac{2}{7} & \frac{3}{7} & \frac{6}{7} \\ \frac{3}{7} & -\frac{6}{7} & \frac{2}{7} \\ \frac{6}{7} & \frac{2}{7} & -\frac{3}{7} \end{bmatrix}$

18. Use the fact stated in Exercise 17 and part (*b*) of Theorem 6.6.2 to show that a composition of rotations can always be accomplished by a single rotation about some appropriate axis.

19. Prove the equivalence of statements (*a*) and (*c*) in Theorem 6.6.1.

Discussion & Discovery

20. A linear operator on R^2 is called ***rigid*** if it does not change the lengths of vectors, and it is called ***angle preserving*** if it does not change the angle between nonzero vectors.

   (a) Name two different types of linear operators that are rigid.

   (b) Name two different types of linear operators that are angle preserving.

   (c) Are there any linear operators on R^2 that are rigid and not angle preserving? Angle preserving and not rigid? Justify your answer.

21. Referring to Exercise 20, what can you say about $\det(A)$ if A is the standard matrix for a rigid linear operator on R^2?

22. Find a, b, and c such that the matrix

$$A = \begin{bmatrix} a & 1/\sqrt{2} & -1/\sqrt{2} \\ b & 1/\sqrt{6} & 1/\sqrt{6} \\ c & 1/\sqrt{3} & 1/\sqrt{3} \end{bmatrix}$$

is orthogonal. Are the values of a, b, and c unique? Explain.

CHAPTER 6
Supplementary Exercises

1. Let R^4 have the Euclidean inner product.

    (a) Find a vector in R^4 that is orthogonal to $\mathbf{u}_1 = (1, 0, 0, 0)$ and $\mathbf{u}_4 = (0, 0, 0, 1)$ and makes equal angles with $\mathbf{u}_2 = (0, 1, 0, 0)$ and $\mathbf{u}_3 = (0, 0, 1, 0)$.

    (b) Find a vector $\mathbf{x} = (x_1, x_2, x_3, x_4)$ of length 1 that is orthogonal to $\mathbf{u}_1$ and $\mathbf{u}_4$ above and such that the cosine of the angle between $\mathbf{x}$ and $\mathbf{u}_2$ is twice the cosine of the angle between $\mathbf{x}$ and $\mathbf{u}_3$.

2. Show that if $\mathbf{x}$ is a nonzero column vector in R^n, then the $n \times n$ matrix

$$A = I_n - \frac{2}{\|\mathbf{x}\|^2} \mathbf{x}\mathbf{x}^T$$

is both orthogonal and symmetric.

3. Let $A\mathbf{x} = \mathbf{0}$ be a system of m equations in n unknowns. Show that

$$\mathbf{x} = \begin{bmatrix} x_1 \\ x_2 \\ \vdots \\ x_n \end{bmatrix}$$

is a solution of the system if and only if the vector $\mathbf{x} = (x_1, x_2, \ldots, x_n)$ is orthogonal to every row vector of A in the Euclidean inner product on R^n.

4. Use the Cauchy–Schwarz inequality to show that if $a_1, a_2, \ldots, a_n$ are positive real numbers, then

$$(a_1 + a_2 + \cdots + a_n)\left(\frac{1}{a_1} + \frac{1}{a_2} + \cdots + \frac{1}{a_n}\right) \geq n^2$$

5. Show that if $\mathbf{x}$ and $\mathbf{y}$ are vectors in an inner product space and c is any scalar, then

$$\|c\mathbf{x} + \mathbf{y}\|^2 = c^2\|\mathbf{x}\|^2 + 2c\langle \mathbf{x}, \mathbf{y} \rangle + \|\mathbf{y}\|^2$$

6. Let R^3 have the Euclidean inner product. Find two vectors of length 1 that are orthogonal to all three of the vectors $\mathbf{u}_1 = (1, 1, -1)$, $\mathbf{u}_2 = (-2, -1, 2)$, and $\mathbf{u}_3 = (-1, 0, 1)$.

7. Find a weighted Euclidean inner product on R^n such that the vectors

$$\mathbf{v}_1 = (1, 0, 0, \ldots, 0)$$
$$\mathbf{v}_2 = (0, \sqrt{2}, 0, \ldots, 0)$$
$$\mathbf{v}_3 = (0, 0, \sqrt{3}, \ldots, 0)$$
$$\vdots$$
$$\mathbf{v}_n = (0, 0, 0, \ldots, \sqrt{n})$$

form an orthonormal set.

8. Is there a weighted Euclidean inner product on R^2 for which the vectors $(1, 2)$ and $(3, -1)$ form an orthonormal set? Justify your answer.

9. Prove: If Q is an orthogonal matrix, then each entry of Q is the same as its cofactor if $\det(Q) = 1$ and is the negative of its cofactor if $\det(Q) = -1$.

10. If **u** and **v** are vectors in an inner product space V, then **u**, **v**, and $\mathbf{u} - \mathbf{v}$ can be regarded as sides of a "triangle" in V (see the accompanying figure). Prove that the law of cosines holds for any such triangle; that is, $\|\mathbf{u} - \mathbf{v}\|^2 = \|\mathbf{u}\|^2 + \|\mathbf{v}\|^2 - 2\|\mathbf{u}\|\|\mathbf{v}\|\cos\theta$, where θ is the angle between **u** and **v**.

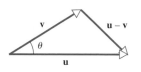

Figure Ex-10

11. (a) In R^3 the vectors $(k, 0, 0)$, $(0, k, 0)$, and $(0, 0, k)$ form the edges of a cube with diagonal (k, k, k) (Figure 3.3.4). Similarly, in R^n the vectors

$$(k, 0, 0, \ldots, 0), \quad (0, k, 0, \ldots, 0), \ldots, \quad (0, 0, 0, \ldots, k)$$

can be regarded as edges of a "cube" with diagonal $(k, k, k, \ldots, k)$. Show that each of the above edges makes an angle of θ with the diagonal, where $\cos\theta = 1/\sqrt{n}$.

(b) **(For Readers Who Have Studied Calculus)** What happens to the angle θ in part (a) as the dimension of R^n approaches $+\infty$?

12. Let **u** and **v** be vectors in an inner product space.

(a) Prove that $\|\mathbf{u}\| = \|\mathbf{v}\|$ if and only if $\mathbf{u} + \mathbf{v}$ and $\mathbf{u} - \mathbf{v}$ are orthogonal.

(b) Give a geometric interpretation of this result in R^2 with the Euclidean inner product.

13. Let **u** be a vector in an inner product space V, and let $\{\mathbf{v}_1, \mathbf{v}_2, \ldots, \mathbf{v}_n\}$ be an orthonormal basis for V. Show that if α_i is the angle between **u** and $\mathbf{v}_i$, then

$$\cos^2\alpha_1 + \cos^2\alpha_2 + \cdots + \cos^2\alpha_n = 1$$

14. Prove: If $\langle\mathbf{u}, \mathbf{v}\rangle_1$ and $\langle\mathbf{u}, \mathbf{v}\rangle_2$ are two inner products on a vector space V, then the quantity $\langle\mathbf{u}, \mathbf{v}\rangle = \langle\mathbf{u}, \mathbf{v}\rangle_1 + \langle\mathbf{u}, \mathbf{v}\rangle_2$ is also an inner product.

15. Show that the inner product on R^n generated by any orthogonal matrix is the Euclidean inner product.

16. Prove part (c) of Theorem 6.2.5.

CHAPTER 6
Technology Exercises

The following exercises are designed to be solved using a technology utility. Typically, this will be MATLAB, *Mathematica*, Maple, Derive, or Mathcad, but it may also be some other type of linear algebra software or a scientific calculator with some linear algebra capabilities. For each exercise you will need to read the relevant documentation for the particular utility you are using. The goal of these exercises is to provide you with a basic proficiency with your technology utility. Once you have mastered the techniques in these exercises, you will be able to use your technology utility to solve many of the problems in the regular exercise sets.

Section 6.1

T1. **(Weighted Euclidean Inner Products)** See if you can program your utility so that it produces the value of a weighted Euclidean inner product when the user enters n, the weights, and the vectors. Check your work by having the program do some specific computations.

T2. **(Inner Product on M_{22})** See if you can program your utility to produce the inner product in Example 7 when the user enters the matrices U and V. Check your work by having the program do some specific computations.

T3. **(Inner Product on $C[a, b]$)** If you are using a CAS or a technology utility that can do numerical integration, see if you can program the utility to compute the inner product given in Example 9 when the user enters a, b, and the functions $f(x)$ and $g(x)$. Check your work by having the program do some specific calculations.

Section 6.3

T1. **(Normalizing a Vector)** See if you can create a program that will normalize a nonzero vector **v** in R^n when the user enters **v**.

T2. **(Gram–Schmidt Process)** Read your documentation on performing the Gram–Schmidt process, and then use your utility to perform the computations in Example 7.

T3. (*QR*-**decomposition**) Read your documentation on performing the Gram–Schmidt process, and then use your utility to perform the computations in Example 8.

Section 6.4 **T1.** (**Least Squares**) Read your documentation on finding least squares solutions of linear systems, and then use your utility to find the least squares solution of the system in Example 1.

T2. (**Orthogonal Projection onto a Subspace**) Use the least squares capability of your technology utility to find the least squares solution $\mathbf{x}$ of the normal system in Example 2, and then complete the computations in the example by computing $A\mathbf{x}$. If you are successful, then see if you can create a program that will produce the orthogonal projection of a vector $\mathbf{u}$ in R^4 onto a subspace W when the user enters $\mathbf{u}$ and a set of vectors that spans W.

Suggestion As the first step, have the program create the matrix A that has the spanning vectors as columns.

Check your work by having your program find the orthogonal projection in Example 2.

Section 6.5 **T1.** (a) Confirm that $B_1 = \{\mathbf{u}_1, \mathbf{u}_2, \mathbf{u}_3, \mathbf{u}_4, \mathbf{u}_5\}$ and $B_2 = \{\mathbf{v}_1, \mathbf{v}_2, \mathbf{v}_3, \mathbf{v}_4, \mathbf{v}_5\}$ are bases for R^5, and find both transition matrices.

$$
\begin{aligned}
\mathbf{u}_1 &= (3, 1, 3, 2, 6) & \mathbf{v}_1 &= (2, 6, 3, 4, 2) \\
\mathbf{u}_2 &= (4, 5, 7, 2, 4) & \mathbf{v}_2 &= (3, 1, 5, 8, 3) \\
\mathbf{u}_3 &= (3, 2, 1, 5, 4) & \mathbf{v}_3 &= (5, 1, 2, 6, 7) \\
\mathbf{u}_4 &= (2, 9, 1, 4, 4) & \mathbf{v}_4 &= (8, 4, 3, 2, 6) \\
\mathbf{u}_5 &= (3, 3, 6, 6, 7) & \mathbf{v}_5 &= (5, 5, 6, 3, 4)
\end{aligned}
$$

(b) Find the coordinate vectors with respect to B_1 and B_2 of $\mathbf{w} = (1, 1, 1, 1, 1)$.

Eigenvalues, Eigenvectors

CHAPTER CONTENTS

INTRODUCTION: If A is an $n \times n$ matrix and **x** is a vector in R^n, then A**x** is also a vector in R^n, but usually there is no simple geometric relationship between **x** and A**x**. However, in the special case where **x** is a nonzero vector and A**x** is a scalar multiple of **x**, a simple geometric relationship occurs. For example, If A is a 2×2 matrix, and if **x** is a nonzero vector such that A**x** is a scalar multiple of **x**, say A**x** $= \lambda$**x**, then each vector on the line through the origin determined by **x** gets mapped back onto the same line under multiplication by A.

Nonzero vectors that get mapped into scalar multiples of themselves under a linear operator arise naturally in the study of vibrations, genetics, population dynamics, quantum mechanics, and economics, as well as in geometry. In this chapter we will study such vectors and their applications.

7.1
EIGENVALUES AND EIGENVECTORS

In Section 2.3 we introduced the concepts of eigenvalue and eigenvector. In this section we will study those ideas in more detail to set the stage for applications of them in later sections.

Review

We begin with a review of some concepts that were mentioned in Sections 2.3 and 4.3.

> **DEFINITION**
>
> If A is an $n \times n$ matrix, then a nonzero vector $\mathbf{x}$ in R^n is called an ***eigenvector*** of A if $A\mathbf{x}$ is a scalar multiple of $\mathbf{x}$; that is, if
>
> $$A\mathbf{x} = \lambda\mathbf{x}$$
>
> for some scalar λ. The scalar λ is called an ***eigenvalue*** of A, and $\mathbf{x}$ is said to be an eigenvector of A ***corresponding*** to λ.

In R^2 and R^3, multiplication by A maps each eigenvector $\mathbf{x}$ of A (if any) onto the same line through the origin as $\mathbf{x}$. Depending on the sign and the magnitude of the eigenvalue λ corresponding to $\mathbf{x}$, the linear operator $A\mathbf{x} = \lambda\mathbf{x}$ compresses or stretches $\mathbf{x}$ by a factor of λ, with a reversal of direction in the case where λ is negative (Figure 7.1.1).

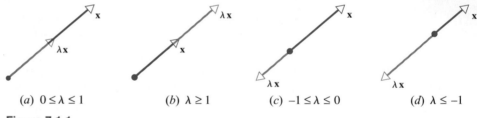

(a) $0 \leq \lambda \leq 1$ (b) $\lambda \geq 1$ (c) $-1 \leq \lambda \leq 0$ (d) $\lambda \leq -1$

Figure 7.1.1

EXAMPLE 1   Eigenvector of a 2 × 2 Matrix

The vector $\mathbf{x} = \begin{bmatrix} 1 \\ 2 \end{bmatrix}$ is an eigenvector of

$$A = \begin{bmatrix} 3 & 0 \\ 8 & -1 \end{bmatrix}$$

corresponding to the eigenvalue $\lambda = 3$, since

$$A\mathbf{x} = \begin{bmatrix} 3 & 0 \\ 8 & -1 \end{bmatrix} \begin{bmatrix} 1 \\ 2 \end{bmatrix} = \begin{bmatrix} 3 \\ 6 \end{bmatrix} = 3\mathbf{x} \quad \blacklozenge$$

To find the eigenvalues of an $n \times n$ matrix A, we rewrite $A\mathbf{x} = \lambda\mathbf{x}$ as

$$A\mathbf{x} = \lambda I \mathbf{x}$$

or, equivalently,

$$(\lambda I - A)\mathbf{x} = \mathbf{0} \tag{1}$$

For λ to be an eigenvalue, there must be a nonzero solution of this equation. By Theorem 6.4.5, Equation (1) has a nonzero solution if and only if

$$\det(\lambda I - A) = 0$$

This is called the ***characteristic equation*** of A; the scalars satisfying this equation are the eigenvalues of A. When expanded, the determinant $\det(\lambda I - A)$ is always a polynomial p in λ, called the ***characteristic polynomial*** of A.

It can be shown (Exercise 15) that if A is an $n \times n$ matrix, then the characteristic polynomial of A has degree n and the coefficient of λ^n is 1; that is, the characteristic polynomial $p(\lambda)$ of an $n \times n$ matrix has the form

$$p(\lambda) = \det(\lambda I - A) = \lambda^n + c_1\lambda^{n-1} + \cdots + c_n$$

It follows from the Fundamental Theorem of Algebra that the characteristic equation

$$\lambda^n + c_1\lambda^{n-1} + \cdots + c_n = 0$$

has at most n distinct solutions, so an $n \times n$ matrix has at most n distinct eigenvalues.

The reader may wish to review Example 6 of Section 2.3, where we found the eigenvalues of a 2×2 matrix by solving the characteristic equation. The following example involves a 3×3 matrix.

EXAMPLE 2 Eigenvalues of a 3 × 3 Matrix

Find the eigenvalues of

$$A = \begin{bmatrix} 0 & 1 & 0 \\ 0 & 0 & 1 \\ 4 & -17 & 8 \end{bmatrix}$$

Solution

The characteristic polynomial of A is

$$\det(\lambda I - A) = \det \begin{bmatrix} \lambda & -1 & 0 \\ 0 & \lambda & -1 \\ -4 & 17 & \lambda - 8 \end{bmatrix} = \lambda^3 - 8\lambda^2 + 17\lambda - 4$$

The eigenvalues of A must therefore satisfy the cubic equation

$$\lambda^3 - 8\lambda^2 + 17\lambda - 4 = 0 \tag{2}$$

To solve this equation, we shall begin by searching for integer solutions. This task can be greatly simplified by exploiting the fact that all integer solutions (if there are any) to a polynomial equation with integer coefficients

$$\lambda^n + c_1\lambda^{n-1} + \cdots + c_n = 0$$

must be divisors of the constant term, c_n. Thus, the only possible integer solutions of (2) are the divisors of -4, that is, $\pm 1, \pm 2, \pm 4$. Successively substituting these values in (2) shows that $\lambda = 4$ is an integer solution. As a consequence, $\lambda - 4$ must be a factor of the left side of (2). Dividing $\lambda - 4$ into $\lambda^3 - 8\lambda^2 + 17\lambda - 4$ shows that (2) can be rewritten as

$$(\lambda - 4)(\lambda^2 - 4\lambda + 1) = 0$$

Thus the remaining solutions of (2) satisfy the quadratic equation

$$\lambda^2 - 4\lambda + 1 = 0$$

which can be solved by the quadratic formula. Thus the eigenvalues of A are

$$\lambda = 4, \quad \lambda = 2 + \sqrt{3}, \quad \text{and} \quad \lambda = 2 - \sqrt{3} \quad \blacklozenge$$

REMARK In practical problems, the matrix A is usually so large that computing the characteristic equation is not practical. As a result, other methods are used to obtain eigenvalues.

EXAMPLE 3 Eigenvalues of an Upper Triangular Matrix

Find the eigenvalues of the upper triangular matrix

$$A = \begin{bmatrix} a_{11} & a_{12} & a_{13} & a_{14} \\ 0 & a_{22} & a_{23} & a_{24} \\ 0 & 0 & a_{33} & a_{34} \\ 0 & 0 & 0 & a_{44} \end{bmatrix}$$

Solution

Recalling that the determinant of a triangular matrix is the product of the entries on the main diagonal (Theorem 2.1.3), we obtain

$$\det(\lambda I - A) = \det \begin{bmatrix} \lambda - a_{11} & -a_{12} & -a_{13} & -a_{14} \\ 0 & \lambda - a_{22} & -a_{23} & -a_{24} \\ 0 & 0 & \lambda - a_{33} & -a_{34} \\ 0 & 0 & 0 & \lambda - a_{44} \end{bmatrix}$$

$$= (\lambda - a_{11})(\lambda - a_{22})(\lambda - a_{33})(\lambda - a_{44})$$

Thus, the characteristic equation is

$$(\lambda - a_{11})(\lambda - a_{22})(\lambda - a_{33})(\lambda - a_{44}) = 0$$

and the eigenvalues are

$$\lambda = a_{11}, \quad \lambda = a_{22}, \quad \lambda = a_{33}, \quad \lambda = a_{44}$$

which are precisely the diagonal entries of A. $\blacklozenge$

The following general theorem should be evident from the computations in the preceding example.

THEOREM 7.1.1

> *If A is an $n \times n$ triangular matrix (upper triangular, lower triangular, or diagonal), then the eigenvalues of A are the entries on the main diagonal of A.*

EXAMPLE 4 Eigenvalues of a Lower Triangular Matrix

By inspection, the eigenvalues of the lower triangular matrix

$$A = \begin{bmatrix} \frac{1}{2} & 0 & 0 \\ -1 & \frac{2}{3} & 0 \\ 5 & -8 & -\frac{1}{4} \end{bmatrix}$$

are $\lambda = \frac{1}{2}$, $\lambda = \frac{2}{3}$, and $\lambda = -\frac{1}{4}$. ◆

Complex Eigenvalues

It is possible for the characteristic equation of a matrix with real entries to have complex solutions. In fact, because the eigenvalues of an $n \times n$ matrix are the roots of a polynomial of precise degree n, every $n \times n$ matrix has exactly n eigenvalues if we count them as we count the roots of a polynomial (meaning that they may be repeated, and may occur in complex conjugate pairs). For example, the characteristic polynomial of the matrix

$$A = \begin{bmatrix} -2 & -1 \\ 5 & 2 \end{bmatrix}$$

is

$$\det(\lambda I - A) = \det \begin{bmatrix} \lambda + 2 & 1 \\ -5 & \lambda - 2 \end{bmatrix} = \lambda^2 + 1$$

so the characteristic equation is $\lambda^2 + 1 = 0$, the solutions of which are the imaginary numbers $\lambda = i$ and $\lambda = -i$. Thus we are forced to consider complex eigenvalues, even for real matrices. This, in turn, leads us to consider the possibility of complex vector spaces—that is, vector spaces in which scalars are allowed to have complex values. Such vector spaces will be considered in Chapter 10. For now, we will allow complex eigenvalues, but we will limit our discussion of eigenvectors to the case of real eigenvalues.

The following theorem summarizes our discussion thus far.

THEOREM 7.1.2

Equivalent Statements

If A is an $n \times n$ matrix and λ is a real number, then the following are equivalent.

(a) λ is an eigenvalue of A.
(b) The system of equations $(\lambda I - A)\mathbf{x} = \mathbf{0}$ has nontrivial solutions.
(c) There is a nonzero vector $\mathbf{x}$ in R^n such that $A\mathbf{x} = \lambda\mathbf{x}$.
(d) λ is a solution of the characteristic equation $\det(\lambda I - A) = 0$.

Finding Eigenvectors and Bases for Eigenspaces

Now that we know how to find eigenvalues, we turn to the problem of finding eigenvectors. The eigenvectors of A corresponding to an eigenvalue λ are the nonzero vectors $\mathbf{x}$ that satisfy $A\mathbf{x} = \lambda\mathbf{x}$. Equivalently, the eigenvectors corresponding to λ are the nonzero vectors in the solution space of $(\lambda I - A)\mathbf{x} = \mathbf{0}$—that is, in the null space of $\lambda I - A$. We call this solution space the **eigenspace** of A corresponding to λ.

EXAMPLE 5 Eigenvectors and Bases for Eigenspaces

Find bases for the eigenspaces of

$$A = \begin{bmatrix} 0 & 0 & -2 \\ 1 & 2 & 1 \\ 1 & 0 & 3 \end{bmatrix}$$

Solution

The characteristic equation of matrix A is $\lambda^3 - 5\lambda^2 + 8\lambda - 4 = 0$, or, in factored form, $(\lambda - 1)(\lambda - 2)^2 = 0$ (verify); thus the eigenvalues of A are $\lambda_1 = 1$ and $\lambda_{2,3} = 2$, so there are two eigenspaces of A.

By definition,

$$\mathbf{x} = \begin{bmatrix} x_1 \\ x_2 \\ x_3 \end{bmatrix}$$

is an eigenvector of A corresponding to λ if and only if $\mathbf{x}$ is a nontrivial solution of $(\lambda I - A)\mathbf{x} = \mathbf{0}$—that is, of

$$\begin{bmatrix} \lambda & 0 & 2 \\ -1 & \lambda - 2 & -1 \\ -1 & 0 & \lambda - 3 \end{bmatrix} \begin{bmatrix} x_1 \\ x_2 \\ x_3 \end{bmatrix} = \begin{bmatrix} 0 \\ 0 \\ 0 \end{bmatrix} \tag{3}$$

If $\lambda = 2$, then (3) becomes

$$\begin{bmatrix} 2 & 0 & 2 \\ -1 & 0 & -1 \\ -1 & 0 & -1 \end{bmatrix} \begin{bmatrix} x_1 \\ x_2 \\ x_3 \end{bmatrix} = \begin{bmatrix} 0 \\ 0 \\ 0 \end{bmatrix}$$

Solving this system using Gaussian elimination yields (verify)

$$x_1 = -s, \qquad x_2 = t, \qquad x_3 = s$$

Thus, the eigenvectors of A corresponding to $\lambda = 2$ are the nonzero vectors of the form

$$\mathbf{x} = \begin{bmatrix} -s \\ t \\ s \end{bmatrix} = \begin{bmatrix} -s \\ 0 \\ s \end{bmatrix} + \begin{bmatrix} 0 \\ t \\ 0 \end{bmatrix} = s \begin{bmatrix} -1 \\ 0 \\ 1 \end{bmatrix} + t \begin{bmatrix} 0 \\ 1 \\ 0 \end{bmatrix}$$

Since

$$\begin{bmatrix} -1 \\ 0 \\ 1 \end{bmatrix} \quad \text{and} \quad \begin{bmatrix} 0 \\ 1 \\ 0 \end{bmatrix}$$

are linearly independent, these vectors form a basis for the eigenspace corresponding to $\lambda = 2$.

If $\lambda = 1$, then (3) becomes

$$\begin{bmatrix} 1 & 0 & 2 \\ -1 & -1 & -1 \\ -1 & 0 & -2 \end{bmatrix} \begin{bmatrix} x_1 \\ x_2 \\ x_3 \end{bmatrix} = \begin{bmatrix} 0 \\ 0 \\ 0 \end{bmatrix}$$

Solving this system yields (verify)

$$x_1 = -2s, \qquad x_2 = s, \qquad x_3 = s$$

Thus the eigenvectors corresponding to $\lambda = 1$ are the nonzero vectors of the form

$$\begin{bmatrix} -2s \\ s \\ s \end{bmatrix} = s \begin{bmatrix} -2 \\ 1 \\ 1 \end{bmatrix} \quad \text{so that} \quad \begin{bmatrix} -2 \\ 1 \\ 1 \end{bmatrix}$$

is a basis for the eigenspace corresponding to $\lambda = 1$. ◆

Notice that the zero vector is in every eigenspace, although it isn't an eigenvector.

Powers of a Matrix

Once the eigenvalues and eigenvectors of a matrix A are found, it is a simple matter to find the eigenvalues and eigenvectors of any positive integer power of A; for example, if λ is an eigenvalue of A and $\mathbf{x}$ is a corresponding eigenvector, then

$$A^2\mathbf{x} = A(A\mathbf{x}) = A(\lambda\mathbf{x}) = \lambda(A\mathbf{x}) = \lambda(\lambda\mathbf{x}) = \lambda^2\mathbf{x}$$

which shows that λ^2 is an eigenvalue of A^2 and that $\mathbf{x}$ is a corresponding eigenvector. In general, we have the following result.

THEOREM 7.1.3

If k is a positive integer, λ is an eigenvalue of a matrix A, and $\mathbf{x}$ is a corresponding eigenvector, then λ^k is an eigenvalue of A^k and $\mathbf{x}$ is a corresponding eigenvector.

EXAMPLE 6 Using Theorem 7.1.3

In Example 5 we showed that the eigenvalues of

$$A = \begin{bmatrix} 0 & 0 & -2 \\ 1 & 2 & 1 \\ 1 & 0 & 3 \end{bmatrix}$$

are $\lambda = 2$ and $\lambda = 1$, so from Theorem 7.1.3, both $\lambda = 2^7 = 128$ and $\lambda = 1^7 = 1$ are eigenvalues of A^7. We also showed that

$$\begin{bmatrix} -1 \\ 0 \\ 1 \end{bmatrix} \quad \text{and} \quad \begin{bmatrix} 0 \\ 1 \\ 0 \end{bmatrix}$$

are eigenvectors of A corresponding to the eigenvalue $\lambda = 2$, so from Theorem 7.1.3, they are also eigenvectors of A^7 corresponding to $\lambda = 2^7 = 128$. Similarly, the eigenvector

$$\begin{bmatrix} -2 \\ 1 \\ 1 \end{bmatrix}$$

of A corresponding to the eigenvalue $\lambda = 1$ is also an eigenvector of A^7 corresponding to $\lambda = 1^7 = 1$. ◆

Eigenvalues and Invertibility

The next theorem establishes a relationship between the eigenvalues and the invertibility of a matrix.

THEOREM 7.1.4

A square matrix A is invertible if and only if $\lambda = 0$ is not an eigenvalue of A.

Proof Assume that A is an $n \times n$ matrix and observe first that $\lambda = 0$ is a solution of the characteristic equation

$$\lambda^n + c_1\lambda^{n-1} + \cdots + c_n = 0$$

if and only if the constant term c_n is zero. Thus it suffices to prove that A is invertible if and only if $c_n \neq 0$. But

$$\det(\lambda I - A) = \lambda^n + c_1\lambda^{n-1} + \cdots + c_n$$

or, on setting $\lambda = 0$,

$$\det(-A) = c_n \quad \text{or} \quad (-1)^n \det(A) = c_n$$

It follows from the last equation that $\det(A) = 0$ if and only if $c_n = 0$, and this in turn implies that A is invertible if and only if $c_n \neq 0$. ∎

EXAMPLE 7 Using Theorem 7.1.4

The matrix A in Example 5 is invertible since it has eigenvalues $\lambda = 1$ and $\lambda = 2$, neither of which is zero. We leave it for the reader to check this conclusion by showing that $\det(A) \neq 0$. ◆

Summary

Theorem 7.1.4 enables us to add an additional result to Theorem 6.4.5.

THEOREM 7.1.5

Equivalent Statements

If A is an $n \times n$ matrix, and if $T_A: R^n \to R^n$ is multiplication by A, then the following are equivalent.

(a) *A is invertible.*
(b) *$A\mathbf{x} = \mathbf{0}$ has only the trivial solution.*
(c) *The reduced row-echelon form of A is I_n.*
(d) *A is expressible as a product of elementary matrices.*
(e) *$A\mathbf{x} = \mathbf{b}$ is consistent for every $n \times 1$ matrix $\mathbf{b}$.*
(f) *$A\mathbf{x} = \mathbf{b}$ has exactly one solution for every $n \times 1$ matrix $\mathbf{b}$.*
(g) *$\det(A) \neq 0$.*
(h) *The range of T_A is R^n.*
(i) *T_A is one-to-one.*
(j) *The column vectors of A are linearly independent.*
(k) *The row vectors of A are linearly independent.*
(l) *The column vectors of A span R^n.*
(m) *The row vectors of A span R^n.*
(n) *The column vectors of A form a basis for R^n.*
(o) *The row vectors of A form a basis for R^n.*
(p) *A has rank n.*
(q) *A has nullity 0.*
(r) *The orthogonal complement of the nullspace of A is R^n.*
(s) *The orthogonal complement of the row space of A is $\{\mathbf{0}\}$.*
(t) *$A^T A$ is invertible.*
(u) *$\lambda = 0$ is not an eigenvalue of A.*

This theorem relates all of the major topics we have studied thus far.

EXERCISE SET
7.1

1. Find the characteristic equations of the following matrices:

(a) $\begin{bmatrix} 3 & 0 \\ 8 & -1 \end{bmatrix}$ 　(b) $\begin{bmatrix} 10 & -9 \\ 4 & -2 \end{bmatrix}$ 　(c) $\begin{bmatrix} 0 & 3 \\ 4 & 0 \end{bmatrix}$

(d) $\begin{bmatrix} -2 & -7 \\ 1 & 2 \end{bmatrix}$ 　(e) $\begin{bmatrix} 0 & 0 \\ 0 & 0 \end{bmatrix}$ 　(f) $\begin{bmatrix} 1 & 0 \\ 0 & 1 \end{bmatrix}$

2. Find the eigenvalues of the matrices in Exercise 1.

3. Find bases for the eigenspaces of the matrices in Exercise 1.

4. Find the characteristic equations of the following matrices:

(a) $\begin{bmatrix} 4 & 0 & 1 \\ -2 & 1 & 0 \\ -2 & 0 & 1 \end{bmatrix}$ 　(b) $\begin{bmatrix} 3 & 0 & -5 \\ \frac{1}{5} & -1 & 0 \\ 1 & 1 & -2 \end{bmatrix}$ 　(c) $\begin{bmatrix} -2 & 0 & 1 \\ -6 & -2 & 0 \\ 19 & 5 & -4 \end{bmatrix}$

(d) $\begin{bmatrix} -1 & 0 & 1 \\ -1 & 3 & 0 \\ -4 & 13 & -1 \end{bmatrix}$ 　(e) $\begin{bmatrix} 5 & 0 & 1 \\ 1 & 1 & 0 \\ -7 & 1 & 0 \end{bmatrix}$ 　(f) $\begin{bmatrix} 5 & 6 & 2 \\ 0 & -1 & -8 \\ 1 & 0 & -2 \end{bmatrix}$

5. Find the eigenvalues of the matrices in Exercise 4.

6. Find bases for the eigenspaces of the matrices in Exercise 4.

7. Find the characteristic equations of the following matrices:

(a) $\begin{bmatrix} 0 & 0 & 2 & 0 \\ 1 & 0 & 1 & 0 \\ 0 & 1 & -2 & 0 \\ 0 & 0 & 0 & 1 \end{bmatrix}$ 　(b) $\begin{bmatrix} 10 & -9 & 0 & 0 \\ 4 & -2 & 0 & 0 \\ 0 & 0 & -2 & -7 \\ 0 & 0 & 1 & 2 \end{bmatrix}$

8. Find the eigenvalues of the matrices in Exercise 7.

9. Find bases for the eigenspaces of the matrices in Exercise 7.

10. By inspection, find the eigenvalues of the following matrices:

(a) $\begin{bmatrix} -1 & 6 \\ 0 & 5 \end{bmatrix}$ 　(b) $\begin{bmatrix} 3 & 0 & 0 \\ -2 & 7 & 0 \\ 4 & 8 & 1 \end{bmatrix}$ 　(c) $\begin{bmatrix} -\frac{1}{3} & 0 & 0 & 0 \\ 0 & -\frac{1}{3} & 0 & 0 \\ 0 & 0 & 1 & 0 \\ 0 & 0 & 0 & \frac{1}{2} \end{bmatrix}$

11. Find the eigenvalues of A^9 for

$$A = \begin{bmatrix} 1 & 3 & 7 & 11 \\ 0 & \frac{1}{2} & 3 & 8 \\ 0 & 0 & 0 & 4 \\ 0 & 0 & 0 & 2 \end{bmatrix}$$

12. Find the eigenvalues and bases for the eigenspaces of A^{25} for

$$A = \begin{bmatrix} -1 & -2 & -2 \\ 1 & 2 & 1 \\ -1 & -1 & 0 \end{bmatrix}$$

13. Let A be a 2×2 matrix, and call a line through the origin of R^2 *invariant* under A if $A\mathbf{x}$ lies on the line when $\mathbf{x}$ does. Find equations for all lines in R^2, if any, that are invariant under the given matrix.

The following theorem shows that the eigenvector problem and the diagonalization problem are equivalent.

THEOREM 7.2.1

> *If A is an n × n matrix, then the following are equivalent.*
>
> (a)   *A is diagonalizable.*
>
> (b)   *A has n linearly independent eigenvectors.*

Proof (a) ⇒ (b)   Since A is assumed diagonalizable, there is an invertible matrix

$$P = \begin{bmatrix} p_{11} & p_{12} & \cdots & p_{1n} \\ p_{21} & p_{22} & \cdots & p_{2n} \\ \vdots & \vdots & & \vdots \\ p_{n1} & p_{n2} & \cdots & p_{nn} \end{bmatrix}$$

such that $P^{-1}AP$ is diagonal, say $P^{-1}AP = D$, where

$$D = \begin{bmatrix} \lambda_1 & 0 & \cdots & 0 \\ 0 & \lambda_2 & \cdots & 0 \\ \vdots & \vdots & & \vdots \\ 0 & 0 & \cdots & \lambda_n \end{bmatrix}$$

It follows from the formula $P^{-1}AP = D$ that $AP = PD$; that is,

$$AP = \begin{bmatrix} p_{11} & p_{12} & \cdots & p_{1n} \\ p_{21} & p_{22} & \cdots & p_{2n} \\ \vdots & \vdots & & \vdots \\ p_{n1} & p_{n2} & \cdots & p_{nn} \end{bmatrix} \begin{bmatrix} \lambda_1 & 0 & \cdots & 0 \\ 0 & \lambda_2 & \cdots & 0 \\ \vdots & \vdots & & \vdots \\ 0 & 0 & \cdots & \lambda_n \end{bmatrix} = \begin{bmatrix} \lambda_1 p_{11} & \lambda_2 p_{12} & \cdots & \lambda_n p_{1n} \\ \lambda_1 p_{21} & \lambda_2 p_{22} & \cdots & \lambda_n p_{2n} \\ \vdots & \vdots & & \vdots \\ \lambda_1 p_{n1} & \lambda_2 p_{n2} & \cdots & \lambda_n p_{nn} \end{bmatrix}$$

$$\tag{1}$$

If we now let $\mathbf{p}_1, \mathbf{p}_2, \ldots, \mathbf{p}_n$ denote the column vectors of P, then from (1), the successive columns of AP are $\lambda_1\mathbf{p}_1, \lambda_2\mathbf{p}_2, \ldots, \lambda_n\mathbf{p}_n$. However, from Formula (6) of Section 1.3, the successive columns of AP are $A\mathbf{p}_1, A\mathbf{p}_2, \ldots, A\mathbf{p}_n$. Thus we must have

$$A\mathbf{p}_1 = \lambda_1\mathbf{p}_1, \quad A\mathbf{p}_2 = \lambda_2\mathbf{p}_2, \ldots, \quad A\mathbf{p}_n = \lambda_n\mathbf{p}_n \tag{2}$$

Since P is invertible, its column vectors are all nonzero; thus, it follows from (2) that $\lambda_1, \lambda_2, \ldots, \lambda_n$ are eigenvalues of A, and $\mathbf{p}_1, \mathbf{p}_2, \ldots, \mathbf{p}_n$ are corresponding eigenvectors. Since P is invertible, it follows from Theorem 7.1.5 that $\mathbf{p}_1, \mathbf{p}_2, \ldots, \mathbf{p}_n$ are linearly independent. Thus A has n linearly independent eigenvectors.

(b) ⇒ (a)   Assume that A has n linearly independent eigenvectors, $\mathbf{p}_1, \mathbf{p}_2, \ldots, \mathbf{p}_n$, with corresponding eigenvalues $\lambda_1, \lambda_2, \ldots, \lambda_n$, and let

$$P = \begin{bmatrix} p_{11} & p_{12} & \cdots & p_{1n} \\ p_{21} & p_{22} & \cdots & p_{2n} \\ \vdots & \vdots & & \vdots \\ p_{n1} & p_{n2} & \cdots & p_{nn} \end{bmatrix}$$

be the matrix whose column vectors are $\mathbf{p}_1, \mathbf{p}_2, \ldots, \mathbf{p}_n$. By Formula (6) of Section 1.3, the column vectors of the product AP are

$$A\mathbf{p}_1, A\mathbf{p}_2, \ldots, A\mathbf{p}_n$$

But

$$A\mathbf{p}_1 = \lambda_1\mathbf{p}_1, \quad A\mathbf{p}_2 = \lambda_2\mathbf{p}_2, \dots, \quad A\mathbf{p}_n = \lambda_n\mathbf{p}_n$$

so

$$AP = \begin{bmatrix} \lambda_1 p_{11} & \lambda_2 p_{12} & \dots & \lambda_n p_{1n} \\ \lambda_1 p_{21} & \lambda_2 p_{22} & \dots & \lambda_n p_{2n} \\ \vdots & \vdots & & \vdots \\ \lambda_1 p_{n1} & \lambda_2 p_{n2} & \dots & \lambda_n p_{nn} \end{bmatrix}$$

$$= \begin{bmatrix} p_{11} & p_{12} & \dots & p_{1n} \\ p_{21} & p_{22} & \dots & p_{2n} \\ \vdots & \vdots & & \vdots \\ p_{n1} & p_{n2} & \dots & p_{nn} \end{bmatrix} \begin{bmatrix} \lambda_1 & 0 & \dots & 0 \\ 0 & \lambda_2 & \dots & 0 \\ \vdots & \vdots & & \vdots \\ 0 & 0 & \dots & \lambda_n \end{bmatrix} = PD \qquad (3)$$

where D is the diagonal matrix having the eigenvalues $\lambda_1, \lambda_2, \dots, \lambda_n$ on the main diagonal. Since the column vectors of P are linearly independent, P is invertible. Thus (3) can be rewritten as $P^{-1}AP = D$; that is, A is diagonalizable. ∎

Procedure for Diagonalizing a Matrix

The preceding theorem guarantees that an $n \times n$ matrix A with n linearly independent eigenvectors is diagonalizable, and the proof provides the following method for diagonalizing A.

Step 1. Find n linearly independent eigenvectors of A, say $\mathbf{p}_1, \mathbf{p}_2, \dots, \mathbf{p}_n$.

Step 2. Form the matrix P having $\mathbf{p}_1, \mathbf{p}_2, \dots, \mathbf{p}_n$ as its column vectors.

Step 3. The matrix $P^{-1}AP$ will then be diagonal with $\lambda_1, \lambda_2, \dots, \lambda_n$ as its successive diagonal entries, where λ_i is the eigenvalue corresponding to $\mathbf{p}_i$, for $i = 1, 2, \dots, n$.

In order to carry out Step 1 of this procedure, one first needs a way of determining whether a given $n \times n$ matrix A has n linearly independent eigenvectors, and then one needs a method for finding them. One can address both problems at the same time by finding bases for the eigenspaces of A. Later in this section, we will show that those basis vectors, as a combined set, are linearly independent, so that if there is a total of n such vectors, then A is diagonalizable, and the n basis vectors can be used as the column vectors of the diagonalizing matrix P. If there are fewer than n basis vectors, then A is not diagonalizable.

EXAMPLE 1   Finding a Matrix *P* That Diagonalizes a Matrix *A*

Find a matrix P that diagonalizes

$$A = \begin{bmatrix} 0 & 0 & -2 \\ 1 & 2 & 1 \\ 1 & 0 & 3 \end{bmatrix}$$

Solution

From Example 5 of the preceding section, we found the characteristic equation of A to be

$$(\lambda - 1)(\lambda - 2)^2 = 0$$

and we found the following bases for the eigenspaces:

$$\lambda = 2: \quad \mathbf{p}_1 = \begin{bmatrix} -1 \\ 0 \\ 1 \end{bmatrix}, \quad \mathbf{p}_2 = \begin{bmatrix} 0 \\ 1 \\ 0 \end{bmatrix} \qquad \lambda = 1: \quad \mathbf{p}_3 = \begin{bmatrix} -2 \\ 1 \\ 1 \end{bmatrix}$$

There are three basis vectors in total, so the matrix A is diagonalizable and

$$P = \begin{bmatrix} -1 & 0 & -2 \\ 0 & 1 & 1 \\ 1 & 0 & 1 \end{bmatrix}$$

diagonalizes A. As a check, the reader should verify that

$$P^{-1}AP = \begin{bmatrix} 1 & 0 & 2 \\ 1 & 1 & 1 \\ -1 & 0 & -1 \end{bmatrix} \begin{bmatrix} 0 & 0 & -2 \\ 1 & 2 & 1 \\ 1 & 0 & 3 \end{bmatrix} \begin{bmatrix} -1 & 0 & -2 \\ 0 & 1 & 1 \\ 1 & 0 & 1 \end{bmatrix} = \begin{bmatrix} 2 & 0 & 0 \\ 0 & 2 & 0 \\ 0 & 0 & 1 \end{bmatrix} \quad \blacklozenge$$

There is no preferred order for the columns of P. Since the ith diagonal entry of $P^{-1}AP$ is an eigenvalue for the ith column vector of P, changing the order of the columns of P just changes the order of the eigenvalues on the diagonal of $P^{-1}AP$. Thus, if we had written

$$P = \begin{bmatrix} -1 & -2 & 0 \\ 0 & 1 & 1 \\ 1 & 1 & 0 \end{bmatrix}$$

in Example 1, we would have obtained

$$P^{-1}AP = \begin{bmatrix} 2 & 0 & 0 \\ 0 & 1 & 0 \\ 0 & 0 & 2 \end{bmatrix}$$

EXAMPLE 2 A Matrix That Is Not Diagonalizable

Find a matrix P that diagonalizes

$$A = \begin{bmatrix} 1 & 0 & 0 \\ 1 & 2 & 0 \\ -3 & 5 & 2 \end{bmatrix}$$

Solution

The characteristic polynomial of A is

$$\det(\lambda I - A) = \begin{vmatrix} \lambda - 1 & 0 & 0 \\ -1 & \lambda - 2 & 0 \\ 3 & -5 & \lambda - 2 \end{vmatrix} = (\lambda - 1)(\lambda - 2)^2$$

so the characteristic equation is

$$(\lambda - 1)(\lambda - 2)^2 = 0$$

Thus the eigenvalues of A are $\lambda_1 = 1$ and $\lambda_{2,3} = 2$. We leave it for the reader to show that bases for the eigenspaces are

$$\lambda = 1: \quad \mathbf{p}_1 = \begin{bmatrix} \frac{1}{8} \\ -\frac{1}{8} \\ 1 \end{bmatrix} \qquad \lambda = 2: \quad \mathbf{p}_2 = \begin{bmatrix} 0 \\ 0 \\ 1 \end{bmatrix}$$

Since A is a 3×3 matrix and there are only two basis vectors in total, A is not diagonalizable.

Alternative Solution

If one is interested only in determining whether a matrix is diagonalizable and is not concerned with actually finding a diagonalizing matrix P, then it is not necessary to compute bases for the eigenspaces; it suffices to find the dimensions of the eigenspaces. For this example, the eigenspace corresponding to $\lambda = 1$ is the solution space of the system

$$\begin{bmatrix} 0 & 0 & 0 \\ -1 & -1 & 0 \\ 3 & -5 & -1 \end{bmatrix} \begin{bmatrix} x_1 \\ x_2 \\ x_3 \end{bmatrix} = \begin{bmatrix} 0 \\ 0 \\ 0 \end{bmatrix}$$

The coefficient matrix has rank 2 (verify). Thus the nullity of this matrix is 1 by Theorem 5.6.3, and hence the solution space is one-dimensional.

The eigenspace corresponding to $\lambda = 2$ is the solution space of the system

$$\begin{bmatrix} 1 & 0 & 0 \\ -1 & 0 & 0 \\ 3 & -5 & 0 \end{bmatrix} \begin{bmatrix} x_1 \\ x_2 \\ x_3 \end{bmatrix} = \begin{bmatrix} 0 \\ 0 \\ 0 \end{bmatrix}$$

This coefficient matrix also has rank 2 and nullity 1 (verify), so the eigenspace corresponding to $\lambda = 2$ is also one-dimensional. Since the eigenspaces produce a total of two basis vectors, the matrix A is not diagonalizable. ◆

There is an assumption in Example 1 that the column vectors of P, which are made up of basis vectors from the various eigenspaces of A, are linearly independent. The following theorem addresses this issue.

THEOREM 7.2.2

> *If $\mathbf{v}_1, \mathbf{v}_2, \ldots, \mathbf{v}_k$ are eigenvectors of A corresponding to distinct eigenvalues $\lambda_1, \lambda_2, \ldots, \lambda_k$, then $\{\mathbf{v}_1, \mathbf{v}_2, \ldots, \mathbf{v}_k\}$ is a linearly independent set.*

Proof Let $\mathbf{v}_1, \mathbf{v}_2, \ldots, \mathbf{v}_k$ be eigenvectors of A corresponding to distinct eigenvalues $\lambda_1, \lambda_2, \ldots, \lambda_k$. We shall assume that $\mathbf{v}_1, \mathbf{v}_2, \ldots, \mathbf{v}_k$ are linearly dependent and obtain a contradiction. We can then conclude that $\mathbf{v}_1, \mathbf{v}_2, \ldots, \mathbf{v}_k$ are linearly independent.

Since an eigenvector is nonzero by definition, $\{\mathbf{v}_1\}$ is linearly independent. Let r be the largest integer such that $\{\mathbf{v}_1, \mathbf{v}_2, \ldots, \mathbf{v}_r\}$ is linearly independent. Since we are assuming that $\{\mathbf{v}_1, \mathbf{v}_2, \ldots, \mathbf{v}_k\}$ is linearly dependent, r satisfies $1 \le r < k$. Moreover, by definition of r, $\{\mathbf{v}_1, \mathbf{v}_2, \ldots, \mathbf{v}_{r+1}\}$ is linearly dependent. Thus there are scalars $c_1, c_2, \ldots, c_{r+1}$, not all zero, such that

$$c_1\mathbf{v}_1 + c_2\mathbf{v}_2 + \cdots + c_{r+1}\mathbf{v}_{r+1} = \mathbf{0} \tag{4}$$

Multiplying both sides of (4) by A and using

$$A\mathbf{v}_1 = \lambda_1\mathbf{v}_1, \quad A\mathbf{v}_2 = \lambda_2\mathbf{v}_2, \ldots, \quad A\mathbf{v}_{r+1} = \lambda_{r+1}\mathbf{v}_{r+1}$$

we obtain

$$c_1\lambda_1\mathbf{v}_1 + c_2\lambda_2\mathbf{v}_2 + \cdots + c_{r+1}\lambda_{r+1}\mathbf{v}_{r+1} = \mathbf{0} \tag{5}$$

Multiplying both sides of (4) by λ_{r+1} and subtracting the resulting equation from (5) yields

$$c_1(\lambda_1 - \lambda_{r+1})\mathbf{v}_1 + c_2(\lambda_2 - \lambda_{r+1})\mathbf{v}_2 + \cdots + c_r(\lambda_r - \lambda_{r+1})\mathbf{v}_r = \mathbf{0}$$

Since $\{\mathbf{v}_1, \mathbf{v}_2, \ldots, \mathbf{v}_r\}$ is a linearly independent set, this equation implies that

$$c_1(\lambda_1 - \lambda_{r+1}) = c_2(\lambda_2 - \lambda_{r+1}) = \cdots = c_r(\lambda_r - \lambda_{r+1}) = 0$$

and since $\lambda_1, \lambda_2, \ldots, \lambda_{r+1}$ are distinct by hypothesis, it follows that

$$c_1 = c_2 = \cdots = c_r = 0 \tag{6}$$

Substituting these values in (4) yields

$$c_{r+1}\mathbf{v}_{r+1} = \mathbf{0}$$

Since the eigenvector $\mathbf{v}_{r+1}$ is nonzero, it follows that

$$c_{r+1} = 0 \tag{7}$$

Equations 6 and 7 contradict the fact that $c_1, c_2, \ldots, c_{r+1}$ are not all zero; this completes the proof. ■

REMARK   Theorem 7.2.2 is a special case of a more general result: Suppose that $\lambda_1, \lambda_2, \ldots, \lambda_k$ are distinct eigenvalues and that we choose a linearly independent set in each of the corresponding eigenspaces. If we then merge all these vectors into a single set, the result will still be a linearly independent set. For example, if we choose three linearly independent vectors from one eigenspace and two linearly independent vectors from another eigenspace, then the five vectors together form a linearly independent set. We omit the proof.

As a consequence of Theorem 7.2.2, we obtain the following important result.

THEOREM 7.2.3

> *If an $n \times n$ matrix A has n distinct eigenvalues, then A is diagonalizable.*

Proof   If $\mathbf{v}_1, \mathbf{v}_2, \ldots, \mathbf{v}_n$ are eigenvectors corresponding to the distinct eigenvalues $\lambda_1, \lambda_2, \ldots, \lambda_n$, then by Theorem 7.2.2, $\mathbf{v}_1, \mathbf{v}_2, \ldots, \mathbf{v}_n$ are linearly independent. Thus A is diagonalizable by Theorem 7.2.1. ■

EXAMPLE 3   Using Theorem 7.2.3

We saw in Example 2 of the preceding section that

$$A = \begin{bmatrix} 0 & 1 & 0 \\ 0 & 0 & 1 \\ 4 & -17 & 8 \end{bmatrix}$$

has three distinct eigenvalues: $\lambda = 4$, $\lambda = 2 + \sqrt{3}$, and $\lambda = 2 - \sqrt{3}$. Therefore, A is diagonalizable. Further,

$$P^{-1}AP = \begin{bmatrix} 4 & 0 & 0 \\ 0 & 2 + \sqrt{3} & 0 \\ 0 & 0 & 2 - \sqrt{3} \end{bmatrix}$$

for some invertible matrix P. If desired, the matrix P can be found using the method shown in Example 1 of this section. ◆

EXAMPLE 4   A Diagonalizable Matrix

From Theorem 7.1.1, the eigenvalues of a triangular matrix are the entries on its main diagonal. Thus, a triangular matrix with distinct entries on the main diagonal is diagonalizable. For example,

$$A = \begin{bmatrix} -1 & 2 & 4 & 0 \\ 0 & 3 & 1 & 7 \\ 0 & 0 & 5 & 8 \\ 0 & 0 & 0 & -2 \end{bmatrix}$$

is a diagonalizable matrix. ◆

EXAMPLE 5   Repeated Eigenvalues and Diagonalizability

It's important to note that Theorem 7.2.3 says only that *if* a matrix has all distinct eigenvalues (whether real or complex), *then* it is diagonalizable; in other words, *only* matrices with repeated eigenvalues *might* be nondiagonalizable. For example, the 3×3 identity matrix

$$I_3 = \begin{bmatrix} 1 & 0 & 0 \\ 0 & 1 & 0 \\ 0 & 0 & 1 \end{bmatrix}$$

has repeated eigenvalues $\lambda_{1,2,3} = 1$ but is diagonalizable since any nonzero vector in R^3 is an eigenvector of the 3×3 identity matrix (verify), and so, in particular, we can find three linearly independent eigenvectors. The matrix

$$J_3 = \begin{bmatrix} 1 & 1 & 0 \\ 0 & 1 & 1 \\ 0 & 0 & 1 \end{bmatrix}$$

also has repeated eigenvalues $\lambda_{1,2,3} = 1$, but solving for its eigenvectors leads to the system

$$(\lambda I - J_3)\mathbf{x} = \begin{bmatrix} 0 & 1 & 0 \\ 0 & 0 & 1 \\ 0 & 0 & 0 \end{bmatrix} \mathbf{x} = \mathbf{0}$$

the solution of which is $x_1 = t, x_2 = 0, x_3 = 0$. Thus every eigenvector of J_3 is a multiple of

$$\begin{bmatrix} 1 \\ 0 \\ 0 \end{bmatrix}$$

which means that the eigenspace has dimension 1 and that J_3 is nondiagonalizable. ◆

Matrices that look like the identity matrix except that the diagonal immediately above the main diagonal also has 1's on it, such as J_3 or

THEOREM 7.3.2

> *If A is a symmetric matrix, then*
>
> (a) *The eigenvalues of A are all real numbers.*
> (b) *Eigenvectors from different eigenspaces are orthogonal.*

Proof (a) The proof of part (*a*), which requires results about complex vector spaces, is discussed in Section 10.6.

Proof (b) Let $\mathbf{v}_1$ and $\mathbf{v}_2$ be eigenvectors corresponding to distinct eigenvalues λ_1 and λ_2 of the matrix A. We want to show that $\mathbf{v}_1 \cdot \mathbf{v}_2 = 0$. The proof of this involves the trick of starting with the expression $A\mathbf{v}_1 \cdot \mathbf{v}_2$. It follows from Formula (8) of Section 4.1 and the symmetry of A that

$$A\mathbf{v}_1 \cdot \mathbf{v}_2 = \mathbf{v}_1 \cdot A^T\mathbf{v}_2 = \mathbf{v}_1 \cdot A\mathbf{v}_2 \tag{3}$$

But $\mathbf{v}_1$ is an eigenvector of A corresponding to λ_1, and $\mathbf{v}_2$ is an eigenvector of A corresponding to λ_2, so (3) yields the relationship

$$\lambda_1 \mathbf{v}_1 \cdot \mathbf{v}_2 = \mathbf{v}_1 \cdot \lambda_2 \mathbf{v}_2$$

which can be rewritten as

$$(\lambda_1 - \lambda_2)(\mathbf{v}_1 \cdot \mathbf{v}_2) = 0 \tag{4}$$

But $\lambda_1 - \lambda_2 \neq 0$, since λ_1 and λ_2 were assumed distinct. Thus it follows from (4) that $\mathbf{v}_1 \cdot \mathbf{v}_2 = 0$. ∎

REMARK We remind the reader that we have assumed to this point that all of our matrices have real entries. Indeed, we shall see in Chapter 10 that part (*a*) of Theorem 7.3.2 is false for matrices with complex entries.

Diagonalization of Symmetric Matrices

As a consequence of the preceding theorem we obtain the following procedure for orthogonally diagonalizing a symmetric matrix.

Step 1. Find a basis for each eigenspace of A.

Step 2. Apply the Gram–Schmidt process to each of these bases to obtain an orthonormal basis for each eigenspace.

Step 3. Form the matrix P whose columns are the basis vectors constructed in Step 2; this matrix orthogonally diagonalizes A.

The justification of this procedure should be clear: Theorem 7.3.2 ensures that eigenvectors from *different* eigenspaces are orthogonal, whereas the application of the Gram–Schmidt process ensures that the eigenvectors obtained within the *same* eigenspace are orthonormal. Therefore, the *entire* set of eigenvectors obtained by this procedure is orthonormal.

EXAMPLE 1 An Orthogonal Matrix *P* That Diagonalizes a Matrix *A*

Find an orthogonal matrix P that diagonalizes

$$A = \begin{bmatrix} 4 & 2 & 2 \\ 2 & 4 & 2 \\ 2 & 2 & 4 \end{bmatrix}$$

Solution

The characteristic equation of A is

$$\det(\lambda I - A) = \det \begin{bmatrix} \lambda - 4 & -2 & -2 \\ -2 & \lambda - 4 & -2 \\ -2 & -2 & \lambda - 4 \end{bmatrix} = (\lambda - 2)^2(\lambda - 8) = 0$$

Thus the eigenvalues of A are $\lambda_{1,2} = 2$ and $\lambda_3 = 8$. By the method used in Example 5 of Section 7.1, it can be shown that

$$\mathbf{u}_1 = \begin{bmatrix} -1 \\ 1 \\ 0 \end{bmatrix} \quad \text{and} \quad \mathbf{u}_2 = \begin{bmatrix} -1 \\ 0 \\ 1 \end{bmatrix} \tag{5}$$

form a basis for the eigenspace corresponding to $\lambda = 2$. Applying the Gram–Schmidt process to $\{\mathbf{u}_1, \mathbf{u}_2\}$ yields the following orthonormal eigenvectors (verify):

$$\mathbf{v}_1 = \begin{bmatrix} -1/\sqrt{2} \\ 1/\sqrt{2} \\ 0 \end{bmatrix} \quad \text{and} \quad \mathbf{v}_2 = \begin{bmatrix} -1/\sqrt{6} \\ -1/\sqrt{6} \\ 2/\sqrt{6} \end{bmatrix} \tag{6}$$

The eigenspace corresponding to $\lambda = 8$ has

$$\mathbf{u}_3 = \begin{bmatrix} 1 \\ 1 \\ 1 \end{bmatrix}$$

as a basis. Applying the Gram–Schmidt process to $\{\mathbf{u}_3\}$ yields

$$\mathbf{v}_3 = \begin{bmatrix} 1/\sqrt{3} \\ 1/\sqrt{3} \\ 1/\sqrt{3} \end{bmatrix}$$

Finally, using $\mathbf{v}_1$, $\mathbf{v}_2$, and $\mathbf{v}_3$ as column vectors, we obtain

$$P = \begin{bmatrix} -1/\sqrt{2} & -1/\sqrt{6} & 1/\sqrt{3} \\ 1/\sqrt{2} & -1/\sqrt{6} & 1/\sqrt{3} \\ 0 & 2/\sqrt{6} & 1/\sqrt{3} \end{bmatrix}$$

which orthogonally diagonalizes A. (As a check, the reader may wish to verify that $P^T AP$ is a diagonal matrix.) ◆

EXERCISE SET
7.3

1. Find the characteristic equation of the given symmetric matrix, and then by inspection determine the dimensions of the eigenspaces.

(a) $\begin{bmatrix} 1 & 2 \\ 2 & 4 \end{bmatrix}$
(b) $\begin{bmatrix} 1 & -4 & 2 \\ -4 & 1 & -2 \\ 2 & -2 & -2 \end{bmatrix}$
(c) $\begin{bmatrix} 1 & 1 & 1 \\ 1 & 1 & 1 \\ 1 & 1 & 1 \end{bmatrix}$

(d) $\begin{bmatrix} 4 & 2 & 2 \\ 2 & 4 & 2 \\ 2 & 2 & 4 \end{bmatrix}$
(e) $\begin{bmatrix} 4 & 4 & 0 & 0 \\ 4 & 4 & 0 & 0 \\ 0 & 0 & 0 & 0 \\ 0 & 0 & 0 & 0 \end{bmatrix}$
(f) $\begin{bmatrix} 2 & -1 & 0 & 0 \\ -1 & 2 & 0 & 0 \\ 0 & 0 & 2 & -1 \\ 0 & 0 & -1 & 2 \end{bmatrix}$

In Exercises 2–9 find a matrix P that orthogonally diagonalizes A, and determine $P^{-1}AP$.

2. $A = \begin{bmatrix} 3 & 1 \\ 1 & 3 \end{bmatrix}$ **3.** $A = \begin{bmatrix} 6 & 2\sqrt{3} \\ 2\sqrt{3} & 7 \end{bmatrix}$ **4.** $A = \begin{bmatrix} 6 & -2 \\ -2 & 3 \end{bmatrix}$

5. $A = \begin{bmatrix} -2 & 0 & -36 \\ 0 & -3 & 0 \\ -36 & 0 & -23 \end{bmatrix}$ **6.** $A = \begin{bmatrix} 1 & 1 & 0 \\ 1 & 1 & 0 \\ 0 & 0 & 0 \end{bmatrix}$ **7.** $A = \begin{bmatrix} 2 & -1 & -1 \\ -1 & 2 & -1 \\ -1 & -1 & 2 \end{bmatrix}$

8. $A = \begin{bmatrix} 3 & 1 & 0 & 0 \\ 1 & 3 & 0 & 0 \\ 0 & 0 & 0 & 0 \\ 0 & 0 & 0 & 0 \end{bmatrix}$ **9.** $A = \begin{bmatrix} -7 & 24 & 0 & 0 \\ 24 & 7 & 0 & 0 \\ 0 & 0 & -7 & 24 \\ 0 & 0 & 24 & 7 \end{bmatrix}$

10. Assuming that $b \neq 0$, find a matrix that orthogonally diagonalizes

$$\begin{bmatrix} a & b \\ b & a \end{bmatrix}$$

11. Prove that if A is any $m \times n$ matrix, then $A^T A$ has an orthonormal set of n eigenvectors.

12. (a) Show that if $\mathbf{v}$ is any $n \times 1$ matrix and I is the $n \times n$ identity matrix, then $I - \mathbf{v}\mathbf{v}^T$ is orthogonally diagonalizable.

(b) Find a matrix P that orthogonally diagonalizes $I - \mathbf{v}\mathbf{v}^T$ if

$$\mathbf{v} = \begin{bmatrix} 1 \\ 0 \\ 1 \end{bmatrix}$$

13. Use the result in Exercise 17 of Section 7.1 to prove Theorem 7.3.2a for 2×2 symmetric matrices.

Discussion & Discovery

14. Indicate whether each statement is always true or sometimes false. Justify your answer by giving a logical argument or a counterexample.

(a) If A is a square matrix, then AA^T and A^TA are orthogonally diagonalizable.

(b) If $\mathbf{v}_1$ and $\mathbf{v}_2$ are eigenvectors from distinct eigenspaces of a symmetric matrix, then $\|\mathbf{v}_1 + \mathbf{v}_2\|^2 = \|\mathbf{v}_1\|^2 + \|\mathbf{v}_2\|^2$.

(c) An orthogonal matrix is orthogonally diagonalizable.

(d) If A is an invertible orthogonally diagonalizable matrix, then A^{-1} is orthogonally diagonalizable.

15. Does there exist a 3×3 symmetric matrix with eigenvalues $\lambda_1 = -1$, $\lambda_2 = 3$, $\lambda_3 = 7$ and corresponding eigenvectors

$$\begin{bmatrix} 0 \\ 1 \\ -1 \end{bmatrix}, \quad \begin{bmatrix} 1 \\ 0 \\ 0 \end{bmatrix}, \quad \begin{bmatrix} 0 \\ 1 \\ 1 \end{bmatrix}?$$

If so, find such a matrix; if not, explain why not.

16. Is the converse of Theorem 7.3.2b true?

CHAPTER 7
Supplementary Exercises

1. (a) Show that if $0 < \theta < \pi$, then

$$A = \begin{bmatrix} \cos\theta & -\sin\theta \\ \sin\theta & \cos\theta \end{bmatrix}$$

has no eigenvalues and consequently no eigenvectors.

(b) Give a geometric explanation of the result in part (a).

2. Find the eigenvalues of

$$A = \begin{bmatrix} 0 & 1 & 0 \\ 0 & 0 & 1 \\ k^3 & -3k^2 & 3k \end{bmatrix}$$

3. (a) Show that if D is a diagonal matrix with nonnegative entries on the main diagonal, then there is a matrix S such that $S^2 = D$.

   (b) Show that if A is a diagonalizable matrix with nonnegative eigenvalues, then there is a matrix S such that $S^2 = A$.

   (c) Find a matrix S such that $S^2 = A$, if

$$A = \begin{bmatrix} 1 & 3 & 1 \\ 0 & 4 & 5 \\ 0 & 0 & 9 \end{bmatrix}$$

4. Prove: If A is a square matrix, then A and A^T have the same characteristic polynomial.

5. Prove: If A is a square matrix and $p(\lambda) = \det(\lambda I - A)$ is the characteristic polynomial of A, then the coefficient of λ^{n-1} in $p(\lambda)$ is the negative of the trace of A.

6. Prove: If $b \neq 0$, then

$$A = \begin{bmatrix} a & b \\ 0 & a \end{bmatrix}$$

   is not diagonalizable.

7. In advanced linear algebra, one proves the **Cayley–Hamilton Theorem**, which states that a square matrix A satisfies its characteristic equation; that is, if

$$c_0 + c_1\lambda + c_2\lambda^2 + \cdots + c_{n-1}\lambda^{n-1} + \lambda^n = 0$$

   is the characteristic equation of A, then

$$c_0 I + c_1 A + c_2 A^2 + \cdots + c_{n-1}A^{n-1} + A^n = 0$$

   Verify this result for

   (a) $A = \begin{bmatrix} 3 & 6 \\ 1 & 2 \end{bmatrix}$ (b) $A = \begin{bmatrix} 0 & 1 & 0 \\ 0 & 0 & 1 \\ 1 & -3 & 3 \end{bmatrix}$

Exercises 8–10 use the Cayley–Hamilton Theorem, stated in Exercise 7.

8. (a) Use Exercise 16 of Section 7.1 to prove the Cayley–Hamilton Theorem for arbitrary 2×2 matrices.

   (b) Prove the Cayley–Hamilton Theorem for $n \times n$ diagonalizable matrices.

9. The Cayley–Hamilton Theorem provides a method for calculating powers of a matrix. For example, if A is a 2×2 matrix with characteristic equation

$$c_0 + c_1\lambda + \lambda^2 = 0$$

   then $c_0 I + c_1 A + A^2 = 0$, so

$$A^2 = -c_1 A - c_0 I$$

   Multiplying through by A yields $A^3 = -c_1 A^2 - c_0 A$, which expresses A^3 in terms of A^2 and A, and multiplying through by A^2 yields $A^4 = -c_1 A^3 - c_0 A^2$, which expresses A^4 in terms of A^3 and A^2. Continuing in this way, we can calculate successive powers of A simply by expressing them in terms of lower powers. Use this procedure to calculate A^2, A^3, A^4, and A^5 for

$$A = \begin{bmatrix} 3 & 6 \\ 1 & 2 \end{bmatrix}$$

10. Use the method of the preceding exercise to calculate A^3 and A^4 for

$$A = \begin{bmatrix} 0 & 1 & 0 \\ 0 & 0 & 1 \\ 1 & -3 & 3 \end{bmatrix}$$

11. Find the eigenvalues of the matrix

$$A = \begin{bmatrix} c_1 & c_2 & \cdots & c_n \\ c_1 & c_2 & \cdots & c_n \\ \vdots & \vdots & & \vdots \\ c_1 & c_2 & \cdots & c_n \end{bmatrix}$$

12. (a) It was shown in Exercise 15 of Section 7.1 that if A is an $n \times n$ matrix, then the coefficient of λ^n in the characteristic polynomial of A is 1. (A polynomial with this property is called **monic**.) Show that the matrix

$$\begin{bmatrix} 0 & 0 & 0 & \cdots & 0 & -c_0 \\ 1 & 0 & 0 & \cdots & 0 & -c_1 \\ 0 & 1 & 0 & \cdots & 0 & -c_2 \\ \vdots & \vdots & \vdots & & \vdots & \vdots \\ 0 & 0 & 0 & \cdots & 1 & -c_{n-1} \end{bmatrix}$$

has characteristic polynomial $p(\lambda) = c_0 + c_1\lambda + \cdots + c_{n-1}\lambda^{n-1} + \lambda^n$. This shows that every monic polynomial is the characteristic polynomial of some matrix. The matrix in this example is called the **companion matrix** of $p(\lambda)$.

Hint Evaluate all determinants in the problem by adding a multiple of the second row to the first to introduce a zero at the top of the first column, and then expanding by cofactors along the first column.

(b) Find a matrix with characteristic polynomial $p(\lambda) = 1 - 2\lambda + \lambda^2 + 3\lambda^3 + \lambda^4$.

13. A square matrix A is called **nilpotent** if $A^n = 0$ for some positive integer n. What can you say about the eigenvalues of a nilpotent matrix?

14. Prove: If A is an $n \times n$ matrix and n is odd, then A has at least one real eigenvalue.

15. Find a 3×3 matrix A that has eigenvalues $\lambda = 0$, 1, and -1 with corresponding eigenvectors

$$\begin{bmatrix} 0 \\ 1 \\ -1 \end{bmatrix}, \quad \begin{bmatrix} 1 \\ -1 \\ 1 \end{bmatrix}, \quad \begin{bmatrix} 0 \\ 1 \\ 1 \end{bmatrix}$$

respectively.

16. Suppose that a 4×4 matrix A has eigenvalues $\lambda_1 = 1$, $\lambda_2 = -2$, $\lambda_3 = 3$, and $\lambda_4 = -3$.

(a) Use the method of Exercise 14 of Section 7.1 to find $\det(A)$.

(b) Use Exercise 5 above to find $\text{tr}(A)$.

17. Let A be a square matrix such that $A^3 = A$. What can you say about the eigenvalues of A?

CHAPTER 7
Technology Exercises

The following exercises are designed to be solved using a technology utility. Typically, this will be MATLAB, *Mathematica*, Maple, Derive, or Mathcad, but it may also be some other type of linear algebra software or a scientific calculator with some linear algebra capabilities. For each exercise you will need to read the relevant documentation for the particular utility you are using. The goal of these exercises is to provide you with a basic proficiency with your technology utility. Once you have mastered the techniques in these exercises, you will be able to use your technology utility to solve many of the problems in the regular exercise sets.

Section 7.1 **T1.** **(Characteristic Polynomial)** Some technology utilities have a specific command for finding characteristic polynomials, and in others you must use the determinant function to compute $\det(\lambda I - A)$. Read your documentation to determine which method you must use, and then use your utility to find $p(\lambda)$ for the matrix in Example 2.

T2. **(Solving the Characteristic Equation)** Depending on the particular characteristic polynomial, your technology utility may or may not be successful in solving the characteristic equation for the eigenvalues. See if your utility can find the eigenvalues in Example 2 by solving the characteristic equation $p(\lambda) = 0$.

T3. (a) Read the statement of the Cayley–Hamilton Theorem in Supplementary Exercise 7 of this chapter, and then use your technology utility to do that exercise.

(b) If you are working with a CAS, use it to prove the Cayley–Hamilton Theorem for 3×3 matrices.

T4. **(Eigenvalues)** Some technology utilities have specific commands for finding the eigenvalues of a matrix directly (though the procedure may not be successful in all cases). If your utility has this capability, read the documentation and then compute the eigenvalues in Example 2 directly.

T5. **(Eigenvectors)** One way to use a technology utility to find eigenvectors corresponding to an eigenvalue λ is to solve the linear system $(\lambda I - A)\mathbf{x} = 0$. Another way is to use a command for finding a basis for the nullspace of $\lambda I - A$ (if available). However, some utilities have specific commands for finding eigenvectors. Read your documentation, and then explore various procedures for finding the eigenvectors in Examples 5 and 6.

Section 7.2 **T1.** **(Diagonalization)** Some technology utilities have specific commands for diagonalizing a matrix. If your utility has this capability, read the documentation and then use your utility to perform the computations in Example 2.

Note Your software may or may not produce the eigenvalues of A and the columns of P in the same order as the example.

Section 7.3 **T1.** **(Orthogonal Diagonalization)** Use your technology utility to check the computations in Example 1.

Linear Transformations

CHAPTER CONTENTS

INTRODUCTION: In Sections 4.2 and 4.3 we studied linear transformations from R^n to R^m. In this chapter we shall define and study linear transformations from an arbitrary vector space V to another arbitrary vector space W. The results we obtain here have important applications in physics, engineering, and various branches of mathematics.

8.1
GENERAL LINEAR TRANSFORMA-TIONS

In Section 4.2 we defined linear transformations from R^n to R^m. In this section we shall extend this idea by defining the more general concept of a linear transformation from one vector space to another.

Definitions and Terminology

Recall that a linear transformation from R^n to R^m was first defined as a function

$$T(x_1, x_2, \ldots, x_n) = (w_1, w_2, \ldots, w_m)$$

for which the equations relating $w_1, w_2, \ldots, w_m$ and $x_1, x_2, \ldots, x_n$ are linear. Subsequently, we showed that a transformation $T: R^n \to R^m$ is linear if and only if the two relationships

$$T(\mathbf{u} + \mathbf{v}) = T(\mathbf{u}) + T(\mathbf{v}) \quad \text{and} \quad T(c\mathbf{u}) = cT(\mathbf{u})$$

hold for all vectors $\mathbf{u}$ and $\mathbf{v}$ in R^n and every scalar c (see Theorem 4.3.2). We shall use these properties as the starting point for general linear transformations.

DEFINITION

If $T: V \to W$ is a function from a vector space V into a vector space W, then T is called a ***linear transformation*** from V to W if, for all vectors $\mathbf{u}$ and $\mathbf{v}$ in V and all scalars c,

(a) $T(\mathbf{u} + \mathbf{v}) = T(\mathbf{u}) + T(\mathbf{v})$ (b) $T(c\mathbf{u}) = cT(\mathbf{u})$

In the special case where $V = W$, the linear transformation $T: V \to V$ is called a ***linear operator*** on V.

EXAMPLE 1 Matrix Transformations

Because the preceding definition of a linear transformation was based on Theorem 4.3.2, linear transformations from R^n to R^m, as defined in Section 4.2, are linear transformations under this more general definition as well. We shall call linear transformations from R^n to R^m ***matrix transformations***, since they can be carried out by matrix multiplication. ◆

EXAMPLE 2 The Zero Transformation

Let V and W be any two vector spaces. The mapping $T: V \to W$ such that $T(\mathbf{v}) = \mathbf{0}$ for every $\mathbf{v}$ in V is a linear transformation called the ***zero transformation***. To see that T is linear, observe that

$$T(\mathbf{u} + \mathbf{v}) = \mathbf{0}, \quad T(\mathbf{u}) = \mathbf{0}, \quad T(\mathbf{v}) = \mathbf{0}, \quad \text{and} \quad T(k\mathbf{u}) = \mathbf{0}$$

Therefore,

$$T(\mathbf{u} + \mathbf{v}) = T(\mathbf{u}) + T(\mathbf{v}) \quad \text{and} \quad T(k\mathbf{u}) = kT(\mathbf{u}) \quad ◆$$

EXAMPLE 3 The Identity Operator

Let V be any vector space. The mapping $I: V \to V$ defined by $I(\mathbf{v}) = \mathbf{v}$ is called the ***identity operator*** on V. The verification that I is linear is left for the reader. ◆

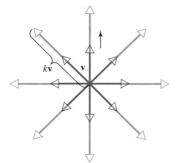

(a) Dilation of *V*

EXAMPLE 4 Dilation and Contraction Operators

Let *V* be any vector space and *k* any fixed scalar. We leave it as an exercise to check that the function $T: V \to V$ defined by

$$T(\mathbf{v}) = k\mathbf{v}$$

is a linear operator on *V*. This linear operator is called a ***dilation*** of *V* with factor *k* if $k > 1$ and is called a ***contraction*** of *V* with factor *k* if $0 < k < 1$. Geometrically, the dilation "stretches" each vector in *V* by a factor of *k*, and the contraction of *V* "compresses" each vector by a factor of *k* (Figure 8.1.1). ◆

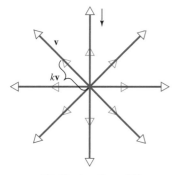

(b) Contraction of *V*

Figure 8.1.1

EXAMPLE 5 Orthogonal Projections

In Section 6.4 we defined the orthogonal projection of R^m onto a subspace *W*. [See Formula (6) and the definition preceding it in that section.] Orthogonal projections can also be defined in general inner product spaces as follows: Suppose that *W* is a finite-dimensional subspace of an inner product space *V*; then the ***orthogonal projection of V onto W*** is the transformation defined by

$$T(\mathbf{v}) = \text{proj}_W \mathbf{v}$$

(Figure 8.1.2). It follows from Theorem 6.3.5 that if

$$S = \{\mathbf{w}_1, \mathbf{w}_2, \ldots, \mathbf{w}_r\}$$

is any orthonormal basis for *W*, then $T(\mathbf{v})$ is given by the formula

$$T(\mathbf{v}) = \text{proj}_W \mathbf{v} = \langle \mathbf{v}, \mathbf{w}_1 \rangle \mathbf{w}_1 + \langle \mathbf{v}, \mathbf{w}_2 \rangle \mathbf{w}_2 + \cdots + \langle \mathbf{v}, \mathbf{w}_r \rangle \mathbf{w}_r$$

The proof that *T* is a linear transformation follows from properties of the inner product. For example,

$$\begin{aligned}
T(\mathbf{u} + \mathbf{v}) &= \langle \mathbf{u} + \mathbf{v}, \mathbf{w}_1 \rangle \mathbf{w}_1 + \langle \mathbf{u} + \mathbf{v}, \mathbf{w}_2 \rangle \mathbf{w}_2 + \cdots + \langle \mathbf{u} + \mathbf{v}, \mathbf{w}_r \rangle \mathbf{w}_r \\
&= \langle \mathbf{u}, \mathbf{w}_1 \rangle \mathbf{w}_1 + \langle \mathbf{u}, \mathbf{w}_2 \rangle \mathbf{w}_2 + \cdots + \langle \mathbf{u}, \mathbf{w}_r \rangle \mathbf{w}_r \\
&\quad + \langle \mathbf{v}, \mathbf{w}_1 \rangle \mathbf{w}_1 + \langle \mathbf{v}, \mathbf{w}_2 \rangle \mathbf{w}_2 + \cdots + \langle \mathbf{v}, \mathbf{w}_r \rangle \mathbf{w}_r \\
&= T(\mathbf{u}) + T(\mathbf{v})
\end{aligned}$$

Similarly, $T(k\mathbf{u}) = kT(\mathbf{u})$. ◆

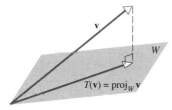

Figure 8.1.2 The orthogonal projection of *V* onto *W*.

EXAMPLE 6 Computing an Orthogonal Projection

As a special case of the preceding example, let $V = R^3$ have the Euclidean inner product. The vectors $\mathbf{w}_1 = (1, 0, 0)$ and $\mathbf{w}_2 = (0, 1, 0)$ form an orthonormal basis for the *xy*-plane. Thus, if $\mathbf{v} = (x, y, z)$ is any vector in R^3, the orthogonal projection of R^3 onto the *xy*-plane is given by

$$\begin{aligned}
T(\mathbf{v}) &= \langle \mathbf{v}, \mathbf{w}_1 \rangle \mathbf{w}_1 + \langle \mathbf{v}, \mathbf{w}_2 \rangle \mathbf{w}_2 \\
&= x(1, 0, 0) + y(0, 1, 0) \\
&= (x, y, 0)
\end{aligned}$$

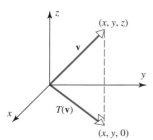

Figure 8.1.3 The orthogonal projection of R^3 onto the *xy*-plane.

(See Figure 8.1.3.) ◆

EXAMPLE 7   A Linear Transformation from a Space V to R^n

Let $S = \{\mathbf{w}_1, \mathbf{w}_2, \ldots, \mathbf{w}_n\}$ be a basis for an n-dimensional vector space V, and let

$$(\mathbf{v})_S = (k_1, k_2, \ldots, k_n)$$

be the coordinate vector relative to S of a vector $\mathbf{v}$ in V; thus

$$\mathbf{v} = k_1\mathbf{w}_1 + k_2\mathbf{w}_2 + \cdots + k_n\mathbf{w}_n$$

Define $T: V \to R^n$ to be the function that maps $\mathbf{v}$ into its coordinate vector relative to S—that is,

$$T(\mathbf{v}) = (\mathbf{v})_S = (k_1, k_2, \ldots, k_n)$$

The function T is a linear transformation. To see that this is so, suppose that $\mathbf{u}$ and $\mathbf{v}$ are vectors in V and that

$$\mathbf{u} = c_1\mathbf{w}_1 + c_2\mathbf{w}_2 + \cdots + c_n\mathbf{w}_n \quad \text{and} \quad \mathbf{v} = d_1\mathbf{w}_1 + d_2\mathbf{w}_2 + \cdots + d_n\mathbf{w}_n$$

Thus

$$(\mathbf{u})_S = (c_1, c_2, \ldots, c_n) \quad \text{and} \quad (\mathbf{v})_S = (d_1, d_2, \ldots, d_n)$$

But

$$\mathbf{u} + \mathbf{v} = (c_1 + d_1)\mathbf{w}_1 + (c_2 + d_2)\mathbf{w}_2 + \cdots + (c_n + d_n)\mathbf{w}_n$$
$$k\mathbf{u} = (kc_1)\mathbf{w}_1 + (kc_2)\mathbf{w}_2 + \cdots + (kc_n)\mathbf{w}_n$$

so

$$(\mathbf{u} + \mathbf{v})_S = (c_1 + d_1, c_2 + d_2, \ldots, c_n + d_n)$$
$$(k\mathbf{u})_S = (kc_1, kc_2, \ldots, kc_n)$$

Therefore,

$$(\mathbf{u} + \mathbf{v})_S = (\mathbf{u})_S + (\mathbf{v})_S \quad \text{and} \quad (k\mathbf{u})_S = k(\mathbf{u})_S$$

Expressing these equations in terms of T, we obtain

$$T(\mathbf{u} + \mathbf{v}) = T(\mathbf{u}) + T(\mathbf{v}) \quad \text{and} \quad T(k\mathbf{u}) = kT(\mathbf{u})$$

which shows that T is a linear transformation. ◆

REMARK   The computations in the preceding example could just as well have been performed using coordinate vectors in column form; that is,

$$[\mathbf{u} + \mathbf{v}]_S = [\mathbf{u}]_S + [\mathbf{v}]_S \quad \text{and} \quad [k\mathbf{u}]_S = k[\mathbf{u}]_S$$

EXAMPLE 8   A Linear Transformation from P_n to P_{n+1}

Let $\mathbf{p} = p(x) = c_0 + c_1 x + \cdots + c_n x^n$ be a polynomial in P_n, and define the function $T: P_n \to P_{n+1}$ by

$$T(\mathbf{p}) = T(p(x)) = xp(x) = c_0 x + c_1 x^2 + \cdots + c_n x^{n+1}$$

The function T is a linear transformation, since for any scalar k and any polynomials $\mathbf{p}_1$ and $\mathbf{p}_2$ in P_n we have

$$T(\mathbf{p}_1 + \mathbf{p}_2) = T(p_1(x) + p_2(x)) = x(p_1(x) + p_2(x))$$
$$= xp_1(x) + xp_2(x) = T(\mathbf{p}_1) + T(\mathbf{p}_2)$$

and

$$T(k\mathbf{p}) = T(kp(x)) = x(kp(x)) = k(xp(x)) = kT(\mathbf{p})$$

(Compare this to Exercise 4 of Section 4.4.) ◆

EXAMPLE 9 A Linear Operator on P_n

Let $\mathbf{p} = p(x) = c_0 + c_1x + \cdots + c_nx^n$ be a polynomial in P_n, and let a and b be any scalars. We leave it as an exercise to show that the function T defined by

$$T(\mathbf{p}) = T(p(x)) = p(ax + b) = c_0 + c_1(ax + b) + \cdots + c_n(ax + b)^n$$

is a linear operator. For example, if $ax + b = 3x - 5$, then $T: P_2 \to P_2$ would be the linear operator given by the formula

$$T(c_0 + c_1x + c_2x^2) = c_0 + c_1(3x - 5) + c_2(3x - 5)^2 \quad \blacklozenge$$

EXAMPLE 10 A Linear Transformation Using an Inner Product

Let V be an inner product space, and let $\mathbf{v}_0$ be any fixed vector in V. Let $T: V \to R$ be the transformation that maps a vector $\mathbf{v}$ into its inner product with $\mathbf{v}_0$—that is,

$$T(\mathbf{v}) = \langle \mathbf{v}, \mathbf{v}_0 \rangle$$

From the properties of an inner product,

$$T(\mathbf{u} + \mathbf{v}) = \langle \mathbf{u} + \mathbf{v}, \mathbf{v}_0 \rangle = \langle \mathbf{u}, \mathbf{v}_0 \rangle + \langle \mathbf{v}, \mathbf{v}_0 \rangle = T(\mathbf{u}) + T(\mathbf{v})$$

and

$$T(k\mathbf{u}) = \langle k\mathbf{u}, \mathbf{v}_0 \rangle = k\langle \mathbf{u}, \mathbf{v}_0 \rangle = kT(\mathbf{u})$$

so T is a linear transformation. $\blacklozenge$

Calculus Required

EXAMPLE 11 A Linear Transformation from $C^1(-\infty, \infty)$ to $F(-\infty, \infty)$

Let $V = C^1(-\infty, \infty)$ be the vector space of functions with continuous first derivatives on $(-\infty, \infty)$, and let $W = F(-\infty, \infty)$ be the vector space of all real-valued functions defined on $(-\infty, \infty)$. Let $D: V \to W$ be the transformation that maps a function $\mathbf{f} = f(x)$ into its derivative—that is,

$$D(\mathbf{f}) = f'(x)$$

From the properties of differentiation, we have

$$D(\mathbf{f} + \mathbf{g}) = D(\mathbf{f}) + D(\mathbf{g}) \quad \text{and} \quad D(k\mathbf{f}) = kD(\mathbf{f})$$

Thus, D is a linear transformation. $\blacklozenge$

Calculus Required

EXAMPLE 12 A Linear Transformation from $C(-\infty, \infty)$ to $C^1(-\infty, \infty)$

Let $V = C(-\infty, \infty)$ be the vector space of continuous functions on $(-\infty, \infty)$, and let $W = C^1(-\infty, \infty)$ be the vector space of functions with continuous first derivatives on

$(-\infty, \infty)$. Let $J: V \to W$ be the transformation that maps $\mathbf{f} = f(x)$ into the integral $\int_0^x f(t)\, dt$. For example, if $\mathbf{f} = x^2$, then

$$J(\mathbf{f}) = \int_0^x t^2\, dt = \left. \frac{t^3}{3} \right|_0^x = \frac{x^3}{3}$$

From the properties of integration, we have

$$J(\mathbf{f} + \mathbf{g}) = \int_0^x (f(t) + g(t))\, dt = \int_0^x f(t)\, dt + \int_0^x g(t)\, dt = J(\mathbf{f}) + J(\mathbf{g})$$

$$J(c\mathbf{f}) = \int_0^x cf(t)\, dt = c \int_0^x f(t)\, dt = cJ(\mathbf{f})$$

so J is a linear transformation. ◆

EXAMPLE 13 A Transformation That Is Not Linear

Let $T: M_{nn} \to R$ be the transformation that maps an $n \times n$ matrix into its determinant:

$$T(A) = \det(A)$$

If $n > 1$, then this transformation does not satisfy either of the properties required of a linear transformation. For example, we saw in Example 1 of Section 2.3 that

$$\det(A_1 + A_2) \neq \det(A_1) + \det(A_2)$$

in general. Moreover, $\det(cA) = c^n \det(A)$, so

$$\det(cA) \neq c \det(A)$$

in general. Thus T is *not* a linear transformation. ◆

Properties of Linear Transformations

If $T: V \to W$ is a linear transformation, then for any vectors $\mathbf{v}_1$ and $\mathbf{v}_2$ in V and any scalars c_1 and c_2, we have

$$T(c_1 \mathbf{v}_1 + c_2 \mathbf{v}_2) = T(c_1 \mathbf{v}_1) + T(c_2 \mathbf{v}_2) = c_1 T(\mathbf{v}_1) + c_2 T(\mathbf{v}_2)$$

and, more generally, if $\mathbf{v}_1, \mathbf{v}_2, \ldots, \mathbf{v}_n$ are vectors in V and $c_1, c_2, \ldots, c_n$ are scalars, then

$$T(c_1 \mathbf{v}_1 + c_2 \mathbf{v}_2 + \cdots + c_n \mathbf{v}_n) = c_1 T(\mathbf{v}_1) + c_2 T(\mathbf{v}_2) + \cdots + c_n T(\mathbf{v}_n) \qquad (1)$$

Formula (1) is sometimes described by saying that *linear transformations preserve linear combinations*.

The following theorem lists three basic properties that are common to all linear transformations.

THEOREM 8.1.1

If $T: V \to W$ is a linear transformation, then

(a) $T(\mathbf{0}) = \mathbf{0}$
(b) $T(-\mathbf{v}) = -T(\mathbf{v})$ *for all $\mathbf{v}$ in V*
(c) $T(\mathbf{v} - \mathbf{w}) = T(\mathbf{v}) - T(\mathbf{w})$ *for all $\mathbf{v}$ and $\mathbf{w}$ in V*

Proof Let $\mathbf{v}$ be any vector in V. Since $0\mathbf{v} = \mathbf{0}$, we have

$$T(\mathbf{0}) = T(0\mathbf{v}) = 0T(\mathbf{v}) = \mathbf{0}$$

which proves (*a*). Also,

$$T(-\mathbf{v}) = T((-1)\mathbf{v}) = (-1)T(\mathbf{v}) = -T(\mathbf{v})$$

which proves (*b*). Finally, $\mathbf{v} - \mathbf{w} = \mathbf{v} + (-1)\mathbf{w}$; thus

$$\begin{aligned}
T(\mathbf{v} - \mathbf{w}) &= T(\mathbf{v} + (-1)\mathbf{w}) \\
&= T(\mathbf{v}) + (-1)T(\mathbf{w}) \\
&= T(\mathbf{v}) - T(\mathbf{w})
\end{aligned}$$

which proves (*c*). ∎

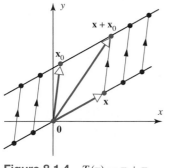

Figure 8.1.4 $T(\mathbf{x}) = \mathbf{x} + \mathbf{x}_0$ translates each point **x** along a line parallel to $\mathbf{x}_0$ through a distance $\|\mathbf{x}_0\|$.

In words, part (*a*) of the preceding theorem states that a linear transformation maps **0** to **0**. This property is useful for identifying transformations that are *not* linear. For example, if $\mathbf{x}_0$ is a fixed nonzero vector in R^2, then the transformation

$$T(\mathbf{x}) = \mathbf{x} + \mathbf{x}_0$$

has the geometric effect of translating each point **x** in a direction parallel to $\mathbf{x}_0$ through a distance of $\|\mathbf{x}_0\|$ (Figure 8.1.4). This cannot be a linear transformation, since $T(\mathbf{0}) = \mathbf{x}_0$, so T does not map **0** to **0**.

Finding Linear Transformations from Images of Basis Vectors

Theorem 4.3.3 shows that if T is a matrix transformation, then the standard matrix for T can be obtained from the images of the standard basis vectors. Stated another way, *a matrix transformation is completely determined by its images of the standard basis vectors*. This is a special case of a more general result: If $T: V \rightarrow W$ is a linear transformation, and if $\{\mathbf{v}_1, \mathbf{v}_2, \ldots, \mathbf{v}_n\}$ is any basis for V, then the image $T(\mathbf{v})$ of any vector **v** in V can be calculated from the images

$$T(\mathbf{v}_1), T(\mathbf{v}_2), \ldots, T(\mathbf{v}_n)$$

of the basis vectors. This can be done by first expressing **v** as a linear combination of the basis vectors, say

$$\mathbf{v} = c_1\mathbf{v}_1 + c_2\mathbf{v}_2 + \cdots + c_n\mathbf{v}_n$$

and then using Formula (1) to write

$$T(\mathbf{v}) = c_1 T(\mathbf{v}_1) + c_2 T(\mathbf{v}_2) + \cdots + c_n T(\mathbf{v}_n)$$

In words, *a linear transformation is completely determined by the images of any set of basis vectors*.

EXAMPLE 14 Computing with Images of Basis Vectors

Consider the basis $S = \{\mathbf{v}_1, \mathbf{v}_2, \mathbf{v}_3\}$ for R^3, where $\mathbf{v}_1 = (1, 1, 1)$, $\mathbf{v}_2 = (1, 1, 0)$, and $\mathbf{v}_3 = (1, 0, 0)$. Let $T: R^3 \rightarrow R^2$ be the linear transformation such that

$$T(\mathbf{v}_1) = (1, 0), \qquad T(\mathbf{v}_2) = (2, -1), \qquad T(\mathbf{v}_3) = (4, 3)$$

Find a formula for $T(x_1, x_2, x_3)$; then use this formula to compute $T(2, -3, 5)$.

Solution

We first express $\mathbf{x} = (x_1, x_2, x_3)$ as a linear combination of $\mathbf{v}_1 = (1, 1, 1)$, $\mathbf{v}_2 = (1, 1, 0)$, and $\mathbf{v}_3 = (1, 0, 0)$. If we write

$$(x_1, x_2, x_3) = c_1(1, 1, 1) + c_2(1, 1, 0) + c_3(1, 0, 0)$$

then on equating corresponding components, we obtain

$$c_1 + c_2 + c_3 = x_1$$
$$c_1 + c_2 \qquad = x_2$$
$$c_1 \qquad\qquad = x_3$$

which yields $c_1 = x_3, c_2 = x_2 - x_3, c_3 = x_1 - x_2$, so

$$(x_1, x_2, x_3) = x_3(1, 1, 1) + (x_2 - x_3)(1, 1, 0) + (x_1 - x_2)(1, 0, 0)$$
$$= x_3\mathbf{v}_1 + (x_2 - x_3)\mathbf{v}_2 + (x_1 - x_2)\mathbf{v}_3$$

Thus

$$T(x_1, x_2, x_3) = x_3 T(\mathbf{v}_1) + (x_2 - x_3)T(\mathbf{v}_2) + (x_1 - x_2)T(\mathbf{v}_3)$$
$$= x_3(1, 0) + (x_2 - x_3)(2, -1) + (x_1 - x_2)(4, 3)$$
$$= (4x_1 - 2x_2 - x_3, 3x_1 - 4x_2 + x_3)$$

From this formula, we obtain

$$T(2, -3, 5) = (9, 23) \quad \blacklozenge$$

In Section 4.2 we defined the composition of matrix transformations. The following definition extends that concept to general linear transformations.

DEFINITION

If $T_1: U \rightarrow V$ and $T_2: V \rightarrow W$ are linear transformations, then the ***composition of T_2 with T_1***, denoted by $T_2 \circ T_1$ (which is read "T_2 circle T_1"), is the function defined by the formula

$$(T_2 \circ T_1)(\mathbf{u}) = T_2(T_1(\mathbf{u})) \tag{2}$$

where $\mathbf{u}$ is a vector in U.

REMARK   Observe that this definition requires that the domain of T_2 (which is V) contain the range of T_1; this is essential for the formula $T_2(T_1(\mathbf{u}))$ to make sense (Figure 8.1.5). The reader should compare (2) to Formula (18) in Section 4.2.

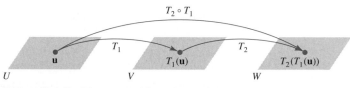

Figure 8.1.5   The composition of T_2 with T_1.

The next result shows that the composition of two linear transformations is itself a linear transformation.

THEOREM 8.1.2

If $T_1: U \rightarrow V$ and $T_2: V \rightarrow W$ are linear transformations, then $(T_2 \circ T_1): U \rightarrow W$ is also a linear transformation.

Proof If **u** and **v** are vectors in U and c is a scalar, then it follows from (2) and the linearity of T_1 and T_2 that

$$(T_2 \circ T_1)(\mathbf{u} + \mathbf{v}) = T_2(T_1(\mathbf{u} + \mathbf{v})) = T_2(T_1(\mathbf{u}) + T_1(\mathbf{v}))$$
$$= T_2(T_1(\mathbf{u})) + T_2(T_1(\mathbf{v}))$$
$$= (T_2 \circ T_1)(\mathbf{u}) + (T_2 \circ T_1)(\mathbf{v})$$

and

$$(T_2 \circ T_1)(c\mathbf{u}) = T_2(T_1(c\mathbf{u})) = T_2(cT_1(\mathbf{u}))$$
$$= cT_2(T_1(\mathbf{u})) = c(T_2 \circ T_1)(\mathbf{u})$$

Thus $T_2 \circ T_1$ satisfies the two requirements of a linear transformation. ∎

EXAMPLE 15 Composition of Linear Transformations

Let $T_1: P_1 \rightarrow P_2$ and $T_2: P_2 \rightarrow P_2$ be the linear transformations given by the formulas

$$T_1(p(x)) = xp(x) \quad \text{and} \quad T_2(p(x)) = p(2x + 4)$$

Then the composition $(T_2 \circ T_1): P_1 \rightarrow P_2$ is given by the formula

$$(T_2 \circ T_1)(p(x)) = T_2(T_1(p(x))) = T_2(xp(x)) = (2x + 4)p(2x + 4)$$

In particular, if $p(x) = c_0 + c_1 x$, then

$$(T_2 \circ T_1)(p(x)) = (T_2 \circ T_1)(c_0 + c_1 x) = (2x + 4)(c_0 + c_1(2x + 4))$$
$$= c_0(2x + 4) + c_1(2x + 4)^2 \quad \blacklozenge$$

EXAMPLE 16 Composition with the Identity Operator

If $T: V \rightarrow V$ is any linear operator, and if $I: V \rightarrow V$ is the identity operator (Example 3), then for all vectors **v** in V, we have

$$(T \circ I)(\mathbf{v}) = T(I(\mathbf{v})) = T(\mathbf{v})$$
$$(I \circ T)(\mathbf{v}) = I(T(\mathbf{v})) = T(\mathbf{v})$$

It follows that $T \circ I$ and $I \circ T$ are the same as T; that is,

$$T \circ I = T \quad \text{and} \quad I \circ T = T \quad \blacklozenge \tag{3}$$

We conclude this section by noting that compositions can be defined for more than two linear transformations. For example, if

$$T_1: U \rightarrow V, \quad T_2: V \rightarrow W, \quad \text{and} \quad T_3: W \rightarrow Y$$

are linear transformations, then the composition $T_3 \circ T_2 \circ T_1$ is defined by

$$(T_3 \circ T_2 \circ T_1)(\mathbf{u}) = T_3(T_2(T_1(\mathbf{u}))) \tag{4}$$

(Figure 8.1.6).

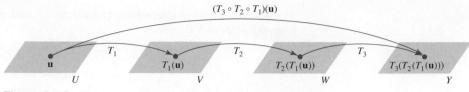

Figure 8.1.6 The composition of three linear transformations.

EXERCISE SET
8.1

1. Use the definition of a linear operator that was given in this section to show that the function $T: R^2 \rightarrow R^2$ given by the formula $T(x_1, x_2) = (x_1 + 2x_2, 3x_1 - x_2)$ is a linear operator.

2. Use the definition of a linear transformation given in this section to show that the function $T: R^3 \rightarrow R^2$ given by the formula $T(x_1, x_2, x_3) = (2x_1 - x_2 + x_3, x_2 - 4x_3)$ is a linear transformation.

In Exercises 3–10 determine whether the function is a linear transformation. Justify your answer.

3. $T: V \rightarrow R$, where V is an inner product space, and $T(\mathbf{u}) = \|\mathbf{u}\|$.

4. $T: R^3 \rightarrow R^3$, where $\mathbf{v}_0$ is a fixed vector in R^3 and $T(\mathbf{u}) = \mathbf{u} \times \mathbf{v}_0$.

5. $T: M_{22} \rightarrow M_{23}$, where B is a fixed 2×3 matrix and $T(A) = AB$.

6. $T: M_{nn} \rightarrow R$, where $T(A) = \text{tr}(A)$. 7. $F: M_{mn} \rightarrow M_{nm}$, where $F(A) = A^T$.

8. $T: M_{22} \rightarrow R$, where

   (a) $T\left(\begin{bmatrix} a & b \\ c & d \end{bmatrix} \right) = 3a - 4b + c - d$ (b) $T\left(\begin{bmatrix} a & b \\ c & d \end{bmatrix} \right) = a^2 + b^2$

9. $T: P_2 \rightarrow P_2$, where

   (a) $T(a_0 + a_1 x + a_2 x^2) = a_0 + a_1(x + 1) + a_2(x + 1)^2$

   (b) $T(a_0 + a_1 x + a_2 x^2) = (a_0 + 1) + (a_1 + 1)x + (a_2 + 1)x^2$

10. $T: F(-\infty, \infty) \rightarrow F(-\infty, \infty)$, where

   (a) $T(f(x)) = 1 + f(x)$ (b) $T(f(x)) = f(x + 1)$

11. Show that the function T in Example 9 is a linear operator.

12. Consider the basis $S = \{\mathbf{v}_1, \mathbf{v}_2\}$ for R^2, where $\mathbf{v}_1 = (1, 1)$ and $\mathbf{v}_2 = (1, 0)$, and let $T: R^2 \rightarrow R^2$ be the linear operator such that

$$T(\mathbf{v}_1) = (1, -2) \quad \text{and} \quad T(\mathbf{v}_2) = (-4, 1)$$

   Find a formula for $T(x_1, x_2)$, and use that formula to find $T(5, -3)$.

13. Consider the basis $S = \{\mathbf{v}_1, \mathbf{v}_2\}$ for R^2, where $\mathbf{v}_1 = (-2, 1)$ and $\mathbf{v}_2 = (1, 3)$, and let $T: R^2 \rightarrow R^3$ be the linear transformation such that

$$T(\mathbf{v}_1) = (-1, 2, 0) \quad \text{and} \quad T(\mathbf{v}_2) = (0, -3, 5)$$

   Find a formula for $T(x_1, x_2)$, and use that formula to find $T(2, -3)$.

14. Consider the basis $S = \{\mathbf{v}_1, \mathbf{v}_2, \mathbf{v}_3\}$ for R^3, where $\mathbf{v}_1 = (1, 1, 1)$, $\mathbf{v}_2 = (1, 1, 0)$, and $\mathbf{v}_3 = (1, 0, 0)$, and let $T: R^3 \rightarrow R^3$ be the linear operator such that

$$T(\mathbf{v}_1) = (2, -1, 4), \qquad T(\mathbf{v}_2) = (3, 0, 1), \qquad T(\mathbf{v}_3) = (-1, 5, 1)$$

   Find a formula for $T(x_1, x_2, x_3)$, and use that formula to find $T(2, 4, -1)$.

15. Consider the basis $S = \{\mathbf{v}_1, \mathbf{v}_2, \mathbf{v}_3\}$ for R^3, where $\mathbf{v}_1 = (1, 2, 1)$, $\mathbf{v}_2 = (2, 9, 0)$, and $\mathbf{v}_3 = (3, 3, 4)$, and let $T: R^3 \rightarrow R^2$ be the linear transformation such that

$$T(\mathbf{v}_1) = (1, 0), \qquad T(\mathbf{v}_2) = (-1, 1), \qquad T(\mathbf{v}_3) = (0, 1)$$

   Find a formula for $T(x_1, x_2, x_3)$, and use that formula to find $T(7, 13, 7)$.

16. Let $\mathbf{v}_1$, $\mathbf{v}_2$, and $\mathbf{v}_3$ be vectors in a vector space V, and let $T: V \rightarrow R^3$ be a linear transformation for which

$$T(\mathbf{v}_1) = (1, -1, 2), \qquad T(\mathbf{v}_2) = (0, 3, 2), \qquad T(\mathbf{v}_3) = (-3, 1, 2)$$

Find $T(2\mathbf{v}_1 - 3\mathbf{v}_2 + 4\mathbf{v}_3)$.

17. Find the domain and codomain of $T_2 \circ T_1$, and find $(T_2 \circ T_1)(x, y)$.

(a) $T_1(x, y) = (2x, 3y)$, $T_2(x, y) = (x - y, x + y)$

(b) $T_1(x, y) = (x - 3y, 0)$, $T_2(x, y) = (4x - 5y, 3x - 6y)$

(c) $T_1(x, y) = (2x, -3y, x + y)$, $T_2(x, y, z) = (x - y, y + z)$

(d) $T_1(x, y) = (x - y, y + z, x - z)$, $T_2(x, y, z) = (0, x + y + z)$

18. Find the domain and codomain of $T_3 \circ T_2 \circ T_1$, and find $(T_3 \circ T_2 \circ T_1)(x, y)$.

(a) $T_1(x, y) = (-2y, 3x, x - 2y)$, $T_2(x, y, z) = (y, z, x)$, $T_3(x, y, z) = (x + z, y - z)$

(b) $T_1(x, y) = (x + y, y, -x)$, $T_2(x, y, z) = (0, x + y + z, 3y)$,
$T_3(x, y, z) = (3x + 2y, 4z - x - 3y)$

19. Let $T_1: M_{22} \rightarrow R$ and $T_2: M_{22} \rightarrow M_{22}$ be the linear transformations given by $T_1(A) = \text{tr}(A)$ and $T_2(A) = A^T$.

(a) Find $(T_1 \circ T_2)(A)$, where $A = \begin{bmatrix} a & b \\ c & d \end{bmatrix}$.

(b) Can you find $(T_2 \circ T_1)(A)$? Explain.

20. Let $T_1: P_n \rightarrow P_n$ and $T_2: P_n \rightarrow P_n$ be the linear operators given by $T_1(p(x)) = p(x - 1)$ and $T_2(p(x)) = p(x + 1)$. Find $(T_1 \circ T_2)(p(x))$ and $(T_2 \circ T_1)(p(x))$.

21. Let $T_1: V \rightarrow V$ be the dilation $T_1(\mathbf{v}) = 4\mathbf{v}$. Find a linear operator $T_2: V \rightarrow V$ such that $T_1 \circ T_2 = I$ and $T_2 \circ T_1 = I$.

22. Suppose that the linear transformations $T_1: P_2 \rightarrow P_2$ and $T_2: P_2 \rightarrow P_3$ are given by the formulas $T_1(p(x)) = p(x + 1)$ and $T_2(p(x)) = xp(x)$. Find $(T_2 \circ T_1)(a_0 + a_1 x + a_2 x^2)$.

23. Let $q_0(x)$ be a fixed polynomial of degree m, and define a function T with domain P_n by the formula $T(p(x)) = p(q_0(x))$.

(a) Show that T is a linear transformation.

(b) What is the codomain of T?

24. Use the definition of $T_3 \circ T_2 \circ T_1$ given by Formula (4) to prove that

(a) $T_3 \circ T_2 \circ T_1$ is a linear transformation

(b) $T_3 \circ T_2 \circ T_1 = (T_3 \circ T_2) \circ T_1$

(c) $T_3 \circ T_2 \circ T_1 = T_3 \circ (T_2 \circ T_1)$

25. Let $T: R^3 \rightarrow R^3$ be the orthogonal projection of R^3 onto the xy-plane. Show that $T \circ T = T$.

26. (a) Let $T: V \rightarrow W$ be a linear transformation, and let k be a scalar. Define the function $(kT): V \rightarrow W$ by $(kT)(\mathbf{v}) = k(T(\mathbf{v}))$. Show that kT is a linear transformation.

(b) Find $(3T)(x_1, x_2)$ if $T: R^2 \rightarrow R^2$ is given by the formula $T(x_1, x_2) = (2x_1 - x_2, x_2 + x_1)$.

27. (a) Let $T_1: V \rightarrow W$ and $T_2: V \rightarrow W$ be linear transformations. Define the functions $(T_1 + T_2): V \rightarrow W$ and $(T_1 - T_2): V \rightarrow W$ by

$$(T_1 + T_2)(\mathbf{v}) = T_1(\mathbf{v}) + T_2(\mathbf{v})$$
$$(T_1 - T_2)(\mathbf{v}) = T_1(\mathbf{v}) - T_2(\mathbf{v})$$

Show that $T_1 + T_2$ and $T_1 - T_2$ are linear transformations.

(b) Find $(T_1 + T_2)(x, y)$ and $(T_1 - T_2)(x, y)$ if $T_1: R^2 \rightarrow R^2$ and $T_2: R^2 \rightarrow R^2$ are given by the formulas $T_1(x, y) = (2y, 3x)$ and $T_2(x, y) = (y, x)$.

28. (a) Prove that if a_1, a_2, b_1, and b_2 are any scalars, then the formula

$$F(x, y) = (a_1 x + b_1 y, a_2 x + b_2 y)$$

defines a linear operator on R^2.

(b) Does the formula $F(x, y) = (a_1 x^2 + b_1 y^2, a_2 x^2 + b_2 y^2)$ define a linear operator on R^2? Explain.

29. Let $\{\mathbf{v}_1, \mathbf{v}_2, \ldots, \mathbf{v}_n\}$ be a basis for a vector space V, and let $T: V \rightarrow W$ be a linear transformation. Show that if $T(\mathbf{v}_1) = T(\mathbf{v}_2) = \cdots = T(\mathbf{v}_n) = \mathbf{0}$, then T is the zero transformation.

30. Let $\{\mathbf{v}_1, \mathbf{v}_2, \ldots, \mathbf{v}_n\}$ be a basis for a vector space V, and let $T: V \rightarrow V$ be a linear operator. Show that if $T(\mathbf{v}_1) = \mathbf{v}_1, T(\mathbf{v}_2) = \mathbf{v}_2, \ldots, T(\mathbf{v}_n) = \mathbf{v}_n$, then T is the identity transformation on V.

31. **(For Readers Who Have Studied Calculus)** Let

$$D(\mathbf{f}) = f'(x) \quad \text{and} \quad J(\mathbf{f}) = \int_0^x f(t)\, dt$$

be the linear transformations in Examples 11 and 12. Find $(J \circ D)(\mathbf{f})$ for

(a) $\mathbf{f}(x) = x^2 + 3x + 2$ (b) $\mathbf{f}(x) = \sin x$ (c) $\mathbf{f}(x) = e^x + 3$

32. **(For Readers Who Have Studied Calculus)** Let $V = C[a, b]$ be the vector space of functions continuous on $[a, b]$, and let $T: V \rightarrow V$ be the transformation defined by

$$T(\mathbf{f}) = 5f(x) + 3 \int_a^x f(t)\, dt$$

Is T a linear operator?

Discussion & Discovery

33. Indicate whether each statement is always true or sometimes false. Justify your answer by giving a logical argument or a counterexample. In each part, V and W are vector spaces.

(a) If $T(c_1 \mathbf{v}_1 + c_2 \mathbf{v}_2) = c_1 T(\mathbf{v}_1) + c_2 T(\mathbf{v}_2)$ for all vectors $\mathbf{v}_1$ and $\mathbf{v}_2$ in V and all scalars c_1 and c_2, then T is a linear transformation.

(b) If $\mathbf{v}$ is a nonzero vector in V, then there is exactly one linear transformation $T: V \rightarrow W$ such that $T(-\mathbf{v}) = -T(\mathbf{v})$.

(c) There is exactly one linear transformation $T: V \rightarrow W$ for which $T(\mathbf{u} + \mathbf{v}) = T(\mathbf{u} - \mathbf{v})$ for all vectors $\mathbf{u}$ and $\mathbf{v}$ in V.

(d) If $\mathbf{v}_0$ is a nonzero vector in V, then the formula $T(\mathbf{v}) = \mathbf{v}_0 + \mathbf{v}$ defines a linear operator on V.

34. If $B = \{\mathbf{v}_1, \mathbf{v}_2, \ldots, \mathbf{v}_n\}$ is a basis for a vector space V, how many different linear operators can be created that map each vector in B back into B? Explain your reasoning.

35. Refer to Section 4.4. Are the transformations from P_n to P_m that correspond to linear transformations from R^{n+1} to R^{m+1} necessarily linear transformations from P_n to P_m?

8.2
KERNEL AND RANGE

In this section we shall develop some basic properties of linear transformations that generalize properties of matrix transformations obtained earlier in the text.

Kernel and Range

Recall that if A is an $m \times n$ matrix, then the nullspace of A consists of all vectors $\mathbf{x}$ in R^n such that $A\mathbf{x} = \mathbf{0}$, and by Theorem 5.5.1 the column space of A consists of all vectors $\mathbf{b}$ in R^m for which there is at least one vector $\mathbf{x}$ in R^n such that $A\mathbf{x} = \mathbf{b}$. From the viewpoint of matrix transformations, the nullspace of A consists of all vectors in R^n that multiplication by A maps into **0**, and the column space of A consists of all vectors in R^m that are images of at least one vector in R^n under multiplication by A. The following definition extends these ideas to general linear transformations.

DEFINITION

If $T: V \rightarrow W$ is a linear transformation, then the set of vectors in V that T maps into **0** is called the *kernel* of T; it is denoted by $\ker(T)$. The set of all vectors in W that are images under T of at least one vector in V is called the *range* of T; it is denoted by $R(T)$.

EXAMPLE 1 Kernel and Range of a Matrix Transformation

If $T_A: R^n \rightarrow R^m$ is multiplication by the $m \times n$ matrix A, then from the discussion preceding the definition above, the kernel of T_A is the nullspace of A, and the range of T_A is the column space of A. ◆

EXAMPLE 2 Kernel and Range of the Zero Transformation

Let $T: V \rightarrow W$ be the zero transformation (Example 2 of Section 8.1). Since T maps *every* vector in V into **0**, it follows that $\ker(T) = V$. Moreover, since **0** is the *only* image under T of vectors in V, we have $R(T) = \{\mathbf{0}\}$. ◆

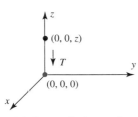

(*a*) $\ker(T)$ is the *z*-axis.

EXAMPLE 3 Kernel and Range of the Identity Operator

Let $I: V \rightarrow V$ be the identity operator (Example 3 of Section 8.1). Since $I(\mathbf{v}) = \mathbf{v}$ for all vectors in V, *every* vector in V is the image of some vector (namely, itself); thus $R(I) = V$. Since the *only* vector that I maps into **0** is **0**, it follows that $\ker(I) = \{\mathbf{0}\}$. ◆

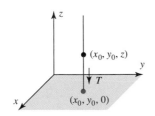

(*b*) $R(T)$ is the entire *xy*-plane.

Figure 8.2.1

EXAMPLE 4 Kernel and Range of an Orthogonal Projection

Let $T: R^3 \rightarrow R^3$ be the orthogonal projection on the *xy*-plane. The kernel of T is the set of points that T maps into $\mathbf{0} = (0, 0, 0)$; these are the points on the *z*-axis (Figure 8.2.1*a*). Since T maps every point in R^3 into the *xy*-plane, the range of T must be some subset of this plane. But every point $(x_0, y_0, 0)$ in the *xy*-plane is the image under T of some point; in fact, it is the image of all points on the vertical line that passes through $(x_0, y_0, 0)$ (Figure 8.2.1*b*). Thus $R(T)$ is the entire *xy*-plane. ◆

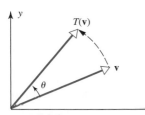

Figure 8.2.2

EXAMPLE 5 Kernel and Range of a Rotation

Let $T: R^2 \rightarrow R^2$ be the linear operator that rotates each vector in the *xy*-plane through the angle θ (Figure 8.2.2). Since *every* vector in the *xy*-plane can be obtained by rotating some vector through the angle θ (why?), we have $R(T) = R^2$. Moreover, the only vector that rotates into **0** is **0**, so ker$(T) = \{\mathbf{0}\}$. ◆

Calculus Required

EXAMPLE 6 Kernel of a Differentiation Transformation

Let $V = C^1(-\infty, \infty)$ be the vector space of functions with continuous first derivatives on $(-\infty, \infty)$, let $W = F(-\infty, \infty)$ be the vector space of all real-valued functions defined on $(-\infty, \infty)$, and let $D: V \rightarrow W$ be the differentiation transformation $D(\mathbf{f}) = f'(x)$. The kernel of D is the set of functions in V with derivative zero. From calculus, this is the set of constant functions on $(-\infty, \infty)$. ◆

Properties of Kernel and Range

In all of the preceding examples, ker(T) and $R(T)$ turned out to be *subspaces*. In Examples 2, 3, and 5 they were either the zero subspace or the entire vector space. In Example 4 the kernel was a line through the origin, and the range was a plane through the origin, both of which are subspaces of R^3. All of this is not accidental; it is a consequence of the following general result.

THEOREM 8.2.1

> *If $T: V \rightarrow W$ is linear transformation, then*
>
> *(a) The kernel of T is a subspace of V.*   *(b) The range of T is a subspace of W.*

Proof (a) To show that ker(T) is a subspace, we must show that it contains at least one vector and is closed under addition and scalar multiplication. By part (a) of Theorem 8.1.1, the vector **0** is in ker(T), so this set contains at least one vector. Let $\mathbf{v}_1$ and $\mathbf{v}_2$ be vectors in ker(T), and let k be any scalar. Then

$$T(\mathbf{v}_1 + \mathbf{v}_2) = T(\mathbf{v}_1) + T(\mathbf{v}_2) = \mathbf{0} + \mathbf{0} = \mathbf{0}$$

so $\mathbf{v}_1 + \mathbf{v}_2$ is in ker(T). Also,

$$T(k\mathbf{v}_1) = kT(\mathbf{v}_1) = k\mathbf{0} = \mathbf{0}$$

so $k\mathbf{v}_1$ is in ker(T).

Proof (b) Since $T(\mathbf{0}) = \mathbf{0}$, there is at least one vector in $R(T)$. Let $\mathbf{w}_1$ and $\mathbf{w}_2$ be vectors in the range of T, and let k be any scalar. To prove this part, we must show that $\mathbf{w}_1 + \mathbf{w}_2$ and $k\mathbf{w}_1$ are in the range of T; that is, we must find vectors **a** and **b** in V such that $T(\mathbf{a}) = \mathbf{w}_1 + \mathbf{w}_2$ and $T(\mathbf{b}) = k\mathbf{w}_1$.

Since $\mathbf{w}_1$ and $\mathbf{w}_2$ are in the range of T, there are vectors $\mathbf{a}_1$ and $\mathbf{a}_2$ in V such that $T(\mathbf{a}_1) = \mathbf{w}_1$ and $T(\mathbf{a}_2) = \mathbf{w}_2$. Let $\mathbf{a} = \mathbf{a}_1 + \mathbf{a}_2$ and $\mathbf{b} = k\mathbf{a}_1$. Then

$$T(\mathbf{a}) = T(\mathbf{a}_1 + \mathbf{a}_2) = T(\mathbf{a}_1) + T(\mathbf{a}_2) = \mathbf{w}_1 + \mathbf{w}_2$$

and

$$T(\mathbf{b}) = T(k\mathbf{a}_1) = kT(\mathbf{a}_1) = k\mathbf{w}_1$$

which completes the proof. ∎

In Section 5.6 we defined the rank of a matrix to be the dimension of its column (or row) space and the nullity to be the dimension of its nullspace. We now extend these definitions to general linear transformations.

If $T: V \to W$ is a linear transformation, then the dimension of the range of T is called the **rank of T** and is denoted by rank(T); the dimension of the kernel is called the **nullity of T** and is denoted by nullity(T).

If A is an $m \times n$ matrix and $T_A: R^n \to R^m$ is multiplication by A, then we know from Example 1 that the kernel of T_A is the nullspace of A and the range of T_A is the column space of A. Thus we have the following relationship between the rank and nullity of a matrix and the rank and nullity of the corresponding matrix transformation.

THEOREM 8.2.2

If A is an $m \times n$ matrix and $T_A: R^n \to R^m$ is multiplication by A, then

(a) nullity(T_A) = nullity(A)   *(b)* rank(T_A) = rank(A)

EXAMPLE 7 Finding Rank and Nullity

Let $T_A: R^6 \to R^4$ be multiplication by

$$A = \begin{bmatrix} -1 & 2 & 0 & 4 & 5 & -3 \\ 3 & -7 & 2 & 0 & 1 & 4 \\ 2 & -5 & 2 & 4 & 6 & 1 \\ 4 & -9 & 2 & -4 & -4 & 7 \end{bmatrix}$$

Find the rank and nullity of T_A.

Solution

In Example 1 of Section 5.6, we showed that rank(A) = 2 and nullity(A) = 4. Thus, from Theorem 8.2.2, we have rank(T_A) = 2 and nullity(T_A) = 4. ◆

EXAMPLE 8 Finding Rank and Nullity

Let $T: R^3 \to R^3$ be the orthogonal projection on the xy-plane. From Example 4, the kernel of T is the z-axis, which is one-dimensional, and the range of T is the xy-plane, which is two-dimensional. Thus

$$\text{nullity}(T) = 1 \quad \text{and} \quad \text{rank}(T) = 2 \quad ◆$$

Dimension Theorem for Linear Transformations

Recall from the Dimension Theorem for Matrices (Theorem 5.6.3) that if A is a matrix with n columns, then

$$\text{rank}(A) + \text{nullity}(A) = n$$

The following theorem, whose proof is deferred to the end of the section, extends this result to general linear transformations.

THEOREM 8.2.3

> ### Dimension Theorem for Linear Transformations
>
> *If $T: V \to W$ is a linear transformation from an n-dimensional vector space V to a vector space W, then*
>
> $$\text{rank}(T) + \text{nullity}(T) = n \qquad (1)$$

In words, this theorem states that *for linear transformations the rank plus the nullity is equal to the dimension of the domain*. This theorem is also known as the Rank Theorem.

REMARK   If A is an $m \times n$ matrix and $T_A: R^n \to R^m$ is multiplication by A, then the domain of T_A has dimension n, so Theorem 8.2.3 agrees with Theorem 5.6.3 in this case.

EXAMPLE 9   Using the Dimension Theorem

Let $T: R^2 \to R^2$ be the linear operator that rotates each vector in the xy-plane through an angle θ. We showed in Example 5 that $\ker(T) = \{\mathbf{0}\}$ and $R(T) = R^2$. Thus

$$\text{rank}(T) + \text{nullity}(T) = 2 + 0 = 2$$

which is consistent with the fact that the domain of T is two-dimensional. ◆

Additional Proof

Proof of Theorem 8.2.3   We must show that

$$\dim(R(T)) + \dim(\ker(T)) = n$$

We shall give the proof for the case where $1 \leq \dim(\ker(T)) < n$. The cases where $\dim(\ker(T)) = 0$ and $\dim(\ker(T)) = n$ are left as exercises. Assume $\dim(\ker(T)) = r$, and let $\mathbf{v}_1, \ldots, \mathbf{v}_r$ be a basis for the kernel. Since $\{\mathbf{v}_1, \ldots, \mathbf{v}_r\}$ is linearly independent, Theorem 5.4.6b states that there are $n - r$ vectors, $\mathbf{v}_{r+1}, \ldots, \mathbf{v}_n$, such that the extended set $\{\mathbf{v}_1, \ldots, \mathbf{v}_r, \mathbf{v}_{r+1}, \ldots, \mathbf{v}_n\}$ is a basis for V. To complete the proof, we shall show that the $n - r$ vectors in the set $S = \{T(\mathbf{v}_{r+1}), \ldots, T(\mathbf{v}_n)\}$ form a basis for the range of T. It will then follow that

$$\dim(R(T)) + \dim(\ker(T)) = (n - r) + r = n$$

First we show that S spans the range of T. If $\mathbf{b}$ is any vector in the range of T, then $\mathbf{b} = T(\mathbf{v})$ for some vector $\mathbf{v}$ in V. Since $\{\mathbf{v}_1, \ldots, \mathbf{v}_r, \mathbf{v}_{r+1}, \ldots, \mathbf{v}_n\}$ is a basis for V, the vector $\mathbf{v}$ can be written in the form

$$\mathbf{v} = c_1 \mathbf{v}_1 + \cdots + c_r \mathbf{v}_r + c_{r+1} \mathbf{v}_{r+1} + \cdots + c_n \mathbf{v}_n$$

Since $\mathbf{v}_1, \ldots, \mathbf{v}_r$ lie in the kernel of T, we have $T(\mathbf{v}_1) = \cdots = T(\mathbf{v}_r) = \mathbf{0}$, so

$$\mathbf{b} = T(\mathbf{v}) = c_{r+1} T(\mathbf{v}_{r+1}) + \cdots + c_n T(\mathbf{v}_n)$$

Thus S spans the range of T.

Finally, we show that S is a linearly independent set and consequently forms a basis for the range of T. Suppose that some linear combination of the vectors in S is zero; that is,

$$k_{r+1} T(\mathbf{v}_{r+1}) + \cdots + k_n T(\mathbf{v}_n) = \mathbf{0} \qquad (2)$$

We must show that $k_{r+1} = \cdots = k_n = 0$. Since T is linear, (2) can be rewritten as

$$T(k_{r+1} \mathbf{v}_{r+1} + \cdots + k_n \mathbf{v}_n) = \mathbf{0}$$

which says that $k_{r+1}\mathbf{v}_{r+1} + \cdots + k_n\mathbf{v}_n$ is in the kernel of T. This vector can therefore be written as a linear combination of the basis vectors $\{\mathbf{v}_1, \ldots, \mathbf{v}_r\}$, say

$$k_{r+1}\mathbf{v}_{r+1} + \cdots + k_n\mathbf{v}_n = k_1\mathbf{v}_1 + \cdots + k_r\mathbf{v}_r$$

Thus,

$$k_1\mathbf{v}_1 + \cdots + k_r\mathbf{v}_r - k_{r+1}\mathbf{v}_{r+1} - \cdots - k_n\mathbf{v}_n = \mathbf{0}$$

Since $\{\mathbf{v}_1, \ldots, \mathbf{v}_n\}$ is linearly independent, all of the k's are zero; in particular, $k_{r+1} = \cdots = k_n = 0$, which completes the proof. ∎

EXERCISE SET 8.2

1. Let $T: R^2 \to R^2$ be the linear operator given by the formula

$$T(x, y) = (2x - y, -8x + 4y)$$

Which of the following vectors are in $R(T)$?

   (a) $(1, -4)$ (b) $(5, 0)$ (c) $(-3, 12)$

2. Let $T: R^2 \to R^2$ be the linear operator in Exercise 1. Which of the following vectors are in $\ker(T)$?

   (a) $(5, 10)$ (b) $(3, 2)$ (c) $(1, 1)$

3. Let $T: R^4 \to R^3$ be the linear transformation given by the formula

$$T(x_1, x_2, x_3, x_4) = (4x_1 + x_2 - 2x_3 - 3x_4, 2x_1 + x_2 + x_3 - 4x_4, 6x_1 - 9x_3 + 9x_4)$$

Which of the following are in $R(T)$?

   (a) $(0, 0, 6)$ (b) $(1, 3, 0)$ (c) $(2, 4, 1)$

4. Let $T: R^4 \to R^3$ be the linear transformation in Exercise 3. Which of the following are in $\ker(T)$?

   (a) $(3, -8, 2, 0)$ (b) $(0, 0, 0, 1)$ (c) $(0, -4, 1, 0)$

5. Let $T: P_2 \to P_3$ be the linear transformation defined by $T(p(x)) = xp(x)$. Which of the following are in $\ker(T)$?

   (a) x^2 (b) 0 (c) $1 + x$

6. Let $T: P_2 \to P_3$ be the linear transformation in Exercise 5. Which of the following are in $R(T)$?

   (a) $x + x^2$ (b) $1 + x$ (c) $3 - x^2$

7. Find a basis for the kernel of

   (a) the linear operator in Exercise 1

   (b) the linear transformation in Exercise 3

   (c) the linear transformation in Exercise 5.

8. Find a basis for the range of

   (a) the linear operator in Exercise 1

   (b) the linear transformation in Exercise 3

   (c) the linear transformation in Exercise 5.

9. Verify Formula (1) of the dimension theorem for

   (a) the linear operator in Exercise 1

   (b) the linear transformation in Exercise 3

   (c) the linear transformation in Exercise 5.

In Exercises 10–13 let T be multiplication by the matrix A. Find

   (a) a basis for the range of T (b) a basis for the kernel of T

   (c) the rank and nullity of T (d) the rank and nullity of A

10. $A = \begin{bmatrix} 1 & -1 & 3 \\ 5 & 6 & -4 \\ 7 & 4 & 2 \end{bmatrix}$ **11.** $A = \begin{bmatrix} 2 & 0 & -1 \\ 4 & 0 & -2 \\ 0 & 0 & 0 \end{bmatrix}$

12. $A = \begin{bmatrix} 4 & 1 & 5 & 2 \\ 1 & 2 & 3 & 0 \end{bmatrix}$ **13.** $A = \begin{bmatrix} 1 & 4 & 5 & 0 & 9 \\ 3 & -2 & 1 & 0 & -1 \\ -1 & 0 & -1 & 0 & -1 \\ 2 & 3 & 5 & 1 & 8 \end{bmatrix}$

14. Describe the kernel and range of

(a) the orthogonal projection on the xz-plane

(b) the orthogonal projection on the yz-plane

(c) the orthogonal projection on the plane defined by the equation $y = x$

15. Let V be any vector space, and let $T: V \to V$ be defined by $T(\mathbf{v}) = 3\mathbf{v}$.

(a) What is the kernel of T?   (b) What is the range of T?

16. In each part, use the given information to find the nullity of T.

(a) $T: R^5 \to R^7$ has rank 3.   (b) $T: P_4 \to P_3$ has rank 1.

(c) The range of $T: R^6 \to R^3$ is R^3.   (d) $T: M_{22} \to M_{22}$ has rank 3.

17. Let A be a 7×6 matrix such that $A\mathbf{x} = \mathbf{0}$ has only the trivial solution, and let $T: R^6 \to R^7$ be multiplication by A. Find the rank and nullity of T.

18. Let A be a 5×7 matrix with rank 4.

(a) What is the dimension of the solution space of $A\mathbf{x} = \mathbf{0}$?

(b) Is $A\mathbf{x} = \mathbf{b}$ consistent for all vectors $\mathbf{b}$ in R^5? Explain.

19. Let $T: R^3 \to V$ be a linear transformation from R^3 to any vector space. Show that the kernel of T is a line through the origin, a plane through the origin, the origin only, or all of R^3.

20. Let $T: V \to R^3$ be a linear transformation from any vector space to R^3. Show that the range of T is a line through the origin, a plane through the origin, the origin only, or all of R^3.

21. Let $T: R^3 \to R^3$ be multiplication by

$$\begin{bmatrix} 1 & 3 & 4 \\ 3 & 4 & 7 \\ -2 & 2 & 0 \end{bmatrix}$$

(a) Show that the kernel of T is a line through the origin, and find parametric equations for it.

(b) Show that the range of T is a plane through the origin, and find an equation for it.

22. Prove: If $\{\mathbf{v}_1, \mathbf{v}_2, \dots, \mathbf{v}_n\}$ is a basis for V and $\mathbf{w}_1, \mathbf{w}_2, \dots, \mathbf{w}_n$ are vectors in W, not necessarily distinct, then there exists a linear transformation $T: V \to W$ such that

$$T(\mathbf{v}_1) = \mathbf{w}_1, \quad T(\mathbf{v}_2) = \mathbf{w}_2, \dots, \quad T(\mathbf{v}_n) = \mathbf{w}_n$$

23. For the positive integer $n > 1$, let $T: M_{nn} \to R$ be the linear transformation defined by $T(A) = \text{tr}(A)$, for A an $n \times n$ matrix with real entries. Determine the dimension of $\ker(T)$.

24. Prove the dimension theorem in the cases

(a) $\dim(\ker(T)) = 0$ (b) $\dim(\ker(T)) = n$

25. **(For Readers Who Have Studied Calculus)** Let $D: P_3 \to P_2$ be the differentiation transformation $D(\mathbf{p}) = p'(x)$. Describe the kernel of D.

26. **(For Readers Who Have Studied Calculus)** Let $J: P_1 \to R$ be the integration transformation $J(\mathbf{p}) = \int_{-1}^{1} p(x)\, dx$. Describe the kernel of J.

27. **(For Readers Who Have Studied Calculus)** Let $D: V \rightarrow W$ be the differentiation transformation $D(\mathbf{f}) = f'(x)$, where $V = C^3(-\infty, \infty)$ and $W = F(-\infty, \infty)$. Describe the kernels of $D \circ D$ and $D \circ D \circ D$.

<table>
<tr><td>

Discussion
& Discovery

</td><td>

28. Fill in the blanks.

   (a) If $T_A: R^n \rightarrow R^m$ is multiplication by A, then the nullspace of A corresponds to the _____ of T_A, and the column space of A corresponds to the _____ of T_A.

   (b) If $T: R^3 \rightarrow R^3$ is the orthogonal projection on the plane $x + y + z = 0$, then the kernel of T is the line through the origin that is parallel to the vector _____.

   (c) If V is a finite-dimensional vector space and $T: V \rightarrow W$ is a linear transformation, then the dimension of the range of T plus the dimension of the kernel of T is _____.

   (d) If $T_A: R^5 \rightarrow R^3$ is multiplication by A, and if $\mathrm{rank}(T_A) = 2$, then the general solution of $A\mathbf{x} = \mathbf{0}$ has _____ (how many?) parameters.

29. (a) If $T: R^3 \rightarrow R^3$ is a linear operator, and if the kernel of T is a line through the origin, then what kind of geometric object is the range of T? Explain your reasoning.

   (b) If $T: R^3 \rightarrow R^3$ is a linear operator, and if the range of T is a plane through the origin, then what kind of geometric object is the kernel of T? Explain your reasoning.

30. **(For Readers Who Have Studied Calculus)** Let V be the vector space of real-valued functions with continuous derivatives of all orders on the interval $(-\infty, \infty)$, and let $W = F(-\infty, \infty)$ be the vector space of real-valued functions defined on $(-\infty, \infty)$.

   (a) Find a linear transformation $T: V \rightarrow W$ whose kernel is P_3.

   (b) Find a linear transformation $T: V \rightarrow W$ whose kernel is P_n.

31. If A is an $m \times n$ matrix, and if the linear system $A\mathbf{x} = \mathbf{b}$ is consistent for every vector $\mathbf{b}$ in R^m, what can you say about the range of $T_A: R^n \rightarrow R^m$?

</td></tr>
</table>

8.3
INVERSE LINEAR TRANSFORMA-TIONS

In Section 4.3 we discussed properties of one-to-one linear transformations from R^n to R^m. In this section we shall extend those ideas to more general kinds of linear transformations.

Recall from Section 4.3 that a linear transformation from R^n to R^m is called *one-to-one* if it maps distinct vectors in R^n into distinct vectors in R^m. The following definition generalizes that idea.

> **DEFINITION**
>
> A linear transformation $T: V \rightarrow W$ is said to be ***one-to-one*** if T maps distinct vectors in V into distinct vectors in W.

EXAMPLE 1   A One-to-One Linear Transformation

Recall from Theorem 4.3.1 that if A is an $n \times n$ matrix and $T_A: R^n \rightarrow R^n$ is multiplication by A, then T_A is one-to-one if and only if A is an invertible matrix. ◆

EXAMPLE 2 A One-to-One Linear Transformation

Let $T: P_n \rightarrow P_{n+1}$ be the linear transformation

$$T(\mathbf{p}) = T(p(x)) = xp(x)$$

discussed in Example 8 of Section 8.1. If

$$\mathbf{p} = p(x) = c_0 + c_1 x + \cdots + c_n x^n \quad \text{and} \quad \mathbf{q} = q(x) = d_0 + d_1 x + \cdots + d_n x^n$$

are distinct polynomials, then they differ in at least one coefficient. Thus,

$$T(\mathbf{p}) = c_0 x + c_1 x^2 + \cdots + c_n x^{n+1} \quad \text{and} \quad T(\mathbf{q}) = d_0 x + d_1 x^2 + \cdots + d_n x^{n+1}$$

also differ in at least one coefficient. Thus T is one-to-one, since it maps distinct polynomials $\mathbf{p}$ and $\mathbf{q}$ into distinct polynomials $T(\mathbf{p})$ and $T(\mathbf{q})$. ◆

Calculus Required

EXAMPLE 3 A Transformation That Is Not One-to-One

Let

$$D: C^1(-\infty, \ \infty) \rightarrow F(-\infty, \ \infty)$$

be the differentiation transformation discussed in Example 11 of Section 8.1. This linear transformation is *not* one-to-one because it maps functions that differ by a constant into the same function. For example,

$$D(x^2) = D(x^2 + 1) = 2x \quad ◆$$

The following theorem establishes a relationship between a one-to-one linear transformation and its kernel.

THEOREM 8.3.1

> **Equivalent Statements**
>
> *If $T: V \rightarrow W$ is a linear transformation, then the following are equivalent.*
>
> (*a*) *T is one-to-one.*
> (*b*) *The kernel of T contains only the zero vector; that is*, $\ker(T) = \{\mathbf{0}\}$.
> (*c*) *nullity$(T) = 0$.*

Proof The equivalence of (*b*) and (*c*) is immediate from the definition of nullity. We shall complete the proof by proving the equivalence of (*a*) and (*b*).

(a) $\Rightarrow$ **(b)** Assume that T is one-to-one, and let $\mathbf{v}$ be any vector in $\ker(T)$. Since $\mathbf{v}$ and $\mathbf{0}$ both lie in $\ker(T)$, we have $T(\mathbf{v}) = \mathbf{0}$ and $T(\mathbf{0}) = \mathbf{0}$, so $T(\mathbf{v}) = T(\mathbf{0})$. But this implies that $\mathbf{v} = \mathbf{0}$, since T is one-to-one; thus $\ker(T)$ contains only the zero vector.

(b) $\Rightarrow$ **(a)** Assume that $\ker(T) = \{\mathbf{0}\}$ and that $\mathbf{v}$ and $\mathbf{w}$ are distinct vectors in V; that is,

$$\mathbf{v} - \mathbf{w} \neq \mathbf{0} \tag{1}$$

To prove that T is one-to-one, we must show that $T(\mathbf{v})$ and $T(\mathbf{w})$ are distinct vectors. But if this were not so, then we would have $T(\mathbf{v}) = T(\mathbf{w})$. Therefore,

$$T(\mathbf{v}) - T(\mathbf{w}) = \mathbf{0} \quad \text{or} \quad T(\mathbf{v} - \mathbf{w}) = \mathbf{0}$$

and so $\mathbf{v} - \mathbf{w}$ is in the kernel of T. Since $\ker(T) = \{\mathbf{0}\}$, this implies that $\mathbf{v} - \mathbf{w} = \mathbf{0}$, which contradicts (1). Thus $T(\mathbf{v})$ and $T(\mathbf{w})$ must be distinct. ∎

EXAMPLE 4 Using Theorem 8.3.1

In each part, determine whether the linear transformation is one-to-one by finding the kernel or the nullity and applying Theorem 8.3.1.

(a) $T: R^2 \rightarrow R^2$ rotates each vector through the angle θ.

(b) $T: R^3 \rightarrow R^3$ is the orthogonal projection on the xy-plane.

(c) $T: R^6 \rightarrow R^4$ is multiplication by the matrix

$$A = \begin{bmatrix} -1 & 2 & 0 & 4 & 5 & -3 \\ 3 & -7 & 2 & 0 & 1 & 4 \\ 2 & -5 & 2 & 4 & 6 & 1 \\ 4 & -9 & 2 & -4 & -4 & 7 \end{bmatrix}$$

Solution (a)

From Example 5 of Section 8.2, $\ker(T) = \{\mathbf{0}\}$, so T is one-to-one.

Solution (b)

From Example 4 of Section 8.2, $\ker(T)$ contains nonzero vectors, so T is not one-to-one.

Solution (c)

From Example 7 of Section 8.2, $\text{nullity}(T) = 4$, so T is not one-to-one. ◆

In the special case where T is a *linear operator* on a *finite-dimensional* vector space, a fourth equivalent statement can be added to those in Theorem 8.3.1.

THEOREM 8.3.2

> *If V is a finite-dimensional vector space, and $T: V \rightarrow V$ is a linear operator, then the following are equivalent.*
>
> *(a)* T is one-to-one.   *(b)* $\ker(T) = \{\mathbf{0}\}$.
> *(c)* $\text{nullity}(T) = 0$.   *(d) The range of T is V; that is, $R(T) = V$.*

Proof We already know that (a), (b), and (c) are equivalent, so we can complete the proof by proving the equivalence of (c) and (d).

(c) $\Rightarrow$ (d) Suppose that $\dim(V) = n$ and $\text{nullity}(T) = 0$. It follows from the Dimension Theorem (Theorem 8.2.3) that

$$\text{rank}(T) = n - \text{nullity}(T) = n$$

By definition, $\text{rank}(T)$ is the dimension of the range of T, so the range of T has dimension n. It now follows from Theorem 5.4.7 that the range of T is V, since the two spaces have the same dimension.

(d) ⇒ (c) Suppose that $\dim(V) = n$ and $R(T) = V$. It follows from these relationships that $\dim(R(T)) = n$, or, equivalently, $\text{rank}(T) = n$. Thus it follows from the Dimension Theorem (Theorem 8.2.3) that

$$\text{nullity}(T) = n - \text{rank}(T) = n - n = 0 \qquad\blacksquare$$

EXAMPLE 5 A Transformation That Is Not One-To-One

Let $T_A: R^4 \to R^4$ be multiplication by

$$A = \begin{bmatrix} 1 & 3 & -2 & 4 \\ 2 & 6 & -4 & 8 \\ 3 & 9 & 1 & 5 \\ 1 & 1 & 4 & 8 \end{bmatrix}$$

Determine whether T_A is one-to-one.

Solution

As noted in Example 1, the given problem is equivalent to determining whether A is invertible. But $\det(A) = 0$, since the first two rows of A are proportional, and consequently, A is not invertible. Thus T_A is not one-to-one. ◆

Inverse Linear Transformations

In Section 4.3 we defined the *inverse* of a one-to-one matrix operator $T_A: R^n \to R^n$ to be $T_{A^{-1}}: R^n \to R^n$, and we showed that if $\mathbf{w}$ is the image of a vector $\mathbf{x}$ under T_A, then $T_{A^{-1}} = T_A^{-1}$ maps $\mathbf{w}$ back into $\mathbf{x}$. We shall now extend these ideas to general linear transformations.

Recall that if $T: V \to W$ is a linear transformation, then the range of T, denoted by $R(T)$, is the subspace of W consisting of all images under T of vectors in V. If T is one-to-one, then each vector $\mathbf{w}$ in $R(T)$ is the image of a unique vector $\mathbf{v}$ in V. This uniqueness allows us to define a new function, called the ***inverse of*** T and denoted by T^{-1}, that maps $\mathbf{w}$ back into $\mathbf{v}$ (Figure 8.3.1).

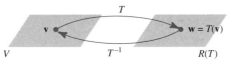

Figure 8.3.1 The inverse of T maps $T(\mathbf{v})$ back into $\mathbf{v}$.

It can be proved (Exercise 19) that $T^{-1}: R(T) \to V$ is a linear transformation. Moreover, it follows from the definition of T^{-1} that

$$T^{-1}(T(\mathbf{v})) = T^{-1}(\mathbf{w}) = \mathbf{v} \tag{2a}$$

$$T(T^{-1}(\mathbf{w})) = T(\mathbf{v}) = \mathbf{w} \tag{2b}$$

so that T and T^{-1}, when applied in succession in either order, cancel the effect of one another.

REMARK   It is important to note that if $T: V \to W$ is a one-to-one linear transformation, then the domain of T^{-1} is the *range* of T. The range may or may not be all of W. However, in the special case where $T: V \to V$ is a one-to-one linear operator, it follows from Theorem 8.3.2 that $R(T) = V$; that is, the domain of T^{-1} is *all* of V.

EXAMPLE 6   An Inverse Transformation

In Example 2 we showed that the linear transformation $T: P_n \to P_{n+1}$ given by

$$T(\mathbf{p}) = T(p(x)) = xp(x)$$

is one-to-one; thus, T has an inverse. Here the range of T is *not* all of P_{n+1}; rather, $R(T)$ is the subspace of P_{n+1} consisting of polynomials with a zero constant term. This is evident from the formula for T:

$$T(c_0 + c_1 x + \cdots + c_n x^n) = c_0 x + c_1 x^2 + \cdots + c_n x^{n+1}$$

It follows that $T^{-1}: R(T) \to P_n$ is given by the formula

$$T^{-1}(c_0 x + c_1 x^2 + \cdots + c_n x^{n+1}) = c_0 + c_1 x + \cdots + c_n x^n$$

For example, in the case where $n \geq 3$,

$$T^{-1}(2x - x^2 + 5x^3 + 3x^4) = 2 - x + 5x^2 + 3x^3 \quad \blacklozenge$$

EXAMPLE 7   An Inverse Transformation

Let $T: R^3 \to R^3$ be the linear operator defined by the formula

$$T(x_1, x_2, x_3) = (3x_1 + x_2, -2x_1 - 4x_2 + 3x_3, 5x_1 + 4x_2 - 2x_3)$$

Determine whether T is one-to-one, if so, find $T^{-1}(x_1, x_2, x_3)$.

Solution

From Theorem 4.3.3, the standard matrix for T is

$$[T] = \begin{bmatrix} 3 & 1 & 0 \\ -2 & -4 & 3 \\ 5 & 4 & -2 \end{bmatrix}$$

(verify). This matrix is invertible, and from Formula (1) of Section 4.3, the standard matrix for T^{-1} is

$$[T^{-1}] = [T]^{-1} = \begin{bmatrix} 4 & -2 & -3 \\ -11 & 6 & 9 \\ -12 & 7 & 10 \end{bmatrix}$$

It follows that

$$T^{-1}\left(\begin{bmatrix} x_1 \\ x_2 \\ x_3 \end{bmatrix}\right) = [T^{-1}]\begin{bmatrix} x_1 \\ x_2 \\ x_3 \end{bmatrix} = \begin{bmatrix} 4 & -2 & -3 \\ -11 & 6 & 9 \\ -12 & 7 & 10 \end{bmatrix}\begin{bmatrix} x_1 \\ x_2 \\ x_3 \end{bmatrix} = \begin{bmatrix} 4x_1 - 2x_2 - 3x_3 \\ -11x_1 + 6x_2 + 9x_3 \\ -12x_1 + 7x_2 + 10x_3 \end{bmatrix}$$

Expressing this result in horizontal notation yields

$$T^{-1}(x_1, x_2, x_3) = (4x_1 - 2x_2 - 3x_3, -11x_1 + 6x_2 + 9x_3, -12x_1 + 7x_2 + 10x_3) \quad \blacklozenge$$

The following theorem shows that a composition of one-to-one linear transformations is one-to-one, and it relates the inverse of the composition to the inverses of the individual linear transformations.

THEOREM 8.3.3

> *If $T_1: U \to V$ and $T_2: V \to W$ are one-to-one linear transformations, then*
>
> (a) $T_2 \circ T_1$ *is one-to-one.*   (b) $(T_2 \circ T_1)^{-1} = T_1^{-1} \circ T_2^{-1}$.

Proof (a) We want to show that $T_2 \circ T_1$ maps distinct vectors in U into distinct vectors in W. But if $\mathbf{u}$ and $\mathbf{v}$ are distinct vectors in U, then $T_1(\mathbf{u})$ and $T_1(\mathbf{v})$ are distinct vectors in V since T_1 is one-to-one. This and the fact that T_2 is one-to-one imply that

$$T_2(T_1(\mathbf{u})) \quad \text{and} \quad T_2(T_1(\mathbf{v}))$$

are also distinct vectors. But these expressions can also be written as

$$(T_2 \circ T_1)(\mathbf{u}) \quad \text{and} \quad (T_2 \circ T_1)(\mathbf{v})$$

so $T_2 \circ T_1$ maps $\mathbf{u}$ and $\mathbf{v}$ into distinct vectors in W.

Proof (b) We want to show that

$$(T_2 \circ T_1)^{-1}(\mathbf{w}) = (T_1^{-1} \circ T_2^{-1})(\mathbf{w})$$

for every vector $\mathbf{w}$ in the range of $T_2 \circ T_1$. For this purpose, let

$$\mathbf{u} = (T_2 \circ T_1)^{-1}(\mathbf{w}) \tag{3}$$

so our goal is to show that

$$\mathbf{u} = (T_1^{-1} \circ T_2^{-1})(\mathbf{w})$$

But it follows from (3) that

$$(T_2 \circ T_1)(\mathbf{u}) = \mathbf{w}$$

or, equivalently,

$$T_2(T_1(\mathbf{u})) = \mathbf{w}$$

Now, taking T_2^{-1} of each side of this equation and then T_1^{-1} of each side of the result and using (2a) yields (verify)

$$\mathbf{u} = T_1^{-1}(T_2^{-1}(\mathbf{w}))$$

or, equivalently,

$$\mathbf{u} = (T_1^{-1} \circ T_2^{-1})(\mathbf{w}) \qquad \blacksquare$$

In words, part (b) of Theorem 8.3.3 states that *the inverse of a composition is the composition of the inverses in the reverse order*. This result can be extended to compositions of three or more linear transformations; for example,

$$(T_3 \circ T_2 \circ T_1)^{-1} = T_1^{-1} \circ T_2^{-1} \circ T_3^{-1} \tag{4}$$

In the case where T_A, T_B, and T_C are matrix operators on R^n, Formula (4) can be written as

$$(T_C \circ T_B \circ T_A)^{-1} = T_A^{-1} \circ T_B^{-1} \circ T_C^{-1}$$

which we might also write as

$$(T_{CBA})^{-1} = T_{A^{-1}B^{-1}C^{-1}} \tag{5}$$

In words, this formula states that *the standard matrix for the inverse of a composition is the product of the inverses of the standard matrices of the individual operators in the reverse order.*

Some problems that use Formula (5) are given in the exercises.

Dimension of Domain and Codomain

In Exercise 16 you are asked to show the important fact that if V and W are finite-dimensional vector spaces with $\dim(W) < \dim(V)$, and if $T: V \rightarrow W$ is a linear transformation, then T cannot be one-to-one. In other words, the dimension of the codomain W must be at least as large as the dimension of the domain V for there to be a one-to-one linear transformation from V to W. This means, for example, that there can be no one-to-one linear transformation from space R^3 to the plane R^2.

EXAMPLE 8 Dimension and One-to-One Linear Transformations

A linear transformation T_A from the plane R^2 to the real line R has a standard matrix

$$A = [a \quad b]$$

If $\mathbf{u} = (x, y)$ is a point in R^2, its image is

$$T_A(\mathbf{u}) = ax + by$$

which is a scalar. But if $ax + by = k$, say, then there are infinitely many other points $\mathbf{v}$ in R^2 that also have $T_A(\mathbf{v}) = k$, since there are infinitely many points on the line

$$ax + by = k$$

This is because if a and b are nonzero, then every point of the form

$$\mathbf{v} = \left(\frac{k - by}{a}, y \right)$$

has $T_A(\mathbf{v}) = k$, whereas if $a = 0$ but b is nonzero, then every point of the form

$$\mathbf{v} = (x, 0)$$

has $T_A(\mathbf{v}) = 0$, and if $b = 0$ but a is nonzero, then every point of the form

$$\mathbf{v} = (0, y)$$

has $T_A(\mathbf{v}) = 0$. Finally, in the degenerate case $a = 0$ and $b = 0$, we have $T_A(\mathbf{v}) = 0$ for every $\mathbf{v}$ in R^2.

In each case, T fails to be one-to-one, so there can be no transformation from the plane to the real line that is both linear and one-to-one. ◆

Of course, even if $\dim(W) \geq \dim(V)$, a linear transformation from V to W might not be one-to-one, as the zero transformation shows.

**EXERCISE SET
8.3**

1. In each part, find $\ker(T)$, and determine whether the linear transformation T is one-to-one.
   (a) $T: R^2 \rightarrow R^2$, where $T(x, y) = (y, x)$
   (b) $T: R^2 \rightarrow R^2$, where $T(x, y) = (0, 2x + 3y)$
   (c) $T: R^2 \rightarrow R^2$, where $T(x, y) = (x + y, x - y)$
   (d) $T: R^2 \rightarrow R^3$, where $T(x, y) = (x, y, x + y)$

(e) $T: R^2 \rightarrow R^3$, where $T(x, y) = (x - y, y - x, 2x - 2y)$

(f) $T: R^3 \rightarrow R^2$, where $T(x, y, z) = (x + y + z, x - y - z)$

2. In each part, let $T: R^2 \rightarrow R^2$ be multiplication by A. Determine whether T has an inverse; if so, find

$$T^{-1}\left(\begin{bmatrix} x_1 \\ x_2 \end{bmatrix}\right)$$

(a) $A = \begin{bmatrix} 5 & 2 \\ 2 & 1 \end{bmatrix}$ (b) $A = \begin{bmatrix} 6 & -3 \\ 4 & -2 \end{bmatrix}$ (c) $A = \begin{bmatrix} 4 & 7 \\ -1 & 3 \end{bmatrix}$

3. In each part, let $T: R^3 \rightarrow R^3$ be multiplication by A. Determine whether T has an inverse; if so, find

$$T^{-1}\left(\begin{bmatrix} x_1 \\ x_2 \\ x_3 \end{bmatrix}\right)$$

(a) $A = \begin{bmatrix} 1 & 5 & 2 \\ 1 & 2 & 1 \\ -1 & 1 & 0 \end{bmatrix}$ (b) $A = \begin{bmatrix} 1 & 4 & -1 \\ 1 & 2 & 1 \\ -1 & 1 & 0 \end{bmatrix}$

(c) $A = \begin{bmatrix} 1 & 0 & 1 \\ 0 & 1 & 1 \\ 1 & 1 & 0 \end{bmatrix}$ (d) $A = \begin{bmatrix} 1 & -1 & 1 \\ 0 & 2 & -1 \\ 2 & 3 & 0 \end{bmatrix}$

4. In each part, determine whether multiplication by A is a one-to-one linear transformation.

(a) $A = \begin{bmatrix} 1 & -2 \\ 2 & -4 \\ -3 & 6 \end{bmatrix}$ (b) $A = \begin{bmatrix} 1 & 3 & 5 & 7 \\ 2 & -1 & 2 & 4 \\ -1 & 3 & 0 & 0 \end{bmatrix}$ (c) $A = \begin{bmatrix} 4 & -2 \\ 1 & 5 \\ 5 & 3 \end{bmatrix}$

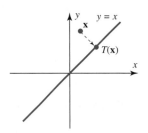

Figure Ex-5

5. As indicated in the accompanying figure, let $T: R^2 \rightarrow R^2$ be the orthogonal projection on the line $y = x$.

(a) Find the kernel of T.

(b) Is T one-to-one? Justify your conclusion.

6. As indicated in the accompanying figure, let $T: R^2 \rightarrow R^2$ be the linear operator that reflects each point about the y-axis.

(a) Find the kernel of T.

(b) Is T one-to-one? Justify your conclusion.

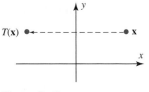

Figure Ex-6

7. In each part, use the given information to determine whether the linear transformation T is one-to-one.

(a) $T: R^m \rightarrow R^m$; nullity$(T) = 0$ (b) $T: R^n \rightarrow R^n$; rank$(T) = n - 1$

(c) $T: R^m \rightarrow R^n$; $n < m$ (d) $T: R^n \rightarrow R^n$; $R(T) = R^n$

8. In each part, determine whether the linear transformation T is one-to-one.

(a) $T: P_2 \rightarrow P_3$, where $T(a_0 + a_1x + a_2x^2) = x(a_0 + a_1x + a_2x^2)$

(b) $T: P_2 \rightarrow P_2$, where $T(p(x)) = p(x + 1)$

9. Let A be a square matrix such that $\det(A) = 0$. Is multiplication by A a one-to-one linear transformation? Justify your conclusion.

10. In each part, determine whether the linear operator $T: R^n \rightarrow R^n$ is one-to-one; if so, find $T^{-1}(x_1, x_2, \ldots, x_n)$.

(a) $T(x_1, x_2, \ldots, x_n) = (0, x_1, x_2, \ldots, x_{n-1})$

(b) $T(x_1, x_2, \ldots, x_n) = (x_n, x_{n-1}, \ldots, x_2, x_1)$

(c) $T(x_1, x_2, \ldots, x_n) = (x_2, x_3, \ldots, x_n, x_1)$

11. Let $T: R^n \to R^n$ be the linear operator defined by the formula

$$T(x_1, x_2, \ldots, x_n) = (a_1 x_1, a_2 x_2, \ldots, a_n x_n)$$

where $a_1, \ldots, a_n$ are constants.

   (a) Under what conditions will T have an inverse?

   (b) Assuming that the conditions determined in part (a) are satisfied, find a formula for $T^{-1}(x_1, x_2, \ldots, x_n)$.

12. Let $T_1: R^2 \to R^2$ and $T_2: R^2 \to R^2$ be the linear operators given by the formulas

$$T_1(x, y) = (x + y, x - y) \quad \text{and} \quad T_2(x, y) = (2x + y, x - 2y)$$

   (a) Show that T_1 and T_2 are one-to-one.

   (b) Find formulas for $T_1^{-1}(x, y)$, and $T_2^{-1}(x, y)$, and $(T_2 \circ T_1)^{-1}(x, y)$.

   (c) Verify that $(T_2 \circ T_1)^{-1} = T_1^{-1} \circ T_2^{-1}$.

13. Let $T_1: P_2 \to P_3$ and $T_2: P_3 \to P_3$ be the linear transformations given by the formulas

$$T_1(p(x)) = xp(x) \quad \text{and} \quad T_2(p(x)) = p(x + 1)$$

   (a) Find formulas for $T_1^{-1}(p(x))$, $T_2^{-1}(p(x))$, and $(T_2 \circ T_1)^{-1}(p(x))$.

   (b) Verify that $(T_2 \circ T_1)^{-1} = T_1^{-1} \circ T_2^{-1}$.

14. Let $T_A: R^3 \to R^3$, $T_B: R^3 \to R^3$, and $T_C: R^3 \to R^3$ be the reflections about the xy-plane, the xz-plane, and the yz-plane, respectively. Verify Formula (5) for these linear operators.

15. Let $T: P_1 \to R^2$ be the function defined by the formula

$$T(p(x)) = (p(0), p(1))$$

   (a) Find $T(1 - 2x)$.

   (b) Show that T is a linear transformation.

   (c) Show that T is one-to-one.

   (d) Find $T^{-1}(2, 3)$, and sketch its graph.

16. Prove: If V and W are finite-dimensional vector spaces such that $\dim W < \dim V$, then there is no one-to-one linear transformation $T: V \to W$.

17. In each part, determine whether the linear operator $T: M_{22} \to M_{22}$ is one-to-one. If so, find

$$T^{-1}\left(\begin{bmatrix} a & b \\ c & d \end{bmatrix}\right)$$

   (a) $T\left(\begin{bmatrix} a & b \\ c & d \end{bmatrix}\right) = \begin{bmatrix} a & 0 \\ 0 & d \end{bmatrix}$ (b) $T\left(\begin{bmatrix} a & b \\ c & d \end{bmatrix}\right) = \begin{bmatrix} a & c \\ b & d \end{bmatrix}$

   (c) $T\left(\begin{bmatrix} a & b \\ c & d \end{bmatrix}\right) = \begin{bmatrix} d & -b \\ -c & a \end{bmatrix}$

18. Let $T: R^2 \to R^2$ be the linear operator given by the formula $T(x, y) = (x + ky, -y)$. Show that T is one-to-one for every real value of k and that $T^{-1} = T$.

19. Prove that if $T: V \to W$ is a one-to-one linear transformation, then $T^{-1}: R(T) \to V$ is a linear transformation.

20. **(For Readers Who Have Studied Calculus)** Let $J: P_1 \to R$ be the integration transformation $J(\mathbf{p}) = \int_{-1}^{1} p(x)\,dx$. Determine whether J is one-to-one. Justify your conclusion.

21. **(For Readers Who Have Studied Calculus)** Let V be the vector space $C^1[0, 1]$ and let $T: V \to R$ be defined by $T(\mathbf{f}) = f(0) + 2f'(0) + 3f'(1)$. Verify that T is a linear transformation. Determine whether T is one-to-one. Justify your conclusion.

In Exercises 22 and 23, determine whether $T_1 \circ T_2 = T_2 \circ T_1$.

22. (a) $T_1: R^2 \to R^2$ is the orthogonal projection on the x-axis, and $T_2: R^2 \to R^2$ is the orthogonal projection on the y-axis.

 (b) $T_1: R^2 \to R^2$ is the rotation about the origin through an angle θ_1, and $T_2: R^2 \to R^2$ is the rotation about the origin through an angle θ_2.

 (c) $T_1: R^3 \to R^3$ is the rotation about the x-axis through an angle θ_1, and $T_2: R^3 \to R^3$ is the rotation about the z-axis through an angle θ_2.

23. (a) $T_1: R^2 \to R^2$ is the reflection about the x-axis, and $T_2: R^2 \to R^2$ is the reflection about the y-axis.

 (b) $T_1: R^2 \to R^2$ is the orthogonal projection on the x-axis, and $T_2: R^2 \to R^2$ is the counterclockwise rotation through an angle θ.

 (c) $T_1: R^3 \to R^3$ is a dilation by a factor k, and $T_2: R^3 \to R^3$ is the counterclockwise rotation about the z-axis through an angle θ.

Discussion & Discovery

24. Indicate whether each statement is always true or sometimes false. Justify your answer by giving a logical argument or a counterexample.

 (a) If $T: R^2 \to R^2$ is the orthogonal projection onto the x-axis, then $T^{-1}: R^2 \to R^2$ maps each point on the x-axis onto a line that is perpendicular to the x-axis.

 (b) If $T_1: U \to V$ and $T_2: V \to W$ are linear transformations, and if T_1 is not one-to-one, then neither is $T_2 \circ T_1$.

 (c) In the xy-plane, a rotation about the origin followed by a reflection about a coordinate axis is one-to-one.

25. Does the formula $T(a, b, c) = ax^2 + bx + c$ define a one-to-one linear transformation from R^3 to P_2? Explain your reasoning.

26. Let E be a fixed 2×2 elementary matrix. Does the formula $T(A) = EA$ define a one-to-one linear operator on M_{22}? Explain your reasoning.

27. Let $\mathbf{a}$ be a fixed vector in R^3. Does the formula $T(\mathbf{v}) = \mathbf{a} \times \mathbf{v}$ define a one-to-one linear operator on R^3? Explain your reasoning.

28. **(For Readers Who Have Studied Calculus)** The Fundamental Theorem of Calculus implies that integration and differentiation reverse the actions of each other in some sense. Define a transformation $D: P_n \to P_{n-1}$ by $D(p(x)) = p'(x)$, and define $J: P_{n-1} \to P_n$ by $J(p(x)) = \int_0^x p(t)\,dt$.

 (a) Show that D and J are linear transformations.

 (b) Explain why J is not the inverse transformation of D.

 (c) Can we restrict the domains and/or codomains of D and J such that they are inverse linear transformations of each other?

8.4
MATRICES OF GENERAL LINEAR TRANSFORMA-TIONS

In this section we shall show that if V and W are finite-dimensional vector spaces (not necessarily R^n and R^m), then with a little ingenuity any linear transformation T: V → W can be regarded as a matrix transformation. The basic idea is to work with coordinate vectors rather than with the vectors themselves.

Matrices of Linear Transformations

Suppose that V is an n-dimensional vector space and W an m-dimensional vector space. If we choose bases B and B' for V and W, respectively, then for each $\mathbf{x}$ in V, the coordinate vector $[\mathbf{x}]_B$ will be a vector in R^n, and the coordinate vector $[T(\mathbf{x})]_{B'}$ will be a vector in R^m (Figure 8.4.1).

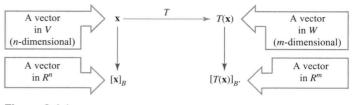

Figure 8.4.1

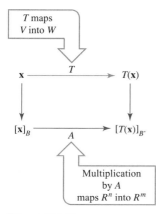

Figure 8.4.2

Suppose $T: V \to W$ is a linear transformation. If, as illustrated in Figure 8.4.2, we complete the rectangle suggested by Figure 8.4.1, we obtain a mapping from R^n to R^m, which can be shown to be a linear transformation. (This is the correspondence discussed in Section 4.4.) If we let A be the standard matrix for this transformation, then

$$A[\mathbf{x}]_B = [T(\mathbf{x})]_{B'} \qquad (1)$$

The matrix A in (1) is called the ***matrix for T with respect to the bases B and B'***.

Later in this section, we shall give some of the uses of the matrix A in (1), but first, let us show how it can be computed. For this purpose, let $B = \{\mathbf{u}_1, \mathbf{u}_2, \ldots, \mathbf{u}_n\}$ be a basis for the n-dimensional space V and $B' = \{\mathbf{v}_1, \mathbf{v}_2, \ldots, \mathbf{v}_m\}$ a basis for the m-dimensional space W. We are looking for an $m \times n$ matrix

$$A = \begin{bmatrix} a_{11} & a_{12} & \cdots & a_{1n} \\ a_{21} & a_{22} & \cdots & a_{2n} \\ \vdots & \vdots & & \vdots \\ a_{m1} & a_{m2} & \cdots & a_{mn} \end{bmatrix}$$

such that (1) holds for all vectors $\mathbf{x}$ in V, meaning that A times the coordinate vector of $\mathbf{x}$ equals the coordinate vector of the image $T(\mathbf{x})$ of $\mathbf{x}$. In particular, we want this equation to hold for the basis vectors $\mathbf{u}_1, \mathbf{u}_2, \ldots, \mathbf{u}_n$; that is,

$$A[\mathbf{u}_1]_B = [T(\mathbf{u}_1)]_{B'}, \quad A[\mathbf{u}_2]_B = [T(\mathbf{u}_2)]_{B'}, \ldots, \quad A[\mathbf{u}_n]_B = [T(\mathbf{u}_n)]_{B'} \qquad (2)$$

But

$$[\mathbf{u}_1]_B = \begin{bmatrix} 1 \\ 0 \\ 0 \\ \vdots \\ 0 \end{bmatrix}, \quad [\mathbf{u}_2]_B = \begin{bmatrix} 0 \\ 1 \\ 0 \\ \vdots \\ 0 \end{bmatrix}, \ldots, \quad [\mathbf{u}_n]_B = \begin{bmatrix} 0 \\ 0 \\ 0 \\ \vdots \\ 1 \end{bmatrix}$$

so

$$A[\mathbf{u}_1]_B = \begin{bmatrix} a_{11} & a_{12} & \cdots & a_{1n} \\ a_{21} & a_{22} & \cdots & a_{2n} \\ \vdots & \vdots & & \vdots \\ a_{m1} & a_{m2} & \cdots & a_{mn} \end{bmatrix} \begin{bmatrix} 1 \\ 0 \\ 0 \\ \vdots \\ 0 \end{bmatrix} = \begin{bmatrix} a_{11} \\ a_{21} \\ \vdots \\ a_{m1} \end{bmatrix}$$

$$A[\mathbf{u}_2]_B = \begin{bmatrix} a_{11} & a_{12} & \cdots & a_{1n} \\ a_{21} & a_{22} & \cdots & a_{2n} \\ \vdots & \vdots & & \vdots \\ a_{m1} & a_{m2} & \cdots & a_{mn} \end{bmatrix} \begin{bmatrix} 0 \\ 1 \\ 0 \\ \vdots \\ 0 \end{bmatrix} = \begin{bmatrix} a_{12} \\ a_{22} \\ \vdots \\ a_{m2} \end{bmatrix}$$

$$\vdots$$

$$A[\mathbf{u}_n]_B = \begin{bmatrix} a_{11} & a_{12} & \cdots & a_{1n} \\ a_{21} & a_{22} & \cdots & a_{2n} \\ \vdots & \vdots & & \vdots \\ a_{m1} & a_{m2} & \cdots & a_{mn} \end{bmatrix} \begin{bmatrix} 0 \\ 0 \\ 0 \\ \vdots \\ 1 \end{bmatrix} = \begin{bmatrix} a_{1n} \\ a_{2n} \\ \vdots \\ a_{mn} \end{bmatrix}$$

Substituting these results into (2) yields

$$\begin{bmatrix} a_{11} \\ a_{21} \\ \vdots \\ a_{m1} \end{bmatrix} = [T(\mathbf{u}_1)]_{B'}, \quad \begin{bmatrix} a_{12} \\ a_{22} \\ \vdots \\ a_{m2} \end{bmatrix} = [T(\mathbf{u}_2)]_{B'}, \ldots, \quad \begin{bmatrix} a_{1n} \\ a_{2n} \\ \vdots \\ a_{mn} \end{bmatrix} = [T(\mathbf{u}_n)]_{B'}$$

which shows that the successive columns of A are the coordinate vectors of

$$T(\mathbf{u}_1), T(\mathbf{u}_2), \ldots, T(\mathbf{u}_n)$$

with respect to the basis B'. Thus the matrix for T with respect to the bases B and B' is

$$A = \begin{bmatrix} [T(\mathbf{u}_1)]_{B'} \mid [T(\mathbf{u}_2)]_{B'} \mid \cdots \mid [T(\mathbf{u}_n)]_{B'} \end{bmatrix} \tag{3}$$

This matrix will be denoted by the symbol

$$[T]_{B',B}$$

so the preceding formula can also be written as

$$[T]_{B',B} = \begin{bmatrix} [T(\mathbf{u}_1)]_{B'} \mid [T(\mathbf{u}_2)]_{B'} \mid \cdots \mid [T(\mathbf{u}_n)]_{B'} \end{bmatrix} \tag{4}$$

and from (1), this matrix has the property

$$[T]_{B',B}[\mathbf{x}]_B = [T(\mathbf{x})]_{B'} \tag{4a}$$

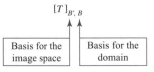

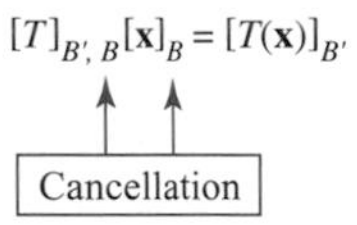

Figure 8.4.3

$[T]_{B',B}[\mathbf{x}]_B = [T(\mathbf{x})]_{B'}$

Cancellation

Figure 8.4.4

REMARK Observe that in the notation $[T]_{B',B}$ the right subscript is a basis for the domain of T, and the left subscript is a basis for the image space of T (Figure 8.4.3). Moreover, observe how the subscript B seems to "cancel out" in Formula (4a) (Figure 8.4.4).

Matrices of Linear Operators

In the special case where $V = W$ (so that $T: V \to V$ is a linear operator), it is usual to take $B = B'$ when constructing a matrix for T. In this case the resulting matrix is called the *matrix for T with respect to the basis B* and is usually denoted by $[T]_B$ rather than $[T]_{B,B}$. If $B = \{\mathbf{u}_1, \mathbf{u}_2, \ldots, \mathbf{u}_n\}$, then Formulas (4) and (4a) become

$$[T]_B = \begin{bmatrix} [T(\mathbf{u}_1)]_B \mid [T(\mathbf{u}_2)]_B \mid \cdots \mid [T(\mathbf{u}_n)]_B \end{bmatrix} \tag{5}$$

and

$$[T]_B[\mathbf{x}]_B = [T(\mathbf{x})]_B \tag{5a}$$

Phrased informally, (4a) and (5a) state that *the matrix for T times the coordinate vector for* **x** *is the coordinate vector for T(**x**).*

EXAMPLE 1 Matrix for a Linear Transformation

Let $T: P_1 \to P_2$ be the linear transformation defined by

$$T(p(x)) = xp(x)$$

Find the matrix for T with respect to the standard bases

$$B = \{\mathbf{u}_1, \mathbf{u}_2\} \quad \text{and} \quad B' = \{\mathbf{v}_1, \mathbf{v}_2, \mathbf{v}_3\}$$

where

$$\mathbf{u}_1 = 1, \quad \mathbf{u}_2 = x; \qquad \mathbf{v}_1 = 1, \quad \mathbf{v}_2 = x, \quad \mathbf{v}_3 = x^2$$

Solution

From the given formula for T we obtain

$$T(\mathbf{u}_1) = T(1) = (x)(1) = x$$
$$T(\mathbf{u}_2) = T(x) = (x)(x) = x^2$$

By inspection, we can determine the coordinate vectors for $T(\mathbf{u}_1)$ and $T(\mathbf{u}_2)$ relative to B'; they are

$$[T(\mathbf{u}_1)]_{B'} = \begin{bmatrix} 0 \\ 1 \\ 0 \end{bmatrix}, \qquad [T(\mathbf{u}_2)]_{B'} = \begin{bmatrix} 0 \\ 0 \\ 1 \end{bmatrix}$$

Thus the matrix for T with respect to B and B' is

$$[T]_{B',B} = \left[[T(\mathbf{u}_1)]_{B'} \mid [T(\mathbf{u}_2)]_{B'}\right] = \begin{bmatrix} 0 & 0 \\ 1 & 0 \\ 0 & 1 \end{bmatrix} \quad \blacklozenge$$

EXAMPLE 2 Verifying Formula (4a)

Let $T: P_1 \to P_2$ be the linear transformation in Example 1. Show that the matrix

$$[T]_{B',B} = \begin{bmatrix} 0 & 0 \\ 1 & 0 \\ 0 & 1 \end{bmatrix}$$

(obtained in Example 1) satisfies (4a) for every vector $\mathbf{x} = a + bx$ in P_1.

Solution

Since $\mathbf{x} = p(x) = a + bx$, we have

$$T(\mathbf{x}) = xp(x) = ax + bx^2$$

For the bases B and B' in Example 1, it follows by inspection that

$$[\mathbf{x}]_B = [ax + b]_B = \begin{bmatrix} a \\ b \end{bmatrix} \quad \text{and} \quad [T(\mathbf{x})]_{B'} = [ax + bx^2] = \begin{bmatrix} 0 \\ a \\ b \end{bmatrix}$$

Thus

$$[T]_{B',B}[\mathbf{x}]_B = \begin{bmatrix} 0 & 0 \\ 1 & 0 \\ 0 & 1 \end{bmatrix} \begin{bmatrix} a \\ b \end{bmatrix} = \begin{bmatrix} 0 \\ a \\ b \end{bmatrix} = [T(\mathbf{x})]_{B'}$$

so (4a) holds. ◆

EXAMPLE 3   Matrix for a Linear Transformation

Let $T: R^2 \to R^3$ be the linear transformation defined by

$$T\left(\begin{bmatrix} x_1 \\ x_2 \end{bmatrix}\right) = \begin{bmatrix} x_2 \\ -5x_1 + 13x_2 \\ -7x_1 + 16x_2 \end{bmatrix}$$

Find the matrix for the transformation T with respect to the bases $B = \{\mathbf{u}_1, \mathbf{u}_2\}$ for R^2 and $B' = \{\mathbf{v}_1, \mathbf{v}_2, \mathbf{v}_3\}$ for R^3, where

$$\mathbf{u}_1 = \begin{bmatrix} 3 \\ 1 \end{bmatrix}, \quad \mathbf{u}_2 = \begin{bmatrix} 5 \\ 2 \end{bmatrix}; \quad \mathbf{v}_1 = \begin{bmatrix} 1 \\ 0 \\ -1 \end{bmatrix}, \quad \mathbf{v}_2 = \begin{bmatrix} -1 \\ 2 \\ 2 \end{bmatrix}, \quad \mathbf{v}_3 = \begin{bmatrix} 0 \\ 1 \\ 2 \end{bmatrix}$$

Solution

From the formula for T,

$$T(\mathbf{u}_1) = \begin{bmatrix} 1 \\ -2 \\ -5 \end{bmatrix}, \qquad T(\mathbf{u}_2) = \begin{bmatrix} 2 \\ 1 \\ -3 \end{bmatrix}$$

Expressing these vectors as linear combinations of $\mathbf{v}_1$, $\mathbf{v}_2$, and $\mathbf{v}_3$, we obtain (verify)

$$T(\mathbf{u}_1) = \mathbf{v}_1 - 2\mathbf{v}_3, \qquad T(\mathbf{u}_2) = 3\mathbf{v}_1 + \mathbf{v}_2 - \mathbf{v}_3$$

Thus

$$[T(\mathbf{u}_1)]_{B'} = \begin{bmatrix} 1 \\ 0 \\ -2 \end{bmatrix}, \qquad [T(\mathbf{u}_2)]_{B'} = \begin{bmatrix} 3 \\ 1 \\ -1 \end{bmatrix}$$

so

$$[T]_{B',B} = \left[[T(\mathbf{u}_1)]_{B'} \mid [T(\mathbf{u}_2)]_{B'}\right] = \begin{bmatrix} 1 & 3 \\ 0 & 1 \\ -2 & -1 \end{bmatrix} \quad ◆$$

EXAMPLE 4   Verifying Formula (5a)

Let $T: R^2 \to R^2$ be the linear operator defined by

$$T\left(\begin{bmatrix} x_1 \\ x_2 \end{bmatrix}\right) = \begin{bmatrix} x_1 + x_2 \\ -2x_1 + 4x_2 \end{bmatrix}$$

and let $B = \{\mathbf{u}_1, \mathbf{u}_2\}$ be the basis, where

$$\mathbf{u}_1 = \begin{bmatrix} 1 \\ 1 \end{bmatrix}, \qquad \mathbf{u}_2 = \begin{bmatrix} 1 \\ 2 \end{bmatrix}$$

(a) Find $[T]_B$.

(b) Verify that (5a) holds for every vector $\mathbf{x}$ in R^2.

Solution (a)

From the given formula for T,

$$T(\mathbf{u}_1) = \begin{bmatrix} 2 \\ 2 \end{bmatrix} = 2\mathbf{u}_1, \qquad T(\mathbf{u}_2) = \begin{bmatrix} 3 \\ 6 \end{bmatrix} = 3\mathbf{u}_2$$

Therefore,

$$[T(\mathbf{u}_1)]_B = \begin{bmatrix} 2 \\ 0 \end{bmatrix} \quad \text{and} \quad [T(\mathbf{u}_2)]_B = \begin{bmatrix} 0 \\ 3 \end{bmatrix}$$

Consequently,

$$[T]_B = \big[[T(\mathbf{u}_1)]_B \mid [T(\mathbf{u}_2)]_B\big] = \begin{bmatrix} 2 & 0 \\ 0 & 3 \end{bmatrix}$$

Solution (b)

If

$$\mathbf{x} = \begin{bmatrix} x_1 \\ x_2 \end{bmatrix} \tag{6}$$

is any vector in R^2, then from the given formula for T,

$$T(\mathbf{x}) = \begin{bmatrix} x_1 + x_2 \\ -2x_1 + 4x_2 \end{bmatrix} \tag{7}$$

To find $[\mathbf{x}]_B$ and $[T(\mathbf{x})]_B$, we must express (6) and (7) as linear combinations of $\mathbf{u}_1$ and $\mathbf{u}_2$. This yields the vector equations

$$\begin{bmatrix} x_1 \\ x_2 \end{bmatrix} = k_1 \begin{bmatrix} 1 \\ 1 \end{bmatrix} + k_2 \begin{bmatrix} 1 \\ 2 \end{bmatrix} \tag{8}$$

$$\begin{bmatrix} x_1 + x_2 \\ -2x_1 + 4x_2 \end{bmatrix} = c_1 \begin{bmatrix} 1 \\ 1 \end{bmatrix} + c_2 \begin{bmatrix} 1 \\ 2 \end{bmatrix} \tag{9}$$

Equating corresponding entries yields the linear systems

$$\begin{aligned} k_1 + k_2 &= x_1 \\ k_1 + 2k_2 &= x_2 \end{aligned} \tag{10}$$

and

$$\begin{aligned} c_1 + c_2 &= x_1 + x_2 \\ c_1 + 2c_2 &= -2x_1 + 4x_2 \end{aligned} \tag{11}$$

Solving (10) for k_1 and k_2 yields

$$k_1 = 2x_1 - x_2, \qquad k_2 = -x_1 + x_2$$

so

$$[\mathbf{x}]_B = \begin{bmatrix} 2x_1 - x_2 \\ -x_1 + x_2 \end{bmatrix}$$

and solving (11) for c_1 and c_2 yields

$$c_1 = 4x_1 - 2x_2, \qquad c_2 = -3x_1 + 3x_2$$

so

$$[T(\mathbf{x})]_B = \begin{bmatrix} 4x_1 - 2x_2 \\ -3x_1 + 3x_2 \end{bmatrix}$$

Thus

$$[T]_B[\mathbf{x}]_B = \begin{bmatrix} 2 & 0 \\ 0 & 3 \end{bmatrix} \begin{bmatrix} 2x_1 - x_2 \\ -x_1 + x_2 \end{bmatrix} = \begin{bmatrix} 4x_1 - 2x_2 \\ -3x_1 + 3x_2 \end{bmatrix} = [T(\mathbf{x})]_B$$

so (5a) holds. ◆

Matrices of Identity Operators

The matrix for the identity operator on V always takes a special form.

EXAMPLE 5   Matrices of Identity Operators

If $B = \{\mathbf{u}_1, \mathbf{u}_2, \ldots, \mathbf{u}_n\}$ is a basis for a finite-dimensional vector space V and $I: V \to V$ is the identity operator on V, then

$$I(\mathbf{u}_1) = \mathbf{u}_1, \quad I(\mathbf{u}_2) = \mathbf{u}_2, \ldots, \quad I(\mathbf{u}_n) = \mathbf{u}_n$$

Therefore,

$$[I(\mathbf{u}_1)]_B = \begin{bmatrix} 1 \\ 0 \\ 0 \\ \vdots \\ 0 \end{bmatrix}, \quad [I(\mathbf{u}_2)]_B = \begin{bmatrix} 0 \\ 1 \\ 0 \\ \vdots \\ 0 \end{bmatrix}, \ldots, \quad [I(\mathbf{u}_n)]_B = \begin{bmatrix} 0 \\ 0 \\ 0 \\ \vdots \\ 1 \end{bmatrix}$$

Thus

$$[I]_B = \begin{bmatrix} 1 & 0 & \cdots & 0 \\ 0 & 1 & \cdots & 0 \\ 0 & 0 & \cdots & 0 \\ \vdots & \vdots & & \vdots \\ 0 & 0 & \cdots & 1 \end{bmatrix} = I$$

Consequently, the matrix of the identity operator with respect to any basis is the $n \times n$ identity matrix. This result could have been anticipated from Formula (5a), since the formula yields

$$[I]_B[\mathbf{x}]_B = [I(\mathbf{x})]_B = [\mathbf{x}]_B$$

which is consistent with the fact that $[I]_B = I$. ◆

We leave it as an exercise to prove the following result.

THEOREM 8.4.1

If $T: R^n \to R^m$ is a linear transformation, and if B and B' are the standard bases for R^n and R^m, respectively, then

$$[T]_{B',B} = [T] \tag{12}$$

This theorem tells us that in the special case where T maps R^n into R^m, the matrix for T with respect to the standard bases is the standard matrix for T. In this special case, Formula (4a) of this section reduces to

$$[T]\mathbf{x} = T(\mathbf{x})$$

Why Matrices of Linear Transformations Are Important

There are two primary reasons for studying matrices for general linear transformations, one theoretical and the other quite practical:

- Answers to theoretical questions about the structure of general linear transformations on finite-dimensional vector spaces can often be obtained by studying just the matrix transformations. Such matters are considered in detail in more advanced linear algebra courses, but we will touch on them in later sections.

- These matrices make it possible to compute images of vectors using matrix multiplication. Such computations can be performed rapidly on computers.

To focus on the latter idea, let $T: V \to W$ be a linear transformation. As shown in Figure 8.4.5, the matrix $[T]_{B',B}$ can be used to calculate $T(\mathbf{x})$ in three steps by the following *indirect* procedure:

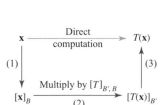

Figure 8.4.5

1. Compute the coordinate vector $[\mathbf{x}]_B$.
2. Multiply $[\mathbf{x}]_B$ on the left by $[T]_{B',B}$ to produce $[T(\mathbf{x})]_{B'}$.
3. Reconstruct $T(\mathbf{x})$ from its coordinate vector $[T(\mathbf{x})]_{B'}$.

EXAMPLE 6   Linear Operator on P_2

Let $T: P_2 \to P_2$ be the linear operator defined by

$$T(p(x)) = p(3x - 5)$$

that is, $T(c_0 + c_1 x + c_2 x^2) = c_0 + c_1(3x - 5) + c_2(3x - 5)^2$.

(a) Find $[T]_B$ with respect to the basis $B = \{1, x, x^2\}$.
(b) Use the indirect procedure to compute $T(1 + 2x + 3x^2)$.
(c) Check the result in (b) by computing $T(1 + 2x + 3x^2)$ directly.

Solution (a)

From the formula for T,

$$T(1) = 1, \qquad T(x) = 3x - 5, \qquad T(x^2) = (3x - 5)^2 = 9x^2 - 30x + 25$$

so

$$[T(1)]_B = \begin{bmatrix} 1 \\ 0 \\ 0 \end{bmatrix}, \qquad [T(x)]_B = \begin{bmatrix} -5 \\ 3 \\ 0 \end{bmatrix}, \qquad [T(x^2)]_B = \begin{bmatrix} 25 \\ -30 \\ 9 \end{bmatrix}$$

Thus

$$[T]_B = \begin{bmatrix} 1 & -5 & 25 \\ 0 & 3 & -30 \\ 0 & 0 & 9 \end{bmatrix}$$

Solution (b)

The coordinate vector relative to B for the vector $\mathbf{p} = 1 + 2x + 3x^2$ is

$$[\mathbf{p}]_B = \begin{bmatrix} 1 \\ 2 \\ 3 \end{bmatrix}$$

Thus, from (5a),

$$[T(1 + 2x + 3x^2)]_B = [T(\mathbf{p})]_B = [T]_B [\mathbf{p}]_B$$

$$= \begin{bmatrix} 1 & -5 & 25 \\ 0 & 3 & -30 \\ 0 & 0 & 9 \end{bmatrix} \begin{bmatrix} 1 \\ 2 \\ 3 \end{bmatrix} = \begin{bmatrix} 66 \\ -84 \\ 27 \end{bmatrix}$$

from which it follows that

$$T(1 + 2x + 3x^2) = 66 - 84x + 27x^2$$

Solution (c)

By direct computation,

$$\begin{aligned} T(1 + 2x + 3x^2) &= 1 + 2(3x - 5) + 3(3x - 5)^2 \\ &= 1 + 6x - 10 + 27x^2 - 90x + 75 \\ &= 66 - 84x + 27x^2 \end{aligned}$$

which agrees with the result in (b). ◆

Matrices of Compositions and Inverse Transformations

We shall now mention two theorems that are generalizations of Formula (21) of Section 4.2 and Formula (1) of Section 4.3. The proofs are omitted.

THEOREM 8.4.2

If $T_1 : U \to V$ and $T_2 : V \to W$ are linear transformations, and if B, B'', and B' are bases for U, V, and W, respectively, then

$$[T_2 \circ T_1]_{B', B} = [T_2]_{B', B''} [T_1]_{B'', B} \tag{13}$$

THEOREM 8.4.3

If $T : V \to V$ is a linear operator, and if B is a basis for V, then the following are equivalent.

(a) T is one-to-one.   *(b) $[T]_B$ is invertible.*

Moreover, when these equivalent conditions hold,

$$[T^{-1}]_B = [T]_B^{-1} \tag{14}$$

REMARK   In (13), observe how the interior subscript B'' (the basis for the intermediate space V) seems to "cancel out," leaving only the bases for the domain and image

$[T_2 \circ T_1]_{B',\,B} = [T_2]_{B',\,B''} \quad [T_1]_{B'',\,B}$

Cancellation

Figure 8.4.6

space of the composition as subscripts (Figure 8.4.6). This cancellation of interior subscripts suggests the following extension of Formula (13) to compositions of three linear transformations (Figure 8.4.7).

$$[T_3 \circ T_2 \circ T_1]_{B',B} = [T_3]_{B',B'''}[T_2]_{B''',B''}[T_1]_{B'',B} \tag{15}$$

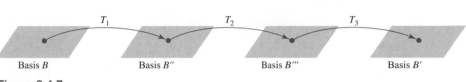

Basis B Basis B'' Basis B''' Basis B'

Figure 8.4.7

The following example illustrates Theorem 8.4.2.

EXAMPLE 7 Using Theorem 8.4.2

Let $T_1 \colon P_1 \to P_2$ be the linear transformation defined by

$$T_1(p(x)) = xp(x)$$

and let $T_2 \colon P_2 \to P_2$ be the linear operator defined by

$$T_2(p(x)) = p(3x - 5)$$

Then the composition $(T_2 \circ T_1) \colon P_1 \to P_2$ is given by

$$(T_2 \circ T_1)(p(x)) = T_2(T_1(p(x))) = T_2(xp(x)) = (3x - 5)p(3x - 5)$$

Thus, if $p(x) = c_0 + c_1 x$, then

$$(T_2 \circ T_1)(c_0 + c_1 x) = (3x - 5)(c_0 + c_1(3x - 5))$$
$$= c_0(3x - 5) + c_1(3x - 5)^2 \tag{16}$$

In this example, P_1 plays the role of U in Theorem 8.4.2, and P_2 plays the roles of both V and W; thus we can take $B' = B''$ in (13) so that the formula simplifies to

$$[T_2 \circ T_1]_{B',B} = [T_2]_{B'}[T_1]_{B',B} \tag{17}$$

Let us choose $B = \{1, x\}$ to be the basis for P_1 and choose $B' = \{1, x, x^2\}$ to be the basis for P_2. We showed in Examples 1 and 6 that

$$[T_1]_{B',B} = \begin{bmatrix} 0 & 0 \\ 1 & 0 \\ 0 & 1 \end{bmatrix} \quad \text{and} \quad [T_2]_{B'} = \begin{bmatrix} 1 & -5 & 25 \\ 0 & 3 & -30 \\ 0 & 0 & 9 \end{bmatrix}$$

Thus it follows from (17) that

$$[T_2 \circ T_1]_{B',B} = \begin{bmatrix} 1 & -5 & 25 \\ 0 & 3 & -30 \\ 0 & 0 & 9 \end{bmatrix} \begin{bmatrix} 0 & 0 \\ 1 & 0 \\ 0 & 1 \end{bmatrix} = \begin{bmatrix} -5 & 25 \\ 3 & -30 \\ 0 & 9 \end{bmatrix} \tag{18}$$

As a check, we will calculate $[T_2 \circ T_1]_{B',B}$ directly from Formula (4). Since $B = \{1, x\}$, it follows from Formula (4) with $\mathbf{u}_1 = 1$ and $\mathbf{u}_2 = x$ that

$$[T_2 \circ T_1]_{B'B} = \big[[(T_2 \circ T_1)(1)]_{B'} \mid [(T_2 \circ T_1)(x)]_{B'} \big] \tag{19}$$

Using (16) yields

$$(T_2 \circ T_1)(1) = 3x - 5 \quad \text{and} \quad (T_2 \circ T_1)(x) = (3x - 5)^2 = 9x^2 - 30x + 25$$

Since $B' = \{1, x, x^2\}$, it follows from this that

$$[(T_2 \circ T_1)(1)]_{B'} = \begin{bmatrix} -5 \\ 3 \\ 0 \end{bmatrix} \quad \text{and} \quad [(T_2 \circ T_1)(x)]_{B'} = \begin{bmatrix} 25 \\ -30 \\ 9 \end{bmatrix}$$

Substituting in (19) yields

$$[T_2 \circ T_1]_{B',B} = \begin{bmatrix} -5 & 25 \\ 3 & -30 \\ 0 & 9 \end{bmatrix}$$

which agrees with (18). ◆

EXERCISE SET
8.4

1. Let $T: P_2 \rightarrow P_3$ be the linear transformation defined by $T(p(x)) = xp(x)$.

   (a) Find the matrix for T with respect to the standard bases

   $$B = \{\mathbf{u}_1, \mathbf{u}_2, \mathbf{u}_3\} \quad \text{and} \quad B' = \{\mathbf{v}_1, \mathbf{v}_2, \mathbf{v}_3, \mathbf{v}_4\}$$

   where

   $$\mathbf{u}_1 = 1, \qquad \mathbf{u}_2 = x, \qquad \mathbf{u}_3 = x^2$$
   $$\mathbf{v}_1 = 1, \qquad \mathbf{v}_2 = x, \qquad \mathbf{v}_3 = x^2, \qquad \mathbf{v}_4 = x^3$$

   (b) Verify that the matrix $[T]_{B',B}$ obtained in part (a) satisfies Formula (4a) for every vector $\mathbf{x} = c_0 + c_1 x + c_2 x^2$ in P_2.

2. Let $T: P_2 \rightarrow P_1$ be the linear transformation defined by

   $$T(a_0 + a_1 x + a_2 x^2) = (a_0 + a_1) - (2a_1 + 3a_2)x$$

   (a) Find the matrix for T with respect to the standard bases $B = \{1, x, x^2\}$ and $B' = \{1, x\}$ for P_2 and P_1.

   (b) Verify that the matrix $[T]_{B',B}$ obtained in part (a) satisfies Formula (4a) for every vector $\mathbf{x} = c_0 + c_1 x + c_2 x^2$ in P_2.

3. Let $T: P_2 \rightarrow P_2$ be the linear operator defined by

   $$T(a_0 + a_1 x + a_2 x^2) = a_0 + a_1(x - 1) + a_2(x - 1)^2$$

   (a) Find the matrix for T with respect to the standard basis $B = \{1, x, x^2\}$ for P_2.

   (b) Verify that the matrix $[T]_B$ obtained in part (a) satisfies Formula (5a) for every vector $\mathbf{x} = a_0 + a_1 x + a_2 x^2$ in P_2.

4. Let $T: R^2 \rightarrow R^2$ be the linear operator defined by

   $$T\left(\begin{bmatrix} x_1 \\ x_2 \end{bmatrix}\right) = \begin{bmatrix} x_1 - x_2 \\ x_1 + x_2 \end{bmatrix}$$

   and let $B = \{\mathbf{u}_1, \mathbf{u}_2\}$ be the basis for which

   $$\mathbf{u}_1 = \begin{bmatrix} 1 \\ 1 \end{bmatrix} \quad \text{and} \quad \mathbf{u}_2 = \begin{bmatrix} -1 \\ 0 \end{bmatrix}$$

   (a) Find $[T]_B$.

   (b) Verify that Formula (5a) holds for every vector $\mathbf{x}$ in R^2.

5. Let $T: R^2 \to R^3$ be defined by

$$T\left(\begin{bmatrix} x_1 \\ x_2 \end{bmatrix}\right) = \begin{bmatrix} x_1 + 2x_2 \\ -x_1 \\ 0 \end{bmatrix}$$

(a) Find the matrix $[T]_{B',B}$ with respect to the bases $B = \{\mathbf{u}_1, \mathbf{u}_2\}$ and $B' = \{\mathbf{v}_1, \mathbf{v}_2, \mathbf{v}_3\}$, where

$$\mathbf{u}_1 = \begin{bmatrix} 1 \\ 3 \end{bmatrix}, \quad \mathbf{u}_2 = \begin{bmatrix} -2 \\ 4 \end{bmatrix}, \quad \mathbf{v}_1 = \begin{bmatrix} 1 \\ 1 \\ 1 \end{bmatrix}, \quad \mathbf{v}_2 = \begin{bmatrix} 2 \\ 2 \\ 0 \end{bmatrix}, \quad \mathbf{v}_3 = \begin{bmatrix} 3 \\ 0 \\ 0 \end{bmatrix}$$

(b) Verify that Formula (4a) holds for every vector

$$\mathbf{x} = \begin{bmatrix} x_1 \\ x_2 \end{bmatrix}$$

in R^2.

6. Let $T: R^3 \to R^3$ be the linear operator defined by

$$T(x_1, x_2, x_3) = (x_1 - x_2, x_2 - x_1, x_1 - x_3).$$

(a) Find the matrix for T with respect to the basis $B = \{\mathbf{v}_1, \mathbf{v}_2, \mathbf{v}_3\}$, where

$$\mathbf{v}_1 = (1, 0, 1), \qquad \mathbf{v}_2 = (0, 1, 1), \qquad \mathbf{v}_3 = (1, 1, 0)$$

(b) Verify that Formula (5a) holds for every vector $\mathbf{x} = (x_1, x_2, x_3)$ in R^3.

(c) Is T one-to-one? If so, find the matrix of T^{-1}.

7. Let $T: P_2 \to P_2$ be the linear operator defined by $T(p(x)) = p(2x + 1)$—that is,

$$T(c_0 + c_1 x + c_2 x^2) = c_0 + c_1(2x + 1) + c_2(2x + 1)^2$$

(a) Find $[T]_B$ with respect to the basis $B = \{1, x, x^2\}$.

(b) Use the indirect procedure illustrated in Figure 8.4.5 to compute $T(2 - 3x + 4x^2)$.

(c) Check the result obtained in part (b) by computing $T(2 - 3x + 4x^2)$ directly.

8. Let $T: P_2 \to P_3$ be the linear transformation defined by $T(p(x)) = xp(x - 3)$—that is,

$$T(c_0 + c_1 x + c_2 x^2) = x(c_0 + c_1(x - 3) + c_2(x - 3)^2)$$

(a) Find $[T]_{B',B}$ with respect to the bases $B = \{1, x, x^2\}$ and $B' = \{1, x, x^2, x^3\}$.

(b) Use the indirect procedure illustrated in Figure 8.4.5 to compute $T(1 + x - x^2)$.

(c) Check the result obtained in part (b) by computing $T(1 + x - x^2)$ directly.

9. Let $\mathbf{v}_1 = \begin{bmatrix} 1 \\ 3 \end{bmatrix}$ and $\mathbf{v}_2 = \begin{bmatrix} -1 \\ 4 \end{bmatrix}$, and let

$$A = \begin{bmatrix} 1 & 3 \\ -2 & 5 \end{bmatrix}$$

be the matrix for $T: R^2 \to R^2$ with respect to the basis $B = \{\mathbf{v}_1, \mathbf{v}_2\}$.

(a) Find $[T(\mathbf{v}_1)]_B$ and $[T(\mathbf{v}_2)]_B$.

(b) Find $T(\mathbf{v}_1)$ and $T(\mathbf{v}_2)$.

(c) Find a formula for $T\left(\begin{bmatrix} x_1 \\ x_2 \end{bmatrix}\right)$.

(d) Use the formula obtained in (c) to compute $T\left(\begin{bmatrix} 1 \\ 1 \end{bmatrix}\right)$.

10. Let $A = \begin{bmatrix} 3 & -2 & 1 & 0 \\ 1 & 6 & 2 & 1 \\ -3 & 0 & 7 & 1 \end{bmatrix}$ be the matrix of $T: R^4 \to R^3$ with respect to the bases

$B = \{\mathbf{v}_1, \mathbf{v}_2, \mathbf{v}_3, \mathbf{v}_4\}$ and $B' = \{\mathbf{w}_1, \mathbf{w}_2, \mathbf{w}_3\}$, where

$$\mathbf{v}_1 = \begin{bmatrix} 0 \\ 1 \\ 1 \\ 1 \end{bmatrix}, \quad \mathbf{v}_2 = \begin{bmatrix} 2 \\ 1 \\ -1 \\ -1 \end{bmatrix}, \quad \mathbf{v}_3 = \begin{bmatrix} 1 \\ 4 \\ -1 \\ 2 \end{bmatrix}, \quad \mathbf{v}_4 = \begin{bmatrix} 6 \\ 9 \\ 4 \\ 2 \end{bmatrix}$$

$$\mathbf{w}_1 = \begin{bmatrix} 0 \\ 8 \\ 8 \end{bmatrix}, \quad \mathbf{w}_2 = \begin{bmatrix} -7 \\ 8 \\ 1 \end{bmatrix}, \quad \mathbf{w}_3 = \begin{bmatrix} -6 \\ 9 \\ 1 \end{bmatrix}$$

(a) Find $[T(\mathbf{v}_1)]_{B'}$, $[T(\mathbf{v}_2)]_{B'}$, $[T(\mathbf{v}_3)]_{B'}$, and $[T(\mathbf{v}_4)]_{B'}$.

(b) Find $T(\mathbf{v}_1)$, $T(\mathbf{v}_2)$, $T(\mathbf{v}_3)$, and $T(\mathbf{v}_4)$.

(c) Find a formula for $T\left(\begin{bmatrix} x_1 \\ x_2 \\ x_3 \\ x_4 \end{bmatrix}\right)$.

(d) Use the formula obtained in (c) to compute $T\left(\begin{bmatrix} 2 \\ 2 \\ 0 \\ 0 \end{bmatrix}\right)$.

11. Let $A = \begin{bmatrix} 1 & 3 & -1 \\ 2 & 0 & 5 \\ 6 & -2 & 4 \end{bmatrix}$ be the matrix of $T: P_2 \to P_2$ with respect to the basis

$B = \{\mathbf{v}_1, \mathbf{v}_2, \mathbf{v}_3\}$, where $\mathbf{v}_1 = 3x + 3x^2$, $\mathbf{v}_2 = -1 + 3x + 2x^2$, $\mathbf{v}_3 = 3 + 7x + 2x^2$.

(a) Find $[T(\mathbf{v}_1)]_B$, $[T(\mathbf{v}_2)]_B$, and $[T(\mathbf{v}_3)]_B$.

(b) Find $T(\mathbf{v}_1)$, $T(\mathbf{v}_2)$, and $T(\mathbf{v}_3)$.

(c) Find a formula for $T(a_0 + a_1x + a_2x^2)$.

(d) Use the formula obtained in (c) to compute $T(1 + x^2)$.

12. Let $T_1: P_1 \to P_2$ be the linear transformation defined by

$$T_1(p(x)) = xp(x)$$

and let $T_2: P_2 \to P_2$ be the linear operator defined by

$$T_2(p(x)) = p(2x + 1)$$

Let $B = \{1, x\}$ and $B' = \{1, x, x^2\}$ be the standard bases for P_1 and P_2.

(a) Find $[T_2 \circ T_1]_{B',B}$, $[T_2]_{B'}$, and $[T_1]_{B',B}$.

(b) State a formula relating the matrices in part (a).

(c) Verify that the matrices in part (a) satisfy the formula you stated in part (b).

13. Let $T_1: P_1 \to P_2$ be the linear transformation defined by

$$T_1(c_0 + c_1x) = 2c_0 - 3c_1x$$

and let $T_2: P_2 \to P_3$ be the linear transformation defined by

$$T_2(c_0 + c_1x + c_2x^2) = 3c_0x + 3c_1x^2 + 3c_2x^3$$

Let $B = \{1, x\}$, $B'' = \{1, x, x^2\}$, and $B' = \{1, x, x^2, x^3\}$.

(a)   Find $[T_2 \circ T_1]_{B',B}$, $[T_2]_{B',B''}$, and $[T_1]_{B'',B}$.

(b)   State a formula relating the matrices in part (a).

(c)   Verify that the matrices in part (a) satisfy the formula you stated in part (b).

14. Show that if $T: V \to W$ is the zero transformation, then the matrix of T with respect to any bases for V and W is a zero matrix.

15. Show that if $T: V \to V$ is a contraction or a dilation of V (Example 4 of Section 8.1), then the matrix of T with respect to any basis for V is a positive scalar multiple of the identity matrix.

16. Let $B = \{\mathbf{v}_1, \mathbf{v}_2, \mathbf{v}_3, \mathbf{v}_4\}$ be a basis for a vector space V. Find the matrix with respect to B of the linear operator $T: V \to V$ defined by $T(\mathbf{v}_1) = \mathbf{v}_2$, $T(\mathbf{v}_2) = \mathbf{v}_3$, $T(\mathbf{v}_3) = \mathbf{v}_4$, $T(\mathbf{v}_4) = \mathbf{v}_1$.

17. Prove that if B and B' are the standard bases for R^n and R^m, respectively, then the matrix for a linear transformation $T: R^n \to R^m$ with respect to the bases B and B' is the standard matrix for T.

18. **(For Readers Who Have Studied Calculus)** Let $D: P_2 \to P_2$ be the differentiation operator $D(\mathbf{p}) = p'(x)$. In parts (a) and (b), find the matrix of D with respect to the basis $B = \{\mathbf{p}_1, \mathbf{p}_2, \mathbf{p}_3\}$.

(a) $\mathbf{p}_1 = 1$, $\mathbf{p}_2 = x$, $\mathbf{p}_3 = x^2$

(b) $\mathbf{p}_1 = 2$, $\mathbf{p}_2 = 2 - 3x$, $\mathbf{p}_3 = 2 - 3x + 8x^2$

(c)   Use the matrix in part (a) to compute $D(6 - 6x + 24x^2)$.

(d)   Repeat the directions for part (c) for the matrix in part (b).

19. **(For Readers Who Have Studied Calculus)** In each part, $B = \{\mathbf{f}_1, \mathbf{f}_2, \mathbf{f}_3\}$ is a basis for a subspace V of the vector space of real-valued functions defined on the real line. Find the matrix with respect to B of the differentiation operator $D: V \to V$.

(a) $\mathbf{f}_1 = 1$, $\mathbf{f}_2 = \sin x$, $\mathbf{f}_3 = \cos x$

(b) $\mathbf{f}_1 = 1$, $\mathbf{f}_2 = e^x$, $\mathbf{f}_3 = e^{2x}$

(c) $\mathbf{f}_1 = e^{2x}$, $\mathbf{f}_2 = xe^{2x}$, $\mathbf{f}_3 = x^2 e^{2x}$

(d)   Use the matrix in part (c) to compute $D(4e^{2x} + 6xe^{2x} - 10x^2 e^{2x})$.

Discussion & Discovery

20. Let V be a four-dimensional vector space with basis B, let W be a seven-dimensional vector space with basis B', and let $T: V \to W$ be a linear transformation. Identify the four vector spaces that contain the vectors at the corners of the accompanying diagram.

Figure Ex-20

21. In each part, fill in the missing part of the equation.

(a) $[T_2 \circ T_1]_{B',B} = [T_2]\_\_?\_\_[T_1]_{B'',B}$

(b) $[T_3 \circ T_2 \circ T_1]_{B',B} = [T_3]\_\_?\_\_[T_2]_{B''',B''}[T_1]_{B'',B}$

22. Give two reasons why matrices for general linear transformations are important.

8.5
SIMILARITY

The matrix of a linear operator $T: V \to V$ depends on the basis selected for V. One of the fundamental problems of linear algebra is to choose a basis for V that makes the matrix for T as simple as possible—a diagonal or a triangular matrix, for example. In this section we shall study this problem.

Simple Matrices for Linear Operators

Standard bases do not necessarily produce the simplest matrices for linear operators. For example, consider the linear operator $T: R^2 \to R^2$ defined by

$$T\left(\begin{bmatrix} x_1 \\ x_2 \end{bmatrix}\right) = \begin{bmatrix} x_1 + x_2 \\ -2x_1 + 4x_2 \end{bmatrix} \tag{1}$$

and the standard basis $B = \{\mathbf{e}_1, \mathbf{e}_2\}$ for R^2, where

$$\mathbf{e}_1 = \begin{bmatrix} 1 \\ 0 \end{bmatrix}, \qquad \mathbf{e}_2 = \begin{bmatrix} 0 \\ 1 \end{bmatrix}$$

By Theorem 8.4.1, the matrix for T with respect to this basis is the standard matrix for T; that is,

$$[T]_B = [T] = [T(\mathbf{e}_1) \mid T(\mathbf{e}_2)]$$

From (1),

$$T(\mathbf{e}_1) = \begin{bmatrix} 1 \\ -2 \end{bmatrix}, \qquad T(\mathbf{e}_2) = \begin{bmatrix} 1 \\ 4 \end{bmatrix}$$

so

$$[T]_B = \begin{bmatrix} 1 & 1 \\ -2 & 4 \end{bmatrix} \tag{2}$$

In comparison, we showed in Example 4 of Section 8.4 that if

$$\mathbf{u}_1 = \begin{bmatrix} 1 \\ 1 \end{bmatrix}, \qquad \mathbf{u}_2 = \begin{bmatrix} 1 \\ 2 \end{bmatrix} \tag{3}$$

then the matrix for T with respect to the basis $B' = \{\mathbf{u}_1, \mathbf{u}_2\}$ is the diagonal matrix

$$[T]_{B'} = \begin{bmatrix} 2 & 0 \\ 0 & 3 \end{bmatrix} \tag{4}$$

This matrix is "simpler" than (2) in the sense that diagonal matrices enjoy special properties that more general matrices do not.

One of the major themes in more advanced linear algebra courses is to determine the "simplest possible form" that can be obtained for the matrix of a linear operator by choosing the basis appropriately. Sometimes it is possible to obtain a diagonal matrix (as above, for example); other times one must settle for a triangular matrix or some other form. We will be able only to touch on this important topic in this text.

The problem of finding a basis that produces the simplest possible matrix for a linear operator $T: V \to V$ can be attacked by first finding a matrix for T relative to *any* basis, say a standard basis, where applicable, and then changing the basis in a manner that simplifies the matrix. Before pursuing this idea, it will be helpful to review some concepts about changing bases.

Recall from Formula (6) in Section 6.5 that if the sets $B = \{\mathbf{u}_1, \mathbf{u}_2, \ldots, \mathbf{u}_n\}$ and $B' = \{\mathbf{u}_1', \mathbf{u}_2', \ldots, \mathbf{u}_n'\}$ are bases for a vector space V, then the *transition matrix* from B' to B is given by the formula

$$P = \left[[\mathbf{u}_1']_B \mid [\mathbf{u}_2']_B \mid \cdots \mid [\mathbf{u}_n']_B\right] \tag{5}$$

This matrix has the property that for every vector **v** in V,

$$P[\mathbf{v}]_{B'} = [\mathbf{v}]_B \tag{6}$$

That is, multiplication by P maps the coordinate matrix for **v** relative to B' into the coordinate matrix for **v** relative to B [see Formula (5) in Section 6.5]. We showed in Theorem 6.5.4 that P is invertible and P^{-1} is the transition matrix from B to B'.

The following theorem gives a useful alternative viewpoint about transition matrices; it shows that the transition matrix from a basis B' to a basis B can be regarded as the matrix of an identity operator.

THEOREM 8.5.1

> *If B and B' are bases for a finite-dimensional vector space V, and if $I: V \to V$ is the identity operator, then the transition matrix from B' to B is $[I]_{B,B'}$.*

Proof Suppose that $B = \{\mathbf{u}_1, \mathbf{u}_2, \ldots, \mathbf{u}_n\}$ and $B' = \{\mathbf{u}'_1, \mathbf{u}'_2, \ldots, \mathbf{u}'_n\}$ are bases for V. Using the fact that $I(\mathbf{v}) = \mathbf{v}$ for all **v** in V, it follows from Formula (4) of Section 8.4 with B and B' *reversed* that

$$[I]_{B,B'} = \left[[I(\mathbf{u}'_1)]_B \mid [I(\mathbf{u}'_2)]_B \mid \cdots \mid [I(\mathbf{u}'_n)]_B\right]$$
$$= \left[[\mathbf{u}'_1]_B \mid [\mathbf{u}'_2]_B \mid \cdots \mid [\mathbf{u}'_n]_B\right]$$

Thus, from (5), we have $[I]_{B,B'} = P$, which shows that $[I]_{B,B'}$ is the transition matrix from B' to B. ∎

The result in this theorem is illustrated in Figure 8.5.1.

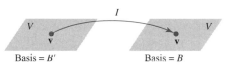

Figure 8.5.1 $[I]_{B,B'}$ is the transition matrix from B' to B.

Effect of Changing Bases on Matrices of Linear Operators

We are now ready to consider the main problem in this section.

Problem: If B and B' are two bases for a finite-dimensional vector space V, and if $T: V \to V$ is a linear operator, what relationship, if any, exists between the matrices $[T]_B$ and $[T]_{B'}$?

The answer to this question can be obtained by considering the composition of the three linear operators on V pictured in Figure 8.5.2.

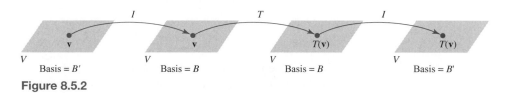

Figure 8.5.2

In this figure, **v** is first mapped into itself by the identity operator, then **v** is mapped into $T(\mathbf{v})$ by T, then $T(\mathbf{v})$ is mapped into itself by the identity operator. All four vector spaces involved in the composition are the same (namely, V); however, the bases for the spaces vary. Since the starting vector is $\mathbf{v}$ and the final vector is $T(\mathbf{v})$, the composition is the same as T; that is,

$$T = I \circ T \circ I \tag{7}$$

If, as illustrated in Figure 8.5.2, the first and last vector spaces are assigned the basis B' and the middle two spaces are assigned the basis B, then it follows from (7) and Formula (15) of Section 8.4 (with an appropriate adjustment in the names of the bases) that

$$[T]_{B',B'} = [I \circ T \circ I]_{B',B'} = [I]_{B',B}[T]_{B,B}[I]_{B,B'} \tag{8}$$

or, in simpler notation,

$$[T]_{B'} = [I]_{B',B}[T]_B[I]_{B,B'} \tag{9}$$

But it follows from Theorem 8.5.1 that $[I]_{B,B'}$ is the transition matrix from B' to B and consequently, $[I]_{B',B}$ is the transition matrix from B to B'. Thus, if we let $P = [I]_{B,B'}$, then $P^{-1} = [I]_{B',B}$, so (9) can be written as

$$[T]_{B'} = P^{-1}[T]_B P$$

In summary, we have the following theorem.

THEOREM 8.5.2

Let $T: V \to V$ be a linear operator on a finite-dimensional vector space V, and let B and B' be bases for V. Then

$$[T]_{B'} = P^{-1}[T]_B P \tag{10}$$

where P is the transition matrix from B' to B.

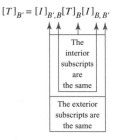

Figure 8.5.3

WARNING When applying Theorem 8.5.2, it is easy to forget whether P is the transition matrix from B to B' (incorrect) or from B' to B (correct). As indicated in Figure 8.5.3, it may help to write (10) in form (9), keeping in mind that the three "interior" subscripts are the same and the two exterior subscripts are the same. Once you master the pattern shown in this figure, you need only remember that $P = [I]_{B,B'}$ is the transition matrix from B' to B and that $P^{-1} = [I]_{B',B}$ is its inverse.

EXAMPLE 1 Using Theorem 8.5.2

Let $T: R^2 \to R^2$ be defined by

$$T\left(\begin{bmatrix} x_1 \\ x_2 \end{bmatrix}\right) = \begin{bmatrix} x_1 + x_2 \\ -2x_1 + 4x_2 \end{bmatrix}$$

Find the matrix of T with respect to the standard basis $B = \{\mathbf{e}_1, \mathbf{e}_2\}$ for R^2; then use Theorem 8.5.2 to find the matrix of T with respect to the basis $B' = \{\mathbf{u}_1', \mathbf{u}_2'\}$, where

$$\mathbf{u}_1' = \begin{bmatrix} 1 \\ 1 \end{bmatrix} \quad \text{and} \quad \mathbf{u}_2' = \begin{bmatrix} 1 \\ 2 \end{bmatrix}$$

Solution

We showed earlier in this section [see (2)] that

$$[T]_B = \begin{bmatrix} 1 & 1 \\ -2 & 4 \end{bmatrix}$$

To find $[T]_{B'}$ from (10), we will need to find the transition matrix

$$P = [I]_{B,B'} = \big[[\mathbf{u}_1']_B \mid [\mathbf{u}_2']_B\big]$$

[see (5)]. By inspection,

$$\mathbf{u}_1' = \mathbf{e}_1 + \mathbf{e}_2$$
$$\mathbf{u}_2' = \mathbf{e}_1 + 2\mathbf{e}_2$$

so

$$[\mathbf{u}_1']_B = \begin{bmatrix} 1 \\ 1 \end{bmatrix} \quad \text{and} \quad [\mathbf{u}_2']_B = \begin{bmatrix} 1 \\ 2 \end{bmatrix}$$

Thus the transition matrix from B' to B is

$$P = \begin{bmatrix} 1 & 1 \\ 1 & 2 \end{bmatrix}$$

The reader can check that

$$P^{-1} = \begin{bmatrix} 2 & -1 \\ -1 & 1 \end{bmatrix}$$

so by Theorem 8.5.2, the matrix of T relative to the basis B' is

$$[T]_{B'} = P^{-1}[T]_B P = \begin{bmatrix} 2 & -1 \\ -1 & 1 \end{bmatrix} \begin{bmatrix} 1 & 1 \\ -2 & 4 \end{bmatrix} \begin{bmatrix} 1 & 1 \\ 1 & 2 \end{bmatrix} = \begin{bmatrix} 2 & 0 \\ 0 & 3 \end{bmatrix}$$

which agrees with (4). ◆

Similarity

The relationship in Formula (10) is of such importance that there is some terminology associated with it.

DEFINITION

If A and B are square matrices, we say that **B is similar to A** if there is an invertible matrix P such that $B = P^{-1}AP$.

REMARK   It is left as an exercise to show that if a matrix B is similar to a matrix A, then necessarily A is similar to B. Therefore, we shall usually simply say that ***A and B are similar***.

Similarity Invariants

Similar matrices often have properties in common; for example, if A and B are similar matrices, then A and B have the same determinant. To see that this is so, suppose that

$$B = P^{-1}AP$$

Then

$$\det(B) = \det(P^{-1}AP) = \det(P^{-1})\det(A)\det(P)$$

$$= \frac{1}{\det(P)}\det(A)\det(P) = \det(A)$$

We make the following definition.

DEFINITION

A property of square matrices is said to be a *similarity invariant* or *invariant under similarity* if that property is shared by any two similar matrices.

In the terminology of this definition, the determinant of a square matrix is a similarity invariant. Table 1 lists some other important similarity invariants. The proofs of some of the results in Table 1 are given in the exercises.

It follows from Theorem 8.5.2 that *two matrices representing the same linear operator* $T: V \rightarrow V$ *with respect to different bases are similar.* Thus, if B is a basis for V, and the matrix $[T]_B$ has some property that is invariant under similarity, then for every basis B', the matrix $[T]_{B'}$ has that same property. For example, for any two bases B and B' we must have

$$\det([T]_B) = \det([T]_{B'})$$

It follows from this equation that the value of the determinant depends on T, but not on the particular basis that is used to obtain the matrix for T. Thus the determinant can be regarded as a property of the linear operator T; indeed, if V is a finite-dimensional vector space, then we can *define* the **determinant of the linear operator** T to be

$$\det(T) = \det([T]_B) \tag{11}$$

where B is any basis for V.

Table 1 Similarity Invariants

Property	Description
Determinant	A and $P^{-1}AP$ have the same determinant.
Invertibility	A is invertible if and only if $P^{-1}AP$ is invertible.
Rank	A and $P^{-1}AP$ have the same rank.
Nullity	A and $P^{-1}AP$ have the same nullity.
Trace	A and $P^{-1}AP$ have the same trace.
Characteristic polynomial	A and $P^{-1}AP$ have the same characteristic polynomial.
Eigenvalues	A and $P^{-1}AP$ have the same eigenvalues.
Eigenspace dimension	If λ is an eigenvalue of A and $P^{-1}AP$, then the eigenspace of A corresponding to λ and the eigenspace of $P^{-1}AP$ corresponding to λ have the same dimension.

EXAMPLE 2 Determinant of a Linear Operator

Let $T: R^2 \to R^2$ be defined by

$$T\left(\begin{bmatrix} x_1 \\ x_2 \end{bmatrix}\right) = \begin{bmatrix} x_1 + x_2 \\ -2x_1 + 4x_2 \end{bmatrix}$$

Find $\det(T)$.

Solution

We can choose any basis B and calculate $\det([T]_B)$. If we take the standard basis, then from Example 1,

$$[T]_B = \begin{bmatrix} 1 & 1 \\ -2 & 4 \end{bmatrix}, \quad \text{so} \quad \det(T) = \begin{vmatrix} 1 & 1 \\ -2 & 4 \end{vmatrix} = 6$$

Had we chosen the basis $B' = \{\mathbf{u}_1, \mathbf{u}_2\}$ of Example 1, then we would have obtained

$$[T]_{B'} = \begin{bmatrix} 2 & 0 \\ 0 & 3 \end{bmatrix}, \quad \text{so} \quad \det(T) = \begin{vmatrix} 2 & 0 \\ 0 & 3 \end{vmatrix} = 6$$

which agrees with the preceding computation. ◆

EXAMPLE 3 Reflection About a Line

Let l be the line in the xy-plane that passes through the origin and makes an angle θ with the positive x-axis, where $0 \le \theta < \pi$. As illustrated in Figure 8.5.4, let $T: R^2 \to R^2$ be the linear operator that maps each vector into its reflection about the line l.

(a) Find the standard matrix for T.

(b) Find the reflection of the vector $\mathbf{x} = (1, 2)$ about the line l through the origin that makes an angle of $\theta = \pi/6$ with the positive x-axis.

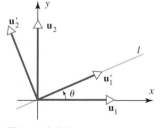

Figure 8.5.4

Solution (a)

We could proceed as in Example (6) of Section 4.3 and try to construct the standard matrix from the formula

$$[T]_B = [T] = [T(\mathbf{e}_1) \mid T(\mathbf{e}_2)]$$

where $B = \{\mathbf{e}_1, \mathbf{e}_2\}$ is the standard basis for R^2. However, it is easier to use a different strategy: Instead of finding $[T]_B$ directly, we shall first find the matrix $[T]_{B'}$, where

$$B' = \{\mathbf{u}_1', \mathbf{u}_2'\}$$

is the basis consisting of a unit vector $\mathbf{u}_1'$ along l and a unit vector $\mathbf{u}_2'$ perpendicular to l (Figure 8.5.5).

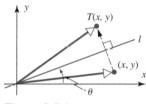

Figure 8.5.5

Once we have found $[T]_{B'}$, we shall perform a change of basis to find $[T]_B$. The computations are as follows:

$$T(\mathbf{u}_1') = \mathbf{u}_1' \quad \text{and} \quad T(\mathbf{u}_2') = -\mathbf{u}_2'$$

so

$$[T(\mathbf{u}_1')]_{B'} = \begin{bmatrix} 1 \\ 0 \end{bmatrix} \quad \text{and} \quad T[(\mathbf{u}_2')]_{B'} = \begin{bmatrix} 0 \\ -1 \end{bmatrix}$$

Thus

$$[T]_{B'} = \begin{bmatrix} 1 & 0 \\ 0 & -1 \end{bmatrix}$$

From the computations in Example 6 of Section 6.5, the transition matrix from B' to B is

$$P = \begin{bmatrix} [\mathbf{u}'_1]_B & | & [\mathbf{u}'_2]_B \end{bmatrix} = \begin{bmatrix} \cos\theta & -\sin\theta \\ \sin\theta & \cos\theta \end{bmatrix} \qquad (12)$$

It follows from Formula (10) that

$$[T]_B = P[T]_{B'} P^{-1}$$

Thus, from (12), the standard matrix for T is

$$[T] = P[T]_{B'} P^{-1} = \begin{bmatrix} \cos\theta & -\sin\theta \\ \sin\theta & \cos\theta \end{bmatrix} \begin{bmatrix} 1 & 0 \\ 0 & -1 \end{bmatrix} \begin{bmatrix} \cos\theta & \sin\theta \\ -\sin\theta & \cos\theta \end{bmatrix}$$

$$= \begin{bmatrix} \cos^2\theta - \sin^2\theta & 2\sin\theta\cos\theta \\ 2\sin\theta\cos\theta & \sin^2\theta - \cos^2\theta \end{bmatrix}$$

$$= \begin{bmatrix} \cos 2\theta & \sin 2\theta \\ \sin 2\theta & -\cos 2\theta \end{bmatrix}$$

Solution (b)

It follows from part (a) that the formula for T in matrix notation is

$$T\left(\begin{bmatrix} x \\ y \end{bmatrix}\right) = \begin{bmatrix} \cos 2\theta & \sin 2\theta \\ \sin 2\theta & -\cos 2\theta \end{bmatrix} \begin{bmatrix} x \\ y \end{bmatrix}$$

Substituting $\theta = \pi/6$ in this formula yields

$$T\left(\begin{bmatrix} x \\ y \end{bmatrix}\right) = \begin{bmatrix} \frac{1}{2} & \frac{\sqrt{3}}{2} \\ \frac{\sqrt{3}}{2} & -\frac{1}{2} \end{bmatrix} \begin{bmatrix} x \\ y \end{bmatrix}$$

so

$$T\left(\begin{bmatrix} 1 \\ 2 \end{bmatrix}\right) = \begin{bmatrix} \frac{1}{2} & \frac{\sqrt{3}}{2} \\ \frac{\sqrt{3}}{2} & -\frac{1}{2} \end{bmatrix} \begin{bmatrix} 1 \\ 2 \end{bmatrix} = \begin{bmatrix} \frac{1}{2} + \sqrt{3} \\ \frac{\sqrt{3}}{2} - 1 \end{bmatrix}$$

Thus $T(1, 2) = \left(\frac{1}{2} + \sqrt{3}, \frac{\sqrt{3}}{2} - 1\right)$. ◆

Eigenvalues of a Linear Operator

Eigenvectors and eigenvalues can be defined for linear operators as well as matrices. A scalar λ is called an *eigenvalue* of a linear operator $T: V \to V$ if there is a nonzero vector $\mathbf{x}$ in V such that $T\mathbf{x} = \lambda\mathbf{x}$. The vector $\mathbf{x}$ is called an *eigenvector* of T corresponding to λ. Equivalently, the eigenvectors of T corresponding to λ are the nonzero vectors in the kernel of $\lambda I - T$ (Exercise 15). This kernel is called the *eigenspace* of T corresponding to λ.

EXAMPLE 4   Eigenvalues of a Linear Operator

Let $V = F(-\infty, \infty)$ and consider the linear operator T on V that maps $f(x)$ to $f(x - 2\pi)$. If $f(x) = \sin(x)$, then $T(\mathbf{f}) = \sin(x - 2\pi) = \sin(x)$, so $\sin(x)$ is an eigenvector of T associated with the eigenvalue 1:

$$T(\sin(x)) = 1 \cdot \sin(x)$$

Other eigenvectors of T associated with the eigenvalue 1 include $\sin(2x)$, $\cos(5x)$, and the constant function 3. ◆

It can be shown that if V is a finite-dimensional vector space, and B is *any* basis for V, then

1. The eigenvalues of T are the same as the eigenvalues of $[T]_B$.
2. A vector $\mathbf{x}$ is an eigenvector of T corresponding to λ if and only if its coordinate matrix $[\mathbf{x}]_B$ is an eigenvector of $[T]_B$ corresponding to λ.

We omit the proofs.

EXAMPLE 5   Eigenvalues and Bases for Eigenspaces

Find the eigenvalues and bases for the eigenspaces of the linear operator $T: P_2 \to P_2$ defined by

$$T(a + bx + cx^2) = -2c + (a + 2b + c)x + (a + 3c)x^2$$

Solution

The matrix for T with respect to the standard basis $B = \{1, x, x^2\}$ is

$$[T]_B = \begin{bmatrix} 0 & 0 & -2 \\ 1 & 2 & 1 \\ 1 & 0 & 3 \end{bmatrix}$$

(verify). The eigenvalues of T are $\lambda = 1$ and $\lambda = 2$ (Example 5 of Section 7.1). Also from that example, the eigenspace of $[T]_B$ corresponding to $\lambda = 2$ has the basis $\{\mathbf{u}_1, \mathbf{u}_2\}$, where

$$\mathbf{u}_1 = \begin{bmatrix} -1 \\ 0 \\ 1 \end{bmatrix}, \qquad \mathbf{u}_2 = \begin{bmatrix} 0 \\ 1 \\ 0 \end{bmatrix}$$

and the eigenspace of $[T]_B$ corresponding to $\lambda = 1$ has the basis $\{\mathbf{u}_3\}$, where

$$\mathbf{u}_3 = \begin{bmatrix} -2 \\ 1 \\ 1 \end{bmatrix}$$

The matrices $\mathbf{u}_1$, $\mathbf{u}_2$, and $\mathbf{u}_3$ are the coordinate matrices relative to B of

$$\mathbf{p}_1 = -1 + x^2, \qquad \mathbf{p}_2 = x, \qquad \mathbf{p}_3 = -2 + x + x^2$$

Thus the eigenspace of T corresponding to $\lambda = 2$ has the basis

$$\{\mathbf{p}_1, \mathbf{p}_2\} = \{-1 + x^2, x\}$$

and that corresponding to $\lambda = 1$ has the basis

$$\{\mathbf{p}_3\} = \{-2 + x + x^2\}$$

As a check, the reader should use the given formula for T to verify that $T(\mathbf{p}_1) = 2\mathbf{p}_1$, $T(\mathbf{p}_2) = 2\mathbf{p}_2$, and $T(\mathbf{p}_3) = \mathbf{p}_3$. ◆

EXAMPLE 6   Diagonal Matrix for a Linear Operator

Let $T: R^3 \to R^3$ be the linear operator given by

$$T\left(\begin{bmatrix} x_1 \\ x_2 \\ x_3 \end{bmatrix}\right) = \begin{bmatrix} -2x_3 \\ x_1 + 2x_2 + x_3 \\ x_1 + 3x_3 \end{bmatrix}$$

Find a basis for R^3 relative to which the matrix for T is diagonal.

Solution

First we will find the standard matrix for T; then we will look for a change of basis that diagonalizes the standard matrix.

   If $B = \{\mathbf{e}_1, \mathbf{e}_2, \mathbf{e}_3\}$ denotes the standard basis for R^3, then

$$T(\mathbf{e}_1) = T\left(\begin{bmatrix} 1 \\ 0 \\ 0 \end{bmatrix}\right) = \begin{bmatrix} 0 \\ 1 \\ 1 \end{bmatrix}, \quad T(\mathbf{e}_2) = T\left(\begin{bmatrix} 0 \\ 1 \\ 0 \end{bmatrix}\right) = \begin{bmatrix} 0 \\ 2 \\ 0 \end{bmatrix}, \quad T(\mathbf{e}_3) = T\left(\begin{bmatrix} 0 \\ 0 \\ 1 \end{bmatrix}\right) = \begin{bmatrix} -2 \\ 1 \\ 3 \end{bmatrix}$$

so the standard matrix for T is

$$[T] = \begin{bmatrix} 0 & 0 & -2 \\ 1 & 2 & 1 \\ 1 & 0 & 3 \end{bmatrix} \tag{13}$$

We now want to change from the standard basis B to a new basis $B' = \{\mathbf{u}_1', \mathbf{u}_2', \mathbf{u}_3'\}$ in order to obtain a diagonal matrix for T. If we let P be the transition matrix from the unknown basis B' to the standard basis B, then by Theorem 8.5.2, the matrices $[T]$ and $[T]_{B'}$ will be related by

$$[T]_{B'} = P^{-1}[T]P \tag{14}$$

In Example 1 of Section 7.2, we found that the matrix in (13) is diagonalized by

$$P = \begin{bmatrix} -1 & 0 & -2 \\ 0 & 1 & 1 \\ 1 & 0 & 1 \end{bmatrix}$$

Since P represents the transition matrix from the basis $B' = \{\mathbf{u}_1', \mathbf{u}_2', \mathbf{u}_3'\}$ to the standard basis $B = \{\mathbf{e}_1, \mathbf{e}_2, \mathbf{e}_3\}$, the columns of P are $[\mathbf{u}_1']_B$, $[\mathbf{u}_2']_B$, and $[\mathbf{u}_3']_B$, so

$$[\mathbf{u}_1']_B = \begin{bmatrix} -1 \\ 0 \\ 1 \end{bmatrix}, \quad [\mathbf{u}_2']_B = \begin{bmatrix} 0 \\ 1 \\ 0 \end{bmatrix}, \quad [\mathbf{u}_3']_B = \begin{bmatrix} -2 \\ 1 \\ 1 \end{bmatrix}$$

Thus

$$\mathbf{u}_1' = (-1)\mathbf{e}_1 + (0)\mathbf{e}_2 + (1)\mathbf{e}_3 = \begin{bmatrix} -1 \\ 0 \\ 1 \end{bmatrix}$$

$$\mathbf{u}_2' = (0)\mathbf{e}_1 + (1)\mathbf{e}_2 + (0)\mathbf{e}_3 = \begin{bmatrix} 0 \\ 1 \\ 0 \end{bmatrix}$$

$$\mathbf{u}_3' = (-2)\mathbf{e}_1 + (1)\mathbf{e}_2 + (1)\mathbf{e}_3 = \begin{bmatrix} -2 \\ 1 \\ 1 \end{bmatrix}$$

are basis vectors that produce a diagonal matrix for $[T]_{B'}$.

As a check, let us compute $[T]_{B'}$ directly. From the given formula for T, we have

$$T(\mathbf{u}_1') = \begin{bmatrix} -2 \\ 0 \\ 2 \end{bmatrix} = 2\mathbf{u}_1', \qquad T(\mathbf{u}_2') = \begin{bmatrix} 0 \\ 2 \\ 0 \end{bmatrix} = 2\mathbf{u}_2', \qquad T(\mathbf{u}_3') = \begin{bmatrix} -2 \\ 1 \\ 1 \end{bmatrix} = \mathbf{u}_3'$$

so that

$$[T(\mathbf{u}_1')]_{B'} = \begin{bmatrix} 2 \\ 0 \\ 0 \end{bmatrix}, \qquad [T(\mathbf{u}_2')]_{B'} = \begin{bmatrix} 0 \\ 2 \\ 0 \end{bmatrix}, \qquad [T(\mathbf{u}_3')]_{B'} = \begin{bmatrix} 0 \\ 0 \\ 1 \end{bmatrix}$$

Thus

$$[T]_{B'} = \left[[T(\mathbf{u}_1')]_{B'} \mid [T(\mathbf{u}_2')]_{B'} \mid [T(\mathbf{u}_3')]_{B'} \right] = \begin{bmatrix} 2 & 0 & 0 \\ 0 & 2 & 0 \\ 0 & 0 & 1 \end{bmatrix}$$

This is consistent with (14) since

$$P^{-1}[T]P = \begin{bmatrix} 1 & 0 & 2 \\ 1 & 1 & 1 \\ -1 & 0 & -1 \end{bmatrix} \begin{bmatrix} 0 & 0 & -2 \\ 1 & 2 & 1 \\ 1 & 0 & 3 \end{bmatrix} \begin{bmatrix} -1 & 0 & -2 \\ 0 & 1 & 1 \\ 1 & 0 & 1 \end{bmatrix}$$

$$= \begin{bmatrix} 2 & 0 & 0 \\ 0 & 2 & 0 \\ 0 & 0 & 1 \end{bmatrix} \blacklozenge$$

We now see that the problem we studied in Section 7.2, that of diagonalizing a matrix A, may be viewed as the problem of finding a diagonal matrix D that is similar to A, or as the problem of finding a basis with respect to which the linear transformation defined by A is diagonal.

**EXERCISE SET
8.5**

In Exercises 1–7 find the matrix of T with respect to the basis B, and use Theorem 8.5.2 to compute the matrix of T with respect to the basis B'.

1. $T: R^2 \rightarrow R^2$ is defined by

$$T\left(\begin{bmatrix} x_1 \\ x_2 \end{bmatrix} \right) = \begin{bmatrix} x_1 - 2x_2 \\ -x_2 \end{bmatrix}$$

$B = \{\mathbf{u}_1, \mathbf{u}_2\}$ and $B' = \{\mathbf{v}_1, \mathbf{v}_2\}$, where

$$\mathbf{u}_1 = \begin{bmatrix} 1 \\ 0 \end{bmatrix}, \quad \mathbf{u}_2 = \begin{bmatrix} 0 \\ 1 \end{bmatrix}, \quad \mathbf{v}_1 = \begin{bmatrix} 2 \\ 1 \end{bmatrix}, \quad \mathbf{v}_2 = \begin{bmatrix} -3 \\ 4 \end{bmatrix}$$

2. $T: R^2 \to R^2$ is defined by

$$T\left(\begin{bmatrix} x_1 \\ x_2 \end{bmatrix} \right) = \begin{bmatrix} x_1 + 7x_2 \\ 3x_1 - 4x_2 \end{bmatrix}$$

$B = \{\mathbf{u}_1, \mathbf{u}_2\}$ and $B' = \{\mathbf{v}_1, \mathbf{v}_2\}$, where

$$\mathbf{u}_1 = \begin{bmatrix} 2 \\ 2 \end{bmatrix}, \quad \mathbf{u}_2 = \begin{bmatrix} 4 \\ -1 \end{bmatrix}, \quad \mathbf{v}_1 = \begin{bmatrix} 1 \\ 3 \end{bmatrix}, \quad \mathbf{v}_2 = \begin{bmatrix} -1 \\ -1 \end{bmatrix}$$

3. $T: R^2 \to R^2$ is the rotation about the origin through $45°$; B and B' are the bases in Exercise 1.

4. $T: R^3 \to R^3$ is defined by

$$T\left(\begin{bmatrix} x_1 \\ x_2 \\ x_3 \end{bmatrix} \right) = \begin{bmatrix} x_1 + 2x_2 - x_3 \\ -x_2 \\ x_1 + 7x_3 \end{bmatrix}$$

B is the standard basis for R^3 and $B' = \{\mathbf{v}_1, \mathbf{v}_2, \mathbf{v}_3\}$, where

$$\mathbf{v}_1 = \begin{bmatrix} 1 \\ 0 \\ 0 \end{bmatrix}, \quad \mathbf{v}_2 = \begin{bmatrix} 1 \\ 1 \\ 0 \end{bmatrix}, \quad \mathbf{v}_3 = \begin{bmatrix} 1 \\ 1 \\ 1 \end{bmatrix}$$

5. $T: R^3 \to R^3$ is the orthogonal projection on the xy-plane; B and B' are as in Exercise 4.

6. $T: R^2 \to R^2$ is defined by $T(\mathbf{x}) = 5\mathbf{x}$; B and B' are the bases in Exercise 2.

7. $T: P_1 \to P_1$ is defined by $T(a_0 + a_1 x) = a_0 + a_1(x + 1)$; $B = \{\mathbf{p}_1, \mathbf{p}_2\}$ and $B' = \{\mathbf{q}_1, \mathbf{q}_2\}$, where $\mathbf{p}_1 = 6 + 3x$, $\mathbf{p}_2 = 10 + 2x$, $\mathbf{q}_1 = 2$, $\mathbf{q}_2 = 3 + 2x$.

8. Find $\det(T)$.

   (a) $T: R^2 \to R^2$, where $T(x_1, x_2) = (3x_1 - 4x_2, -x_1 + 7x_2)$

   (b) $T: R^3 \to R^3$, where $T(x_1, x_2, x_3) = (x_1 - x_2, x_2 - x_3, x_3 - x_1)$

   (c) $T: P_2 \to P_2$, where $T(p(x)) = p(x - 1)$

9. Prove that the following are similarity invariants:

   (a) rank   (b) nullity   (c) invertibility

10. Let $T: P_4 \to P_4$ be the linear operator given by the formula $T(p(x)) = p(2x + 1)$.

    (a) Find a matrix for T with respect to some convenient basis; then use Theorem 8.2.2 to find the rank and nullity of T.

    (b) Use the result in part (a) to determine whether T is one-to-one.

11. In each part, find a basis for R^2 relative to which the matrix for T is diagonal.

    (a) $T\left(\begin{bmatrix} x_1 \\ x_2 \end{bmatrix} \right) = \begin{bmatrix} x_1 - x_2 \\ 2x_1 + 4x_2 \end{bmatrix}$ (b) $T\left(\begin{bmatrix} x_1 \\ x_2 \end{bmatrix} \right) = \begin{bmatrix} 4x_1 - x_2 \\ -3x_1 + x_2 \end{bmatrix}$

12. In each part, find a basis for R^3 relative to which the matrix for T is diagonal.

    (a) $T\left(\begin{bmatrix} x_1 \\ x_2 \\ x_3 \end{bmatrix} \right) = \begin{bmatrix} -2x_1 + x_2 - x_3 \\ x_1 - 2x_2 - x_3 \\ -x_1 - x_2 - 2x_3 \end{bmatrix}$

    (b) $T\left(\begin{bmatrix} x_1 \\ x_2 \\ x_3 \end{bmatrix} \right) = \begin{bmatrix} -x_2 + x_3 \\ -x_1 + x_3 \\ x_1 + x_2 \end{bmatrix}$

(c) $T\left(\begin{bmatrix} x_1 \\ x_2 \\ x_3 \end{bmatrix}\right) = \begin{bmatrix} 4x_1 + x_3 \\ 2x_1 + 3x_2 + 2x_3 \\ x_1 + 4x_3 \end{bmatrix}$

13. Let $T: P_2 \rightarrow P_2$ be defined by

$$T(a_0 + a_1x + a_2x^2) = (5a_0 + 6a_1 + 2a_2) - (a_1 + 8a_2)x + (a_0 - 2a_2)x^2$$

   (a) Find the eigenvalues of T.

   (b) Find bases for the eigenspaces of T.

14. Let $T: M_{22} \rightarrow M_{22}$ be defined by

$$T\left(\begin{bmatrix} a & b \\ c & d \end{bmatrix}\right) = \begin{bmatrix} 2c & a+c \\ b-2c & d \end{bmatrix}$$

   (a) Find the eigenvalues of T.

   (b) Find bases for the eigenspaces of T.

15. Let λ be an eigenvalue of a linear operator $T: V \rightarrow V$. Prove that the eigenvectors of T corresponding to λ are the nonzero vectors in the kernel of $\lambda I - T$.

16. (a) Prove that if A and B are similar matrices, then A^2 and B^2 are also similar. More generally, prove that A^k and B^k are similar, where k is any positive integer.

   (b) If A^2 and B^2 are similar, must A and B be similar?

17. Let C and D be $m \times n$ matrices, and let $B = \{\mathbf{v}_1, \mathbf{v}_2, \ldots, \mathbf{v}_n\}$ be a basis for a vector space V. Show that if $C[\mathbf{x}]_B = D[\mathbf{x}]_B$ for all $\mathbf{x}$ in V, then $C = D$.

18. Let l be a line in the xy-plane that passes through the origin and makes an angle θ with the positive x-axis. As illustrated in the accompanying figure, let $T: R^2 \rightarrow R^2$ be the orthogonal projection of R^2 onto l. Use the method of Example 3 to show that

$$T\left(\begin{bmatrix} x \\ y \end{bmatrix}\right) = \begin{bmatrix} \cos^2\theta & \sin\theta\cos\theta \\ \sin\theta\cos\theta & \sin^2\theta \end{bmatrix}\begin{bmatrix} x \\ y \end{bmatrix}$$

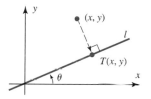

Figure Ex-18

Note   See Example 6 of Section 4.3.

Discussion & Discovery

19. Indicate whether each statement is always true or sometimes false. Justify your answer by giving a logical argument or a counterexample.

   (a) A matrix cannot be similar to itself.

   (b) If A is similar to B, and B is similar to C, then A is similar to C.

   (c) If A and B are similar and B is singular, then A is singular.

   (d) If A and B are invertible and similar, then A^{-1} and B^{-1} are similar.

20. Find two nonzero 2×2 matrices that are not similar, and explain why they are not.

21. Complete the proof by filling in the blanks with an appropriate justification.

   Hypothesis: A and B are similar matrices.

   Conclusion: A and B have the same characteristic polynomial (and hence the same eigenvalues).

   Proof: (1) $\det(\lambda I - B) = \det(\lambda I - P^{-1}AP)$ _____

   (2) $\qquad\qquad\quad = \det(\lambda P^{-1}P - P^{-1}AP)$ _____

   (3) $\qquad\qquad\quad = \det(P^{-1}(\lambda I - A)P)$ _____

   (4) $\qquad\qquad\quad = \det(P^{-1})\det(\lambda I - A)\det(P)$ _____

   (5) $\qquad\qquad\quad = \det(P^{-1})\det(P)\det(\lambda I - A)$ _____

   (6) $\qquad\qquad\quad = \det(\lambda I - A)$ _____

22. If A and B are similar matrices, say $B = P^{-1}AP$, then Exercise 21 shows that A and B have the same eigenvalues. Suppose that λ is one of the common eigenvalues and $\mathbf{x}$ is a corresponding eigenvector for A. See if you can find an eigenvector of B corresponding to λ, expressed in terms of λ, $\mathbf{x}$, and P.

23. Since the standard basis for R^n is so simple, why would one want to represent a linear operator on R^n in another basis?

24. Characterize the eigenspace of $\lambda = 1$ in Example 4.

25. Prove that the trace is a similarity invariant.

8.6
ISOMORPHISM

Our previous work shows that every real vector space of dimension n can be related to R^n through coordinate vectors and that every linear transformation from a real vector space of dimension n to one of dimension m can be related to R^n and R^m through transition matrices. In this section we shall further strengthen the connection between a real vector space of dimension n and R^n.

Onto Transformations

Let V and W be real vector spaces. We say that the linear transformation $T: V \to W$ is *onto* if the range of T is W—that is, if for every $\mathbf{w}$ in W, there is a $\mathbf{v}$ in V such that

$$T(\mathbf{v}) = \mathbf{w}$$

An onto transformation is also said to be *surjective* or to be a *surjection*. For a surjective mapping, then, the range and the codomain coincide.

EXAMPLE 1   Onto Transformations

Consider the projection $P: R^3 \to R^2$ defined by $P(x, y, z) = (x, y)$. This is an onto mapping, because if $\mathbf{w} = (x, y)$ is a point in R^2, then $\mathbf{v} = (x, y, 0)$ is mapped to it. (Of course, so are infinitely many other points in R^3.)

Consider the transformation $Q: R^3 \to R^3$ defined by $P(x, y, z) = (x, y, 0)$. This is essentially the same as P except that we consider the result to be a vector in R^3 rather than a vector in R^2. This mapping is not onto, because, for example, the point $(1, 1, 1)$ in the codomain is not the image of any $\mathbf{v}$ in the domain. ◆

If a transformation $T: V \to W$ is both one-to-one (also called *injective* or an *injection*) and onto, then it is a one-to-one mapping to its range W and so has an inverse $T^{-1}: W \to V$. A transformation that is one-to-one and onto is also said to be *bijective* or to be a *bijection* between V and W. In the exercises, you'll be asked to show that the inverse of a bijection is also a bijection.

In Section 8.3 it was stated that if V and W are finite-dimensional vector spaces, then the dimension of the codomain W must be at least as large as the dimension of the domain V for there to exist a one-to-one linear transformation from V to W. That is, there can be an injective linear transformation from V to W only if $\dim(V) \leq \dim(W)$. Similarly, there can be a surjective linear transformation from V to W only if $\dim(V) \geq \dim(W)$. Theorem 8.6.1 follows immediately.

THEOREM 8.6.1

Bijective Linear Transformations

Let V and W be finite-dimensional vector spaces. If $\dim(V) \neq \dim(W)$, *then there can be no bijective linear transformation from V to W.*

Isomorphisms

Bijective linear transformations between vector spaces are sufficiently important that they have their own name.

DEFINITION

An *isomorphism* between V and W is a bijective linear transformation from V to W.

Note that if T is an isomorphism between V and W, then T^{-1} exists and is an isomorphism between W and V. For this reason, we say that V and W are *isomorphic* if there is an isomorphism from V to W. The term *isomorphic* means "same shape," so isomorphic vector spaces have the same form or structure.

Theorem 8.6.1 does not guarantee that if $\dim(V) = \dim(W)$, then there is an isomorphism from V to W. However, every real vector space V of dimension n admits at least one bijective linear transformation to R^n: the transformation $T(\mathbf{v}) = (\mathbf{v})_S$ that takes a vector in V to its coordinate vector in R^n with respect to the standard basis for R^n.

THEOREM 8.6.2

Isomorphism Theorem

Let V be a finite-dimensional real vector space. If $\dim(V) = n$, *then there is an isomorphism from V to* R^n.

We leave the proof of Theorem 8.6.2 as an exercise.

EXAMPLE 2   An Isomorphism between P_3 and R^4

The vector space P_3 is isomorphic to R^4, because the transformation

$$T(a + xb + cx^2 + dx^3) = (a, b, c, d)$$

is one-to-one, onto, and linear (verify). ◆

EXAMPLE 3   An Isomorphism between M_{22} and R^4

The vector space M_{22} is isomorphic to R^4, because the transformation

$$T\left(\begin{bmatrix} a & b \\ c & d \end{bmatrix}\right) = (a, b, c, d)$$

is one-to-one, onto, and linear (verify). ◆

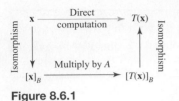

Figure 8.6.1

The significance of the Isomorphism Theorem is this: It is a formal statement of the fact, represented in Figure 8.4.5 and repeated here as Figure 8.6.1 for the case $V = W$, that any computation involving a linear operator T on V is equivalent to a computation involving a linear operator on R^n; that is, any computation involving a linear operator on V is equivalent to matrix multiplication. Operations on V are effectively the same as those on R^n.

If $\dim(V) = n$, then we say that V and R^n have the same *algebraic structure*. This means that although the names conventionally given to the vectors and corresponding operations in V may differ from the corresponding traditional names in R^n, as vector spaces they really are the same.

Isomorphisms between Vector Spaces

It is easy to show that compositions of bijective linear transformations are themselves bijective linear transformations. (See the exercises.) This leads to the following theorem.

THEOREM 8.6.3

Isomorphism of Finite-Dimensional Vector Spaces

Let V and W be finite-dimensional vector spaces. If $\dim(V) = \dim(W)$, *then V and W are isomorphic.*

Proof We must show that there is an isomorphism from V to W. Let n be the common dimension of V and W. Then there is an isomorphism $T: V \to R^n$ by Theorem 8.6.2. Similarly, there is an isomorphism $S: W \to R^n$. Let $R = S^{-1}$. Then $R \circ T$ is an isomorphism from V to W, so V and W are isomorphic. ∎

EXAMPLE 4 An Isomorphism between P_3 and M_{22}

Because $\dim(P_3) = 4$ and $\dim(M_{22}) = 4$, these spaces are isomorphic. We can find an isomorphism T between them by identifying the natural bases for these spaces under $T: P_3 \to M_{22}$:

$$T(1) = \begin{bmatrix} 1 & 0 \\ 0 & 0 \end{bmatrix}$$

$$T(x) = \begin{bmatrix} 0 & 1 \\ 0 & 0 \end{bmatrix}$$

$$T(x^2) = \begin{bmatrix} 0 & 0 \\ 1 & 0 \end{bmatrix}$$

$$T(x^3) = \begin{bmatrix} 0 & 0 \\ 0 & 1 \end{bmatrix}$$

If $p(x) = a + xb + cx^2 + dx^3$ is in P_3, then by linearity,

$$T(p(x)) = a \begin{bmatrix} 1 & 0 \\ 0 & 0 \end{bmatrix} + b \begin{bmatrix} 0 & 1 \\ 0 & 0 \end{bmatrix} + c \begin{bmatrix} 0 & 0 \\ 1 & 0 \end{bmatrix} + d \begin{bmatrix} 0 & 0 \\ 0 & 1 \end{bmatrix}$$

$$= \begin{bmatrix} a & b \\ c & d \end{bmatrix}$$

This is one-to-one and onto linear transformation (verify), so it is an isomorphism between P_3 and M_{22}. ◆

In the sense of isomorphism, then, there is only one real vector space of dimension n, with many different names. We take R^n as the canonical example of a real vector space of dimension n because of the importance of coordinate vectors. Coordinate vectors are vectors in R^n because they are the vectors of the coefficients in linear combinations

$$a_1\mathbf{v}_1 + \cdots + a_n\mathbf{v}_n$$

and since our scalars a_i are real, the coefficients $(a_1, \ldots, a_n)$ are real n-tuples.

Think for a moment about the practical import of this result. If you want to program a computer to perform linear operations, such as the basic operations of the calculus on polynomials, you can do it using matrix multiplication. If you want to do video game graphics requiring rotations and reflections, you can do it using matrix multiplication. (Indeed, the special architectures of high-end video game consoles are designed to optimize the speed of matrix–matrix and matrix–vector calculations for computing new positions of objects and for lighting and rendering them. Supercomputer clusters have been created from these devices!) This is why every high-level computer programming language has facilities for arrays (vectors and matrices). Isomorphism ensures that any linear operation on vector spaces can be done using just those capabilities, and most operations of interest either will be linear or may be approximated by a linear operator.

EXERCISE SET 8.6

1. Which of the transformations in Exercise 1 of Section 8.3 are onto?

2. Let A be an $n \times n$ matrix. When is $T_A : R^n \to R^n$ not onto?

3. Which of the transformations in Exercise 3 of Section 8.3 are onto?

4. Which of the transformations in Exercise 4 of Section 8.3 are onto?

5. Which of the following transformations are bijections?
   (a) $T: P_2(x) \to P_3(x),\ T(p(x)) = xp(x)$
   (b) $T: M_{22} \to M_{22},\ T(A) = A^T$
   (c) $T: R^4 \to R^3,\ T(x, y, z, w) = (x, y, 0)$
   (d) $T: P_3 \to R^3,\ T(a + bx + cx^2 + dx^3) = (b, c, d)$

6. Show that the inverse of a bijective transformation from V to W is a bijective transformation from W to V. Also, show that the inverse of a bijective *linear* transformation is a bijective *linear* transformation.

7. Prove: There can be a surjective linear transformation from V to W only if $\dim(V) \geq \dim(W)$.

8. (a) Find an isomorphism between the vector space of all 3×3 symmetric matrices and R^6.
   (b) Find two different isomorphisms between the vector space of all 2×2 matrices and R^4.
   (c) Find an isomorphism between the vector space of all polynomials of degree at most 3 such that $p(0) = 0$ and R^3.
   (d) Find an isomorphism between the vector space span$\{1, \sin(x), \cos(x)\}$ and R^3.

9. Let S be the standard basis for R^n. Prove Theorem 8.6.2 by showing that the linear transformation $T: V \to R^n$ that maps $\mathbf{v} \in V$ to its coordinate vector $(\mathbf{v})_S$ in R^n is an isomorphism.

10. Show that if T_1, T_2 are bijective linear transformations, then the composition $T_2 \circ T_1$ is a bijective linear transformation.

11. **(For Readers Who Have Studied Calculus)** How could differentiation of functions in the vector space span$\{1, \sin(x), \cos(x), \sin(2x), \cos(2x)\}$ be computed by matrix multiplication in R^5? Use your method to find the derivative of $3 - 4\sin(x) + \sin(2x) + 5\cos(2x)$.

Discussion & Discovery

12. Isomorphisms preserve the algebraic structure of vector spaces. The geometric structure depends on notions of angle and distance and so, ultimately, on the inner product. If V and W are finite-dimensional inner product spaces, then we say that $T: V \to W$ is an *inner product space isomorphism* if it is an isomorphism between V and W, and furthermore,

$$\langle \mathbf{u}, \mathbf{v} \rangle_V = \langle T(\mathbf{u}), T(\mathbf{v}) \rangle_W$$

That is, the inner product of $\mathbf{u}$ and $\mathbf{v}$ in V is equal to the inner product of their images in W.

(a) Prove that an inner product space isomorphism preserves angles and distances—that is, the angle between $\mathbf{u}$ and $\mathbf{v}$ in V is equal to the angle between $T(\mathbf{u})$ and $T(\mathbf{v})$ in W, and $\|\mathbf{u} - \mathbf{v}\|_V = \|T(\mathbf{u}) - T(\mathbf{v})\|_W$.

(b) Prove that such a T maps orthonormal sets in V to orthonormal sets in W. Is this true for an isomorphism in general?

(c) Prove that if W is Euclidean n-space and if $\dim(V) = n$, then there is an inner product space isomorphism between V and W.

(d) Use the result of part (c) to prove that if $\dim(V) = \dim(W)$, then there is an inner product space isomorphism between V and W.

(e) Find an inner product space isomorphism between P_5 and M_{23}.

CHAPTER 8
Supplementary Exercises

1. Let A be an $n \times n$ matrix, B a nonzero $n \times 1$ matrix, and $\mathbf{x}$ a vector in R^n expressed in matrix notation. Is $T(\mathbf{x}) = A\mathbf{x} + B$ a linear operator on R^n? Justify your answer.

2. Let
$$A = \begin{bmatrix} \cos\theta & -\sin\theta \\ \sin\theta & \cos\theta \end{bmatrix}$$

(a) Show that
$$A^2 = \begin{bmatrix} \cos 2\theta & -\sin 2\theta \\ \sin 2\theta & \cos 2\theta \end{bmatrix} \quad \text{and} \quad A^3 = \begin{bmatrix} \cos 3\theta & -\sin 3\theta \\ \sin 3\theta & \cos 3\theta \end{bmatrix}$$

(b) Guess the form of the matrix A^n for any positive integer n.

(c) By considering the geometric effect of $T: R^2 \to R^2$, where T is multiplication by A, obtain the result in (b) geometrically.

3. Let $\mathbf{v}_0$ be a fixed vector in an inner product space V, and let $T: V \to V$ be defined by $T(\mathbf{v}) = \langle \mathbf{v}, \mathbf{v}_0 \rangle \mathbf{v}_0$. Show that T is a linear operator on V.

4. Let $\mathbf{v}_1, \mathbf{v}_2, \ldots, \mathbf{v}_m$ be fixed vectors in R^n, and let $T: R^n \to R^m$ be the function defined by $T(\mathbf{x}) = (\mathbf{x} \cdot \mathbf{v}_1, \mathbf{x} \cdot \mathbf{v}_2, \ldots, \mathbf{x} \cdot \mathbf{v}_m)$, where $\mathbf{x} \cdot \mathbf{v}_i$ is the Euclidean inner product on R^n.

(a) Show that T is a linear transformation.

(b) Show that the matrix with row vectors $\mathbf{v}_1, \mathbf{v}_2, \ldots, \mathbf{v}_m$ is the standard matrix for T.

5. Let $\{\mathbf{e}_1, \mathbf{e}_2, \mathbf{e}_3, \mathbf{e}_4\}$ be the standard basis for R^4, and let $T: R^4 \to R^3$ be the linear transformation for which
$$T(\mathbf{e}_1) = (1, 2, 1), \qquad T(\mathbf{e}_2) = (0, 1, 0),$$
$$T(\mathbf{e}_3) = (1, 3, 0), \qquad T(\mathbf{e}_4) = (1, 1, 1)$$

(a) Find bases for the range and kernel of T.

(b) Find the rank and nullity of T.

6. Suppose that vectors in R^3 are denoted by 1×3 matrices, and define $T: R^3 \to R^3$ by

$$T([x_1 \quad x_2 \quad x_3]) = [x_1 \quad x_2 \quad x_3] \begin{bmatrix} -1 & 2 & 4 \\ 3 & 0 & 1 \\ 2 & 2 & 5 \end{bmatrix}$$

(a) Find a basis for the kernel of T.

(b) Find a basis for the range of T.

7. Let $B = \{v_1, v_2, v_3, v_4\}$ be a basis for a vector space V, and let $T: V \to V$ be the linear operator for which

$$T(v_1) = v_1 + v_2 + v_3 + 3v_4$$
$$T(v_2) = v_1 - v_2 + 2v_3 + 2v_4$$
$$T(v_3) = 2v_1 - 4v_2 + 5v_3 + 3v_4$$
$$T(v_4) = -2v_1 + 6v_2 - 6v_3 - 2v_4$$

(a) Find the rank and nullity of T.

(b) Determine whether T is one-to-one.

8. Let V and W be vector spaces, let T, T_1, and T_2 be linear transformations from V to W, and let k be a scalar. Define new transformations, $T_1 + T_2$ and kT, by the formulas

$$(T_1 + T_2)(x) = T_1(x) + T_2(x)$$
$$(kT)(x) = k(T(x))$$

(a) Show that $(T_1 + T_2): V \to W$ and $kT: V \to W$ are linear transformations.

(b) Show that the set of all linear transformations from V to W with the operations in part (a) forms a vector space.

9. Let A and B be similar matrices. Prove:

(a) A^T and B^T are similar.

(b) If A and B are invertible, then A^{-1} and B^{-1} are similar.

10. **(Fredholm Alternative Theorem)** Let $T: V \to V$ be a linear operator on an n-dimensional vector space. Prove that exactly one of the following statements holds:

(i) The equation $T(x) = b$ has a solution for all vectors b in V.

(ii) Nullity of $T > 0$.

11. Let $T: M_{22} \to M_{22}$ be the linear operator defined by

$$T(X) = \begin{bmatrix} 1 & 1 \\ 0 & 0 \end{bmatrix} X + X \begin{bmatrix} 0 & 0 \\ 1 & 1 \end{bmatrix}$$

Find the rank and nullity of T.

12. Prove: If A and B are similar matrices, and if B and C are similar matrices, then A and C are similar matrices.

13. Let $L: M_{22} \to M_{22}$ be the linear operator defined by $L(M) = M^T$. Find the matrix for L with respect to the standard basis for M_{22}.

14. Let $B = \{u_1, u_2, u_3\}$ and $B' = \{v_1, v_2, v_3\}$ be bases for a vector space V, and let

$$P = \begin{bmatrix} 2 & -1 & 3 \\ 1 & 1 & 4 \\ 0 & 1 & 2 \end{bmatrix}$$

be the transition matrix from B' to B.

(a) Express v_1, v_2, v_3 as linear combinations of u_1, u_2, u_3.

(b) Express u_1, u_2, u_3 as linear combinations of v_1, v_2, v_3.

15. Let $B = \{\mathbf{u}_1, \mathbf{u}_2, \mathbf{u}_3\}$ be a basis for a vector space V, and let $T: V \to V$ be a linear operator such that

$$[T]_B = \begin{bmatrix} -3 & 4 & 7 \\ 1 & 0 & -2 \\ 0 & 1 & 0 \end{bmatrix}$$

Find $[T]_{B'}$, where $B' = \{\mathbf{v}_1, \mathbf{v}_2, \mathbf{v}_3\}$ is the basis for V defined by

$$\mathbf{v}_1 = \mathbf{u}_1, \qquad \mathbf{v}_2 = \mathbf{u}_1 + \mathbf{u}_2, \qquad \mathbf{v}_3 = \mathbf{u}_1 + \mathbf{u}_2 + \mathbf{u}_3$$

16. Show that the matrices

$$\begin{bmatrix} 1 & 1 \\ -1 & 4 \end{bmatrix} \quad \text{and} \quad \begin{bmatrix} 2 & 1 \\ 1 & 3 \end{bmatrix}$$

are similar but that

$$\begin{bmatrix} 3 & 1 \\ -6 & -2 \end{bmatrix} \quad \text{and} \quad \begin{bmatrix} -1 & 2 \\ 1 & 0 \end{bmatrix}$$

are not.

17. Suppose that $T: V \to V$ is a linear operator and B is a basis for V such that for any vector $\mathbf{x}$ in V,

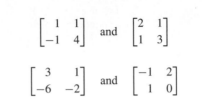

$$[T(\mathbf{x})]_B = \begin{bmatrix} x_1 - x_2 + x_3 \\ x_2 \\ x_1 - x_3 \end{bmatrix} \quad \text{if} \quad [\mathbf{x}]_B = \begin{bmatrix} x_1 \\ x_2 \\ x_3 \end{bmatrix}$$

Find $[T]_B$.

18. Let $T: V \to V$ be a linear operator. Prove that T is one-to-one if and only if $\det(T) \neq 0$.

19. **(For Readers Who Have Studied Calculus)**

(a) Show that if $\mathbf{f} = f(x)$, then the function $D: C^2(-\infty, \infty) \to F(-\infty, \infty)$ defined by $D(\mathbf{f}) = f''(x)$ is a linear transformation.

(b) Find a basis for the kernel of D.

(c) Show that the functions satisfying the equation $D(\mathbf{f}) = f(x)$ form a two-dimensional subspace of $C^2(-\infty, \infty)$, and find a basis for this subspace.

20. Let $T: P_2 \to R^3$ be the function defined by the formula

$$T(p(x)) = \begin{bmatrix} p(-1) \\ p(0) \\ p(1) \end{bmatrix}$$

(a) Find $T(x^2 + 5x + 6)$.

(b) Show that T is a linear transformation.

(c) Show that T is one-to-one.

(d) Find

$$T^{-1}\left(\begin{bmatrix} 0 \\ 3 \\ 0 \end{bmatrix} \right)$$

(e) Sketch the graph of the polynomial in part (d).

21. Let x_1, x_2, and x_3 be distinct real numbers such that $x_1 < x_2 < x_3$, and let $T: P_2 \to R^3$ be the function defined by the formula

$$T(p(x)) = \begin{bmatrix} p(x_1) \\ p(x_2) \\ p(x_3) \end{bmatrix}$$

(a) Show that T is a linear transformation.

(b)   Show that T is one-to-one.

(c)   Verify that if a_1, a_2, and a_3 are any real numbers, then

$$T^{-1}\left(\begin{bmatrix} a_1 \\ a_2 \\ a_3 \end{bmatrix}\right) = a_1 P_1(x) + a_2 P_2(x) + a_3 P_3(x)$$

where

$$P_1(x) = \frac{(x - x_2)(x - x_3)}{(x_1 - x_2)(x_1 - x_3)}, \quad P_2(x) = \frac{(x - x_1)(x - x_3)}{(x_2 - x_1)(x_2 - x_3)}, \quad P_3(x) = \frac{(x - x_1)(x - x_2)}{(x_3 - x_1)(x_3 - x_2)}$$

(d)   What relationship exists between the graph of the function

$$a_1 P_1(x) + a_2 P_2(x) + a_3 P_3(x)$$

and the points (x_1, a_1), (x_2, a_2), and (x_3, a_3)?

22.   **(For Readers Who Have Studied Calculus)** Let $p(x)$ and $q(x)$ be continuous functions, and let V be the subspace of $C(-\infty, +\infty)$ consisting of all twice differentiable functions. Define $L: V \to V$ by

$$L(y(x)) = y''(x) + p(x)y'(x) + q(x)y(x)$$

(a)   Show that L is a linear transformation.

(b)   Consider the special case where $p(x) = 0$ and $q(x) = 1$. Show that the function $\phi(x) = c_1 \sin x + c_2 \cos x$ is in the nullspace of L for all real values of c_1 and c_2.

23.   **(For Readers Who Have Studied Calculus)** Let $D: P_n \to P_n$ be the differentiation operator $D(\mathbf{p}) = \mathbf{p}'$. Show that the matrix for D with respect to the basis $B = \{1, x, x^2, \ldots, x^n\}$ is

$$\begin{bmatrix} 0 & 1 & 0 & 0 & \cdots & 0 \\ 0 & 0 & 2 & 0 & \cdots & 0 \\ 0 & 0 & 0 & 3 & \cdots & 0 \\ \vdots & \vdots & \vdots & \vdots & & \vdots \\ 0 & 0 & 0 & 0 & \cdots & n \\ 0 & 0 & 0 & 0 & \cdots & 0 \end{bmatrix}$$

24.   **(For Readers Who Have Studied Calculus)** It can be shown that for any real number c, the vectors

$$1, x - c, \frac{(x - c)^2}{2!}, \ldots, \frac{(x - c)^n}{n!}$$

form a basis for P_n. Find the matrix for the differentiation operator of Exercise 23 with respect to this basis.

25.   **(For Readers Who Have Studied Calculus)** Let $J: P_n \to P_{n+1}$ be the integration transformation defined by

$$J(\mathbf{p}) = \int_0^x (a_0 + a_1 x + \cdots + a_n x^n)\, dx = a_0 x + \frac{a_1}{2} x^2 + \cdots + \frac{a_n}{n+1} x^{n+1}$$

where $\mathbf{p} = a_0 + a_1 x + \cdots + a_n x^n$. Find the matrix for T with respect to the standard bases for P_n and P_{n+1}.

CHAPTER 8

**Technology
Exercises**

The following exercise is designed to be solved using a technology utility. Typically, this will be MATLAB, *Mathematica*, Maple, Derive, or Mathcad, but it may also be some other type of linear algebra software or a scientific calculator with some linear algebra capabilities. For this exercise you will need to read the relevant documentation for the particular utility you are using. The goal of this exercise is to provide you with a basic proficiency with your technology utility. Once you have mastered the techniques in this exercise, you will be able to use your technology utility to solve many of the problems in the regular exercise sets.

Section 8.3   **T1.**   **(Transition Matrices)** Use your technology utility to verify Formula (5).

Section 8.5   **T1.**   **(Similarity Invariants)** Choose a nonzero 3×3 matrix A and an invertible 3×3 matrix P. Compute $P^{-1}AP$ and confirm the statements in Table 1.

Additional Topics

CHAPTER CONTENTS

INTRODUCTION: In this chapter we shall see how some of the topics that we have studied in earlier chapters can be applied to other areas of mathematics, such as differential equations, analytic geometry, curve fitting, and Fourier series. The chapter concludes by returning once again to the fundamental problem of solving systems of linear equations $A\mathbf{x} = \mathbf{b}$. This time we solve a system not by another elimination procedure but by factoring the coefficient matrix into two different triangular matrices. This is the method that is generally used in computer programs for solving linear systems in real-world applications.

9.1

APPLICATION TO DIFFERENTIAL EQUATIONS

Many laws of physics, chemistry, biology, engineering, and economics are described in terms of differential equations—that is, equations involving functions and their derivatives. The purpose of this section is to illustrate one way in which linear algebra can be applied to certain systems of differential equations. The scope of this section is narrow, but it illustrates an important area of application of linear algebra.

Terminology

One of the simplest differential equations is

$$y' = ay \tag{1}$$

where $y = f(x)$ is an unknown function to be determined, $y' = dy/dx$ is its derivative, and a is a constant. Like most differential equations, (1) has infinitely many solutions; they are the functions of the form

$$y = ce^{ax} \tag{2}$$

where c is an arbitrary constant. Each function of this form is a solution of $y' = ay$ since

$$y' = cae^{ax} = ay$$

Conversely, every solution of $y' = ay$ must be a function of the form ce^{ax} (Exercise 5), so (2) describes all solutions of $y' = ay$. We call (2) the ***general solution*** of $y' = ay$.

Sometimes the physical problem that generates a differential equation imposes some added conditions that enable us to isolate one ***particular solution*** from the general solution. For example, if we require that the solution of $y' = ay$ satisfy the added condition

$$y(0) = 3 \tag{3}$$

that is, $y = 3$ when $x = 0$, then on substituting these values in the general solution $y = ce^{ax}$ we obtain a value for c—namely, $3 = ce^0 = c$. Thus

$$y = 3e^{ax}$$

is the only solution of $y' = ay$ that satisfies the added condition. A condition such as (3), which specifies the value of the solution at a point, is called an ***initial condition***, and the problem of solving a differential equation subject to an initial condition is called an ***initial-value problem***.

Linear Systems of First-Order Equations

In this section we will be concerned with solving systems of differential equations having the form

$$
\begin{aligned}
y_1' &= a_{11}y_1 + a_{12}y_2 + \cdots + a_{1n}y_n \\
y_2' &= a_{21}y_1 + a_{22}y_2 + \cdots + a_{2n}y_n \\
&\vdots \qquad \vdots \qquad \vdots \qquad\qquad \vdots \\
y_n' &= a_{n1}y_1 + a_{n2}y_2 + \cdots + a_{nn}y_n
\end{aligned} \tag{4}
$$

where $y_1 = f_1(x)$, $y_2 = f_2(x), \ldots, y_n = f_n(x)$ are functions to be determined, and the a_{ij}'s are constants. In matrix notation, (4) can be written as

$$
\begin{bmatrix} y_1' \\ y_2' \\ \vdots \\ y_n' \end{bmatrix} = \begin{bmatrix} a_{11} & a_{12} & \cdots & a_{1n} \\ a_{21} & a_{22} & \cdots & a_{2n} \\ \vdots & \vdots & & \vdots \\ a_{n1} & a_{n2} & \cdots & a_{nn} \end{bmatrix} \begin{bmatrix} y_1 \\ y_2 \\ \vdots \\ y_n \end{bmatrix}
$$

or, more briefly,

$$\mathbf{y}' = A\mathbf{y}$$

EXAMPLE 1 Solution of a System with Initial Conditions

(a) Write the following system in matrix form:

$$
\begin{aligned}
y_1' &= 3y_1 \\
y_2' &= -2y_2 \\
y_3' &= 5y_3
\end{aligned}
$$

(b) Solve the system.

(c) Find a solution of the system that satisfies the initial conditions $y_1(0) = 1$, $y_2(0) = 4$, and $y_3(0) = -2$.

Solution (a)

$$
\begin{bmatrix} y_1' \\ y_2' \\ y_3' \end{bmatrix} = \begin{bmatrix} 3 & 0 & 0 \\ 0 & -2 & 0 \\ 0 & 0 & 5 \end{bmatrix} \begin{bmatrix} y_1 \\ y_2 \\ y_3 \end{bmatrix} \tag{5}
$$

or

$$
\mathbf{y}' = \begin{bmatrix} 3 & 0 & 0 \\ 0 & -2 & 0 \\ 0 & 0 & 5 \end{bmatrix} \mathbf{y}
$$

Solution (b)

Because each equation involves only one unknown function, we can solve the equations individually. From (2), we obtain

$$
\begin{aligned}
y_1 &= c_1 e^{3x} \\
y_2 &= c_2 e^{-2x} \\
y_3 &= c_3 e^{5x}
\end{aligned}
$$

or, in matrix notation,

$$
\mathbf{y} = \begin{bmatrix} y_1 \\ y_2 \\ y_3 \end{bmatrix} = \begin{bmatrix} c_1 e^{3x} \\ c_2 e^{-2x} \\ c_3 e^{5x} \end{bmatrix} \tag{6}
$$

Solution (c)

From the given initial conditions, we obtain

$$
\begin{aligned}
1 &= y_1(0) = c_1 e^0 = c_1 \\
4 &= y_2(0) = c_2 e^0 = c_2 \\
-2 &= y_3(0) = c_3 e^0 = c_3
\end{aligned}
$$

so the solution satisfying the initial conditions is

$$
y_1 = e^{3x}, \qquad y_2 = 4e^{-2x}, \qquad y_3 = -2e^{5x}
$$

or, in matrix notation,

$$
\mathbf{y} = \begin{bmatrix} y_1 \\ y_2 \\ y_3 \end{bmatrix} = \begin{bmatrix} e^{3x} \\ 4e^{-2x} \\ -2e^{5x} \end{bmatrix} \quad \blacklozenge
$$

The system in the preceding example is easy to solve because each equation involves only one unknown function, and this is the case because the matrix of coefficients for the system in (5) is diagonal. But how do we handle a system $\mathbf{y}' = A\mathbf{y}$ in which the matrix A is not diagonal? The idea is simple: Try to make a substitution for $\mathbf{y}$ that will yield a new system with a diagonal coefficient matrix; solve this new simpler system, and then use this solution to determine the solution of the original system.

The kind of substitution we have in mind is

$$
\begin{aligned}
y_1 &= p_{11}u_1 + p_{12}u_2 + \cdots + p_{1n}u_n \\
y_2 &= p_{21}u_1 + p_{22}u_2 + \cdots + p_{2n}u_n \\
&\vdots \qquad \vdots \qquad \vdots \qquad\qquad \vdots \\
y_n &= p_{n1}u_1 + p_{n2}u_2 + \cdots + p_{nn}u_n
\end{aligned}
\tag{7}
$$

or, in matrix notation,

$$
\begin{bmatrix} y_1 \\ y_2 \\ \vdots \\ y_n \end{bmatrix} = \begin{bmatrix} p_{11} & p_{12} & \cdots & p_{1n} \\ p_{21} & p_{22} & \cdots & p_{2n} \\ \vdots & \vdots & & \vdots \\ p_{n1} & p_{n2} & \cdots & p_{nn} \end{bmatrix} \begin{bmatrix} u_1 \\ u_2 \\ \vdots \\ u_n \end{bmatrix} \quad \text{or, more briefly,} \quad \mathbf{y} = P\mathbf{u}
$$

In this substitution, the p_{ij}'s are constants to be determined in such a way that the new system involving the unknown functions $u_1, u_2, \ldots, u_n$ has a diagonal coefficient matrix. We leave it for the reader to differentiate each equation in (7) and deduce

$$\mathbf{y}' = P\mathbf{u}'$$

If we make the substitutions $\mathbf{y} = P\mathbf{u}$ and $\mathbf{y}' = P\mathbf{u}'$ in the original system

$$\mathbf{y}' = A\mathbf{y}$$

and if we assume P to be invertible, then we obtain

$$P\mathbf{u}' = A(P\mathbf{u})$$

or

$$\mathbf{u}' = (P^{-1}AP)\mathbf{u} \quad \text{or} \quad \mathbf{u}' = D\mathbf{u}$$

where $D = P^{-1}AP$. The choice for P is now clear; if we want the new coefficient matrix D to be diagonal, we must choose P to be a matrix that diagonalizes A.

Solution by Diagonalization

The preceding discussion suggests the following procedure for solving a system $\mathbf{y}' = A\mathbf{y}$ with a diagonalizable coefficient matrix A.

Step 1. Find a matrix P that diagonalizes A.

Step 2. Make the substitutions $\mathbf{y} = P\mathbf{u}$ and $\mathbf{y}' = P\mathbf{u}'$ to obtain a new "diagonal system" $\mathbf{u}' = D\mathbf{u}$, where $D = P^{-1}AP$.

Step 3. Solve $\mathbf{u}' = D\mathbf{u}$.

Step 4. Determine $\mathbf{y}$ from the equation $\mathbf{y} = P\mathbf{u}$.

EXAMPLE 2 Solution Using Diagonalization

(a) Solve the system

$$
\begin{aligned}
y_1' &= y_1 + y_2 \\
y_2' &= 4y_1 - 2y_2
\end{aligned}
$$

(b) Find the solution that satisfies the initial conditions $y_1(0) = 1$, $y_2(0) = 6$.

Solution (a)

The coefficient matrix for the system is

$$A = \begin{bmatrix} 1 & 1 \\ 4 & -2 \end{bmatrix}$$

As discussed in Section 7.2, A will be diagonalized by any matrix P whose columns are linearly independent eigenvectors of A. Since

$$\det(\lambda I - A) = \begin{vmatrix} \lambda - 1 & -1 \\ -4 & \lambda + 2 \end{vmatrix} = \lambda^2 + \lambda - 6 = (\lambda + 3)(\lambda - 2)$$

the eigenvalues of A are $\lambda = 2$, $\lambda = -3$. By definition,

$$\mathbf{x} = \begin{bmatrix} x_1 \\ x_2 \end{bmatrix}$$

is an eigenvector of A corresponding to λ if and only if $\mathbf{x}$ is a nontrivial solution of $(\lambda I - A)\mathbf{x} = \mathbf{0}$—that is, of

$$\begin{bmatrix} \lambda - 1 & -1 \\ -4 & \lambda + 2 \end{bmatrix} \begin{bmatrix} x_1 \\ x_2 \end{bmatrix} = \begin{bmatrix} 0 \\ 0 \end{bmatrix}$$

If $\lambda = 2$, this system becomes

$$\begin{bmatrix} 1 & -1 \\ -4 & 4 \end{bmatrix} \begin{bmatrix} x_1 \\ x_2 \end{bmatrix} = \begin{bmatrix} 0 \\ 0 \end{bmatrix}$$

Solving this system yields $x_1 = t$, $x_2 = t$, so

$$\begin{bmatrix} x_1 \\ x_2 \end{bmatrix} = \begin{bmatrix} t \\ t \end{bmatrix} = t \begin{bmatrix} 1 \\ 1 \end{bmatrix}$$

Thus

$$\mathbf{p}_1 = \begin{bmatrix} 1 \\ 1 \end{bmatrix}$$

is a basis for the eigenspace corresponding to $\lambda = 2$. Similarly, the reader can show that

$$\mathbf{p}_2 = \begin{bmatrix} -\frac{1}{4} \\ 1 \end{bmatrix}$$

is a basis for the eigenspace corresponding to $\lambda = -3$. Thus

$$P = \begin{bmatrix} 1 & -\frac{1}{4} \\ 1 & 1 \end{bmatrix}$$

diagonalizes A, and

$$D = P^{-1}AP = \begin{bmatrix} 2 & 0 \\ 0 & -3 \end{bmatrix}$$

Therefore, the substitution

$$\mathbf{y} = P\mathbf{u} \quad \text{and} \quad \mathbf{y}' = P\mathbf{u}'$$

yields the new "diagonal system"

$$\mathbf{u}' = D\mathbf{u} = \begin{bmatrix} 2 & 0 \\ 0 & -3 \end{bmatrix} \mathbf{u} \quad \text{or} \quad \begin{array}{l} u_1' = 2u_1 \\ u_2' = -3u_2 \end{array}$$

From (2) the solution of this system is

$$u_1 = c_1 e^{2x} \quad \text{or} \quad \mathbf{u} = \begin{bmatrix} c_1 e^{2x} \\ c_2 e^{-3x} \end{bmatrix}$$
$$u_2 = c_2 e^{-3x}$$

so the equation $\mathbf{y} = P\mathbf{u}$ yields, as the solution for $\mathbf{y}$,

$$\mathbf{y} = \begin{bmatrix} y_1 \\ y_2 \end{bmatrix} = \begin{bmatrix} 1 & -\frac{1}{4} \\ 1 & 1 \end{bmatrix} \begin{bmatrix} c_1 e^{2x} \\ c_2 e^{-3x} \end{bmatrix} = \begin{bmatrix} c_1 e^{2x} - \frac{1}{4} c_2 e^{-3x} \\ c_1 e^{2x} + c_2 e^{-3x} \end{bmatrix}$$

or

$$y_1 = c_1 e^{2x} - \tfrac{1}{4} c_2 e^{-3x}$$
$$y_2 = c_1 e^{2x} + c_2 e^{-3x} \tag{8}$$

Solution (b)

If we substitute the given initial conditions in (8), we obtain

$$c_1 - \tfrac{1}{4} c_2 = 1$$
$$c_1 + c_2 = 6$$

Solving this system, we obtain $c_1 = 2$, $c_2 = 4$, so from (8), the solution satisfying the initial conditions is

$$y_1 = 2e^{2x} - e^{-3x}$$
$$y_2 = 2e^{2x} + 4e^{-3x} \quad \blacklozenge$$

We have assumed in this section that the coefficient matrix of $\mathbf{y}' = A\mathbf{y}$ is diagonalizable. If this is not the case, other methods must be used to solve the system. Such methods are discussed in more advanced texts.

EXERCISE SET 9.1

1. (a) Solve the system
$$y_1' = y_1 + 4y_2$$
$$y_2' = 2y_1 + 3y_2$$
   (b) Find the solution that satisfies the initial conditions $y_1(0) = 0$, $y_2(0) = 0$.

2. (a) Solve the system
$$y_1' = y_1 + 3y_2$$
$$y_2' = 4y_1 + 5y_2$$
   (b) Find the solution that satisfies the conditions $y_1(0) = 2$, $y_2'(0) = 1$.

3. (a) Solve the system
$$y_1' = 4y_1 + y_3$$
$$y_2' = -2y_1 + y_2$$
$$y_3' = -2y_1 + y_3$$
   (b) Find the solution that satisfies the initial conditions $y_1(0) = -1$, $y_2(0) = 1$, $y_3(0) = 0$.

4. Solve the system
$$y_1' = 4y_1 + 2y_2 + 2y_3$$
$$y_2' = 2y_1 + 4y_2 + 2y_3$$
$$y_3' = 2y_1 + 2y_2 + 4y_3$$

5. Show that every solution of $y' = ay$ has the form $y = ce^{ax}$.

   Hint Let $y = f(x)$ be a solution of the equation, and show that $f(x)e^{-ax}$ is constant.

6. Show that if A is diagonalizable and

$$\mathbf{y} = \begin{bmatrix} y_1 \\ y_2 \\ \vdots \\ y_n \end{bmatrix}$$

satisfies $\mathbf{y}' = A\mathbf{y}$, then each y_i is a linear combination of $e^{\lambda_1 x}, e^{\lambda_2 x}, \ldots, e^{\lambda_n x}$, where $\lambda_1, \lambda_2, \ldots, \lambda_n$ are the eigenvalues of A.

7. It is possible to solve a single differential equation by expressing the equation as a system and then using the method of this section. For the differential equation $y'' - y' - 6y = 0$, show that the substitutions $y_1 = y$ and $y_2 = y'$ lead to the system

$$y_1' = y_2$$
$$y_2' = 6y_1 + y_2$$

Solve this system and then solve the original differential equation.

8. Use the procedure in Exercise 7 to solve $y'' + y' - 12y = 0$.

9. Discuss: How can the procedure in Exercise 7 be used to solve $y''' - 6y'' + 11y' - 6y = 0$? Carry out your ideas.

Discussion & Discovery

10. (a) By rewriting (8) in matrix form, show that the solution of the system in Example 2 can be expressed as

$$\mathbf{y} = c_1 e^{2x} \begin{bmatrix} 1 \\ 1 \end{bmatrix} + c_2 e^{-3x} \begin{bmatrix} -\frac{1}{4} \\ 1 \end{bmatrix}$$

This is called the *general solution* of the system.

(b) Note that in part (a), the vector in the first term is an eigenvector corresponding to the eigenvalue $\lambda_1 = 2$ and the vector in the second term is an eigenvector corresponding to the eigenvalue $\lambda_2 = -3$. This is a special case of the following general result:

THEOREM

> *If the coefficient matrix A of the system $\mathbf{y}' = A\mathbf{y}$ in (4) is diagonalizable, then the general solution of the system can be expressed as*
>
> $$\mathbf{y} = c_1 e^{\lambda_1 x} \mathbf{x}_1 + c_2 e^{\lambda_2 x} \mathbf{x}_2 + \cdots + c_n e^{\lambda_n x} \mathbf{x}_n$$
>
> *where $\lambda_1, \lambda_2, \ldots, \lambda_n$ are the eigenvalues of A, and $\mathbf{x}_i$ is an eigenvector of A corresponding to λ_i.*

Prove this result by tracing through the four-step procedure discussed in the section with

$$D = \begin{bmatrix} \lambda_1 & 0 & \cdots & 0 \\ 0 & \lambda_2 & \cdots & 0 \\ \vdots & \vdots & & \vdots \\ 0 & 0 & \cdots & \lambda_n \end{bmatrix} \quad \text{and} \quad P = [\mathbf{x}_1 \mid \mathbf{x}_2 \mid \cdots \mid \mathbf{x}_n]$$

11. Consider the system of differential equations $\mathbf{y}' = A\mathbf{y}$ where A is a 2×2 matrix. For what values of $a_{11}, a_{12}, a_{21}, a_{22}$ do the component solutions $y_1(t), y_2(t)$ tend to zero as $t \to \infty$? In particular, what must be true about the determinant and the trace of A for this to happen?

12. Solve the nondiagonalizable system $y_1' = y_1 + y_2, y_2' = y_2$.

13. Use diagonalization to solve the system $y_1' = 2y_1 + y_2 + t, y_2' = y_1 + 2y_2 + 2t$ by first writing it in the form $\mathbf{y}' = A\mathbf{y} + \mathbf{f}$. Note the presence of a forcing function in each equation.

14. Use diagonalization to solve the system $y_1' = y_1 + y_2 + e^t$, $y_2' = y_1 - y_2 + e^{-t}$ by first writing it in the form $\mathbf{y}' = A\mathbf{y} + \mathbf{f}$. Note the presence of a forcing function in each equation.

9.2
GEOMETRY OF LINEAR OPERATORS ON R^2

In Section 4.2 we studied some of the geometric properties of linear operators on R^2 and R^3. In this section we shall study linear operators on R^2 in a little more depth. Some of the ideas that will be developed here have important applications to the field of computer graphics.

Vectors or Points

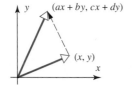

(a) T maps vectors to vectors

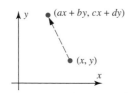

(b) T maps points to points

Figure 9.2.1

If $T: R^2 \to R^2$ is the matrix operator whose standard matrix is

$$A = \begin{bmatrix} a & b \\ c & d \end{bmatrix}$$

then

$$T\left(\begin{bmatrix} x \\ y \end{bmatrix}\right) = \begin{bmatrix} a & b \\ c & d \end{bmatrix}\begin{bmatrix} x \\ y \end{bmatrix} = \begin{bmatrix} ax + by \\ cx + dy \end{bmatrix} \tag{1}$$

There are two equally good geometric interpretations of this formula. We may view the entries in the matrices

$$\begin{bmatrix} x \\ y \end{bmatrix} \quad \text{and} \quad \begin{bmatrix} ax + by \\ cx + dy \end{bmatrix}$$

either as components of vectors or as coordinates of points. With the first interpretation, T maps arrows to arrows, and with the second, points to points (Figure 9.2.1). The choice is a matter of taste.

In this section we shall view linear operators on R^2 as mapping points to points. One useful device for visualizing the behavior of a linear operator is to observe its effect on the points of simple figures in the plane. For example, Table 1 shows the effect of some basic linear operators on a unit square that has been partially colored.

In Section 4.2 we discussed reflections, projections, rotations, contractions, and dilations of R^2. We shall now consider some other basic linear operators on R^2.

Compressions and Expansions

If the x-coordinate of each point in the plane is multiplied by a positive constant k, then the effect is to compress or expand each plane figure in the x-direction. If $0 < k < 1$, the result is a compression, and if $k > 1$, it is an expansion (Figure 9.2.2). We call such an operator a ***compression*** (or an ***expansion***) ***in the x-direction with factor k***. Similarly, if the y-coordinate of each point is multiplied by a positive constant k, we obtain a ***compression*** (or ***expansion***) ***in the y-direction with factor k***. It can be shown that compressions and expansions along the coordinate axes are linear transformations.

If $T: R^2 \to R^2$ is a compression or expansion in the x-direction with factor k, then

$$T(\mathbf{e}_1) = T\left(\begin{bmatrix} 1 \\ 0 \end{bmatrix}\right) = \begin{bmatrix} k \\ 0 \end{bmatrix}, \qquad T(\mathbf{e}_2) = T\left(\begin{bmatrix} 0 \\ 1 \end{bmatrix}\right) = \begin{bmatrix} 0 \\ 1 \end{bmatrix}$$

so the standard matrix for T is

$$\begin{bmatrix} k & 0 \\ 0 & 1 \end{bmatrix}$$

Similarly, the standard matrix for a compression or expansion in the y-direction is

$$\begin{bmatrix} 1 & 0 \\ 0 & k \end{bmatrix}$$

Table 1

Operator	Standard Matrix	Effect on the Unit Square
Reflection about the y-axis	$\begin{bmatrix} -1 & 0 \\ 0 & 1 \end{bmatrix}$	
Reflection about the x-axis	$\begin{bmatrix} 1 & 0 \\ 0 & -1 \end{bmatrix}$	
Reflection about the line $y = x$	$\begin{bmatrix} 0 & 1 \\ 1 & 0 \end{bmatrix}$	
Counterclockwise rotation through an angle θ	$\begin{bmatrix} \cos \theta & -\sin \theta \\ \sin \theta & \cos \theta \end{bmatrix}$	

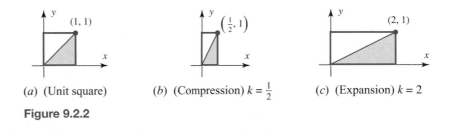

(a) (Unit square)   *(b)* (Compression) $k = \frac{1}{2}$ *(c)* (Expansion) $k = 2$

Figure 9.2.2

EXAMPLE 1   Operating with Diagonal Matrices

Suppose that the xy-plane first is compressed or expanded by a factor of k_1 in the x-direction and then is compressed or expanded by a factor of k_2 in the y-direction. Find a single matrix operator that performs both operations.

Solution

The standard matrices for the two operations are

$$\begin{bmatrix} k_1 & 0 \\ 0 & 1 \end{bmatrix} \qquad \begin{bmatrix} 1 & 0 \\ 0 & k_2 \end{bmatrix}$$

x-compression (expansion)   **y-compression (expansion)**

Thus the standard matrix for the composition of the x-operation followed by the y-operation is

$$A = \begin{bmatrix} 1 & 0 \\ 0 & k_2 \end{bmatrix} \begin{bmatrix} k_1 & 0 \\ 0 & 1 \end{bmatrix} = \begin{bmatrix} k_1 & 0 \\ 0 & k_2 \end{bmatrix} \tag{2}$$

This shows that multiplication by a diagonal 2×2 matrix compresses or expands the plane in the x-direction and also in the y-direction. In the special case where k_1 and k_2 are the same, say $k_1 = k_2 = k$, note that (2) simplifies to

$$A = \begin{bmatrix} k & 0 \\ 0 & k \end{bmatrix}$$

which is a contraction or a dilation (Table 8 of Section 4.2). ◆

Shears

A **shear in the x-direction with factor k** is a transformation that moves each point (x, y) parallel to the x-axis by an amount ky to the new position $(x + ky, y)$. Under such a transformation, points on the x-axis are unmoved since $y = 0$. However, as we progress away from the x-axis, the magnitude of y increases, so points farther from the x-axis move a greater distance than those closer (Figure 9.2.3).

A **shear in the y-direction with factor k** is a transformation that moves each point (x, y) parallel to the y-axis by an amount kx to the new position $(x, y + kx)$. Under such a transformation, points on the y-axis remain fixed, and points farther from the y-axis move a greater distance than those that are closer.

It can be shown that shears are linear transformations. If $T: R^2 \to R^2$ is a shear with factor k in the x-direction, then

$$T(\mathbf{e}_1) = T\left(\begin{bmatrix} 1 \\ 0 \end{bmatrix}\right) = \begin{bmatrix} 1 \\ 0 \end{bmatrix}, \qquad T(\mathbf{e}_2) = T\left(\begin{bmatrix} 0 \\ 1 \end{bmatrix}\right) = \begin{bmatrix} k \\ 1 \end{bmatrix}$$

so the standard matrix for T is

$$\begin{bmatrix} 1 & k \\ 0 & 1 \end{bmatrix}$$

Similarly, the standard matrix for a shear in the y-direction with factor k is

$$\begin{bmatrix} 1 & 0 \\ k & 1 \end{bmatrix}$$

REMARK Multiplication by the 2×2 identity matrix is the identity operator on R^2. This operator can be viewed as a rotation through $0°$, or as a shear along either axis with $k = 0$, or as a compression or expansion along either axis with factor $k = 1$.

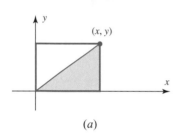

(a)

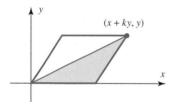

(b) Shear in x-direction with factor $k > 0$

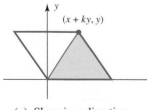

(c) Shear in x-direction with factor $k < 0$

Figure 9.2.3

EXAMPLE 2 Finding Matrix Transformations

(a) Find a matrix transformation from R^2 to R^2 that first shears by a factor of 2 in the x-direction and then reflects about $y = x$.

(b) Find a matrix transformation from R^2 to R^2 that first reflects about $y = x$ and then shears by a factor of 2 in the x-direction.

Solution (a)

The standard matrix for the shear is

$$A_1 = \begin{bmatrix} 1 & 2 \\ 0 & 1 \end{bmatrix}$$

and for the reflection is

$$A_2 = \begin{bmatrix} 0 & 1 \\ 1 & 0 \end{bmatrix}$$

Thus the standard matrix for the shear followed by the reflection is

$$A_2 A_1 = \begin{bmatrix} 0 & 1 \\ 1 & 0 \end{bmatrix} \begin{bmatrix} 1 & 2 \\ 0 & 1 \end{bmatrix} = \begin{bmatrix} 0 & 1 \\ 1 & 2 \end{bmatrix}$$

Solution (b)

The reflection followed by the shear is represented by

$$A_1 A_2 = \begin{bmatrix} 1 & 2 \\ 0 & 1 \end{bmatrix} \begin{bmatrix} 0 & 1 \\ 1 & 0 \end{bmatrix} = \begin{bmatrix} 2 & 1 \\ 1 & 0 \end{bmatrix} \blacklozenge$$

In the last example, note that $A_1 A_2 \neq A_2 A_1$, so the effect of shearing and then reflecting is different from the effect of reflecting and then shearing. This is illustrated

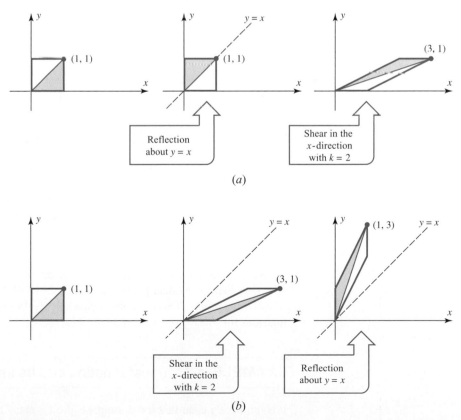

Figure 9.2.4

geometrically in Figure 9.2.4, where we show the effects of the transformations on a unit square.

EXAMPLE 3 Transformations Using Elementary Matrices

Show that if $T: R^2 \rightarrow R^2$ is multiplication by an *elementary matrix*, then the transformation is one of the following:

(a) a shear along a coordinate axis

(b) a reflection about $y = x$

(c) a compression along a coordinate axis

(d) an expansion along a coordinate axis

(e) a reflection about a coordinate axis

(f) a compression or expansion along a coordinate axis followed by a reflection about a coordinate axis.

Solution

Because a 2×2 elementary matrix results from performing a single elementary row operation on the 2×2 identity matrix, it must have one of the following forms (verify):

$$\begin{bmatrix} 1 & 0 \\ k & 1 \end{bmatrix}, \quad \begin{bmatrix} 1 & k \\ 0 & 1 \end{bmatrix}, \quad \begin{bmatrix} 0 & 1 \\ 1 & 0 \end{bmatrix}, \quad \begin{bmatrix} k & 0 \\ 0 & 1 \end{bmatrix}, \quad \begin{bmatrix} 1 & 0 \\ 0 & k \end{bmatrix}$$

The first two matrices represent shears along coordinate axes; the third represents a reflection about $y = x$. If $k > 0$, the last two matrices represent compressions or expansions along coordinate axes, depending on whether $0 \leq k \leq 1$ or $k \geq 1$. If $k < 0$, and if we express k in the form $k = -k_1$, where $k_1 > 0$, then the last two matrices can be written as

$$\begin{bmatrix} k & 0 \\ 0 & 1 \end{bmatrix} = \begin{bmatrix} -k_1 & 0 \\ 0 & 1 \end{bmatrix} = \begin{bmatrix} -1 & 0 \\ 0 & 1 \end{bmatrix} \begin{bmatrix} k_1 & 0 \\ 0 & 1 \end{bmatrix} \tag{3}$$

$$\begin{bmatrix} 1 & 0 \\ 0 & k \end{bmatrix} = \begin{bmatrix} 1 & 0 \\ 0 & -k_1 \end{bmatrix} = \begin{bmatrix} 1 & 0 \\ 0 & -1 \end{bmatrix} \begin{bmatrix} 1 & 0 \\ 0 & k_1 \end{bmatrix} \tag{4}$$

Since $k_1 > 0$, the product in (3) represents a compression or expansion along the x-axis followed by a reflection about the y-axis, and (4) represents a compression or expansion along the y-axis followed by a reflection about the x-axis. In the case where $k = -1$, transformations (3) and (4) are simply reflections about the y-axis and x-axis, respectively. ◆

Reflections, rotations, compressions, expansions, and shears are all one-to-one linear operators. This is evident geometrically, since all of those operators map distinct points into distinct points. This can also be checked algebraically by verifying that the standard matrices for those operators are invertible.

EXAMPLE 4 A Transformation and Its Inverse

It is intuitively clear that if we compress the xy-plane by a factor of $\frac{1}{2}$ in the y-direction, then we must expand the xy-plane by a factor of 2 in the y-direction to move each point back to its original position. This is indeed the case, since

$$A = \begin{bmatrix} 1 & 0 \\ 0 & \frac{1}{2} \end{bmatrix}$$

represents a compression of factor $\frac{1}{2}$ in the y-direction, and

$$A^{-1} = \begin{bmatrix} 1 & 0 \\ 0 & 2 \end{bmatrix}$$

is an expansion of factor 2 in the y-direction. ◆

Geometric Properties of Linear Operators on R^2

We conclude this section with two theorems that provide some insight into the geometric properties of linear operators on R^2.

THEOREM 9.2.1

If $T: R^2 \to R^2$ is multiplication by an invertible matrix A, then the geometric effect of T is the same as an appropriate succession of shears, compressions, expansions, and reflections.

Proof Since A is invertible, it can be reduced to the identity by a finite sequence of elementary row operations. An elementary row operation can be performed by multiplying on the left by an elementary matrix, and so there exist elementary matrices $E_1, E_2, \ldots, E_k$ such that

$$E_k \cdots E_2 E_1 A = I$$

Solving for A yields

$$A = E_1^{-1} E_2^{-1} \cdots E_k^{-1} I$$

or, equivalently,

$$A = E_1^{-1} E_2^{-1} \cdots E_k^{-1} \tag{5}$$

This equation expresses A as a product of elementary matrices (since the inverse of an elementary matrix is also elementary by Theorem 1.5.2). The result now follows from Example 3. ∎

EXAMPLE 5 Geometric Effect of Multiplication by a Matrix

Assuming that k_1 and k_2 are positive, express the diagonal matrix

$$A = \begin{bmatrix} k_1 & 0 \\ 0 & k_2 \end{bmatrix}$$

as a product of elementary matrices, and describe the geometric effect of multiplication by A in terms of compressions and expansions.

Solution

From Example 1 we have

$$A = \begin{bmatrix} k_1 & 0 \\ 0 & k_2 \end{bmatrix} = \begin{bmatrix} 1 & 0 \\ 0 & k_2 \end{bmatrix}\begin{bmatrix} k_1 & 0 \\ 0 & 1 \end{bmatrix}$$

which shows that multiplication by A has the geometric effect of compressing or expanding by a factor of k_1 in the x-direction and then compressing or expanding by a factor of k_2 in the y-direction. ◆

EXAMPLE 6 Analyzing the Geometric Effect of a Matrix Operator

---

Express

$$A = \begin{bmatrix} 1 & 2 \\ 3 & 4 \end{bmatrix}$$

as a product of elementary matrices, and then describe the geometric effect of multiplication by A in terms of shears, compressions, expansions, and reflections.

Solution

A can be reduced to I as follows:

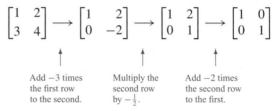

The three successive row operations can be performed by multiplying on the left successively by

$$E_1 = \begin{bmatrix} 1 & 0 \\ -3 & 1 \end{bmatrix}, \qquad E_2 = \begin{bmatrix} 1 & 0 \\ 0 & -\frac{1}{2} \end{bmatrix}, \qquad E_3 = \begin{bmatrix} 1 & -2 \\ 0 & 1 \end{bmatrix}$$

Inverting these matrices and using (5) yields

$$A = E_1^{-1} E_2^{-1} E_3^{-1} = \begin{bmatrix} 1 & 0 \\ 3 & 1 \end{bmatrix} \begin{bmatrix} 1 & 0 \\ 0 & -2 \end{bmatrix} \begin{bmatrix} 1 & 2 \\ 0 & 1 \end{bmatrix}$$

Reading from right to left and noting that

$$\begin{bmatrix} 1 & 0 \\ 0 & -2 \end{bmatrix} = \begin{bmatrix} 1 & 0 \\ 0 & -1 \end{bmatrix} \begin{bmatrix} 1 & 0 \\ 0 & 2 \end{bmatrix}$$

it follows that the effect of multiplying by A is equivalent to

1. shearing by a factor of 2 in the x-direction,
2. then expanding by a factor of 2 in the y-direction,
3. then reflecting about the x-axis,
4. then shearing by a factor of 3 in the y-direction. ◆

The proofs for parts of the following theorem are discussed in the exercises.

THEOREM 9.2.2

Images of Lines

If $T: R^2 \to R^2$ is multiplication by an invertible matrix, then

(a) *The image of a straight line is a straight line.*
(b) *The image of a straight line through the origin is a straight line through the origin.*
(c) *The images of parallel straight lines are parallel straight lines.*

(d) *The image of the line segment joining points P and Q is the line segment joining the images of P and Q.*

(e) *The images of three points lie on a line if and only if the points themselves lie on some line.*

REMARK   It follows from parts (c), (d), and (e) that multiplication by an invertible 2×2 matrix A maps triangles into triangles and parallelograms into parallelograms.

EXAMPLE 7   Image of a Square

The square with vertices $P_1(0, 0)$, $P_2(1, 0)$, $P_3(1, 1)$, and $P_4(0, 1)$ is called the **unit square**. Sketch the image of the unit square under multiplication by

$$A = \begin{bmatrix} -1 & 2 \\ 2 & -1 \end{bmatrix}$$

Solution

Since

$$\begin{bmatrix} -1 & 2 \\ 2 & -1 \end{bmatrix}\begin{bmatrix} 0 \\ 0 \end{bmatrix} = \begin{bmatrix} 0 \\ 0 \end{bmatrix} \qquad \begin{bmatrix} -1 & 2 \\ 2 & -1 \end{bmatrix}\begin{bmatrix} 1 \\ 0 \end{bmatrix} = \begin{bmatrix} -1 \\ 2 \end{bmatrix}$$

$$\begin{bmatrix} -1 & 2 \\ 2 & -1 \end{bmatrix}\begin{bmatrix} 0 \\ 1 \end{bmatrix} = \begin{bmatrix} 2 \\ -1 \end{bmatrix} \qquad \begin{bmatrix} -1 & 2 \\ 2 & -1 \end{bmatrix}\begin{bmatrix} 1 \\ 1 \end{bmatrix} = \begin{bmatrix} 1 \\ 1 \end{bmatrix}$$

the image of the square is a parallelogram with vertices $(0, 0)$, $(-1, 2)$, $(2, -1)$, and $(1, 1)$ (Figure 9.2.5). ◆

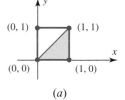

(a)

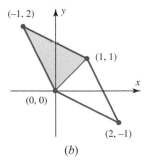

(b)

Figure 9.2.5

EXAMPLE 8   Image of a Line

According to Theorem 2, the invertible matrix

$$A = \begin{bmatrix} 3 & 1 \\ 2 & 1 \end{bmatrix}$$

maps the line $y = 2x + 1$ into another line. Find its equation.

Solution

Let (x, y) be a point on the line $y = 2x + 1$, and let (x', y') be its image under multiplication by A. Then

$$\begin{bmatrix} x' \\ y' \end{bmatrix} = \begin{bmatrix} 3 & 1 \\ 2 & 1 \end{bmatrix}\begin{bmatrix} x \\ y \end{bmatrix} \quad \text{and} \quad \begin{bmatrix} x \\ y \end{bmatrix} = \begin{bmatrix} 3 & 1 \\ 2 & 1 \end{bmatrix}^{-1}\begin{bmatrix} x' \\ y' \end{bmatrix} = \begin{bmatrix} 1 & -1 \\ -2 & 3 \end{bmatrix}\begin{bmatrix} x' \\ y' \end{bmatrix}$$

so

$$\begin{aligned} x &= x' - y' \\ y &= -2x' + 3y' \end{aligned}$$

Substituting in $y = 2x + 1$ yields

$$-2x' + 3y' = 2(x' - y') + 1 \quad \text{or, equivalently,} \quad y' = \tfrac{4}{5}x' + \tfrac{1}{5}$$

Thus (x', y') satisfies

$$y = \tfrac{4}{5}x + \tfrac{1}{5}$$

which is the equation we want. ◆

EXERCISE SET
9.2

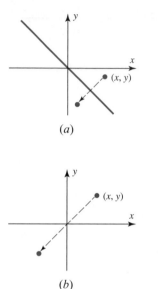

(a)

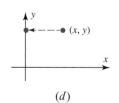

(b)

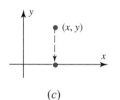

(c)

(d)

Figure Ex-1

1. Find the standard matrix for the linear operator $T: R^2 \rightarrow R^2$ that maps a point (x, y) into (see the accompanying figure)

   (a) its reflection about the line $y = -x$

   (b) its reflection through the origin

   (c) its orthogonal projection on the x-axis

   (d) its orthogonal projection on the y-axis

2. For each part of Exercise 1, use the matrix you have obtained to compute $T(2, 1)$. Check your answers geometrically by plotting the points $(2, 1)$ and $T(2, 1)$.

3. Find the standard matrix for the linear operator $T: R^3 \rightarrow R^3$ that maps a point (x, y, z) into

   (a) its reflection through the xy-plane

   (b) its reflection through the xz-plane

   (c) its reflection through the yz-plane

4. For each part of Exercise 3, use the matrix you have obtained to compute $T(1, 1, 1)$. Check your answers geometrically by sketching the vectors $(1, 1, 1)$ and $T(1, 1, 1)$.

5. Find the standard matrix for the linear operator $T: R^3 \rightarrow R^3$ that

   (a) rotates each vector 90° counterclockwise about the z-axis (looking along the positive z-axis toward the origin)

   (b) rotates each vector 90° counterclockwise about the x-axis (looking along the positive x-axis toward the origin)

   (c) rotates each vector 90° counterclockwise about the y-axis (looking along the positive y-axis toward the origin)

6. Sketch the image of the rectangle with vertices $(0, 0)$, $(1, 0)$, $(1, 2)$, and $(0, 2)$ under

   (a) a reflection about the x-axis

   (b) a reflection about the y-axis

   (c) a compression of factor $k = \tfrac{1}{4}$ in the y-direction

   (d) an expansion of factor $k = 2$ in the x-direction

   (e) a shear of factor $k = 3$ in the x-direction

   (f) a shear of factor $k = 2$ in the y-direction

7. Sketch the image of the square with vertices $(0, 0)$, $(1, 0)$, $(0, 1)$, and $(1, 1)$ under multiplication by

   $$A = \begin{bmatrix} -3 & 0 \\ 0 & 1 \end{bmatrix}$$

8. Find the matrix that rotates a point (x, y) about the origin through

   (a) 45° (b) 90° (c) 180° (d) 270° (e) −30°

9. Find the matrix that shears by

   (a) a factor of $k = 4$ in the y-direction

   (b) a factor of $k = -2$ in the x-direction

10. Find the matrix that compresses or expands by

    (a) a factor of $\tfrac{1}{3}$ in the y-direction

    (b) a factor of 6 in the x-direction

11. In each part, describe the geometric effect of multiplication by the given matrix.

    (a) $\begin{bmatrix} 3 & 0 \\ 0 & 1 \end{bmatrix}$ (b) $\begin{bmatrix} 1 & 0 \\ 0 & -5 \end{bmatrix}$ (c) $\begin{bmatrix} 1 & 4 \\ 0 & 1 \end{bmatrix}$

12. Express the matrix as a product of elementary matrices, and then describe the effect of multiplication by the given matrix in terms of compressions, expansions, reflections, and shears.

    (a) $\begin{bmatrix} 2 & 0 \\ 0 & 3 \end{bmatrix}$ (b) $\begin{bmatrix} 1 & 4 \\ 2 & 9 \end{bmatrix}$ (c) $\begin{bmatrix} 0 & -2 \\ 4 & 0 \end{bmatrix}$ (d) $\begin{bmatrix} 1 & -3 \\ 4 & 6 \end{bmatrix}$

13. In each part, find a single matrix that performs the indicated succession of operations:

    (a) compresses by a factor of $\frac{1}{2}$ in the x-direction, then expands by a factor of 5 in the y-direction

    (b) expands by a factor of 5 in the y-direction, then shears by a factor of 2 in the y-direction

    (c) reflects about $y = x$, then rotates through an angle of 180° about the origin

14. In each part, find a single matrix that performs the indicated succession of operations:

    (a) reflects about the y-axis, then expands by a factor of 5 in the x-direction, and then reflects about $y = x$

    (b) rotates through 30° about the origin, then shears by a factor of -2 in the y-direction, and then expands by a factor of 3 in the y-direction

15. By matrix inversion, show the following:

    (a) The inverse transformation for a reflection about $y = x$ is a reflection about $y = x$.

    (b) The inverse transformation for a compression along an axis is an expansion along that axis.

    (c) The inverse transformation for a reflection about a coordinate axis is a reflection about that axis.

    (d) The inverse transformation for a shear along a coordinate axis is a shear along that axis.

16. Find the equation of the image of the line $y = -4x + 3$ under multiplication by

    $$A = \begin{bmatrix} 4 & -3 \\ 3 & -2 \end{bmatrix}$$

17. In parts (a) through (e), find the equation of the image of the line $y = 2x$ under

    (a) a shear of factor 3 in the x-direction

    (b) a compression of factor $\frac{1}{2}$ in the y-direction

    (c) a reflection about $y = x$

    (d) a reflection about the y-axis

    (e) a rotation of 60° about the origin

18. Find the matrix for a shear in the x-direction that transforms the triangle with vertices $(0, 0)$, $(2, 1)$, and $(3, 0)$ into a right triangle with the right angle at the origin.

19. (a) Show that multiplication by

    $$A = \begin{bmatrix} 3 & 1 \\ 6 & 2 \end{bmatrix}$$

    maps every point in the plane onto the line $y = 2x$.

    (b) It follows from part (a) that the noncollinear points $(1, 0)$, $(0, 1)$, $(-1, 0)$ are mapped on a line. Does this violate part (e) of Theorem 2?

20. Prove part (a) of Theorem 2.

    Hint A line in the plane has an equation of the form $Ax + By + C = 0$, where A and B are not both zero. Use the method of Example 8 to show that the image of this line under multiplication by the invertible matrix

$$\begin{bmatrix} a & b \\ c & d \end{bmatrix}$$

has the equation $A'x + B'y + C = 0$, where

$$A' = (dA - cB)/(ad - bc) \quad \text{and} \quad B' = (-bA + aB)/(ad - bc)$$

Then show that A' and B' are not both zero to conclude that the image is a line.

21. Use the hint in Exercise 20 to prove parts (b) and (c) of Theorem 2.

22. In each part, find the standard matrix for the linear operator $T: R^3 \to R^3$ described by the accompanying figure.

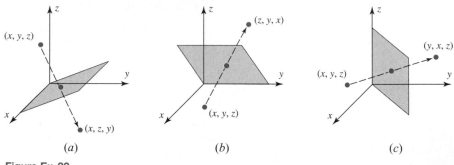

(a)   (b)   (c)

Figure Ex-22

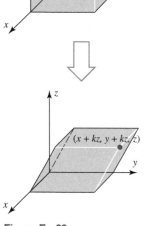

Figure Ex-23

23. In R^3 the ***shear in the xy-direction with factor k*** is the linear transformation that moves each point (x, y, z) parallel to the xy-plane to the new position $(x + kz, y + kz, z)$. (See the accompanying figure.)

(a) Find the standard matrix for the shear in the xy-direction with factor k.

(b) How would you define the shear in the xz-direction with factor k and the shear in the yz-direction with factor k? Find the standard matrices for these linear transformations.

24. In each part, find as many linearly independent eigenvectors as you can by inspection (by visualizing the geometric effect of the transformation on R^2). For each of your eigenvectors find the corresponding eigenvalue by inspection; then check your results by computing the eigenvalues and bases for the eigenspaces from the standard matrix for the transformation.

(a) reflection about the x-axis

(b) reflection about the y-axis

(c) reflection about $y = x$

(d) shear in the x-direction with factor k

(e) shear in the y-direction with factor k

(f) rotation through the angle θ

9.3

LEAST SQUARES FITTING TO DATA

In this section we shall use results about orthogonal projections in inner product spaces to obtain a technique for fitting a line or other polynomial curve to a set of experimentally determined points in the plane.

Fitting a Curve to Data

A common problem in experimental work is to obtain a mathematical relationship $y = f(x)$ between two variables x and y by "fitting" a curve to points in the plane corresponding to various experimentally determined values of x and y, say

$$(x_1, y_1), (x_2, y_2), \ldots, (x_n, y_n)$$

On the basis of theoretical considerations or simply by the pattern of the points, one decides on the general form of the curve $y = f(x)$ to be fitted. Some possibilities are (Figure 9.3.1)

(*a*) A straight line: $y = a + bx$

(*b*) A quadratic polynomial: $y = a + bx + cx^2$

(*c*) A cubic polynomial: $y = a + bx + cx^2 + dx^3$

Because the points are obtained experimentally, there is usually some measurement "error" in the data, making it impossible to find a curve of the desired form that passes through all the points. Thus, the idea is to choose the curve (by determining its coefficients) that "best" fits the data. We begin with the simplest and most common case: fitting a straight line to the data points.

Least Squares Fit of a Straight Line

(*a*) $y = a + bx$

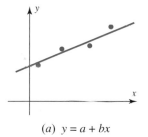

(*b*) $y = a + bx + cx^2$

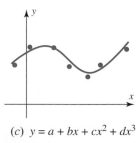

(*c*) $y = a + bx + cx^2 + dx^3$

Figure 9.3.1

Suppose we want to fit a straight line $y = a + bx$ to the experimentally determined points

$$(x_1, y_1), (x_2, y_2), \ldots, (x_n, y_n)$$

If the data points were collinear, the line would pass through all n points, and so the unknown coefficients a and b would satisfy

$$y_1 = a + bx_1$$
$$y_2 = a + bx_2$$
$$\vdots$$
$$y_n = a + bx_n$$

We can write this system in matrix form as

$$\begin{bmatrix} 1 & x_1 \\ 1 & x_2 \\ \vdots & \vdots \\ 1 & x_n \end{bmatrix} \begin{bmatrix} a \\ b \end{bmatrix} = \begin{bmatrix} y_1 \\ y_2 \\ \vdots \\ y_n \end{bmatrix}$$

or, more compactly, as

$$M\mathbf{v} = \mathbf{y} \tag{1}$$

where

$$\mathbf{y} = \begin{bmatrix} y_1 \\ y_2 \\ \vdots \\ y_n \end{bmatrix}, \qquad M = \begin{bmatrix} 1 & x_1 \\ 1 & x_2 \\ \vdots & \vdots \\ 1 & x_n \end{bmatrix}, \qquad \mathbf{v} = \begin{bmatrix} a \\ b \end{bmatrix} \tag{2}$$

If the data points are not collinear, then it is impossible to find coefficients a and b that satisfy system (1) exactly; that is, the system is inconsistent. In this case we shall look for a least squares solution

$$\mathbf{v} = \mathbf{v}^* = \begin{bmatrix} a^* \\ b^* \end{bmatrix}$$

We call a line $y = a^* + b^*x$ whose coefficients come from a least squares solution a *regression line* or a *least squares straight line fit* to the data. To explain this terminology, recall that a least squares solution of (1) minimizes

$$\|\mathbf{y} - M\mathbf{v}\| \tag{3}$$

If we express the square of (3) in terms of components, we obtain

$$\|\mathbf{y} - M\mathbf{v}\|^2 = (y_1 - a - bx_1)^2 + (y_2 - a - bx_2)^2 + \cdots + (y_n - a - bx_n)^2 \tag{4}$$

If we now let

$$d_1 = |y_1 - a - bx_1|, \quad d_2 = |y_2 - a - bx_2|, \ldots, \quad d_n = |y_n - a - bx_n|$$

then (4) can be written as

$$\|\mathbf{y} - M\mathbf{v}\|^2 = d_1^2 + d_2^2 + \cdots + d_n^2 \tag{5}$$

As illustrated in Figure 9.3.2, d_i can be interpreted as the vertical distance between the line $y = a + bx$ and the data point (x_i, y_i). This distance is a measure of the "error" at the point (x_i, y_i) resulting from the inexact fit of $y = a + bx$ to the data points. The assumption is that the x_i are known exactly and that all the error is in the measurement of the y_i. We model the error in the y_i as an additive error—that is, the measured y_i is equal to $a + bx_i + d_i$ for some unknown error d_i. Since (3) and (5) are minimized by the same vector $\mathbf{v}^*$, the least squares straight line fit minimizes the sum of the squares of the estimated errors d_i, hence the name *least squares straight line fit*.

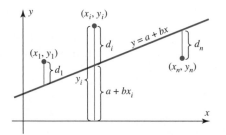

Figure 9.3.2 d_i measures the vertical error in the least squares straight line.

Normal Equations

Recall from Theorem 6.4.2 that the least squares solutions of (1) can be obtained by solving the associated normal system

$$M^T M \mathbf{v} = M^T \mathbf{y}$$

the equations of which are called the **normal equations**.

In the exercises it will be shown that the column vectors of M are linearly independent if and only if the n data points do not lie on a vertical line in the xy-plane. In this case it follows from Theorem 6.4.4 that the least squares solution is unique and is given by

$$\mathbf{v}^* = (M^T M)^{-1} M^T \mathbf{y}$$

In summary, we have the following theorem.

THEOREM 9.3.1

Least Squares Solution

Let $(x_1, y_1), (x_2, y_2), \ldots, (x_n, y_n)$ be a set of two or more data points, not all lying on a vertical line, and let

$$M = \begin{bmatrix} 1 & x_1 \\ 1 & x_2 \\ \vdots & \vdots \\ 1 & x_n \end{bmatrix} \quad and \quad \mathbf{y} = \begin{bmatrix} y_1 \\ y_2 \\ \vdots \\ y_n \end{bmatrix}$$

Then there is a unique least squares straight line fit

$$y = a^* + b^* x$$

to the data points. Moreover,

$$\mathbf{v}^* = \begin{bmatrix} a^* \\ b^* \end{bmatrix}$$

is given by the formula

$$\mathbf{v}^* = (M^T M)^{-1} M^T \mathbf{y} \tag{6}$$

which expresses the fact that $\mathbf{v} = \mathbf{v}^*$ is the unique solution of the normal equations

$$M^T M \mathbf{v} = M^T \mathbf{y} \tag{7}$$

EXAMPLE 1   Least Squares Line: Using Formula (6)

Find the least squares straight line fit to the four points $(0, 1)$, $(1, 3)$, $(2, 4)$, and $(3, 4)$. (See Figure 9.3.3.)

Solution

We have

$$M = \begin{bmatrix} 1 & 0 \\ 1 & 1 \\ 1 & 2 \\ 1 & 3 \end{bmatrix}, \quad M^T M = \begin{bmatrix} 4 & 6 \\ 6 & 14 \end{bmatrix}, \quad \text{and} \quad (M^T M)^{-1} = \frac{1}{10} \begin{bmatrix} 7 & -3 \\ -3 & 2 \end{bmatrix}$$

$$\mathbf{v}^* = (M^T M)^{-1} M^T \mathbf{y} = \frac{1}{10} \begin{bmatrix} 7 & -3 \\ -3 & 2 \end{bmatrix} \begin{bmatrix} 1 & 1 & 1 & 1 \\ 0 & 1 & 2 & 3 \end{bmatrix} \begin{bmatrix} 1 \\ 3 \\ 4 \\ 4 \end{bmatrix} = \begin{bmatrix} 1.5 \\ 1 \end{bmatrix}$$

so the desired line is $y = 1.5 + x$. ◆

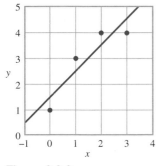

Figure 9.3.3

EXAMPLE 2   Spring Constant

Hooke's law in physics states that the length x of a uniform spring is a linear function of the force y applied to it. If we write $y = a + bx$, then the coefficient b is called the spring constant. Suppose a particular unstretched spring has a measured length of 6.1 inches (i.e., $x = 6.1$ when $y = 0$). Forces of 2 pounds, 4 pounds, and 6 pounds are then applied to the spring, and the corresponding lengths are found to be 7.6 inches, 8.7 inches, and 10.4 inches (see Figure 9.3.4). Find the spring constant of this spring.

Solution

We have

$$M = \begin{bmatrix} 1 & 6.1 \\ 1 & 7.6 \\ 1 & 8.7 \\ 1 & 10.4 \end{bmatrix}, \quad \mathbf{y} = \begin{bmatrix} 0 \\ 2 \\ 4 \\ 6 \end{bmatrix},$$

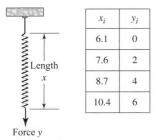

x_i	y_i
6.1	0
7.6	2
8.7	4
10.4	6

Figure 9.3.4

and

$$\mathbf{v}^* = \begin{bmatrix} a^* \\ b^* \end{bmatrix} = (M^T M)^{-1} M^T \mathbf{y} \doteq \begin{bmatrix} -8.6 \\ 1.4 \end{bmatrix}$$

where the numerical values have been rounded to one decimal place. Thus the estimated value of the spring constant is $b^* \doteq 1.4$ pounds/inch. ◆

Least Squares Fit of a Polynomial

The technique described for fitting a straight line to data points generalizes easily to fitting a polynomial of any specified degree to data points. Let us attempt to fit a polynomial of fixed degree m

$$y = a_0 + a_1 x + \cdots + a_m x^m \tag{8}$$

to n points

$$(x_1, y_1), (x_2, y_2), \ldots, (x_n, y_n)$$

Substituting these n values of x and y into (8) yields the n equations

$$\begin{aligned}
y_1 &= a_0 + a_1 x_1 + \cdots + a_m x_1^m \\
y_2 &= a_0 + a_1 x_2 + \cdots + a_m x_2^m \\
&\ \ \vdots \qquad \vdots \qquad \vdots \qquad\qquad \vdots \\
y_n &= a_0 + a_1 x_n + \cdots + a_m x_n^m
\end{aligned}$$

or, in matrix form,

$$M\mathbf{v} = \mathbf{y} \tag{9}$$

where

$$\mathbf{y} = \begin{bmatrix} y_1 \\ y_2 \\ \vdots \\ y_n \end{bmatrix}, \qquad M = \begin{bmatrix} 1 & x_1 & x_1^2 & \cdots & x_1^m \\ 1 & x_2 & x_2^2 & \cdots & x_2^m \\ \vdots & \vdots & \vdots & & \vdots \\ 1 & x_n & x_n^2 & \cdots & x_n^m \end{bmatrix}, \qquad \mathbf{v} = \begin{bmatrix} a_0 \\ a_1 \\ \vdots \\ a_m \end{bmatrix} \tag{10}$$

As before, the solutions of the normal equations

$$M^T M \mathbf{v} = M^T \mathbf{y}$$

determine the coefficients of the polynomial. The vector $\mathbf{v}$ minimizes

$$\|\mathbf{y} - M\mathbf{v}\|$$

Conditions that guarantee the invertibility of $M^T M$ are discussed in the exercises. If $M^T M$ is invertible, then the normal equations have a unique solution $\mathbf{v} = \mathbf{v}^*$, which is given by

$$\mathbf{v}^* = (M^T M)^{-1} M^T \mathbf{y} \tag{11}$$

Space Exploration

On October 5, 1991 the *Magellan* spacecraft entered the atmosphere of Venus and transmitted the temperature T in kelvins (K) versus the altitude h in kilometers (km) until its signal was lost at an altitude of about 34 km. Discounting the initial erratic signal, the data strongly suggested a linear relationship, so a least squares straight line fit was used on the linear part of the data to obtain the equation

$$T = 737.5 - 8.125h$$

By setting $h = 0$ in this equation, the surface temperature of Venus was estimated at $T \approx 737.5$ K.

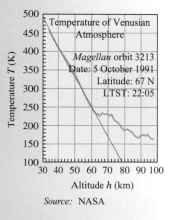

Temperature of Venusian Atmosphere

Magellan orbit 3213
Date: 5 October 1991
Latitude: 67 N
LTST: 22:05

Source: NASA

EXAMPLE 3 Fitting a Quadratic Curve to Data

According to Newton's second law of motion, a body near the earth's surface falls vertically downward according to the equation

$$s = s_0 + v_0 t + \tfrac{1}{2} g t^2 \tag{12}$$

where

s = vertical displacement downward relative to some fixed point

s_0 = initial displacement at time $t = 0$

v_0 = initial velocity at time $t = 0$

g = acceleration of gravity at the earth's surface

Suppose that a laboratory experiment is performed to evaluate g using this equation. A weight is released with unknown initial displacement and velocity, and at certain times the distances fallen relative to some fixed reference point are measured. In particular, suppose it is found that at times $t = .1, .2, .3, .4,$ and $.5$ seconds, the weight has fallen $s = -0.18, 0.31, 1.03, 2.48,$ and 3.73 feet, respectively, from the reference point. Find an approximate value of g using these data.

Solution

The mathematical problem is to fit a quadratic curve

$$s = a_0 + a_1 t + a_2 t^2 \tag{13}$$

to the five data points:

$$(.1, -0.18), \quad (.2, 0.31), \quad (.3, 1.03), \quad (.4, 2.48), \quad (.5, 3.73)$$

With the appropriate adjustments in notation, the matrices M and $\mathbf{y}$ in (10) are

$$M = \begin{bmatrix} 1 & t_1 & t_1^2 \\ 1 & t_2 & t_2^2 \\ 1 & t_3 & t_3^2 \\ 1 & t_4 & t_4^2 \\ 1 & t_5 & t_5^2 \end{bmatrix} = \begin{bmatrix} 1 & .1 & .01 \\ 1 & .2 & .04 \\ 1 & .3 & .09 \\ 1 & .4 & .16 \\ 1 & .5 & .25 \end{bmatrix}, \qquad \mathbf{y} = \begin{bmatrix} s_1 \\ s_2 \\ s_3 \\ s_4 \\ s_5 \end{bmatrix} = \begin{bmatrix} -0.18 \\ 0.31 \\ 1.03 \\ 2.48 \\ 3.73 \end{bmatrix}$$

Thus, from (11),

$$\mathbf{v}^* = \begin{bmatrix} a_0^* \\ a_1^* \\ a_2^* \end{bmatrix} = (M^T M)^{-1} M^T \mathbf{y} \doteq \begin{bmatrix} -0.40 \\ 0.35 \\ 16.1 \end{bmatrix}$$

From (12) and (13), we have $a_2 = \frac{1}{2}g$, so the estimated value of g is

$$g = 2a_2^* = 2(16.1) = 32.2 \text{ feet/second}^2$$

If desired, we can also estimate the initial displacement and initial velocity of the weight:

$$s_0 = a_0^* = -0.40 \text{ feet}$$
$$v_0 = a_1^* = 0.35 \text{ feet/second}$$

In Figure 9.3.5 we have plotted the five data points and the approximating polynomial. ◆

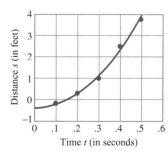

Figure 9.3.5

EXERCISE SET
9.3

1. Find the least squares straight line fit to the three points $(0, 0)$, $(1, 2)$, and $(2, 7)$.

2. Find the least squares straight line fit to the four points $(0, 1)$, $(2, 0)$, $(3, 1)$, and $(3, 2)$.

3. Find the quadratic polynomial that best fits the four points $(2, 0)$, $(3, -10)$, $(5, -48)$, and $(6, -76)$.

4. Find the cubic polynomial that best fits the five points $(-1, -14)$, $(0, -5)$, $(1, -4)$, $(2, 1)$, and $(3, 22)$.

5. Show that the matrix M in Equation (2) has linearly independent columns if and only if at least two of the numbers $x_1, x_2, \ldots, x_n$ are distinct.

6. Show that the columns of the $n \times (m + 1)$ matrix M in Equation (10) are linearly independent if $n > m$ and at least $m + 1$ of the numbers $x_1, x_2, \ldots, x_n$ are distinct.

    Hint   A nonzero polynomial of degree m has at most m distinct roots.

7. Let M be the matrix in Equation (10). Using Exercise 6, show that a sufficient condition for the matrix $M^T M$ to be invertible is that $n > m$ and that at least $m + 1$ of the numbers $x_1, x_2, \ldots, x_n$ are distinct.

8. The owner of a rapidly expanding business finds that for the first five months of the year the sales (in thousands) are \$4.0, \$4.4, \$5.2, \$6.4, and \$8.0. The owner plots these figures on a graph and conjectures that for the rest of the year, the sales curve can be approximated by a quadratic polynomial. Find the least squares quadratic polynomial fit to the sales curve, and use it to project the sales for the twelfth month of the year.

9. A corporation obtains the following data relating the number of sales representatives on its staff to annual sales:

Number of Sales Representatives	5	10	15	20	25	30
Annual Sales (millions)	3.4	4.3	5.2	6.1	7.2	8.3

    Explain how you might use least squares methods to estimate the annual sales with 45 representatives, and discuss the assumptions that you are making. (You need not perform the actual computations.)

10. Find a curve of the form $y = a + (b/x)$ that best fits the data points $(1, 7)$, $(3, 3)$, $(6, 1)$ by making the substitution $X = 1/x$. Draw the curve and plot the data points in the same coordinate system.

9.4
APPROXIMATION PROBLEMS; FOURIER SERIES

In this section we shall use results about orthogonal projections in inner product spaces to solve problems that involve approximating a given function by simpler functions. Such problems arise in a variety of engineering and scientific applications.

Best Approximations

All of the problems that we will study in this section will be special cases of the following general problem.

Approximation Problem: Given a function f that is continuous on an interval $[a, b]$, find the "best possible approximation" to f using only functions from a specified subspace W of $C[a, b]$.

Here are some examples of such problems:

(a) Find the best possible approximation to e^x over $[0, 1]$ by a polynomial of the form $a_0 + a_1 x + a_2 x^2$.

(b) Find the best possible approximation to $\sin \pi x$ over $[-1, 1]$ by a function of the form $a_0 + a_1 e^x + a_2 e^{2x} + a_3 e^{3x}$.

(c) Find the best possible approximation to x over $[0, 2\pi]$ by a function of the form $a_0 + a_1 \sin x + a_2 \sin 2x + b_1 \cos x + b_2 \cos 2x$.

In the first example W is the subspace of $C[0, 1]$ spanned by 1, x, and x^2; in the second example W is the subspace of $C[-1, 1]$ spanned by 1, e^x, e^{2x}, and e^{3x}; and in the third example W is the subspace of $C[0, 2\pi]$ spanned by 1, $\sin x$, $\sin 2x$, $\cos x$, and $\cos 2x$.

Measurements of Error

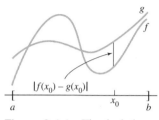

Figure 9.4.1 The deviation between f and g at x_0.

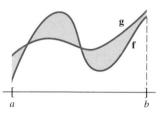

Figure 9.4.2 The area between the graphs of **f** and **g** over $[a, b]$ measures the error in approximating f by g over $[a, b]$.

To solve approximation problems of the preceding types, we must make the phrase "best approximation over $[a, b]$" mathematically precise; to do this, we need a precise way of measuring the error that results when one continuous function is approximated by another over $[a, b]$. If we were concerned only with approximating $f(x)$ at a single point x_0, then the error at x_0 by an approximation $g(x)$ would be simply

$$\text{error} = |f(x_0) - g(x_0)|$$

sometimes called the ***deviation*** between f and g at x_0 (Figure 9.4.1). However, we are concerned with approximation over the entire interval $[a, b]$, not at a single point. Consequently, in one part of the interval an approximation g_1 to f may have smaller deviations from f than an approximation g_2 to f, and in another part of the interval it might be the other way around. How do we decide which is the better overall approximation? What we need is some way of measuring the overall error in an approximation $g(x)$. One possible measure of overall error is obtained by integrating the deviation $|f(x) - g(x)|$ over the entire interval $[a, b]$; that is,

$$\text{error} = \int_a^b |f(x) - g(x)|\, dx \tag{1}$$

Geometrically, (1) is the area between the graphs of $f(x)$ and $g(x)$ over the interval $[a, b]$ (Figure 9.4.2); the greater the area, the greater the overall error.

Although (1) is natural and appealing geometrically, most mathematicians and scientists generally favor the following alternative measure of error, called the ***mean square error***.

$$\text{mean square error} = \int_a^b [f(x) - g(x)]^2\, dx$$

Mean square error emphasizes the effect of larger errors because of the squaring and has the added advantage that it allows us to bring to bear the theory of inner product spaces. To see how, suppose that **f** is a continuous function on $[a, b]$ that we want to approximate by a function **g** from a subspace W of $C[a, b]$, and suppose that $C[a, b]$ is given the inner product

$$\langle \mathbf{f}, \mathbf{g} \rangle = \int_a^b f(x)g(x)\, dx$$

It follows that

$$\|\mathbf{f} - \mathbf{g}\|^2 = \langle \mathbf{f} - \mathbf{g}, \mathbf{f} - \mathbf{g} \rangle = \int_a^b [f(x) - g(x)]^2\, dx = \text{mean square error}$$

so minimizing the mean square error is the same as minimizing $\|\mathbf{f} - \mathbf{g}\|^2$. Thus the approximation problem posed informally at the beginning of this section can be restated more precisely as follows:

Least Squares Approximation

Least Squares Approximation Problem: Let **f** be a function that is continuous on an interval $[a, b]$, let $C[a, b]$ have the inner product

$$\langle \mathbf{f}, \mathbf{g} \rangle = \int_a^b f(x)g(x)\, dx$$

and let W be a finite-dimensional subspace of $C[a, b]$. Find a function $\mathbf{g}$ in W that minimizes

$$\|\mathbf{f} - \mathbf{g}\|^2 = \int_a^b [f(x) - g(x)]^2 \, dx$$

Since $\|\mathbf{f} - \mathbf{g}\|^2$ and $\|\mathbf{f} - \mathbf{g}\|$ are minimized by the same function $\mathbf{g}$, the preceding problem is equivalent to looking for a function $\mathbf{g}$ in W that is closest to $\mathbf{f}$. But we know from Theorem 6.4.1 that $\mathbf{g} = \text{proj}_W \mathbf{f}$ is such a function (Figure 9.4.3). Thus we have the following result.

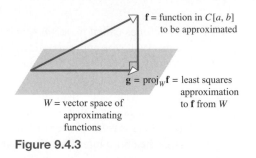

$\mathbf{f} =$ function in $C[a, b]$ to be approximated

$\mathbf{g} = \text{proj}_W \mathbf{f} =$ least squares approximation to $\mathbf{f}$ from W

$W =$ vector space of approximating functions

Figure 9.4.3

Solution of the Least Squares Approximation Problem: If $\mathbf{f}$ is a continuous function on $[a, b]$, and W is a finite-dimensional subspace of $C[a, b]$, then the function $\mathbf{g}$ in W that minimizes the mean square error

$$\int_a^b [f(x) - g(x)]^2 \, dx$$

is $\mathbf{g} = \text{proj}_W \mathbf{f}$, where the orthogonal projection is relative to the inner product

$$\langle \mathbf{f}, \mathbf{g} \rangle = \int_a^b f(x)g(x) \, dx$$

The function $\mathbf{g} = \text{proj}_W \mathbf{f}$ is called the ***least squares approximation*** to $\mathbf{f}$ from W.

Fourier Series

A function of the form

$$\tau(x) = c_0 + c_1 \cos x + c_2 \cos 2x + \cdots + c_n \cos nx$$
$$+ \, d_1 \sin x + d_2 \sin 2x + \cdots + d_n \sin nx \tag{2}$$

is called a ***trigonometric polynomial***; if c_n and d_n are not both zero, then $\tau(x)$ is said to have ***order n***. For example,

$$\tau(x) = 2 + \cos x - 3 \cos 2x + 7 \sin 4x$$

is a trigonometric polynomial with

$$c_0 = 2, \quad c_1 = 1, \quad c_2 = -3, \quad c_3 = 0, \quad c_4 = 0, \quad d_1 = 0, \quad d_2 = 0, \quad d_3 = 0, \quad d_4 = 7$$

The order of $\tau(x)$ is 4.

It is evident from (2) that the trigonometric polynomials of order n or less are the various possible linear combinations of

$$1, \quad \cos x, \quad \cos 2x, \ldots, \quad \cos nx, \quad \sin x, \quad \sin 2x, \ldots, \quad \sin nx \tag{3}$$

It can be shown that these $2n + 1$ functions are linearly independent and that consequently, for any interval $[a, b]$, they form a basis for a $(2n + 1)$-dimensional subspace of $C[a, b]$.

Let us now consider the problem of finding the least squares approximation of a continuous function $f(x)$ over the interval $[0, 2\pi]$ by a trigonometric polynomial of order n or less. As noted above, the least squares approximation to $\mathbf{f}$ from W is the orthogonal projection of $\mathbf{f}$ on W. To find this orthogonal projection, we must find an orthonormal basis $\mathbf{g}_0, \mathbf{g}_1, \ldots, \mathbf{g}_{2n}$ for W, after which we can compute the orthogonal projection on W from the formula

$$\text{proj}_W \mathbf{f} = \langle \mathbf{f}, \mathbf{g}_0 \rangle \mathbf{g}_0 + \langle \mathbf{f}, \mathbf{g}_1 \rangle \mathbf{g}_1 + \cdots + \langle \mathbf{f}, \mathbf{g}_{2n} \rangle \mathbf{g}_{2n} \qquad (4)$$

[see Theorem 6.3.5]. An orthonormal basis for W can be obtained by applying the Gram–Schmidt process to the basis (3), using the inner product

$$\langle \mathbf{f}, \mathbf{g} \rangle = \int_0^{2\pi} f(x) g(x) \, dx$$

This yields (Exercise 6) the orthonormal basis

$$\mathbf{g}_0 = \frac{1}{\sqrt{2\pi}}, \quad \mathbf{g}_1 = \frac{1}{\sqrt{\pi}} \cos x, \ldots, \quad \mathbf{g}_n = \frac{1}{\sqrt{\pi}} \cos nx,$$

$$\mathbf{g}_{n+1} = \frac{1}{\sqrt{\pi}} \sin x, \ldots, \quad \mathbf{g}_{2n} = \frac{1}{\sqrt{\pi}} \sin nx \qquad (5)$$

If we introduce the notation

$$a_0 = \frac{2}{\sqrt{2\pi}} \langle \mathbf{f}, \mathbf{g}_0 \rangle, \quad a_1 = \frac{1}{\sqrt{\pi}} \langle \mathbf{f}, \mathbf{g}_1 \rangle, \ldots, \quad a_n = \frac{1}{\sqrt{\pi}} \langle \mathbf{f}, \mathbf{g}_n \rangle$$

$$b_1 = \frac{1}{\sqrt{\pi}} \langle \mathbf{f}, \mathbf{g}_{n+1} \rangle, \ldots, \quad b_n = \frac{1}{\sqrt{\pi}} \langle \mathbf{f}, \mathbf{g}_{2n} \rangle \qquad (6)$$

then on substituting (5) in (4), we obtain

$$\text{proj}_W \mathbf{f} = \frac{a_0}{2} + [a_1 \cos x + \cdots + a_n \cos nx] + [b_1 \sin x + \cdots + b_n \sin nx] \qquad (7)$$

where

$$a_0 = \frac{2}{\sqrt{2\pi}} \langle \mathbf{f}, \mathbf{g}_0 \rangle = \frac{2}{\sqrt{2\pi}} \int_0^{2\pi} f(x) \frac{1}{\sqrt{2\pi}} \, dx = \frac{1}{\pi} \int_0^{2\pi} f(x) \, dx$$

$$a_1 = \frac{1}{\sqrt{\pi}} \langle \mathbf{f}, \mathbf{g}_1 \rangle = \frac{1}{\sqrt{\pi}} \int_0^{2\pi} f(x) \frac{1}{\sqrt{\pi}} \cos x \, dx = \frac{1}{\pi} \int_0^{2\pi} f(x) \cos x \, dx$$

$$\vdots$$

$$a_n = \frac{1}{\sqrt{\pi}} \langle \mathbf{f}, \mathbf{g}_n \rangle = \frac{1}{\sqrt{\pi}} \int_0^{2\pi} f(x) \frac{1}{\sqrt{\pi}} \cos nx \, dx = \frac{1}{\pi} \int_0^{2\pi} f(x) \cos nx \, dx$$

$$b_1 = \frac{1}{\sqrt{\pi}} \langle \mathbf{f}, \mathbf{g}_{n+1} \rangle = \frac{1}{\sqrt{\pi}} \int_0^{2\pi} f(x) \frac{1}{\sqrt{\pi}} \sin x \, dx = \frac{1}{\pi} \int_0^{2\pi} f(x) \sin x \, dx$$

$$\vdots$$

$$b_n = \frac{1}{\sqrt{\pi}} \langle \mathbf{f}, \mathbf{g}_{2n} \rangle = \frac{1}{\sqrt{\pi}} \int_0^{2\pi} f(x) \frac{1}{\sqrt{\pi}} \sin nx \, dx = \frac{1}{\pi} \int_0^{2\pi} f(x) \sin nx \, dx$$

In short,

$$a_k = \frac{1}{\pi} \int_0^{2\pi} f(x) \cos kx \, dx, \qquad b_k = \frac{1}{\pi} \int_0^{2\pi} f(x) \sin kx \, dx \qquad (8)$$

The numbers $a_0, a_1, \ldots, a_n, b_1, \ldots, b_n$ are called the ***Fourier coefficients*** of $\mathbf{f}$.

Jean Baptiste Joseph Fourier *(1768–1830)* was a French mathematician and physicist who discovered the Fourier series and related ideas while working on problems of heat diffusion. This discovery was one of the most influential in the history of mathematics; it is the cornerstone of many fields of mathematical research and a basic tool in many branches of engineering. Fourier, a political activist during the French revolution, spent time in jail for his defense of many victims during the Terror. He later became a favorite of Napoleon and was named a baron.

EXAMPLE 1   Least Squares Approximations

Find the least squares approximation of $f(x) = x$ on $[0, 2\pi]$ by

(a)   a trigonometric polynomial of order 2 or less;

(b)   a trigonometric polynomial of order n or less.

Solution (a)

$$a_0 = \frac{1}{\pi} \int_0^{2\pi} f(x)\, dx = \frac{1}{\pi} \int_0^{2\pi} x\, dx = 2\pi \tag{9a}$$

For $k = 1, 2, \ldots$, integration by parts yields (verify)

$$a_k = \frac{1}{\pi} \int_0^{2\pi} f(x) \cos kx\, dx = \frac{1}{\pi} \int_0^{2\pi} x \cos kx\, dx = 0 \tag{9b}$$

$$b_k = \frac{1}{\pi} \int_0^{2\pi} f(x) \sin kx\, dx = \frac{1}{\pi} \int_0^{2\pi} x \sin kx\, dx = -\frac{2}{k} \tag{9c}$$

Thus the least squares approximation to x on $[0, 2\pi]$ by a trigonometric polynomial of order 2 or less is

$$x \simeq \frac{a_0}{2} + a_1 \cos x + a_2 \cos 2x + b_1 \sin x + b_2 \sin 2x$$

or, from (9a), (9b), and (9c),

$$x \simeq \pi - 2 \sin x - \sin 2x$$

Solution (b)

The least squares approximation to x on $[0, 2\pi]$ by a trigonometric polynomial of order n or less is

$$x \simeq \frac{a_0}{2} + [a_1 \cos x + \cdots + a_n \cos nx] + [b_1 \sin x + \cdots + b_n \sin nx]$$

or, from (9a), (9b), and (9c),

$$x \simeq \pi - 2 \left(\sin x + \frac{\sin 2x}{2} + \frac{\sin 3x}{3} + \cdots + \frac{\sin nx}{n} \right)$$

The graphs of $y = x$ and some of these approximations are shown in Figure 9.4.4.

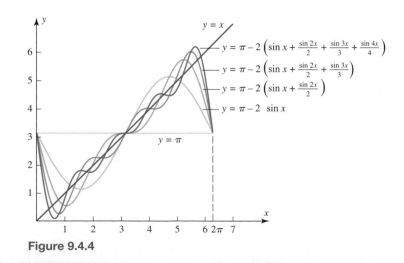

Figure 9.4.4

It is natural to expect that the mean square error will diminish as the number of terms in the least squares approximation

$$f(x) \simeq \frac{a_0}{2} + \sum_{k=1}^{n}(a_k \cos kx + b_k \sin kx)$$

increases. It can be proved that for functions f in $C[0, 2\pi]$, the mean square error approaches zero as $n \to +\infty$; this is denoted by writing

$$f(x) = \frac{a_0}{2} + \sum_{k=1}^{\infty}(a_k \cos kx + b_k \sin kx)$$

The right side of this equation is called the **Fourier series** for f over the interval $[0, 2\pi]$. Such series are of major importance in engineering, science, and mathematics. ◆

EXERCISE SET 9.4

1. Find the least squares approximation of $f(x) = 1 + x$ over the interval $[0, 2\pi]$ by
   (a) a trigonometric polynomial of order 2 or less
   (b) a trigonometric polynomial of order n or less

2. Find the least squares approximation of $f(x) = x^2$ over the interval $[0, 2\pi]$ by
   (a) a trigonometric polynomial of order 3 or less
   (b) a trigonometric polynomial of order n or less

3. (a) Find the least squares approximation of x over the interval $[0, 1]$ by a function of the form $a + be^x$.
   (b) Find the mean square error of the approximation.

4. (a) Find the least squares approximation of e^x over the interval $[0, 1]$ by a polynomial of the form $a_0 + a_1 x$.
   (b) Find the mean square error of the approximation.

5. (a) Find the least squares approximation of $\sin \pi x$ over the interval $[-1, 1]$ by a polynomial of the form $a_0 + a_1 x + a_2 x^2$.
   (b) Find the mean square error of the approximation.

6. Use the Gram–Schmidt process to obtain the orthonormal basis (5) from the basis (3).

7. Carry out the integrations in (9a), (9b), and (9c).

8. Find the Fourier series of $f(x) = \pi - x$ over the interval $[0, 2\pi]$.

9. Find the Fourier series of $f(x) = 1, 0 < x < \pi$ and $f(x) = 0, \pi \leq x \leq 2\pi$ over the interval $[0, 2\pi]$.

10. What is the Fourier series of $\sin(3x)$?

9.5 QUADRATIC FORMS

In this section we shall study functions in which the terms are squares of variables or products of two variables. Such functions arise in a variety of applications, including geometry, vibrations of mechanical systems, statistics, and electrical engineering.

Quadratic Forms

Up to now, we have been interested primarily in linear equations—that is, in equations of the form

$$a_1 x_1 + a_2 x_2 + \cdots + a_n x_n = b$$

The expression on the left side of this equation,

$$a_1 x_1 + a_2 x_2 + \cdots + a_n x_n$$

is a function of n variables, called a ***linear form***. In a linear form, all variables occur to the first power, and there are no products of variables in the expression. Here, we will be concerned with ***quadratic forms***, which are functions of the form

$$a_1 x_1^2 + a_2 x_2^2 + \cdots + a_n x_n^2 + \text{ (all possible terms of the form } a_k x_i x_j \text{ for } i < j) \quad (1)$$

For example, the most general quadratic form in the variables x_1 and x_2 is

$$a_1 x_1^2 + a_2 x_2^2 + a_3 x_1 x_2 \quad (2)$$

and the most general quadratic form in the variables x_1, x_2, and x_3 is

$$a_1 x_1^2 + a_2 x_2^2 + a_3 x_3^2 + a_4 x_1 x_2 + a_5 x_1 x_3 + a_6 x_2 x_3 \quad (3)$$

The terms in a quadratic form that involve products of different variables are called the ***cross-product terms***. Thus, in (2) the last term is a cross-product term, and in (3) the last three terms are cross-product terms.

If we follow the convention of omitting brackets on the resulting 1×1 matrices, then (2) can be written in matrix form as

$$\begin{bmatrix} x_1 & x_2 \end{bmatrix} \begin{bmatrix} a_1 & a_3/2 \\ a_3/2 & a_2 \end{bmatrix} \begin{bmatrix} x_1 \\ x_2 \end{bmatrix} \quad (4)$$

and (3) can be written as

$$\begin{bmatrix} x_1 & x_2 & x_3 \end{bmatrix} \begin{bmatrix} a_1 & a_4/2 & a_5/2 \\ a_4/2 & a_2 & a_6/2 \\ a_5/2 & a_6/2 & a_3 \end{bmatrix} \begin{bmatrix} x_1 \\ x_2 \\ x_3 \end{bmatrix} \quad (5)$$

(verify by multiplying out). Note that the products in (4) and (5) are both of the form $\mathbf{x}^T A \mathbf{x}$, where $\mathbf{x}$ is the column vector of variables, and A is a symmetric matrix whose diagonal entries are the coefficients of the squared terms and whose entries off the main diagonal are half the coefficients of the cross-product terms. More precisely, the diagonal entry in row i and column i is the coefficient of x_i^2, and the off-diagonal entry in row i and column j is half the coefficient of the product $x_i x_j$. Here are some examples.

EXAMPLE 1 Matrix Representation of Quadratic Forms

$$2x^2 + 6xy - 7y^2 = \begin{bmatrix} x & y \end{bmatrix} \begin{bmatrix} 2 & 3 \\ 3 & -7 \end{bmatrix} \begin{bmatrix} x \\ y \end{bmatrix}$$

$$4x^2 - 5y^2 = \begin{bmatrix} x & y \end{bmatrix} \begin{bmatrix} 4 & 0 \\ 0 & -5 \end{bmatrix} \begin{bmatrix} x \\ y \end{bmatrix}$$

$$xy = \begin{bmatrix} x & y \end{bmatrix} \begin{bmatrix} 0 & \frac{1}{2} \\ \frac{1}{2} & 0 \end{bmatrix} \begin{bmatrix} x \\ y \end{bmatrix}$$

$$x_1^2 + 7x_2^2 - 3x_3^2 + 4x_1 x_2 - 2x_1 x_3 + 6x_2 x_3 = \begin{bmatrix} x_1 & x_2 & x_3 \end{bmatrix} \begin{bmatrix} 1 & 2 & -1 \\ 2 & 7 & 3 \\ -1 & 3 & -3 \end{bmatrix} \begin{bmatrix} x_1 \\ x_2 \\ x_3 \end{bmatrix}$$

◆

Symmetric matrices are useful, but not essential, for representing quadratic forms. For example, the quadratic form $2x^2 + 6xy - 7y^2$, which we represented in Example 1 as $\mathbf{x}^T A \mathbf{x}$ with a symmetric matrix A, can also be written as

$$2x^2 + 6xy - 7y^2 = [x \quad y] \begin{bmatrix} 2 & 5 \\ 1 & -7 \end{bmatrix} \begin{bmatrix} x \\ y \end{bmatrix}$$

where the coefficient 6 of the cross-product term has been split as $5 + 1$ rather than $3 + 3$, as in the symmetric representation. However, symmetric matrices are usually more convenient to work with, so when we write a quadratic form as $\mathbf{x}^T A \mathbf{x}$, it will always be understood, even if it is not stated explicitly, that A is symmetric. When convenient, we can use Formula (7) of Section 4.1 to express a quadratic form $\mathbf{x}^T A \mathbf{x}$ in terms of the Euclidean inner product as

$$\mathbf{x}^T A \mathbf{x} = A \mathbf{x} \cdot \mathbf{x} \quad \text{or by symmetry of the dot product,} \quad \mathbf{x}^T A \mathbf{x} = \mathbf{x} \cdot A \mathbf{x}$$

If preferred, we can use the notation $\mathbf{u} \cdot \mathbf{v} = \langle \mathbf{u}, \mathbf{v} \rangle$ for the dot product and write these expressions as

$$\mathbf{x}^T A \mathbf{x} = \mathbf{x}^T (A \mathbf{x}) = \langle A \mathbf{x}, \mathbf{x} \rangle = \langle \mathbf{x}, A \mathbf{x} \rangle \tag{6}$$

Problems Involving Quadratic Forms

The study of quadratic forms is an extensive topic that we can only touch on in this section. The following are some of the important mathematical problems that involve quadratic forms.

- Find the maximum and minimum values of the quadratic form $\mathbf{x}^T A \mathbf{x}$ if $\mathbf{x}$ is constrained so that
$$\|\mathbf{x}\| = (x_1^2 + x_2^2 + \cdots + x_n^2)^{1/2} = 1$$

- What conditions must A satisfy in order for a quadratic form to satisfy the inequality $\mathbf{x}^T A \mathbf{x} > 0$ for all $\mathbf{x} \neq \mathbf{0}$?

- If $\mathbf{x}^T A \mathbf{x}$ is a quadratic form in two or three variables and c is a constant, what does the graph of the equation $\mathbf{x}^T A \mathbf{x} = c$ look like?

- If P is an orthogonal matrix, the change of variables $\mathbf{x} = P\mathbf{y}$ converts the quadratic form $\mathbf{x}^T A \mathbf{x}$ to $(P\mathbf{y})^T A (P\mathbf{y}) = \mathbf{y}^T (P^T A P)\mathbf{y}$. But $P^T A P$ is a symmetric matrix if A is (verify), so $\mathbf{y}^T (P^T A P)\mathbf{y}$ is a new quadratic form in the variables of $\mathbf{y}$. It is important to know whether P can be chosen such that this new quadratic form has no cross-product terms.

In this section we shall study the first two problems, and in the following sections we shall study the last two. The following theorem provides a solution to the first problem. The proof is deferred to the end of this section.

THEOREM 9.5.1

Let A be a symmetric $n \times n$ matrix with eigenvalues $\lambda_1 \geq \lambda_2 \geq \cdots \geq \lambda_n$. If $\mathbf{x}$ is constrained so that $\|\mathbf{x}\| = 1$, then

(a) $\lambda_1 \geq \mathbf{x}^T A \mathbf{x} \geq \lambda_n$.

(b) $\mathbf{x}^T A \mathbf{x} = \lambda_n$ if $\mathbf{x}$ is an eigenvector of A corresponding to λ_n and $\mathbf{x}^T A \mathbf{x} = \lambda_1$ if $\mathbf{x}$ is an eigenvector of A corresponding to λ_1.

It follows from this theorem that subject to the constraint

$$\|\mathbf{x}\| = (x_1^2 + x_2^2 + \cdots + x_n^2)^{1/2} = 1$$

the quadratic form $\mathbf{x}^T A \mathbf{x}$ has a maximum value of λ_1 (the largest eigenvalue) and a minimum value of λ_n (the smallest eigenvalue).

To bring the conic into standard position, the $x'y'$ axes must be translated. Proceeding as in Example 5, we rewrite (11) as

$$4(x'^2 - 2x') + 9(y'^2 - 4y') = -4$$

Completing the squares yields

$$4(x'^2 - 2x' + 1) + 9(y'^2 - 4y' + 4) = -4 + 4 + 36$$

or

$$4(x' - 1)^2 + 9(y' - 2)^2 = 36 \qquad (12)$$

If we translate the coordinate axes by means of the translation equations

$$x'' = x' - 1, \qquad y'' = y' - 2$$

then (12) becomes

$$4x''^2 + 9y''^2 = 36 \quad \text{or} \quad \frac{x''^2}{9} + \frac{y''^2}{4} = 1$$

which is the equation of the ellipse sketched in Figure 9.6.5. In that figure, the vectors $\mathbf{v}_1$ and $\mathbf{v}_2$ are the column vectors of P—that is, the eigenvectors of A. ◆

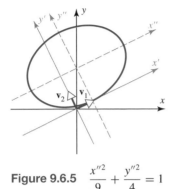

Figure 9.6.5 $\quad \dfrac{x''^2}{9} + \dfrac{y''^2}{4} = 1$

EXERCISE SET 9.6

1. In each part, find a change of variables that reduces the quadratic form to a sum or difference of squares, and express the quadratic form in terms of the new variables.
   (a) $2x_1^2 + 2x_2^2 - 2x_1x_2$
   (b) $5x_1^2 + 2x_2^2 + 4x_1x_2$
   (c) $2x_1x_2$
   (d) $-3x_1^2 + 5x_2^2 + 2x_1x_2$

2. In each part, find a change of variables that reduces the quadratic form to a sum or difference of squares, and express the quadratic form in terms of the new variables.
   (a) $3x_1^2 + 4x_2^2 + 5x_3^2 + 4x_1x_2 - 4x_2x_3$
   (b) $2x_1^2 + 5x_2^2 + 5x_3^2 + 4x_1x_2 - 4x_1x_3 - 8x_2x_3$
   (c) $-5x_1^2 + x_2^2 - x_3^2 + 6x_1x_3 + 4x_1x_2$
   (d) $2x_1x_3 + 6x_2x_3$

3. Find the quadratic forms associated with the following quadratic equations.
   (a) $2x^2 - 3xy + 4y^2 - 7x + 2y + 7 = 0$
   (b) $x^2 - xy + 5x + 8y - 3 = 0$
   (c) $5xy = 8$
   (d) $4x^2 - 2y^2 = 7$
   (e) $y^2 + 7x - 8y - 5 = 0$

4. Find the matrices of the quadratic forms in Exercise 3.

5. Express each of the quadratic equations in Exercise 3 in the matrix form $\mathbf{x}^T A \mathbf{x} + K\mathbf{x} + f = 0$.

6. Name the following conics.
   (a) $2x^2 + 5y^2 = 20$
   (b) $4x^2 + 9y^2 = 1$
   (c) $x^2 - y^2 - 8 = 0$
   (d) $4y^2 - 5x^2 = 20$
   (e) $x^2 + y^2 - 25 = 0$
   (f) $7y^2 - 2x = 0$
   (g) $-x^2 = 2y$
   (h) $3x - 11y^2 = 0$
   (i) $y - x^2 = 0$
   (j) $x^2 - 3 = -y^2$

7. In each part, a translation will put the conic in standard position. Name the conic and give its equation in the translated coordinate system.
   (a) $9x^2 + 4y^2 - 36x - 24y + 36 = 0$
   (b) $x^2 - 16y^2 + 8x + 128y = 256$
   (c) $y^2 - 8x - 14y + 49 = 0$
   (d) $x^2 + y^2 + 6x - 10y + 18 = 0$
   (e) $2x^2 - 3y^2 + 6x + 20y = -41$
   (f) $x^2 + 10x + 7y = -32$

8. The following nondegenerate conics are rotated out of standard position. In each part, rotate the coordinate axes to remove the xy-term. Name the conic and give its equation in the rotated coordinate system.

   (a) $2x^2 - 4xy - y^2 + 8 = 0$ (b) $5x^2 + 4xy + 5y^2 = 9$

   (c) $11x^2 + 24xy + 4y^2 - 15 = 0$

In Exercises 9–14 translate and rotate the coordinate axes, if necessary, to put the conic in standard position. Name the conic and give its equation in the final coordinate system.

9. $9x^2 - 4xy + 6y^2 - 10x - 20y = 5$ 10. $3x^2 - 8xy - 12y^2 - 30x - 64y = 0$

11. $2x^2 - 4xy - y^2 - 4x - 8y = -14$ 12. $21x^2 + 6xy + 13y^2 - 114x + 34y + 73 = 0$

13. $x^2 - 6xy - 7y^2 + 10x + 2y + 9 = 0$ 14. $4x^2 - 20xy + 25y^2 - 15x - 6y = 0$

15. The graph of a quadratic equation in x and y can, in certain cases, be a point, a line, or a pair of lines. These are called ***degenerate*** conics. It is also possible that the equation is not satisfied by any real values of x and y. In such cases the equation has no graph; it is said to represent an ***imaginary conic***. Each of the following represents a degenerate or imaginary conic. Where possible, sketch the graph.

   (a) $x^2 - y^2 = 0$ (b) $x^2 + 3y^2 + 7 = 0$

   (c) $8x^2 + 7y^2 = 0$ (d) $x^2 - 2xy + y^2 = 0$

   (e) $9x^2 + 12xy + 4y^2 - 52 = 0$ (f) $x^2 + y^2 - 2x - 4y = -5$

16. Prove: If $b \neq 0$, then the cross-product term can be eliminated from the quadratic form $ax^2 + 2bxy + cy^2$ by rotating the coordinate axes through an angle θ that satisfies the equation

$$\cot 2\theta = \frac{a-c}{2b}$$

9.7
QUADRIC SURFACES

In this section we shall apply the diagonalization techniques developed in the preceding section to quadratic equations in three variables, and we shall use our results to study quadric surfaces.

In Section 9.6 we looked at quadratic equations in two variables.

Quadric Surfaces

An equation of the form

$$ax^2 + by^2 + cz^2 + 2dxy + 2exz + 2fyz + gx + hy + iz + j = 0 \qquad (1)$$

where $a, b, \ldots, f$ are not all zero, is called a ***quadratic equation in x, y, and z***; the expression

$$ax^2 + by^2 + cz^2 + 2dxy + 2exz + 2fyz$$

is called the ***associated quadratic form***, which now involves three variables: x, y, and z.

Equation (1) can be written in the matrix form

$$[x \quad y \quad z]\begin{bmatrix} a & d & e \\ d & b & f \\ e & f & c \end{bmatrix}\begin{bmatrix} x \\ y \\ z \end{bmatrix} + [g \quad h \quad i]\begin{bmatrix} x \\ y \\ z \end{bmatrix} + j = 0$$

or

$$\mathbf{x}^T A \mathbf{x} + K\mathbf{x} + j = 0$$

where

$$\mathbf{x} = \begin{bmatrix} x \\ y \\ z \end{bmatrix}, \qquad A = \begin{bmatrix} a & d & e \\ d & b & f \\ e & f & c \end{bmatrix}, \qquad K = [g \quad h \quad i]$$

EXAMPLE 1 Associated Quadratic Form

The quadratic form associated with the quadratic equation

$$3x^2 + 2y^2 - z^2 + 4xy + 3xz - 8yz + 7x + 2y + 3z - 7 = 0$$

is

$$3x^2 + 2y^2 - z^2 + 4xy + 3xz - 8yz \quad \blacklozenge$$

Graphs of quadratic equations in x, y, and z are called **quadrics** or **quadric surfaces**. The simplest equations for quadric surfaces occur when those surfaces are placed in certain **standard positions** relative to the coordinate axes. Figure 9.7.1 shows the six basic quadric surfaces and the equations for those surfaces when the surfaces are in the standard positions shown in the figure. If a quadric surface is cut by a plane, then the curve of intersection is called the **trace** of the plane on the surface. To help visualize the quadric surfaces in Figure 9.7.1, we have shown and described the traces made by planes parallel to the coordinate planes. The presence of one or more of the cross-product terms xy, xz, and yz in the equation of a quadric indicates that the quadric is rotated out of standard position; the presence of both x^2 and x terms, y^2 and y terms, or z^2 and z terms in a quadric with no cross-product term indicates the quadric is translated out of standard position.

EXAMPLE 2 Identifying a Quadric Surface

Describe the quadric surface whose equation is

$$4x^2 + 36y^2 - 9z^2 - 16x - 216y + 304 = 0$$

Solution

Rearranging terms gives

$$4(x^2 - 4x) + 36(y^2 - 6y) - 9z^2 = -304$$

Completing the squares yields

$$4(x^2 - 4x + 4) + 36(y^2 - 6y + 9) - 9z^2 = -304 + 16 + 324$$

or

$$4(x - 2)^2 + 36(y - 3)^2 - 9z^2 = 36$$

or

$$\frac{(x - 2)^2}{9} + (y - 3)^2 - \frac{z^2}{4} = 1$$

Translating the axes by means of the translation equations

$$x' = x - 2, \qquad y' = y - 3, \qquad z' = z$$

yields

$$\frac{x'^2}{9} + y'^2 - \frac{z'^2}{4} = 1$$

which is the equation of a hyperboloid of one sheet. $\blacklozenge$

Eliminating Cross-Product Terms

The procedure for identifying quadrics that are rotated out of standard position is similar to the procedure for conics. Let Q be a quadric surface whose equation in xyz-coordinates is

$$\mathbf{x}^T A \mathbf{x} + K\mathbf{x} + j = 0 \tag{2}$$

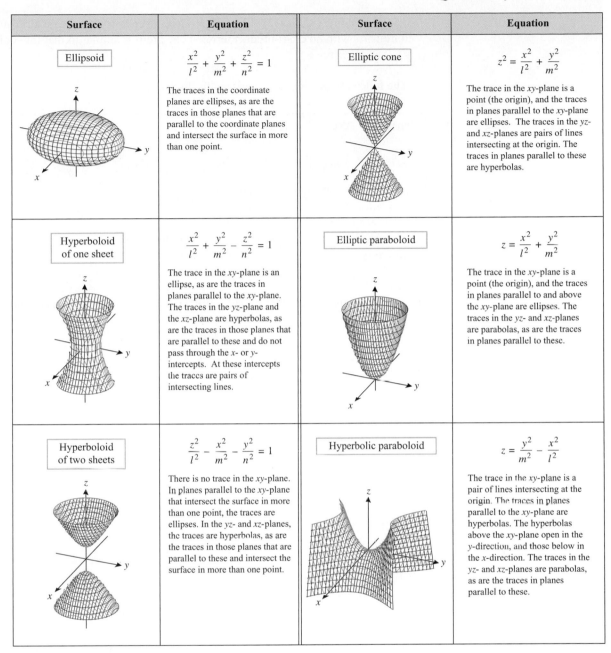

Surface	Equation	Surface	Equation
Ellipsoid	$\dfrac{x^2}{l^2} + \dfrac{y^2}{m^2} + \dfrac{z^2}{n^2} = 1$ The traces in the coordinate planes are ellipses, as are the traces in those planes that are parallel to the coordinate planes and intersect the surface in more than one point.	Elliptic cone	$z^2 = \dfrac{x^2}{l^2} + \dfrac{y^2}{m^2}$ The trace in the xy-plane is a point (the origin), and the traces in planes parallel to the xy-plane are ellipses. The traces in the yz- and xz-planes are pairs of lines intersecting at the origin. The traces in planes parallel to these are hyperbolas.
Hyperboloid of one sheet	$\dfrac{x^2}{l^2} + \dfrac{y^2}{m^2} - \dfrac{z^2}{n^2} = 1$ The trace in the xy-plane is an ellipse, as are the traces in planes parallel to the xy-plane. The traces in the yz-plane and the xz-plane are hyperbolas, as are the traces in those planes that are parallel to these and do not pass through the x- or y-intercepts. At these intercepts the traces are pairs of intersecting lines.	Elliptic paraboloid	$z = \dfrac{x^2}{l^2} + \dfrac{y^2}{m^2}$ The trace in the xy-plane is a point (the origin), and the traces in planes parallel to and above the xy-plane are ellipses. The traces in the yz- and xz-planes are parabolas, as are the traces in planes parallel to these.
Hyperboloid of two sheets	$\dfrac{z^2}{l^2} - \dfrac{x^2}{m^2} - \dfrac{y^2}{n^2} = 1$ There is no trace in the xy-plane. In planes parallel to the xy-plane that intersect the surface in more than one point, the traces are ellipses. In the yz- and xz-planes, the traces are hyperbolas, as are the traces in those planes that are parallel to these and intersect the surface in more than one point.	Hyperbolic paraboloid	$z = \dfrac{y^2}{m^2} - \dfrac{x^2}{l^2}$ The trace in the xy-plane is a pair of lines intersecting at the origin. The traces in planes parallel to the xy-plane are hyperbolas. The hyperbolas above the xy-plane open in the y-direction, and those below in the x-direction. The traces in the yz- and xz-planes are parabolas, as are the traces in planes parallel to these.

Figure 9.7.1

We want to rotate the xyz-coordinate axes so that the equation of the quadric in the new $x'y'z'$-coordinate system has no cross-product terms. This can be done as follows:

Step 1. Find a matrix P that orthogonally diagonalizes $\mathbf{x}^T A \mathbf{x}$.

Step 2. Interchange two columns of P, if necessary, to make $\det(P) = 1$. This ensures that the orthogonal coordinate transformation

$$\mathbf{x} = P\mathbf{x}', \quad \text{that is,} \quad \begin{bmatrix} x \\ y \\ z \end{bmatrix} = P \begin{bmatrix} x' \\ y' \\ z' \end{bmatrix} \tag{3}$$

is a rotation.

Step 3. Substitute (3) into (2). This will produce an equation for the quadric in $x'y'z'$-coordinates with no cross-product terms. (The proof is similar to that for conics and is left as an exercise.)

The following theorem summarizes this discussion.

THEOREM 9.7.1

Principal Axes Theorem for R^3

Let

$$ax^2 + by^2 + cz^2 + 2dxy + 2exz + 2fyz + gx + hy + iz + j = 0$$

be the equation of a quadric Q, and let

$$\mathbf{x}^T A \mathbf{x} = ax^2 + by^2 + cz^2 + 2dxy + 2exz + 2fyz$$

be the associated quadratic form. The coordinate axes can be rotated so that the equation of Q in the $x'y'z'$-coordinate system has the form

$$\lambda_1 x'^2 + \lambda_2 y'^2 + \lambda_3 z'^2 + g'x' + h'y' + i'z' + j = 0$$

where λ_1, λ_2, and λ_3 are the eigenvalues of A. The rotation can be accomplished by the substitution

$$\mathbf{x} = P\mathbf{x}'$$

where P orthogonally diagonalizes A and $\det(P) = 1$.

EXAMPLE 3 Eliminating Cross-Product Terms

Describe the quadric surface whose equation is

$$4x^2 + 4y^2 + 4z^2 + 4xy + 4xz + 4yz - 3 = 0$$

Solution

The matrix form of the above quadratic equation is

$$\mathbf{x}^T A \mathbf{x} - 3 = 0 \tag{4}$$

where

$$A = \begin{bmatrix} 4 & 2 & 2 \\ 2 & 4 & 2 \\ 2 & 2 & 4 \end{bmatrix}$$

As shown in Example 1 of Section 7.3, the eigenvalues of A are $\lambda = 2$ and $\lambda = 8$, and A is orthogonally diagonalized by the matrix

$$P = \begin{bmatrix} -1/\sqrt{2} & -1/\sqrt{6} & 1/\sqrt{3} \\ 1/\sqrt{2} & -1/\sqrt{6} & 1/\sqrt{3} \\ 0 & 2/\sqrt{6} & 1/\sqrt{3} \end{bmatrix}$$

where the first two column vectors in P are eigenvectors corresponding to $\lambda = 2$, and the third column vector is an eigenvector corresponding to $\lambda = 8$.

Since $\det(P) = 1$ (verify), the orthogonal coordinate transformation $\mathbf{x} = P\mathbf{x}'$ is a rotation. Substituting this expression in (4) yields

$$(P\mathbf{x}')^T A (P\mathbf{x}') - 3 = 0$$

or, equivalently,

$$(\mathbf{x}')^T(P^TAP)\mathbf{x}' - 3 = 0 \tag{5}$$

But

$$P^TAP = \begin{bmatrix} 2 & 0 & 0 \\ 0 & 2 & 0 \\ 0 & 0 & 8 \end{bmatrix}$$

so (5) becomes

$$[x' \quad y' \quad z'] \begin{bmatrix} 2 & 0 & 0 \\ 0 & 2 & 0 \\ 0 & 0 & 8 \end{bmatrix} \begin{bmatrix} x' \\ y' \\ z' \end{bmatrix} - 3 = 0$$

or

$$2x'^2 + 2y'^2 + 8z'^2 = 3$$

This can be rewritten as

$$\frac{x'^2}{3/2} + \frac{y'^2}{3/2} + \frac{z'^2}{3/8} = 1$$

which is the equation of an ellipsoid. ◆

EXERCISE SET 9.7

1. Find the quadratic forms associated with the following quadratic equations.
   (a) $x^2 + 2y^2 - z^2 + 4xy - 5yz + 7x + 2z = 3$
   (b) $3x^2 + 7z^2 + 2xy - 3xz + 4yz - 3x = 4$
   (c) $xy + xz + yz = 1$
   (d) $x^2 + y^2 - z^2 = 7$
   (e) $3z^2 + 3xz - 14y + 9 = 0$
   (f) $2z^2 + 2xz + y^2 + 2x - y + 3z = 0$

2. Find the matrices of the quadratic forms in Exercise 1.

3. Express each of the quadratic equations given in Exercise 1 in the matrix form $\mathbf{x}^TA\mathbf{x} + K\mathbf{x} + j = 0$.

4. Name the following quadrics.
   (a) $36x^2 + 9y^2 + 4z^2 - 36 = 0$ (b) $2x^2 + 6y^2 - 3z^2 = 18$
   (c) $6x^2 - 3y^2 - 2z^2 - 6 = 0$ (d) $9x^2 + 4y^2 - z^2 = 0$
   (e) $16x^2 + y^2 = 16z$ (f) $7x^2 - 3y^2 + z = 0$
   (g) $x^2 + y^2 + z^2 = 25$

5. In Exercise 4, identify the trace in the plane $z = 1$ in each case.

6. Find the matrices of the quadratic forms in Exercise 4. Express each of the quadratic equations in the matrix form $\mathbf{x}^TA\mathbf{x} + K\mathbf{x} + j = 0$.

7. In each part, determine the translation equations that will put the quadric in standard position, and find the equation of the quadric in the translated coordinate system. Name the quadric.
   (a) $9x^2 + 36y^2 + 4z^2 - 18x - 144y - 24z + 153 = 0$
   (b) $6x^2 + 3y^2 - 2z^2 + 12x - 18y - 8z = -7$
   (c) $3x^2 - 3y^2 - z^2 + 42x + 144 = 0$
   (d) $4x^2 + 9y^2 - z^2 - 54y - 50z = 544$
   (e) $x^2 + 16y^2 + 2x - 32y - 16z - 15 = 0$

(f) $7x^2 - 3y^2 + 126x + 72y + z + 135 = 0$

(g) $x^2 + y^2 + z^2 - 2x + 4y - 6z = 11$

8. In each part, find a rotation $\mathbf{x} = P\mathbf{x}'$ that removes the cross-product terms, and give its equation in the $x'y'z'$-system. Name the quadric.

(a) $2x^2 + 3y^2 + 23z^2 + 72xz + 150 = 0$

(b) $4x^2 + 4y^2 + 4z^2 + 4xy + 4xz + 4yz - 5 = 0$

(c) $144x^2 + 100y^2 + 81z^2 - 216xz - 540x - 720z = 0$

(d) $2xy + z = 0$

In Exercises 7–10 translate and rotate the coordinate axes to put the quadric in standard position. Name the quadric and give its equation in the final coordinate system.

9. $2xy + 2xz + 2yz - 6x - 6y - 4z = -9$

10. $7x^2 + 7y^2 + 10z^2 - 2xy - 4xz + 4yz - 12x + 12y + 60z = 24$

11. $2xy - 6x + 10y + z - 31 = 0$

12. $2x^2 + 2y^2 + 5z^2 - 4xy - 2xz + 2yz + 10x - 26y - 2z = 0$

13. Prove Theorem 9.7.1.

9.8

COMPARISON OF PROCEDURES FOR SOLVING LINEAR SYSTEMS

In this section we shall discuss some practical aspects of solving systems of linear equations, inverting matrices, and finding eigenvalues. Although we have previously discussed methods for performing these computations, we now consider their suitability for the computer solution of the large-scale problems that arise in real-world applications.

Counting Operations

Since computers are limited in the number of decimal places they can carry, they round off or truncate most numerical quantities. For example, a computer designed to store eight decimal places might record $\frac{2}{3}$ as either .66666667 (rounded off) or .66666666 (truncated).[†] In either case, an error is introduced that we shall call ***roundoff error*** or ***rounding error***.

The main practical considerations in solving linear algebra problems on digital computers are minimizing the computer time (and thus cost) needed to obtain the solution, and minimizing inaccuracies due to roundoff errors. Thus, a good computer algorithm uses as few operations and memory accesses as possible, and performs the operations in a way that minimizes the effect of roundoff errors.

In this text we have studied four methods for solving a linear system, $A\mathbf{x} = \mathbf{b}$, of n equations in n unknowns:

1. Gaussian elimination with back-substitution

2. Gauss–Jordan elimination

3. Computing A^{-1}, then forming $\mathbf{x} = A^{-1}\mathbf{b}$

4. Cramer's rule

[†]Although many calculators perform calculations in base 10, most computers work in base 2. Hence they carry finitely many bits, not decimal places. The general effects described in this section are the same in either case, however.

To determine how these methods compare as computational tools, we need to know how many arithmetic operations each requires. It is usual to group divisions and multiplications together and to group additions and subtractions together. Divisions and multiplications are considerably slower than additions and subtractions, in general. We shall refer to either multiplications or divisions as "multiplications" and to additions or subtractions as "additions."

In Table 1 we list the number of operations required to solve a linear system $A\mathbf{x} = \mathbf{b}$ of n equations in n unknowns by each of the four methods discussed in the text, as well as the number of operations required to invert A or to compute its determinant by row reduction.

Table 1 Operation Counts for an Invertible $n \times n$ Matrix A

Method	Number of Additions	Number of Multiplications
Solve $A\mathbf{x} = \mathbf{b}$ by Gauss–Jordan elimination	$\frac{1}{3}n^3 + \frac{1}{2}n^2 - \frac{5}{6}n$	$\frac{1}{3}n^3 + n^2 - \frac{1}{3}n$
Solve $A\mathbf{x} = \mathbf{b}$ by Gaussian elimination	$\frac{1}{3}n^3 + \frac{1}{2}n^2 - \frac{5}{6}n$	$\frac{1}{3}n^3 + n^2 - \frac{1}{3}n$
Find A^{-1} by reducing $[A \mid I]$ to $[I \mid A^{-1}]$	$n^3 - 2n^2 + n$	n^3
Solve $A\mathbf{x} = \mathbf{b}$ as $\mathbf{x} = A^{-1}\mathbf{b}$	$n^3 - n^2$	$n^3 + n^2$
Find $\det(A)$ by row reduction	$\frac{1}{3}n^3 - \frac{1}{2}n^2 + \frac{1}{6}n$	$\frac{1}{3}n^3 + \frac{2}{3}n - 1$
Solve $A\mathbf{x} = \mathbf{b}$ by Cramer's rule	$\frac{1}{3}n^4 - \frac{1}{6}n^3 - \frac{1}{3}n^2 + \frac{1}{6}n$	$\frac{1}{3}n^4 + \frac{1}{3}n^3 + \frac{2}{3}n^2 + \frac{2}{3}n - 1$

Note that the text methods of Gauss–Jordan elimination and Gaussian elimination have the same operation counts. It is not hard to see why this is so. Both methods begin by reducing the augmented matrix to row-echelon form. This is called the *forward phase* or *forward pass*. Then the solution is completed by back-substitution in Gaussian elimination and by continued reduction to reduced row-echelon form in Gauss–Jordan elimination. This is called the **backward phase** or **backward pass**. It turns out that the number of operations required for the backward phase is the same whether one uses back-substitution or continued reduction to reduced row-echelon form. Thus the text method of Gaussian elimination and the text method of Gauss–Jordan elimination have the same operation counts.

REMARK There is a common variation of Gauss–Jordan elimination that is less efficient than the one presented in this text. In our method the augmented matrix is first reduced to reduced row-echelon form by introducing zeros *below* the leading 1's; then the reduction is completed by introducing zeros above the leading 1's. An alternative procedure is to introduce zeros *above* and *below* a leading 1 as soon as it is obtained. This method requires

$$\frac{n^3}{2} - \frac{n}{2} \quad \text{additions} \quad \text{and} \quad \frac{n^3}{2} + \frac{n^2}{2} \quad \text{multiplications}$$

both of which are larger than our values for all $n \geq 3$.

To illustrate how the results in Table 1 are computed, we shall derive the operation counts for Gauss–Jordan elimination. For this discussion we need the following formulas for the sum of the first n positive integers and the sum of the squares of the first n positive integers:

$$1 + 2 + 3 + \cdots + n = \frac{n(n+1)}{2} \tag{1}$$

$$1^2 + 2^2 + 3^2 + \cdots + n^2 = \frac{n(n+1)(2n+1)}{6} \tag{2}$$

Derivations of these formulas are discussed in the exercises. We also need formulas for the sum of the first $n - 1$ positive integers and the sum of the squares of the first $n - 1$ positive integers. These can be obtained by substituting $n - 1$ for n in (1) and (2).

$$1 + 2 + 3 + \cdots + (n-1) = \frac{(n-1)n}{2} \tag{3}$$

$$1^2 + 2^2 + 3^2 + \cdots + (n-1)^2 = \frac{(n-1)n(2n-1)}{6} \tag{4}$$

Operation Count for Gauss–Jordan Elimination

Let $A\mathbf{x} = \mathbf{b}$ be a system of n linear equations in n unknowns, and assume that A is invertible, so that the system has a unique solution. Also assume, for simplicity, that no row interchanges are required to put the augmented matrix $[A \mid \mathbf{b}]$ in reduced row-echelon form. This assumption is justified by the fact that row interchanges are performed as bookkeeping operations on a computer (that is, they are simulated, not actually performed) and so require much less time than arithmetic operations.

Since no row interchanges are required, the first step in the Gauss–Jordan elimination process is to introduce a leading 1 in the first row by multiplying the elements in that row by the reciprocal of the leftmost entry in the row. We shall represent this step schematically as follows:

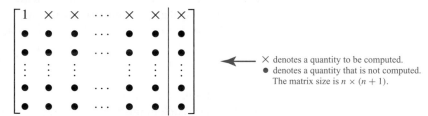

$\times$ denotes a quantity to be computed.
$\bullet$ denotes a quantity that is not computed.
The matrix size is $n \times (n + 1)$.

Note that the leading 1 is simply recorded and requires no computation; only the remaining n entries in the first row must be computed.

The following is a schematic description of the steps and the number of operations required to reduce $[A \mid \mathbf{b}]$ to row-echelon form.

Step 1.

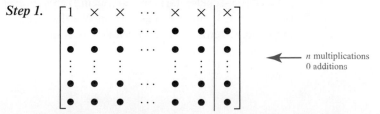

n multiplications
0 additions

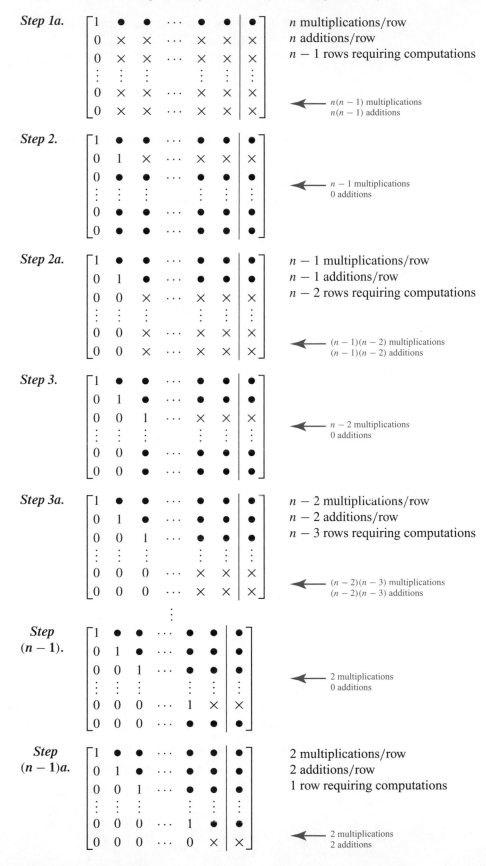

Step 1a.

n multiplications/row
n additions/row
$n - 1$ rows requiring computations

← $n(n - 1)$ multiplications
$n(n - 1)$ additions

Step 2.

← $n - 1$ multiplications
0 additions

Step 2a.

$n - 1$ multiplications/row
$n - 1$ additions/row
$n - 2$ rows requiring computations

← $(n - 1)(n - 2)$ multiplications
$(n - 1)(n - 2)$ additions

Step 3.

← $n - 2$ multiplications
0 additions

Step 3a.

$n - 2$ multiplications/row
$n - 2$ additions/row
$n - 3$ rows requiring computations

← $(n - 2)(n - 3)$ multiplications
$(n - 2)(n - 3)$ additions

Step $(n - 1)$.

← 2 multiplications
0 additions

Step $(n - 1)a$.

2 multiplications/row
2 additions/row
1 row requiring computations

← 2 multiplications
2 additions

Step n.

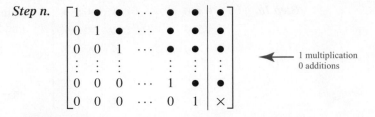

1 multiplication
0 additions

Thus, the number of operations required to complete successive steps is as follows:

Steps 1 and 1a.
   Multiplications: $n + n(n-1) = n^2$
   Additions: $n(n-1) = n^2 - n$

Steps 2 and 2a.
   Multiplications: $(n-1) + (n-1)(n-2) = (n-1)^2$
   Additions: $(n-1)(n-2) = (n-1)^2 - (n-1)$

Steps 3 and 3a.
   Multiplications: $(n-2) + (n-2)(n-3) = (n-2)^2$
   Additions: $(n-2)(n-3) = (n-2)^2 - (n-2)$
   $\vdots$

Steps $(n-1)$ and $(n-1)a$.
   Multiplications: $4\ (= 2^2)$
   Additions: $2\ (= 2^2 - 2)$

Step n.
   Multiplications: $1\ (= 1^2)$
   Additions: $0\ (= 1^2 - 1)$

   Therefore, the total number of operations required to reduce $[A \mid \mathbf{b}]$ to row-echelon form is

   Multiplications: $n^2 + (n-1)^2 + (n-2)^2 + \cdots + 1^2$
   Additions: $[n^2 + (n-1)^2 + (n-2)^2 + \cdots + 1^2]$
   $$- [n + (n-1) + (n-2) + \cdots + 1]$$

or, on applying Formulas (1) and (2),

$$\text{Multiplications:} \quad \frac{n(n+1)(2n+1)}{6} = \frac{n^3}{3} + \frac{n^2}{2} + \frac{n}{6} \tag{5}$$

$$\text{Additions:} \quad \frac{n(n+1)(2n+1)}{6} - \frac{n(n+1)}{2} = \frac{n^3}{3} - \frac{n}{3} \tag{6}$$

This completes the operation count for the forward phase. For the backward phase we must put the row-echelon form of $[A \mid \mathbf{b}]$ into reduced row-echelon form by introducing zeros above the leading 1's. The operations are as follows:

Step 1.

$$\left[\begin{array}{ccccccc|c} 1 & \bullet & \bullet & \cdots & \bullet & 0 & & \times \\ 0 & 1 & \bullet & \cdots & \bullet & 0 & & \times \\ 0 & 0 & 1 & \cdots & \bullet & 0 & & \times \\ \vdots & \vdots & \vdots & & \vdots & \vdots & & \vdots \\ 0 & 0 & 0 & \cdots & 1 & 0 & & \times \\ 0 & 0 & 0 & \cdots & 0 & 1 & & \bullet \end{array}\right]$$

$n - 1$ multiplications
$n - 1$ additions

Step 2.

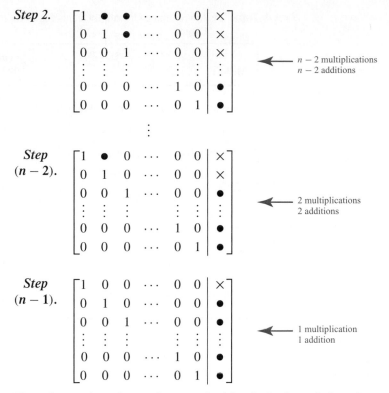

Thus, the number of operations required for the backward phase is

Multiplications: $(n-1)+(n-2)+\cdots+2+1$
Additions: $(n-1)+(n-2)+\cdots+2+1$

or, on applying Formula (3),

$$\text{Multiplications:}\quad \frac{(n-1)n}{2}=\frac{n^2}{2}-\frac{n}{2} \qquad (7)$$

$$\text{Additions:}\quad \frac{(n-1)n}{2}=\frac{n^2}{2}-\frac{n}{2} \qquad (8)$$

Thus, from (5), (6), (7), and (8), the total operation count for Gauss–Jordan elimination is

$$\text{Multiplications:}\quad \left(\frac{n^3}{3}+\frac{n^2}{2}+\frac{n}{6}\right)+\left(\frac{n^2}{2}-\frac{n}{2}\right)=\frac{n^3}{3}+n^2-\frac{n}{3} \qquad (9)$$

$$\text{Additions:}\quad \left(\frac{n^3}{3}-\frac{n}{3}\right)+\left(\frac{n^2}{2}-\frac{n}{2}\right)=\frac{n^3}{3}+\frac{n^2}{2}-\frac{5n}{6} \qquad (10)$$

Comparison of Methods for Solving Linear Systems

In practical applications it is common to encounter linear systems with thousands of equations in thousands of unknowns. Thus we shall be interested in Table 1 for large values of n. It is a fact about polynomials that for large values of the variable, a polynomial can be approximated well by its term of highest degree; that is, if $a_k \neq 0$, then

$$a_0+a_1x+\cdots+a_kx^k \approx a_kx^k \quad \text{for large } x$$

(Exercise 12). Thus, for large values of n, the operation counts in Table 1 can be approximated as shown in Table 2.

It follows from Table 2 that for large n, the best of these methods for solving $A\mathbf{x}=\mathbf{b}$ are Gaussian elimination and Gauss–Jordan elimination. The method of multiplying

Table 2 Approximate Operation Counts for an Invertible $n \times n$ Matrix A for Large n

Method	Number of Additions	Number of Multiplications
Solve $Ax = \mathbf{b}$ by Gauss–Jordan elimination	$\approx \dfrac{n^3}{3}$	$\approx \dfrac{n^3}{3}$
Solve $Ax = \mathbf{b}$ by Gaussian elimination	$\approx \dfrac{n^3}{3}$	$\approx \dfrac{n^3}{3}$
Find A^{-1} by reducing $[A \mid I]$ to $[I \mid A^{-1}]$	$\approx n^3$	$\approx n^3$
Solve $Ax = \mathbf{b}$ as $\mathbf{x} = A^{-1}\mathbf{b}$	$\approx n^3$	$\approx n^3$
Find $\det(A)$ by row reduction	$\approx \dfrac{n^3}{3}$	$\approx \dfrac{n^3}{3}$
Solve $Ax = \mathbf{b}$ by Cramer's rule	$\approx \dfrac{n^4}{3}$	$\approx \dfrac{n^4}{3}$

by A^{-1} is much worse than these (it requires three times as many operations), and the poorest of the four methods is Cramer's rule.

REMARK We observed in the remark following Table 1 that if Gauss–Jordan elimination is performed by introducing zeros above and below leading 1's as soon as they are obtained, then the operation count is

$$\frac{n^3}{2} - \frac{n}{2} \quad \text{additions} \quad \text{and} \quad \frac{n^3}{2} + \frac{n^2}{2} \quad \text{multiplications}$$

Thus, for large n, this procedure requires $\approx n^3/2$ multiplications, which is 50% greater than the $n^3/3$ multiplications required by the text method. Similarly for additions.

It is reasonable to ask if it is possible to devise other methods for solving linear systems that might require significantly fewer than the $\approx n^3/3$ additions and multiplications needed in Gaussian elimination and Gauss–Jordan elimination. The answer is a qualified "yes." In recent years, methods have been devised that require $\approx Cn^q$ multiplications, where q is slightly larger than 2.3. However, these methods have little practical value because the programming is complicated, the constant C is very large, and the number of additions required is excessive. In short, there is currently no practical method for the direct solution of general linear systems that significantly improves on the operation counts for Gaussian elimination and the text method of Gauss–Jordan elimination.

Operation counts are not the only criterion by which to judge a method for the computer solution of a linear system. As the speed of computers has increased, the time it takes to move entries of the matrix from memory to the processing unit has become increasingly important. For very large matrices, the time for memory accesses greatly exceeds the time required to do the actual computations! Despite this, the conclusion above still stands: Except for extremely large matrices, Gaussian elimination or a variant thereof is nearly always the method of choice for solving $Ax = \mathbf{b}$. It is almost never necessary to compute A^{-1}, and we should avoid doing so whenever possible. Solving $Ax = \mathbf{b}$ by Cramer's rule would be senseless for numerical purposes, despite its theoretical value.

EXAMPLE 1 Avoiding the Inverse

Suppose we needed to compute the product $AB^{-1}C\mathbf{x}$. The result is a vector $\mathbf{y}$. Rather than computing $\mathbf{y} = A(B^{-1}(C\mathbf{x}))$ as given, it would be more efficient to write this as

$$\mathbf{z} = B^{-1}C\mathbf{x}$$
$$\mathbf{y} = A\mathbf{z}$$

that is, as

$$B\mathbf{z} = C\mathbf{x}$$
$$\mathbf{y} = A\mathbf{z}$$

and to compute the result as follows: First, compute the vector $\mathbf{w} = C\mathbf{x}$; second, solve $B\mathbf{z} = \mathbf{w}$ for $\mathbf{z}$ using Gaussian elimination; third, compute the vector $\mathbf{y} = A\mathbf{z}$. ◆

For extremely large matrices, such as the ones that occur in numerical weather prediction, approximate methods for solving $A\mathbf{x} = \mathbf{b}$ are often employed. In such cases, the matrix is typically *sparse*; that is, it has very few nonzero entries. These techniques are beyond the scope of this text.

EXERCISE SET 9.8

1. Find the number of additions and multiplications required to compute AB if A is an $m \times n$ matrix and B is an $n \times p$ matrix.

2. Use the result in Exercise 1 to find the number of additions and multiplications required to compute A^k by direct multiplication if A is an $n \times n$ matrix.

3. Assuming A to be an $n \times n$ matrix, use the formulas in Table 1 to determine the number of operations required for the procedures in Table 3.

Table 3

	$n = 5$ +	$n = 5$ ×	$n = 10$ +	$n = 10$ ×	$n = 100$ +	$n = 100$ ×	$n = 1000$ +	$n = 1000$ ×
Solve $A\mathbf{x} = \mathbf{b}$ by Gauss–Jordan elimination								
Solve $A\mathbf{x} = \mathbf{b}$ by Gaussian elimination								
Find A^{-1} by reducing $[A \mid I]$ to $[I \mid A^{-1}]$								
Solve $A\mathbf{x} = \mathbf{b}$ as $\mathbf{x} = A^{-1}\mathbf{b}$								
Find $\det(A)$ by row reduction								
Solve $A\mathbf{x} = \mathbf{b}$ by Cramer's rule								

4. Assuming for simplicity a computer execution time of 2.0 microseconds for multiplications and 0.5 microsecond for additions, use the results in Exercise 3 to fill in the execution times in seconds for the procedures in Table 4.

Table 4

	$n = 5$ Execution Time (sec)	$n = 10$ Execution Time (sec)	$n = 100$ Execution Time (sec)	$n = 1000$ Execution Time (sec)
Solve $A\mathbf{x} = \mathbf{b}$ by Gauss–Jordan elimination				
Solve $A\mathbf{x} = \mathbf{b}$ by Gaussian elimination				
Find A^{-1} by reducing $[A \mid I]$ to $[I \mid A^{-1}]$				
Solve $A\mathbf{x} = \mathbf{b}$ as $\mathbf{x} = A^{-1}\mathbf{b}$				
Find $\det(A)$ by row reduction				
Solve $A\mathbf{x} = \mathbf{b}$ by Cramer's rule				

5. Derive the formula
$$1 + 2 + 3 + \cdots + n = \frac{n(n + 1)}{2}$$

Hint Let $S_n = 1 + 2 + 3 + \cdots + n$. Write the terms of S_n in reverse order and add the two expressions for S_n.

6. Use the result in Exercise 5 to show that
$$1 + 2 + 3 + \cdots + (n - 1) = \frac{(n - 1)n}{2}$$

7. Derive the formula
$$1^2 + 2^2 + 3^2 + \cdots + n^2 = \frac{n(n + 1)(2n + 1)}{6}$$

using the following steps.

(a) Show that $(k + 1)^3 - k^3 = 3k^2 + 3k + 1$.

(b) Show that
$$[2^3 - 1^3] + [3^3 - 2^3] + [4^3 - 3^3] + \cdots + [(n + 1)^3 - n^3] = (n + 1)^3 - 1$$

(c) Apply (a) to each term on the left side of (b) to show that
$$(n + 1)^3 - 1 = 3[1^2 + 2^2 + 3^2 + \cdots + n^2] + 3[1 + 2 + 3 + \cdots + n] + n$$

(d) Solve the equation in (c) for $1^2 + 2^2 + 3^2 + \cdots + n^2$, use the result of Exercise 5, and then simplify.

8. Use the result in Exercise 7 to show that
$$1^2 + 2^2 + 3^2 + \cdots + (n - 1)^2 = \frac{(n - 1)n(2n - 1)}{6}$$

9. Let R be a row-echelon form of an invertible $n \times n$ matrix. Show that solving the linear system $R\mathbf{x} = \mathbf{b}$ by back-substitution requires
$$\frac{n^2}{2} - \frac{n}{2} \quad \text{multiplications} \quad \text{and} \quad \frac{n^2}{2} - \frac{n}{2} \quad \text{additions}$$

10. Show that to reduce an invertible $n \times n$ matrix to I_n by the text method requires
$$\frac{n^3}{3} - \frac{n}{3} \quad \text{multiplications} \quad \text{and} \quad \frac{n^3}{3} - \frac{n^2}{2} + \frac{n}{6} \quad \text{additions}$$

Note Assume that no row interchanges are required.

11. Consider the variation of Gauss–Jordan elimination in which zeros are introduced above and below a leading 1 as soon as it is obtained, and let A be an invertible $n \times n$ matrix. Show that to solve a linear system $A\mathbf{x} = \mathbf{b}$ using this version of Gauss–Jordan elimination requires

$$\frac{n^3}{2} + \frac{n^2}{2} \quad \text{multiplications} \qquad \text{and} \qquad \frac{n^3}{2} - \frac{n}{2} \quad \text{additions}$$

   Note Assume that no row interchanges are required.

12. **(For Readers Who Have Studied Calculus)** Show that if $p(x) = a_0 + a_1 x + \cdots + a_k x^k$, where $a_k \neq 0$, then

$$\lim_{x \to +\infty} \frac{p(x)}{a_k x^k} = 1$$

   This result justifies the approximation $a_0 + a_1 x + \cdots + a_k x^k \approx a_k x^k$ for large values of x.

13. (a) Why is $\mathbf{y} = ((AB^{-1})C)\mathbf{x}$ an even less efficient way to find $\mathbf{y}$ in Example 1?

   (b) Use the result of Exercise 1 to find the operation count for this approach and for $\mathbf{y} = A(B^{-1}(C\mathbf{x}))$.

9.9
LU-DECOMPOSI-TIONS

With Gaussian elimination and Gauss–Jordan elimination, a linear system is solved by operating systematically on an augmented matrix. In this section we shall discuss a different organization of this approach, one based on factoring the coefficient matrix into a product of lower and upper triangular matrices. This method is well suited for computers and is the basis for many practical computer programs.[†]

Solving Linear Systems by Factoring

We shall proceed in two stages. First, we shall show how a linear system $A\mathbf{x} = \mathbf{b}$ can be solved very easily once the coefficient matrix A is factored into a product of lower and upper triangular matrices. Second, we shall show how to construct such factorizations.

If an $n \times n$ matrix A can be factored into a product of $n \times n$ matrices as

$$A = LU$$

where L is lower triangular and U is upper triangular, then the linear system $A\mathbf{x} = \mathbf{b}$ can be solved as follows:

Step 1. Rewrite the system $A\mathbf{x} = \mathbf{b}$ as

$$LU\mathbf{x} = \mathbf{b} \tag{1}$$

Step 2. Define a new $n \times 1$ matrix $\mathbf{y}$ by

$$U\mathbf{x} = \mathbf{y} \tag{2}$$

Step 3. Use (2) to rewrite (1) as $L\mathbf{y} = \mathbf{b}$ and solve this system for $\mathbf{y}$.

Step 4. Substitute $\mathbf{y}$ in (2) and solve for $\mathbf{x}$.

Although this procedure replaces the problem of solving the single system $A\mathbf{x} = \mathbf{b}$ by the problem of solving the two systems $L\mathbf{y} = \mathbf{b}$ and $U\mathbf{x} = \mathbf{y}$, the latter systems are easy to solve because the coefficient matrices are triangular. The following example illustrates this procedure.

[†] In 1979 an important library of machine-independent linear algebra programs called LINPACK was developed at Argonne National Laboratories. Many of the programs in that library use the factorization methods that we will study in this section. Variations of the LINPACK routines, including the LAPACK routines, are used in many computer programs, including MATLAB, *Mathematica*, and Maple.

EXAMPLE 1   Solving a System by Factorization

Later in this section we will derive the factorization

$$
\begin{bmatrix} 2 & 6 & 2 \\ -3 & -8 & 0 \\ 4 & 9 & 2 \end{bmatrix} = \begin{bmatrix} 2 & 0 & 0 \\ -3 & 1 & 0 \\ 4 & -3 & 7 \end{bmatrix} \begin{bmatrix} 1 & 3 & 1 \\ 0 & 1 & 3 \\ 0 & 0 & 1 \end{bmatrix}
$$

Use this result and the method described above to solve the system

$$
\begin{bmatrix} 2 & 6 & 2 \\ -3 & -8 & 0 \\ 4 & 9 & 2 \end{bmatrix} \begin{bmatrix} x_1 \\ x_2 \\ x_3 \end{bmatrix} = \begin{bmatrix} 2 \\ 2 \\ 3 \end{bmatrix} \tag{3}
$$

Solution

Rewrite (3) as

$$
\begin{bmatrix} 2 & 0 & 0 \\ -3 & 1 & 0 \\ 4 & -3 & 7 \end{bmatrix} \begin{bmatrix} 1 & 3 & 1 \\ 0 & 1 & 3 \\ 0 & 0 & 1 \end{bmatrix} \begin{bmatrix} x_1 \\ x_2 \\ x_3 \end{bmatrix} = \begin{bmatrix} 2 \\ 2 \\ 3 \end{bmatrix} \tag{4}
$$

As specified in Step 2 above, define y_1, y_2, and y_3 by the equation

$$
\begin{bmatrix} 1 & 3 & 1 \\ 0 & 1 & 3 \\ 0 & 0 & 1 \end{bmatrix} \begin{bmatrix} x_1 \\ x_2 \\ x_3 \end{bmatrix} = \begin{bmatrix} y_1 \\ y_2 \\ y_3 \end{bmatrix} \tag{5}
$$

so (4) can be rewritten as

$$
\begin{bmatrix} 2 & 0 & 0 \\ -3 & 1 & 0 \\ 4 & -3 & 7 \end{bmatrix} \begin{bmatrix} y_1 \\ y_2 \\ y_3 \end{bmatrix} = \begin{bmatrix} 2 \\ 2 \\ 3 \end{bmatrix}
$$

or, equivalently,

$$
\begin{aligned}
2y_1 \qquad\qquad &= 2 \\
-3y_1 + \ y_2 \qquad &= 2 \\
4y_1 - 3y_2 + 7y_3 &= 3
\end{aligned}
$$

The procedure for solving this system is similar to back-substitution except that the equations are solved from the top down instead of from the bottom up. This procedure, which is called forward-substitution, yields

$$
y_1 = 1, \qquad y_2 = 5, \qquad y_3 = 2
$$

(verify). Substituting these values in (5) yields the linear system

$$
\begin{bmatrix} 1 & 3 & 1 \\ 0 & 1 & 3 \\ 0 & 0 & 1 \end{bmatrix} \begin{bmatrix} x_1 \\ x_2 \\ x_3 \end{bmatrix} = \begin{bmatrix} 1 \\ 5 \\ 2 \end{bmatrix}
$$

or, equivalently,

$$
\begin{aligned}
x_1 + 3x_2 + \ x_3 &= 1 \\
x_2 + 3x_3 &= 5 \\
x_3 &= 2
\end{aligned}
$$

Solving this system by back-substitution yields the solution

$$x_1 = 2, \qquad x_2 = -1, \qquad x_3 = 2$$

(verify). ◆

LU-Decompositions

Now that we have seen how a linear system of n equations in n unknowns can be solved by factoring the coefficient matrix, we shall turn to the problem of constructing such factorizations. To motivate the method, suppose that an $n \times n$ matrix A has been reduced to a row-echelon form U by Gaussian elimination—that is, by a certain sequence of elementary row operations. By Theorem 1.5.1 each of these operations can be accomplished by multiplying on the left by an appropriate elementary matrix. Thus there are elementary matrices $E_1, E_2, \ldots, E_k$ such that

$$E_k \cdots E_2 E_1 A = U \tag{6}$$

By Theorem 1.5.2, $E_1, E_2, \ldots, E_k$ are invertible, so we can multiply both sides of Equation (6) on the left successively by

$$E_k^{-1}, \ldots, E_2^{-1}, E_1^{-1}$$

to obtain

$$A = E_1^{-1} E_2^{-1} \cdots E_k^{-1} U \tag{7}$$

In Exercise 15 we will help the reader to show that the matrix L defined by

$$L = E_1^{-1} E_2^{-1} \cdots E_k^{-1} \tag{8}$$

is lower triangular provided that *no row interchanges are used in reducing A to U*. Assuming this to be so, substituting (8) into (7) yields

$$A = LU$$

which is a factorization of A into a product of a lower triangular matrix and an upper triangular matrix.

The following theorem summarizes the above result.

THEOREM 9.9.1

> *If A is a square matrix that can be reduced to a row-echelon form U by Gaussian elimination without row interchanges, then A can be factored as $A = LU$, where L is a lower triangular matrix.*

> **DEFINITION**
>
> A factorization of a square matrix A as $A = LU$, where L is lower triangular and U is upper triangular, is called an ***LU-decomposition*** or ***triangular decomposition*** of the matrix A.

EXAMPLE 2   An *LU*-Decomposition

Find an LU-decomposition of

$$A = \begin{bmatrix} 2 & 6 & 2 \\ -3 & -8 & 0 \\ 4 & 9 & 2 \end{bmatrix}$$

Solution

To obtain an LU-decomposition, $A = LU$, we shall reduce A to a row-echelon form U using Gaussian elimination and then calculate L from (8). The steps are as follows:

	Reduction to Row-Echelon Form	Elementary Matrix Corresponding to the Row Operation	Inverse of the Elementary Matrix
	$\begin{bmatrix} 2 & 6 & 2 \\ -3 & -8 & 0 \\ 4 & 9 & 2 \end{bmatrix}$		
Step 1		$E_1 = \begin{bmatrix} \frac{1}{2} & 0 & 0 \\ 0 & 1 & 0 \\ 0 & 0 & 1 \end{bmatrix}$	$E_1^{-1} = \begin{bmatrix} 2 & 0 & 0 \\ 0 & 1 & 0 \\ 0 & 0 & 1 \end{bmatrix}$
	$\begin{bmatrix} 1 & 3 & 1 \\ -3 & -8 & 0 \\ 4 & 9 & 2 \end{bmatrix}$		
Step 2		$E_2 = \begin{bmatrix} 1 & 0 & 0 \\ 3 & 1 & 0 \\ 0 & 0 & 1 \end{bmatrix}$	$E_2^{-1} = \begin{bmatrix} 1 & 0 & 0 \\ -3 & 1 & 0 \\ 0 & 0 & 1 \end{bmatrix}$
	$\begin{bmatrix} 1 & 3 & 1 \\ 0 & 1 & 3 \\ 4 & 9 & 2 \end{bmatrix}$		
Step 3		$E_3 = \begin{bmatrix} 1 & 0 & 0 \\ 0 & 1 & 0 \\ -4 & 0 & 1 \end{bmatrix}$	$E_3^{-1} = \begin{bmatrix} 1 & 0 & 0 \\ 0 & 1 & 0 \\ 4 & 0 & 1 \end{bmatrix}$
	$\begin{bmatrix} 1 & 3 & 1 \\ 0 & 1 & 3 \\ 0 & -3 & -2 \end{bmatrix}$		
Step 4		$E_4 = \begin{bmatrix} 1 & 0 & 0 \\ 0 & 1 & 0 \\ 0 & 3 & 1 \end{bmatrix}$	$E_4^{-1} = \begin{bmatrix} 1 & 0 & 0 \\ 0 & 1 & 0 \\ 0 & -3 & 1 \end{bmatrix}$
	$\begin{bmatrix} 1 & 3 & 1 \\ 0 & 1 & 3 \\ 0 & 0 & 7 \end{bmatrix}$		
Step 5		$E_5 = \begin{bmatrix} 1 & 0 & 0 \\ 0 & 1 & 0 \\ 0 & 0 & \frac{1}{7} \end{bmatrix}$	$E_5^{-1} = \begin{bmatrix} 1 & 0 & 0 \\ 0 & 1 & 0 \\ 0 & 0 & 7 \end{bmatrix}$
	$\begin{bmatrix} 1 & 3 & 1 \\ 0 & 1 & 3 \\ 0 & 0 & 1 \end{bmatrix}$		

Thus

$$U = \begin{bmatrix} 1 & 3 & 1 \\ 0 & 1 & 3 \\ 0 & 0 & 1 \end{bmatrix}$$

and, from (8),

$$L = \begin{bmatrix} 2 & 0 & 0 \\ 0 & 1 & 0 \\ 0 & 0 & 1 \end{bmatrix} \begin{bmatrix} 1 & 0 & 0 \\ -3 & 1 & 0 \\ 0 & 0 & 1 \end{bmatrix} \begin{bmatrix} 1 & 0 & 0 \\ 0 & 1 & 0 \\ 4 & 0 & 1 \end{bmatrix} \begin{bmatrix} 1 & 0 & 0 \\ 0 & 1 & 0 \\ 0 & -3 & 1 \end{bmatrix} \begin{bmatrix} 1 & 0 & 0 \\ 0 & 1 & 0 \\ 0 & 0 & 7 \end{bmatrix}$$

$$= \begin{bmatrix} 2 & 0 & 0 \\ -3 & 1 & 0 \\ 4 & -3 & 7 \end{bmatrix}$$

so

$$\begin{bmatrix} 2 & 6 & 2 \\ -3 & -8 & 0 \\ 4 & 9 & 2 \end{bmatrix} = \begin{bmatrix} 2 & 0 & 0 \\ -3 & 1 & 0 \\ 4 & -3 & 7 \end{bmatrix} \begin{bmatrix} 1 & 3 & 1 \\ 0 & 1 & 3 \\ 0 & 0 & 1 \end{bmatrix}$$

is an *LU*-decomposition of *A*. ◆

Procedure for Finding *LU*-Decompositions

As this example shows, most of the work in constructing an *LU*-decomposition is expended in the calculation of *L*. However, *all* this work can be eliminated by some careful bookkeeping of the operations used to reduce *A* to *U*. Because we are assuming that no row interchanges are required to reduce *A* to *U*, there are only two types of operations involved: multiplying a row by a nonzero constant, and adding a multiple of one row to another. The first operation is used to introduce the leading 1's and the second to introduce zeros below the leading 1's.

In Example 2, the multipliers needed to introduce the leading 1's in successive rows were as follows:

$$\tfrac{1}{2} \text{ for the first row}$$

$$1 \text{ for the second row}$$

$$\tfrac{1}{7} \text{ for the third row}$$

Note that in (9), the successive diagonal entries in *L* were precisely the reciprocals of these multipliers:

$$L = \begin{bmatrix} ② & 0 & 0 \\ -3 & ① & 0 \\ 4 & -3 & ⑦ \end{bmatrix} \qquad (9)$$

Next, observe that to introduce zeros below the leading 1 in the first row, we used the operations

add 3 times the first row to the second

add −4 times the first row to the third

and to introduce the zero below the leading 1 in the second row, we used the operation

add 3 times the second row to the third

Now note in (10) that in each position below the main diagonal of *L*, the entry is the *negative* of the multiplier in the operation that introduced the zero in that position in *U*:

$$L = \begin{bmatrix} 2 & 0 & 0 \\ -3 & 1 & 0 \\ 4 & -3 & 7 \end{bmatrix} \qquad (10)$$

We state without proof that the same happens in the general case. Therefore, we have the following procedure for constructing an *LU*-decomposition of a square matrix *A* provided that *A* can be reduced to row-echelon form without row interchanges.

Step 1. Reduce *A* to a row-echelon form *U* by Gaussian elimination without row interchanges, keeping track of the multipliers used to introduce the leading 1's and the multipliers used to introduce the zeros below the leading 1's.

Step 2. In each position along the main diagonal of *L*, place the reciprocal of the multiplier that introduced the leading 1 in that position in *U*.

Step 3. In each position below the main diagonal of *L*, place the negative of the multiplier used to introduce the zero in that position in *U*.

EXAMPLE 3   Finding an *LU*-Decomposition

Find an *LU* -decomposition of

$$L = \begin{bmatrix} 6 & -2 & 0 \\ 9 & -1 & 1 \\ 3 & 7 & 5 \end{bmatrix}$$

Solution

We begin by reducing *A* to row-echelon form, keeping track of all multipliers.

$$\begin{bmatrix} 6 & -2 & 0 \\ 9 & -1 & 1 \\ 3 & 7 & 5 \end{bmatrix}$$

$$\begin{bmatrix} ① & -\frac{1}{3} & 0 \\ 9 & -1 & 1 \\ 3 & 7 & 5 \end{bmatrix} \longleftarrow \text{multiplier} = \frac{1}{6}$$

$$\begin{bmatrix} 1 & -\frac{1}{3} & 0 \\ ⓪ & 2 & 1 \\ ⓪ & 8 & 5 \end{bmatrix} \begin{matrix} \\ \longleftarrow \text{multiplier} = -9 \\ \longleftarrow \text{multiplier} = -3 \end{matrix}$$

$$\begin{bmatrix} 1 & -\frac{1}{3} & 0 \\ 0 & ① & \frac{1}{2} \\ 0 & 8 & 5 \end{bmatrix} \longleftarrow \text{multiplier} = \frac{1}{2}$$

$$\begin{bmatrix} 1 & -\frac{1}{3} & 0 \\ 0 & 1 & \frac{1}{2} \\ 0 & ⓪ & 1 \end{bmatrix} \longleftarrow \text{multiplier} = -8$$

$$\begin{bmatrix} 1 & -\frac{1}{3} & 0 \\ 0 & 1 & \frac{1}{2} \\ 0 & 0 & ① \end{bmatrix} \longleftarrow \text{multiplier} = 1$$

← No actual operation is performed here, since there is already a leading 1 in the third row.

Constructing L from the multipliers yields the LU-decomposition.

$$A = LU = \begin{bmatrix} 6 & 0 & 0 \\ 9 & 2 & 0 \\ 3 & 8 & 1 \end{bmatrix} \begin{bmatrix} 1 & -\frac{1}{3} & 0 \\ 0 & 1 & \frac{1}{2} \\ 0 & 0 & 1 \end{bmatrix} \blacklozenge$$

We conclude this section by briefly discussing two fundamental questions about LU-decompositions:

1. Does every square matrix have an LU-decomposition?

2. Can a square matrix have more than one LU-decomposition?

We already know that if a square matrix A can be reduced to row-echelon form by Gaussian elimination without row interchanges, then A has an LU-decomposition. In general, if row interchanges are required to reduce matrix A to row-echelon form, then there is no LU-decomposition of A. However, in such cases it is possible to factor A in the form of a ***PLU-decomposition***

$$A = PLU$$

where L is lower triangular, U is upper triangular, and P is a matrix obtained by interchanging the rows of I_n appropriately (see Exercise 17). Any matrix that is equal to the identity matrix with the order of its rows changed is called a ***permutation matrix***.

In the absence of additional restrictions, LU-decompositions are not unique. For example, if

$$A = LU = \begin{bmatrix} l_{11} & 0 & 0 \\ l_{21} & l_{22} & 0 \\ l_{31} & l_{32} & l_{33} \end{bmatrix} \begin{bmatrix} 1 & u_{12} & u_{13} \\ 0 & 1 & u_{23} \\ 0 & 0 & 1 \end{bmatrix}$$

and L has nonzero diagonal entries, then we can shift the diagonal entries from the left factor to the right factor by writing

$$A = \begin{bmatrix} 1 & 0 & 0 \\ \dfrac{l_{21}}{l_{11}} & 1 & 0 \\ \dfrac{l_{31}}{l_{11}} & \dfrac{l_{32}}{l_{22}} & 1 \end{bmatrix} \begin{bmatrix} l_{11} & 0 & 0 \\ 0 & l_{22} & 0 \\ 0 & 0 & l_{33} \end{bmatrix} \begin{bmatrix} 1 & u_{12} & u_{13} \\ 0 & 1 & u_{23} \\ 0 & 0 & 1 \end{bmatrix}$$

$$= \begin{bmatrix} 1 & 0 & 0 \\ \dfrac{l_{21}}{l_{11}} & 1 & 0 \\ \dfrac{l_{31}}{l_{11}} & \dfrac{l_{32}}{l_{22}} & 1 \end{bmatrix} \begin{bmatrix} l_{11} & l_{11}u_{12} & l_{11}u_{13} \\ 0 & l_{22} & l_{22}u_{23} \\ 0 & 0 & l_{33} \end{bmatrix}$$

which is another triangular decomposition of A.

EXERCISE SET 9.9

1. Use the method of Example 1 and the LU-decomposition

$$\begin{bmatrix} 3 & -6 \\ -2 & 5 \end{bmatrix} = \begin{bmatrix} 3 & 0 \\ -2 & 1 \end{bmatrix} \begin{bmatrix} 1 & -2 \\ 0 & 1 \end{bmatrix}$$

to solve the system

$$3x_1 - 6x_2 = 0$$
$$-2x_1 + 5x_2 = 1$$

2. Use the method of Example 1 and the LU-decomposition

$$\begin{bmatrix} 3 & -6 & -3 \\ 2 & 0 & 6 \\ -4 & 7 & 4 \end{bmatrix} = \begin{bmatrix} 3 & 0 & 0 \\ 2 & 4 & 0 \\ -4 & -1 & 2 \end{bmatrix} \begin{bmatrix} 1 & -2 & -1 \\ 0 & 1 & 2 \\ 0 & 0 & 1 \end{bmatrix}$$

to solve the system

$$3x_1 - 6x_2 - 3x_3 = -3$$
$$2x_1 \qquad + 6x_3 = -22$$
$$-4x_1 + 7x_2 + 4x_3 = \quad 3$$

In Exercises 3–10 find an LU-decomposition of the coefficient matrix; then use the method of Example 1 to solve the system.

3. $\begin{bmatrix} 2 & 8 \\ -1 & -1 \end{bmatrix} \begin{bmatrix} x_1 \\ x_2 \end{bmatrix} = \begin{bmatrix} -2 \\ -2 \end{bmatrix}$ 4. $\begin{bmatrix} -5 & -10 \\ 6 & 5 \end{bmatrix} \begin{bmatrix} x_1 \\ x_2 \end{bmatrix} = \begin{bmatrix} -10 \\ 19 \end{bmatrix}$

5. $\begin{bmatrix} 2 & -2 & -2 \\ 0 & -2 & 2 \\ -1 & 5 & 2 \end{bmatrix} \begin{bmatrix} x_1 \\ x_2 \\ x_3 \end{bmatrix} = \begin{bmatrix} -4 \\ -2 \\ 6 \end{bmatrix}$ 6. $\begin{bmatrix} -3 & 12 & -6 \\ 1 & -2 & 2 \\ 0 & 1 & 1 \end{bmatrix} \begin{bmatrix} x_1 \\ x_2 \\ x_3 \end{bmatrix} = \begin{bmatrix} -33 \\ 7 \\ -1 \end{bmatrix}$

7. $\begin{bmatrix} 5 & 5 & 10 \\ -8 & -7 & -9 \\ 0 & 4 & 26 \end{bmatrix} \begin{bmatrix} x_1 \\ x_2 \\ x_3 \end{bmatrix} = \begin{bmatrix} 0 \\ 1 \\ 4 \end{bmatrix}$ 8. $\begin{bmatrix} -1 & -3 & -4 \\ 3 & 10 & -10 \\ -2 & -4 & 11 \end{bmatrix} \begin{bmatrix} x_1 \\ x_2 \\ x_3 \end{bmatrix} = \begin{bmatrix} -6 \\ -3 \\ 9 \end{bmatrix}$

9. $\begin{bmatrix} -1 & 0 & 1 & 0 \\ 2 & 3 & -2 & 6 \\ 0 & -1 & 2 & 0 \\ 0 & 0 & 1 & 5 \end{bmatrix} \begin{bmatrix} x_1 \\ x_2 \\ x_3 \\ x_4 \end{bmatrix} = \begin{bmatrix} 5 \\ -1 \\ 3 \\ 7 \end{bmatrix}$ 10. $\begin{bmatrix} 2 & -4 & 0 & 0 \\ 1 & 2 & 12 & 0 \\ 0 & -1 & -4 & -5 \\ 0 & 0 & 2 & 11 \end{bmatrix} \begin{bmatrix} x_1 \\ x_2 \\ x_3 \\ x_4 \end{bmatrix} = \begin{bmatrix} 8 \\ 0 \\ 1 \\ 0 \end{bmatrix}$

11. Let

$$A = \begin{bmatrix} 2 & 1 & -1 \\ -2 & -1 & 2 \\ 2 & 1 & 0 \end{bmatrix}$$

   (a) Find an LU-decomposition of A.

   (b) Express A in the form $A = L_1 D U_1$, where L_1 is lower triangular with 1's along the main diagonal, U_1 is upper triangular, and D is a diagonal matrix.

   (c) Express A in the form $A = L_2 U_2$, where L_2 is lower triangular with 1's along the main diagonal and U_2 is upper triangular.

12. (a) Show that the matrix

$$\begin{bmatrix} 0 & 1 \\ 1 & 0 \end{bmatrix}$$

   has no LU-decomposition.

   (b) Find a PLU-decomposition of this matrix.

13. Let

$$A = \begin{bmatrix} a & b \\ c & d \end{bmatrix}$$

   (a) Prove: If $a \neq 0$, then A has a unique LU-decomposition with 1's along the main diagonal of L.

   (b) Find the LU-decomposition described in part (a).

14. Let $A\mathbf{x} = \mathbf{b}$ be a linear system of n equations in n unknowns, and assume that A is an invertible matrix that can be reduced to row-echelon form without row interchanges. How many additions and multiplications are required to solve the system by the method of Example 1?

    Note Count subtractions as additions and divisions as multiplications.

15. Recall from Theorem 1.7.1*b* that a product of lower triangular matrices is lower triangular. Use this fact to prove that the matrix L in (8) is lower triangular.

16. Use the result in Exercise 15 to prove that a product of finitely many upper triangular matrices is upper triangular.

17. Prove: If A is any $n \times n$ matrix, then A can be factored as $A = PLU$, where L is lower triangular, U is upper triangular, and P can be obtained by interchanging the rows of I_n appropriately.

    Hint Let U be a row-echelon form of A, and let all row interchanges required in the reduction of A to U be performed first.

18. Factor

$$A = \begin{bmatrix} 3 & -1 & 0 \\ 3 & -1 & 1 \\ 0 & 2 & 1 \end{bmatrix}$$

    as $A = PLU$, where P is a permutation matrix obtained from I_3 by interchanging rows appropriately, L is lower triangular, and U is upper triangular.

19. Show that if $A = PLU$, then $A\mathbf{x} = \mathbf{b}$ may be solved by a two-step process similar to the process in Example 1. Use this method to solve $A\mathbf{x} = \mathbf{b}$, where A is the matrix in Exercise 18 and $\mathbf{b} = \mathbf{e}_2$.

CHAPTER 9

Technology Exercises

The following exercises are designed to be solved using a technology utility. Typically, this will be MATLAB, *Mathematica*, Maple, Derive, or Mathcad, but it may also be some other type of linear algebra software or a scientific calculator with some linear algebra capabilities. For each exercise you will need to read the relevant documentation for the particular utility you are using. The goal of these exercises is to provide you with a basic proficiency with your technology utility. Once you have mastered the techniques in these exercises, you will be able to use your technology utility to solve many of the problems in the regular exercise sets.

Section 9.1 **T1.** (a) Find a general solution of the system

$$\begin{aligned} y_1' &= 3y_1 + 2y_2 + 2y_3 \\ y_2' &= y_1 + 4y_2 + y_3 \\ y_3' &= -2y_1 - 4y_2 - y_3 \end{aligned}$$

by computing appropriate eigenvalues and eigenvectors.

(b) Find the solution that satisfies the initial conditions $y_1(0) = 0$, $y_2(0) = 1$, $y_3(0) = -3$.

T2. The electrical circuit in the accompanying figure, called a ***parallel LRC circuit***, contains a resistor with resistance R ohms (Ω), an inductor with inductance L henries (H), and a capacitor with capacitance C farads (F). It is shown in electrical circuit theory that the current I in amperes (A) through the inductor and the voltage drop V in volts (V) across the capacitor satisfy the system of differential equations

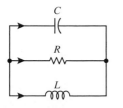

Figure Ex-T2

$$\frac{dI}{dt} = \frac{V}{L}$$

$$\frac{dV}{dt} = -\frac{I}{C} - \frac{V}{RC}$$

where the derivatives are with respect to the time t. Find I and V as functions of t if $L = 0.5$ H, $C = 0.2$ F, $R = 2 \, \Omega$, and the initial values of V and I are $V(0) = 1$ V and $I(0) = 2$ A.

Section 9.3   **T1.** **(Least Squares Straight Line Fit)** Read your documentation on finding the least squares straight line fit to a set of data points, and then use your utility to find the line of best fit to the data in Example 1. Do not imitate the method in the example; rather, use the command provided by your utility.

T2. **(Least Squares Polynomial Fit)** Read your documentation on fitting polynomials to a set of data points by least squares, and then use your utility to find the polynomial fit to the data in Example 3. Do not imitate the method in the example; rather, use the command provided by your utility.

Section 9.7   **T1.** **(Quadric Surfaces)** Use your technology utility to perform the computations in Example 3.

Section 9.9   **T1.** **(LU-decomposition)** Technology utilities vary widely in how they handle LU- and PLU-decompositions. For example, most programs perform row interchanges to reduce roundoff error and hence produce a PLU-decomposition, even when asked for an LU-decomposition. Read your documentation, and then see what happens when you use your utility to find an LU-decomposition of the matrix A in Example 3.

Complex Vector Spaces

CHAPTER CONTENTS

INTRODUCTION: Up to now we have considered only vector spaces for which the scalars are real numbers. However, for many important applications of vectors, it is desirable to allow the scalars to be complex numbers. For example, in problems involving systems of differential equations, complex eigenvalues are often the case of greatest interest.

In the first three sections of this chapter we will review some of the basic properties of complex numbers, and in subsequent sections we will discuss vector spaces in which scalars can be complex numbers.

10.1
COMPLEX NUMBERS

In this section we shall review the definition of a complex number and discuss the addition, subtraction, and multiplication of such numbers. We will also consider matrices with complex entries and explain how addition and subtraction of complex numbers can be viewed as operations on vectors.

Complex Numbers

Since $x^2 \geq 0$ for every real number x, the equation $x^2 = -1$ has no real solutions. To deal with this problem, mathematicians of the eighteenth century introduced the "imaginary" number,

$$i = \sqrt{-1}$$

which they assumed had the property

$$i^2 = (\sqrt{-1})^2 = -1$$

but which otherwise could be treated like an ordinary number. Expressions of the form

$$a + bi \tag{1}$$

where a and b are real numbers, were called "complex numbers," and these were manipulated according to the standard rules of arithmetic with the added property that $i^2 = -1$.

By the beginning of the nineteenth century it was recognized that a complex number (1) could be regarded as an alternative symbol for the ordered pair

$$(a, b)$$

of real numbers, and that operations of addition, subtraction, multiplication, and division could be defined on these ordered pairs so that the familiar laws of arithmetic hold and $i^2 = -1$. This is the approach we will follow.

> **DEFINITION**
>
> A *complex number* is an ordered pair of real numbers, denoted either by (a, b) or by $a + bi$, where $i^2 = -1$.

EXAMPLE 1   Two Notations for a Complex Number

Some examples of complex numbers in both notations are as follows:

Ordered Pair	Equivalent Notation
(3, 4)	$3 + 4i$
(−1, 2)	$-1 + 2i$
(0, 1)	$0 + i$
(2, 0)	$2 + 0i$
(4, −2)	$4 + (-2)i$

For simplicity, the last three complex numbers would usually be abbreviated as

$$0 + i = i, \qquad 2 + 0i = 2, \qquad 4 + (-2)i = 4 - 2i \quad \blacklozenge$$

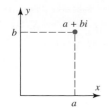

(*a*) Complex number as a point

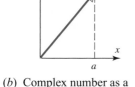

(*b*) Complex number as a vector

Figure 10.1.1

Geometrically, a complex number can be viewed as either a point or a vector in the xy-plane (Figure 10.1.1).

EXAMPLE 2 Complex Numbers as Points and as Vectors

Some complex numbers are shown as points in Figure 10.1.2*a* and as vectors in Figure 10.1.2*b*. ◆

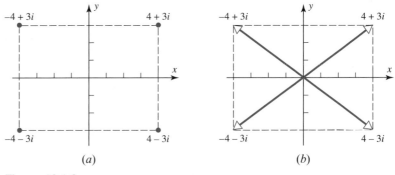

(*a*)   (*b*)

Figure 10.1.2

The Complex Plane

Sometimes it is convenient to use a single letter, such as z, to denote a complex number. Thus we might write

$$z = a + bi$$

The real number a is called the ***real part of z***, and the real number b is called the ***imaginary part of z***. These numbers are denoted by $\mathrm{Re}(z)$ and $\mathrm{Im}(z)$, respectively. Thus

$$\mathrm{Re}(4 - 3i) = 4 \quad \text{and} \quad \mathrm{Im}(4 - 3i) = -3$$

When complex numbers are represented geometrically in an xy-coordinate system, the x-axis is called the ***real axis***, the y-axis is called the ***imaginary axis***, and the plane is called the ***complex plane*** (Figure 10.1.3). The resulting plot is called an ***Argand diagram***.

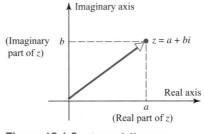

Figure 10.1.3   Argand diagram.

Operations on Complex Numbers

Just as two vectors in R^2 are defined to be equal if they have the same components, so we define two complex numbers to be equal if their real parts are equal and their imaginary parts are equal:

Two complex numbers, $a + bi$ and $c + di$, are defined to be **equal**, written

$$a + bi = c + di$$

if $a = c$ and $b = d$.

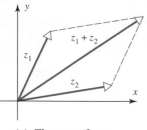

(a) The sum of two complex numbers

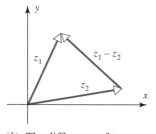

(b) The difference of two complex numbers

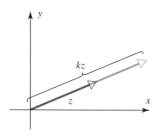

(c) The product of a complex number z and a positive real number k

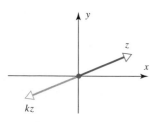

(d) The product of a complex number z and a negative real number k

Figure 10.1.4

If $b = 0$, then the complex number $a + bi$ reduces to $a + 0i$, which we write simply as a. Thus, for any real number a,

$$a = a + 0i$$

so the real numbers can be regarded as complex numbers with an imaginary part of zero. Geometrically, the real numbers correspond to points on the real axis. If we have $a = 0$, then $a + bi$ reduces to $0 + bi$, which we usually write as bi. These complex numbers, which correspond to points on the imaginary axis, are called **pure imaginary numbers**.

Just as vectors in R^2 are added by adding corresponding components, so complex numbers are added by adding their real parts and adding their imaginary parts:

$$(a + bi) + (c + di) = (a + c) + (b + d)i \qquad (2)$$

The operations of subtraction and multiplication by a *real* number are also similar to the corresponding vector operations in R^2:

$$(a + bi) - (c + di) = (a - c) + (b - d)i \qquad (3)$$

$$k(a + bi) = (ka) + (kb)i, \quad k \text{ real} \qquad (4)$$

Because the operations of addition, subtraction, and multiplication of a complex number by a real number parallel the corresponding operations for vectors in R^2, the familiar geometric interpretations of these operations hold for complex numbers (see Figure 10.1.4).

It follows from (4) that $(-1)z + z = 0$ (verify), so we denote $(-1)z$ as $-z$ and call it the **negative of** z.

EXAMPLE 3   Adding, Subtracting, and Multiplying by Real Numbers

---

If $z_1 = 4 - 5i$ and $z_2 = -1 + 6i$, find $z_1 + z_2$, $z_1 - z_2$, $3z_1$, and $-z_2$.

Solution

$$z_1 + z_2 = (4 - 5i) + (-1 + 6i) = (4 - 1) + (-5 + 6)i = 3 + i$$
$$z_1 - z_2 = (4 - 5i) - (-1 + 6i) = (4 + 1) + (-5 - 6)i = 5 - 11i$$
$$3z_1 = 3(4 - 5i) = 12 - 15i$$
$$-z_2 = (-1)z_2 = (-1)(-1 + 6i) = 1 - 6i \quad \blacklozenge$$

So far, there has been a parallel between complex numbers and vectors in R^2. However, we now define multiplication of complex numbers, an operation with no vector analog in R^2. To motivate the definition, we expand the product

$$(a + bi)(c + di)$$

following the usual rules of algebra but treating i^2 as -1. This yields

$$(a + bi)(c + di) = ac + bdi^2 + adi + bci$$
$$= (ac - bd) + (ad + bc)i$$

which suggests the following definition:

$$(a + bi)(c + di) = (ac - bd) + (ad + bc)i \tag{5}$$

EXAMPLE 4 Multiplying Complex Numbers

$$(3 + 2i)(4 + 5i) = (3 \cdot 4 - 2 \cdot 5) + (3 \cdot 5 + 2 \cdot 4)i$$
$$= 2 + 23i$$
$$(4 - i)(2 - 3i) = [4 \cdot 2 - (-1)(-3)] + [(4)(-3) + (-1)(2)]i$$
$$= 5 - 14i$$
$$i^2 = (0 + i)(0 + i) = (0 \cdot 0 - 1 \cdot 1) + (0 \cdot 1 + 1 \cdot 0)i = -1 \quad \blacklozenge$$

We leave it as an exercise to verify the following rules of complex arithmetic:

$$z_1 + z_2 = z_2 + z_1$$
$$z_1 z_2 = z_2 z_1$$
$$z_1 + (z_2 + z_3) = (z_1 + z_2) + z_3$$
$$z_1(z_2 z_3) = (z_1 z_2)z_3$$
$$z_1(z_2 + z_3) = z_1 z_2 + z_1 z_3$$
$$0 + z = z$$
$$z + (-z) = 0$$
$$1 \cdot z = z$$

These rules make it possible to multiply complex numbers without using Formula (5) directly. Following the procedure used to motivate this formula, we can simply multiply each term of $a + bi$ by each term of $c + di$, set $i^2 = -1$, and simplify.

EXAMPLE 5 Multiplication of Complex Numbers

$$(3 + 2i)(4 + i) = 12 + 3i + 8i + 2i^2 = 12 + 11i - 2 = 10 + 11i$$
$$\left(5 - \tfrac{1}{2}i\right)(2 + 3i) = 10 + 15i - i - \tfrac{3}{2}i^2 = 10 + 14i + \tfrac{3}{2} = \tfrac{23}{2} + 14i$$
$$i(1 + i)(1 - 2i) = i(1 - 2i + i - 2i^2) = i(3 - i) = 3i - i^2 = 1 + 3i \quad \blacklozenge$$

REMARK Unlike the real numbers, there is no size ordering for the complex numbers. Thus, the order symbols $<$, $\leq$, $>$, and $\geq$ are not used with complex numbers.

Now that we have defined addition, subtraction, and multiplication of complex numbers, it is possible to add, subtract, and multiply matrices with complex entries and to multiply a matrix by a complex number. Without going into detail, we note that the matrix operations and terminology discussed in Chapter 1 carry over without change to matrices with complex entries.

EXAMPLE 6 Matrices with Complex Entries

If

$$A = \begin{bmatrix} 1 & -i \\ 1+i & 4-i \end{bmatrix} \quad \text{and} \quad B = \begin{bmatrix} i & 1-i \\ 2-3i & 4 \end{bmatrix}$$

then

$$A+B = \begin{bmatrix} 1+i & 1-2i \\ 3-2i & 8-i \end{bmatrix}, \quad A-B = \begin{bmatrix} 1-i & -1 \\ -1+4i & -i \end{bmatrix}$$

$$iA = \begin{bmatrix} i & -i^2 \\ i+i^2 & 4i-i^2 \end{bmatrix} = \begin{bmatrix} i & 1 \\ -1+i & 1+4i \end{bmatrix}$$

$$AB = \begin{bmatrix} 1 & -i \\ 1+i & 4-i \end{bmatrix}\begin{bmatrix} i & 1-i \\ 2-3i & 4 \end{bmatrix}$$

$$= \begin{bmatrix} 1\cdot i+(-i)\cdot(2-3i) & 1\cdot(1-i)+(-i)\cdot 4 \\ (1+i)\cdot i+(4-i)\cdot(2-3i) & (1+i)\cdot(1-i)+(4-i)\cdot 4 \end{bmatrix}$$

$$= \begin{bmatrix} -3-i & 1-5i \\ 4-13i & 18-4i \end{bmatrix} \blacklozenge$$

EXERCISE SET 10.1

1. In each part, plot the point and sketch the vector that corresponds to the given complex number.
   (a) $2+3i$ (b) -4 (c) $-3-2i$ (d) $-5i$

2. Express each complex number in Exercise 1 as an ordered pair of real numbers.

3. In each part, use the given information to find the real numbers x and y.
   (a) $x-iy = -2+3i$ (b) $(x+y)+(x-y)i = 3+i$

4. Given that $z_1 = 1-2i$ and $z_2 = 4+5i$, find
   (a) z_1+z_2 (b) z_1-z_2 (c) $4z_1$
   (d) $-z_2$ (e) $3z_1+4z_2$ (f) $\frac{1}{2}z_1-\frac{3}{2}z_2$

5. In each part, solve for z.
   (a) $z+(1-i) = 3+2i$ (b) $-5z = 5+10i$
   (c) $(i-z)+(2z-3i) = -2+7i$

6. In each part, sketch the vectors z_1, z_2, z_1+z_2, and z_1-z_2.
   (a) $z_1 = 3+i, z_2 = 1+4i$ (b) $z_1 = -2+2i, z_2 = 4+5i$

7. In each part, sketch the vectors z and kz.
   (a) $z = 1+i, k = 2$ (b) $z = -3-4i, k = -2$ (c) $z = 4+6i, k = \frac{1}{2}$

8. In each part, find real numbers k_1 and k_2 that satisfy the equation.
   (a) $k_1 i+k_2(1+i) = 3-2i$ (b) $k_1(2+3i)+k_2(1-4i) = 7+5i$

9. In each part, find $z_1 z_2, z_1^2$, and z_2^2.
   (a) $z_1 = 3i, z_2 = 1-i$ (b) $z_1 = 4+6i, z_2 = 2-3i$
   (c) $z_1 = \frac{1}{3}(2+4i), z_2 = \frac{1}{2}(1-5i)$

10. Given that $z_1 = 2-5i$ and $z_2 = -1-i$, find
   (a) $z_1-z_1 z_2$ (b) $(z_1+3z_2)^2$ (c) $[z_1+(1+z_2)]^2$ (d) $iz_2-z_1^2$

In Exercises 11–18 perform the calculations and express the result in the form $a+bi$.

11. $(1+2i)(4-6i)^2$ **12.** $(2-i)(3+i)(4-2i)$

13. $(1 - 3i)^3$

14. $i(1 + 7i) - 3i(4 + 2i)$

15. $\left[(2 + i) \left(\frac{1}{2} + \frac{3}{4}i \right) \right]^2$

16. $(\sqrt{2} + i) - i\sqrt{2}(1 + \sqrt{2}i)$

17. $(1 + i + i^2 + i^3)^{100}$

18. $(3 - 2i)^2 - (3 + 2i)^2$

19. Let

$$A = \begin{bmatrix} 1 & i \\ -i & 3 \end{bmatrix}, \qquad B = \begin{bmatrix} 2 & 2 + i \\ 3 - i & 4 \end{bmatrix}$$

Find

(a) $A + 3iB$
(b) BA
(c) AB
(d) $B^2 - A^2$

20. Let

$$A = \begin{bmatrix} 3 + 2i & 0 \\ -i & 2 \\ 1 + i & 1 - i \end{bmatrix}, \qquad B = \begin{bmatrix} -i & 2 \\ 0 & i \end{bmatrix}, \qquad C = \begin{bmatrix} -1 - i & 0 & -i \\ 3 & 2i & -5 \end{bmatrix}$$

Find

(a) $A(BC)$
(b) $(BC)A$
(c) $(CA)B^2$
(d) $(1 + i)(AB) + (3 - 4i)A$

21. Show that

(a) $\text{Im}(iz) = \text{Re}(z)$
(b) $\text{Re}(iz) = -\text{Im}(z)$

22. In each part, solve the equation by the quadratic formula and check your results by substituting the solutions into the given equation.

(a) $z^2 + 2z + 2 = 0$
(b) $z^2 - z + 1 = 0$

23. (a) Show that if n is a positive integer, then the only possible values for i^n are $1, -1, i$, and $-i$.

(b) Find i^{2509}.

24. Prove: If $z_1 z_2 = 0$, then $z_1 = 0$ or $z_2 = 0$.

25. Use the result of Exercise 24 to prove: If $zz_1 = zz_2$ and $z \neq 0$, then $z_1 = z_2$.

26. Prove that for all complex numbers z_1, z_2, and z_3,

(a) $z_1 + z_2 = z_2 + z_1$
(b) $z_1 + (z_2 + z_3) = (z_1 + z_2) + z_3$

27. Prove that for all complex numbers z_1, z_2, and z_3,

(a) $z_1 z_2 = z_2 z_1$
(b) $z_1 (z_2 z_3) = (z_1 z_2) z_3$

28. Prove that $z_1 (z_2 + z_3) = z_1 z_2 + z_1 z_3$ for all complex numbers z_1, z_2, and z_3.

29. In quantum mechanics the **_Dirac matrices_** are

$$\beta = \begin{bmatrix} 1 & 0 & 0 & 0 \\ 0 & 1 & 0 & 0 \\ 0 & 0 & -1 & 0 \\ 0 & 0 & 0 & -1 \end{bmatrix}, \qquad \alpha_x = \begin{bmatrix} 0 & 0 & 0 & 1 \\ 0 & 0 & 1 & 0 \\ 0 & 1 & 0 & 0 \\ 1 & 0 & 0 & 0 \end{bmatrix},$$

$$\alpha_y = \begin{bmatrix} 0 & 0 & 0 & -i \\ 0 & 0 & i & 0 \\ 0 & -i & 0 & 0 \\ i & 0 & 0 & 0 \end{bmatrix}, \qquad \alpha_z = \begin{bmatrix} 0 & 0 & 1 & 0 \\ 0 & 0 & 0 & -1 \\ 1 & 0 & 0 & 0 \\ 0 & -1 & 0 & 0 \end{bmatrix}$$

(a) Prove that $\beta^2 = \alpha_x^2 = \alpha_y^2 = \alpha_z^2 = I$.

(b) Two matrices A and B are called **_anticommutative_** if $AB = -BA$. Prove that any two distinct Dirac matrices are anticommutative.

30. Describe the set of all complex numbers $z = a + bi$ such that $a^2 + b^2 = 1$. Show that if z_1, z_2 are such numbers, then so is $z_1 z_2$.

10.2
DIVISION OF COMPLEX NUMBERS

In the last section we defined multiplication of complex numbers. In this section we shall define division of complex numbers as the inverse of multiplication.

We begin with some preliminary ideas.

Complex Conjugates

If $z = a + bi$ is any complex number, then the **complex conjugate** of z (also called the **conjugate** of z) is denoted by the symbol $\bar{z}$ (read "z bar" or "z conjugate") and is defined by

$$\bar{z} = a - bi$$

In words, $\bar{z}$ is obtained by reversing the sign of the imaginary part of z. Geometrically, $\bar{z}$ is the reflection of z about the real axis (Figure 10.2.1).

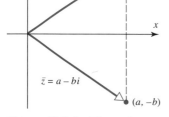

Figure 10.2.1 The conjugate of a complex number.

EXAMPLE 1 Examples of Conjugates

$$
\begin{aligned}
z &= 3 + 2i & \bar{z} &= 3 - 2i \\
z &= -4 - 2i & \bar{z} &= -4 + 2i \\
z &= i & \bar{z} &= -i \\
z &= 4 & \bar{z} &= 4 \quad \blacklozenge
\end{aligned}
$$

REMARK The last line in Example 1 illustrates the fact that a real number is the same as its conjugate. More precisely, it can be shown (Exercise 22) that $z = \bar{z}$ if and only if z is a real number.

If a complex number z is viewed as a vector in R^2, then the norm or length of the vector is called the modulus of z. More precisely:

Paul Adrien Maurice Dirac *(1902–1984) was a British theoretical physicist who devised a new form of quantum mechanics and a theory that predicted electron spin and the existence of a fundamental atomic particle called a positron. He received the Nobel Prize for physics in 1933 and the medal of the Royal Society in 1939.*

> **DEFINITION**
>
> The **modulus** of a complex number $z = a + bi$, denoted by $|z|$, is defined by
>
> $$|z| = \sqrt{a^2 + b^2} \qquad (1)$$

If $b = 0$, then $z = a$ is a real number, and

$$|z| = \sqrt{a^2 + 0^2} = \sqrt{a^2} = |a|$$

so the modulus of a real number is simply its absolute value. Thus the modulus of z is also called the **absolute value** of z.

EXAMPLE 2 Modulus of a Complex Number

Find $|z|$ if $z = 3 - 4i$.

Solution

From (1), with $a = 3$ and $b = -4$, $|z| = \sqrt{(3)^2 + (-4)^2} = \sqrt{25} = 5$. $\blacklozenge$

The following theorem establishes a basic relationship between $\bar{z}$ and $|z|$.

THEOREM 10.2.1

> For any complex number z,
> $$z\bar{z} = |z|^2$$

Proof If $z = a + bi$, then

$$z\bar{z} = (a + bi)(a - bi) = a^2 - abi + bai - b^2i^2 = a^2 + b^2 = |z|^2 \qquad \blacksquare$$

Division of Complex Numbers

We now turn to the division of complex numbers. Our objective is to define division as the inverse of multiplication. Thus, if $z_2 \neq 0$, then our definition of $z = z_1/z_2$ should be such that

$$z_1 = z_2 z \qquad (2)$$

Our procedure will be to prove that (2) has a unique solution for z if $z_2 \neq 0$, and then to define z_1/z_2 to be this value of z. As with real numbers, division by zero is not allowed.

THEOREM 10.2.2

> If $z_2 \neq 0$, then Equation (2) has a unique solution, which is
> $$z = \frac{1}{|z_2|^2} z_1 \bar{z}_2 \qquad (3)$$

Proof Let $z = x + iy$, $z_1 = x_1 + iy_1$, and $z_2 = x_2 + iy_2$. Then (2) can be written as

$$x_1 + iy_1 = (x_2 + iy_2)(x + iy)$$

or

$$x_1 + iy_1 = (x_2 x - y_2 y) + i(y_2 x + x_2 y)$$

or, on equating real and imaginary parts,

$$x_2 x - y_2 y = x_1$$
$$y_2 x + x_2 y = y_1$$

or

$$\begin{bmatrix} x_2 & -y_2 \\ y_2 & x_2 \end{bmatrix} \begin{bmatrix} x \\ y \end{bmatrix} = \begin{bmatrix} x_1 \\ y_1 \end{bmatrix} \qquad (4)$$

Since $z_2 = x_2 + iy_2 \neq 0$, it follows that x_2 and y_2 are not both zero, so

$$\begin{vmatrix} x_2 & -y_2 \\ y_2 & x_2 \end{vmatrix} = x_2^2 + y_2^2 \neq 0$$

Thus, by Cramer's rule (Theorem 2.1.4), system (4) has the unique solution

$$x = \frac{\begin{vmatrix} x_1 & -y_2 \\ y_1 & x_2 \end{vmatrix}}{\begin{vmatrix} x_2 & -y_2 \\ y_2 & x_2 \end{vmatrix}} = \frac{x_1 x_2 + y_1 y_2}{x_2^2 + y_2^2} = \frac{x_1 x_2 + y_1 y_2}{|z_2|^2}$$

$$y = \frac{\begin{vmatrix} x_2 & x_1 \\ y_2 & y_1 \end{vmatrix}}{\begin{vmatrix} x_2 & -y_2 \\ y_2 & x_2 \end{vmatrix}} = \frac{y_1 x_2 - x_1 y_2}{x_2^2 + y_2^2} = \frac{y_1 x_2 - x_1 y_2}{|z_2|^2}$$

Therefore,

$$z = x + iy = \frac{1}{|z_2|^2} \left[(x_1 x_2 + y_1 y_2) + i(y_1 x_2 - x_1 y_2) \right]$$

$$= \frac{1}{|z_2|^2} (x_1 + iy_1)(x_2 - iy_2) = \frac{1}{|z_2|^2} z_1 \bar{z}_2 \qquad \blacksquare$$

Thus, for $z_2 \neq 0$, we define

$$\frac{z_1}{z_2} = \frac{1}{|z_2|^2} z_1 \bar{z}_2 \qquad (5)$$

REMARK   To remember this formula, multiply the numerator and denominator of z_1/z_2 by $\bar{z}_2$:

$$\frac{z_1}{z_2} = \frac{z_1 \bar{z}_2}{z_2 \bar{z}_2} = \frac{z_1 \bar{z}_2}{|z_2|^2} = \frac{1}{|z_2|^2} z_1 \bar{z}_2$$

EXAMPLE 3   Quotient in the Form $a + bi$

Express

$$\frac{3 + 4i}{1 - 2i}$$

in the form $a + bi$.

Solution

From (5) with $z_1 = 3 + 4i$ and $z_2 = 1 - 2i$,

$$\frac{3 + 4i}{1 - 2i} = \frac{1}{|1 - 2i|^2}(3 + 4i)(\overline{1 - 2i}) = \frac{1}{5}(3 + 4i)(1 + 2i)$$

$$= \frac{1}{5}(-5 + 10i) = -1 + 2i$$

Alternative Solution

As in the remark above, multiply numerator and denominator by the conjugate of the denominator:

$$\frac{3 + 4i}{1 - 2i} = \frac{3 + 4i}{1 - 2i} \cdot \frac{1 + 2i}{1 + 2i} = \frac{-5 + 10i}{5} = -1 + 2i \quad \blacklozenge$$

Systems of linear equations with complex coefficients arise in various applications. Without going into detail, we note that all the results about linear systems studied in Chapters 1 and 2 carry over without change to systems with complex coefficients. Note, however, that a few results studied in other chapters *will* change for complex matrices.

EXAMPLE 4   A Linear System with Complex Coefficients

Use Cramer's rule to solve

$$ix + 2y = 1 - 2i$$
$$4x - iy = -1 + 3i$$

Solution

$$x = \frac{\begin{vmatrix} 1-2i & 2 \\ -1+3i & -i \end{vmatrix}}{\begin{vmatrix} i & 2 \\ 4 & -i \end{vmatrix}} = \frac{(-i)(1-2i) - 2(-1+3i)}{i(-i) - 2(4)} = \frac{-7i}{-7} = i$$

$$y = \frac{\begin{vmatrix} i & 1-2i \\ 4 & -1+3i \end{vmatrix}}{\begin{vmatrix} i & 2 \\ 4 & -i \end{vmatrix}} = \frac{(i)(-1+3i) - 4(1-2i)}{i(-i) - 2(4)} = \frac{-7+7i}{-7} = 1 - i$$

Thus the solution is $x = i$, $y = 1 - i$. ◆

We conclude this section by listing some properties of the complex conjugate that will be useful in later sections.

THEOREM 10.2.3

> ## Properties of the Conjugate
>
> *For any complex numbers z, z_1, and z_2:*
>
> (a) $\overline{z_1 + z_2} = \overline{z}_1 + \overline{z}_2$ $\qquad$ (b) $\overline{z_1 - z_2} = \overline{z}_1 - \overline{z}_2$
>
> (c) $\overline{z_1 z_2} = \overline{z}_1 \overline{z}_2$ $\qquad$ (d) $\overline{(z_1/z_2)} = \overline{z}_1 / \overline{z}_2$
>
> (e) $\overline{\overline{z}} = z$

We prove (a) and leave the rest as exercises.

Proof (a) Let $z_1 = a_1 + b_1 i$ and $z_2 = a_2 + b_2 i$; then

$$\begin{aligned} \overline{z_1 + z_2} &= \overline{(a_1 + a_2) + (b_1 + b_2)i} \\ &= (a_1 + a_2) - (b_1 + b_2)i \\ &= (a_1 - b_1 i) + (a_2 - b_2 i) \\ &= \overline{z}_1 + \overline{z}_2 \end{aligned}$$ ∎

REMARK It is possible to extend part (a) of Theorem 10.2.3 to n terms and part (c) to n factors. More precisely,

$$\overline{z_1 + z_2 + \cdots + z_n} = \overline{z}_1 + \overline{z}_2 + \cdots + \overline{z}_n$$
$$\overline{z_1 z_2 \cdots z_n} = \overline{z}_1 \overline{z}_2 \cdots \overline{z}_n$$

EXERCISE SET
10.2

1. In each part, find $\overline{z}$.

   (a) $z = 2 + 7i$ $\qquad$ (b) $z = -3 - 5i$ $\qquad$ (c) $z = 5i$

   (d) $z = -i$ $\qquad$ (e) $z = -9$ $\qquad$ (f) $z = 0$

2. In each part, find $|z|$.

   (a) $z = i$ $\qquad$ (b) $z = -7i$ $\qquad$ (c) $z = -3 - 4i$

   (d) $z = 1 + i$ $\qquad$ (e) $z = -8$ $\qquad$ (f) $z = 0$

3. Verify that $z\overline{z} = |z|^2$ for

   (a) $z = 2 - 4i$ $\qquad$ (b) $z = -3 + 5i$ $\qquad$ (c) $z = \sqrt{2} - \sqrt{2}i$

4. Given that $z_1 = 1 - 5i$ and $z_2 = 3 + 4i$, find

   (a) z_1/z_2 (b) $\bar{z}_1/z_2$ (c) $z_1/\bar{z}_2$

   (d) $\overline{(z_1/z_2)}$ (e) $z_1/|z_2|$ (f) $|z_1/z_2|$

5. In each part, find $1/z$.

   (a) $z = i$ (b) $z = 1 - 5i$ (c) $z = \dfrac{-i}{7}$

6. Given that $z_1 = 1 + i$ and $z_2 = 1 - 2i$, find

   (a) $z_1 - \left(\dfrac{z_1}{z_2}\right)$ (b) $\dfrac{z_1 - 1}{z_2}$ (c) $z_1^2 - \left(\dfrac{iz_1}{z_2}\right)$ (d) $\dfrac{z_1}{iz_2}$

In Exercises 7–14 perform the calculations and express the result in the form $a + bi$.

7. $\dfrac{i}{1+i}$ 8. $\dfrac{2}{(1-i)(3+i)}$ 9. $\dfrac{1}{(3+4i)^2}$

10. $\dfrac{2+i}{i(-3+4i)}$ 11. $\dfrac{\sqrt{3}+i}{(1-i)(\sqrt{3}-i)}$ 12. $\dfrac{1}{i(3-2i)(1+i)}$

13. $\dfrac{i}{(1-i)(1-2i)(1+2i)}$ 14. $\dfrac{1-2i}{3+4i} - \dfrac{2+i}{5i}$

15. In each part, solve for z.

   (a) $iz = 2 - i$ (b) $(4 - 3i)\bar{z} = i$

16. Use Theorem 10.2.3 to prove the following identities:

   (a) $\overline{\bar{z} + 5i} = z - 5i$ (b) $\overline{i\bar{z}} = -i\bar{z}$ (c) $\dfrac{\overline{i + \bar{z}}}{i - z} = -1$

17. In each part, sketch the set of points in the complex plane that satisfies the equation.

   (a) $|z| = 2$ (b) $|z - (1+i)| = 1$ (c) $|z - i| = |z + i|$ (d) $\text{Im}(\bar{z} + i) = 3$

18. In each part, sketch the set of points in the complex plane that satisfies the given condition(s).

   (a) $|z + i| \leq 1$ (b) $1 < |z| < 2$ (c) $|2z - 4i| < 1$ (d) $|z| \leq |z + i|$

19. Given that $z = x + iy$, find

   (a) $\text{Re}(\overline{iz})$ (b) $\text{Im}(\overline{iz})$ (c) $\text{Re}(i\bar{z})$ (d) $\text{Im}(i\bar{z})$

20. (a) Show that if n is a positive integer, then the only possible values for $(1/i)^n$ are $1, -1, i,$ and $-i$.

   (b) Find $(1/i)^{2509}$.

   Hint See Exercise 23(b) of Section 10.1.

21. Prove:

   (a) $\dfrac{1}{2}(z + \bar{z}) = \text{Re}(z)$ (b) $\dfrac{1}{2i}(z - \bar{z}) = \text{Im}(z)$

22. Prove: $z = \bar{z}$ if and only if z is a real number.

23. Given that $z_1 = x_1 + iy_1$ and $z_2 = x_2 + iy_2 \neq 0$, find

   (a) $\text{Re}\left(\dfrac{z_1}{z_2}\right)$ (b) $\text{Im}\left(\dfrac{z_1}{z_2}\right)$

24. Prove: If $(\bar{z})^2 = z^2$, then z is either real or pure imaginary.

25. Prove that $|z| = |\bar{z}|$.

26. Prove:

   (a) $\overline{z_1 - z_2} = \bar{z}_1 - \bar{z}_2$ (b) $\overline{z_1 z_2} = \bar{z}_1 \bar{z}_2$ (c) $\overline{(z_1/z_2)} = \bar{z}_1/\bar{z}_2$ (d) $\bar{\bar{z}} = z$

27. (a) Prove that $\overline{z^2} = (\bar{z})^2$.

   (b) Prove that if n is a positive integer, then $\overline{z^n} = (\bar{z})^n$.

   (c) Is the result in part (b) true if n is a negative integer? Explain.

In Exercises 28–31 solve the system of linear equations by Cramer's rule.

28. $ix_1 - ix_2 = -2$
$2x_1 + x_2 = i$

29. $x_1 + x_2 = 2$
$x_1 - x_2 = 2i$

30. $x_1 + x_2 + x_3 = 3$
$x_1 + x_2 - x_3 = 2 + 2i$
$x_1 - x_2 + x_3 = -1$

31. $ix_1 + 3x_2 + (1 + i)x_3 = -i$
$x_1 + ix_2 + 3x_3 = -2i$
$x_1 + x_2 + x_3 = 0$

In Exercises 32 and 33 solve the system of linear equations by Gauss–Jordan elimination.

32. $\begin{bmatrix} -1 & -1-i \\ -1+i & -2 \end{bmatrix} \begin{bmatrix} x_1 \\ x_2 \end{bmatrix} = \begin{bmatrix} 0 \\ 0 \end{bmatrix}$

33. $\begin{bmatrix} 2 & -1-i \\ -1+i & 1 \end{bmatrix} \begin{bmatrix} x_1 \\ x_2 \end{bmatrix} = \begin{bmatrix} 0 \\ 0 \end{bmatrix}$

34. Solve the following system of linear equations by Gauss–Jordan elimination.

$$x_1 + ix_2 - ix_3 = 0$$
$$-x_1 + (1-i)x_2 + 2ix_3 = 0$$
$$2x_1 + (-1+2i)x_2 - 3ix_3 = 0$$

35. In each part, use the formula in Theorem 1.4.5 to compute the inverse of the matrix, and check your result by showing that $AA^{-1} = A^{-1}A = I$.

(a) $A = \begin{bmatrix} i & -2 \\ 1 & i \end{bmatrix}$
(b) $A = \begin{bmatrix} 2 & i \\ 1 & 0 \end{bmatrix}$

36. Let $p(x) = a_0 + a_1x + a_2x^2 + \cdots + a_nx^n$ be a polynomial for which the coefficients $a_0, a_1, a_2, \ldots, a_n$ are real. Prove that if z is a solution of the equation $p(z) = 0$, then so is $\bar{z}$.

37. Prove: For any complex number z, $|\text{Re}(z)| \leq |z|$ and $|\text{Im}(z)| \leq |z|$.

38. Prove that

$$\frac{|\text{Re}(z)| + |\text{Im}(z)|}{\sqrt{2}} \leq |z|$$

Hint Let $z = x + iy$ and use the fact that $(|x| - |y|)^2 \geq 0$.

39. In each part, use the method of Example 4 in Section 1.5 to find A^{-1}, and check your result by showing that $AA^{-1} = A^{-1}A = I$.

(a) $A = \begin{bmatrix} 1 & 1+i & 0 \\ 0 & 1 & i \\ -i & 1-2i & 2 \end{bmatrix}$
(b) $A = \begin{bmatrix} i & 0 & -i \\ 0 & 1 & -1-4i \\ 2-i & i & 3 \end{bmatrix}$

40. Show that $|z - 1| = |\bar{z} - 1|$. Discuss the geometric interpretation of the result.

41. (a) If $z_1 = a_1 + b_1i$ and $z_2 = a_2 + b_2i$, find $|z_1 - z_2|$ and interpret the result geometrically.

(b) Use part (a) to show that the complex numbers $12, 6 + 2i$, and $8 + 8i$ are vertices of a right triangle.

42. Use Theorem 10.2.3 to show that if the coefficients a, b, and c in a quadratic polynomial are real, then the solutions of the equation $az^2 + bz + c = 0$ are complex conjugates. What can you conclude if a, b, and c are complex?

10.3
POLAR FORM OF A COMPLEX NUMBER

In this section we shall discuss a way to represent complex numbers using trigonometric properties. Our work will lead to an important formula for powers of complex numbers and to a method for finding nth roots of complex numbers.

Polar Form

If $z = x + iy$ is a nonzero complex number, $r = |z|$, and θ measures the angle from the positive real axis to the vector z, then, as suggested by Figure 10.3.1,

$$x = r\cos\theta, \qquad y = r\sin\theta \qquad (1)$$

so that $z = x + iy$ can be written as $z = r\cos\theta + ir\sin\theta$ or

$$z = r(\cos\theta + i\sin\theta) \tag{2}$$

This is called a ***polar form of*** z.

Argument of a Complex Number

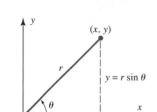

Figure 10.3.1

The angle θ is called an ***argument of*** z and is denoted by

$$\theta = \arg z$$

The argument of z is not uniquely determined because we can add or subtract any multiple of 2π from θ to produce another value of the argument. However, there is only one value of the argument in radians that satisfies

$$-\pi < \theta \leq \pi$$

This is called the ***principal argument of*** z and is denoted by

$$\theta = \text{Arg } z$$

EXAMPLE 1 Polar Forms

Express the following complex numbers in polar form using their principal arguments:

$$\text{(a) } z = 1 + \sqrt{3}i \qquad \text{(b) } z = -1 - i$$

Solution (a)

The value of r is

$$r = |z| = \sqrt{1^2 + (\sqrt{3})^2} = \sqrt{4} = 2$$

and since $x = 1$ and $y = \sqrt{3}$, it follows from (1) that

$$1 = 2\cos\theta \quad \text{and} \quad \sqrt{3} = 2\sin\theta$$

so $\cos\theta = 1/2$ and $\sin\theta = \sqrt{3}/2$. The only value of θ that satisfies these relations and meets the requirement $-\pi < \theta \leq \pi$ is $\theta = \pi/3 \,(= 60°)$ (see Figure 10.3.2*a*). Thus a polar form of z is

$$z = 2\left(\cos\frac{\pi}{3} + i\sin\frac{\pi}{3}\right)$$

Solution (b)

The value of r is

$$r = |z| = \sqrt{(-1)^2 + (-1)^2} = \sqrt{2}$$

and since $x = -1$, $y = -1$, it follows from (1) that

$$-1 = \sqrt{2}\cos\theta \quad \text{and} \quad -1 = \sqrt{2}\sin\theta$$

so $\cos\theta = -1/\sqrt{2}$ and $\sin\theta = -1/\sqrt{2}$. The only value of θ that satisfies these relations and meets the requirement $-\pi < \theta \leq \pi$ is $\theta = -3\pi/4 \,(= -135°)$ (Figure 10.3.2*b*). Thus, a polar form of z is

$$z = \sqrt{2}\left(\cos\frac{-3\pi}{4} + i\sin\frac{-3\pi}{4}\right) \quad \blacklozenge$$

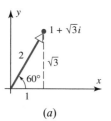

(a)

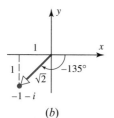

(b)

Figure 10.3.2

Multiplication and Division Interpreted Geometrically

We now show how polar forms can be used to give geometric interpretations of multiplication and division of complex numbers. Let

$$z_1 = r_1(\cos\theta_1 + i\sin\theta_1) \quad \text{and} \quad z_2 = r_2(\cos\theta_2 + i\sin\theta_2)$$

Multiplying, we obtain

$$z_1 z_2 = r_1 r_2[(\cos\theta_1\cos\theta_2 - \sin\theta_1\sin\theta_2) + i(\sin\theta_1\cos\theta_2 + \cos\theta_1\sin\theta_2)]$$

Recalling the trigonometric identities

$$\cos(\theta_1 + \theta_2) = \cos\theta_1\cos\theta_2 - \sin\theta_1\sin\theta_2$$
$$\sin(\theta_1 + \theta_2) = \sin\theta_1\cos\theta_2 + \cos\theta_1\sin\theta_2$$

we obtain

$$z_1 z_2 = r_1 r_2[\cos(\theta_1 + \theta_2) + i\sin(\theta_1 + \theta_2)] \tag{3}$$

which is a polar form of the complex number with modulus $r_1 r_2$ and argument $\theta_1 + \theta_2$. Thus we have shown that

$$|z_1 z_2| = |z_1||z_2| \tag{4}$$

and

$$\arg(z_1 z_2) = \arg z_1 + \arg z_2$$

(Why?) In words, *the product of two complex numbers is obtained by multiplying their moduli and adding their arguments* (Figure 10.3.3).

We leave it as an exercise to show that if $z_2 \neq 0$, then

$$\frac{z_1}{z_2} = \frac{r_1}{r_2}[\cos(\theta_1 - \theta_2) + i\sin(\theta_1 - \theta_2)] \tag{5}$$

from which it follows that

$$\left|\frac{z_1}{z_2}\right| = \frac{|z_1|}{|z_2|} \quad \text{if } z_2 \neq 0$$

and

$$\arg\left(\frac{z_1}{z_2}\right) = \arg z_1 - \arg z_2$$

In words, *the quotient of two complex numbers is obtained by dividing their moduli and subtracting their arguments* (*in the appropriate order*).

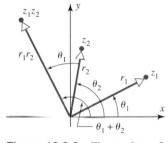

Figure 10.3.3 The product of two complex numbers.

EXAMPLE 2 A Quotient Using Polar Forms

Let

$$z_1 = 1 + \sqrt{3}i \quad \text{and} \quad z_2 = \sqrt{3} + i$$

Polar forms of these complex numbers are

$$z_1 = 2\left(\cos\frac{\pi}{3} + i\sin\frac{\pi}{3}\right) \quad \text{and} \quad z_2 = 2\left(\cos\frac{\pi}{6} + i\sin\frac{\pi}{6}\right)$$

(verify) so that from (3),

$$z_1 z_2 = 4\left[\cos\left(\frac{\pi}{3} + \frac{\pi}{6}\right) + i\sin\left(\frac{\pi}{3} + \frac{\pi}{6}\right)\right]$$

$$= 4\left[\cos\frac{\pi}{2} + i\sin\frac{\pi}{2}\right] = 4[0 + i] = 4i$$

and from (5),

$$\frac{z_1}{z_2} = 1 \cdot \left[\cos\left(\frac{\pi}{3} - \frac{\pi}{6}\right) + i \sin\left(\frac{\pi}{3} - \frac{\pi}{6}\right) \right]$$

$$= \cos\frac{\pi}{6} + i \sin\frac{\pi}{6} = \frac{\sqrt{3}}{2} + \frac{1}{2}i$$

As a check, we calculate $z_1 z_2$ and z_1/z_2 directly without using polar forms for z_1 and z_2:

$$z_1 z_2 = (1 + \sqrt{3}i)(\sqrt{3} + i) = (\sqrt{3} - \sqrt{3}) + (3 + 1)i = 4i$$

$$\frac{z_1}{z_2} = \frac{1 + \sqrt{3}i}{\sqrt{3} + i} \cdot \frac{\sqrt{3} - i}{\sqrt{3} - i} = \frac{(\sqrt{3} + \sqrt{3}) + (-i + 3i)}{4} = \frac{\sqrt{3}}{2} + \frac{1}{2}i$$

which agrees with our previous results. ◆

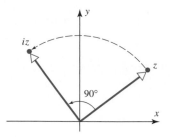

Figure 10.3.4 Multiplying by i rotates z counterclockwise by 90°.

The complex number i has a modulus of 1 and an argument of $\pi/2$ ($= 90°$), so the product iz has the same modulus as z, but its argument is 90° greater than that of z. In short, *multiplying z by i rotates z counterclockwise by 90°* (Figure 10.3.4).

DeMoivre's Formula

If n is a positive integer and $z = r(\cos\theta + i\sin\theta)$, then from Formula (3),

$$z^n = \underbrace{z \cdot z \cdot z \cdots z}_{n\text{-factors}} = r^n[\cos\underbrace{(\theta + \theta + \cdots + \theta)}_{n\text{-terms}} + i\sin\underbrace{(\theta + \theta + \cdots + \theta)}_{n\text{-terms}}]$$

or

$$z^n = r^n(\cos n\theta + i\sin n\theta) \tag{6}$$

Moreover, (6) also holds for negative integers if $z \neq 0$ (see Exercise 23).

In the special case where $r = 1$, we have $z = \cos\theta + i\sin\theta$, so (6) becomes

$$(\cos\theta + i\sin\theta)^n = \cos n\theta + i\sin n\theta \tag{7}$$

which is called **DeMoivre's formula**. Although we derived (7) assuming n to be a positive integer, it will be shown in the exercises that this formula is valid for all integers n.

Finding nth Roots

We now show how DeMoivre's formula can be used to obtain roots of complex numbers. If n is a positive integer and z is any complex number, then we define an **nth root of z** to be any complex number w that satisfies the equation

$$w^n = z \tag{8}$$

We denote an nth root of z by $z^{1/n}$. If $z \neq 0$, then we can derive formulas for the nth roots of z as follows. Let

$$w = \rho(\cos\alpha + i\sin\alpha) \quad \text{and} \quad z = r(\cos\theta + i\sin\theta)$$

If we assume that w satisfies (8), then it follows from (6) that

$$\rho^n(\cos n\alpha + i\sin n\alpha) = r(\cos\theta + i\sin\theta) \tag{9}$$

Comparing the moduli of the two sides, we see that $\rho^n = r$ or

$$\rho = \sqrt[n]{r}$$

where $\sqrt[n]{r}$ denotes the real positive nth root of r. Moreover, in order to have the equalities $\cos n\alpha = \cos\theta$ and $\sin n\alpha = \sin\theta$ in (9), the angles $n\alpha$ and θ must either be equal or differ by a multiple of 2π. That is,

$$n\alpha = \theta + 2k\pi \quad \text{or} \quad \alpha = \frac{\theta}{n} + \frac{2k\pi}{n}, \qquad k = 0, \pm1, \pm2, \ldots$$

Thus the values of $w = \rho(\cos\alpha + i\sin\alpha)$ that satisfy (8) are given by

$$w = \sqrt[n]{r}\left[\cos\left(\frac{\theta}{n} + \frac{2k\pi}{n}\right) + i\sin\left(\frac{\theta}{n} + \frac{2k\pi}{n}\right)\right], \qquad k = 0, \pm1, \pm2, \ldots$$

Although there are infinitely many values of k, it can be shown (see Exercise 16) that $k = 0, 1, 2, \ldots, n-1$ produce distinct values of w satisfying (8) but all other choices of k yield duplicates of these. Therefore, there are exactly n different nth roots of $z = r(\cos\theta + i\sin\theta)$, and these are given by

$$z^{1/n} = \sqrt[n]{r}\left[\cos\left(\frac{\theta}{n} + \frac{2k\pi}{n}\right) + i\sin\left(\frac{\theta}{n} + \frac{2k\pi}{n}\right)\right], \qquad k = 0, 1, 2, \ldots, n-1$$

(10)

Abraham DeMoivre
(1667–1754) was a French mathematician who made important contributions to probability, statistics, and trigonometry. He developed the concept of statistically independent events, wrote a major and influential treatise on probability, and helped transform trigonometry from a branch of geometry into a branch of analysis through his use of complex numbers. In spite of his important work, he barely managed to eke out a living as a tutor and a consultant on gambling and insurance.

EXAMPLE 3   Cube Roots of a Complex Number

Find all cube roots of -8.

Solution

Since -8 lies on the negative real axis, we can use $\theta = \pi$ as an argument. Moreover, $r = |z| = |-8| = 8$, so a polar form of -8 is

$$-8 = 8(\cos\pi + i\sin\pi)$$

From (10) with $n = 3$, it follows that

$$(-8)^{1/3} = \sqrt[3]{8}\left[\cos\left(\frac{\pi}{3} + \frac{2k\pi}{3}\right) + i\sin\left(\frac{\pi}{3} + \frac{2k\pi}{3}\right)\right], \qquad k = 0, 1, 2$$

Thus the cube roots of -8 are

$$2\left(\cos\frac{\pi}{3} + i\sin\frac{\pi}{3}\right) = 2\left(\frac{1}{2} + \frac{\sqrt{3}}{2}i\right) = 1 + \sqrt{3}i$$

$$2(\cos\pi + i\sin\pi) = 2(-1) = -2$$

$$2\cos\left(\frac{5\pi}{3} + i\sin\frac{5\pi}{3}\right) = 2\left(\frac{1}{2} - \frac{\sqrt{3}}{2}i\right) = 1 - \sqrt{3}i \quad \blacklozenge$$

As shown in Figure 10.3.5, the three cube roots of -8 obtained in Example 3 are equally spaced $\pi/3$ radians ($= 120°$) apart around the circle of radius 2 centered at the origin. This is not accidental. In general, it follows from Formula (10) that the nth roots of z lie on the circle of radius $\sqrt[n]{r}$ ($= \sqrt[n]{|z|}$) and are equally spaced $2\pi/n$ radians apart. (Can you see why?) Thus, once one nth root of z is found, the remaining $n-1$ roots can be generated by rotating this root successively through increments of $2\pi/n$ radians.

Figure 10.3.5   The cube roots of -8.

EXAMPLE 4   Fourth Roots of a Complex Number

Find all fourth roots of 1.

Solution

We could apply Formula (10). Instead, we observe that $w = 1$ is one fourth root of 1, so the remaining three roots can be generated by rotating this root through increments of

$2\pi/4 = \pi/2$ radians ($= 90°$). From Figure 10.3.6, we see that the fourth roots of 1 are

$$1, \quad i, \quad -1, \quad -i \quad \blacklozenge$$

Complex Exponents

We conclude this section with some comments on notation.

In more detailed studies of complex numbers, complex exponents are defined, and it is shown that

$$\cos\theta + i\sin\theta = e^{i\theta} \tag{11}$$

where e is an irrational real number given approximately by $e \approx 2.71828\ldots$ (For readers who have studied calculus, a proof of this result is given in Exercise 18.)

It follows from (11) that the polar form

$$z = r(\cos\theta + i\sin\theta)$$

can be written more briefly as

$$z = re^{i\theta} \tag{12}$$

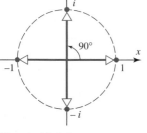

Figure 10.3.6 The fourth roots of 1.

EXAMPLE 5 Expressing a Complex Number in Form (12)

In Example 1 it was shown that

$$1 + \sqrt{3}i = 2\left(\cos\frac{\pi}{3} + i\sin\frac{\pi}{3}\right)$$

From (12) this can also be written as

$$1 + \sqrt{3}i = 2e^{i\pi/3} \quad \blacklozenge$$

It can be proved that complex exponents follow the same laws as real exponents, so if

$$z_1 = r_1 e^{i\theta_1} \quad \text{and} \quad z_2 = r_2 e^{i\theta_2}$$

are nonzero complex numbers, then

$$z_1 z_2 = r_1 r_2 e^{i\theta_1 + i\theta_2} = r_1 r_2 e^{i(\theta_1 + \theta_2)}$$

$$\frac{z_1}{z_2} = \frac{r_1}{r_2} e^{i\theta_1 - i\theta_2} = \frac{r_1}{r_2} e^{i(\theta_1 - \theta_2)}$$

But these are just Formulas (3) and (5) in a different notation.

We conclude this section with a useful formula for $\bar{z}$ in polar notation. If

$$z = re^{i\theta} = r(\cos\theta + i\sin\theta)$$

then

$$\bar{z} = r(\cos\theta - i\sin\theta) \tag{13}$$

Recalling the trigonometric identities

$$\sin(-\theta) = -\sin\theta \quad \text{and} \quad \cos(-\theta) = \cos\theta$$

we can rewrite (13) as

$$\bar{z} = r[\cos(-\theta) + i\sin(-\theta)] = re^{i(-\theta)}$$

or, equivalently,

$$\bar{z} = re^{-i\theta} \tag{14}$$

In the special case where $r = 1$, the polar form of z is $z = e^{i\theta}$, and (14) yields the formula

$$\overline{e^{i\theta}} = e^{-i\theta} \tag{15}$$

EXERCISE SET
10.3

1. In each part, find the principal argument of z.

    (a) $z = 1$ (b) $z = i$ (c) $z = -i$

    (d) $z = 1 + i$ (e) $z = -1 + \sqrt{3}i$ (f) $z = 1 - i$

2. In each part, find the value of $\theta = \arg(1 - \sqrt{3}i)$ that satisfies the given condition.

    (a) $0 < \theta \le 2\pi$ (b) $-\pi < \theta \le \pi$ (c) $-\dfrac{\pi}{6} \le \theta < \dfrac{11\pi}{6}$

3. In each part, express the complex number in polar form using its principal argument.

    (a) $2i$ (b) -4 (c) $5 + 5i$

    (d) $-6 + 6\sqrt{3}i$ (e) $-3 - 3i$ (f) $2\sqrt{3} - 2i$

4. Given that $z_1 = 2(\cos \pi/4 + i \sin \pi/4)$ and $z_2 = 3(\cos \pi/6 + i \sin \pi/6)$, find a polar form of

    (a) $z_1 z_2$ (b) $\dfrac{z_1}{z_2}$ (c) $\dfrac{z_2}{z_1}$ (d) $\dfrac{z_1^5}{z_2^2}$

5. Express $z_1 = i$, $z_2 = 1 - \sqrt{3}i$, and $z_3 = \sqrt{3} + i$ in polar form, and use your results to find $z_1 z_2 / z_3$. Check your results by performing the calculations without using polar forms.

6. Use Formula (6) to find

    (a) $(1 + i)^{12}$ (b) $\left(\dfrac{1}{\sqrt{2}} - \dfrac{1}{\sqrt{2}}i\right)^{-6}$ (c) $(\sqrt{3} + i)^7$ (d) $(1 - i\sqrt{3})^{-10}$

7. In each part, find all the roots and sketch them as vectors in the complex plane.

    (a) $(-i)^{1/2}$ (b) $(1 + \sqrt{3}i)^{1/2}$ (c) $(-27)^{1/3}$

    (d) $(i)^{1/3}$ (e) $(-1)^{1/4}$ (f) $(-8 + 8\sqrt{3}i)^{1/4}$

8. Use the method of Example 4 to find all cube roots of 1.

9. Use the method of Example 4 to find all sixth roots of 1.

10. Find all square roots of $1 + i$ and express your results in polar form.

11. Find all solutions of the equation $z^4 - 16 = 0$.

12. Find all solutions of the equation $z^4 + 8 = 0$ and use your results to factor $z^4 + 8$ into two quadratic factors with real coefficients.

13. It was shown in the text that multiplying z by i rotates z counterclockwise by 90°. What is the geometric effect of dividing z by i?

14. In each part, use (6) to calculate the given power.

    (a) $(1 + i)^8$ (b) $(-2\sqrt{3} + 2i)^{-9}$

15. In each part, find $\operatorname{Re}(z)$ and $\operatorname{Im}(z)$.

    (a) $z = 3e^{i\pi}$ (b) $z = 3e^{-i\pi}$ (c) $\bar{z} = \sqrt{2}e^{\pi i/2}$ (d) $\bar{z} = -3e^{-2\pi i}$

16. (a) Show that the values of $z^{1/n}$ in Formula (10) are all different.

    (b) Show that integer values of k other than $k = 0, 1, 2, \ldots, n - 1$ produce values of $z^{1/n}$ that are duplicates of those in Formula (10).

17. Show that Formula (7) is valid if $n = 0$ or n is a negative integer.

18. **(For Readers Who Have Studied Calculus)** To prove Formula (11), recall that the Maclaurin series for e^x is

$$e^x = 1 + x + \frac{x^2}{2!} + \cdots + \frac{x^n}{n!} + \cdots$$

   (a) By substituting $x = i\theta$ in this series and simplifying, show that

   $$e^{i\theta} = \left(1 - \frac{\theta^2}{2!} + \frac{\theta^4}{4!} - \frac{\theta^6}{6!} + \cdots\right) + i\left(\theta - \frac{\theta^3}{3!} + \frac{\theta^5}{5!} - \frac{\theta^7}{7!} + \cdots\right)$$

   (b) Use the result in part (a) to obtain Formula (11).

19. Derive Formula (5).

20. When $n = 2$ and $n = 3$, Equation (7) gives

$$(\cos\theta + i\sin\theta)^2 = \cos 2\theta + i\sin 2\theta$$
$$(\cos\theta + i\sin\theta)^3 = \cos 3\theta + i\sin 3\theta$$

   Use these two equations to obtain trigonometric identities for $\cos 2\theta$, $\sin 2\theta$, $\cos 3\theta$, and $\sin 3\theta$.

21. Use Formula (11) to show that

$$\cos\theta = \frac{e^{i\theta} + e^{-i\theta}}{2} \quad \text{and} \quad \sin\theta = \frac{e^{i\theta} - e^{-i\theta}}{2i}$$

22. Show that if $(a + bi)^3 = 8$, then $a^2 + b^2 = 4$.

23. Show that Formula (6) is valid for negative integer exponents if $z \neq 0$.

10.4
COMPLEX VECTOR SPACES

In this section we shall develop the basic properties of vector spaces with complex scalars and discuss some of the ways in which they differ from real vector spaces. However, before going farther, the reader should review the vector space axioms given in Section 5.1.

Basic Properties

Recall that a vector space in which the scalars are allowed to be complex numbers is called a ***complex vector space***. Linear combinations of vectors in a complex vector space are defined exactly as in a real vector space except that the scalars are allowed to be complex numbers. More precisely, a vector $\mathbf{w}$ is called a ***linear combination*** of the vectors of $\mathbf{v}_1, \mathbf{v}_2, \ldots, \mathbf{v}_r$ if $\mathbf{w}$ can be expressed in the form

$$\mathbf{w} = k_1\mathbf{v}_1 + k_2\mathbf{v}_2 + \cdots + k_r\mathbf{v}_r$$

where $k_1, k_2, \ldots, k_r$ are complex numbers.

The notions of ***linear independence***, ***spanning***, ***basis***, ***dimension***, and ***subspace*** carry over without change to complex vector spaces, and the theorems developed in Chapter 5 continue to hold with R^n changed to C^n.

Among the real vector spaces the most important one is R^n, the space of n-tuples of real numbers, with addition and scalar multiplication performed coordinatewise. Among the complex vector spaces the most important one is C^n, the space of n-tuples of complex numbers, with addition and scalar multiplication performed coordinatewise. A vector $\mathbf{u}$ in C^n can be written either in vector notation,

$$\mathbf{u} = (u_1, u_2, \ldots, u_n)$$

or in matrix notation,

$$\mathbf{u} = \begin{bmatrix} u_1 \\ u_2 \\ \vdots \\ u_n \end{bmatrix}$$

where

$$u_1 = a_1 + b_1 i, \quad u_2 = a_2 + b_2 i, \dots, \quad u_n = a_n + b_n i$$

EXAMPLE 1 Vector Addition and Scalar Multiplication

If

$$\mathbf{u} = (i, 1 + i, -2) \quad \text{and} \quad \mathbf{v} = (2 + i, 1 - i, 3 + 2i)$$

then

$$\mathbf{u} + \mathbf{v} = (i, 1 + i, -2) + (2 + i, 1 - i, 3 + 2i) = (2 + 2i, 2, 1 + 2i)$$

and

$$i\mathbf{u} = i(i, 1 + i, -2) = (i^2, i + i^2, -2i) = (-1, -1 + i, -2i) \quad \blacklozenge$$

In C^n as in R^n, the vectors

$$\mathbf{e}_1 = (1, 0, 0, \dots, 0), \quad \mathbf{e}_2 = (0, 1, 0, \dots, 0), \dots, \quad \mathbf{e}_n = (0, 0, 0, \dots, 1)$$

form a basis. It is called the ***standard basis*** for C^n. Since there are n vectors in this basis, C^n is an n-dimensional vector space.

REMARK Do not confuse the complex number $i = \sqrt{-1}$ with the vector $\mathbf{i} = (1, 0, 0)$ from the standard basis for R^3 (see Example 3, Section 3.4). The complex number i will always be set in lightface type and the vector $\mathbf{i}$ in boldface.

EXAMPLE 2 Complex M_{mn}

In Example 3 of Section 5.1 we defined the vector space M_{mn} of $m \times n$ matrices with real entries. The complex analog of this space is the vector space of $m \times n$ matrices with complex entries and the operations of matrix addition and scalar multiplication. We refer to this space as complex M_{mn}. $\blacklozenge$

EXAMPLE 3 Complex-Valued Function of a Real Variable

If $f_1(x)$ and $f_2(x)$ are real-valued functions of the real variable x, then the expression

$$f(x) = f_1(x) + i f_2(x)$$

is called a ***complex-valued function of the real variable x***. Two examples are

$$f(x) = 2x + ix^3 \quad \text{and} \quad g(x) = 2\sin x + i\cos x \tag{1}$$

Let V be the set of all complex-valued functions that are defined on the entire line. If $\mathbf{f} = f_1(x) + i f_2(x)$ and $\mathbf{g} = g_1(x) + i g_2(x)$ are two such functions and k is any complex number, then we define the ***sum*** function $\mathbf{f} + \mathbf{g}$ and the ***scalar multiple*** $k\mathbf{f}$ by

$$(\mathbf{f} + \mathbf{g})(x) = [f_1(x) + g_1(x)] + i[f_2(x) + g_2(x)]$$
$$(k\mathbf{f})(x) = kf_1(x) + ikf_2(x)$$

For example, if $\mathbf{f} = f(x)$ and $\mathbf{g} = g(x)$ are the functions in (1), then

$$(\mathbf{f} + \mathbf{g})(x) = (2x + 2\sin x) + i(x^3 + \cos x)$$
$$(i\mathbf{f})(x) = 2xi + i^2 x^3 = -x^3 + 2xi$$
$$((1 + i)\mathbf{g})(x) = (1 + i)(2\sin x + i\cos x) = (2\sin x - \cos x) + i(2\sin x + \cos x)$$

It can be shown that V together with the stated operations is a complex vector space. It is the complex analog of the vector space $F(-\infty, \infty)$ of real-valued functions discussed in Example 4 of Section 5.1. ◆

Calculus Required

EXAMPLE 4 Complex C$(-\infty, \infty)$

If $f(x) = f_1(x) + if_2(x)$ is a complex-valued function of the real variable x, then f is said to be ***continuous*** if $f_1(x)$ and $f_2(x)$ are continuous. We leave it as an exercise to show that the set of all continuous complex-valued functions of a real variable x is a subspace of the vector space of all complex-valued functions of x. This space is the complex analog of the vector space $C(-\infty, \infty)$ discussed in Example 6 of Section 5.2 and is called complex $C(-\infty, \infty)$. A closely related example is complex $C[a, b]$, the vector space of all complex-valued functions that are continuous on the closed interval $[a, b]$.

◆

Recall that in R^n the Euclidean inner product of two vectors

$$\mathbf{u} = (u_1, u_2, \ldots, u_n) \quad \text{and} \quad \mathbf{v} = (v_1, v_2, \ldots, v_n)$$

was defined as

$$\mathbf{u} \cdot \mathbf{v} = u_1 v_1 + u_2 v_2 + \cdots + u_n v_n \tag{2}$$

and the Euclidean norm (or length) of $\mathbf{u}$ as

$$\|\mathbf{u}\| = (\mathbf{u} \cdot \mathbf{u})^{1/2} = \sqrt{u_1^2 + u_2^2 + \cdots + u_n^2} \tag{3}$$

Unfortunately, these definitions are not appropriate for vectors in C^n. For example, if (3) were applied to the vector $\mathbf{u} = (i, 1)$ in C^2, we would obtain

$$\|\mathbf{u}\| = \sqrt{i^2 + 1} = \sqrt{0} = 0$$

so $\mathbf{u}$ would be a *nonzero* vector with zero length—a situation that is clearly unsatisfactory.

To extend the notions of norm, distance, and angle to C^n properly, we must modify the inner product slightly.

DEFINITION

If $\mathbf{u} = (u_1, u_2, \ldots, u_n)$ and $\mathbf{v} = (v_1, v_2, \ldots, v_n)$ are vectors in C^n, then their ***complex Euclidean inner product*** $\mathbf{u} \cdot \mathbf{v}$ is defined by

$$\mathbf{u} \cdot \mathbf{v} = u_1 \bar{v}_1 + u_2 \bar{v}_2 + \cdots + u_n \bar{v}_n$$

where $\bar{v}_1, \bar{v}_2, \ldots, \bar{v}_n$ are the conjugates of $v_1, v_2, \ldots, v_n$.

REMARK Observe that the Euclidean inner product of vectors in C^n is a complex number, whereas the Euclidean inner product of vectors in R^n is a real number.

EXAMPLE 5 Complex Inner Product

The complex Euclidean inner product of the vectors

$$\mathbf{u} = (-i, 2, 1 + 3i) \quad \text{and} \quad \mathbf{v} = (1 - i, 0, 1 + 3i)$$

is

$$\mathbf{u} \cdot \mathbf{v} = (-i)\overline{(1-i)} + (2)(\overline{0}) + (1+3i)\overline{(1+3i)}$$
$$= (-i)(1+i) + (2)(0) + (1+3i)(1-3i)$$
$$= -i - i^2 + 1 - 9i^2 = 11 - i \quad \blacklozenge$$

Theorem 4.1.2 listed the four main properties of the Euclidean inner product on R^n. The following theorem is the corresponding result for the complex Euclidean inner product on C^n.

THEOREM 10.4.1

> **Properties of the Complex Inner Product**
>
> *If* $\mathbf{u}$, $\mathbf{v}$, *and* $\mathbf{w}$ *are vectors in* C^n, *and k is any complex number, then*
>
> (*a*) $\mathbf{u} \cdot \mathbf{v} = \overline{\mathbf{v} \cdot \mathbf{u}}$
> (*b*) $(\mathbf{u} + \mathbf{v}) \cdot \mathbf{w} = \mathbf{u} \cdot \mathbf{w} + \mathbf{v} \cdot \mathbf{w}$
> (*c*) $(k\mathbf{u}) \cdot \mathbf{v} = k(\mathbf{u} \cdot \mathbf{v})$
> (*d*) $\mathbf{v} \cdot \mathbf{v} \geq 0$. *Further,* $\mathbf{v} \cdot \mathbf{v} = 0$ *if and only if* $\mathbf{v} = \mathbf{0}$.

Note the difference between part (*a*) of this theorem and part (*a*) of Theorem 4.1.2. We will prove parts (*a*) and (*d*) and leave the rest as exercises.

Proof (a) Let $\mathbf{u} = (u_1, u_2, \ldots, u_n)$ and $\mathbf{v} = (v_1, v_2, \ldots, v_n)$. Then

$$\mathbf{u} \cdot \mathbf{v} = u_1 \overline{v}_1 + u_2 \overline{v}_2 + \cdots + u_n \overline{v}_n$$

and

$$\mathbf{v} \cdot \mathbf{u} = v_1 \overline{u}_1 + v_2 \overline{u}_2 + \cdots + v_n \overline{u}_n$$

so

$$\overline{\mathbf{v} \cdot \mathbf{u}} = \overline{v_1 \overline{u}_1 + v_2 \overline{u}_2 + \cdots + v_n \overline{u}_n}$$
$$= \overline{v}_1 \overline{\overline{u}}_1 + \overline{v}_2 \overline{\overline{u}}_2 + \cdots + \overline{v}_n \overline{\overline{u}}_n \quad \text{[Theorem 10.2.3, parts (a) and (c)]}$$
$$= \overline{v}_1 u_1 + \overline{v}_2 u_2 + \cdots + \overline{v}_n u_n \quad \text{[Theorem 10.2.3, part (e)]}$$
$$= u_1 \overline{v}_1 + u_2 \overline{v}_2 + \cdots + u_n \overline{v}_n$$
$$= \mathbf{u} \cdot \mathbf{v}$$

Proof (d)

$$\mathbf{v} \cdot \mathbf{v} = v_1 \overline{v}_1 + v_2 \overline{v}_2 + \cdots + v_n \overline{v}_n = |v_1|^2 + |v_2|^2 + \cdots + |v_n|^2 \geq 0$$

Moreover, equality holds if and only if $|v_1| = |v_2| = \cdots = |v_n| = 0$. But this is true if and only if $v_1 = v_2 = \cdots = v_n = 0$; that is, it is true if and only if $\mathbf{v} = \mathbf{0}$. $\blacksquare$

REMARK We leave it as an exercise to prove that

$$\mathbf{u} \cdot (k\mathbf{v}) = \overline{k}(\mathbf{u} \cdot \mathbf{v})$$

for vectors in C^n. Compare this to the corresponding formula

$$\mathbf{u} \cdot (k\mathbf{v}) = k(\mathbf{u} \cdot \mathbf{v})$$

for vectors in R^n.

Norm and Distance in C^n

By analogy with (3), we define the ***Euclidean norm*** (or ***Euclidean length***) of a vector $\mathbf{u} = (u_1, u_2, \ldots, u_n)$ in C^n by

$$\|\mathbf{u}\| = (\mathbf{u} \cdot \mathbf{u})^{1/2} = \sqrt{|u_1|^2 + |u_2|^2 + \cdots + |u_n|^2}$$

and we define the ***Euclidean distance*** between the points $\mathbf{u} = (u_1, u_2, \ldots, u_n)$ and $\mathbf{v} = (v_1, v_2, \ldots, v_n)$ by

$$d(\mathbf{u}, \mathbf{v}) = \|\mathbf{u} - \mathbf{v}\| = \sqrt{|u_1 - v_1|^2 + |u_2 - v_2|^2 + \cdots + |u_n - v_n|^2}$$

EXAMPLE 6   Norm and Distance

If $\mathbf{u} = (i, 1 + i, 3)$ and $\mathbf{v} = (1 - i, 2, 4i)$, then

$$\|\mathbf{u}\| = \sqrt{|i|^2 + |1 + i|^2 + |3|^2} = \sqrt{1 + 2 + 9} = \sqrt{12} = 2\sqrt{3}$$

and

$$d(\mathbf{u}, \mathbf{v}) = \sqrt{|i - (1 - i)|^2 + |(1 + i) - 2|^2 + |3 - 4i|^2}$$

$$= \sqrt{|-1 + 2i|^2 + |-1 + i|^2 + |3 - 4i|^2}$$

$$= \sqrt{5 + 2 + 25} = \sqrt{32} = 4\sqrt{2} \quad \blacklozenge$$

The complex vector space C^n with norm and inner product defined above is called ***complex Euclidean n-space***.

EXERCISE SET 10.4

1. Let $\mathbf{u} = (2i, 0, -1, 3)$, $\mathbf{v} = (-i, i, 1 + i, -1)$, and $\mathbf{w} = (1 + i, -i, -1 + 2i, 0)$. Find
   - (a) $\mathbf{u} - \mathbf{v}$
   - (b) $i\mathbf{v} + 2\mathbf{w}$
   - (c) $-\mathbf{w} + \mathbf{v}$
   - (d) $3(\mathbf{u} - (1 + i)\mathbf{v})$
   - (e) $-i\mathbf{v} + 2i\mathbf{w}$
   - (f) $2\mathbf{v} - (\mathbf{u} + \mathbf{w})$

2. Let $\mathbf{u}$, $\mathbf{v}$, and $\mathbf{w}$ be the vectors in Exercise 1. Find the vector $\mathbf{x}$ that satisfies $\mathbf{u} - \mathbf{v} + i\mathbf{x} = 2i\mathbf{x} + \mathbf{w}$.

3. Let $\mathbf{u}_1 = (1 - i, i, 0)$, $\mathbf{u}_2 = (2i, 1 + i, 1)$, and $\mathbf{u}_3 = (0, 2i, 2 - i)$. Find scalars c_1, c_2, and c_3 such that $c_1\mathbf{u}_1 + c_2\mathbf{u}_2 + c_3\mathbf{u}_3 = (-3 + i, 3 + 2i, 3 - 4i)$.

4. Show that there do not exist scalars c_1, c_2, and c_3 such that

$$c_1(i, 2 - i, 2 + i) + c_2(1 + i, -2i, 2) + c_3(3, i, 6 + i) = (i, i, i)$$

5. Find the Euclidean norm of $\mathbf{v}$ if
   - (a) $\mathbf{v} = (1, i)$
   - (b) $\mathbf{v} = (1 + i, 3i, 1)$
   - (c) $\mathbf{v} = (2i, 0, 2i + 1, -1)$
   - (d) $\mathbf{v} = (-i, i, i, 3, 3 + 4i)$

6. Let $\mathbf{u} = (3i, 0, -i)$, $\mathbf{v} = (0, 3 + 4i, -2i)$, and $\mathbf{w} = (1 + i, 2i, 0)$. Find
   - (a) $\|\mathbf{u} + \mathbf{v}\|$
   - (b) $\|\mathbf{u}\| + \|\mathbf{v}\|$
   - (c) $\|-i\mathbf{u}\| + i\|\mathbf{u}\|$
   - (d) $\|3\mathbf{u} - 5\mathbf{v} + \mathbf{w}\|$
   - (e) $\dfrac{1}{\|\mathbf{w}\|}\mathbf{w}$
   - (f) $\left\|\dfrac{1}{\|\mathbf{w}\|}\mathbf{w}\right\|$

7. Show that if $\mathbf{v}$ is a nonzero vector in C^n, then $(1/\|\mathbf{v}\|)\mathbf{v}$ has Euclidean norm 1.

8. Find all scalars k such that $\|k\mathbf{v}\| = 1$, where $\mathbf{v} = (3i, 4i)$.

9. Find the Euclidean inner product $\mathbf{u} \cdot \mathbf{v}$ if
   - (a) $\mathbf{u} = (-i, 3i)$, $\mathbf{v} = (3i, 2i)$
   - (b) $\mathbf{u} = (3 - 4i, 2 + i, -6i)$, $\mathbf{v} = (1 + i, 2 - i, 4)$
   - (c) $\mathbf{u} = (1 - i, 1 + i, 2i, 3)$, $\mathbf{v} = (4 + 6i, -5i, -1 + i, i)$

In Exercises 10 and 11 a set of objects is given, together with operations of addition and scalar multiplication. Determine which sets are complex vector spaces under the given operations. For those that are not, list all axioms that fail to hold.

10. The set of all triples of complex numbers (z_1, z_2, z_3) with the operations

$$(z_1, z_2, z_3) + (z_1', z_2', z_3') = (z_1 + z_1', z_2 + z_2', z_3 + z_3')$$

and

$$k(z_1, z_2, z_3) = (\bar{k}z_1, \bar{k}z_2, \bar{k}z_3)$$

11. The set of all complex 2×2 matrices of the form

$$\begin{bmatrix} z & 0 \\ 0 & \bar{z} \end{bmatrix}$$

with the standard matrix operations of addition and scalar multiplication.

12. Use Theorem 5.2.1 to determine which of the following sets are subspaces of C^3:

(a) all vectors of the form $(z, 0, 0)$

(b) all vectors of the form (z, i, i)

(c) all vectors of the form (z_1, z_2, z_3), where $z_3 = \bar{z}_1 + \bar{z}_2$

(d) all vectors of the form (z_1, z_2, z_3), where $z_3 = z_1 + z_2 + i$

13. Let $T: C^3 \to C^3$ be a linear operator defined by $T(\mathbf{x}) = A\mathbf{x}$, where

$$A = \begin{bmatrix} i & -i & -1 \\ 1 & -i & 1+i \\ 0 & 1-i & 1 \end{bmatrix}$$

Find the kernel and nullity of T.

14. Use Theorem 5.2.1 to determine which of the following are subspaces of complex M_{22}:

(a) All complex matrices of the form

$$\begin{bmatrix} z_1 & z_2 \\ z_3 & z_4 \end{bmatrix}$$

where z_1 and z_2 are real.

(b) All complex matrices of the form

$$\begin{bmatrix} z_1 & z_2 \\ z_3 & z_4 \end{bmatrix}$$

where $z_1 + z_4 = 0$.

(c) All 2×2 complex matrices A such that $(\bar{A})^T = A$, where $\bar{A}$ is the matrix whose entries are the conjugates of the corresponding entries in A.

15. Use Theorem 5.2.1 to determine which of the following are subspaces of the vector space of complex-valued functions of the real variable x:

(a) all f such that $f(1) = 0$

(b) all f such that $f(0) = i$

(c) all f such that $f(-x) = \overline{f(x)}$

(d) all f of the form $k_1 + k_2 e^{ix}$, where k_1 and k_2 are complex numbers

16. Which of the following are linear combinations of $\mathbf{u} = (i, -i, 3i)$ and $\mathbf{v} = (2i, 4i, 0)$?

(a) $(3i, 3i, 3i)$ (b) $(4i, 2i, 6i)$ (c) $(i, 5i, 6i)$ (d) $(0, 0, 0)$

17. Express the following as linear combinations of $\mathbf{u} = (1, 0, -i)$, $\mathbf{v} = (1 + i, 1, 1 - 2i)$, and $\mathbf{w} = (0, i, 2)$.

    (a) $(1, 1, 1)$ (b) $(i, 0, -i)$ (c) $(0, 0, 0)$ (d) $(2 - i, 1, 1 + i)$

18. In each part, determine whether the given vectors span C^3.

    (a) $\mathbf{v}_1 = (i, i, i)$, $\mathbf{v}_2 = (2i, 2i, 0)$, $\mathbf{v}_3 = (3i, 0, 0)$

    (b) $\mathbf{v}_1 = (1 + i, 2 - i, 3 + i)$, $\mathbf{v}_2 = (2 + 3i, 0, 1 - i)$

    (c) $\mathbf{v}_1 = (1, 0, -i)$, $\mathbf{v}_2 = (1 + i, 1, 1 - 2i)$, $\mathbf{v}_3 = (0, i, 2)$

    (d) $\mathbf{v}_1 = (1, i, 0)$, $\mathbf{v}_2 = (0, -i, 1)$, $\mathbf{v}_3 = (1, 0, 1)$

19. Determine which of the following lie in the space spanned by

    $$\mathbf{f} = e^{ix} \quad \text{and} \quad \mathbf{g} = e^{-ix}$$

    (a) $\cos x$ (b) $\sin x$ (c) $\cos x + 3i \sin x$

20. Explain why the following are linearly dependent sets of vectors. (Solve this problem by inspection.)

    (a) $\mathbf{u}_1 = (1 - i, i)$ and $\mathbf{u}_2 = (1 + i, -1)$ in C^2

    (b) $\mathbf{u}_1 = (1, -i)$, $\mathbf{u}_2 = (2 + i, -1)$, $\mathbf{u}_3 = (4, 0)$ in C^2

    (c) $A = \begin{bmatrix} i & 3i \\ 2i & 0 \end{bmatrix}$ and $B = \begin{bmatrix} 1 & 3 \\ 2 & 0 \end{bmatrix}$ in complex M_{22}

21. Which of the following sets of vectors in C^3 are linearly independent?

    (a) $\mathbf{u}_1 = (1 - i, 1, 0)$, $\mathbf{u}_2 = (2, 1 + i, 0)$, $\mathbf{u}_3 = (1 + i, i, 0)$

    (b) $\mathbf{u}_1 = (1, 0, -i)$, $\mathbf{u}_2 = (1 + i, 1, 1 - 2i)$, $\mathbf{u}_3 = (0, i, 2)$

    (c) $\mathbf{u}_1 = (i, 0, 2 - i)$, $\mathbf{u}_2 = (0, 1, i)$, $\mathbf{u}_3 = (-i, -1 - 4i, 3)$

22. Given the vectors $\mathbf{v} = [a, b + ci]$ and $\mathbf{w} = [b - ci, a]$, for what real values of a, b, and c are the vectors $\mathbf{v}$ and $\mathbf{w}$ linearly dependent?

23. Let V be the vector space of all complex-valued functions of the real variable x. Show that the following vectors are linearly dependent.

    $$\mathbf{f} = 3 + 3i \cos 2x, \qquad \mathbf{g} = \sin^2 x + i \cos^2 x, \qquad \mathbf{h} = \cos^2 x - i \sin^2 x$$

24. Explain why the following sets of vectors are not bases for the indicated vector spaces. (Solve this problem by inspection.)

    (a) $\mathbf{u}_1 = (i, 2i)$, $\mathbf{u}_2 = (0, 3i)$, $\mathbf{u}_3 = (1, 7i)$ for C^2

    (b) $\mathbf{u}_1 = (-1 + i, 0, 2 - i)$, $\mathbf{u}_2 = (1, -i, 1 + i)$ for C^3

25. Which of the following sets of vectors are bases for C^2?

    (a) $(2i, -i)$, $(4i, 0)$ (b) $(1 + i, 1)$, $(1 + i, i)$

    (c) $(0, 0)$, $(1 + i, 1 - i)$ (d) $(2 - 3i, i)$, $(3 + 2i, -1)$

26. Which of the following sets of vectors are bases for C^3?

    (a) $(i, 0, 0)$, $(i, i, 0)$, (i, i, i)

    (b) $(1, 0, -i)$, $(1 + i, 1, 1 - 2i)$, $(0, i, 2)$

    (c) $(i, 0, 2 - i)$, $(0, 1, i)$, $(-i, -1 - 4i, 3)$

    (d) $(1, 0, i)$, $(2 - i, 1, 2 + i)$, $(0, 3i, 3i)$

In Exercises 27–30, determine the dimension of and a basis for the solution space of the system.

27.
$$x_1 + (1 + i)x_2 = 0$$
$$(1 - i)x_1 + \qquad 2x_2 = 0$$

28.
$$2x_1 - (1 + i)x_2 = 0$$
$$(-1 + i)x_1 + \qquad x_2 = 0$$

29.
$$x_1 + (2 - i)x_2 \qquad\qquad = 0$$
$$x_2 + 3ix_3 = 0$$
$$ix_1 + (2 + 2i)x_2 + 3ix_3 = 0$$

30.
$$x_1 + ix_2 - 2ix_3 + \quad x_4 = 0$$
$$ix_1 + 3x_2 + 4x_3 - 2ix_4 = 0$$

31. Prove: If **u** and **v** are vectors in complex Euclidean *n*-space, then

$$\mathbf{u} \cdot (k\mathbf{v}) = \overline{k}(\mathbf{u} \cdot \mathbf{v})$$

32. (a) Prove part (*b*) of Theorem 10.4.1.

(b) Prove part (*c*) of Theorem 10.4.1.

33. Establish the identity

$$\mathbf{u} \cdot \mathbf{v} = \frac{1}{4}\|\mathbf{u}+\mathbf{v}\|^2 - \frac{1}{4}\|\mathbf{u}-\mathbf{v}\|^2 + \frac{i}{4}\|\mathbf{u}+i\mathbf{v}\|^2 - \frac{i}{4}\|\mathbf{u}-i\mathbf{v}\|^2$$

for vectors in complex Euclidean *n*-space.

34. **(For Readers Who Have Studied Calculus)** Prove that complex $C(-\infty, \infty)$ is a subspace of the vector space of complex-valued functions of a real variable.

10.5
COMPLEX INNER PRODUCT SPACES

In Section 6.1 we defined the notion of an inner product on a real vector space by using the basic properties of the Euclidean inner product on R^n as axioms. In this section we shall define inner products on complex vector spaces by using the properties of the Euclidean inner product on C^n as axioms.

Complex Inner Product Spaces

Motivated by Theorem 10.4.1, we make the following definition.

> **DEFINITION**
>
> An *inner product on a complex vector space* V is a function that associates a complex number $\langle \mathbf{u}, \mathbf{v} \rangle$ with each pair of vectors **u** and **v** in V in such a way that the following axioms are satisfied for all vectors **u**, **v**, and **w** in V and all scalars k.
>
> 1. $\langle \mathbf{u}, \mathbf{v} \rangle = \overline{\langle \mathbf{v}, \mathbf{u} \rangle}$
> 2. $\langle \mathbf{u}+\mathbf{v}, \mathbf{w} \rangle = \langle \mathbf{u}, \mathbf{w} \rangle + \langle \mathbf{v}, \mathbf{w} \rangle$
> 3. $\langle k\mathbf{u}, \mathbf{v} \rangle = k\langle \mathbf{u}, \mathbf{v} \rangle$
> 4. $\langle \mathbf{v}, \mathbf{v} \rangle \geq 0$ and $\langle \mathbf{v}, \mathbf{v} \rangle = 0$ if and only if $\mathbf{v} = \mathbf{0}$

A complex vector space with an inner product is called a *complex inner product space*.

The following additional properties follow immediately from the four inner product axioms:

(i) $\langle \mathbf{0}, \mathbf{v} \rangle = \langle \mathbf{v}, \mathbf{0} \rangle = 0$

(ii) $\langle \mathbf{u}, \mathbf{v}+\mathbf{w} \rangle = \langle \mathbf{u}, \mathbf{v} \rangle + \langle \mathbf{u}, \mathbf{w} \rangle$

(iii) $\langle \mathbf{u}, k\mathbf{v} \rangle = \overline{k}\langle \mathbf{u}, \mathbf{v} \rangle$

Since only (iii) differs from the corresponding results for real inner products, we will prove it and leave the other proofs as exercises.

$$\begin{aligned} \langle \mathbf{u}, k\mathbf{v} \rangle &= \overline{\langle k\mathbf{v}, \mathbf{u} \rangle} && \text{[Axiom 1]} \\ &= \overline{k\langle \mathbf{v}, \mathbf{u} \rangle} && \text{[Axiom 3]} \\ &= \overline{k}\,\overline{\langle \mathbf{v}, \mathbf{u} \rangle} && \text{[Property of conjugates]} \\ &= \overline{k}\langle \mathbf{u}, \mathbf{v} \rangle && \text{[Axiom 1]} \end{aligned}$$

EXAMPLE 1 Inner Product on C^n

Let $\mathbf{u} = (u_1, u_2, \ldots, u_n)$ and $\mathbf{v} = (v_1, v_2, \ldots, v_n)$ be vectors in C^n. The Euclidean inner product $\langle \mathbf{u}, \mathbf{v} \rangle = \mathbf{u} \cdot \mathbf{v} = u_1\bar{v}_1 + u_2\bar{v}_2 + \cdots + u_n\bar{v}_n$ satisfies all the inner product axioms by Theorem 10.4.1. ◆

EXAMPLE 2 Inner Product on Complex M_{22}

If

$$U = \begin{bmatrix} u_1 & u_2 \\ u_3 & u_4 \end{bmatrix} \quad \text{and} \quad V = \begin{bmatrix} v_1 & v_2 \\ v_3 & v_4 \end{bmatrix}$$

are any 2×2 matrices with complex entries, then the following formula defines a complex inner product on complex M_{22} (verify):

$$\langle U, V \rangle = u_1\bar{v}_1 + u_2\bar{v}_2 + u_3\bar{v}_3 + u_4\bar{v}_4$$

For example, if

$$U = \begin{bmatrix} 0 & i \\ 1 & 1+i \end{bmatrix} \quad \text{and} \quad V = \begin{bmatrix} 1 & -i \\ 0 & 2i \end{bmatrix}$$

then

$$
\begin{aligned}
\langle U, V \rangle &= (0)(\bar{1}) + i(\overline{-i}) + (1)(\bar{0}) + (1+i)(\overline{2i}) \\
&= (0)(1) + i(i) + (1)(0) + (1+i)(-2i) \\
&= 0 + i^2 + 0 - 2i - 2i^2 \\
&= 1 - 2i \quad ◆
\end{aligned}
$$

Calculus Required   ### EXAMPLE 3 Inner Product on Complex $C[a, b]$

If $f(x) = f_1(x) + if_2(x)$ is a complex-valued function of the real variable x, and if $f_1(x)$ and $f_2(x)$ are continuous on $[a, b]$, then we define

$$\int_a^b f(x)\, dx = \int_a^b [f_1(x) + if_2(x)]\, dx = \int_a^b f_1(x)\, dx + i\int_a^b f_2(x)\, dx$$

In words, *the integral of $f(x)$ is the integral of the real part of f plus i times the integral of the imaginary part of f*.

We leave it as an exercise to show that if the functions $\mathbf{f} = f_1(x) + if_2(x)$ and $\mathbf{g} = g_1(x) + ig_2(x)$ are vectors in complex $C[a, b]$, then the following formula defines an inner product on complex $C[a, b]$:

$$
\begin{aligned}
\langle \mathbf{f}, \mathbf{g} \rangle &= \int_a^b [f_1(x) + if_2(x)]\overline{[g_1(x) + ig_2(x)]}\, dx \\
&= \int_a^b [f_1(x) + if_2(x)][g_1(x) - ig_2(x)]\, dx \\
&= \int_a^b [f_1(x)g_1(x) + f_2(x)g_2(x)]\, dx + i\int_a^b [f_2(x)g_1(x) - f_1(x)g_2(x)]\, dx \quad ◆
\end{aligned}
$$

In complex inner product spaces, as in real inner product spaces, the ***norm*** (or ***length***) of a vector **u** is defined by

$$\|\mathbf{u}\| = \langle \mathbf{u}, \mathbf{u} \rangle^{1/2}$$

and the ***distance*** between two vectors **u** and **v** is defined by

$$d(\mathbf{u}, \mathbf{v}) = \|\mathbf{u} - \mathbf{v}\|$$

It can be shown that with these definitions, Theorems 6.2.2 and 6.2.3 remain true in complex inner product spaces (Exercise 35).

EXAMPLE 4   Norm and Distance in C^n

If $\mathbf{u} = (u_1, u_2, \ldots, u_n)$ and $\mathbf{v} = (v_1, v_2, \ldots, v_n)$ are vectors in C^n with the Euclidean inner product, then

$$\|\mathbf{u}\| = \langle \mathbf{u}, \mathbf{u} \rangle^{1/2} = \sqrt{|u_1|^2 + |u_2|^2 + \cdots + |u_n|^2}$$

and

$$d(\mathbf{u}, \mathbf{v}) = \|\mathbf{u} - \mathbf{v}\| = \langle \mathbf{u} - \mathbf{v}, \mathbf{u} - \mathbf{v} \rangle^{1/2}$$
$$= \sqrt{|u_1 - v_1|^2 + |u_2 - v_2|^2 + \cdots + |u_n - v_n|^2}$$

Observe that these are just the formulas for the Euclidean norm and distance discussed in Section 10.4. ◆

Calculus Required   ### EXAMPLE 5   Norm of a Function in Complex $C[0, 2\pi]$

If complex $C[0, 2\pi]$ has the inner product of Example 3, and if $\mathbf{f} = e^{imx}$, where m is any integer, then with the help of Formula (15) of Section 10.3, we obtain

$$\|\mathbf{f}\| = \langle \mathbf{f}, \mathbf{f} \rangle^{1/2} = \left[\int_0^{2\pi} e^{imx}\overline{e^{imx}}\, dx \right]^{1/2}$$
$$= \left[\int_0^{2\pi} e^{imx}e^{-imx}\, dx \right]^{1/2} = \left[\int_0^{2\pi} dx \right]^{1/2} = \sqrt{2\pi} \;\blacklozenge$$

Orthogonal Sets

The definitions of such terms as ***orthogonal vectors***, ***orthogonal set***, ***orthonormal set***, and ***orthonormal basis*** carry over to complex inner product spaces without change. Moreover, Theorems 6.2.4, 6.3.1, 6.3.3, 6.3.4, 6.3.5, 6.3.6, and 6.5.1 remain valid in complex inner product spaces, and the Gram–Schmidt process can be used to convert an arbitrary basis for a complex inner product space into an orthonormal basis.

EXAMPLE 6   Orthogonal Vectors in C^2

The vectors

$$\mathbf{u} = (i, 1) \quad \text{and} \quad \mathbf{v} = (1, i)$$

in C^2 are orthogonal with respect to the Euclidean inner product, since

$$\mathbf{u} \cdot \mathbf{v} = (i)(\overline{1}) + (1)(\overline{i}) = (i)(1) + (1)(-i) = 0 \quad \blacklozenge$$

EXAMPLE 7 Constructing an Orthonormal Basis for C^3

Consider the vector space C^3 with the Euclidean inner product. Apply the Gram–Schmidt process to transform the basis vectors $\mathbf{u}_1 = (i, i, i)$, $\mathbf{u}_2 = (0, i, i)$, $\mathbf{u}_3 = (0, 0, i)$ into an orthonormal basis.

Solution

Step 1. $\mathbf{v}_1 = \mathbf{u}_1 = (i, i, i)$

Step 2. $\mathbf{v}_2 = \mathbf{u}_2 - \mathrm{proj}_{W_1} \mathbf{u}_2 = \mathbf{u}_2 - \dfrac{\langle \mathbf{u}_2, \mathbf{v}_1 \rangle}{\|\mathbf{v}_1\|^2} \mathbf{v}_1$

$$= (0, i, i) - \frac{2}{3}(i, i, i) = \left(-\frac{2}{3}i, \frac{1}{3}i, \frac{1}{3}i \right)$$

Step 3. $\mathbf{v}_3 = \mathbf{u}_3 - \mathrm{proj}_{W_2} \mathbf{u}_3 = \mathbf{u}_3 - \dfrac{\langle \mathbf{u}_3, \mathbf{v}_1 \rangle}{\|\mathbf{v}_1\|^2} \mathbf{v}_1 - \dfrac{\langle \mathbf{u}_3, \mathbf{v}_2 \rangle}{\|\mathbf{v}_2\|^2} \mathbf{v}_2$

$$= (0, 0, i) - \frac{1}{3}(i, i, i) - \frac{1/3}{2/3} \left(-\frac{2}{3}i, \frac{1}{3}i, \frac{1}{3}i \right)$$

$$= \left(0, -\frac{1}{2}i, \frac{1}{2}i \right)$$

Thus

$$\mathbf{v}_1 = (i, i, i), \qquad \mathbf{v}_2 = \left(-\frac{2}{3}i, \frac{1}{3}i, \frac{1}{3}i \right), \qquad \mathbf{v}_3 = \left(0, -\frac{1}{2}i, \frac{1}{2}i \right)$$

form an orthogonal basis for C^3. The norms of these vectors are

$$\|\mathbf{v}_1\| = \sqrt{3}, \qquad \|\mathbf{v}_2\| = \frac{\sqrt{6}}{3}, \qquad \|\mathbf{v}_3\| = \frac{1}{\sqrt{2}}$$

so an orthonormal basis for C^3 is

$$\frac{\mathbf{v}_1}{\|\mathbf{v}_1\|} = \left(\frac{i}{\sqrt{3}}, \frac{i}{\sqrt{3}}, \frac{i}{\sqrt{3}} \right), \qquad \frac{\mathbf{v}_2}{\|\mathbf{v}_2\|} = \left(-\frac{2i}{\sqrt{6}}, \frac{i}{\sqrt{6}}, \frac{i}{\sqrt{6}} \right),$$

$$\frac{\mathbf{v}_3}{\|\mathbf{v}_3\|} = \left(0, -\frac{i}{\sqrt{2}}, \frac{i}{\sqrt{2}} \right) \quad \blacklozenge$$

Calculus Required ### EXAMPLE 8 Orthonormal Set in Complex $C[0, 2\pi]$

Let complex $C[0, 2\pi]$ have the inner product of Example 3, and let W be the set of vectors in $C[0, 2\pi]$ of the form

$$e^{imx} = \cos mx + i \sin mx$$

where m is an integer. The set W is orthogonal because if

$$\mathbf{f} = e^{ikx} \quad \text{and} \quad \mathbf{g} = e^{ilx}$$

are distinct vectors in W, then

$$\langle \mathbf{f}, \mathbf{g} \rangle = \int_0^{2\pi} e^{ikx} \overline{e^{ilx}} \, dx = \int_0^{2\pi} e^{ikx} e^{-ilx} \, dx = \int_0^{2\pi} e^{i(k-l)x} \, dx$$

$$= \int_0^{2\pi} \cos(k-l)x \, dx + i \int_0^{2\pi} \sin(k-l)x \, dx$$

$$= \left[\frac{1}{k-l} \sin(k-l)x \right]_0^{2\pi} - i \left[\frac{1}{k-l} \cos(k-l)x \right]_0^{2\pi}$$

$$= (0) - i(0) = 0$$

If we normalize each vector in the orthogonal set W, we obtain an orthonormal set. But in Example 5 we showed that each vector in W has norm $\sqrt{2\pi}$, so the vectors

$$\frac{1}{\sqrt{2\pi}} e^{imx}, \qquad m = 0, \pm 1, \pm 2, \ldots$$

form an orthonormal set in complex $C[0, 2\pi]$. ◆

EXERCISE SET
10.5

1. Let $\mathbf{u} = (u_1, u_2)$ and $\mathbf{v} = (v_1, v_2)$. Show that $\langle \mathbf{u}, \mathbf{v} \rangle = 3u_1 \bar{v}_1 + 2u_2 \bar{v}_2$ defines an inner product on C^2.

2. Compute $\langle \mathbf{u}, \mathbf{v} \rangle$ using the inner product in Exercise 1.

   (a) $\mathbf{u} = (2i, -i)$, $\mathbf{v} = (-i, 3i)$ (b) $\mathbf{u} = (0, 0)$, $\mathbf{v} = (1 - i, 7 - 5i)$

   (c) $\mathbf{u} = (1 + i, 1 - i)$, $\mathbf{v} = (1 - i, 1 + i)$ (d) $\mathbf{u} = (3i, \quad 1 + 2i)$, $\mathbf{v} = (3i, -1 + 2i)$

3. Let $\mathbf{u} = (u_1, u_2)$ and $\mathbf{v} = (v_1, v_2)$. Show that

   $$\langle \mathbf{u}, \mathbf{v} \rangle = u_1 \bar{v}_1 + (1 + i)u_1 \bar{v}_2 + (1 - i)u_2 \bar{v}_1 + 3u_2 \bar{v}_2$$

   defines an inner product on C^2.

4. Compute $\langle \mathbf{u}, \mathbf{v} \rangle$ using the inner product in Exercise 3.

   (a) $\mathbf{u} = (2i, -i)$, $\mathbf{v} = (-i, 3i)$ (b) $\mathbf{u} = (0, 0)$, $\mathbf{v} = (1 - i, 7 - 5i)$

   (c) $\mathbf{u} = (1 + i, 1 - i)$, $\mathbf{v} = (1 - i, 1 + i)$ (d) $\mathbf{u} = (3i, -1 + 2i)$, $\mathbf{v} = (3i, -1 + 2i)$

5. Let $\mathbf{u} = (u_1, u_2)$ and $\mathbf{v} = (v_1, v_2)$. Determine which of the following are inner products on C^2. For those that are not, list the axioms that do not hold.

   (a) $\langle \mathbf{u}, \mathbf{v} \rangle = u_1 \bar{v}_1$

   (b) $\langle \mathbf{u}, \mathbf{v} \rangle = u_1 \bar{v}_1 - u_2 \bar{v}_2$

   (c) $\langle \mathbf{u}, \mathbf{v} \rangle = |u_1|^2 |v_1|^2 + |u_2|^2 |v_2|^2$

   (d) $\langle \mathbf{u}, \mathbf{v} \rangle = 2u_1 \bar{v}_1 + iu_1 \bar{v}_2 + iu_2 \bar{v}_1 + 2u_2 \bar{v}_2$

   (e) $\langle \mathbf{u}, \mathbf{v} \rangle = 2u_1 \bar{v}_1 + iu_1 \bar{v}_2 - iu_2 \bar{v}_1 + 2u_2 \bar{v}_2$

6. Use the inner product of Example 2 to find $\langle U, V \rangle$ if

   $$U = \begin{bmatrix} -i & 1 + i \\ 1 - i & i \end{bmatrix} \quad \text{and} \quad V = \begin{bmatrix} 3 & -2 - 3i \\ 4i & 1 \end{bmatrix}$$

7. Let $\mathbf{u} = (u_1, u_2, u_3)$ and $\mathbf{v} = (v_1, v_2, v_3)$. Does $\langle \mathbf{u}, \mathbf{v} \rangle = u_1 \bar{v}_1 + u_2 \bar{v}_2 + u_3 \bar{v}_3 - iu_3 \bar{v}_1$ define an inner product on C^3? If not, list all axioms that fail to hold.

8. Let V be the vector space of complex-valued functions of the real variable x, and let $\mathbf{f} = f_1(x) + if_2(x)$ and $\mathbf{g} = g_1(x) + ig_2(x)$ be vectors in V. Does

   $$\langle \mathbf{f}, \mathbf{g} \rangle = (f_1(0) + if_2(0))\overline{(g_1(0) + ig_2(0))}$$

   define an inner product on V? If not, list all axioms that fail to hold.

9. Let C^2 have the inner product of Exercise 1. Find $\|\mathbf{w}\|$ if
   (a) $\mathbf{w} = (-i, 3i)$ (b) $\mathbf{w} = (1 - i, 1 + i)$
   (c) $\mathbf{w} = (0, 2 - i)$ (d) $\mathbf{w} = (0, 0)$

10. For each vector in Exercise 9, use the Euclidean inner product to find $\|\mathbf{w}\|$.

11. Use the inner product of Exercise 3 to find $\|\mathbf{w}\|$ if
    (a) $\mathbf{w} = (1, -i)$ (b) $\mathbf{w} = (1 - i, 1 + i)$
    (c) $\mathbf{w} = (3 - 4i, 0)$ (d) $\mathbf{w} = (0, 0)$

12. Use the inner product of Example 2 to find $\|A\|$ if

   (a) $A = \begin{bmatrix} -i & 7i \\ 6i & 2i \end{bmatrix}$ (b) $A = \begin{bmatrix} -1 & 1+i \\ 1-i & 3 \end{bmatrix}$

13. Let C^2 have the inner product of Exercise 1. Find $d(\mathbf{x}, \mathbf{y})$ if
    (a) $\mathbf{x} = (1, 1)$, $\mathbf{y} = (i, -i)$ (b) $\mathbf{x} = (1 - i, 3 + 2i)$, $\mathbf{y} = (1 + i, 3)$

14. Repeat the directions of Exercise 13 using the Euclidean inner product on C^2.

15. Repeat the directions of Exercise 13 using the inner product of Exercise 3.

16. Let complex M_{22} have the inner product of Example 2. Find $d(A, B)$ if

   (a) $A = \begin{bmatrix} i & 5i \\ 8i & 3i \end{bmatrix}$ and $B = \begin{bmatrix} -5i & 0 \\ 7i & -3i \end{bmatrix}$

   (b) $A = \begin{bmatrix} -1 & 1-i \\ 1+i & 2 \end{bmatrix}$ and $B = \begin{bmatrix} 2i & 2-3i \\ i & 1 \end{bmatrix}$

17. Let C^3 have the Euclidean inner product. For which complex values of k are $\mathbf{u}$ and $\mathbf{v}$ orthogonal?
    (a) $\mathbf{u} = (2i, i, 3i)$, $\mathbf{v} = (i, 6i, k)$ (b) $\mathbf{u} = (k, k, 1 + i)$, $\mathbf{v} = (1, -1, 1 - i)$

18. Let complex M_{22} have the inner product of Example 2. Determine which of the following are orthogonal to

$$A = \begin{bmatrix} 2i & i \\ -i & 3i \end{bmatrix}$$

   (a) $\begin{bmatrix} -3 & 1-i \\ 1-i & 2 \end{bmatrix}$ (b) $\begin{bmatrix} 1 & 1 \\ 0 & -1 \end{bmatrix}$ (c) $\begin{bmatrix} 0 & 0 \\ 0 & 0 \end{bmatrix}$ (d) $\begin{bmatrix} 0 & 1 \\ 3-i & 0 \end{bmatrix}$

19. Let C^3 have the Euclidean inner product. Show that for all values of the variable θ, the vector
    $$\mathbf{x} = e^{i\theta} \left(\frac{i}{\sqrt{3}}, \frac{1}{\sqrt{3}}, \frac{1}{\sqrt{3}} \right)$$ has norm 1 and is orthogonal to both $(1, i, 0)$ and $(0, i, -i)$.

20. Let C^2 have the Euclidean inner product. Which of the following form orthonormal sets?
    (a) $(i, 0), (0, 1 - i)$ (b) $\left(\frac{i}{\sqrt{2}}, -\frac{i}{\sqrt{2}} \right), \left(\frac{i}{\sqrt{2}}, \frac{i}{\sqrt{2}} \right)$
    (c) $\left(\frac{i}{\sqrt{2}}, \frac{i}{\sqrt{2}} \right), \left(-\frac{i}{\sqrt{2}}, -\frac{i}{\sqrt{2}} \right)$ (d) $(i, 0), (0, 0)$

21. Let C^3 have the Euclidean inner product. Which of the following form orthonormal sets?
    (a) $\left(\frac{i}{\sqrt{2}}, 0, \frac{i}{\sqrt{2}} \right), \left(\frac{i}{\sqrt{3}}, \frac{i}{\sqrt{3}}, -\frac{i}{\sqrt{3}} \right), \left(-\frac{i}{\sqrt{2}}, 0, \frac{i}{\sqrt{2}} \right)$
    (b) $\left(\frac{2}{3}i, -\frac{2}{3}i, \frac{1}{3}i \right), \left(\frac{2}{3}i, \frac{1}{3}i, -\frac{2}{3}i \right), \left(\frac{1}{3}i, \frac{2}{3}i, \frac{2}{3}i \right)$
    (c) $\left(\frac{i}{\sqrt{6}}, \frac{i}{\sqrt{6}}, -\frac{2i}{\sqrt{6}} \right), \left(\frac{i}{\sqrt{2}}, -\frac{i}{\sqrt{2}}, 0 \right)$

22. Let

$$\mathbf{x} = \left(\frac{i}{\sqrt{5}}, -\frac{i}{\sqrt{5}} \right) \quad \text{and} \quad \mathbf{y} = \left(\frac{2i}{\sqrt{30}}, \frac{3i}{\sqrt{30}} \right)$$

Show that $\{\mathbf{x}, \mathbf{y}\}$ is an orthonormal set if C^2 has the inner product

$$\langle \mathbf{u}, \mathbf{v} \rangle = 3u_1 \bar{v}_1 + 2u_2 \bar{v}_2$$

but is not orthonormal if C^2 has the Euclidean inner product.

23. Show that

$$\mathbf{u}_1 = (i, 0, 0, i), \quad \mathbf{u}_2 = (-i, 0, 2i, i), \quad \mathbf{u}_3 = (2i, 3i, 2i, -2i), \quad \mathbf{u}_4 = (-i, 2i, -i, i)$$

is an orthogonal set in C^4 with the Euclidean inner product. By normalizing each of these vectors, obtain an orthonormal set.

24. Let C^2 have the Euclidean inner product. Use the Gram–Schmidt process to transform the basis $\{\mathbf{u}_1, \mathbf{u}_2\}$ into an orthonormal basis.

(a) $\mathbf{u}_1 = (i, -3i)$, $\mathbf{u}_2 = (2i, 2i)$ (b) $\mathbf{u}_1 = (i, 0)$, $\mathbf{u}_2 = (3i, -5i)$

25. Let C^3 have the Euclidean inner product. Use the Gram–Schmidt process to transform the basis $\{\mathbf{u}_1, \mathbf{u}_2, \mathbf{u}_3\}$ into an orthonormal basis.

(a) $\mathbf{u}_1 = (i, i, i)$, $\mathbf{u}_2 = (-i, i, 0)$, $\mathbf{u}_3 = (i, 2i, i)$

(b) $\mathbf{u}_1 = (i, 0, 0)$, $\mathbf{u}_2 = (3i, 7i, -2i)$, $\mathbf{u}_3 = (0, 4i, i)$

26. Let C^4 have the Euclidean inner product. Use the Gram–Schmidt process to transform the basis $\{\mathbf{u}_1, \mathbf{u}_2, \mathbf{u}_3, \mathbf{u}_4\}$ into an orthonormal basis.

$$\mathbf{u}_1 = (0, 2i, i, 0), \quad \mathbf{u}_2 = (i, -i, 0, 0), \quad \mathbf{u}_3 = (i, 2i, 0, -i), \quad \mathbf{u}_4 = (i, 0, i, i)$$

27. Let C^3 have the Euclidean inner product. Find an orthonormal basis for the subspace spanned by $(0, i, 1 - i)$ and $(-i, 0, 1 + i)$.

28. Let C^4 have the Euclidean inner product. Express the vector $\mathbf{w} = (-i, 2i, 6i, 0)$ in the form $\mathbf{w} = \mathbf{w}_1 + \mathbf{w}_2$, where the vector $\mathbf{w}_1$ is in the space W spanned by $\mathbf{u}_1 = (-i, 0, i, 2i)$ and $\mathbf{u}_2 = (0, i, 0, i)$, and $\mathbf{w}_2$ is orthogonal to W.

29. (a) Prove: If k is a complex number and $\langle \mathbf{u}, \mathbf{v} \rangle$ is an inner product on a complex vector space, then $\langle \mathbf{u} - k\mathbf{v}, \mathbf{u} - k\mathbf{v} \rangle = \langle \mathbf{u}, \mathbf{u} \rangle - \bar{k}\langle \mathbf{u}, \mathbf{v} \rangle - k\overline{\langle \mathbf{u}, \mathbf{v} \rangle} + k\bar{k}\langle \mathbf{v}, \mathbf{v} \rangle$.

(b) Use the result in part (a) to prove that $0 \le \langle \mathbf{u}, \mathbf{u} \rangle - \bar{k}\langle \mathbf{u}, \mathbf{v} \rangle - k\overline{\langle \mathbf{u}, \mathbf{v} \rangle} + k\bar{k}\langle \mathbf{v}, \mathbf{v} \rangle$.

30. Prove that if $\mathbf{u}$ and $\mathbf{v}$ are vectors in a complex inner product space, then

$$|\langle \mathbf{u}, \mathbf{v} \rangle|^2 \le \langle \mathbf{u}, \mathbf{u} \rangle \langle \mathbf{v}, \mathbf{v} \rangle$$

This result, called the *Cauchy–Schwarz inequality for complex inner product spaces*, differs from its real analog (Theorem 6.2.1) in that an absolute value sign must be included on the left side.

Hint   Let $k = \langle \mathbf{u}, \mathbf{v} \rangle / \langle \mathbf{v}, \mathbf{v} \rangle$ in the inequality of Exercise 29(b).

31. Prove: If $\mathbf{u} = (u_1, u_2, \ldots, u_n)$ and $\mathbf{v} = (v_1, v_2, \ldots, v_n)$ are vectors in C^n, then

$$|u_1 \bar{v}_1 + u_2 \bar{v}_2 + \cdots + u_n \bar{v}_n| \le (|u_1|^2 + |u_2|^2 + \cdots + |u_n|^2)^{1/2}(|v_1|^2 + |v_2|^2 + \cdots + |v_n|^2)^{1/2}$$

This is the complex version of Formula (4) in Theorem 4.1.3.

Hint   Use Exercise 30.

32. Prove that equality holds in the Cauchy–Schwarz inequality for complex vector spaces if and only if $\mathbf{u}$ and $\mathbf{v}$ are linearly dependent.

33. Prove that if $\langle \mathbf{u}, \mathbf{v} \rangle$ is an inner product on a complex vector space, then

$$\langle \mathbf{0}, \mathbf{v} \rangle = \langle \mathbf{v}, \mathbf{0} \rangle = 0$$

34. Prove that if $\langle \mathbf{u}, \mathbf{v} \rangle$ is an inner product on a complex vector space, then

$$\langle \mathbf{u}, \mathbf{v} + \mathbf{w} \rangle = \langle \mathbf{u}, \mathbf{v} \rangle + \langle \mathbf{u}, \mathbf{w} \rangle$$

Charles Hermite
(1822–1901) was a French mathematician who made fundamental contributions to algebra, matrix theory, and various branches of analysis. He is noted for using integrals to solve a general fifth-degree polynomial equation. He also proved that the number e (the base for natural logarithms) is a transcendental number—that is, a number that is not the root of any polynomial equation with rational coefficients.

EXAMPLE 2 A 2 x 2 Unitary Matrix

The matrix

$$A = \begin{bmatrix} \dfrac{1+i}{2} & \dfrac{1+i}{2} \\ \dfrac{1-i}{2} & \dfrac{-1+i}{2} \end{bmatrix} \tag{1}$$

has row vectors

$$\mathbf{r}_1 = \left(\frac{1+i}{2}, \frac{1+i}{2} \right), \qquad \mathbf{r}_2 = \left(\frac{1-i}{2}, \frac{-1+i}{2} \right)$$

Relative to the Euclidean inner product on C^n, we have

$$\|\mathbf{r}_1\| = \sqrt{ \left| \frac{1+i}{2} \right|^2 + \left| \frac{1+i}{2} \right|^2 } = \sqrt{ \frac{1}{2} + \frac{1}{2} } = 1$$

$$\|\mathbf{r}_2\| = \sqrt{ \left| \frac{1-i}{2} \right|^2 + \left| \frac{-1+i}{2} \right|^2 } = \sqrt{ \frac{1}{2} + \frac{1}{2} } = 1$$

and

$$\mathbf{r}_1 \cdot \mathbf{r}_2 = \left(\frac{1+i}{2} \right) \overline{\left(\frac{1-i}{2} \right)} + \left(\frac{1+i}{2} \right) \overline{\left(\frac{-1+i}{2} \right)}$$

$$= \left(\frac{1+i}{2} \right) \left(\frac{1+i}{2} \right) + \left(\frac{1+i}{2} \right) \left(\frac{-1-i}{2} \right)$$

$$= \frac{i}{2} - \frac{i}{2} = 0$$

so the row vectors form an orthonormal set in C^2. Thus A is unitary and

$$A^{-1} = A^* = \begin{bmatrix} \dfrac{1-i}{2} & \dfrac{1+i}{2} \\ \dfrac{1-i}{2} & \dfrac{-1-i}{2} \end{bmatrix} \tag{2}$$

The reader should verify that matrix (2) is the inverse of matrix (1) by showing that $AA^* = A^*A = I$. ◆

Recall that a square matrix A with real entries is called *orthogonally diagonalizable* if there is an orthogonal matrix P such that $P^{-1}AP \ (= P^TAP)$ is diagonal. For complex matrices we have an analogous concept.

DEFINITION

A square matrix A with complex entries is called ***unitarily diagonalizable*** if there is a unitary P such that $P^{-1}AP \ (= P^*AP)$ is diagonal; the matrix P is said to ***unitarily diagonalize*** A.

We have two questions to consider:

- Which matrices are unitarily diagonalizable?
- How do we find a unitary matrix P to carry out the diagonalization?

Before pursuing these questions, we note that our earlier definitions of the terms *eigenvector*, *eigenvalue*, *eigenspace*, *characteristic equation*, and *characteristic polynomial* carry over without change to complex vector spaces.

Hermitian Matrices

In Section 7.3 we saw that the problem of orthogonally diagonalizing a matrix with real entries led to consideration of the symmetric matrices. The most natural complex analogs of the real symmetric matrices are the *Hermitian* matrices, which are defined as follows:

DEFINITION

A square matrix A with complex entries is called ***Hermitian*** if

$$A = A^*$$

EXAMPLE 3 A 3 x 3 Hermitian Matrix

If

$$A = \begin{bmatrix} 1 & i & 1+i \\ -i & -5 & 2-i \\ 1-i & 2+i & 3 \end{bmatrix}$$

then

$$\overline{A} = \begin{bmatrix} 1 & -i & 1-i \\ i & -5 & 2+i \\ 1+i & 2-i & 3 \end{bmatrix}, \quad \text{so} \quad A^* = \overline{A}^T = \begin{bmatrix} 1 & i & 1+i \\ -i & -5 & 2-i \\ 1-i & 2+i & 3 \end{bmatrix} = A$$

which means that A is Hermitian. ◆

It is easy to recognize Hermitian matrices by inspection: As seen in (3), the entries on the main diagonal are real numbers, and the "mirror image" of each entry across the main diagonal is its complex conjugate.

$$\begin{bmatrix} 1 & i & 1+i \\ -i & -5 & 2-i \\ 1-i & 2+i & 3 \end{bmatrix} \tag{3}$$

Normal Matrices

Hermitian matrices enjoy many but not all of the properties of real symmetric matrices. For example, just as the real symmetric matrices are orthogonally diagonalizable, so we shall see that the Hermitian matrices are unitarily diagonalizable. However, whereas the real symmetric matrices are the only matrices with real entries that are orthogonally diagonalizable (Theorem 7.3.1), the Hermitian matrices do not constitute the entire class of unitarily diagonalizable matrices; that is, there are unitarily diagonalizable matrices that are not Hermitian. To explain why this is so, we shall need the following definition:

DEFINITION

A square matrix A with complex entries is called ***normal*** if

$$AA^* = A^*A$$

EXAMPLE 4 Hermitian and Unitary Matrices

Every Hermitian matrix A is normal since $AA^* = AA = A^*A$, and every unitary matrix A is normal since $AA^* = I = A^*A$. ◆

The following two theorems are the complex analogs of Theorems 7.3.1 and 7.3.2. The proofs will be omitted.

THEOREM 10.6.3

Equivalent Statements

If A is a square matrix with complex entries, then the following are equivalent:

(a) A is unitarily diagonalizable.
(b) A has an orthonormal set of n eigenvectors.
(c) A is normal.

THEOREM 10.6.4

If A is a normal matrix, then eigenvectors from different eigenspaces of A are orthogonal.

Theorem 10.6.3 tells us that a square matrix A with complex entries is unitarily diagonalizable if and only if it is normal. Theorem 10.6.4 will be the key to constructing a matrix that unitarily diagonalizes a normal matrix.

Diagonalization Procedure

We saw in Section 7.3 that a symmetric matrix A is orthogonally diagonalized by any orthogonal matrix whose column vectors are eigenvectors of A. Similarly, a normal matrix A is diagonalized by any unitary matrix whose column vectors are eigenvectors of A. The procedure for diagonalizing a normal matrix is as follows:

Step 1. Find a basis for each eigenspace of A.

Step 2. Apply the Gram–Schmidt process to each of these bases to obtain an orthonormal basis for each eigenspace.

Step 3. Form the matrix P whose columns are the basis vectors constructed in Step 2. This matrix unitarily diagonalizes A.

The justification of this procedure should be clear. Theorem 10.6.4 ensures that eigenvectors from *different* eigenspaces are orthogonal, and the application of the Gram–Schmidt process ensures that the eigenvectors within the *same* eigenspace are orthonormal. Thus the *entire* set of eigenvectors obtained by this procedure is orthonormal. Theorem 10.6.3 ensures that this orthonormal set of eigenvectors is a basis.

EXAMPLE 5 Unitary Diagonalization

The matrix

$$A = \begin{bmatrix} 2 & 1+i \\ 1-i & 3 \end{bmatrix}$$

is unitarily diagonalizable because it is Hermitian and therefore normal. Find a matrix P that unitarily diagonalizes A.

Solution

The characteristic polynomial of A is

$$\det(\lambda I - A) = \det \begin{bmatrix} \lambda - 2 & -1 - i \\ -1 + i & \lambda - 3 \end{bmatrix} = (\lambda - 2)(\lambda - 3) - 2 = \lambda^2 - 5\lambda + 4$$

so the characteristic equation is

$$\lambda^2 - 5\lambda + 4 = (\lambda - 1)(\lambda - 4) = 0$$

and the eigenvalues are $\lambda = 1$ and $\lambda = 4$.

By definition,

$$\mathbf{x} = \begin{bmatrix} x_1 \\ x_2 \end{bmatrix}$$

will be an eigenvector of A corresponding to λ if and only if $\mathbf{x}$ is a nontrivial solution of

$$\begin{bmatrix} \lambda - 2 & -1 - i \\ -1 + i & \lambda - 3 \end{bmatrix} \begin{bmatrix} x_1 \\ x_2 \end{bmatrix} = \begin{bmatrix} 0 \\ 0 \end{bmatrix} \tag{4}$$

To find the eigenvectors corresponding to $\lambda = 1$, we substitute this value in (4):

$$\begin{bmatrix} -1 & -1 - i \\ -1 + i & -2 \end{bmatrix} \begin{bmatrix} x_1 \\ x_2 \end{bmatrix} = \begin{bmatrix} 0 \\ 0 \end{bmatrix}$$

Solving this system by Gauss–Jordan elimination yields (verify)

$$x_1 = (-1 - i)s, \qquad x_2 = s$$

Thus the eigenvectors of A corresponding to $\lambda = 1$ are the nonzero vectors in C^2 of the form

$$\mathbf{x} = \begin{bmatrix} (-1 - i)s \\ s \end{bmatrix} = s \begin{bmatrix} -1 - i \\ 1 \end{bmatrix}$$

Thus this eigenspace is one-dimensional with basis

$$\mathbf{u} = \begin{bmatrix} -1 - i \\ 1 \end{bmatrix} \tag{5}$$

In this case the Gram–Schmidt process involves only one step: normalizing this vector. Since

$$\|\mathbf{u}\| = \sqrt{|-1 - i|^2 + |1|^2} = \sqrt{2 + 1} = \sqrt{3}$$

the vector

$$\mathbf{p}_1 = \frac{\mathbf{u}}{\|\mathbf{u}\|} = \begin{bmatrix} \dfrac{-1 - i}{\sqrt{3}} \\ \dfrac{1}{\sqrt{3}} \end{bmatrix}$$

is an orthonormal basis for the eigenspace corresponding to $\lambda = 1$.

To find the eigenvectors corresponding to $\lambda = 4$, we substitute this value in (4):

$$\begin{bmatrix} 2 & -1 - i \\ -1 + i & 1 \end{bmatrix} \begin{bmatrix} x_1 \\ x_2 \end{bmatrix} = \begin{bmatrix} 0 \\ 0 \end{bmatrix}$$

Solving this system by Gauss–Jordan elimination yields (verify)

$$x_1 = \left(\frac{1 + i}{2}\right)s, \qquad x_2 = s$$

so the eigenvectors of A corresponding to $\lambda = 4$ are the nonzero vectors in C^2 of the form

$$\mathbf{x} = \begin{bmatrix} \left(\dfrac{1+i}{2}\right)s \\ s \end{bmatrix} = s\begin{bmatrix} \dfrac{1+i}{2} \\ 1 \end{bmatrix}$$

Thus the eigenspace is one-dimensional with basis

$$\mathbf{u} = \begin{bmatrix} \dfrac{1+i}{2} \\ 1 \end{bmatrix}$$

Applying the Gram–Schmidt process (that is, normalizing this vector) yields

$$\mathbf{p}_2 = \frac{\mathbf{u}}{\|\mathbf{u}\|} = \begin{bmatrix} \dfrac{1+i}{\sqrt{6}} \\ \dfrac{2}{\sqrt{6}} \end{bmatrix}$$

Thus

$$P = [\mathbf{p}_1 \mid \mathbf{p}_2] = \begin{bmatrix} \dfrac{-1-i}{\sqrt{3}} & \dfrac{1+i}{\sqrt{6}} \\ \dfrac{1}{\sqrt{3}} & \dfrac{2}{\sqrt{6}} \end{bmatrix}$$

diagonalizes A and

$$P^{-1}AP = \begin{bmatrix} 1 & 0 \\ 0 & 4 \end{bmatrix} \quad \blacklozenge$$

Eigenvalues of Hermitian and Symmetric Matrices

In Theorem 7.3.2 it was stated that the eigenvalues of a symmetric matrix with real entries are real numbers. This important result is a corollary of the following more general theorem.

THEOREM 10.6.5

> *The eigenvalues of a Hermitian matrix are real numbers.*

Proof If λ is an eigenvalue and $\mathbf{v}$ a corresponding eigenvector of an $n \times n$ Hermitian matrix A, then

$$A\mathbf{v} = \lambda\mathbf{v}$$

If we multiply each side of this equation on the left by $\mathbf{v}^*$ and then use the remark following Theorem 10.6.1 to write $\mathbf{v}^*\mathbf{v} = \|\mathbf{v}\|^2$ (with the Euclidean inner product on C^n), then we obtain

$$\mathbf{v}^*A\mathbf{v} = \mathbf{v}^*(\lambda\mathbf{v}) = \lambda\mathbf{v}^*\mathbf{v} = \lambda\|\mathbf{v}\|^2$$

But if we agree not to distinguish between the 1×1 matrix $\mathbf{v}^*A\mathbf{v}$ and its entry, and if we use the fact that eigenvectors are nonzero, then we can express λ as

$$\lambda = \frac{\mathbf{v}^*A\mathbf{v}}{\|\mathbf{v}\|^2} \tag{6}$$

Thus, to show that λ is a real number, it suffices to show that the entry of $\mathbf{v}^*A\mathbf{v}$ is real. One way to do this is to show that the matrix $\mathbf{v}^*A\mathbf{v}$ is Hermitian, since we know that Hermitian matrices have real numbers on the main diagonal. However,

$$(\mathbf{v}^*A\mathbf{v})^* = \mathbf{v}^*A^*(\mathbf{v}^*)^* = \mathbf{v}^*A\mathbf{v}$$

which shows that $\mathbf{v}^*A\mathbf{v}$ is Hermitian and completes the proof. ■

The proof of the following theorem is an immediate consequence of Theorem 10.6.5 and is left as an exercise.

THEOREM 10.6.6

The eigenvalues of a symmetric matrix with real entries are real numbers.

EXERCISE SET 10.6

1. In each part, find A^*.

(a) $A = \begin{bmatrix} 2i & 1-i \\ 4 & 3+i \\ 5+i & 0 \end{bmatrix}$

(b) $A = \begin{bmatrix} 2i & 1-i & -1+i \\ 4 & 5-7i & -i \\ i & 3 & 1 \end{bmatrix}$

(c) $A = [7i \quad 0 \quad -3i]$

(d) $A = \begin{bmatrix} a_{11} & a_{12} & a_{13} \\ a_{21} & a_{22} & a_{23} \end{bmatrix}$

2. Which of the following are Hermitian matrices?

(a) $\begin{bmatrix} 0 & i \\ i & 2 \end{bmatrix}$

(b) $\begin{bmatrix} 1 & 1+i \\ 1-i & -3 \end{bmatrix}$

(c) $\begin{bmatrix} i & i \\ -i & i \end{bmatrix}$

(d) $\begin{bmatrix} -2 & 1-i & -1+i \\ 1+i & 0 & 3 \\ -1-i & 3 & 5 \end{bmatrix}$

(e) $\begin{bmatrix} 1 & 0 & 0 \\ 0 & 1 & 0 \\ 0 & 0 & 1 \end{bmatrix}$

3. Find k, l, and m to make A a Hermitian matrix.

$$A = \begin{bmatrix} -1 & k & -i \\ 3-5i & 0 & m \\ l & 2+4i & 2 \end{bmatrix}$$

4. Use Theorem 10.6.2 to determine which of the following are unitary matrices.

(a) $\begin{bmatrix} i & 0 \\ 0 & i \end{bmatrix}$

(b) $\begin{bmatrix} \dfrac{i}{\sqrt{2}} & \dfrac{1}{\sqrt{2}} \\ -\dfrac{i}{\sqrt{2}} & \dfrac{1}{\sqrt{2}} \end{bmatrix}$

(c) $\begin{bmatrix} 1+i & 1+i \\ 1-i & -1+i \end{bmatrix}$

(d) $\begin{bmatrix} -\dfrac{i}{\sqrt{2}} & \dfrac{i}{\sqrt{6}} & \dfrac{i}{\sqrt{3}} \\ 0 & -\dfrac{i}{\sqrt{6}} & \dfrac{i}{\sqrt{3}} \\ \dfrac{i}{\sqrt{2}} & \dfrac{i}{\sqrt{6}} & \dfrac{i}{\sqrt{3}} \end{bmatrix}$

5. In each part, verify that the matrix is unitary and find its inverse.

(a) $\begin{bmatrix} \dfrac{3}{5} & \dfrac{4}{5}i \\ -\dfrac{4}{5} & \dfrac{3}{5}i \end{bmatrix}$

(b) $\begin{bmatrix} \dfrac{1}{\sqrt{2}} & \dfrac{1}{\sqrt{2}} \\ -\dfrac{1+i}{2} & \dfrac{1+i}{2} \end{bmatrix}$

(c) $\begin{bmatrix} \dfrac{1}{2\sqrt{2}}(\sqrt{3}+i) & \dfrac{1}{2\sqrt{2}}(1-i\sqrt{3}) \\[3mm] \dfrac{1}{2\sqrt{2}}(1+i\sqrt{3}) & \dfrac{1}{2\sqrt{2}}(i-\sqrt{3}) \end{bmatrix}$
(d) $\begin{bmatrix} \dfrac{1+i}{2} & -\dfrac{1}{2} & \dfrac{1}{2} \\[3mm] \dfrac{i}{\sqrt{3}} & \dfrac{1}{\sqrt{3}} & -\dfrac{i}{\sqrt{3}} \\[3mm] \dfrac{3+i}{2\sqrt{15}} & \dfrac{4+3i}{2\sqrt{15}} & \dfrac{5i}{2\sqrt{15}} \end{bmatrix}$

6. Show that the matrix

$$\frac{1}{\sqrt{2}} \begin{bmatrix} e^{i\theta} & e^{-i\theta} \\ ie^{i\theta} & -ie^{-i\theta} \end{bmatrix}$$

is unitary for every real value of θ.

In Exercises 7–12 find a unitary matrix P that diagonalizes A, and determine $P^{-1}AP$.

7. $A = \begin{bmatrix} 4 & 1-i \\ 1+i & 5 \end{bmatrix}$
8. $A = \begin{bmatrix} 3 & -i \\ i & 3 \end{bmatrix}$

9. $A = \begin{bmatrix} 6 & 2+2i \\ 2-2i & 4 \end{bmatrix}$
10. $\begin{bmatrix} 0 & 3+i \\ 3-i & -3 \end{bmatrix}$

11. $A = \begin{bmatrix} 5 & 0 & 0 \\ 0 & -1 & -1+i \\ 0 & -1-i & 0 \end{bmatrix}$
12. $A = \begin{bmatrix} 2 & \dfrac{i}{\sqrt{2}} & -\dfrac{i}{\sqrt{2}} \\[3mm] -\dfrac{i}{\sqrt{2}} & 2 & 0 \\[3mm] \dfrac{i}{\sqrt{2}} & 0 & 2 \end{bmatrix}$

13. Show that the eigenvalues of the symmetric matrix

$$A = \begin{bmatrix} 1 & 4i \\ 4i & 3 \end{bmatrix}$$

are not real. Does this violate Theorem 10.6.6?

14. (a) Find a 2×2 matrix that is both Hermitian and unitary and whose entries are not all real numbers.

(b) What can you say about the inverse of a matrix that is both Hermitian and unitary?

15. Prove: If A is an $n \times n$ matrix with complex entries, then $\det(\overline{A}) = \overline{\det(A)}$.

   Hint First show that the signed elementary products from $\overline{A}$ are the conjugates of the signed elementary products from A.

16. (a) Use the result of Exercise 15 to prove that if A is an $n \times n$ matrix with complex entries, then $\det(A^*) = \overline{\det(A)}$.

(b) Prove: If A is Hermitian, then $\det(A)$ is real.

(c) Prove: If A is unitary, then $|\det(A)| = 1$.

17. Prove that the entries on the main diagonal of a Hermitian matrix are real numbers.

18. Let

$$A = \begin{bmatrix} a_{11} & a_{12} & a_{13} \\ a_{21} & a_{22} & a_{23} \\ a_{31} & a_{32} & a_{33} \end{bmatrix} \quad \text{and} \quad B = \begin{bmatrix} b_{11} & b_{12} & b_{13} \\ b_{21} & b_{22} & b_{23} \\ b_{31} & b_{32} & b_{33} \end{bmatrix}$$

be matrices with complex entries. Show that

(a) $(A^*)^* = A$
(b) $(A+B)^* = A^* + B^*$

(c) $(kA)^* = \overline{k}A^*$
(d) $(AB)^* = B^*A^*$

19. Prove: If A is invertible, then so is A^*, in which case $(A^*)^{-1} = (A^{-1})^*$.

20. Show that if A is a unitary matrix, then A^* is also unitary.

21. Prove that an $n \times n$ matrix with complex entries is unitary if and only if its rows form an orthonormal set in C^n with the Euclidean inner product.

22. Use Exercises 20 and 21 to show that an $n \times n$ matrix is unitary if and only if its columns form an orthonormal set in C^n with the Euclidean inner product.

23. Let λ and μ be distinct eigenvalues of a Hermitian matrix A.

    (a) Prove that if $\mathbf{x}$ is an eigenvector corresponding to λ and $\mathbf{y}$ an eigenvector corresponding to μ, then $\mathbf{x}^* A \mathbf{y} = \lambda \mathbf{x}^* \mathbf{y}$ and $\mathbf{x}^* A \mathbf{y} = \mu \mathbf{x}^* \mathbf{y}$.

    (b) Prove Theorem 10.6.4.

CHAPTER 10

Supplementary Exercises

1. Let $\mathbf{u} = (u_1, u_2, \ldots, u_n)$ and $\mathbf{v} = (v_1, v_2, \ldots, v_n)$ be vectors in C^n, and let $\overline{\mathbf{u}} = (\overline{u}_1, \overline{u}_2, \ldots, \overline{u}_n)$ and $\overline{\mathbf{v}} = (\overline{v}_1, \overline{v}_2, \ldots, \overline{v}_n)$.

    (a) Prove: $\overline{\mathbf{u} \cdot \mathbf{v}} = \overline{\mathbf{u}} \cdot \overline{\mathbf{v}}$.

    (b) Prove: $\mathbf{u}$ and $\mathbf{v}$ are orthogonal if and only if $\overline{\mathbf{u}}$ and $\overline{\mathbf{v}}$ are orthogonal.

2. Show that if the matrix

$$\begin{bmatrix} a & b \\ -\overline{b} & \overline{a} \end{bmatrix}$$

is nonzero, then it is invertible.

3. Find a basis for the solution space of the system

$$\begin{bmatrix} -1 & -i & 1 \\ -i & 1 & i \\ 1 & i & -1 \end{bmatrix} \begin{bmatrix} x_1 \\ x_2 \\ x_3 \end{bmatrix} = \begin{bmatrix} 0 \\ 0 \\ 0 \end{bmatrix}$$

4. Prove: If a and b are complex numbers such that $|a|^2 + |b|^2 = 1$, and if θ is a real number, then

$$A = \begin{bmatrix} a & b \\ -e^{i\theta}\overline{b} & e^{i\theta}\overline{a} \end{bmatrix}$$

is a unitary matrix.

5. Find the eigenvalues of the matrix

$$\begin{bmatrix} 0 & 0 & 1 \\ 1 & 0 & \omega + 1 + \dfrac{1}{\omega} \\ 0 & 1 & -\omega - 1 - \dfrac{1}{\omega} \end{bmatrix}$$

where $\omega = e^{2\pi i/3}$.

6. Consider the relation between the complex number $z = a + ib$ and the corresponding 2×2 matrix with real entries $Z = \begin{bmatrix} a & -b \\ b & a \end{bmatrix}$.

    (a) How are the eigenvalues of Z related to z?

    (b) How is the complex number z related to the determinant of Z?

    (c) Show that z^{-1} corresponds to Z^{-1}.

    (d) Show that the product $(a + bi)(c + di)$ corresponds to the matrix that is the product of the matrices corresponding to $a + bi$ and $c + di$.

7. (a) Prove that if z is a complex number other than 1, then

$$1 + z + z^2 + \cdots + z^n = \frac{1 - z^{n+1}}{1 - z}$$

    Hint Let S be the sum on the left side of the equation and consider the quantity $S - zS$.

(b) Use the result in part (a) to prove that if $z^n = 1$ and $z \neq 1$, then $1 + z + z^2 + \cdots + z^{n-1} = 0$.

(c) Use the result in part (a) to obtain Lagrange's trigonometric identity

$$1 + \cos\theta + \cos 2\theta + \cdots + \cos n\theta = \frac{1}{2} + \frac{\sin\left[\left(n + \frac{1}{2}\right)\theta\right]}{2\sin(\theta/2)}$$

for $0 < \theta < 2\pi$.

Hint Let $z = \cos\theta + i\sin\theta$.

8. Let $\omega = e^{2\pi i/3}$. Show that the vectors $\mathbf{v}_1 = (1/\sqrt{3})(1, 1, 1)$, $\mathbf{v}_2 = (1/\sqrt{3})(1, \omega, \omega^2)$, and $\mathbf{v}_3 = (1/\sqrt{3})(1, \omega^2, \omega^4)$ form an orthonormal set in C^3.

Hint Use part (b) of Exercise 7.

9. Show that if U is an $n \times n$ unitary matrix and $|z_1| = |z_2| = \cdots = |z_n| = 1$, then the product

$$U \begin{bmatrix} z_1 & 0 & 0 & \cdots & 0 \\ 0 & z_2 & 0 & \cdots & 0 \\ \vdots & \vdots & \vdots & & \vdots \\ 0 & 0 & 0 & \cdots & z_n \end{bmatrix}$$

is also unitary.

10. Suppose that $A^* = -A$.

(a) Show that iA is Hermitian.

(b) Show that A is unitarily diagonalizable and has pure imaginary eigenvalues.

11. Show that the eigenvalues of a unitary matrix have modulus 1.

12. Under what conditions is the following matrix normal?

$$A = \begin{bmatrix} a & 0 & 0 \\ 0 & 0 & c \\ 0 & b & 0 \end{bmatrix}$$

13. Show that if $\mathbf{u}$ is a nonzero vector in C^n that is expressed in column form, then $P = \mathbf{u}\mathbf{u}^*$ is Hermitian and has rank 1.

14. Show that if $\mathbf{u}$ is a unit vector in C^n that is expressed in column form, then $H = I - 2\mathbf{u}\mathbf{u}^*$ is unitary and Hermitian. This is called a *Householder matrix*.

15. What geometric interpretations might you reasonably give to multiplication by the matrices $P = \mathbf{u}\mathbf{u}^*$ and $H = I - 2\mathbf{u}\mathbf{u}^*$ in Exercises 13 and 14?

CHAPTER 10
Technology Exercises

The following exercises are designed to be solved using a technology utility. Typically, this will be MATLAB, *Mathematica*, Maple, Derive, or Mathcad, but it may also be some other type of linear algebra software or a scientific calculator with some linear algebra capabilities. For each exercise you will need to read the relevant documentation for the particular utility you are using. The goal of these exercises is to provide you with a basic proficiency with your technology utility. Once you have mastered the techniques in these exercises, you will be able to use your technology utility to solve many of the problems in the regular exercise sets.

Sections 10.1 and 10.2   **T1.** **(Complex Numbers and Numerical Operations)** Read your documentation on entering and displaying complex numbers and for performing the basic arithmetic operations of addition, subtraction, multiplication, and division. Experiment with numbers of your own choosing until you feel you have mastered the operations.

T2. **(Matrices with Complex Entries)** For most technology utilities the procedures for adding, subtracting, multiplying, and inverting matrices with complex entries are the same as for matrices with real entries. Experiment with these operations on some matrices of your own choosing, and then try using your utility to solve some of the exercises in Sections 10.1 and 10.2.

T3. **(Complex Conjugate)** Read your documentation on finding the conjugate of a complex number, and then use your utility to perform the computations in Example 1 of Section 10.2.

Section 10.3 **T1.** **(Modulus and Argument)** Read your documentation on finding the modulus and argument of a complex number, and then use your utility to perform the computations in Example 1.

Section 10.6 **T1.** **(Conjugate Transpose)** Read your documentation on finding the conjugate transpose of a matrix with complex entries, and then use your utility to perform the computations in Examples 1 and 3.

T2. **(Unitary Diagonalization)** Use your technology utility to diagonalize the matrix A in Example 5 and to find a matrix P that unitarily diagonalizes A. (See Technology Exercise T1 of Section 7.2.)

ANSWERS TO EXERCISES

Exercise Set 1.1
(page 6)

1. (a), (c), (f)

3. (a) $x = \frac{3}{7} + \frac{5}{7}t$
   $y = t$

   (b) $x_1 = \frac{5}{3}s - \frac{4}{3}t + \frac{7}{3}$ $x_1 = \frac{1}{4}r - \frac{5}{8}s + \frac{3}{4}t - \frac{1}{8}$ $v = \frac{8}{3}q - \frac{2}{3}r + \frac{1}{3}s - \frac{4}{3}t$
         $x_2 = s$ $x_2 = r$ $w = q$
         $x_3 = t$ $x_3 = s$ $x = r$
                            $x_4 = t$ $y = s$
                                                 $z = t$

4. (a) $\begin{bmatrix} 3 & -2 & -1 \\ 4 & 5 & 3 \\ 7 & 3 & 2 \end{bmatrix}$ (b) $\begin{bmatrix} 2 & 0 & 2 & 1 \\ 3 & -1 & 4 & 7 \\ 6 & 1 & -1 & 0 \end{bmatrix}$

    (c) $\begin{bmatrix} 1 & 2 & 0 & -1 & 1 & 1 \\ 0 & 3 & 1 & 0 & -1 & 2 \\ 0 & 0 & 1 & 7 & 0 & 1 \end{bmatrix}$ (d) $\begin{bmatrix} 1 & 0 & 0 & 1 \\ 0 & 1 & 0 & 2 \\ 0 & 0 & 1 & 3 \end{bmatrix}$

5. (a) $\begin{aligned} 2x_1 &= 0 \\ 3x_1 - 4x_2 &= 0 \\ x_2 &= 1 \end{aligned}$ (b) $\begin{aligned} 3x_1 \quad\quad - 2x_3 &= 5 \\ 7x_1 + x_2 + 4x_3 &= -3 \\ -2x_2 + x_3 &= 7 \end{aligned}$

    (c) $\begin{aligned} 7x_1 + 2x_2 + x_3 - 3x_4 &= 5 \\ x_1 + 2x_2 + 4x_3 \quad\quad &= 1 \end{aligned}$ (d) $\begin{aligned} x_1 \quad\quad\quad\quad &= 7 \\ x_2 \quad\quad &= -2 \\ x_3 \quad &= 3 \\ x_4 &= 4 \end{aligned}$

6. (a) $x - 2y = 5$ (b) Let $x = t$; then $t - 2y = 5$. Solving for y yields $y = \frac{1}{2}t - \frac{5}{2}$.

12. (a) The lines have no common point of intersection.
    (b) The lines intersect in exactly one point.   (c) The three lines coincide.

Exercise Set 1.2
(page 19)

1. (a), (b), (c), (d), (h), (i), (j)

3. (a) Both   (b) Neither   (c) Both
   (d) Row-echelon   (e) Neither   (f) Both

4. (a) $x_1 = -3,\ x_2 = 0,\ x_3 = 7$
   (b) $x_1 = 7t + 8,\ x_2 = -3t + 2,\ x_3 = -t - 5,\ x_4 = t$
   (c) $x_1 = 6s - 3t - 2,\ x_2 = s,\ x_3 = -4t + 7,\ x_4 = -5t + 8,\ x_5 = t$
   (d) Inconsistent

6. (a) $x_1 = 3,\ x_2 = 1,\ x_3 = 2$ (b) $x_1 = -\frac{1}{7} - \frac{3}{7}t,\ x_2 = \frac{1}{7} - \frac{4}{7},\ x_3 = t$
   (c) $x = t - 1,\ y = 2s,\ z = s,\ w = t$ (d) Inconsistent

8. (a) Inconsistent   (b) $x_1 = -4,\ x_2 = 2,\ x_3 = 7$
   (c) $x_1 = 3 + 2t,\ x_2 = t$ (d) $x = \frac{8}{5} - \frac{3}{5}t - \frac{3}{5}s,\ y = \frac{1}{10} + \frac{2}{5}t - \frac{1}{10}s,\ z = t,\ w = s$

12. (a), (c), (d)

13. (a) $x_1 = 0,\ x_2 = 0,\ x_3 = 0$ (b) $x_1 = -s,\ x_2 = -t - s,\ x_3 = 4s,\ x_4 = t$
    (c) $w = t,\ x = -t,\ y = t,\ z = 0$

14. **(a)** Only the trivial solution   **(b)** $u = 7s - 5t,\ v = -6s + 4t,\ w = 2s,\ x = 2t$
(c) Only the trivial solution

15. **(a)** $I_1 = -1,\ I_2 = 0,\ I_3 = 1,\ I_4 = 2$
(b) $Z_1 = -s - t,\ Z_2 = s,\ Z_3 = -t,\ Z_4 = 0,\ Z_5 = t$

19. $\begin{bmatrix} 1 & 3 \\ 0 & 1 \end{bmatrix}$ and $\begin{bmatrix} 1 & 0 \\ 0 & 1 \end{bmatrix}$ are possible answers.   **20.** $\alpha = \pi/2,\ \beta = \pi,\ \gamma = 0$

23. If $\lambda = 1$, then $x_1 = x_2 = -\frac{1}{2}s,\ x_3 = s$
If $\lambda = 2$, then $x_1 = -\frac{1}{2}s,\ x_2 = 0,\ x_3 = s$

24. $x = -13/7,\ y = 91/54,\ z = -91/8$ **25.** $a = 1,\ b = -6,\ c = 2,\ d = 10$

30. **(a)** Three lines, at least two of which are distinct   **(b)** Three identical lines

32. **(a)** False   **(b)** False   **(c)** False   **(d)** False

Exercise Set 1.3
(page 34)

1. **(a)** Undefined   **(b)** 4×2 **(c)** Undefined   **(d)** Undefined
(e) 5×5 **(f)** 5×2 **(g)** Undefined   **(h)** 5×2

2. $a = 5,\ b = -3,\ c = 4,\ d = 1$

4. **(a)** $\begin{bmatrix} 7 & 2 & 4 \\ 3 & 5 & 7 \end{bmatrix}$ **(b)** $\begin{bmatrix} -5 & 0 & -1 \\ 4 & -1 & 1 \\ -1 & -1 & 1 \end{bmatrix}$ **(c)** $\begin{bmatrix} -5 & 0 & -1 \\ 4 & -1 & 1 \\ -1 & -1 & 1 \end{bmatrix}$

(d) Undefined   **(e)** $\begin{bmatrix} -\frac{1}{4} & \frac{3}{2} \\ \frac{9}{4} & 0 \\ \frac{3}{4} & \frac{9}{4} \end{bmatrix}$ **(f)** $\begin{bmatrix} 0 & -1 \\ 1 & 0 \end{bmatrix}$

(g) $\begin{bmatrix} 9 & 1 & -1 \\ -13 & 2 & -4 \\ 0 & 1 & -6 \end{bmatrix}$ **(h)** $\begin{bmatrix} 9 & -13 & 0 \\ 1 & 2 & 1 \\ -1 & -4 & -6 \end{bmatrix}$

5. **(a)** $\begin{bmatrix} 12 & -3 \\ -4 & 5 \\ 4 & 1 \end{bmatrix}$ **(b)** Undefined   **(c)** $\begin{bmatrix} 42 & 108 & 75 \\ 12 & -3 & 21 \\ 36 & 78 & 63 \end{bmatrix}$

(d) $\begin{bmatrix} 3 & 45 & 9 \\ 11 & -11 & 17 \\ 7 & 17 & 13 \end{bmatrix}$ **(e)** $\begin{bmatrix} 3 & 45 & 9 \\ 11 & -11 & 17 \\ 7 & 17 & 13 \end{bmatrix}$ **(f)** $\begin{bmatrix} 21 & 17 \\ 17 & 35 \end{bmatrix}$

(g) $\begin{bmatrix} 0 & -2 & 11 \\ 12 & 1 & 8 \end{bmatrix}$ **(h)** $\begin{bmatrix} 12 & 6 & 9 \\ 48 & -20 & 14 \\ 24 & 8 & 16 \end{bmatrix}$ **(i)** 61   **(j)** 35   **(k)** (28)

8. **(a)** $\begin{bmatrix} 67 \\ 64 \\ 63 \end{bmatrix} = 6\begin{bmatrix} 3 \\ 6 \\ 0 \end{bmatrix} + 0\begin{bmatrix} -2 \\ 5 \\ 4 \end{bmatrix} + 7\begin{bmatrix} 7 \\ 4 \\ 9 \end{bmatrix}$ **(b)** $\begin{bmatrix} 6 \\ 6 \\ 63 \end{bmatrix} = 3\begin{bmatrix} 6 \\ 0 \\ 7 \end{bmatrix} + 6\begin{bmatrix} -2 \\ 1 \\ 7 \end{bmatrix} + 0\begin{bmatrix} 4 \\ 3 \\ 5 \end{bmatrix}$

$\begin{bmatrix} 41 \\ 21 \\ 67 \end{bmatrix} = -2\begin{bmatrix} 3 \\ 6 \\ 0 \end{bmatrix} + 1\begin{bmatrix} -2 \\ 5 \\ 4 \end{bmatrix} + 7\begin{bmatrix} 7 \\ 4 \\ 9 \end{bmatrix}$ $\begin{bmatrix} -6 \\ 17 \\ 41 \end{bmatrix} = -2\begin{bmatrix} 6 \\ 0 \\ 7 \end{bmatrix} + 5\begin{bmatrix} -2 \\ 1 \\ 7 \end{bmatrix} + 4\begin{bmatrix} 4 \\ 3 \\ 5 \end{bmatrix}$

$\begin{bmatrix} 41 \\ 59 \\ 57 \end{bmatrix} = 4\begin{bmatrix} 3 \\ 6 \\ 0 \end{bmatrix} + 3\begin{bmatrix} -2 \\ 5 \\ 4 \end{bmatrix} + 5\begin{bmatrix} 7 \\ 4 \\ 9 \end{bmatrix}$ $\begin{bmatrix} 70 \\ 31 \\ 122 \end{bmatrix} = 7\begin{bmatrix} 6 \\ 0 \\ 7 \end{bmatrix} + 4\begin{bmatrix} -2 \\ 1 \\ 7 \end{bmatrix} + 9\begin{bmatrix} 4 \\ 3 \\ 5 \end{bmatrix}$

13. **(a)** $A = \begin{bmatrix} 2 & -3 & 5 \\ 9 & -1 & 1 \\ 1 & 5 & 4 \end{bmatrix}$, $\mathbf{x} = \begin{bmatrix} x_1 \\ x_2 \\ x_3 \end{bmatrix}$, $\mathbf{b} = \begin{bmatrix} 7 \\ -1 \\ 0 \end{bmatrix}$

(b) $A = \begin{bmatrix} 4 & 0 & -3 & 1 \\ 5 & 1 & 0 & -8 \\ 2 & -5 & 9 & -1 \\ 0 & 3 & -1 & 7 \end{bmatrix}$, $\mathbf{x} = \begin{bmatrix} x_1 \\ x_2 \\ x_3 \\ x_4 \end{bmatrix}$, $\mathbf{b} = \begin{bmatrix} 1 \\ 3 \\ 0 \\ 2 \end{bmatrix}$

16. **(a)** $\begin{bmatrix} -3 & -15 & -11 \\ 21 & -15 & 44 \end{bmatrix}$ **(b)** $\begin{bmatrix} 4 & -7 & -19 & -43 \\ 2 & 2 & 18 & 17 \\ 0 & 5 & 25 & 35 \\ 2 & 3 & 23 & 24 \end{bmatrix}$ **(c)** $\begin{bmatrix} 3 & 3 \\ -1 & 4 \\ 1 & 5 \\ 4 & -4 \\ 0 & 14 \end{bmatrix}$

17. **(a)** A_{11} is a 2×3 matrix and B_{11} is a 2×2 matrix. $A_{11}B_{11}$ does not exist.

(b) $\begin{bmatrix} -1 & 23 & -10 \\ 37 & -13 & 8 \\ 29 & 23 & 41 \end{bmatrix}$

21. **(a)** $\begin{bmatrix} a_{11} & 0 & 0 & 0 & 0 & 0 \\ 0 & a_{22} & 0 & 0 & 0 & 0 \\ 0 & 0 & a_{33} & 0 & 0 & 0 \\ 0 & 0 & 0 & a_{44} & 0 & 0 \\ 0 & 0 & 0 & 0 & a_{55} & 0 \\ 0 & 0 & 0 & 0 & 0 & a_{66} \end{bmatrix}$ **(b)** $\begin{bmatrix} a_{11} & a_{12} & a_{13} & a_{14} & a_{15} & a_{16} \\ 0 & a_{22} & a_{23} & a_{24} & a_{25} & a_{26} \\ 0 & 0 & a_{33} & a_{34} & a_{35} & a_{36} \\ 0 & 0 & 0 & a_{44} & a_{45} & a_{46} \\ 0 & 0 & 0 & 0 & a_{55} & a_{56} \\ 0 & 0 & 0 & 0 & 0 & a_{66} \end{bmatrix}$

(c) $\begin{bmatrix} a_{11} & 0 & 0 & 0 & 0 & 0 \\ a_{21} & a_{22} & 0 & 0 & 0 & 0 \\ a_{31} & a_{32} & a_{33} & 0 & 0 & 0 \\ a_{41} & a_{42} & a_{43} & a_{44} & 0 & 0 \\ a_{51} & a_{52} & a_{53} & a_{54} & a_{55} & 0 \\ a_{61} & a_{62} & a_{63} & a_{64} & a_{65} & a_{66} \end{bmatrix}$ **(d)** $\begin{bmatrix} a_{11} & a_{12} & 0 & 0 & 0 & 0 \\ a_{21} & a_{22} & a_{23} & 0 & 0 & 0 \\ 0 & a_{32} & a_{33} & a_{34} & 0 & 0 \\ 0 & 0 & a_{43} & a_{44} & a_{45} & 0 \\ 0 & 0 & 0 & a_{54} & a_{55} & a_{56} \\ 0 & 0 & 0 & 0 & a_{65} & a_{66} \end{bmatrix}$

27. One; namely, $A = \begin{bmatrix} 1 & 1 & 0 \\ 1 & -1 & 0 \\ 0 & 0 & 0 \end{bmatrix}$

30. **(a)** Yes; for example, $\begin{bmatrix} 0 & 1 \\ 0 & 0 \end{bmatrix}$ **(b)** Yes; for example, $\begin{bmatrix} 1 & 0 \\ 0 & 0 \end{bmatrix}$

32. **(a)** True **(b)** False; for example, $A = \begin{bmatrix} 1 & -1 \\ 1 & -1 \end{bmatrix}$ **(c)** True **(d)** True

Exercise Set 1.4 (page 48)

4. $A^{-1} = \begin{bmatrix} 2 & -1 \\ -5 & 3 \end{bmatrix}$, $B^{-1} = \begin{bmatrix} \frac{1}{5} & \frac{3}{20} \\ -\frac{1}{5} & \frac{1}{10} \end{bmatrix}$, $C^{-1} = \begin{bmatrix} -\frac{1}{2} & -2 \\ 1 & 3 \end{bmatrix}$, $D^{-1} = \begin{bmatrix} \frac{1}{2} & 0 \\ 0 & \frac{1}{3} \end{bmatrix}$

7. **(a)** $A = \begin{bmatrix} \frac{5}{13} & \frac{1}{13} \\ -\frac{3}{13} & \frac{2}{13} \end{bmatrix}$ **(b)** $A = \begin{bmatrix} \frac{2}{7} & 1 \\ \frac{1}{7} & \frac{3}{7} \end{bmatrix}$

(c) $A = \begin{bmatrix} -\frac{2}{5} & 1 \\ -\frac{1}{5} & \frac{3}{5} \end{bmatrix}$ **(d)** $A = \begin{bmatrix} -\frac{9}{13} & \frac{1}{13} \\ \frac{2}{13} & -\frac{6}{13} \end{bmatrix}$

9. (a) $p(A) = \begin{bmatrix} 1 & 1 \\ 2 & -1 \end{bmatrix}$ (b) $p(A) = \begin{bmatrix} 20 & 7 \\ 14 & 6 \end{bmatrix}$ (c) $p(A) = \begin{bmatrix} 39 & 13 \\ 26 & 13 \end{bmatrix}$

11. $\begin{bmatrix} \cos\theta & -\sin\theta \\ \sin\theta & \cos\theta \end{bmatrix}$ 13. $A^{-1} = \begin{bmatrix} \frac{1}{a_{11}} & 0 & \cdots & 0 \\ 0 & \frac{1}{a_{22}} & \cdots & 0 \\ \vdots & \vdots & & \vdots \\ 0 & 0 & \cdots & \frac{1}{a_{nn}} \end{bmatrix}$ 18. $C = -A^{-1}BA^{-1}$

19. (a) $\begin{bmatrix} \frac{1}{2} & -\frac{1}{2} & 0 & 0 \\ \frac{1}{2} & \frac{1}{2} & 0 & 0 \\ 0 & 0 & \frac{1}{2} & -\frac{1}{2} \\ -1 & 0 & \frac{1}{2} & \frac{1}{2} \end{bmatrix}$ (b) $\begin{bmatrix} 1 & -1 & 0 & 0 \\ 0 & 1 & 0 & 0 \\ 0 & 0 & 1 & -1 \\ 0 & 0 & 0 & 1 \end{bmatrix}$

20. (a) One example is $\begin{bmatrix} 1 & 2 & 3 \\ 2 & 1 & 4 \\ 3 & 4 & 5 \end{bmatrix}$. (b) One example is $\begin{bmatrix} 0 & -1 & -1 \\ 1 & 0 & -1 \\ 1 & 1 & 0 \end{bmatrix}$.

22. Yes 23. $A^{-1} = \begin{bmatrix} \frac{1}{2} & \frac{1}{2} & -\frac{1}{2} \\ -\frac{1}{2} & \frac{1}{2} & \frac{1}{2} \\ \frac{1}{2} & -\frac{1}{2} & \frac{1}{2} \end{bmatrix}$ 33. $\begin{bmatrix} \pm 1 & 0 & 0 \\ 0 & \pm 1 & 0 \\ 0 & 0 & \pm 1 \end{bmatrix}$

34. (a) If A is invertible, then A^T is invertible. (b) True

**Exercise Set 1.5
(page 57)**

1. (a), (c), (d), (f)

3. (a) $\begin{bmatrix} 0 & 0 & 1 \\ 0 & 1 & 0 \\ 1 & 0 & 0 \end{bmatrix}$ (b) $\begin{bmatrix} 0 & 0 & 1 \\ 0 & 1 & 0 \\ 1 & 0 & 0 \end{bmatrix}$ (c) $\begin{bmatrix} 1 & 0 & 0 \\ 0 & 1 & 0 \\ -2 & 0 & 1 \end{bmatrix}$ (d) $\begin{bmatrix} 1 & 0 & 0 \\ 0 & 1 & 0 \\ 2 & 0 & 1 \end{bmatrix}$

6. (a) $\begin{bmatrix} -7 & 4 \\ 2 & -1 \end{bmatrix}$ (b) $\begin{bmatrix} -\frac{5}{39} & \frac{2}{13} \\ \frac{4}{39} & \frac{1}{13} \end{bmatrix}$ (c) Not invertible

8. (a) $\begin{bmatrix} 1 & 3 & 1 \\ 0 & 1 & -1 \\ -2 & 2 & 0 \end{bmatrix}$ (b) $\begin{bmatrix} \frac{\sqrt{2}}{26} & \frac{-3\sqrt{2}}{26} & 0 \\ \frac{4\sqrt{2}}{26} & \frac{\sqrt{2}}{26} & 0 \\ 0 & 0 & 1 \end{bmatrix}$ (c) $\begin{bmatrix} 1 & 0 & 0 & 0 \\ -\frac{1}{3} & \frac{1}{3} & 0 & 0 \\ 0 & -\frac{1}{5} & \frac{1}{5} & 0 \\ 0 & 0 & -\frac{1}{7} & \frac{1}{7} \end{bmatrix}$

(d) Not invertible (e) $\begin{bmatrix} -\frac{4}{5} & \frac{3}{5} & \frac{1}{5} & \frac{1}{5} \\ \frac{3}{2} & 0 & -1 & 0 \\ \frac{1}{2} & 0 & 0 & 0 \\ \frac{4}{5} & \frac{2}{5} & -\frac{1}{5} & -\frac{1}{5} \end{bmatrix}$

10. (a) $E_1 = \begin{bmatrix} 1 & 0 \\ 5 & 1 \end{bmatrix}$, $E_2 = \begin{bmatrix} 1 & 0 \\ 0 & \frac{1}{2} \end{bmatrix}$ (b) $A^{-1} = E_2 E_1$ (c) $A = E_1^{-1} E_2^{-1}$

11. (a) $\begin{bmatrix} 1 & -4 & 7 \\ 4 & 5 & -3 \\ 2 & -1 & 0 \end{bmatrix}$ (b) $\begin{bmatrix} 2 & -1 & 0 \\ \frac{4}{3} & \frac{5}{3} & -1 \\ 1 & -4 & 7 \end{bmatrix}$ (c) $\begin{bmatrix} 10 & 9 & -6 \\ 4 & 5 & -3 \\ 1 & -4 & 7 \end{bmatrix}$

14. $\begin{bmatrix} 0 & 1 & 0 \\ 1 & 0 & 0 \\ 0 & 0 & 1 \end{bmatrix} \begin{bmatrix} 1 & 0 & 0 \\ 0 & 1 & 0 \\ -2 & 0 & 1 \end{bmatrix} \begin{bmatrix} 1 & 0 & 0 \\ 0 & 1 & 0 \\ 0 & 1 & 1 \end{bmatrix} \begin{bmatrix} 1 & 3 & 3 & 8 \\ 0 & 1 & 7 & 8 \\ 0 & 0 & 0 & 0 \end{bmatrix}$

19. **(b)** Add -1 times the first row to the second row.
Add -1 times the first row to the third row.
Add -1 times the second row to the first row.
Add the second row to the third row.

24. In general, no. Try $b = 1$, $a = c = d = 0$.

**Exercise Set 1.6
(page 66)**

1. $x_1 = 3$, $x_2 = -1$ **4.** $x_1 = 1$, $x_2 = -11$, $x_3 = 16$

6. $w = -6$, $x = 1$, $y = 10$, $z = -7$

9. **(a)** $x_1 = \frac{16}{3}$, $x_2 = -\frac{4}{3}$, $x_3 = -\frac{11}{3}$ **(b)** $x_1 = -\frac{5}{3}$, $x_2 = \frac{5}{3}$, $x_3 = \frac{10}{3}$
(c) $x_1 = 3$, $x_2 = 0$, $x_3 = -4$

11. **(a)** $x_1 = \frac{22}{17}$, $x_2 = \frac{1}{17}$ **(b)** $x_1 = \frac{21}{17}$, $x_2 = \frac{11}{17}$

13. **(a)** $x_1 = \frac{7}{15}$, $x_2 = \frac{4}{15}$ **(b)** $x_1 = \frac{34}{15}$, $x_2 = \frac{28}{15}$
(c) $x_1 = \frac{19}{15}$, $x_2 = \frac{13}{15}$ **(d)** $x_1 = -\frac{1}{5}$, $x_2 = \frac{3}{5}$

15. **(a)** $x_1 = -12 - 3t$, $x_2 = -5 - t$, $x_3 = t$ **(b)** $x_1 = 7 - 3t$, $x_2 = 3 - t$, $x_3 = t$

19. $b_1 = b_3 + b_4$, $b_2 = 2b_3 + b_4$ **21.** $X = \begin{bmatrix} 11 & 12 & -3 & 27 & 26 \\ -6 & -8 & 1 & -18 & -17 \\ -15 & -21 & 9 & -38 & -35 \end{bmatrix}$

22. **(a)** Only the trivial solution $x_1 = x_2 = x_3 = x_4 = 0$; invertible
(b) Infinitely many solutions; not invertible

28. **(a)** $I - A$ is invertible.   **(b)** $\mathbf{x} = (I - A)^{-1}\mathbf{b}$

30. Yes, for nonsquare matrices

**Exercise Set 1.7
(page 73)**

1. **(a)** $\begin{bmatrix} \frac{1}{2} & 0 \\ 0 & -\frac{1}{5} \end{bmatrix}$ **(b)** Not invertible   **(c)** $\begin{bmatrix} -1 & 0 & 0 \\ 0 & \frac{1}{2} & 0 \\ 0 & 0 & 3 \end{bmatrix}$

3. **(a)** $A^2 = \begin{bmatrix} 1 & 0 \\ 0 & 4 \end{bmatrix}$, $A^{-2} = \begin{bmatrix} 1 & 0 \\ 0 & \frac{1}{4} \end{bmatrix}$, $A^{-k} = \begin{bmatrix} 1 & 0 \\ 0 & 1/(-2)^k \end{bmatrix}$

(b) $A^2 = \begin{bmatrix} \frac{1}{4} & 0 & 0 \\ 0 & \frac{1}{9} & 0 \\ 0 & 0 & \frac{1}{16} \end{bmatrix}$, $A^{-2} = \begin{bmatrix} 4 & 0 & 0 \\ 0 & 9 & 0 \\ 0 & 0 & 16 \end{bmatrix}$, $A^{-k} = \begin{bmatrix} 2^k & 0 & 0 \\ 0 & 3^k & 0 \\ 0 & 0 & 4^k \end{bmatrix}$

5. **(a)**   **7.** $a = 2$, $b = -1$

10. **(a)** $\begin{bmatrix} 1 & 0 & 0 \\ 0 & -1 & 0 \\ 0 & 0 & -1 \end{bmatrix}$ **(b)** $\begin{bmatrix} \pm\frac{1}{3} & 0 & 0 \\ 0 & \pm\frac{1}{2} & 0 \\ 0 & 0 & \pm 1 \end{bmatrix}$

11. **(a)** $\begin{bmatrix} a_{11} & a_{12} & a_{13} \\ a_{21} & a_{22} & a_{23} \\ a_{31} & a_{32} & a_{33} \end{bmatrix}\begin{bmatrix} 3 & 0 & 0 \\ 0 & 5 & 0 \\ 0 & 0 & 7 \end{bmatrix}$ **(b)** No

16. **(b)** Yes   **17.** Yes

19. $\begin{bmatrix} 4 & 0 & 0 \\ 0 & 4 & 0 \\ 0 & 0 & 4 \end{bmatrix}$, $\begin{bmatrix} 4 & 0 & 0 \\ 0 & 4 & 0 \\ 0 & 0 & -1 \end{bmatrix}$, $\begin{bmatrix} 4 & 0 & 0 \\ 0 & -1 & 0 \\ 0 & 0 & 4 \end{bmatrix}$, $\begin{bmatrix} -1 & 0 & 0 \\ 0 & 4 & 0 \\ 0 & 0 & 4 \end{bmatrix}$,
$\begin{bmatrix} -1 & 0 & 0 \\ 0 & -1 & 0 \\ 0 & 0 & 4 \end{bmatrix}$, $\begin{bmatrix} -1 & 0 & 0 \\ 0 & 4 & 0 \\ 0 & 0 & -1 \end{bmatrix}$, $\begin{bmatrix} 4 & 0 & 0 \\ 0 & -1 & 0 \\ 0 & 0 & -1 \end{bmatrix}$, $\begin{bmatrix} -1 & 0 & 0 \\ 0 & -1 & 0 \\ 0 & 0 & -1 \end{bmatrix}$

20. **(a)** Yes   **(b)** No (unless $n = 1$)   **(c)** Yes   **(d)** No (unless $n = 1$)

24. **(a)** $x_1 = \frac{7}{4}$, $x_2 = 1$, $x_3 = -\frac{1}{2}$ **(b)** $x_1 = -8$, $x_2 = -4$, $x_3 = 3$

25. $A = \begin{bmatrix} 1 & 10 \\ 0 & -2 \end{bmatrix}$ **26.** $\frac{n}{2}(1+n)$

Supplementary Exercises (page 76)

1. $x' = \frac{3}{5}x + \frac{4}{5}y, \; y' = -\frac{4}{5}x + \frac{3}{5}y$

3. One possible answer is
$$x_1 - 2x_2 - x_3 - x_4 = 0$$
$$x_1 + 5x_2 + 2x_4 \quad\quad = 0$$

5. $x = 4, \; y = 2, \; z = 3$

7. **(a)** $a \neq 0, \; b \neq 2$ **(b)** $a \neq 0, \; b = 2$ **(c)** $a = 0, \; b = 2$ **(d)** $a = 0, \; b \neq 2$

9. $K = \begin{bmatrix} 0 & 2 \\ 1 & 1 \end{bmatrix}$

11. **(a)** $X = \begin{bmatrix} -1 & 3 & -1 \\ 6 & 0 & 1 \end{bmatrix}$ **(b)** $X = \begin{bmatrix} 1 & -2 \\ 3 & 1 \end{bmatrix}$ **(c)** $X = \begin{bmatrix} -\frac{113}{37} & -\frac{160}{37} \\ -\frac{20}{37} & -\frac{46}{37} \end{bmatrix}$

13. mpn multiplications and $mp(n-1)$ additions   **15.** $a = 1, \; b = -2, \; c = 3$

16. $a = 1, \; b = -4, \; c = -5$ **26.** $A = -\frac{7}{5}, \; B = \frac{4}{5}, \; C = \frac{3}{5}$

29. **(b)** $\begin{bmatrix} a^n & 0 & 0 \\ 0 & b^n & 0 \\ d & 0 & c^n \end{bmatrix}$, where $d = \begin{cases} \dfrac{a^n - c^n}{a - c} & \text{if } a \neq c \\ na^{n-1} & \text{if } a = c \end{cases}$

Exercise Set 2.1 (page 94)

1. **(a)** $M_{11} = 29, \; M_{12} = 21, \; M_{13} = 27, \; M_{21} = -11, \; M_{22} = 13,$
$M_{23} = -5, \; M_{31} = -19, \; M_{32} = -19, \; M_{33} = 19$

(b) $C_{11} = 29, \; C_{12} = -21, \; C_{13} = 27, \; C_{21} = 11, \; C_{22} = 13,$
$C_{23} = 5, \; C_{31} = -19, \; C_{32} = 19, \; C_{33} = 19$

3. 152

4. **(a)** $\text{adj}(A) = \begin{bmatrix} 29 & 11 & -19 \\ -21 & 13 & 19 \\ 27 & 5 & 19 \end{bmatrix}$ **(b)** $A^{-1} = \begin{bmatrix} \frac{29}{152} & \frac{11}{152} & -\frac{19}{152} \\ -\frac{21}{152} & \frac{13}{152} & \frac{19}{152} \\ \frac{27}{152} & \frac{5}{152} & \frac{19}{152} \end{bmatrix}$

6. -66 **8.** $k^3 - 8k^2 - 10k + 95$ **11.** $A^{-1} = \begin{bmatrix} 3 & -5 & -5 \\ -3 & 4 & 5 \\ 2 & -2 & -3 \end{bmatrix}$

13. $A^{-1} = \begin{bmatrix} \frac{1}{2} & \frac{3}{2} & 1 \\ 0 & 1 & \frac{3}{2} \\ 0 & 0 & \frac{1}{2} \end{bmatrix}$ **15.** $A^{-1} = \begin{bmatrix} -4 & 3 & 0 & -1 \\ 2 & -1 & 0 & 0 \\ -7 & 0 & -1 & 8 \\ 6 & 0 & 1 & -7 \end{bmatrix}$

16. $x_1 = 1, \; x_2 = 2$ **18.** $x = -\frac{144}{55}, \; y = -\frac{61}{55}, \; z = \frac{46}{11}$

21. Cramer's rule does not apply.   **22.** $A^{-1} = \begin{bmatrix} \cos\theta & -\sin\theta & 0 \\ \sin\theta & \cos\theta & 0 \\ 0 & 0 & 1 \end{bmatrix}$

24. $x = 1, \; y = 0, \; z = 2, \; w = 0$ **31.** $\det(A) = 10 \times (-108) = -1080$ **34.** One

Exercise Set 2.2 (page 101)

2. **(a)** -30 **(b)** -2 **(c)** 0 **(d)** 0

4. 30   **6.** -17 **8.** 39   **11.** -2

12. **(a)** -6 **(b)** 72   **(c)** -6 **(d)** 18

16. **(a)** $\det(A) = -1$ **(b)** $\det(A) = 1$ **18.** $x = 0, \; -1, \; \frac{1}{2}$

Exercise Set 2.3
(page 109)

1. **(a)** $\det(2A) = -40 = 2^2 \det(A)$ **(b)** $\det(-2A) = -448 = (-2)^3 \det(A)$

4. **(a)** Invertible **(b)** Not invertible **(c)** Not invertible **(d)** Not invertible

6. If $x = 0$, the first and third rows are proportional.
   If $x = 2$, the first and second rows are proportional.

12. **(a)** $k = \dfrac{5 \pm \sqrt{17}}{2}$ **(b)** $k = -1$

14. **(a)** $\begin{bmatrix} \lambda - 1 & -2 \\ -2 & \lambda - 1 \end{bmatrix} \begin{bmatrix} x_1 \\ x_2 \end{bmatrix} = \begin{bmatrix} 0 \\ 0 \end{bmatrix}$ **(b)** $\begin{bmatrix} \lambda - 2 & -3 \\ -4 & \lambda - 3 \end{bmatrix} \begin{bmatrix} x_1 \\ x_2 \end{bmatrix} = \begin{bmatrix} 0 \\ 0 \end{bmatrix}$

    (c) $\begin{bmatrix} \lambda - 3 & -1 \\ 5 & \lambda + 3 \end{bmatrix} \begin{bmatrix} x_1 \\ x_2 \end{bmatrix} = \begin{bmatrix} 0 \\ 0 \end{bmatrix}$

15. (i) $\lambda^2 - 2\lambda - 3 = 0$ (ii) $\lambda = -1$, $\lambda = 3$ (iii) $\begin{bmatrix} -t \\ t \end{bmatrix}$, $\begin{bmatrix} t \\ t \end{bmatrix}$

    (i) $\lambda^2 - 5\lambda - 6 = 0$ (ii) $\lambda = -1$, $\lambda = 6$ (iii) $\begin{bmatrix} -t \\ t \end{bmatrix}$, $\begin{bmatrix} \frac{3}{4}t \\ t \end{bmatrix}$

    (i) $\lambda^2 - 4 = 0$ (ii) $\lambda = -2$, $\lambda = 2$ (iii) $\begin{bmatrix} -\frac{t}{5} \\ t \end{bmatrix}$, $\begin{bmatrix} -t \\ t \end{bmatrix}$

20. No 21. AB is singular.

22. **(a)** False **(b)** True **(c)** False **(d)** True

23. **(a)** True **(b)** True **(c)** False **(d)** True

Exercise Set 2.4
(page 117)

1. **(a)** 5 **(b)** 9 **(c)** 6 **(d)** 10 **(e)** 0 **(f)** 2

3. 22 5. 52 7. $a^2 - 5a + 21$ 9. -65 11. -123

13. **(a)** $\lambda = 1$, $\lambda = -3$ **(b)** $\lambda = -2$, $\lambda = 3$, $\lambda = 4$ 16. 275

17. **(a)** $= -120$ **(b)** $= -120$ 18. $x = \dfrac{3 \pm \sqrt{33}}{4}$ 22. Equals 0 if $n > 1$

Supplementary
Exercises
(page 118)

1. $x' = \frac{3}{5}x + \frac{4}{5}y$, $y' = -\frac{4}{5}x + \frac{3}{5}y$ 4. 2

5. $\cos \beta = \dfrac{c^2 + a^2 - b^2}{2ac}$, $\cos \gamma = \dfrac{a^2 + b^2 - c^2}{2ab}$ 12. $\det(B) = (-1)^{n(n-1)/2} \det(A)$

13. **(a)** The ith and jth columns will be interchanged.
    (b) The ith column will be divided by c.
    (c) $-c$ times the jth column will be added to the ith column.

15. **(a)** $\lambda^3 + (-a_{11} - a_{22} - a_{33})\lambda^2$
    $+ (a_{11}a_{22} + a_{11}a_{33} + a_{22}a_{33} - a_{12}a_{21} - a_{13}a_{31} - a_{23}a_{32})\lambda$
    $+ (a_{11}a_{23}a_{32} + a_{12}a_{21}a_{33} + a_{13}a_{22}a_{31} - a_{11}a_{22}a_{33} - a_{12}a_{23}a_{31} - a_{13}a_{21}a_{32})$

18. **(a)** $\lambda = -5$, $\lambda = 2$, $\lambda = 4$; $\begin{bmatrix} -2t \\ t \\ t \end{bmatrix}$, $\begin{bmatrix} 5t \\ t \\ t \end{bmatrix}$, $\begin{bmatrix} 7t \\ 19t \\ t \end{bmatrix}$ **(b)** $\lambda = 1$; $\begin{bmatrix} \frac{1}{2}t \\ -\frac{1}{2}t \\ t \end{bmatrix}$

Exercise Set 3.1
(page 130)

3. **(a)** $\overrightarrow{P_1 P_2} = (-1, -1)$ **(b)** $\overrightarrow{P_1 P_2} = (-7, -2)$ **(c)** $\overrightarrow{P_1 P_2} = (2, 1)$
   (d) $\overrightarrow{P_1 P_2} = (a, b)$ **(e)** $\overrightarrow{P_1 P_2} = (-5, 12, -6)$ **(f)** $\overrightarrow{P_1 P_2} = (1, -1, -2)$
   (g) $\overrightarrow{P_1 P_2} = (-a, -b, -c)$ **(h)** $\overrightarrow{P_1 P_2} = (a, b, c)$

5. **(a)** $P(-1, 2, -4)$ is one possible answer. **(b)** $P(7, -2, -6)$ is one possible answer.

6. **(a)** $(-2, 1, -4)$ **(b)** $(-10, 6, 4)$ **(c)** $(-7, 1, 10)$
   (d) $(80, -20, -80)$ **(e)** $(132, -24, -72)$ **(f)** $(-77, 8, 94)$

8. $c_1 = 2$, $c_2 = -1$, $c_3 = 2$ 10. $c_1 = c_2 = c_3 = 0$

12. (a) $x' = 5$, $y' = 8$ (b) $x = -1$, $y = 3$

15. $\mathbf{u} = \left(\frac{\sqrt{3}}{2}, \frac{1}{2}\right)$, $\mathbf{v} = \left(-\frac{1}{2}, -\frac{\sqrt{3}}{2}\right)$,

$\mathbf{u} + \mathbf{v} = \left(\frac{\sqrt{3}-1}{2}, \frac{1-\sqrt{3}}{2}\right)$, $\mathbf{u} - \mathbf{v} = \left(\frac{\sqrt{3}+1}{2}, \frac{\sqrt{3}+1}{2}\right)$

Exercise Set 3.2
(page 134)

1. (a) 5   (b) $\sqrt{13}$ (c) 5   (d) $2\sqrt{3}$ (e) $3\sqrt{6}$ (f) 6

3. (a) $\sqrt{83}$ (b) $\sqrt{17} + \sqrt{26}$ (c) $4\sqrt{17}$ (d) $\sqrt{466}$

(e) $\left(\frac{3}{\sqrt{61}}, \frac{6}{\sqrt{61}}, -\frac{4}{\sqrt{61}}\right)$ (f) 1

9. (b) $\left(\frac{3}{5}, \frac{4}{5}\right)$ (c) $\left(\frac{2}{7}, -\frac{3}{7}, \frac{6}{7}\right)$

10. A sphere of radius 1 centered at (x_0, y_0, z_0)

16. (a) $a = c = 0$ (b) At least one of a or c is not zero, that is, $a^2 + c^2 > 0$

17. (a) The distance from x to the origin is less than 1.   (b) $\|x - x_0\| > 1$

Exercise Set 3.3
(page 142)

1. (a) -11 (b) -24 (c) 0   (d) 0

3. (a) Orthogonal   (b) Obtuse   (c) Acute   (d) Obtuse

5. (a) $(6, 2)$ (b) $\left(-\frac{21}{13}, -\frac{14}{13}\right)$ (c) $\left(\frac{55}{13}, 1, -\frac{11}{13}\right)$ (d) $\left(\frac{73}{89}, -\frac{12}{89}, -\frac{32}{89}\right)$

8. (b) $(3k, 2k)$ for any scalar k (c) $\left(\frac{4}{5}, \frac{3}{5}\right), \left(-\frac{4}{5}, -\frac{3}{5}\right)$

11. $\cos\theta_1 = \frac{\sqrt{10}}{10}$, $\cos\theta_2 = \frac{3\sqrt{10}}{10}$, $\cos\theta_3 = 0$ 13. $\pm(1/\sqrt{3}, 1/\sqrt{3}, -1/\sqrt{3})$

16. (a) $\frac{10}{3}$ (b) $-\frac{6}{5}$ (c) $\frac{-60+34\sqrt{3}}{33}$ (d) $\frac{1}{2}$

20. $\cos^{-1}\left(\frac{2}{\sqrt{6}}\right)$ 21. (b) $\cos\beta = \frac{b}{\|\mathbf{v}\|}$, $\cos\gamma = \frac{c}{\|\mathbf{v}\|}$

27. (a) The vector $\mathbf{u}$ is dotted with a scalar.   (b) A scalar is added to the vector $\mathbf{w}$.
(c) Scalars do not have norms.   (d) The scalar k is dotted with a vector.

29. No; it merely says that $\mathbf{u}$ is orthogonal to $\mathbf{v} - \mathbf{w}$.

30. $\mathbf{r} = (\mathbf{u} \cdot \mathbf{r})\frac{\mathbf{u}}{\|\mathbf{u}\|^2} + (\mathbf{v} \cdot \mathbf{r})\frac{\mathbf{v}}{\|\mathbf{v}\|^2} + (\mathbf{w} \cdot \mathbf{r})\frac{\mathbf{w}}{\|\mathbf{w}\|^2}$ 31. Theorem of Pythagoras

Exercise Set 3.4
(page 153)

1. (a) $(32, -6, -4)$ (b) $(-14, -20, -82)$ (c) $(27, 40, -42)$
(d) $(0, 176, -264)$ (e) $(-44, 55, -22)$ (f) $(-8, -3, -8)$

3. (a) $\sqrt{59}$ (b) $\sqrt{101}$ (c) 0

7. For example, $(1, 1, 1) \times (2, -3, 5) = (8, -3, -5)$

9. (a) -3 (b) 3   (c) 3   (d) -3 (e) -3 (f) 0

11. (a) No   (b) Yes   (c) No   13. $\left(\frac{6}{\sqrt{61}}, -\frac{3}{\sqrt{61}}, \frac{4}{\sqrt{61}}\right), \left(-\frac{6}{\sqrt{61}}, \frac{3}{\sqrt{61}}, -\frac{4}{\sqrt{61}}\right)$

15. $2(\mathbf{v} \times \mathbf{u})$ 17. (a) $\frac{\sqrt{26}}{2}$ (b) $\frac{\sqrt{26}}{3}$ 21. (a) $\sqrt{122}$ (b) $\theta \approx 40°19''$

23. (a) $\mathbf{m} = (0, 1, 0)$ and $\mathbf{n} = (1, 0, 0)$ (b) $(-1, 0, 0)$ (c) $(0, 0, -1)$

28. $(-8, 0, -8)$ 31. (a) $\frac{2}{3}$ (b) $\frac{1}{2}$ 35. (b) $\mathbf{u} \cdot \mathbf{w} \neq 0$, $\mathbf{v} \cdot \mathbf{w} = 0$

36. No, the equation is equivalent to $\mathbf{u} \times (\mathbf{v} - \mathbf{w}) = 0$ and hence to $\mathbf{v} - \mathbf{w} = k\mathbf{u}$ for some scalar k.

38. They are collinear.

Exercise Set 3.5
(page 162)

1. (a) $-2(x + 1) + (y - 3) - (z + 2) = 0$ (b) $(x - 1) + 9(y - 1) + 8(z - 4) = 0$
(c) $2z = 0$ (d) $x + 2y + 3z = 0$

3. (a) $(0, 0, 5)$ is a point in the plane and $\mathbf{n} = (-3, 7, 2)$ is a normal vector so that
$-3(x - 0) + 7(y - 0) + 2(z - 5) = 0$ is a point-normal form; other points and normals
yield other correct answers.
(b) $(x - 0) + 0(y - 0) - 4(z - 0) = 0$ is a possibility

5.   (a) Not parallel   (b) Parallel   (c) Parallel

9.   (a) $x = 3 + 2t$, $y = -1 + t$, $z = 2 + 3t$ (b) $x = -2 + 6t$, $y = 3 - 6t$, $z = -3 - 2t$
     (c) $x = 2$, $y = 2 + t$, $z = 6$ (d) $x = t$, $y = -2t$, $z = 3t$

11.   (a) $x = -12 - 7t$, $y = -41 - 23t$, $z = t$ (b) $x = \frac{5}{2}t$, $y = 0$, $z = t$

13.   (a) Parallel   (b) Not parallel   17. $2x + 3y - 5z + 36 = 0$

19.   (a) $z - z_0 = 0$ (b) $x - x_0 = 0$ (c) $y - y_0 = 0$ 21. $5x - 2y + z - 34 = 0$

23. $y + 2z - 9 = 0$ 27. $x + 5y + 3z - 18 = 0$

29. $4x + 13y - z - 17 = 0$ 31. $3x - y - z - 2 = 0$

37.   (a) $x = \frac{11}{23} + \frac{7}{23}t$, $y = -\frac{41}{23} - \frac{1}{23}t$, $z = t$ (b) $x = -\frac{2}{5}t$, $y = 0$, $z = t$

39.   (a) $\frac{5}{3}$ (b) $\frac{1}{\sqrt{29}}$ (c) $\frac{4}{\sqrt{3}}$

43.   (a) $\dfrac{x - 3}{2} = y + 1 = \dfrac{z - 2}{3}$ (b) $\dfrac{x + 2}{6} = -\dfrac{y - 3}{6} = -\dfrac{z + 3}{2}$

44.   (a) $x - 2y - 17 = 0$ and $x + 4z - 27 = 0$ is one possible answer.
      (b) $x - 2y = 0$ and $-7y + 2z = 0$ is one possible answer.

45.   (a) $\theta \approx 35°$ (b) $\theta \approx 79°$ 47.   They are identical.

Exercise Set 4.1
(page 178)

1.   (a) $(-1, 9, -11, 1)$ (b) $(22, 53, -19, 14)$ (c) $(-13, 13, -36, -2)$
     (d) $(-90, -114, 60, -36)$ (e) $(-9, -5, -5, -3)$ (f) $(27, 29, -27, 9)$

3. $c_1 = 1$, $c_2 = 1$, $c_3 = -1$, $c_4 = 1$ 5.   (a) $\sqrt{29}$ (b) 3   (c) 13   (d) $\sqrt{31}$

8. $k = \pm\frac{5}{7}$ 10.   (a) $\left(\frac{1}{\sqrt{10}}, \frac{3}{\sqrt{10}}\right)$, $\left(-\frac{1}{\sqrt{10}}, -\frac{3}{\sqrt{10}}\right)$

14.   (a) Yes   (b) No   (c) Yes   (d) No   (e) No   (f) Yes

15.   (a) $k = -3$ (b) $k = -2$, $k = -3$ 19. $x_1 = 1$, $x_2 = -1$, $x_3 = 2$

22.   The component in the **a** direction is $\text{proj}_\mathbf{a}\mathbf{u} = \frac{4}{15}(-1, 1, 2, 3)$; the orthogonal component
      is $\frac{1}{15}(34, 11, 52, -27)$.

23.   They do not intersect.

33.   (a) Euclidean measure of "box" in R^n: $a_1 a_2 \cdots a_n$

      (b) Length of diagonal: $\sqrt{a_1^2 + a_2^2 + \cdots + a_n^2}$

35.   (a) $d(\mathbf{u}, \mathbf{v}) = \sqrt{2}$

37.   (a) True   (b) True   (c) False   (d) True   (e) True, unless $\mathbf{u} = \mathbf{0}$

Exercise Set 4.2
(page 193)

1.   (a) Linear; $R^3 \rightarrow R^2$ (b) Nonlinear; $R^2 \rightarrow R^3$ (c) Linear; $R^3 \rightarrow R^3$
     (d) Nonlinear; $R^4 \rightarrow R^2$

3. $\begin{bmatrix} 3 & 5 & -1 \\ 4 & -1 & 1 \\ 3 & 2 & -1 \end{bmatrix}$; $T(-1, 2, 4) = (3, -2, -3)$

5.   (a) $\begin{bmatrix} 0 & 1 \\ -1 & 0 \\ 1 & 3 \\ 1 & -1 \end{bmatrix}$ (b) $\begin{bmatrix} 7 & 2 & -1 & 1 \\ 0 & 1 & 1 & 0 \\ -1 & 0 & 0 & 0 \end{bmatrix}$ (c) $\begin{bmatrix} 0 & 0 & 0 \\ 0 & 0 & 0 \\ 0 & 0 & 0 \\ 0 & 0 & 0 \\ 0 & 0 & 0 \end{bmatrix}$

     (d) $\begin{bmatrix} 0 & 0 & 0 & 1 \\ 1 & 0 & 0 & 0 \\ 0 & 0 & 1 & 0 \\ 0 & 1 & 0 & 0 \\ 1 & 0 & -1 & 0 \end{bmatrix}$

7.   (a) $T(-1, 4) = (5, 4)$ (b) $T(2, 1, -3) = (0, -2, 0)$

9. (a) $(2, -5, -3)$ (b) $(2, 5, 3)$ (c) $(-2, -5, 3)$

13. (a) $\left(-2, \frac{\sqrt{3}-2}{2}, \frac{1+2\sqrt{3}}{2}\right)$ (b) $(0, 1, 2\sqrt{2})$ (c) $(-1, -2, 2)$

15. (a) $\left(-2, \frac{\sqrt{3}+2}{2}, \frac{-1+2\sqrt{3}}{2}\right)$ (b) $(-2\sqrt{2}, 1, 0)$ (c) $(1, 2, 2)$

17. (a) $\begin{bmatrix} 0 & 0 \\ 1/2 & -\sqrt{3}/2 \end{bmatrix}$ (b) $\begin{bmatrix} -\sqrt{2} & \sqrt{2} \\ \sqrt{2} & \sqrt{2} \end{bmatrix}$ (c) $\begin{bmatrix} -1 & 0 \\ 0 & -1 \end{bmatrix}$

19. (a) $\begin{bmatrix} \sqrt{3}/8 & -\sqrt{3}/16 & 1/16 \\ 1/8 & 3/16 & -\sqrt{3}/16 \\ 0 & 1/8 & \sqrt{3}/8 \end{bmatrix}$ (b) $\begin{bmatrix} 0 & 0 & 0 \\ 0 & -1 & 0 \\ 0 & 0 & -1 \end{bmatrix}$

(c) $\begin{bmatrix} 0 & 1 & 0 \\ 0 & 0 & -1 \\ -1 & 0 & 0 \end{bmatrix}$

21. (a) Yes   (b) No

24. $\begin{bmatrix} \frac{1}{3}(1-\cos\theta)+\cos\theta & \frac{1}{3}(1-\cos\theta)-\frac{1}{\sqrt{3}}\sin\theta & \frac{1}{3}(1-\cos\theta)-\frac{1}{\sqrt{3}}\sin\theta \\ \frac{1}{3}(1-\cos\theta)-\frac{1}{\sqrt{3}}\sin\theta & \frac{1}{3}(1-\cos\theta)+\cos\theta & \frac{1}{3}(1-\cos\theta)-\frac{1}{\sqrt{3}}\sin\theta \\ \frac{1}{3}(1-\cos\theta)-\frac{1}{\sqrt{3}}\sin\theta & \frac{1}{3}(1-\cos\theta)-\frac{1}{\sqrt{3}}\sin\theta & \frac{1}{3}(1-\cos\theta)+\cos\theta \end{bmatrix}$

28. (c) $90°$

29. (a) Twice the orthogonal projection on the x-axis
 (b) Twice the reflection about the x-axis

30. (a) The x-coordinate is stretched by a factor of 2 and the y-coordinate is stretched by a factor of 3.
 (b) Rotation through $30°$

31. Rotation through the angle 2θ **34.** Only if $b = 0$.

Exercise Set 4.3
(page 206)

1. (a) Not one-to-one   (b) One-to-one   (c) One-to-one   (d) One-to-one
 (e) One-to-one   (f) One-to-one   (g) One-to-one

3. For example, the vector $(1, 3)$ is not in the range.

5. (a) One-to-one; $\begin{bmatrix} \frac{1}{3} & -\frac{2}{3} \\ \frac{1}{3} & \frac{1}{3} \end{bmatrix}$; $T^{-1}(w_1, w_2) = \left(\frac{1}{3}w_1 - \frac{2}{3}w_2, \frac{1}{3}w_1 + \frac{1}{3}w_2\right)$
 (b) Not one-to-one
 (c) One-to-one; $\begin{bmatrix} 0 & -1 \\ -1 & 0 \end{bmatrix}$; $T^{-1}(w_1, w_2) = (-w_2, -w_1)$ (d) Not one-to-one

7. (a) Reflection about the x-axis   (b) Rotation through the angle $-\pi/4$
 (c) Contraction by a factor of $\frac{1}{3}$ (d) Reflection about the yz-plane
 (e) Dilation by a factor of 5

9. (a) Linear   (b) Nonlinear   (c) Linear   (d) Nonlinear

12. (a) For a reflection about the y-axis, $T(\mathbf{e}_1) = \begin{bmatrix} -1 \\ 0 \end{bmatrix}$ and $T(\mathbf{e}_2) = \begin{bmatrix} 0 \\ 1 \end{bmatrix}$.

 Thus, $T = \begin{bmatrix} -1 & 0 \\ 0 & 1 \end{bmatrix}$.

 (b) For a reflection about the xz-plane, $T(\mathbf{e}_1) = \begin{bmatrix} 1 \\ 0 \\ 0 \end{bmatrix}$, $T(\mathbf{e}_2) = \begin{bmatrix} 0 \\ -1 \\ 0 \end{bmatrix}$, and

 $T(\mathbf{e}_3) = \begin{bmatrix} 0 \\ 0 \\ 1 \end{bmatrix}$. Thus, $T = \begin{bmatrix} 1 & 0 & 0 \\ 0 & -1 & 0 \\ 0 & 0 & 1 \end{bmatrix}$.

(c) For an orthogonal projection on the x-axis, $T(\mathbf{e}_1) = \begin{bmatrix} 1 \\ 0 \end{bmatrix}$ and $T(\mathbf{e}_2) = \begin{bmatrix} 0 \\ 0 \end{bmatrix}$.

Thus, $T = \begin{bmatrix} 1 & 0 \\ 0 & 0 \end{bmatrix}$.

(d) For an orthogonal projection on the yz-plane, $T(\mathbf{e}_1) = \begin{bmatrix} 0 \\ 0 \\ 0 \end{bmatrix}$, $T(\mathbf{e}_2) = \begin{bmatrix} 0 \\ 1 \\ 0 \end{bmatrix}$, and

$T(\mathbf{e}_3) = \begin{bmatrix} 0 \\ 0 \\ 1 \end{bmatrix}$. Thus, $T = \begin{bmatrix} 0 & 0 & 0 \\ 0 & 1 & 0 \\ 0 & 0 & 1 \end{bmatrix}$.

(e) For a rotation through a positive angle θ, $T(\mathbf{e}_1) = \begin{bmatrix} \cos\theta \\ \sin\theta \end{bmatrix}$ and $T(\mathbf{e}_2) = \begin{bmatrix} -\sin\theta \\ \cos\theta \end{bmatrix}$.

Thus, $T = \begin{bmatrix} \cos\theta & -\sin\theta \\ \sin\theta & \cos\theta \end{bmatrix}$.

(f) For a dilation by a factor $k \geq 1$, $T(\mathbf{e}_1) = \begin{bmatrix} k \\ 0 \\ 0 \end{bmatrix}$, $T(\mathbf{e}_2) = \begin{bmatrix} 0 \\ k \\ 0 \end{bmatrix}$, and $T(\mathbf{e}_3) = \begin{bmatrix} 0 \\ 0 \\ k \end{bmatrix}$.

Thus, $T = \begin{bmatrix} k & 0 & 0 \\ 0 & k & 0 \\ 0 & 0 & k \end{bmatrix}$.

13. (a) $T(\mathbf{e}_1) = \begin{bmatrix} -1 \\ 0 \end{bmatrix}$ and $T(\mathbf{e}_2) = \begin{bmatrix} 0 \\ 0 \end{bmatrix}$. Thus, $T = \begin{bmatrix} -1 & 0 \\ 0 & 0 \end{bmatrix}$.

(b) $T(\mathbf{e}_1) = \begin{bmatrix} 0 \\ -1 \end{bmatrix}$ and $T(\mathbf{e}_2) = \begin{bmatrix} 1 \\ 0 \end{bmatrix}$. Thus, $T = \begin{bmatrix} 0 & 1 \\ -1 & 0 \end{bmatrix}$.

(c) $T(\mathbf{e}_1) = \begin{bmatrix} 0 \\ 3 \end{bmatrix}$ and $T(\mathbf{e}_2) = \begin{bmatrix} 0 \\ 0 \end{bmatrix}$. Thus, $T = \begin{bmatrix} 0 & 0 \\ 3 & 0 \end{bmatrix}$.

16. (a) Linear transformation from $R^2 \to R^3$; one-to-one
    (b) Linear transformation from $R^3 \to R^2$; not one-to-one

17. (a) $\left(\frac{1}{2}, \frac{1}{2}\right)$ (b) $\left(\frac{3}{4}, \frac{\sqrt{3}}{4}\right)$ (c) $\left(\frac{1-5\sqrt{3}}{4}, \frac{15-\sqrt{3}}{4}\right)$

19. (a) $\lambda = 1$; $\begin{bmatrix} 0 \\ s \\ t \end{bmatrix}$ $\lambda = -1$; $\begin{bmatrix} t \\ 0 \\ 0 \end{bmatrix}$ (b) $\lambda = 1$; $\begin{bmatrix} s \\ 0 \\ t \end{bmatrix}$ $\lambda = 0$; $\begin{bmatrix} 0 \\ t \\ 0 \end{bmatrix}$

(c) $\lambda = 2$; all vectors in R^3 are eigenvectors (d) $\lambda = 1$; $\begin{bmatrix} 0 \\ 0 \\ t \end{bmatrix}$

23. (a) $\begin{bmatrix} \cos 2\theta & \sin 2\theta \\ \sin 2\theta & -\cos 2\theta \end{bmatrix}$ (b) $\left(\frac{1+5\sqrt{3}}{2}, \frac{\sqrt{3}-5}{2}\right)$

27. (a) The range of T is a proper subset of R^n.
    (b) T must map infinitely many vectors to 0.

Exercise Set 4.4
(page 217)

1. (a) $x^2 + 2x - 1 - 2(3x^2 + 2) = -5x^2 + 2x - 5$ 4. (a) Yes; $A = \begin{bmatrix} 1 & 0 & 0 \\ 0 & 1 & 0 \\ 0 & 0 & 1 \\ 0 & 0 & 0 \end{bmatrix}$

7. (a) $L: P_1 \to P_1$ where L maps $ax + b$ to $(a+b)x + a - b$

9. (a) $3e^t + 3e^{-t} = 6\cosh(t)$ (b) Yes

12. $y = 2x^2$ **14. (a)** $y = x^3 - x$ **15. (a)** $y = 2x^3 - 2x + 2$

18. (a) No, because of the arbitrary constant of integration
(b) No (except for P_0)

21. (a) Each $L_i(x)$ is a polynomial of degree at most n and hence so is the sum
$y_0 L(x) + \cdots + y_n L(x)$; also, $p(x_i) = 0 + 0 + \cdots + 0 + y_i \cdot L_i(x_i) + 0 + \cdots + 0$
$+ 0 = y_i$, showing that this function is an interpolant of degree at most n.
(b) It is $I_{n+1}\mathbf{c} = \mathbf{y}$ where $\mathbf{c}$ is the vector of c_i values and $\mathbf{y}$ is the vector of y-values.

Exercise Set 5.1
(page 226)

1. Not a vector space. Axiom 8 fails.

3. Not a vector space. Axioms 9 and 10 fail.

5. The set is a vector space under the given operations.

7. The set is a vector space under the given operations.

9. Not a vector space. Axioms 1, 4, 5, and 6 fail.

11. The set is a vector space under the given operations.

13. The set is a vector space under the given operations.

25. No. A vector space must have a zero element.

26. No. Axioms 1, 4, and 6 will fail.

29. (1) Axiom 7   (2) Axiom 4   (3) Axiom 5   (4) Follows from statement 2
(5) Axiom 3   (6) Axiom 5   (7) Axiom 4

32. No; $\mathbf{0}_1 = \mathbf{0}_1 + \mathbf{0}_2 = \mathbf{0}_2$

Exercise Set 5.2
(page 238)

1. (a), (c)   **3.** (a), (b), (d)   **5.** (a), (b), (d)

6. (a) Line; $x = -\frac{1}{2}t$, $y = -\frac{3}{2}t$, $z = t$ **(b)** Line; $x = 2t$, $y = t$, $z = 0$ **(c)** Origin
(d) Origin   **(e)** Line; $x = -3t$, $y = -2t$, $z = t$ **(f)** Plane; $x - 3y + z = 0$

9. (a) $-9 - 7x - 15x^2 = -2\mathbf{p}_1 + \mathbf{p}_2 - 2\mathbf{p}_3$ **(b)** $6 + 11x + 6x^2 = 4\mathbf{p}_1 - 5\mathbf{p}_2 + \mathbf{p}_3$
(c) $0 = 0\mathbf{p}_1 + 0\mathbf{p}_2 + 0\mathbf{p}_3$ **(d)** $7 + 8x + 9x^2 = 0\mathbf{p}_1 - 2\mathbf{p}_2 + 3\mathbf{p}_3$

11. (a) The vectors span.   **(b)** The vectors do not span.
(c) The vectors do not span.   **(d)** The vectors span.

12. (a), (c), (e)   **15.** $y = z$

24. (a) They span a line if they are collinear and not both 0. They span a plane if they are not collinear.
(b) If $\mathbf{u} = a\mathbf{v}$ and $\mathbf{v} = b\mathbf{u}$ for some real numbers a, b
(c) We must have $\mathbf{b} = \mathbf{0}$ since a subspace must contain $\mathbf{x} = \mathbf{0}$ and then $\mathbf{b} = A\mathbf{0} = \mathbf{0}$.

26. (a) For example, $\begin{bmatrix} 1 & 0 \\ 0 & 0 \end{bmatrix}$, $\begin{bmatrix} 0 & 1 \\ 0 & 0 \end{bmatrix}$, $\begin{bmatrix} 0 & 0 \\ 1 & 0 \end{bmatrix}$, $\begin{bmatrix} 0 & 0 \\ 0 & 1 \end{bmatrix}$
(b) The set of matrices having one entry equal to 1 and all other entries equal to 0

Exercise Set 5.3
(page 248)

1. (a) $\mathbf{u}_2$ is a scalar multiple of $\mathbf{u}_1$.
(b) The vectors are linearly dependent by Theorem 5.3.3.
(c) $\mathbf{p}_2$ is a scalar multiple of $\mathbf{p}_1$.   **(d)** B is a scalar multiple of A.

3. None   **5. (a)** They do not lie in a plane.   **(b)** They do lie in a plane.

7. (b) $\mathbf{v}_1 = \frac{2}{7}\mathbf{v}_2 - \frac{3}{7}\mathbf{v}_3$, $\mathbf{v}_2 = \frac{7}{2}\mathbf{v}_1 + \frac{3}{2}\mathbf{v}_3$, $\mathbf{v}_3 = -\frac{7}{3}\mathbf{v}_1 + \frac{2}{3}\mathbf{v}_2$ **9.** $\lambda = -\frac{1}{2}$, $\lambda = 1$

18. If and only if the vector is not zero

19. (a) They are linearly independent since $\mathbf{v}_1$, $\mathbf{v}_2$, and $\mathbf{v}_3$ do not lie in the same plane when they are placed with their initial points at the origin.
(b) They are not linearly independent since $\mathbf{v}_1$, $\mathbf{v}_2$, and $\mathbf{v}_3$ lie in the same plane when they are placed with their initial points at the origin.

20. (a), (d), (e), (f)

24. (a) False   **(b)** False   **(c)** True   **(d)** False   **27. (a)** Yes

Exercise Set 5.4
(page 263)

1. **(a)** A basis for R^2 has two linearly independent vectors.
   (b) A basis for R^3 has three linearly independent vectors.
   (c) A basis for P_2 has three linearly independent vectors.
   (d) A basis for M_{22} has four linearly independent vectors.

3. (a), (b)   7. **(a)** $(\mathbf{w})_S = (3, -7)$ **(b)** $(\mathbf{w})_S = \left(\frac{5}{28}, \frac{3}{14}\right)$ **(c)** $(\mathbf{w})_S = \left(a, \frac{b-a}{2}\right)$

9. **(a)** $(\mathbf{v})_S = (3, -2, 1)$ **(b)** $(\mathbf{v})_S = (-2, 0, 1)$ 11. $(A)_S = (-1, 1, -1, 3)$

13. Basis: $\left(-\frac{1}{4}, -\frac{1}{4}, 1, 0\right)$, $(0, -1, 0, 1)$; dimension $= 2$

15. Basis: $(3, 1, 0)$, $(-1, 0, 1)$; dimension $= 2$

19. **(a)** 3-dimensional   **(b)** 2-dimensional   **(c)** 1-dimensional

20. 3-dimensional

21. **(a)** $\{\mathbf{v}_1, \mathbf{v}_2, \mathbf{e}_1\}$ or $\{\mathbf{v}_1, \mathbf{v}_2, \mathbf{e}_2\}$ **(b)** $\{\mathbf{v}_1, \mathbf{v}_2, \mathbf{e}_1\}$ or $\{\mathbf{v}_1, \mathbf{v}_2, \mathbf{e}_2\}$ or $\{\mathbf{v}_1, \mathbf{v}_2, \mathbf{e}_3\}$

27. **(a)** One possible answer is $\{-1 + x - 2x^2, 3 + 3x + 6x^2, 9\}$.
    (b) One possible answer is $\{1 + x, x^2, -2 + 2x^2\}$.
    (c) One possible answer is $\{1 + x - 3x^2\}$.

29. **(a)** $(2, 0)$ **(b)** $\left(\frac{2}{\sqrt{3}}, -\frac{1}{\sqrt{3}}\right)$ **(c)** $(0, 1)$ **(d)** $\left(\frac{2}{\sqrt{3}}a, b - \frac{a}{\sqrt{3}}\right)$

31. Yes; for example, $\begin{bmatrix} 1 & 0 \\ 0 & \pm 1 \end{bmatrix}$, $\begin{bmatrix} 0 & 1 \\ \pm 1 & 0 \end{bmatrix}$

32. **(a)** n **(b)** $n(n+1)/2$ **(c)** $n(n+1)/2$

35. **(a)** The dimension is $n - 1$.
    (b) $(1, 0, 0, \ldots, 0, -1)$, $(0, 1, 0, \ldots, 0, -1)$, $(0, 0, 1, \ldots, 0, -1)$, $\ldots$, $(0, 0, 0, \ldots, 1, -1)$ is a basis of size $n - 1$.

Exercise Set 5.5
(page 276)

1. $\mathbf{r}_1 = (2, -1, 0, 1)$, $\mathbf{r}_2 = (3, 5, 7, -1)$, $\mathbf{r}_3 = (1, 4, 2, 7)$;

   $$\mathbf{c}_1 = \begin{bmatrix} 2 \\ 3 \\ 1 \end{bmatrix}, \quad \mathbf{c}_2 = \begin{bmatrix} -1 \\ 5 \\ 4 \end{bmatrix}, \quad \mathbf{c}_3 = \begin{bmatrix} 0 \\ 7 \\ 2 \end{bmatrix}, \quad \mathbf{c}_4 = \begin{bmatrix} 1 \\ -1 \\ 7 \end{bmatrix}$$

3. **(a)** $\begin{bmatrix} -2 \\ 10 \end{bmatrix} = \begin{bmatrix} 1 \\ 4 \end{bmatrix} - \begin{bmatrix} 3 \\ -6 \end{bmatrix}$ **(b)** $\mathbf{b}$ is not in the column space of A.

   (c) $\begin{bmatrix} 1 \\ 9 \\ 1 \end{bmatrix} - 3\begin{bmatrix} -1 \\ 3 \\ 1 \end{bmatrix} + \begin{bmatrix} 1 \\ 1 \\ 1 \end{bmatrix} = \begin{bmatrix} 5 \\ 1 \\ -1 \end{bmatrix}$

   (d) $\begin{bmatrix} 2 \\ 0 \\ 0 \end{bmatrix} = \begin{bmatrix} 1 \\ 1 \\ -1 \end{bmatrix} + (t-1)\begin{bmatrix} -1 \\ 1 \\ -1 \end{bmatrix} + t\begin{bmatrix} 1 \\ -1 \\ 1 \end{bmatrix}$

   (e) $\begin{bmatrix} 4 \\ 3 \\ 5 \\ 7 \end{bmatrix} = -26\begin{bmatrix} 1 \\ 0 \\ 1 \\ 0 \end{bmatrix} + 13\begin{bmatrix} 2 \\ 1 \\ 2 \\ 1 \end{bmatrix} - 7\begin{bmatrix} 0 \\ 2 \\ 1 \\ 2 \end{bmatrix} + 4\begin{bmatrix} 1 \\ 1 \\ 3 \\ 2 \end{bmatrix}$

5. **(a)** $\begin{bmatrix} 1 \\ 0 \end{bmatrix} + t\begin{bmatrix} 3 \\ 1 \end{bmatrix}$; $t\begin{bmatrix} 3 \\ 1 \end{bmatrix}$ **(b)** $\begin{bmatrix} -2 \\ 7 \\ 0 \end{bmatrix} + t\begin{bmatrix} -1 \\ -1 \\ 1 \end{bmatrix}$; $t\begin{bmatrix} -1 \\ -1 \\ 1 \end{bmatrix}$

(c) $\begin{bmatrix} -1 \\ 0 \\ 0 \\ 0 \end{bmatrix} + r\begin{bmatrix} 2 \\ 1 \\ 0 \\ 0 \end{bmatrix} + s\begin{bmatrix} -1 \\ 0 \\ 1 \\ 0 \end{bmatrix} + t\begin{bmatrix} -2 \\ 0 \\ 0 \\ 1 \end{bmatrix}$; $r\begin{bmatrix} 2 \\ 1 \\ 0 \\ 0 \end{bmatrix} + s\begin{bmatrix} -1 \\ 0 \\ 1 \\ 0 \end{bmatrix} + t\begin{bmatrix} -2 \\ 0 \\ 0 \\ 1 \end{bmatrix}$

(d) $\begin{bmatrix} \frac{6}{5} \\ \frac{7}{5} \\ 0 \\ 0 \end{bmatrix} + s\begin{bmatrix} \frac{7}{5} \\ \frac{4}{5} \\ 1 \\ 0 \end{bmatrix} + t\begin{bmatrix} \frac{1}{5} \\ -\frac{3}{5} \\ 0 \\ 1 \end{bmatrix}$; $s\begin{bmatrix} \frac{7}{5} \\ \frac{4}{5} \\ 1 \\ 0 \end{bmatrix} + t\begin{bmatrix} \frac{1}{5} \\ -\frac{3}{5} \\ 0 \\ 1 \end{bmatrix}$

7. **(a)** $\mathbf{r}_1 = [1 \quad 0 \quad 2]$, $\mathbf{r}_2 = [0 \quad 0 \quad 1]$, $\mathbf{c} = \begin{bmatrix} 1 \\ 0 \\ 0 \end{bmatrix}$, $\mathbf{c}_2 = \begin{bmatrix} 2 \\ 1 \\ 0 \end{bmatrix}$

(b) $\mathbf{r}_1 = [1 \quad -3 \quad 0 \quad 0]$, $\mathbf{r}_2 = [0 \quad 1 \quad 0 \quad 0]$, $\mathbf{c}_1 = \begin{bmatrix} 1 \\ 0 \\ 0 \\ 0 \end{bmatrix}$, $\mathbf{c}_2 = \begin{bmatrix} -3 \\ 1 \\ 0 \\ 0 \end{bmatrix}$

(c) $\mathbf{r}_1 = [1 \quad 2 \quad 4 \quad 5]$, $\mathbf{r}_2 = [0 \quad 1 \quad -3 \quad 0]$,
$\mathbf{r}_3 = [0 \quad 0 \quad 1 \quad -3]$, $\mathbf{r}_4 = [0 \quad 0 \quad 0 \quad 1]$,

$\mathbf{c}_1 = \begin{bmatrix} 1 \\ 0 \\ 0 \\ 0 \\ 0 \end{bmatrix}$, $\mathbf{c}_2 = \begin{bmatrix} 2 \\ 1 \\ 0 \\ 0 \\ 0 \end{bmatrix}$, $\mathbf{c}_3 = \begin{bmatrix} 4 \\ -3 \\ 1 \\ 0 \\ 0 \end{bmatrix}$, $\mathbf{c}_4 = \begin{bmatrix} 5 \\ 0 \\ -3 \\ 1 \\ 0 \end{bmatrix}$

(d) $\mathbf{r}_1 = [1 \quad 2 \quad -1 \quad 5]$, $\mathbf{r}_2 = [0 \quad 1 \quad 4 \quad 3]$,
$\mathbf{r}_3 = [0 \quad 0 \quad 1 \quad -7]$, $\mathbf{r}_4 = [0 \quad 0 \quad 0 \quad 1]$,

$\mathbf{c}_1 = \begin{bmatrix} 1 \\ 0 \\ 0 \\ 0 \end{bmatrix}$, $\mathbf{c}_2 = \begin{bmatrix} 2 \\ 1 \\ 0 \\ 0 \end{bmatrix}$, $\mathbf{c}_3 = \begin{bmatrix} -1 \\ 4 \\ 1 \\ 0 \end{bmatrix}$, $\mathbf{c}_4 = \begin{bmatrix} 5 \\ 3 \\ -7 \\ 1 \end{bmatrix}$

9. **(a)** $\begin{bmatrix} 1 \\ 5 \\ 7 \end{bmatrix}$, $\begin{bmatrix} -1 \\ -4 \\ -6 \end{bmatrix}$ **(b)** $\begin{bmatrix} 2 \\ 4 \\ 0 \end{bmatrix}$ **(c)** $\begin{bmatrix} 1 \\ 2 \\ -1 \end{bmatrix}$, $\begin{bmatrix} 4 \\ 1 \\ 3 \end{bmatrix}$

(d) $\begin{bmatrix} 1 \\ 3 \\ -1 \\ 2 \end{bmatrix}$, $\begin{bmatrix} 4 \\ -2 \\ 0 \\ 3 \end{bmatrix}$ **(e)** $\begin{bmatrix} 1 \\ 0 \\ 2 \\ 3 \\ -2 \end{bmatrix}$, $\begin{bmatrix} -3 \\ 3 \\ -3 \\ -6 \\ 9 \end{bmatrix}$, $\begin{bmatrix} 2 \\ 6 \\ -2 \\ 0 \\ 2 \end{bmatrix}$

11. **(a)** $(1, 1, -4, -3)$, $(0, 1, -5, -2)$, $\left(0, 0, 1, -\frac{1}{2}\right)$
(b) $\left(1, -1, 2, 0\right)$, $(0, 1, 0, 0)$, $\left(0, 0, 1, -\frac{1}{6}\right)$
(c) $(1, 1, 0, 0)$, $(0, 1, 1, 1)$, $(0, 0, 1, 1)$, $(0, 0, 0, 1)$

14. **(b)** $\begin{bmatrix} 0 & 0 & 0 \\ 0 & 1 & 0 \\ 0 & 0 & 1 \end{bmatrix}$ **17.** $\begin{bmatrix} 3a & -5a \\ 3b & -5b \end{bmatrix}$ for all real numbers a, b not both 0.

Exercise Set 5.6
(page 288)

1. Rank $(A) = $ rank$(A^T) = 2$
3. **(a)** $2; 1$ **(b)** $1; 2$ **(c)** $2; 2$ **(d)** $2; 3$ **(e)** $3; 2$
5. **(a)** Rank $= 4$, nullity $= 0$ **(b)** Rank $= 3$, nullity $= 2$ **(c)** Rank $= 3$, nullity $= 0$

7. **(a)** Yes, 0   **(b)** No   **(c)** Yes, 2   **(d)** Yes, 7   **(e)** No
(f) Yes, 4   **(g)** Yes, 0

9. $b_1 = r$, $b_2 = s$, $b_3 = 4s - 3r$, $b_4 = 2r - s$, $b_5 = 8s - 7r$ **11.** No

13. Rank is 2 if $r = 2$ and $s = 1$; the rank is never 1.

16. **(a)** $\begin{bmatrix} 1 & 0 & 0 \\ 0 & 1 & 0 \\ 0 & 0 & 0 \end{bmatrix}$ **(b)** A line through the origin   **(c)** A plane through the origin

(d) The nullspace is a line through the origin and the row space is a plane through the origin.

19. **(a)** 3   **(b)** 5   **(c)** 3   **(d)** 3

Supplementary Exercises (page 290)

1. **(a)** All of R^3 **(b)** Plane: $2x - 3y + z = 0$
(c) Line: $x = 2t$, $y = t$, $z = 0$ **(d)** The origin: $(0, 0, 0)$

3. **(a)** $a(4, 1, 1) + b(0, -1, 2)$ **(b)** $(a + c)(3, -1, 2) + b(1, 4, 1)$
(c) $a(2, 3, 0) + b(-1, 0, 4) + c(4, -1, 1)$

5. **(a)** $\mathbf{v} = (-1 + r)\mathbf{v}_1 + \left(\frac{2}{3} - r\right)\mathbf{v}_2 + r\mathbf{v}_3$; r arbitrary   **7.** No

9. **(a)** Rank $= 2$, nullity $= 1$ **(b)** Rank $= 3$, nullity $= 2$
(c) Rank $= n + 1$, nullity $= n$

11. $\{1, x^2, x^3, x^4, x^5, x^6, \ldots, x^n\}$ **13.** **(a)** 2   **(b)** 1   **(c)** 2   **(d)** 3

Exercise Set 6.1 (page 304)

1. **(a)** 2   **(b)** 11   **(c)** -13 **(d)** -8 **(e)** 0   **3.** **(a)** 3   **(b)** 56

5. **(b)** 29   **7.** **(a)** $\begin{bmatrix} \sqrt{3} & 0 \\ 0 & \sqrt{5} \end{bmatrix}$ **(b)** $\begin{bmatrix} 2 & 0 \\ 0 & \sqrt{6} \end{bmatrix}$

9. **(a)** No. Axiom 4 fails.   **(b)** No. Axioms 2 and 3 fail.
(c) Yes   **(d)** No. Axiom 4 fails.

11. **(a)** $3\sqrt{2}$ **(b)** $3\sqrt{5}$ **(c)** $3\sqrt{13}$ **13.** **(a)** $\sqrt{74}$ **(b)** 0

15. **(a)** $\sqrt{105}$ **(b)** $\sqrt{47}$ **17.** **(a)** $\sqrt{2}$, $\frac{1}{3}\sqrt{6}$, $\frac{1}{5}\sqrt{10}$ **(b)** $\frac{2}{3}\sqrt{6}$

19. $\langle \mathbf{u}, \mathbf{v} \rangle = \frac{1}{9}u_1 v_1 + u_2 v_2$

23. No for P_3, since $\mathbf{p} = x\left(x - \frac{1}{2}\right)(x - 1)$ satisfies $\langle \mathbf{p}, \mathbf{p} \rangle = 0$

27. **(a)** $-\frac{28}{15}$ **(b)** 0   **34.** $a = 1/25$, $b = 1/16$

Exercise Set 6.2 (page 315)

1. **(a)** Yes   **(b)** No   **(c)** Yes   **(d)** No   **(e)** No   **(f)** Yes

5. **(a)** $-\frac{1}{\sqrt{2}}$ **(b)** $-\frac{3}{\sqrt{73}}$ **(c)** 0   **(d)** $-\frac{20}{9\sqrt{10}}$ **(e)** $-\frac{1}{\sqrt{2}}$ **(f)** $\frac{2}{\sqrt{55}}$

9. **(a)** Orthogonal   **(b)** Orthogonal   **(c)** Orthogonal   **(d)** Not orthogonal

11. $\pm\frac{1}{57}(-34, 44, -6, 11)$

15. **(a)** $x = t$, $y = -2t$, $z = -3t$ **(b)** $2x - 5y + 4z = 0$ **(c)** $x - z = 0$

17. **(a)** $\begin{bmatrix} 1 \\ 3 \\ 1 \end{bmatrix}$, $\begin{bmatrix} 0 \\ 1 \\ 1 \end{bmatrix}$; $\begin{bmatrix} 2 \\ -1 \\ 1 \end{bmatrix}$

18. **(a)** $(16, 19, 1)$ **(b)** $(0, 1, 0)$, $\left(\frac{1}{2}, 0, 1\right)$ **(c)** $(-1, -1, 1, 0)$, $\left(\frac{2}{7}, -\frac{4}{7}, 0, 1\right)$
(d) $(-1, -1, 1, 0, 0)$, $(-2, -1, 0, 1, 0)$, $(-1, -2, 0, 0, 1)$

32. $\langle \mathbf{u}, \mathbf{v} \rangle = \frac{1}{2}u_1 v_1 + \frac{1}{6}u_2 v_2$

35. **(a)** The line $y = -x$ **(b)** The xz-plane   **(c)** The x-axis

37. **(a)** False   **(b)** True   **(c)** True   **(d)** False

Exercise Set 6.3 (page 328)

1. (a), (b), (d)   **3.** (b), (d)   **5.** (a)

7. **(a)** $\left(-\frac{1}{\sqrt{5}}, \frac{2}{\sqrt{5}}\right)$, $\left(\frac{2}{\sqrt{5}}, \frac{1}{\sqrt{5}}\right)$ **(b)** $\left(\frac{1}{\sqrt{2}}, 0, -\frac{1}{\sqrt{2}}\right)$, $\left(\frac{1}{\sqrt{2}}, 0, \frac{1}{\sqrt{2}}\right)$, $(0, 1, 0)$

(c) $\left(\frac{1}{\sqrt{3}}, \frac{1}{\sqrt{3}}, \frac{1}{\sqrt{3}}\right)$, $\left(-\frac{1}{\sqrt{2}}, \frac{1}{\sqrt{2}}, 0\right)$, $\left(\frac{1}{\sqrt{6}}, \frac{1}{\sqrt{6}}, -\frac{2}{\sqrt{6}}\right)$

9. **(a)** $-\frac{7}{5}\mathbf{v}_1 + \frac{1}{5}\mathbf{v}_2 + 2\mathbf{v}_3$ **(b)** $-\frac{37}{5}\mathbf{v}_1 - \frac{9}{5}\mathbf{v}_2 + 4\mathbf{v}_3$ **(c)** $-\frac{3}{7}\mathbf{v}_1 - \frac{1}{7}\mathbf{v}_2 + \frac{5}{7}\mathbf{v}_3$

11. **(a)** $(\mathbf{w})_S = (-2\sqrt{2}, 5\sqrt{2})$ **(b)** $(\mathbf{w})_S = (0, -2, 1)$

13. **(a)** $\mathbf{u} = \left(1, \frac{14}{5}, -\frac{2}{5}\right)$, $\mathbf{v} = \left(0, -\frac{17}{5}, \frac{6}{5}\right)$, $\mathbf{w} = \left(-4, -\frac{11}{5}, \frac{23}{5}\right)$

(b) $\|\mathbf{v}\| = \sqrt{13}$, $d(\mathbf{u}, \mathbf{v}) = 5\sqrt{3}$, $\langle \mathbf{w}, \mathbf{v} \rangle = 13$

15. **(b)** $\mathbf{u} = -\frac{4}{5}\mathbf{v}_1 - \frac{11}{10}\mathbf{v}_2 + 0\mathbf{v}_3 + \frac{1}{2}\mathbf{v}_4$

17. **(a)** $\left(\frac{1}{\sqrt{3}}, \frac{1}{\sqrt{3}}, \frac{1}{\sqrt{3}}\right)$, $\left(-\frac{1}{\sqrt{2}}, \frac{1}{\sqrt{2}}, 0\right)$, $\left(\frac{1}{\sqrt{6}}, \frac{1}{\sqrt{6}}, -\frac{2}{\sqrt{6}}\right)$

(b) $(1, 0, 0)$, $\left(0, \frac{7}{\sqrt{53}}, -\frac{2}{\sqrt{53}}\right)$, $\left(0, \frac{2}{\sqrt{53}}, \frac{7}{\sqrt{53}}\right)$

19. $\left(0, \frac{1}{\sqrt{5}}, \frac{2}{\sqrt{5}}\right)$, $\left(-\frac{\sqrt{5}}{\sqrt{6}}, -\frac{2}{\sqrt{30}}, \frac{1}{\sqrt{30}}\right)$ 21. $\mathbf{w}_1 = \left(-\frac{4}{5}, 2, \frac{3}{5}\right)$, $\mathbf{w}_2 = \left(\frac{9}{5}, 0, \frac{12}{5}\right)$

24. **(a)** $\begin{bmatrix} \frac{1}{\sqrt{5}} & -\frac{2}{\sqrt{5}} \\ \frac{2}{\sqrt{5}} & \frac{1}{\sqrt{5}} \end{bmatrix} \begin{bmatrix} \sqrt{5} & \sqrt{5} \\ 0 & \sqrt{5} \end{bmatrix}$ **(b)** $\begin{bmatrix} \frac{1}{\sqrt{2}} & -\frac{1}{\sqrt{3}} \\ 0 & \frac{1}{\sqrt{3}} \\ \frac{1}{\sqrt{2}} & \frac{1}{\sqrt{3}} \end{bmatrix} \begin{bmatrix} \sqrt{2} & 3\sqrt{2} \\ 0 & \sqrt{3} \end{bmatrix}$

(c) $\begin{bmatrix} \frac{1}{3} & \frac{8}{\sqrt{234}} \\ -\frac{2}{3} & \frac{11}{\sqrt{234}} \\ \frac{2}{3} & \frac{7}{\sqrt{234}} \end{bmatrix} \begin{bmatrix} 3 & \frac{1}{3} \\ 0 & \frac{\sqrt{26}}{3} \end{bmatrix}$ **(d)** $\begin{bmatrix} \frac{1}{\sqrt{2}} & -\frac{1}{\sqrt{3}} & \frac{1}{\sqrt{6}} \\ 0 & \frac{1}{\sqrt{3}} & \frac{2}{\sqrt{6}} \\ \frac{1}{\sqrt{2}} & \frac{1}{\sqrt{3}} & -\frac{1}{\sqrt{6}} \end{bmatrix} \begin{bmatrix} \sqrt{2} & \sqrt{2} & \sqrt{2} \\ 0 & \sqrt{3} & -\frac{1}{\sqrt{3}} \\ 0 & 0 & \frac{4}{\sqrt{6}} \end{bmatrix}$

(e) $\begin{bmatrix} \frac{1}{\sqrt{2}} & \frac{\sqrt{2}}{2\sqrt{19}} & -\frac{3}{\sqrt{19}} \\ \frac{1}{\sqrt{2}} & -\frac{\sqrt{2}}{2\sqrt{19}} & \frac{3}{\sqrt{19}} \\ 0 & \frac{3\sqrt{2}}{\sqrt{19}} & \frac{1}{\sqrt{19}} \end{bmatrix} \begin{bmatrix} \sqrt{2} & \frac{3}{\sqrt{2}} & \sqrt{2} \\ 0 & \frac{\sqrt{19}}{\sqrt{2}} & \frac{3\sqrt{2}}{\sqrt{19}} \\ 0 & 0 & \frac{1}{\sqrt{19}} \end{bmatrix}$

(f) Columns not linearly independent

29. $\mathbf{v}_1 = \frac{1}{\sqrt{2}}$, $\mathbf{v}_2 = \sqrt{\frac{3}{2}}x$, $\mathbf{v}_3 = \frac{\sqrt{5}}{2\sqrt{2}}(3x^2 - 1)$

31. $\mathbf{v}_1 = 1$, $\mathbf{v}_2 = \sqrt{3}(2x - 1)$, $\mathbf{v}_3 = \sqrt{5}(6x^2 - 6x + 1)$

35. $(1/\sqrt{5}, 1/\sqrt{5})$, $(2/\sqrt{30}, -3/\sqrt{30})$

Exercise Set 6.4
(page 339)

1. **(a)** $\begin{bmatrix} 21 & 25 \\ 25 & 35 \end{bmatrix} \begin{bmatrix} x_1 \\ x_2 \end{bmatrix} = \begin{bmatrix} 20 \\ 20 \end{bmatrix}$ **(b)** $\begin{bmatrix} 15 & -1 & 5 \\ -1 & 22 & 30 \\ 5 & 30 & 45 \end{bmatrix} \begin{bmatrix} x_1 \\ x_2 \\ x_3 \end{bmatrix} = \begin{bmatrix} -1 \\ 9 \\ 13 \end{bmatrix}$

3. **(a)** $x_1 = 5$, $x_2 = \frac{1}{2}$; $\begin{bmatrix} \frac{11}{2} \\ -\frac{9}{2} \\ -4 \end{bmatrix}$ **(b)** $x_1 = \frac{3}{7}$, $x_2 = -\frac{2}{3}$; $\begin{bmatrix} \frac{46}{21} \\ -\frac{5}{21} \\ \frac{13}{21} \end{bmatrix}$

(c) $x_1 = 12$, $x_2 = -3$, $x_3 = 9$; $\begin{bmatrix} 3 \\ 3 \\ 9 \\ 0 \end{bmatrix}$ **(d)** $x_1 = 14$, $x_2 = 30$, $x_3 = 26$; $\begin{bmatrix} 2 \\ 6 \\ -2 \\ 4 \end{bmatrix}$

5. **(a)** $(7, 2, 9, 5)$ **(b)** $\left(-\frac{12}{5}, -\frac{4}{5}, \frac{12}{5}, \frac{16}{5}\right)$

7. **(a)** $\begin{bmatrix} 1 & 0 \\ 0 & 0 \end{bmatrix}$ **(b)** $\begin{bmatrix} 0 & 0 \\ 0 & 1 \end{bmatrix}$

11. **(a)** $\mathbf{v}_1 = (2, -1, 4)$ **(b)** $\begin{bmatrix} \frac{4}{21} & -\frac{2}{21} & \frac{8}{21} \\ -\frac{2}{21} & \frac{1}{21} & -\frac{4}{21} \\ \frac{8}{21} & -\frac{4}{21} & \frac{16}{21} \end{bmatrix}$ **(c)** $\begin{bmatrix} \frac{4}{21}x_0 - \frac{2}{21}y_0 + \frac{8}{21}z_0 \\ -\frac{2}{21}x_0 + \frac{1}{21}y_0 - \frac{4}{21}z_0 \\ \frac{8}{21}x_0 - \frac{4}{21}y_0 + \frac{16}{21}z_0 \end{bmatrix}$

   (d) $\frac{\sqrt{497}}{7}$

17. $[P] = A^T(AA^T)^{-1}A$

18. (1) Since $A^T\mathbf{0} = \mathbf{0}$ (2) Since A^TA is invertible

   (3) Since the nullspace of A is nonzero if and only if the columns of A are dependent

Exercise Set 6.5
(page 345)

1. **(a)** $[\mathbf{w}]_S = \begin{bmatrix} 3 \\ -7 \end{bmatrix}$ **(b)** $[\mathbf{w}]_S = \begin{bmatrix} \frac{5}{28} \\ \frac{3}{14} \end{bmatrix}$ **(c)** $[\mathbf{w}]_S = \begin{bmatrix} a \\ b-a \\ \frac{}{2} \end{bmatrix}$

3. **(a)** $(\mathbf{p})_S = (4, -3, 1), \ [\mathbf{p}]_S = \begin{bmatrix} 4 \\ -3 \\ 1 \end{bmatrix}$ **(b)** $(\mathbf{p})_S = (0, 2, -1), \ [\mathbf{p}]_S = \begin{bmatrix} 0 \\ 2 \\ -1 \end{bmatrix}$

5. **(a)** $\mathbf{w} = (16, 10, 12)$ **(b)** $\mathbf{q} = 3 + 4x^2$ **(c)** $B = \begin{bmatrix} 15 & -1 \\ 6 & 3 \end{bmatrix}$

7. **(a)** $\begin{bmatrix} \frac{13}{10} & -\frac{1}{2} \\ -\frac{2}{5} & 0 \end{bmatrix}$ **(b)** $\begin{bmatrix} 0 & -\frac{5}{2} \\ -2 & -\frac{13}{2} \end{bmatrix}$ **(c)** $[\mathbf{w}]_B = \begin{bmatrix} -\frac{17}{10} \\ \frac{8}{5} \end{bmatrix}, \ [\mathbf{w}]_{B'} = \begin{bmatrix} -4 \\ -7 \end{bmatrix}$

9. **(a)** $\begin{bmatrix} 3 & 2 & \frac{5}{2} \\ -2 & -3 & -\frac{1}{2} \\ 5 & 1 & 6 \end{bmatrix}$ **(b)** $\begin{bmatrix} -\frac{7}{2} \\ \frac{23}{2} \\ 6 \end{bmatrix}$

11. **(b)** $\begin{bmatrix} 2 & 0 \\ 1 & 3 \end{bmatrix}$ **(c)** $\begin{bmatrix} \frac{1}{2} & 0 \\ -\frac{1}{6} & \frac{1}{3} \end{bmatrix}$ **(d)** $[\mathbf{h}]_B = \begin{bmatrix} 2 \\ -5 \end{bmatrix}, \ [\mathbf{h}]_{B'} = \begin{bmatrix} 1 \\ -2 \end{bmatrix}$

Exercise Set 6.6
(page 353)

1. **(b)** $\begin{bmatrix} \frac{4}{5} & -\frac{9}{25} & \frac{12}{25} \\ 0 & \frac{4}{5} & \frac{3}{5} \\ -\frac{3}{5} & -\frac{12}{25} & \frac{16}{25} \end{bmatrix}$

3. **(a)** $\begin{bmatrix} 1 & 0 \\ 0 & 1 \end{bmatrix}$ **(b)** $\begin{bmatrix} \frac{1}{\sqrt{2}} & \frac{1}{\sqrt{2}} \\ -\frac{1}{\sqrt{2}} & \frac{1}{\sqrt{2}} \end{bmatrix}$ **(d)** $\begin{bmatrix} -\frac{1}{\sqrt{2}} & 0 & \frac{1}{\sqrt{2}} \\ \frac{1}{\sqrt{6}} & -\frac{2}{\sqrt{6}} & \frac{1}{\sqrt{6}} \\ \frac{1}{\sqrt{3}} & \frac{1}{\sqrt{3}} & \frac{1}{\sqrt{3}} \end{bmatrix}$

   (e) $\begin{bmatrix} \frac{1}{2} & \frac{1}{2} & \frac{1}{2} & \frac{1}{2} \\ \frac{1}{2} & -\frac{5}{6} & \frac{1}{6} & \frac{1}{6} \\ \frac{1}{2} & \frac{1}{6} & \frac{1}{6} & -\frac{5}{6} \\ \frac{1}{2} & \frac{1}{6} & -\frac{5}{6} & \frac{1}{6} \end{bmatrix}$

7. **(a)** $(-1 + 3\sqrt{3}, 3 + \sqrt{3})$ **(b)** $\left(\frac{5}{2} - \sqrt{3}, \frac{5}{2}\sqrt{3} + 1\right)$

9. **(a)** $\left(-\frac{1}{2} - \frac{5}{2}\sqrt{3}, 2, \frac{5}{2} - \frac{1}{2}\sqrt{3}\right)$ **(b)** $\left(\frac{1}{2} - \frac{3}{2}\sqrt{3}, 6, -\frac{3}{2} - \frac{1}{2}\sqrt{3}\right)$

11. **(a)** $A = \begin{bmatrix} \cos\theta & 0 & -\sin\theta \\ 0 & 1 & 0 \\ \sin\theta & 0 & \cos\theta \end{bmatrix}$ **(b)** $A = \begin{bmatrix} 1 & 0 & 0 \\ 0 & \cos\theta & \sin\theta \\ 0 & -\sin\theta & \cos\theta \end{bmatrix}$

12. $\begin{bmatrix} \frac{\sqrt{2}}{4} & \frac{\sqrt{6}}{4} & -\frac{\sqrt{2}}{2} \\ -\frac{\sqrt{3}}{2} & \frac{1}{2} & 0 \\ \frac{\sqrt{2}}{4} & \frac{\sqrt{6}}{4} & \frac{\sqrt{2}}{2} \end{bmatrix}$

16. (a) Rotation **(b)** Rotation followed by a reflection

20. (a) Rotation and reflection **(b)** Rotation and dilation
(c) Any rigid operator is angle preserving. Any dilation or contraction with $k \neq 0, 1$ is angle preserving but not rigid.

22. $a = 0, \quad b = \sqrt{2/3}, \quad c = -\sqrt{1/3}$ or $a = 0, \quad b = -\sqrt{2/3}, \quad c = \sqrt{1/3}$

Supplementary Exercises (page 356)

1. (a) $(0, a, a, 0)$ with $a \neq 0$ **(b)** $\pm\left(0, \frac{2}{\sqrt{5}}, \frac{1}{\sqrt{5}}, 0\right)$ **6.** $\pm\left(\frac{1}{\sqrt{2}}, 0, \frac{1}{\sqrt{2}}\right)$

7. $w_k = \dfrac{1}{k}, \quad k = 1, 2, \ldots, n$ **11. (b)** θ approaches $\dfrac{\pi}{2}$

12. (b) The diagonals of a parallelogram are perpendicular if and only if its sides have the same length.

Exercise Set 7.1 (page 367)

1. (a) $\lambda^2 - 2\lambda - 3 = 0$ **(b)** $\lambda^2 - 8\lambda + 16 = 0$ **(c)** $\lambda^2 - 12 = 0$
(d) $\lambda^2 + 3 = 0$ **(e)** $\lambda^2 = 0$ **(f)** $\lambda^2 - 2\lambda + 1 = 0$

3. (a) Basis for eigenspace corresponding to $\lambda = 3$: $\begin{bmatrix} \frac{1}{2} \\ 1 \end{bmatrix}$; basis for eigenspace corresponding to $\lambda = -1$: $\begin{bmatrix} 0 \\ 1 \end{bmatrix}$.

(b) Basis for eigenspace corresponding to $\lambda = 4$: $\begin{bmatrix} \frac{3}{2} \\ 1 \end{bmatrix}$

(c) Basis for eigenspace corresponding to $\lambda = \sqrt{12}$: $\begin{bmatrix} \frac{3}{\sqrt{12}} \\ 1 \end{bmatrix}$; basis for eigenspace corresponding to $\lambda = -\sqrt{12}$: $\begin{bmatrix} -\frac{3}{\sqrt{12}} \\ 1 \end{bmatrix}$

(d) There are no eigenspaces.

(e) Basis for eigenspace corresponding to $\lambda = 0$: $\begin{bmatrix} 1 \\ 0 \end{bmatrix}$, $\begin{bmatrix} 0 \\ 1 \end{bmatrix}$

(f) Basis for eigenspace corresponding to $\lambda = 1$: $\begin{bmatrix} 1 \\ 0 \end{bmatrix}$, $\begin{bmatrix} 0 \\ 1 \end{bmatrix}$

6. (a) $\lambda = 1$: basis $\begin{bmatrix} 0 \\ 1 \\ 0 \end{bmatrix}$; $\lambda = 2$: basis $\begin{bmatrix} -\frac{1}{2} \\ 1 \\ 1 \end{bmatrix}$; $\lambda = 3$: basis $\begin{bmatrix} -1 \\ 1 \\ 1 \end{bmatrix}$

(b) $\lambda = 0$: basis $\begin{bmatrix} \frac{5}{3} \\ \frac{1}{3} \\ 1 \end{bmatrix}$; $\lambda = \sqrt{2}$: basis $\begin{bmatrix} \frac{1}{7}(15 + 5\sqrt{2}) \\ \frac{1}{7}(-1 + 2\sqrt{2}) \\ 1 \end{bmatrix}$;

$\lambda = -\sqrt{2}$: basis $\begin{bmatrix} \frac{1}{7}(15 - 5\sqrt{2}) \\ \frac{1}{7}(-1 - 2\sqrt{2}) \\ 1 \end{bmatrix}$

(c) $\lambda = -8$: basis $\begin{bmatrix} -\frac{1}{6} \\ -\frac{1}{6} \\ 1 \end{bmatrix}$ **(d)** $\lambda = 2$: basis $\begin{bmatrix} \frac{1}{3} \\ \frac{1}{3} \\ 1 \end{bmatrix}$ **(e)** $\lambda = 2$: basis $\begin{bmatrix} -\frac{1}{3} \\ -\frac{1}{3} \\ 1 \end{bmatrix}$

(f) $\lambda = -4$: basis $\begin{bmatrix} -2 \\ \frac{8}{3} \\ 1 \end{bmatrix}$; $\lambda = 3$: basis $\begin{bmatrix} 5 \\ -2 \\ 1 \end{bmatrix}$

8. **(a)** $\lambda = 1$, $\lambda = -2$, $\lambda = -1$ **(b)** $\lambda = 4$

10. **(a)** $\lambda = -1$, $\lambda = 5$ **(b)** $\lambda = 3$, $\lambda = 7$, $\lambda = 1$ **(c)** $\lambda = -\frac{1}{3}$, $\lambda = 1$, $\lambda = \frac{1}{2}$

13. **(a)** $y = x$ and $y = 2x$ **(b)** No lines   **(c)** $y = 0$ **14.** **(a)** -5 **(b)** 7

22. **(a)** $\lambda_1 = 1: \begin{bmatrix} 1 \\ 0 \\ 1 \end{bmatrix}$; $\lambda_2 = \frac{1}{2}: \begin{bmatrix} \frac{1}{2} \\ 1 \\ 0 \end{bmatrix}$; $\lambda_3 = \frac{1}{3}: \begin{bmatrix} 1 \\ 1 \\ 1 \end{bmatrix}$

     (b) $\lambda_1 = -2: \begin{bmatrix} 1 \\ 0 \\ 1 \end{bmatrix}$; $\lambda_2 = -1: \begin{bmatrix} \frac{1}{2} \\ 1 \\ 0 \end{bmatrix}$; $\lambda_3 = 0: \begin{bmatrix} 1 \\ 1 \\ 1 \end{bmatrix}$

     (c) $\lambda_1 = 3: \begin{bmatrix} 1 \\ 0 \\ 1 \end{bmatrix}$; $\lambda_2 = 4: \begin{bmatrix} \frac{1}{2} \\ 1 \\ 0 \end{bmatrix}$; $\lambda_3 = 5: \begin{bmatrix} 1 \\ 1 \\ 1 \end{bmatrix}$

25. **(a)** A is 6×6.   **(b)** A is invertible.   **(c)** A has three eigenspaces.

Exercise Set 7.2
(page 378)

1. $\lambda = 0 : 1$ or 2; $\lambda = 1 : 1$; $\lambda = 2 : 1$, 2, or 3   3. Not diagonalizable

5. Not diagonalizable   7. Not diagonalizable

9. $P = \begin{bmatrix} \frac{1}{3} & 0 \\ 1 & 1 \end{bmatrix}$; $P^{-1}AP = \begin{bmatrix} 1 & 0 \\ 0 & -1 \end{bmatrix}$

11. $P = \begin{bmatrix} -2 & 0 & 1 \\ 0 & 1 & 0 \\ 1 & 0 & 0 \end{bmatrix}$; $P^{-1}AP = \begin{bmatrix} 3 & 0 & 0 \\ 0 & 3 & 0 \\ 0 & 0 & 2 \end{bmatrix}$

13. $P = \begin{bmatrix} 1 & 2 & 1 \\ 1 & 3 & 3 \\ 1 & 3 & 4 \end{bmatrix}$; $P^{-1}AP = \begin{bmatrix} 1 & 0 & 0 \\ 0 & 2 & 0 \\ 0 & 0 & 3 \end{bmatrix}$

16. Not diagonalizable

17. $P = \begin{bmatrix} 1 & 1 & 0 & 0 \\ 0 & 1 & 1 & 0 \\ 0 & 0 & 1 & 1 \\ 0 & 0 & 0 & 1 \end{bmatrix}$; $P^{-1}AP = \begin{bmatrix} -2 & 0 & 0 & 0 \\ 0 & -2 & 0 & 0 \\ 0 & 0 & 3 & 0 \\ 0 & 0 & 0 & 3 \end{bmatrix}$

19. $\begin{bmatrix} -1 & 10237 & -2047 \\ 0 & 1 & 0 \\ 0 & 10245 & -2048 \end{bmatrix}$

21. $A^n = PD^nP^{-1} = \begin{bmatrix} 1 & 1 & 1 \\ 2 & 0 & -1 \\ 1 & -1 & 1 \end{bmatrix} \begin{bmatrix} 1^n & 0 & 0 \\ 0 & 3^n & 0 \\ 0 & 0 & 4^n \end{bmatrix} \begin{bmatrix} \frac{1}{6} & \frac{1}{3} & \frac{1}{6} \\ \frac{1}{2} & 0 & -\frac{1}{2} \\ \frac{1}{3} & -\frac{1}{3} & \frac{1}{3} \end{bmatrix}$

     One possibility is $P = \begin{bmatrix} -b & -b \\ a - \lambda_1 & a - \lambda_2 \end{bmatrix}$ where λ_1 and λ_2 are as in Exercise 18 of Section 7.1.

25. **(a)** False   **(b)** False   **(c)** True   **(d)** True   **(e)** True

27. **(a)** Eigenvalues λ must satisfy $-1 < \lambda \le 1$.
     (b) If $A = PDP^{-1}$ with D diagonal, then $\lim_{k \to +\infty} A^k = PD'P^{-1}$, where D' is obtained from D by setting all diagonal entries that are not 1 to 0.

Exercise Set 7.3
(page 383)

1. **(a)** $\lambda^2 - 5\lambda = 0$; $\lambda = 0$: one-dimensional; $\lambda = 5$: one-dimensional
     (b) $\lambda^3 - 27\lambda - 54 = 0$; $\lambda = 6$: one-dimensional; $\lambda = -3$: two-dimensional
     (c) $\lambda^3 - 3\lambda^2 = 0$; $\lambda = 3$: one-dimensional; $\lambda = 0$: two-dimensional
     (d) $\lambda^3 - 12\lambda^2 + 36\lambda - 32 = 0$; $\lambda = 2$: two-dimensional; $\lambda = 8$: one-dimensional

(e) $\lambda^4 - 8\lambda^3 = 0$; $\lambda = 0$: three-dimensional; $\lambda = 8$: one-dimensional
(f) $\lambda^4 - 8\lambda^3 + 22\lambda^2 - 24\lambda + 9 = 0$; $\lambda = 1$: two-dimensional;
$\lambda = 3$: two-dimensional

3. $P = \begin{bmatrix} -\frac{2}{\sqrt{7}} & \frac{\sqrt{3}}{\sqrt{7}} \\ \frac{\sqrt{3}}{\sqrt{7}} & \frac{2}{\sqrt{7}} \end{bmatrix}$; $P^{-1}AP = \begin{bmatrix} 3 & 0 \\ 0 & 10 \end{bmatrix}$

5. $P = \begin{bmatrix} -\frac{4}{5} & 0 & \frac{3}{5} \\ 0 & 1 & 0 \\ \frac{3}{5} & 0 & \frac{4}{5} \end{bmatrix}$; $P^{-1}AP = \begin{bmatrix} 25 & 0 & 0 \\ 0 & -3 & 0 \\ 0 & 0 & -50 \end{bmatrix}$

7. $P = \begin{bmatrix} \frac{1}{\sqrt{3}} & \frac{1}{\sqrt{6}} & \frac{1}{\sqrt{2}} \\ \frac{1}{\sqrt{3}} & -\frac{2}{\sqrt{6}} & 0 \\ \frac{1}{\sqrt{3}} & \frac{1}{\sqrt{6}} & -\frac{1}{\sqrt{2}} \end{bmatrix}$ $\begin{bmatrix} 0 & 0 & 0 \\ 0 & 3 & 0 \\ 0 & 0 & 3 \end{bmatrix}$

9. $P = \begin{bmatrix} -\frac{4}{5} & \frac{3}{5} & 0 & 0 \\ \frac{3}{5} & \frac{4}{5} & 0 & 0 \\ 0 & 0 & -\frac{4}{5} & \frac{3}{5} \\ 0 & 0 & \frac{3}{5} & \frac{4}{5} \end{bmatrix}$; $P^{-1}AP = \begin{bmatrix} -25 & 0 & 0 & 0 \\ 0 & 25 & 0 & 0 \\ 0 & 0 & -25 & 0 \\ 0 & 0 & 0 & 25 \end{bmatrix}$

12. (b) $\begin{bmatrix} \frac{1}{\sqrt{2}} & 0 & \frac{1}{\sqrt{2}} \\ 0 & 1 & 0 \\ -\frac{1}{\sqrt{2}} & 0 & \frac{1}{\sqrt{2}} \end{bmatrix}$ 15. Yes; take $A = \begin{bmatrix} 3 & 0 & 0 \\ 0 & 3 & 4 \\ 0 & 4 & 3 \end{bmatrix}$.

Supplementary Exercises (page 384)

1. (b) The transformation rotates vectors through the angle θ; therefore, if $0 < \theta < \pi$, then no nonzero vector is transformed into a vector in the same or opposite direction.

3. (c) $\begin{bmatrix} 1 & 1 & 0 \\ 0 & 2 & 1 \\ 0 & 0 & 3 \end{bmatrix}$

9. $A^2 = \begin{bmatrix} 15 & 30 \\ 5 & 10 \end{bmatrix}$, $A^3 = \begin{bmatrix} 75 & 150 \\ 25 & 50 \end{bmatrix}$, $A^4 = \begin{bmatrix} 375 & 750 \\ 125 & 250 \end{bmatrix}$, $A^5 = \begin{bmatrix} 1875 & 3750 \\ 625 & 1250 \end{bmatrix}$

12. (b) $\begin{bmatrix} 0 & 0 & 0 & -1 \\ 1 & 0 & 0 & 2 \\ 0 & 1 & 0 & -1 \\ 0 & 0 & 1 & -3 \end{bmatrix}$ 15. $\begin{bmatrix} 1 & 0 & 0 \\ -1 & -\frac{1}{2} & -\frac{1}{2} \\ 1 & -\frac{1}{2} & -\frac{1}{2} \end{bmatrix}$

13. They are all 0, 1, or -1.

Exercise Set 8.1 (page 398)

3. Nonlinear   5. Linear   9. (a) Linear   (b) Nonlinear

13. $T(x_1, x_2) = \frac{1}{7}(3x_1 - x_2, -9x_1 - 4x_2, 5x_1 + 10x_2)$; $T(2, -3) = \left(\frac{9}{7}, -\frac{6}{7}, -\frac{20}{7}\right)$

15. $T(x_1, x_2, x_3) = (-41x_1 + 9x_2 + 24x_3, 14x_1 - 3x_2 - 8x_3)$; $T(7, 13, 7) = (-2, 3)$

17. (a) Domain: R^2; codomain: R^2; $(T_2 \circ T_1)(x, y) = (2x - 3y, 2x + 3y)$
(b) Domain: R^2; codomain: R^2; $(T_2 \circ T_1)(x, y) = (4x - 12y, 3x - 9y)$
(c) Domain: R^2; codomain: R^2; $(T_2 \circ T_1)(x, y) = (2x + 3y, x - 2y)$
(d) Domain: R^2; codomain: R^2; $(T_2 \circ T_1)(x, y) = (0, 2x)$

19. (a) $a + d$ (b) $(T_2 \circ T_1)(A)$ does not exist since $T_1(A)$ is not a 2×2 matrix.

22. $(T_2 \circ T_1)(a_0 + a_1x + a_2x^2) = (a_0 + a_1 + a_2)x + (a_1 + 2a_2)x^2 + a_2x^3$

26. (b) $(3T)(x_1, x_2) = (6x_1 - 3x_2, 3x_2 + 3x_1)$

28. **(b)** No   31. **(a)** $x^2 + 3x$ **(b)** $\sin x$ **(c)** $e^x - 1$

Exercise Set 8.2
(page 405)

1. (a), (c)   3. (a), (b), (c)   5. (b)

7. **(a)** $\left(\frac{1}{2}, 1\right)$ **(b)** $\left(\frac{3}{2}, -4, 1, 0\right)$ **(c)** No basis exists.

11. **(a)** $\begin{bmatrix} 1 \\ 2 \\ 0 \end{bmatrix}$ **(b)** $\begin{bmatrix} \frac{1}{2} \\ 0 \\ 1 \end{bmatrix}$, $\begin{bmatrix} 0 \\ 1 \\ 0 \end{bmatrix}$ **(c)** Rank$(T) = 1$, nullity$(T) = 2$

(d) Rank$(A) = 1$, nullity$(A) = 2$

13. **(a)** $\begin{bmatrix} 1 \\ 3 \\ -1 \\ 2 \end{bmatrix}$, $\begin{bmatrix} 0 \\ 1 \\ -\frac{2}{7} \\ \frac{5}{14} \end{bmatrix}$, $\begin{bmatrix} 0 \\ 0 \\ 0 \\ 1 \end{bmatrix}$ **(b)** $\begin{bmatrix} -1 \\ -1 \\ 1 \\ 0 \\ 0 \end{bmatrix}$, $\begin{bmatrix} -1 \\ -2 \\ 0 \\ 0 \\ 1 \end{bmatrix}$

(c) Rank$(T) = 3$, nullity$(T) = 2$ **(d)** Rank$(A) = 3$, nullity$(A) = 2$

15. ker$(T) = \{\mathbf{0}\}$; $R(T) = V$ 17. Nullity$(T) = 0$, rank$(T) = 6$

21. **(a)** $x = -t$, $y = -t$, $z = t$, $-\infty < t < +\infty$ **(b)** $14x - 8y - 5z = 0$

25. ker(D) consists of all constant polynomials.

27. ker$(D \circ D)$ consists of all functions of the form $ax + b$; ker$(D \circ D \circ D)$ consists of all functions of the form $ax^2 + bx + c$.

30. **(a)** $D \circ D \circ D \circ D$, where D is differentiation   **(b)** $D \circ D \circ \cdots \circ D$ ($n + 1$ times)

Exercise Set 8.3
(page 413)

1. **(a)** ker$(T) = \{\mathbf{0}\}$; T is one-to-one.
   (b) ker$(T) = \left\{k\left(-\frac{3}{2}, 1\right)\right\}$; T is not one-to-one.
   (c) ker$(T) = \{\mathbf{0}\}$; T is one-to-one.
   (d) ker$(T) = \{\mathbf{0}\}$; T is one-to-one.
   (e) ker$(T) = \{k(1, 1)\}$; T is not one-to-one.
   (f) ker$(T) = \{k(0, 1, -1)\}$; T is not one-to-one.

3. **(a)** T has no inverse.   **(b)** $T^{-1}\begin{bmatrix} x_1 \\ x_2 \\ x_3 \end{bmatrix} = \begin{bmatrix} \frac{1}{8}x_1 + \frac{1}{8}x_2 - \frac{3}{4}x_3 \\ \frac{1}{8}x_1 + \frac{1}{8}x_2 + \frac{1}{4}x_3 \\ -\frac{3}{8}x_1 + \frac{5}{8}x_2 + \frac{1}{4}x_3 \end{bmatrix}$

(c) $T^{-1}\begin{bmatrix} x_1 \\ x_2 \\ x_3 \end{bmatrix} = \begin{bmatrix} \frac{1}{2}x_1 - \frac{1}{2}x_2 + \frac{1}{2}x_3 \\ -\frac{1}{2}x_1 + \frac{1}{2}x_2 + \frac{1}{2}x_3 \\ \frac{1}{2}x_1 + \frac{1}{2}x_2 - \frac{1}{2}x_3 \end{bmatrix}$ **(d)** $T^{-1}\begin{bmatrix} x_1 \\ x_2 \\ x_3 \end{bmatrix} = \begin{bmatrix} 3x_1 + 3x_2 - x_3 \\ -2x_1 - 2x_2 + x_3 \\ -4x_1 - 5x_2 + 2x_3 \end{bmatrix}$

5. **(a)** ker$(T) = \{k(-1, 1)\}$ **(b)** T is not one-to-one since ker$(T) \neq \{\mathbf{0}\}$.

7. **(a)** T is one-to-one.   **(b)** T is not one-to-one.
   (c) T is not one-to-one.   **(d)** T is one-to-one.

11. **(a)** $a_i \neq 0$ for $i = 1, 2, 3, \ldots, n$
    (b) $T^{-1}(x_1, x_2, x_3, \ldots, x_n) = \left(\dfrac{1}{a_1}x_1, \dfrac{1}{a_2}x_2, \dfrac{1}{a_3}x_3, \ldots, \dfrac{1}{a_n}x_n\right)$

13. **(a)** $T_1^{-1}(p(x)) = \dfrac{p(x)}{x}$; $T_2^{-1}(p(x)) = p(x - 1)$; $(T_2 \circ T_1)^{-1}(p(x)) = \dfrac{1}{x}p(x - 1)$

15. **(a)** $(1, -1)$ **(d)** $T^{-1}(2, 3) = 2 + x$

17. **(a)** T is not one-to-one.   **(b)** T is one-to-one. $T^{-1}\begin{bmatrix} a & b \\ c & d \end{bmatrix} = \begin{bmatrix} a & c \\ b & d \end{bmatrix}$

(c) T is one-to-one. $T^{-1}\begin{bmatrix} a & b \\ c & d \end{bmatrix} = \begin{bmatrix} d & -b \\ -c & a \end{bmatrix}$

21. T is not one-to-one since, for example, $f(x) = x^2(x-1)^2$ is in its kernel.

25. Yes; it is one-to-one.

Exercise Set 8.4
(page 426)

1. (a) $\begin{bmatrix} 0 & 0 & 0 \\ 1 & 0 & 0 \\ 0 & 1 & 0 \\ 0 & 0 & 1 \end{bmatrix}$ **3. (a)** $\begin{bmatrix} 1 & -1 & 1 \\ 0 & 1 & -2 \\ 0 & 0 & 1 \end{bmatrix}$

5. (a) $\begin{bmatrix} 0 & 0 \\ -\frac{1}{2} & 1 \\ \frac{8}{3} & \frac{4}{3} \end{bmatrix}$ **7. (a)** $\begin{bmatrix} 1 & 1 & 1 \\ 0 & 2 & 4 \\ 0 & 0 & 4 \end{bmatrix}$ **(b)** $3 + 10x + 16x^2$

9. (a) $[T(\mathbf{v}_1)]_B = \begin{bmatrix} 1 \\ -2 \end{bmatrix}$, $[T(\mathbf{v}_2)]_B = \begin{bmatrix} 3 \\ 5 \end{bmatrix}$ **(b)** $T(\mathbf{v}_1) = \begin{bmatrix} 3 \\ -5 \end{bmatrix}$, $T(\mathbf{v}_2) = \begin{bmatrix} -2 \\ 29 \end{bmatrix}$

(c) $T\left(\begin{bmatrix} x_1 \\ x_2 \end{bmatrix}\right) = \begin{bmatrix} \frac{18}{7} & \frac{1}{7} \\ -\frac{107}{7} & \frac{24}{7} \end{bmatrix}\begin{bmatrix} x_1 \\ x_2 \end{bmatrix}$ **(d)** $\begin{bmatrix} \frac{19}{7} \\ -\frac{83}{7} \end{bmatrix}$

11. (a) $[T(\mathbf{v}_1)]_B = \begin{bmatrix} 1 \\ 2 \\ 6 \end{bmatrix}$, $[T(\mathbf{v}_2)]_B = \begin{bmatrix} 3 \\ 0 \\ -2 \end{bmatrix}$, $[T(\mathbf{v}_3)]_B = \begin{bmatrix} -1 \\ 5 \\ 4 \end{bmatrix}$

(b) $T(\mathbf{v}_1) = 16 + 51x + 19x^2$, $T(\mathbf{v}_2) = -6 - 5x + 5x^2$, $T(\mathbf{v}_3) = 7 + 40x + 15x^2$

(c) $T(a_0 + a_1x + a_2x^2) = \dfrac{239a_0 - 161a_1 + 289a_2}{24}$

$\qquad\qquad + \dfrac{201a_0 - 111a_1 + 247a_2}{8}x + \dfrac{61a_0 - 31a_1 + 107a_2}{12}x^2$

(d) $T(1 + x^2) = 22 + 56x + 14x^2$

13. (a) $[T_2 \circ T_1]_{B',B} = \begin{bmatrix} 0 & 0 \\ 6 & 0 \\ 0 & 0 \\ 0 & -9 \end{bmatrix}$, $[T_2]_{B',B''} = \begin{bmatrix} 0 & 0 & 0 \\ 3 & 0 & 0 \\ 0 & 3 & 0 \\ 0 & 0 & 3 \end{bmatrix}$, $[T_1]_{B'',B} = \begin{bmatrix} 2 & 0 \\ 0 & 0 \\ 0 & -3 \end{bmatrix}$

(b) $[T_2 \circ T_1]_{B',B} = [T_2]_{B',B''}[T_1]_{B'',B}$

19. (a) $\begin{bmatrix} 0 & 0 & 0 \\ 0 & 0 & -1 \\ 0 & 1 & 0 \end{bmatrix}$ **(b)** $\begin{bmatrix} 0 & 0 & 0 \\ 0 & 1 & 0 \\ 0 & 0 & 2 \end{bmatrix}$ **(c)** $\begin{bmatrix} 2 & 1 & 0 \\ 0 & 2 & 2 \\ 0 & 0 & 2 \end{bmatrix}$

(d) $14e^{2x} - 8xe^{2x} - 20x^2e^{2x}$ since $\begin{bmatrix} 2 & 1 & 0 \\ 0 & 2 & 2 \\ 0 & 0 & 2 \end{bmatrix}\begin{bmatrix} 4 \\ 6 \\ -10 \end{bmatrix} = \begin{bmatrix} 14 \\ -8 \\ -20 \end{bmatrix}$

21. (a) B', B'' **(b)** B', B'''

22. We can easily compute kernels, ranges, and compositions of linear transformations.

Exercise Set 8.5
(page 439)

1. $[T]_B = \begin{bmatrix} 1 & -2 \\ 0 & -1 \end{bmatrix}$, $[T]_{B'} = \begin{bmatrix} -\frac{3}{11} & -\frac{56}{11} \\ -\frac{2}{11} & \frac{3}{11} \end{bmatrix}$

3. $[T]_B = \begin{bmatrix} \frac{1}{\sqrt{2}} & -\frac{1}{\sqrt{2}} \\ \frac{1}{\sqrt{2}} & \frac{1}{\sqrt{2}} \end{bmatrix}$, $[T]_{B'} = \begin{bmatrix} \frac{13}{11\sqrt{2}} & -\frac{25}{11\sqrt{2}} \\ \frac{5}{11\sqrt{2}} & \frac{9}{11\sqrt{2}} \end{bmatrix}$

5. $[T]_B = \begin{bmatrix} 1 & 0 & 0 \\ 0 & 1 & 0 \\ 0 & 0 & 0 \end{bmatrix}$, $[T]_{B'} = \begin{bmatrix} 1 & 0 & 0 \\ 0 & 1 & 1 \\ 0 & 0 & 0 \end{bmatrix}$

8. (a) $\det(T) = 17$ **(b)** $\det(T) = 0$ **(c)** $\det(T) = 1$

10. **(a)** $[T]_B = \begin{bmatrix} 1 & 1 & 1 & 1 & 1 \\ 0 & 2 & 4 & 6 & 8 \\ 0 & 0 & 4 & 12 & 24 \\ 0 & 0 & 0 & 8 & 32 \\ 0 & 0 & 0 & 0 & 16 \end{bmatrix}$, where B is the standard basis for P_4;

rank $(T) = 5$ and nullity $(T) = 0$.   **(b)** T is one-to-one.

12. **(a)** $\mathbf{u}_1' = \begin{bmatrix} -1 \\ 1 \\ 0 \end{bmatrix}$, $\mathbf{u}_2' = \begin{bmatrix} 1 \\ 0 \\ 1 \end{bmatrix}$, $\mathbf{u}_3' = \begin{bmatrix} -1 \\ -1 \\ 1 \end{bmatrix}$

    (b) $\mathbf{u}_1' = \begin{bmatrix} -1 \\ 1 \\ 0 \end{bmatrix}$, $\mathbf{u}_2' = \begin{bmatrix} 1 \\ 0 \\ 1 \end{bmatrix}$, $\mathbf{u}_3' = \begin{bmatrix} -1 \\ -1 \\ 1 \end{bmatrix}$

    (c) $\mathbf{u}_1' = \begin{bmatrix} 1 \\ 2 \\ 1 \end{bmatrix}$, $\mathbf{u}_2' = \begin{bmatrix} 0 \\ 1 \\ 0 \end{bmatrix}$, $\mathbf{u}_3' = \begin{bmatrix} -1 \\ 0 \\ 1 \end{bmatrix}$

14. **(a)** $\lambda = 1$, $\lambda = -2$, $\lambda = -1$

    (b) Basis for eigenspace corresponding to $\lambda = 1$: $\begin{bmatrix} 0 & 0 \\ 0 & 1 \end{bmatrix}$ and $\begin{bmatrix} 2 & 3 \\ 1 & 0 \end{bmatrix}$;

        basis for eigenspace corresponding to $\lambda = -2$: $\begin{bmatrix} -1 & 0 \\ 1 & 0 \end{bmatrix}$;

        basis for eigenspace corresponding to $\lambda = -1$: $\begin{bmatrix} -2 & 1 \\ 1 & 0 \end{bmatrix}$

21. (1) $B = P^{-1}AP$ is similar to A.

    (2) $I = P^{-1}P$

    (3) The distributive law for matrices

    (4) The determinant of a product is the product of the determinants.

    (5) The commutative law for real multiplication   (6) $\det(P^{-1}) = 1/\det(P)$

23. The choice of an appropriate basis can yield a better understanding of the linear operator.

Exercise Set 8.6 (page 445)

2. When A is noninvertible.

5. **(a)** No (not onto)   **(b)** Yes   **(c)** No (not one-to-one)   **(d)** No (not one-to-one)

11. The matrix is $\begin{bmatrix} 0 & 0 & 0 & 0 & 0 \\ 0 & 0 & -1 & 0 & 0 \\ 0 & 1 & 0 & 0 & 0 \\ 0 & 0 & 0 & 0 & -2 \\ 0 & 0 & 0 & 2 & 0 \end{bmatrix}$.

Supplementary Exercises (page 446)

1. No. $T(\mathbf{x}_1 + \mathbf{x}_2) = A(\mathbf{x}_1 + \mathbf{x}_2) + B \neq (A\mathbf{x}_1 + B) + (A\mathbf{x}_2 + B) = T(\mathbf{x}_1) + T(\mathbf{x}_2)$, and if $c \neq 1$, then $T(c\mathbf{x}) = cA\mathbf{x} + B \neq c(A\mathbf{x} + B) = cT(\mathbf{x})$.

5. **(a)** $T(\mathbf{e}_3)$ and any two of $T(\mathbf{e}_1)$, $T(\mathbf{e}_2)$, and $T(\mathbf{e}_4)$ form bases for the range; $(-1, 1, 0, 1)$ is a basis for the kernel.

    (b) Rank $= 3$, nullity $= 1$

7. **(a)** Rank $(T) = 2$ and nullity $(T) = 2$ **(b)** T is not one-to-one.

11. Rank $= 3$, nullity $= 1$ 13. $\begin{bmatrix} 1 & 0 & 0 & 0 \\ 0 & 0 & 1 & 0 \\ 0 & 1 & 0 & 0 \\ 0 & 0 & 0 & 1 \end{bmatrix}$

14. **(a)** $\mathbf{v}_1 = 2\mathbf{u}_1 + \mathbf{u}_2$, $\mathbf{v}_2 = -\mathbf{u}_1 + \mathbf{u}_2 + \mathbf{u}_3$, $\mathbf{v}_3 = 3\mathbf{u}_1 + 4\mathbf{u}_2 + 2\mathbf{u}_3$

    (b) $\mathbf{u}_1 = -2\mathbf{v}_1 - 2\mathbf{v}_2 + \mathbf{v}_3$, $\mathbf{u}_2 = 5\mathbf{v}_1 + 4\mathbf{v}_2 - 2\mathbf{v}_3$, $\mathbf{u}_3 = -7\mathbf{v}_1 - 5\mathbf{v}_2 + 3\mathbf{v}_3$

17. $[T]_B = \begin{bmatrix} 1 & -1 & 1 \\ 0 & 1 & 0 \\ 1 & 0 & -1 \end{bmatrix}$

20. **(a)** $\begin{bmatrix} 2 \\ 6 \\ 12 \end{bmatrix}$ **(d)** $-3x^2 + 3$ **(e)**

21. The points are on the graph. 24. $\begin{bmatrix} 0 & 1 & 0 & 0 & \cdots & 0 \\ 0 & 0 & 1 & 0 & \cdots & 0 \\ 0 & 0 & 0 & 1 & \cdots & 0 \\ \vdots & \vdots & \vdots & \vdots & & \vdots \\ 0 & 0 & 0 & 0 & \cdots & 1 \\ 0 & 0 & 0 & 0 & \cdots & 0 \end{bmatrix}$

Exercise Set 9.1
(page 456)

1. **(a)** $y_1 = c_1 e^{5x} - 2c_2 e^{-x}$ **(b)** $y_1 = 0$
   $y_2 = c_1 e^{5x} + c_2 e^{-x}$ $\quad\quad y_2 = 0$

3. **(a)** $y_1 = -c_2 e^{2x} + c_3 e^{3x}$ **(b)** $y_1 = e^{2x} - 2e^{3x}$
   $y_2 = c_1 e^x + 2c_2 e^{2x} - c_3 e^{3x}$ $\quad\quad y_2 = e^x - 2e^{2x} + 2e^{3x}$
   $y_3 = 2c_2 e^{2x} - c_3 e^{3x}$ $\quad\quad y_3 = -2e^{2x} + 2e^{3x}$

7. $y = c_1 e^{3x} + c_2 e^{-2x}$ 9. $y = c_1 e^x + c_2 e^{2x} + c_3 e^{3x}$

Exercise Set 9.2
(Page 466)

1. **(a)** $\begin{bmatrix} 0 & -1 \\ -1 & 0 \end{bmatrix}$ **(b)** $\begin{bmatrix} -1 & 0 \\ 0 & -1 \end{bmatrix}$ **(c)** $\begin{bmatrix} 1 & 0 \\ 0 & 0 \end{bmatrix}$ **(d)** $\begin{bmatrix} 0 & 0 \\ 0 & 1 \end{bmatrix}$

3. **(a)** $\begin{bmatrix} 1 & 0 & 0 \\ 0 & 1 & 0 \\ 0 & 0 & -1 \end{bmatrix}$ **(b)** $\begin{bmatrix} 1 & 0 & 0 \\ 0 & -1 & 0 \\ 0 & 0 & 1 \end{bmatrix}$ **(c)** $\begin{bmatrix} -1 & 0 & 0 \\ 0 & 1 & 0 \\ 0 & 0 & 1 \end{bmatrix}$

5. **(a)** $\begin{bmatrix} 0 & -1 & 0 \\ 1 & 0 & 0 \\ 0 & 0 & 1 \end{bmatrix}$ **(b)** $\begin{bmatrix} 1 & 0 & 0 \\ 0 & 0 & -1 \\ 0 & 1 & 0 \end{bmatrix}$ **(c)** $\begin{bmatrix} 0 & 0 & 1 \\ 0 & 1 & 0 \\ -1 & 0 & 0 \end{bmatrix}$

7. Rectangle with vertices at $(0, 0)$, $(-3, 0)$, $(0, 1)$, $(-3, 1)$

9. **(a)** $\begin{bmatrix} 1 & 0 \\ 4 & 1 \end{bmatrix}$ **(b)** $\begin{bmatrix} 1 & -2 \\ 0 & 1 \end{bmatrix}$ 10. **(a)** $\begin{bmatrix} 1 & 0 \\ 0 & \frac{1}{3} \end{bmatrix}$ **(b)** $\begin{bmatrix} 6 & 0 \\ 0 & 1 \end{bmatrix}$

12. **(a)** $\begin{bmatrix} 2 & 0 \\ 0 & 1 \end{bmatrix}\begin{bmatrix} 3 & 0 \\ 0 & 1 \end{bmatrix}$; expansion in the y-direction by a factor of 3, then expansion in the x-direction by a factor of 2

(b) $\begin{bmatrix} 1 & 0 \\ 2 & 1 \end{bmatrix}\begin{bmatrix} 1 & 4 \\ 0 & 1 \end{bmatrix}$; shear in the x-direction by a factor of 4, then shear in the y-direction by a factor of 2

(c) $\begin{bmatrix} 0 & 1 \\ 1 & 0 \end{bmatrix}\begin{bmatrix} 4 & 0 \\ 0 & 1 \end{bmatrix}\begin{bmatrix} 1 & 0 \\ 0 & -2 \end{bmatrix}$; expansion in the y-direction by a factor of -2, then expansion in the x-direction by a factor of 4, then reflection about $y = x$

(d) $\begin{bmatrix} 1 & 0 \\ 4 & 1 \end{bmatrix}\begin{bmatrix} 1 & 0 \\ 1 & 18 \end{bmatrix}\begin{bmatrix} 1 & -3 \\ 0 & 1 \end{bmatrix}$; shear in the x-direction by a factor of -3, then expansion in the y-direction by a factor of 18, then shear in the y-direction by a factor of 4

14. **(a)** $\begin{bmatrix} 0 & 1 \\ -5 & 0 \end{bmatrix}$　　**(b)** $\dfrac{1}{2}\begin{bmatrix} \sqrt{3} & -1 \\ -6\sqrt{3}+3 & 6+3\sqrt{3} \end{bmatrix}$

17. **(a)** $y = \frac{2}{7}x$　　**(b)** $y = x$　　**(c)** $y = \frac{1}{2}x$　　**(d)** $y = -2x$

22. **(a)** $\begin{bmatrix} 1 & 0 & 0 \\ 0 & 0 & 1 \\ 0 & 1 & 0 \end{bmatrix}$　　**(b)** $\begin{bmatrix} 0 & 0 & 1 \\ 0 & 1 & 0 \\ 1 & 0 & 0 \end{bmatrix}$　　**(c)** $\begin{bmatrix} 0 & 1 & 0 \\ 1 & 0 & 0 \\ 0 & 0 & 1 \end{bmatrix}$

24. **(a)** $\lambda_1 = 1: \begin{bmatrix} 1 \\ 0 \end{bmatrix}$; $\lambda_2 = -1: \begin{bmatrix} 0 \\ 1 \end{bmatrix}$　　**(b)** $\lambda_1 = 1: \begin{bmatrix} 0 \\ 1 \end{bmatrix}$; $\lambda_2 = -1: \begin{bmatrix} 1 \\ 0 \end{bmatrix}$

(c) $\lambda_1 = 1: \begin{bmatrix} 1 \\ 1 \end{bmatrix}$; $\lambda_2 = -1: \begin{bmatrix} -1 \\ 1 \end{bmatrix}$　　**(d)** $\lambda = 1: \begin{bmatrix} 1 \\ 0 \end{bmatrix}$

(e) $\lambda = 1: \begin{bmatrix} 0 \\ 1 \end{bmatrix}$　　　**(f)** (θ an odd integer multiple of π) $\lambda = -1: (1, 0), (0, 1)$
(θ an even integer multiple of π) $\lambda = 1: (1, 0), (0, 1)$
(θ not an integer multiple of π) no real eigenvalues

Exercise Set 9.3
(page 473)

1. $y = -\frac{1}{2} + \frac{7}{2}x$　　**3.** $y = 2 + 5x - 3x^2$

8. $y = 4 - .2x + .2x^2$; if $x = 12$, then $y = 30.4$ ($\$30.4$ thousand)

Exercise Set 9.4
(page 479)

1. **(a)** $(1 + \pi) - 2\sin x - \sin 2x$

(b) $(1 + \pi) - 2\left[\sin x + \dfrac{\sin 2x}{2} + \dfrac{\sin 3x}{3} + \cdots + \dfrac{\sin nx}{n}\right]$

3. **(a)** $-\dfrac{1}{2} + \dfrac{1}{e-1}e^x$　　**(b)** $\dfrac{1}{12} - \dfrac{3-e}{2e-2}$

5. **(a)** $\dfrac{3}{\pi}x$　　**(b)** $1 - \dfrac{6}{\pi^2}$　　**8.** $\displaystyle\sum_{k=1}^{\infty} \dfrac{2}{k}\sin(kx)$

Exercise Set 9.5
(page 485)

1. (a), (c), (e), (g), (h)

3. **(a)** $A = \begin{bmatrix} 9 & 3 & -4 \\ 3 & -1 & \frac{1}{2} \\ -4 & \frac{1}{2} & 4 \end{bmatrix}$　　**(b)** $\begin{bmatrix} 1 & -\frac{5}{2} & \frac{9}{2} \\ -\frac{5}{2} & 1 & 0 \\ \frac{9}{2} & 0 & -3 \end{bmatrix}$　　**(c)** $A = \begin{bmatrix} 0 & \frac{1}{2} & \frac{1}{2} \\ \frac{1}{2} & 0 & \frac{1}{2} \\ \frac{1}{2} & \frac{1}{2} & 0 \end{bmatrix}$

(d) $A = \begin{bmatrix} \sqrt{2} & \sqrt{2} & -4\sqrt{3} \\ \sqrt{2} & 0 & 0 \\ -4\sqrt{3} & 0 & -\sqrt{3} \end{bmatrix}$　　**(e)** $A = \begin{bmatrix} 1 & 1 & 0 & -5 \\ 1 & 1 & 0 & 0 \\ 0 & 0 & -1 & 2 \\ -5 & 0 & 2 & -1 \end{bmatrix}$

5. **(a)** max value $= 5$ at $\pm(1, 0)$; min value $= -1$ at $\pm(0, 1)$

(b) max value $= \dfrac{11+\sqrt{10}}{2}$ at $\pm\left(\dfrac{1}{\sqrt{20-6\sqrt{10}}}, \dfrac{1}{\sqrt{20+6\sqrt{10}}}\right)$;

min value $= \dfrac{11-\sqrt{10}}{2}$ at $\pm\left(\dfrac{-1}{\sqrt{20+6\sqrt{10}}}, \dfrac{1}{\sqrt{20-6\sqrt{10}}}\right)$

(c) max value $= \frac{7+\sqrt{10}}{2}$ at $\pm\left(\frac{1}{\sqrt{20-6\sqrt{10}}}, \frac{-1}{\sqrt{20-6\sqrt{10}}}\right)$;

min value $= \frac{7-\sqrt{10}}{2}$ at $\pm\left(\frac{1}{\sqrt{20+6\sqrt{10}}}, \frac{1}{\sqrt{20-6\sqrt{10}}}\right)$

(d) max value $= \frac{3+\sqrt{10}}{2}$ at $\pm\left(\frac{3}{\sqrt{20-2\sqrt{10}}}, \frac{3}{\sqrt{20+2\sqrt{10}}}\right)$;

min value $= \frac{3-\sqrt{10}}{2}$ at $\pm\left(\frac{3}{\sqrt{20+2\sqrt{10}}}, \frac{-3}{\sqrt{20-2\sqrt{10}}}\right)$

7. (b)   **9. (a)**

11. (a) Positive definite   **(b)** Negative definite   **(c)** Positive semidefinite
(d) Negative semidefinite   **(e)** Indefinite   **(f)** Indefinite

13. (c)

16. (a) $A = \begin{bmatrix} \frac{1}{n} & \frac{-1}{n(n-1)} & \frac{-1}{n(n-1)} & \cdots & \frac{-1}{n(n-1)} \\ \frac{-1}{n(n-1)} & \frac{1}{n} & \frac{-1}{n(n-1)} & \cdots & \frac{-1}{n(n-1)} \\ \vdots & \vdots & \vdots & & \vdots \\ \frac{-1}{n(n-1)} & \frac{-1}{n(n-1)} & \frac{-1}{n(n-1)} & \cdots & \frac{1}{n} \end{bmatrix}$

(b) Positive semidefinite

Exercise Set 9.6
(page 496)

1. (a) $\begin{bmatrix} x_1 \\ x_2 \end{bmatrix} = \begin{bmatrix} \frac{1}{\sqrt{2}} & \frac{1}{\sqrt{2}} \\ \frac{1}{\sqrt{2}} & -\frac{1}{\sqrt{2}} \end{bmatrix}\begin{bmatrix} y_1 \\ y_2 \end{bmatrix}$; $y_1^2 + 3y_2^2$

(b) $\begin{bmatrix} x_1 \\ x_2 \end{bmatrix} = \begin{bmatrix} \frac{1}{\sqrt{5}} & \frac{2}{\sqrt{5}} \\ -\frac{2}{\sqrt{5}} & \frac{1}{\sqrt{5}} \end{bmatrix}\begin{bmatrix} y_1 \\ y_2 \end{bmatrix}$; $y_1^2 + 6y_2^2$

(c) $\begin{bmatrix} x_1 \\ x_2 \end{bmatrix} = \begin{bmatrix} \frac{1}{\sqrt{2}} & \frac{1}{\sqrt{2}} \\ \frac{1}{\sqrt{2}} & -\frac{1}{\sqrt{2}} \end{bmatrix}\begin{bmatrix} y_1 \\ y_2 \end{bmatrix}$; $y_1^2 - y_2^2$

(d) $\begin{bmatrix} x_1 \\ x_2 \end{bmatrix} = \begin{bmatrix} \frac{\sqrt{17}-4}{\sqrt{34-8\sqrt{17}}} & \frac{\sqrt{17}+4}{\sqrt{34+8\sqrt{17}}} \\ \frac{1}{\sqrt{34-8\sqrt{17}}} & \frac{-1}{\sqrt{34+8\sqrt{17}}} \end{bmatrix}\begin{bmatrix} y_1 \\ y_2 \end{bmatrix}$; $(1+\sqrt{17})y_1^2 + (1-\sqrt{17})y_2^2$

3. (a) $2x^2 - 3xy + 4y^2$ **(b)** $x^2 - xy$ **(c)** $5xy$ **(d)** $4x^2 - 2y^2$ **(e)** y^2

5. (a) $[x \ \ y]\begin{bmatrix} 2 & -\frac{3}{2} \\ -\frac{3}{2} & 4 \end{bmatrix}\begin{bmatrix} x \\ y \end{bmatrix} + [-7 \ \ 2]\begin{bmatrix} x \\ y \end{bmatrix} + 7 = 0$

(b) $[x \ \ y]\begin{bmatrix} 1 & -\frac{1}{2} \\ -\frac{1}{2} & 0 \end{bmatrix}\begin{bmatrix} x \\ y \end{bmatrix} + [5 \ \ 8]\begin{bmatrix} x \\ y \end{bmatrix} - 3 = 0$

(c) $[x \ \ y]\begin{bmatrix} 0 & \frac{5}{2} \\ \frac{5}{2} & 0 \end{bmatrix}\begin{bmatrix} x \\ y \end{bmatrix} - 8 = 0$ **(d)** $[x \ \ y]\begin{bmatrix} 4 & 0 \\ 0 & -2 \end{bmatrix}\begin{bmatrix} x \\ y \end{bmatrix} - 7 = 0$

(e) $[x \ \ y]\begin{bmatrix} 0 & 0 \\ 0 & 1 \end{bmatrix}\begin{bmatrix} x \\ y \end{bmatrix} + [7 \ \ -8]\begin{bmatrix} x \\ y \end{bmatrix} - 5 = 0$

7. (a) $9x'^2 + 4y'^2 = 36$, ellipse   **(b)** $x'^2 - 16y'^2 = 16$, hyperbola
(c) $y'^2 = 8x'$, parabola   **(d)** $x'^2 + y'^2 = 16$, circle

(e) $18y'^2 - 12x'^2 = 419$, hyperbola　　**(f)** $y' = -\frac{1}{7}x'^2$, parabola

9. $2x''^2 + y''^2 = 6$, ellipse　　**11.** $2x''^2 - 3y''^2 = 24$, hyperbola

15. (a) Two intersecting lines, $y = x$ and $y = -x$
(b) No graph
(c) The graph is the single point $(0, 0)$.
(d) The graph is the line $y = x$.
(e) The graph consists of two parallel lines $\frac{3}{\sqrt{13}}x + \frac{2}{\sqrt{13}}y = \pm 2$.
(f) The graph is the single point $(1, 2)$.

**Exercise Set 9.7
(page 501)**

1. (a) $x^2 + 2y^2 - z^2 + 4xy - 5yz$　　**(b)** $3x^2 + 7z^2 + 2xy - 3xz + 4yz$
(c) $xy + xz + yz$　　**(d)** $x^2 + y^2 - z^2$
(e) $3z^2 + 3xz$　　**(f)** $2z^2 + 2xz + y^2$

3. (a) $\begin{bmatrix} x & y & z \end{bmatrix} \begin{bmatrix} 1 & 2 & 0 \\ 2 & 2 & -\frac{5}{2} \\ 0 & -\frac{5}{2} & -1 \end{bmatrix} \begin{bmatrix} x \\ y \\ z \end{bmatrix} + \begin{bmatrix} 7 & 0 & 2 \end{bmatrix} \begin{bmatrix} x \\ y \\ z \end{bmatrix} - 3 = 0$

(b) $\begin{bmatrix} x & y & z \end{bmatrix} \begin{bmatrix} 3 & 1 & -\frac{3}{2} \\ 1 & 0 & 2 \\ -\frac{3}{2} & 2 & 7 \end{bmatrix} \begin{bmatrix} x \\ y \\ z \end{bmatrix} + \begin{bmatrix} -3 & 0 & 0 \end{bmatrix} \begin{bmatrix} x \\ y \\ z \end{bmatrix} - 4 = 0$

(c) $\begin{bmatrix} x & y & z \end{bmatrix} \begin{bmatrix} 0 & \frac{1}{2} & \frac{1}{2} \\ \frac{1}{2} & 0 & \frac{1}{2} \\ \frac{1}{2} & \frac{1}{2} & 0 \end{bmatrix} \begin{bmatrix} x \\ y \\ z \end{bmatrix} - 1 = 0$

(d) $\begin{bmatrix} x & y & z \end{bmatrix} \begin{bmatrix} 1 & 0 & 0 \\ 0 & 1 & 0 \\ 0 & 0 & -1 \end{bmatrix} \begin{bmatrix} x \\ y \\ z \end{bmatrix} - 7 = 0$

(e) $\begin{bmatrix} x & y & z \end{bmatrix} \begin{bmatrix} 0 & 0 & \frac{3}{2} \\ 0 & 0 & 0 \\ \frac{3}{2} & 0 & 3 \end{bmatrix} \begin{bmatrix} x \\ y \\ z \end{bmatrix} + \begin{bmatrix} 0 & -14 & 0 \end{bmatrix} \begin{bmatrix} x \\ y \\ z \end{bmatrix} + 9 = 0$

(f) $\begin{bmatrix} x & y & z \end{bmatrix} \begin{bmatrix} 0 & 0 & 1 \\ 0 & 1 & 0 \\ 1 & 0 & 2 \end{bmatrix} \begin{bmatrix} x \\ y \\ z \end{bmatrix} + \begin{bmatrix} 2 & -1 & 3 \end{bmatrix} \begin{bmatrix} x \\ y \\ z \end{bmatrix} = 0$

7. (a) $9x'^2 + 36y'^2 + 4z'^2 = 36$, ellipsoid
(b) $6x'^2 + 3y'^2 - 2z'^2 = 18$, hyperboloid of one sheet
(c) $3x'^2 - 3y'^2 - z'^2 = 3$, hyperboloid of two sheets
(d) $4x'^2 + 9y'^2 - z'^2 = 0$, elliptic cone
(e) $x'^2 + 16y'^2 - 16z' = 0$, elliptic paraboloid
(f) $7x'^2 - 3y'^2 + z' = 0$, hyperbolic paraboloid
(g) $x'^2 + y'^2 + z'^2 = 25$, sphere

9. $x''^2 + y''^2 - 2z''^2 = -1$, hyperboloid of two sheets

11. $x''^2 - y''^2 + z'' = 0$, hyperbolic paraboloid

**Exercise Set 9.8
(page 509)**

1. Multiplications: mpn; additions: $mp(n - 1)$

3.

	$n = 5$	$n = 10$	$n = 100$	$n = 1000$
Solve $A\mathbf{x} = \mathbf{b}$ by Gauss–Jordan elimination	+: 50 ×: 65	+: 375 ×: 430	+: 383,250 ×: 343,300	+: 333,283,500 ×: 334,333,000
Solve $A\mathbf{x} = \mathbf{b}$ by Gaussian elimination	+: 50 ×: 65	+: 375 ×: 430	+: 383,250 ×: 343,300	+: 333,283,500 ×: 334,333,000
Find A^{-1} by reducing $[A \mid I]$ to $[I \mid A^{-1}]$	+: 80 ×: 125	+: 810 ×: 1000	+: 980,100 ×: 1,000,000	+: 998,001,000 ×: 1,000,000,000
Solve $A\mathbf{x} = \mathbf{b}$ as $\mathbf{x} = A^{-1}\mathbf{b}$	+: 100 ×: 150	+: 900 ×: 1100	+: 990,000 ×: 1,010,000	+: 999,000,000 ×: 1,001,000,000
Find $\det(A)$ by row reduction	+: 30 ×: 44	+: 285 ×: 339	+: 328,350 ×: 333,399	+: 332,833,500 ×: 333,333,999
Solve $A\mathbf{x} = \mathbf{b}$ by Cramer's Rule	+: 180 ×: 264	+: 3135 ×: 3729	+: 33,163,350 ×: 33,673,399	+: $33,316,633 \times 10^4$ ×: $33,366,733 \times 10^4$

4.

	$n = 5$ Execution Time (sec)	$n = 10$ Execution Time (sec)	$n = 100$ Execution Time (sec)	$n = 1000$ Execution Time (sec)
Solve $A\mathbf{x} = \mathbf{b}$ by Gauss–Jordan elimination	1.55×10^{-4}	1.05×10^{-3}	.878	836
Solve $A\mathbf{x} = \mathbf{b}$ by Gaussian elimination	1.55×10^{-4}	1.05×10^{-3}	.878	836
Find A^{-1} by reducing $[A \mid I]$ to $[I \mid A^{-1}]$	2.84×10^{-4}	2.41×10^{-3}	2.49	2499
Solve $A\mathbf{x} = \mathbf{b}$ as $\mathbf{x} = A^{-1}\mathbf{b}$	3.50×10^{-4}	2.65×10^{-3}	2.52	2502
Find $\det(A)$ by row reduction	1.03×10^{-4}	8.21×10^{-4}	.831	833
Solve $A\mathbf{x} = \mathbf{b}$ by Cramer's Rule	6.18×10^{-4}	90.3×10^{-4}	83.9	834×10^3

**Exercise Set 9.9
(page 517)**

1. $x_1 = 2,\ x_2 = 1$ **3.** $x_1 = 3,\ x_2 = -1$ **5.** $x_1 = -1,\ x_2 = 1,\ x_3 = 0$

7. $x_1 = -1,\ x_2 = 1,\ x_3 = 0$ **9.** $x_1 = -3,\ x_2 = 1,\ x_3 = 2,\ x_4 = 1$

11. (a) $A = LU = \begin{bmatrix} 2 & 0 & 0 \\ -2 & 1 & 0 \\ 2 & 1 & 1 \end{bmatrix} \begin{bmatrix} 1 & \frac{1}{2} & -\frac{1}{2} \\ 0 & 0 & 1 \\ 0 & 0 & 0 \end{bmatrix}$

(b) $A = L_1 DU = \begin{bmatrix} 1 & 0 & 0 \\ -1 & 1 & 0 \\ 1 & 1 & 1 \end{bmatrix} \begin{bmatrix} 2 & 0 & 0 \\ 0 & 1 & 0 \\ 0 & 0 & 1 \end{bmatrix} \begin{bmatrix} 1 & \frac{1}{2} & -\frac{1}{2} \\ 0 & 0 & 1 \\ 0 & 0 & 0 \end{bmatrix}$

(c) $A = L_2 U_2 = \begin{bmatrix} 1 & 0 & 0 \\ -1 & 1 & 0 \\ 1 & 1 & 1 \end{bmatrix} \begin{bmatrix} 2 & 1 & -1 \\ 0 & 0 & 1 \\ 0 & 0 & 0 \end{bmatrix}$

13. (b) $\begin{bmatrix} a & b \\ c & d \end{bmatrix} = \begin{bmatrix} 1 & 0 \\ \dfrac{c}{a} & 1 \end{bmatrix} \begin{bmatrix} a & b \\ 0 & \dfrac{ad - bc}{a} \end{bmatrix}$

18. $A = PLU = \begin{bmatrix} 1 & 0 & 0 \\ 0 & 0 & 1 \\ 0 & 1 & 0 \end{bmatrix} \begin{bmatrix} 3 & 0 & 0 \\ 0 & 2 & 0 \\ 3 & 0 & 1 \end{bmatrix} \begin{bmatrix} 1 & -\frac{1}{3} & 0 \\ 0 & 1 & \frac{1}{2} \\ 0 & 0 & 1 \end{bmatrix}$

Exercise Set 10.1
(page 526)

1. (a–d)

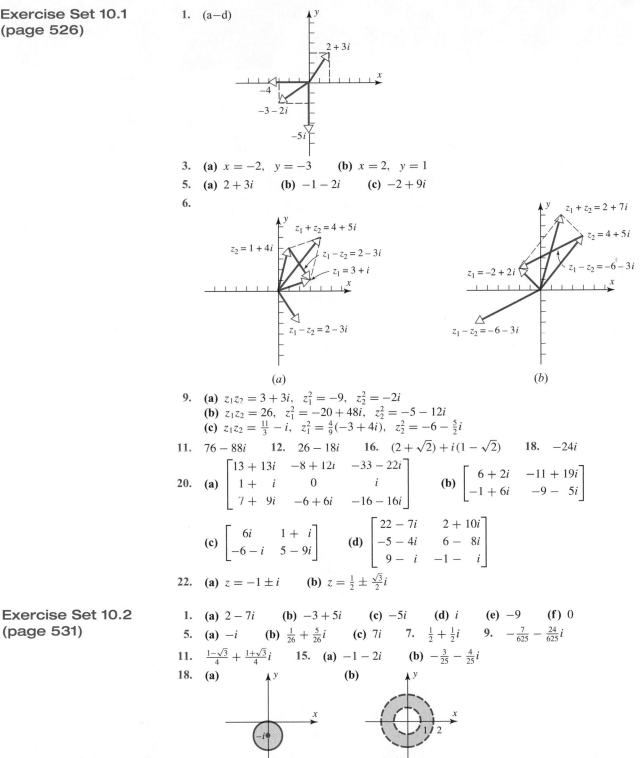

3. **(a)** $x = -2,\ y = -3$ **(b)** $x = 2,\ y = 1$

5. **(a)** $2 + 3i$ **(b)** $-1 - 2i$ **(c)** $-2 + 9i$

6.

9. **(a)** $z_1z_2 = 3 + 3i,\ z_1^2 = -9,\ z_2^2 = -2i$
   (b) $z_1z_2 = 26,\ z_1^2 = -20 + 48i,\ z_2^2 = -5 - 12i$
   (c) $z_1z_2 = \frac{11}{3} - i,\ z_1^2 = \frac{4}{9}(-3 + 4i),\ z_2^2 = -6 - \frac{5}{2}i$

11. $76 - 88i$ **12.** $26 - 18i$ **16.** $(2 + \sqrt{2}) + i(1 - \sqrt{2})$ **18.** $-24i$

20. **(a)** $\begin{bmatrix} 13 + 13i & -8 + 12i & -33 - 22i \\ 1 + i & 0 & i \\ 7 + 9i & -6 + 6i & -16 - 16i \end{bmatrix}$ **(b)** $\begin{bmatrix} 6 + 2i & -11 + 19i \\ -1 + 6i & -9 - 5i \end{bmatrix}$

     (c) $\begin{bmatrix} 6i & 1 + i \\ -6 - i & 5 - 9i \end{bmatrix}$ **(d)** $\begin{bmatrix} 22 - 7i & 2 + 10i \\ -5 - 4i & 6 - 8i \\ 9 - i & -1 - i \end{bmatrix}$

22. **(a)** $z = -1 \pm i$ **(b)** $z = \frac{1}{2} \pm \frac{\sqrt{3}}{2}i$

Exercise Set 10.2
(page 531)

1. **(a)** $2 - 7i$ **(b)** $-3 + 5i$ **(c)** $-5i$ **(d)** i **(e)** -9 **(f)** 0

5. **(a)** $-i$ **(b)** $\frac{1}{26} + \frac{5}{26}i$ **(c)** $7i$ **7.** $\frac{1}{2} + \frac{1}{2}i$ **9.** $-\frac{7}{625} - \frac{24}{625}i$

11. $\frac{1 - \sqrt{3}}{4} + \frac{1 + \sqrt{3}}{4}i$ **15.** **(a)** $-1 - 2i$ **(b)** $-\frac{3}{25} - \frac{4}{25}i$

18. **(a)**   **(b)**

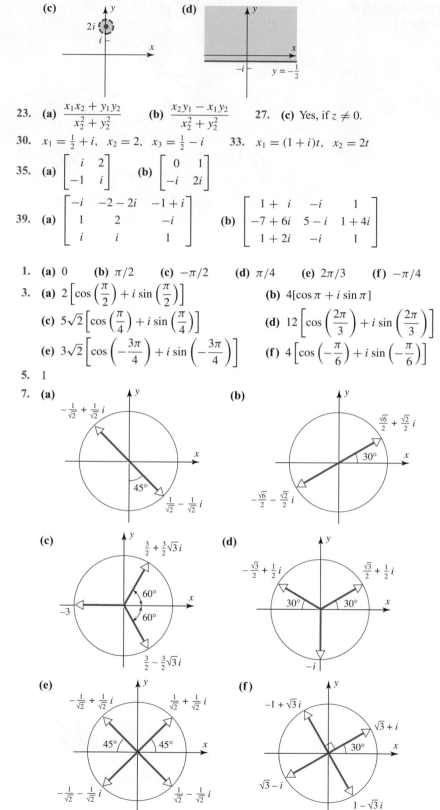

(c) $2i$; i

(d) $-i$; $y = -\frac{1}{2}$

23. **(a)** $\dfrac{x_1 x_2 + y_1 y_2}{x_2^2 + y_2^2}$ **(b)** $\dfrac{x_2 y_1 - x_1 y_2}{x_2^2 + y_2^2}$ 27. **(c)** Yes, if $z \neq 0$.

30. $x_1 = \frac{1}{2} + i$, $x_2 = 2$, $x_3 = \frac{1}{2} - i$ 33. $x_1 = (1 + i)t$, $x_2 = 2t$

35. **(a)** $\begin{bmatrix} i & 2 \\ -1 & i \end{bmatrix}$ **(b)** $\begin{bmatrix} 0 & 1 \\ -i & 2i \end{bmatrix}$

39. **(a)** $\begin{bmatrix} -i & -2-2i & -1+i \\ 1 & 2 & -i \\ i & i & 1 \end{bmatrix}$ **(b)** $\begin{bmatrix} 1+i & -i & 1 \\ -7+6i & 5-i & 1+4i \\ 1+2i & -i & 1 \end{bmatrix}$

Exercise Set 10.3
(page 539)

1. **(a)** 0   **(b)** $\pi/2$ **(c)** $-\pi/2$ **(d)** $\pi/4$ **(e)** $2\pi/3$ **(f)** $-\pi/4$

3. **(a)** $2\left[\cos\left(\dfrac{\pi}{2}\right) + i\sin\left(\dfrac{\pi}{2}\right)\right]$ **(b)** $4[\cos\pi + i\sin\pi]$

   (c) $5\sqrt{2}\left[\cos\left(\dfrac{\pi}{4}\right) + i\sin\left(\dfrac{\pi}{4}\right)\right]$ **(d)** $12\left[\cos\left(\dfrac{2\pi}{3}\right) + i\sin\left(\dfrac{2\pi}{3}\right)\right]$

   (e) $3\sqrt{2}\left[\cos\left(-\dfrac{3\pi}{4}\right) + i\sin\left(-\dfrac{3\pi}{4}\right)\right]$ **(f)** $4\left[\cos\left(-\dfrac{\pi}{6}\right) + i\sin\left(-\dfrac{\pi}{6}\right)\right]$

5. 1

7. **(a)**, **(b)**, **(c)**, **(d)**, **(e)**, **(f)**

10. $\sqrt[4]{2}\left[\cos\left(\dfrac{\pi}{8}\right) + i\sin\left(\dfrac{\pi}{8}\right)\right]$, $\sqrt[4]{2}\left[\cos\left(\dfrac{9\pi}{8}\right) + i\sin\left(\dfrac{9\pi}{8}\right)\right]$

12. The roots are $\pm(2^{1/4} + 2^{1/4}i)$, $\pm(2^{1/4} - 2^{1/4}i)$ and the factorization is
$z^4 + 8 = (z^2 - 2^{5/4}z + 2^{3/2})(z^2 + 2^{5/4}z + 2^{3/2})$.

15. (a) $\text{Re}(z) = -3$, $\text{Im}(z) = 0$ (b) $\text{Re}(z) = -3$, $\text{Im}(z) = 0$
(c) $\text{Re}(z) = 0$, $\text{Im}(z) = -\sqrt{2}$ (d) $\text{Re}(z) = -3$, $\text{Im}(z) = 0$

20. $\cos 2\theta = \cos^2\theta - \sin^2\theta$, $\sin 2\theta = 2\sin\theta\cos\theta$
$\cos 3\theta = \cos^3\theta - 3\sin^2\theta\cos\theta$, $\sin 3\theta = 3\sin\theta\cos^2\theta - \sin^3\theta$

Exercise Set 10.4
(page 544)

1. (a) $(3i, -i, -2 - i, 4)$ (b) $(3 + 2i, -1 - 2i, -3 + 5i, -i)$
(c) $(-1 - 2i, 2i, 2 - i, -1)$ (d) $(-3 + 9i, 3 - 3i, -3 - 6i, 12 + 3i)$
(e) $(-3 + 2i, 3, -3 - 3i, i)$ (f) $(-1 - 5i, 3i, 4, -5)$

5. (a) $\sqrt{2}$ (b) $2\sqrt{3}$ (c) $\sqrt{10}$ (d) $\sqrt{37}$

9. (a) 3   (b) $2 - 27i$ (c) $-5 - 10i$

11. Not a vector space. Axiom 6 fails; that is, the set is not closed under scalar multiplication. (Multiply by i, for example.)

13. ker T is all multiples of $\begin{bmatrix} 1 + 3i \\ 1 + i \\ -2 \end{bmatrix}$; nullity of $T = 1$

17. (a) $(-3 - 2i)\mathbf{u} + (3 - i)\mathbf{v} + (1 + 2i)\mathbf{w}$ (b) $(2 + i)\mathbf{u} + (-1 + i)\mathbf{v} + (-1 - i)\mathbf{w}$
(c) $0\mathbf{u} + 0\mathbf{v} + 0\mathbf{w}$ (d) $(-5 - 4i)\mathbf{u} + (5 + 2i)\mathbf{v} + (2 + 4i)\mathbf{w}$

19. (a), (b), (c)   21. (b), (c)   23. $\mathbf{f} - 3\mathbf{g} - 3\mathbf{h} = \mathbf{0}$ 25. (a), (b)

27. $(-1 - i, 1)$; dimension $= 1$ 30. $\left(\frac{5}{2}i, -\frac{1}{2}, 1, 0\right)$, $\left(-\frac{1}{4}, \frac{3}{4}i, 0, 1\right)$; dimension $= 2$

Exercise Set 10.5
(page 551)

2. (a) -12 (b) 0   (c) $2i$ (d) 37

4. (a) $-4 + 5i$ (b) 0   (c) $4 - 4i$ (d) 42   6. $-9 - 5i$

8. No. Axiom 4 fails.   10. (a) $\sqrt{10}$ (b) 2   (c) $\sqrt{5}$ (d) 0

12. (a) $3\sqrt{10}$ (b) $\sqrt{14}$ 14. (a) 2   (b) $2\sqrt{2}$

16. (a) $7\sqrt{2}$ (b) $2\sqrt{3}$ 20. (b)

23. $\left(\dfrac{i}{\sqrt{2}}, 0, 0, \dfrac{i}{\sqrt{2}}\right)$, $\left(-\dfrac{i}{\sqrt{6}}, 0, \dfrac{2i}{\sqrt{6}}, \dfrac{i}{\sqrt{6}}\right)$, $\left(\dfrac{2i}{\sqrt{21}}, \dfrac{3i}{\sqrt{21}}, \dfrac{2i}{\sqrt{21}}, \dfrac{-2i}{\sqrt{21}}\right)$,
$\left(-\dfrac{i}{\sqrt{7}}, \dfrac{2i}{\sqrt{7}}, -\dfrac{i}{\sqrt{7}}, \dfrac{i}{\sqrt{7}}\right)$

25. (a) $\mathbf{v}_1 = \left(\dfrac{i}{\sqrt{3}}, \dfrac{i}{\sqrt{3}}, \dfrac{i}{\sqrt{3}}\right)$, $\mathbf{v}_2 = \left(-\dfrac{i}{\sqrt{2}}, \dfrac{i}{\sqrt{2}}, 0\right)$, $\mathbf{v}_3 = \left(\dfrac{i}{\sqrt{6}}, \dfrac{i}{\sqrt{6}}, -\dfrac{2i}{\sqrt{6}}\right)$

     (b) $\mathbf{v}_1 = (i, 0, 0)$, $\mathbf{v}_2 = \left(0, \dfrac{7i}{\sqrt{53}}, \dfrac{-2i}{\sqrt{53}}\right)$, $\mathbf{v}_3 = \left(0, \dfrac{2i}{\sqrt{53}}, \dfrac{7i}{\sqrt{53}}\right)$

27. $\mathbf{v}_1 = \left(0, \dfrac{i}{\sqrt{3}}, \dfrac{1 - i}{\sqrt{3}}\right)$, $\mathbf{v}_2 = \left(-\dfrac{3i}{\sqrt{15}}, \dfrac{2}{\sqrt{15}}, \dfrac{1 + i}{\sqrt{15}}\right)$

36. $\mathbf{u} = -\sqrt{3}i\mathbf{v}_1 + \dfrac{3}{\sqrt{6}}\mathbf{v}_2 - \dfrac{1}{\sqrt{2}}\mathbf{v}_3$

Exercise Set 10.6
(page 561)

1. (a) $\begin{bmatrix} -2i & 4 & 5 - i \\ 1 + i & 3 - i & 0 \end{bmatrix}$ (b) $\begin{bmatrix} -2i & 4 & -i \\ 1 + i & 5 + 7i & 3 \\ -1 - i & i & 1 \end{bmatrix}$

(c) $\begin{bmatrix} -7i \\ 0 \\ 3i \end{bmatrix}$ (d) $\begin{bmatrix} \bar{a}_{11} & \bar{a}_{21} \\ \bar{a}_{12} & \bar{a}_{22} \\ \bar{a}_{13} & \bar{a}_{23} \end{bmatrix}$

3. $k = 3 + 5i$, $l = i$, $m = 2 - 4i$ 4. (a), (b)

5. **(a)** $A^{-1} = \begin{bmatrix} \frac{3}{5} & -\frac{4}{5} \\ -\frac{4}{5}i & -\frac{3}{5}i \end{bmatrix}$ **(b)** $A^{-1} = \begin{bmatrix} \dfrac{1}{\sqrt{2}} & \dfrac{-1+i}{2} \\ \dfrac{1}{\sqrt{2}} & \dfrac{1-i}{2} \end{bmatrix}$

(c) $A^{-1} = \begin{bmatrix} \frac{1}{2\sqrt{2}}(\sqrt{3}-i) & \frac{1}{2\sqrt{2}}(1-\sqrt{3}i) \\ \frac{1}{2\sqrt{2}}(1+\sqrt{3}i) & \frac{1}{2\sqrt{2}}(-\sqrt{3}-i) \end{bmatrix}$

(d) $A^{-1} = \begin{bmatrix} \dfrac{1-i}{2} & -\dfrac{i}{\sqrt{3}} & \dfrac{3-i}{2\sqrt{15}} \\ -\dfrac{1}{2} & \dfrac{1}{\sqrt{3}} & \dfrac{4-3i}{2\sqrt{15}} \\ \dfrac{1}{2} & \dfrac{i}{\sqrt{3}} & -\dfrac{5i}{2\sqrt{15}} \end{bmatrix}$

7. $P = \begin{bmatrix} \dfrac{-1+i}{\sqrt{3}} & \dfrac{1-i}{\sqrt{6}} \\ \dfrac{1}{\sqrt{3}} & \dfrac{2}{\sqrt{6}} \end{bmatrix}$; $P^{-1}AP = \begin{bmatrix} 3 & 0 \\ 0 & 6 \end{bmatrix}$

9. $P = \begin{bmatrix} -\dfrac{1+i}{\sqrt{6}} & \dfrac{1+i}{\sqrt{3}} \\ \dfrac{2}{\sqrt{6}} & \dfrac{1}{\sqrt{3}} \end{bmatrix}$; $P^{-1}AP = \begin{bmatrix} 2 & 0 \\ 0 & 8 \end{bmatrix}$

11. $P = \begin{bmatrix} 0 & 1 & 0 \\ -\dfrac{1-i}{\sqrt{6}} & 0 & \dfrac{1-i}{\sqrt{3}} \\ \dfrac{2}{\sqrt{6}} & 0 & \dfrac{1}{\sqrt{3}} \end{bmatrix}$; $P^{-1}AP = \begin{bmatrix} 1 & 0 & 0 \\ 0 & 5 & 0 \\ 0 & 0 & -2 \end{bmatrix}$

14. **(a)** $\begin{bmatrix} 0 & i \\ -i & 0 \end{bmatrix}$ is one possibility.

**Supplementary
Exercises
(page 563)**

3. $\begin{bmatrix} -i \\ 1 \\ 0 \end{bmatrix}$, $\begin{bmatrix} 1 \\ 0 \\ 1 \end{bmatrix}$ is one possibility.   5. $\lambda = 1,\ \omega,\ \omega^2\ (= \overline{\omega})$